Salinity Tolerance in Plants

Salinity Tolerance in Plants

Special Issue Editor

Jose Antonio Hernández Cortés

MDPI • Basel • Beijing • Wuhan • Barcelona • Belgrade

MDPI

Special Issue Editor
Jose Antonio Hernández Cortés
Group of Fruit Biotechnology
Department of Fruit Breeding
Spain

Editorial Office
MDPI
St. Alban-Anlage 66
4052 Basel, Switzerland

This is a reprint of articles from the Special Issue published online in the open access journal *International Journal of Molecular Sciences* (ISSN 1422-0067) from 2018 to 2019 (available at: https://www.mdpi.com/journal/ijms/special_issues/plant_salinity_tolerance)

For citation purposes, cite each article independently as indicated on the article page online and as indicated below:

LastName, A.A.; LastName, B.B.; LastName, C.C. Article Title. *Journal Name* **Year**, *Article Number*, Page Range.

ISBN 978-3-03921-026-8 (Pbk)
ISBN 978-3-03921-027-5 (PDF)

Cover image courtesy of Jose Antonio Hernández Cortés.

Contents

About the Special Issue Editor

Jose Antonio Hernández Cortés is a Senior Researcher at the Spanish National Research Council (CSIC) in Murcia (CEBAS, Spain). He has worked in CEBAS-CSIC since 1988, and received a permanent position in 2000. During the years 1992–1995 he made a stay at the ENSAT (Toulouse, France) under the supervision of Dr. Claudine Balagué, and he worked in the isolation and characterization of genes encoding for ACC oxidase in melon plants. Later, in 1997, he worked in John Innes Centre (Norwich, England) under the supervision of Prof. Phil Mullineaux on the effect of high light stress in antioxidant metabolism in peas plants. He works in the fields of Plant Physiology and Biochemistry in different subjects, such as salinity, seed biology, antioxidant metabolism, and redox signaling. Since 2008, Dr. Hernandez has been part of the Fruit Biotechnology Group of the Plant Breeding Department at CEBAS-CSIC. In this group, he has studied the response of fruit trees to abiotic (salinity and drought) and biotic stresses (plum pox virus infection) in fruit trees. In the past 2–3 years, Dr. Hernandez has been working in the physiological and biochemical characterization of peach bud dormancy, studying the evolution of sugar and starch contents, hormonal profiles, and antioxidant metabolism.

International Journal of
Molecular Sciences

MDPI

Editorial

Salinity Tolerance in Plants: Trends and Perspectives

Jose Antonio Hernández

Group of Fruit Trees Biotechnology, Dept. Plant Breeding, CEBAS-CSIC, Campus Universitario de Espinardo, 25, 30100 Murcia, Spain; jahernan@cebas.csic.es; Tel.: +34-968-396-200

Received: 30 April 2019; Accepted: 14 May 2019; Published: 15 May 2019

Salinity stress is one of the more prevailing abiotic stresses which results in significant losses in agricultural crop production, particularly in arid and semi-arid areas. According to FAO, around 800 million hectares of land are affected by salinity worldwide. Therefore, it is of vital importance to know the mechanisms of salinity tolerance in order to obtain plants with a better response to this abiotic stress. At the same time, it is necessary to achieve these objectives with sustainable agricultural practices that allow obtaining more productive crops under a future scenario of climate change.

The first symptoms from the early hours until a few days later associated with saline stress are displayed in the roots by suffering an osmotic stress associated with the accumulation of phytotoxic ions. In the long term, salinity induces ion toxicity due to a nutrient imbalance in the cytosol. In addition, salt stress is also manifested as an oxidative stress at the subcellular level, mediated by reactive oxygen species (ROS) [1]. All these responses to salinity contribute to deleterious effects on plants, although there are tolerant plants to NaCl that can implement a series of adaptations to acclimate to salinity that can help their survival. These adaptation mechanisms include morphological, physiological, biochemical, and molecular changes [1].

The majority of research on salt tolerance in plants in this Special Issue is focused on determining which genes are involved in the molecular mechanisms of tolerance. Likewise, there are also an important number of works using transgenic plants in order to get a better response to salinity.

1. Transcriptomic and Genomic Approaches

Transcriptome sequencing may provide a functional view of the plant resistance mechanisms to salt stress. Wang et al. [2] performed a transcriptome analysis of short-term acclimation (for 24 h) in the algae *Chlamydomonas reinhardtii* to salt stress (200 mM NaCl) [2]. The authors identified 10,635 unigenes as differentially expressed in *C. reinhardtii* under salt stress by RNA-seq, including 5920 that were up-regulated and 4715 that were down-regulated. Using GO (gene onthology) terms, MapMan, and KEGG (Kyoto Encyclopedia of Genes and Genomes) functional enrichment analyses, the potential mechanisms for responses to salt stress were identified [2]. These analyses reported that lipid homeostasis and the regulation of phosphatidic acid acetate levels had a key role in improving tolerance to salt stress, and use as an alternative source of energy for solving the impairment of photosynthesis and the enhancement of glycolysis metabolism [2].

By using also *C. reinhardtii* as an experimental organism, [3] evaluated the role of the basic leucine-region zipper (bZIP) transcription factors (TFs) in response to salt stress [3]. They identified, using a genome-wide analysis, 17 *C. reinhardtii* bZIP (*CrebZIP*) TFs containing typical bZIP structure, as well as the CrebZIP gene structures and their chromosomal assignment were also analysed [3]. The expression profiling of *CrebZIP* genes by qRT-PCR indicated that six *CrebZIPs* might be involved in stress response and lipid accumulation. The authors also concluded that CrebZIPs TFs may play important roles in mediating photosynthesis, as suggested by the reported reduced chlorophyll content and Fv/Fm, and the increased NPQ, carotenoid, and oil contents which could be interpreted as adaptive mechanisms to salt stress [3].

Wu et al. [4] sequenced the flax (*Linum usitatissimum* L.) transcriptome to identify differentially-expressed unigenes (DEUs) under NaCl stress [4]. After the results of the flax transcriptome were confirmed using qRT-PCR, a large-scale analysis of expressed sequence tag-derived simple sequence repeat (EST-SSRs) markers was conducted using public resources in order to understand the functions of the identified genes. The authors identified 33,774 significant DEUs (18,040 up-regulated and 15,734 down-regulated) [4]. The functional categories of the DEUs were mostly assigned as signal transduction of plant hormones, photosynthesis-antenna proteins, and biosynthesis of amino acids, which are important in flax responses to NaCl exposure [4]. They also identified a number of DEUs homologous to known plant transcription factors that regulate abiotic stress responses, such as *bZIP, HD-ZIP, NAC, MYB, GATA, CAMTA,* and *B3.* The authors also suggested an important role of the bZIP TFs in the response to salinity [4].

Grapevine (*Vitis vinifera*) is an economically important fruit crop. This fact makes the search for salt-tolerant genotypes a relevant issue. Recently, next-generation sequencing (NGS) technology based on high throughput RNA-Seq technology has been extensively used to unveil and compare the transcriptome profile under abiotic stresses [5], which provides large-scale data to identify and characterize the differentially-expressed genes DEGs. Guan et al. [6] carried out the transcriptomic sequencing of cDNA generated from both control and salt-treated grapevine leaf samples [6]. The GO and KEGG analyses of DEGs in response to salt stress suggested that many genes were involved in various defense-related biological pathways, including ROS scavenging, ion transport, heat shock proteins (HSPs), pathogenesis-related proteins (PRs), and hormone signaling. Furthermore, many DEGs encoded for TFs and essential regulatory proteins involved in signal transduction by regulating genes associated with salinity in grapevine [6]. The authors also observed that salinity negatively affected all gas exchange parameters. The analysis of antioxidant enzymes showed that at short-term (up to 60 h) salinity significantly increased the superoxide dismutase, peroxidase, catalase, and glutathione S-transferase activities in grapevine leaves, suggesting that salt stress induced an oxidative stress. Regarding the ion contents, they showed that Cl^- accumulated in the roots more than in the leaves [6], and this response can be considered as an adaptive mechanism limiting the accumulation of these ions in the canopy [1]. Finally, the authors proposed four genes as candidates as potential markers of salt stress (nonspecific lipid-transfer protein, *LTP*; proline-rich cell wall protein-like, *PRPs*; glutathione transferase-like, *GST*; photosystem II reaction center protein, with the aim of explore new approaches to applying the gene information in genetic engineering and breeding purposes [6].

Yu et al. [7] performed a genome-wide association study (GWAS) of salt-tolerance-related phenotypes in rice during the germination stage, in order to identify candidate genes related to salt tolerance [7]. In that regard, they characterized 17 genes that may contribute to salt tolerance during the seed germination stage, which contained highly associated SNPs (single-nucleotide polymorphisms)/indels whiting the coding region. Among these genes, *OSMADS31* is involved in floral organ specification as well as in plant growth and development. *OSHAK11* coded for a high-affinity K^+ uptake transporter, significantly induced by salt-stress and K^+ starvation. *AGOs* play important roles in the regulation of development and stress responses. *OsPIN* is encoded for an auxin efflux carrier protein, and is involved in the root elongation growth and lateral root formation. *Germin family proteins,* is involved in plant cell defense and diseases. Finally, *SAP* (A20/AN1 zinc finger containing proteins) and zinc-induced facilitator-like (*ZIFL*) family genes are involved in the regulation of stress signaling in plants [7].

Using the same approach (GWAS mapping), Dilnur et al. [8] revealed nine SNP-rich regions significantly associated with plant performance parameters [the relative fresh weight (RFW), the relative stem length (RSL), the relative water content (RWC)] and the comprehensive index of salt tolerance (CIST) in 215 accessions of Asiatic cotton (*Gossypium arboretum* L.) grown in the absence or the presence of 150 mM NaCl [8]. The analysis showed that most of SNPs which related positively to salt tolerance indicators (RFW, RSL, and RWC) were on chromosome 7. Moreover, most of SNPs related negatively to salt tolerance indicators (relative electric conductivity and relative methylene dioxyamphetamine content,) were on chromosome 3 [8]. In the mentioned SNP-rich region, the authors identified candidate genes possibly associated with salt tolerance in *G. arboreum*, suggesting that this information could provide essential knowledge which would be very useful to the breeder in order to produce new salt tolerant cotton cultivars [8].

2. Transgenic Strategy

The use of transgenic plants to study the response to salinity is a strategy widely addressed in numerous research groups for many years. The homolog of *More Axillary Branches 2* (*MAX2*) encodes for a key component in the strigolactones (SL) signalling pathway. The overexpression of *MAX2* from *Sapium sebiferum* (*SsMAX2*) in Arabidopsis plants significantly promoted resistance to different abiotic stresses (drought, osmotic, and salinity) [9]. The authors showed that the protein MAX2 potentially influences chlorophyll metabolism, anthocyanin biosynthesis, soluble sugars, and proline accumulation. The physiological and biochemical analyses demonstrated that SsMAX2 plays a pivotal role in the regulation of redox homeostasis via the regulation of antioxidative enzymes. In this sense, a potential interaction between SL and abscisic acid (ABA) in the adaptation to abiotic was suggested [9].

Rice nucleolin protein (OsNUC1), consisting of two isoforms, OsNUC1-L and OsNUC1-S, is a multifunctional protein involved in salt-stress tolerance. The overexpression of *OsNUC1-S* gene improved rice productivity under saline conditions, which correlated with a better behaviour of gas exchange parameters (net photosynthesis, stomatal conductance, and transpiration rate) and higher carotenoids contents [10].

Rho-like GTPases (ROPs) from plants are a subfamily of small GTP-binding proteins crucial for plant survival when subjected to abiotic stress. Miao et al. [11] described a novel *ROP* gene from banana (*MaROP5g*) whose overexpression increased the tolerance to salt stress in transgenic *Arabidopsis thaliana* plants. This response was related to minor injury in the plasmatic membrane, as has been shown by reduced lipid peroxidation and electrolyte leakage values. It was also related to an increase in the cytosolic K^+/Na^+ ratio and the Ca^{2+} concentration. In addition, the *MaROP5g* overexpression up-regulated the salt overly sensitive (SOS)-pathway genes and several genes encoding calcium-signalling pathway proteins, including calcineurin B-like (CBL) proteins, CBL-interacting protein kinases (CIPKs), and calcium-dependent protein kinases (CDPKs) [12].

Bernal-Vicente et al. [13] studied the response to salt stress of a transgenic plum line (J8-1) harbouring four copies of the cytosolic ascorbate peroxidase gene (*cytapx*) from pea. The authors reported that this plum line was more tolerant to salinity stress in terms of plant growth, chlorophyll contents, chlorophyll fluorescence parameters, and root water contents. In addition, they proposed a connection between the salicylic acid and cyanogenic glucoside (CNglcs) biosynthetic pathways under salt-stress conditions [13].

The overexpression of the cucumber *TGase* (transglutaminase) gene from *Cucumis sativus* L. (*CsTGase*) in tobacco effectively ameliorated salt-induced photoinhibition by increasing the levels of polyamines (PAs) in the chloroplast as well as gas exchange and chlorophyll fluorescence parameters, along with greater abundance of D1 and D2 proteins under saline conditions [14]. In addition, *TGase* overexpression resulted in chloroplasts that showed more quantity and size of grana compared with wild type plants, suggesting a role of *TGase* in the chloroplast development. Thus, overexpression of *TGase* may be an effective strategy for enhancing resistance to salt stress in crops especially sensitive for agronomic production [14].

3. Physiological and Biochemical Mechanisms

Zhang et al. [15] presented a review about the physiological and molecular responses of *Populus* sp. to salinity. Poplars are used as a model species to study physiological and molecular responses of trees to NaCl stress, taken into account that salinity is one of the limiting factors of afforestation programs. The authors compared the response of salt-tolerant and salt-sensitive *Populus* species in terms of salinity injury (plant growth, photosynthesis) and primary salt-tolerance mechanisms (ion homeostasis, accumulation of soluble osmolytes), with reactive oxygen species (ROS) and reactive nitrogen species (RNS) metabolism and signaling networks induced by salinity, and they identified candidate genes for improving salt tolerance in the *Populus* sp.

4. Biostimulants and Salt-Stress Response

4.1. Biostimulants

The use of biostimulants is another strategy addressed to overcome the negative effects of salinity. Zhan et al. [16] presented an excellent review about the effect of melatonin in the plant response to salinity. They described the effects of exogenous melatonin in the modulation of the expression of genes involved in melatonin metabolism, the increase of the transcript levels of different stress-responsive genes, and transcription factors involved in the ROS scavenging and of the genes responsible for the maintenance of ion homeostasis. Melatonin also regulates hormone metabolism by up-regulation of gibberellic acid (GA) biosynthesis and abscisic acid (ABA) catabolism genes [16]. Finally, the authors described the identification of a plant melatonin receptor in Arabidopsis [17]. These finding opens new perspectives of research on the role of melatonin in response to abiotic stresses in general, and to salinity in particular.

Related to the previous revision, Zhao et al. [18] described that the treatment of *Brassica napus* L. seedlings with melatonin and NO-releasing compounds such as sodium nitroprusside (SNP), diethylamine NONOate (NONOate), and S-nitrosoglutathione (GSNO) produced synergistic effects that counteracted the seedling growth inhibition induced by NaCl exposure. At the same time, such treatments re-established the redox and ion homeostasis, by decreasing the ROS and lipid peroxidation accumulation as well as the Na^+/K^+ ratio. The addition of PTIO (a NO-scavenger) impaired the coupled response of melatonin and SNP, suggesting that NO is required to potentiate the effects of melatonin in protecting plants from salt stress [18]. Chen et al. (2018) [19] studied the salt-stress response of *Apocynum venetum* L plants, used in traditional Chinese medicine. The authors studied the changes in photosynthetic pigments, osmolytes, lipid peroxidation, some antioxidant enzymes, and ascorbic acid. By using UFLC-QTRAP-MS/MS technology a total of 43 bioactive constituents, including amino acids, nucleosides, organic acids, and flavonoids were successfully identified to change in response to salt stress. They applied a multivariate statistical analysis to evaluate the quality of *Apocynum venetum* L plants grown under saline conditions [19].

4.2. Plant Hormones

The role of root ABA (including ABA translocation from root to leaf) in the protection of photosystems and photosynthesis against salt stress was studied in Jerusalem artichoke [20]. In this study, the pretreatment of Jerusalem artichoke plants with sodium tungstate (a specific ABA synthesis inhibitor) followed by exposure to salt stress (150 mM NaCl) induced a drastic overaccumulation of Na^+ in leaves. Moreover, a decline in net photosynthesis, ØPSII (actual photochemical efficiency of photosystem II) and Fv/Fm (the maximal photosystem II (PSII) quantum yield) was produced, indicating photoinhibition of PSII, along with the establishment of an oxidative stress due to an increase in H_2O_2 and lipid peroxidation levels. These results suggest that root ABA can participate in protecting PSII against photoinhibition in Jerusalem artichoke under salt stress, likely via a reduction of Na^+ toxicity. In that regard, it has been reported that Na^+ can irreversibly inactivate PSII and PSI by inducing secondary oxidative injury or through direct damage on photosynthetic proteins [21,22]. This finding was corroborated by immunoblotting analysis, where a decline in the PSII reaction center protein (PsbA) abundance was observed [20].

4.3. Protein Kinases, ROS, and Ion Homeostasis

Szymanska et al. [23] and Zhang et al. [24] described the involvement of different protein kinases families in the regulation of plant adaptation to salt stress [23,24]. Specifically, Szymanska et al. [23]

showed that the SNF1-related protein kinases (SnRK2.4 and SnRK2.10) have a role in the modulation of ROS homeostasis in response to salinity by regulating the expression of several genes related to ROS generation and scavenging in Arabidopsis.

Zhang et al. [24] described the importance of CDPKs (Ca^{2+}-dependent protein kinases) in the adaptation of Arabidopsis to salt stress. In that regard, they reported that the *CPK12*-RNA interference (RNAi) mutant was more sensitive to salinity than the wild-type plants in terms of seedling growth. This response seemed to be related to the accumulation of phytotoxic ions in the roots as well as the overgeneration of H_2O_2 in the *CPK12*-RNAi mutants [24].

Regarding the effect of salt stress in ion homeostasis, Ali et al. [25] provided a brief overview of the role of the high-affinity potassium-type transporter 1 (HKT1) and their importance in different plant species under salt stress. HKT1-type transporters play a crucial role in Na^+ homeostasis, being of pivotal importance to maintain an optimal K^+/Na^+ balance in the cytoplasm in response to salt stress for plant survival. The authors described the role of HKT1-type transporters and their functional differences in glycophytes and halophytes [25].

Luo et al. [26] showed that Arabidopsis plants overexpressing a SKn-type dehydrin from *Capsicum annuum* L. (*CaDHN5*) resulted an increased tolerance to salt and osmotic stress, suggesting an important role for *CaDHN5* in response to the mentioned abiotic stresses [26]. In addition, using VIGS (virus-induced gene silencing) technique, the authors reported that knockdown of the *CaDHN5* gen suppressed the expression of manganese superoxide dismutase (*MnSOD*) and peroxidase (*POD*) genes in transformed pepper plants [26]. These changes caused a higher oxidative stress in the VIGS lines than in control pepper plants under NaCl or osmotic stress conditions, as observed by the data of some stress-oxidative parameters (superoxide accumulation, lipid peroxidation, and electrolyte leakage), chlorophyll levels, and the rate of water loss. The results demonstrated an important role for the *CaDHN5* gene in the tolerance of plants to salt and osmotic stresses as well as in the salt and osmotic stress signalling pathways [26]. The results also indicated that *CaDHN5* positively regulates the expression of the *MnSOD* and *POD* genes, but also other stress-related genes, including *AtSOD1* (encoding a H^+/Na^+ plasma membrane antiporter), *AtDREB2A* (a transcription factor in the ABA signalling pathway), and *AtRSA1* and *AtRITF1* genes that regulate the transcription of several ROS scavenging-related genes and the *AtSOS1* gene [26].

5. Proteomic Approach

The isobaric tags for relative and absolute quantitation (iTRAQ)-based proteomic technique was used to identify the differentially-expressed proteins in leaves of two rice genotypes that differ in their tolerance to salt stress [27]. The iTRAQ protein profiling identified in both rice genotypes revealed that the differentially-expressed proteins were mainly involved in the regulation of salt-stress responses, in oxidation-reduction responses, in photosynthesis, and in carbohydrate metabolism. Regarding their subcellular localization, most of them were predicted to localize in cytoplasm and chloroplasts (67.2% of the total up-regulated proteins) [27].

6. Conclusions and Outlook

Salinity is one of the major factors that limits geographical distribution of plants and adversely affects crop productivity and quality worldwide. Salinization affects about 30% of the irrigated land of the world, increasing this area approximately 1–2% per year due to salt-affected land surfaces (FAO, 2014). In Europe, about 3 million hectares of the land are affected by salinization. Unfortunately, this situation will worsen in a context of climate change, where there will be an overall increase in temperature and a decrease in average annual rainfall worldwide.

Although an important part of the studies on response to salinity are carried out with Arabidopsis plants, nowadays the use of other species with agronomic interest is also remarkable, including woody plants.

Studies on salinity tolerance have focused on different points of view: agronomic, physiological, biochemical, and molecular. However, in recent years, the number of works that address tolerance to salinity from a molecular point of view has increased considerably, in order to search for candidate genes that may be useful to the search for resistant genotypes. The identification of the candidate genes would provide valuable information about the molecular and genetic mechanisms involved in the salt tolerance response, and it would also supply important resources to the breeding programs in order to look for salt tolerance in crop plants. Therefore, obtaining salt-tolerant species is one of the goals for breeders, and probably, the use of transformed plants could improve the salt response in crop plants. In this way, transformed plants with enhanced antioxidant defenses have been obtained in different laboratories, and, in most cases, these plants displayed an improved salt-tolerance response. The overexpression of certain proteins can afford protection against salt stress in plants. In this Special Issue, the author shows that the overexpression of certain transgenes improved the response to salinity in plants in terms of photosynthesis rate, improved the gas exchange parameters, and increased photosynthetic pigments, antioxidant mechanisms, and accumulation of anthocyanins, as well as improved ion homeostasis responses, up-regulation of ABA biosynthesis genes, and plant hormone signaling. However, the use of transgenic plants for agricultural purposes still has a high level of rejection by consumers, for example in the European Union, motivated by its agricultural policy.

Another feasible strategy to mitigate salinity impacts on crop production would be breeding salt tolerant cultivars for the production of new varieties which can thrive in more extreme environmental conditions. In this sense, crop wild relatives may contain genes of potential value for plant salinity tolerance. Despite the vast pool of resources that exists, much of the crop germplasm richness found in gene banks is underutilized.

In addition, cultivation of halophytic plants at the same time or prior to the cultivation of crop plants (intercropping) would allow the desalination of the soil favoring crop yield and/or, alternatively, the use of saline irrigation water. Complementarily, the use of biostimulants, such as antioxidant compounds, melatonin, plant hormones or NO-releasing compounds can improve the response of plants to salinity.

The proteomic approach to study the response to salt stress can provide relevant information in order to know the physiological and biochemical processes affected by salinity. This information can also support the breeding programs to attempt selection of salt-tolerant plants.

Finally, the development of plant metabolomics techniques can supply relevant information about the effect of salt stress on cell metabolism. In addition, these techniques may allow for the identification of new metabolites that can be used as markers to better understand the salt tolerance response and help breeders select new tolerant species.

Conflicts of Interest: The authors declare no conflict of interest.

References

1. Acosta-Motos, J.; Ortuño, M.; Bernal-Vicente, A.; Diaz-Vivancos, P.; Sanchez-Blanco, M.; Hernandez, J. Plant Responses to Salt Stress: Adaptive Mechanisms. *Agronomy* **2017**, *7*, 18. [CrossRef]
2. Wang, N.; Qian, Z.; Luo, M.; Fan, S.; Zhang, X.; Zhang, L. Identification of salt stress responding genes using transcriptome analysis in green alga *Chlamydomonas reinhardtii*. *Int. J. Mol. Sci.* **2018**, *19*, 3359. [CrossRef] [PubMed]
3. Ji, C.; Mao, X.; Hao, J.; Wang, X.; Xue, J.; Cui, H.; Li, R. Analysis of bZIP transcription factor family and their expressions under salt stress in *Chlamydomonas reinhardtii*. *Int. J. Mol. Sci.* **2018**, *19*, 2800. [CrossRef] [PubMed]
4. Wu, J.; Zhao, Q.; Wu, G.; Yuan, H.; Ma, Y.; Lin, H.; Pan, L.; Li, S.; Sun, D. Comprehensive analysis of differentially expressed unigenes under nacl stress in flax (*Linum usitatissimum* L.) using RNA-Seq. *Int. J. Mol. Sci.* **2019**, *20*, 369. [CrossRef]

5. Haider, M.S.; Kurjogi, M.M.; Khalil-Ur-Rehman, M.; Fiaz, M.; Pervaiz, T.; Jiu, S.; Haifeng, J.; Chen, W.; Fang, J. Grapevine immune signaling network in response to drought stress as revealed by transcriptomic analysis. *Plant Physiol. Biochem.* **2017**, *121*, 187–195. [CrossRef] [PubMed]

6. Guan, L.; Haider, M.S.; Khan, N.; Nasim, M.; Jiu, S.; Fiaz, M.; Zhu, X.; Zhang, K.; Fang, J. Transcriptome sequence analysis elaborates a complex defensive mechanism of grapevine (*Vitis vinifera* L.) in response to salt stress. *Int. J. Mol. Sci.* **2018**, *19*, 4019. [CrossRef]

7. Yu, J.; Zhao, W.; Tong, W.; He, Q.; Yoon, M.Y.; Li, F.P.; Choi, B.; Heo, E.B.; Kim, K.W.; Park, Y.J. A Genome-wide association study reveals candidate genes related to salt tolerance in rice (*Oryza sativa*) at the germination stage. *Int. J. Mol. Sci.* **2018**, *19*, 3145. [CrossRef]

8. Dilnur, T.; Peng, Z.; Pan, Z.; Palanga, K.K.; Jia, Y.; Gong, W.; Du, X. Association analysis of salt tolerance in Asiatic cotton (*Gossypium arboretum*) with SNP markers. *Int. J. Mol. Sci.* **2019**, *20*, 2168. [CrossRef]

9. Wang, Q.; Ni, J.; Shah, F.; Liu, W.; Wang, D.; Yao, Y.; Hu, H.; Huang, S.; Hou, J.; Fu, S.; et al. Overexpression of the Stress-Inducible SsMAX2 Promotes Drought and Salt Resistance via the Regulation of Redox Homeostasis in Arabidopsis. *Int. J. Mol. Sci.* **2019**, *20*, 837. [CrossRef]

10. Boonchai, C.; Udomchalothorn, T.; Sripinyowanich, S.; Comai, L.; Buaboocha, T.; Chadchawan, S. Rice overexpressing OsNUC1-S reveals differential gene expression leading to yield loss reduction after salt stress at the booting stage. *Int. J. Mol. Sci.* **2018**, *19*, 3936. [CrossRef]

11. Miao, H.; Sun, P.; Liu, J.; Wang, J.; Xu, B.; Jin, Z. Overexpression of a novel ROP gene from the banana (MaROP5g) confers increased salt stress tolerance. *Int. J. Mol. Sci.* **2018**, *19*, 3108. [CrossRef] [PubMed]

12. Miranda, R.S.; Alvarez-Pizarro, J.C.; Costa, J.H.; Paula, S.O.; Pirsco, J.T.; Gomes-Filho, E. Putative role of glutamine in the activation of CBL/CIPK signaling pathways during salt stress in sorghum. *Plant Signal. Behav.* **2017**, *12*, e1361075. [CrossRef]

13. Bernal-Vicente, A.; Cantabella, D.; Petri, C.; Hernández, J.A.; Diaz-Vivancos, P. The Salt-Stress Response of the Transgenic Plum Line J8-1 and Its Interaction with the Salicylic Acid Biosynthetic Pathway from Mandelonitrile. *Int. J. Mol. Sci.* **2018**, *19*, 3519. [CrossRef]

14. Zhong, M.; Wang, Y.; Zhang, Y.; Shu, S.; Sun, J.; Guo, S. Overexpression of transglutaminase from cucumber in tobacco increases salt tolerance through regulation of photosynthesis. *Int. J. Mol. Sci.* **2019**, *20*, 894. [CrossRef] [PubMed]

15. Zhang, Y.; Liu, L.; Chen, B.; Qin, Z.; Xiao, Y.; Zhang, Y.; Yao, R.; Liu, H.; Yang, H. Progress in understanding the physiological and molecular responses of *Populus* to salt stress. *Int. J. Mol. Sci.* **2019**, *20*, 1312. [CrossRef]

16. Zhan, H.; Nie, X.; Zhang, T.; Li, S.; Wang, X.; Du, X.; Tong, W.; Song, W. Melatonin: A small molecule but important for salt stress tolerance in plants. *Int. J. Mol. Sci.* **2019**, *20*, 709. [CrossRef]

17. Wei, J.; Li, D.-X.; Zhang, J.-R.; Shan, C.; Rengel, Z.; Song, Z.-B.; Chen, Q. Phytomelatonin receptor PMTR1-mediated signaling regulates stomatal closure in Arabidopsis thaliana. *J. Pineal Res.* **2018**, *65*, e12500. [CrossRef]

18. Zhao, G.; Zhao, Y.; Yu, X.; Kiprotich, F.; Han, H.; Guan, R.; Wang, R.; Shen, W. Nitric oxide is required for melatonin-enhanced tolerance against salinity stress in rapeseed (*Brassica napus* L.) seedlings. *Int. J. Mol. Sci.* **2018**, *19*, 1912. [CrossRef]

19. Chen, C.; Wang, C.; Liu, Z.; Liu, X.; Zou, L.; Shi, J.; Chen, S.; Chen, J.; Tan, M. Variations in physiology and multiple bioactive constituents under salt stress provide insight into the quality evaluation of Apocyni Veneti Folium. *Int. J. Mol. Sci.* **2018**, *19*, 3042. [CrossRef]

20. Yan, K.; Bian, T.; He, W.; Han, G.; Lv, M.; Guo, M.; Lu, M. Root abscisic acid contributes to defending photoinibition in jerusalem artichoke (*Helianthus tuberosus* L.) under salt stress. *Int. J. Mol. Sci.* **2018**, *19*, 3934. [CrossRef] [PubMed]

21. Murata, N.; Takahashi, S.; Nishiyama, Y.; Allakhverdiev, S.I. Photoinhibition of photosystem II under environmental stress. *Biochim. Biophys. Acta* **2007**, *1767*, 414–421. [CrossRef] [PubMed]

22. Yang, C.; Zhang, Z.S.; Gao, H.Y.; Fan, X.L.; Liu, M.J.; Li, X.D. The mechanism by which NaCl treatment alleviates PSI photoinhibition under chilling-light treatment. *J. Photochem. Photobiol. B Biol.* **2014**, *140*, 286–291. [CrossRef]

23. Szymanska, K.P.; Polkowska-Kowalczyk, L.; Lichocka, M.; Maszkowska, J.; Dobrowolska, G. SNF1-related protein kinases SnRK2.4 and SnRK2.10 modulate ROS homeostasis in plant response to salt stress. *Int. J. Mol. Sci.* **2019**, *20*, 143. [CrossRef] [PubMed]

24. Zhang, H.; Zhang, Y.; Deng, C.; Deng, S.; Li, N.; Zhao, C.; Zhao, R.; Liang, S.; Chen, S. The Arabidopsis Ca2+-dependent protein kinase CPK12 is involved in plant response to salt stress. *Int. J. Mol. Sci.* **2018**, *19*, 4062. [CrossRef]
25. Ali, A.; Maggio, A.; Bressan, R.A.; Yun, D.J. Role and functional differences of HKT1-type transporters in plants under salt stress. *Int. J. Mol. Sci.* **2019**, *20*, 1059. [CrossRef]
26. Luo, D.; Hou, X.; Zhang, Y.; Meng, Y.; Zhang, H.; Liu, S.; Wang, X.; Chen, R. *CaDHN5*, a dehydrin gene from pepper, plays an important role in salt and osmotic stress responses. *Int. J. Mol. Sci.* **2019**, *20*, 1989. [CrossRef]
27. Hussain, S.; Zhu, C.; Bai, Z.; Huang, J.; Zhu, L.; Cao, X.; Nanda, S.; Hussain, S.; Riaz, A.; Liang, Q.; et al. iTRAQ-based protein profiling and biochemical analysis of two contrasting rice genotypes revealed their differential responses to salt stress. *Int. J. Mol. Sci.* **2019**, *20*, 547. [CrossRef]

International Journal of
Molecular Sciences

MDPI

Article

Identification of Salt Stress Responding Genes Using Transcriptome Analysis in Green Alga *Chlamydomonas reinhardtii*

Ning Wang [†], Zhixin Qian [†], Manwei Luo, Shoujin Fan, Xuejie Zhang and Luoyan Zhang *

Key Lab of Plant Stress Research, College of Life Science, Shandong Normal University,
No. 88 Wenhuadong Road, Jinan 250014, China; wangning_sdnu@163.com (N.W.);
qianzhixin_sdnu@163.com (Z.Q.); luomanwei_sdnu@163.com (M.L.);
fansj@sdnu.edu.cn (S.F.); zxjpublic@sohu.com (X.Z.)
* Correspondence: zhangluoyan@sdnu.edu.cn; Tel.: +86-531-86180718
† These authors contributed equally to this work.

Received: 16 September 2018; Accepted: 24 October 2018; Published: 26 October 2018

Abstract: Salinity is one of the most important abiotic stresses threatening plant growth and agricultural productivity worldwide. In green alga *Chlamydomonas reinhardtii*, physiological evidence indicates that saline stress increases intracellular peroxide levels and inhibits photosynthetic-electron flow. However, understanding the genetic underpinnings of salt-responding traits in plantae remains a daunting challenge. In this study, the transcriptome analysis of short-term acclimation to salt stress (200 mM NaCl for 24 h) was performed in *C. reinhardtii*. A total of 10,635 unigenes were identified as being differently expressed by RNA-seq, including 5920 up- and 4715 down-regulated unigenes. A series of molecular cues were screened for salt stress response, including maintaining the lipid homeostasis by regulating phosphatidic acid, acetate being used as an alternative source of energy for solving impairment of photosynthesis, and enhancement of glycolysis metabolism to decrease the carbohydrate accumulation in cells. Our results may help understand the molecular and genetic underpinnings of salt stress responses in green alga *C. reinhardtii*.

Keywords: *Chlamydomonas reinhardtii*; salt stress; transcriptome analysis; impairment of photosynthesis; underpinnings of salt stress responses

---

1. Introduction

Salinity is one of the most important abiotic stresses threatening agricultural productivity worldwide. Although plants have gradually evolved a series of adaptive molecular, physiology and biochemistry processes to respond to salinity stress, it could threaten 30% of cultivable soils by 2050 [1,2]. Understanding the molecular machineries of salt stress response in model plants of basal taxa, such as green algae, may contribute to finding the evolutionary cues of abiotic stress response in plants and developing salt-resistant crops with additional salt-responding traits [2–9].

Salt stress causes diverse impacts on plant growth by disturbing the osmotic/ionic balance and eliciting Na^+ toxicity [9,10]. Under aquatic saline stress, a series of physical and biochemical processes are recruited by algae to respond to the damage caused by osmotic and ionic stresses, such as photosynthesis inhibition, macromolecular compound synthesis and homeostasis adjustment [6,10–14]. It has been reported that salt stress leads to decreased photosynthetic efficiency [15,16] which influences chlorophyll content in plant leaves [17,18]. In green algae, salt stress remarkably influences the structure and functions of the photosynthetic apparatus in *Scenedesmus obliquus* [19] and reduces the maximum quantum yield of photosystem II (PSII) in *Dunaliella salina* [20]. In alga *Botryococcus braunii*, metabolism of lutein was significantly enhanced under stress conditions [12].

Chlamydomonas reinhardtii is a free-living freshwater alga with unicellular vegetative cell. Previous studies exposed the *C. reinhardtii* strain 21 gr and CC-503 to salt stress and demonstrated the physiological and metabolic processes impacted by ionic toxicity and osmotic stress caused by salt damage [6–8,16,21]. Vega [22] demonstrated that 200 mM NaCl in the culture medium was highly toxic for *C. reinhardtii* productivity. The addition of NaCl immediately blocked the photosynthetic activity of the alga which partially recovered, after 1 h of treatment, remaining high during the following 24 h. However, after 24 h treatment with NaCl 200 mM, the intracellular catalase activity of the alga reached a 20-fold higher level than in the control cells. The physiological data indicate that saline stress induces in *C. reinhardtii* an increase of intracellular peroxide, which parallels a significant inhibition of the photosynthetic-electron flow. However, the related machineries of up-stream regulating and the triggering of appropriate cellular and physiological responses to cope with stress circumstances are still largely unknown.

Transcriptome sequencing is an effective strategy for detecting potential participants of stress response on a genome-wide scale. Hundreds of studies about salt stress responses in model plant *Arabidopsis thaliana* [23–26], crops *Oryza sativa* [23,27] and *Glycine max* [28], and in some halophytes (plants able to complete their life cycles under saline environments) have been widely conducted using sequencing technologies [7,29–41]. The integrations of genes' spatio-temporal expression patterns and responding traits have helped to identify a large number of salt stress-related differentially expressed genes (DEGs) and mechanisms.

Keeping this in mind, the work presented here was carried out to explore the saline stress-responding mechanisms of *C. reinhardtii* by transcriptome sequencing of strains GY-D55 wild type. The aim of this study was to identify dys-regulated genes in *C. reinhardtii* cells under salt stress by RNA-seq, screen physiological and biochemical cues by gene ontology (GO) terms and MapMan functional enrichment analyses, and investigate the physiological adaptions and cellular regulatory networks for salt stress responding.

2. Results

2.1. Transcriptome Profiling of C. reinhardtii

After sequencing with the Illumina HiSeq X platform, a total of 56,438,218, 72,853,712, 47,551,786, 56,962,722, 52,926,804 and 55,998,748 high-quality pair-end reads were obtained from three control and three salt stress treated samples of *C. reinhardtii* (Table 1), respectively. *De novo* transcriptome assembly generated 91,242 unigenes, with an average length of 2691 nt and N50 of 4554. On average, 90.66% of the reads from six samples were mapped to the reference genome (Table 1). The assembled transcriptome information of *C. reinhardtii* is shown in Supplementary Figure S1.

Table 1. Summary of mapping transcriptome reads to reference sequence.

Sample Name	Sample Description	Total Reads	Total Mapped	Ratio of Mapped Reads
C_0_1	Control replication 1	56,438,218	51,454,456	91.17%
C_0_2	Control replication 2	72,853,712	66,008,290	90.60%
C_0_3	Control replication 3	47,551,786	43,268,544	90.99%
S_200_1	Salt stress replication 1	56,962,722	51,633,614	90.64%
S_200_2	Salt stress replication 2	52,926,804	47,815,814	90.34%
S_200_3	Salt stress replication 3	55,998,748	50,507,824	90.19%

2.2. Functional Annotations of Unigenes

Similarity searches were performed to annotate unigenes against different databases using BLASTX. For *C. reinhardtii*, 65,679 (71.98%) unigenes were annotated in at least one database (Figure 1C and Supplementary Figure S1). A total of 52,884 (57.96%) and 58,062 (63.63%) unigenes showed similarity to sequences in NR and PFAM databases with an E-value threshold of 1×10^{-5}.

About 58,651 (64.28%) unigenes were annotated in the GO database by Blast2GO v2.5 with an E-value cutoff of 1×10^{-6} (Figure 1C and Supplementary Figure S1). Unigenes of the *C. reinhardtii* were assigned to *C. reinhardtii* and *A. thaliana* gene IDs for GO annotation mapping and TFs/PKs perdition. By sequence alignment, a total of 48,158 unigenes were aligned to *C. reinhardtii* PLAZA genome genes. A total of 54,509 unigenes were assigned to TAIR10 locus IDs by BLASTP with an E-value cutoff of 1×10^{-5} and classified into GO categories for GO analysis (Supplementary Table S1).

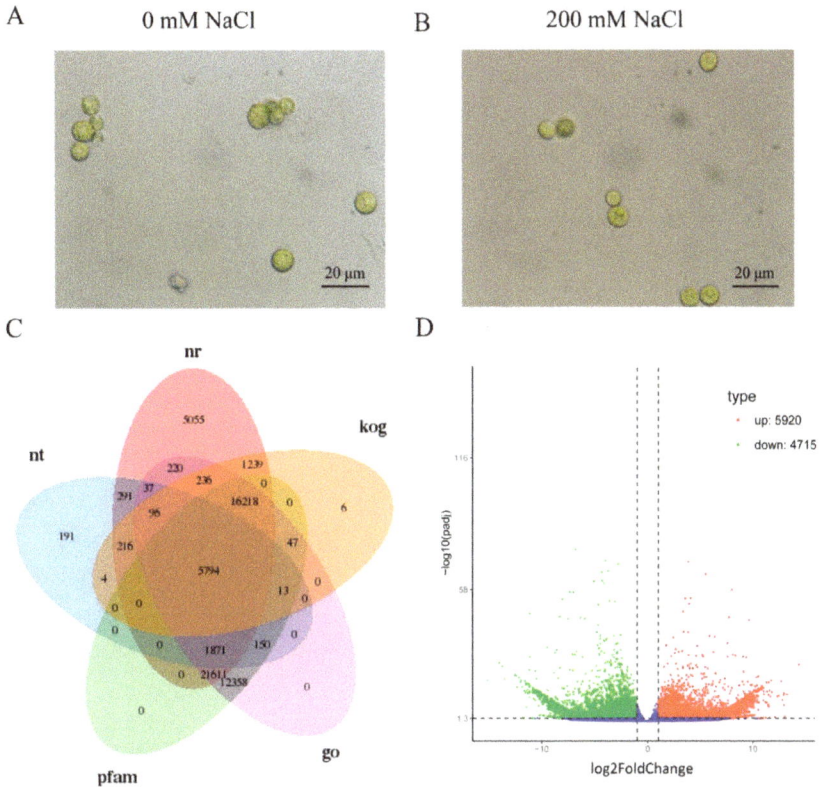

Figure 1. (**A**) The morphology of *C. reinhardtii* cells without addition of NaCl. (**B**) The morphology of *C. reinhardtii* cells under 200 mM NaCl treatment. (**C**) Venn diagram of functional annotations of unigenes in nt (NCBI non-redundant protein sequences), nr (NCBI non-redundant protein sequences), kog (Clusters of Orthologous Groups of proteins), go (Gene Ontology) and pfam (Protein family) databases. (**D**) Expression patterns of differentially expressed genes (DEGs) identified between 200 mM NaCl treated and control. S_200 indicated cells under 200 mM NaCl stressed condition for 24 h; C_0 indicated cells cultured under control condition. Red and green dots represent DEGs, blue dots indicate genes that were not differentially expressed. In total, 10,635 unigenes were identified as DEGs (padj < 0.05) between S_200 and C_0, including 5920 upregulated genes and 4715 downregulated genes.

2.3. Differently Expressed Genes (DEGs) Calculation

To evaluate the relative level of gene expression in *C. reinhardtii* under control or salt stress treatment, the FPKM values were calculated based on the uniquely mapped reads. The FPKM distributions of unigenes in six samples are shown in Supplementary Figure S2. The FPKM value for genes detected in six samples ranged from 0 to 40,486.05, with mean value of 7.08. By comparative analysis, a part of the genes was observed to be differently expressed in 200 Mm NaCl treated samples: 5920 unigenes were calculated as up-regulated in salt treated samples and 4715 filtered as down-regulated genes with the cutoff of padj < 0.05 and | log2(foldchange) | > 1 (Supplementary Table S2).

The most significantly dysregulated 30 genes are recorded in Table 2. The most significantly upregulated unigenes included RNA recognition motif containing gene Cluster-2749.47186 (log2FoldChange [L_2fc] = 3.894), "transcription, DNA-templated" participating gene Cluster-2749.64181 (L_2fc = 5.573) and "potassium ion transport" gene Cluster-2749.61362 (L_2fc = 8.112) (Table 2 and Supplementary Table S2). Downregulated unigenes, included "chlorophyll metabolic process" related gene Cluster-2749.44503 (L_2fc = −8.623) with the lowest *p*-value, "proteolysis" related gene Cluster-2749.61923 (L_2fc = −6.748) and "regulation of transcription, DNA-templated" participating gene Cluster-2749.45379 (L_2fc = −3.663).

Table 2. Top30 dysregulated genes in *C. reinhardtii* under 200 mM NaCl treated and control conditions.

Gene_ID	L_2fc	*p*val	BP Description
		Up-regulated	
Cluster-2749.47186	3.894	3.77×10^{-75}	
Cluster-2749.64181	5.573	1.55×10^{-69}	transcription, DNA-templated
Cluster-2749.61362	8.112	1.95×10^{-62}	potassium ion transport
Cluster-2749.33332	4.129	1.19×10^{-58}	signal transduction
Cluster-2749.48242	3.610	1.64×10^{-58}	
Cluster-2749.21356	3.975	4.34×10^{-56}	
Cluster-2749.37168	3.413	1.00×10^{-52}	
Cluster-2749.23874	7.849	5.01×10^{-50}	lipid metabolic process
Cluster-2749.57700	9.756	9.76×10^{-49}	iron-sulfur cluster assembly
Cluster-2749.59287	3.459	1.42×10^{-43}	cell adhesion
Cluster-2749.53252	3.877	2.36×10^{-43}	pathogenesis
Cluster-2749.49912	5.957	1.07×10^{-41}	lipoprotein metabolic process
Cluster-2749.84953	6.468	2.29×10^{-41}	
Cluster-2749.82821	2.504	5.20×10^{-41}	regulation of protein kinase activity
Cluster-2749.3203	7.706	1.83×10^{-38}	
		Down-regulated	
Cluster-2749.44503	−8.623	4.01×10^{-178}	chlorophyll metabolic process
Cluster-2749.61923	−6.748	6.07×10^{-81}	proteolysis
Cluster-2749.38883	−3.906	7.54×10^{-76}	
Cluster-2749.44595	−2.699	3.50×10^{-74}	metabolic process
Cluster-2749.45379	−3.663	6.53×10^{-71}	regulation of transcription, DNA-templated
Cluster-2749.49076	−4.268	2.29×10^{-70}	chlorophyll biosynthetic process
Cluster-2749.44117	−4.239	1.30×10^{-66}	oxidation-reduction process
Cluster-2749.42573	−5.023	3.04×10^{-66}	protein glycosylation
Cluster-2749.32226	−4.043	2.67×10^{-65}	proteolysis
Cluster-2749.45636	−7.283	1.98×10^{-61}	
Cluster-2749.44732	−6.934	2.08×10^{-61}	
Cluster-2749.49721	−7.951	3.18×10^{-58}	
Cluster-2749.65261	−3.524	1.91×10^{-57}	
Cluster-2749.36258	−2.996	5.32×10^{-57}	
Cluster-2749.43872	−4.589	1.10×10^{-55}	cell adhesion

Note: Top30 dysregulated genes with the lowest *p*-value (*p*val) are represented; L_2fc indicates the log2FoldChange of genes differently expressed in 200 mM NaCl treated samples and control samples; BP Description means descriptions of genes' potential participating biological process predicted by sequence similarity search.

2.4. GO Enrichment of DEGs

For uncovering the differences of molecular mechanisms of *C. reinhardtii* under salt stress, the DEGs were then characterized with GO databases. A total of 353 biological processes (BP) terms were enriched by the 5920 up-regulated unigenes, like "oxidation-reduction process" (GO:0055114), "response to cadmium ion" (GO:0046686) and "response to salt stress" (GO:0009651) (Table 3; Supplementary Table S3). The 4715 down-regulated genes were calculated enriched in 313 BP terms, as "photosynthesis, light harvesting in photosystem I" (GO:0009768), "chlorophyll biosynthetic process" (GO:0015995) and "isoleucine biosynthetic process" (GO:0009097) (Table 3; Supplementary Table S3).

Table 3. Top30 biological processes enriched by the up- and down-regulated genes.

GO ID	GO Term	Annotated Gene Number	Enriched Gene Number	*p*-Value
	Up-Regulated			
GO:0008150	biological process	33682	2820	1.00×10^{-30}
GO:0055114	oxidation-reduction process	3653	385	2.90×10^{-27}
GO:0046686	response to cadmium ion	1317	159	3.40×10^{-18}
GO:0042542	response to hydrogen peroxide	189	41	1.10×10^{-15}
GO:0009408	response to heat	717	122	1.40×10^{-15}
GO:0051259	protein oligomerization	109	25	7.50×10^{-12}
GO:0010090	trichome morphogenesis	131	26	4.60×10^{-10}
GO:0009414	response to water deprivation	668	79	6.70×10^{-10}
GO:0009651	response to salt stress	1488	143	3.90×10^{-09}
GO:0043335	protein unfolding	39	14	1.80×10^{-08}
GO:0016036	cellular response to phosphate starvation	262	40	2.40×10^{-08}
GO:0010030	positive regulation of seed germination	85	20	6.50×10^{-08}
GO:0030866	cortical actin cytoskeleton organization	31	12	7.20×10^{-08}
GO:0016477	cell migration	31	12	7.20×10^{-08}
GO:0045010	actin nucleation	31	12	7.20×10^{-08}
	Down-Regulated			
GO:0008150	biological process	33682	2018	1.00×10^{-30}
GO:0009768	photosynthesis, light harvesting in photosystem I	87	46	1.00×10^{-30}
GO:0009645	response to low light intensity stimulus	72	37	1.00×10^{-30}
GO:0015995	chlorophyll biosynthetic process	242	54	4.40×10^{-29}
GO:0009644	response to high light intensity	393	71	6.70×10^{-22}
GO:0006412	translation	1779	179	3.80×10^{-16}
GO:0009409	response to cold	978	103	5.30×10^{-16}
GO:0009269	response to desiccation	41	18	1.00×10^{-14}
GO:0009769	photosynthesis, light harvesting in photosystem II	36	17	1.30×10^{-14}
GO:0010218	response to far red light	101	25	8.70×10^{-14}
GO:0006364	rRNA processing	742	89	2.10×10^{-12}
GO:0010114	response to red light	159	28	5.90×10^{-11}
GO:0015979	photosynthesis	853	137	2.40×10^{-10}
GO:0009097	isoleucine biosynthetic process	53	16	2.60×10^{-10}
GO:0009099	valine biosynthetic process	43	14	1.10×10^{-09}

2.5. MapMan Enrichment of DEGs

A more specific comparison of metabolic and regulatory pathways was conducted using MapMan. A total of 5920 up- and 4715 down-regulated genes were assigned to 1334 and 1050 homologs in *Arabidopsis thaliana*, respectively. Consequently, these uniquely expressed genes were mapped to 797 pathways by MapMan, of which, 22 pathways were filtered enriched by the dysregulated genes with the cutoff *p*-value < 0.05 (Figure 3A; Supplementary Table S4). The expression of genes implicated in "TCA/org. transformation.TCA", "Tetrapyrrole synthesis", "Starch" and "Sucrose" were over-expressed in *C. reinhardtii*, while those genes involved in "PS.lightreaction", "PS.lightreaction.photosystem I" and "PS.lightreaction.photosystem I.LHC-I" were down-regulated in *C. reinhardtii* during salt stress responding (Figure 3A).

2.6. KEGG Enrichment of DEGs

To gain a deeper insight into the regulation of photosynthesis underlying salt stress response, down-regulated unigenes involved in "photosynthesis" KEGG pathways (ko00195) were mapped and shown in Figure 2B. Orthologs of 44 genes annotated in this pathway were filtered as down-regulated in the NaCl treated samples in the green alga, such as, photosystem II oxygen-evolving enhancer protein PSBO Cluster-2749.35825 (L_2fc = −2.5458) and Cluster-2749.43661 (L_2fc = −2.1558), cytochrome b6-f complex iron-sulfur subunit PETC, Cluster-2749.42943 (L_2fc = −2.7088), and F-type H+-transporting ATPase subunit ATPF0A, Cluster-2966.0 (L_2fc = −3.1245) (Figure 2B; Supplementary Table S2).

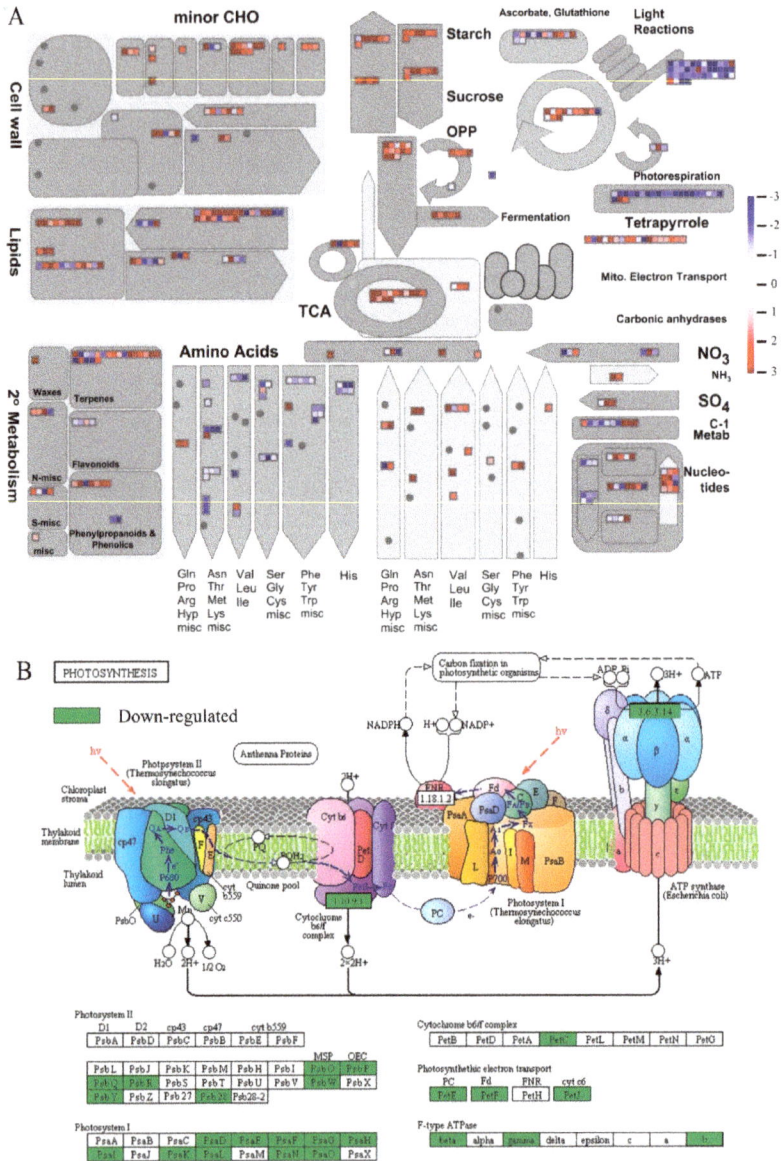

Figure 2. (**A**) Global view of differently expressed genes (DEGs) involved in diverse metabolic pathways. DEGs genes were selected for the metabolic pathways analysis using the MapMan software (3.5.1 R2). The colored boxes indicate the Log$_2$ of expression ratio of DEGs genes. The dys-regulated unigenes were assigned to 1334 and 1050 homologs in Arabidopsis, respectively. These genes were mapped to 797 pathways by MapMan, of which, 22 pathways were filtered enriched by the dys-regulated genes with the cutoff p-value < 0.05. (**B**) The KEGG pathways (ko00195) "photosynthesis" mapped with 44 down-regulated unigenes. The down-regulated genes are marked by a green frame. The black solid line with a black arrow means molecular interaction or relation; the black dash line with a black arrow means indirect link or unknown reaction; the red dash line with a red arrow stands for the light quanta.

2.7. The Differentially Expressed TFs and PKs

Among the expressed unigenes, 2050 and 1624 sequences were assigned to 45 TF families and 78 PK families, respectively (Supplementary Table S5). Of the TF families, MYB family had the largest number of upregulated genes (16 unigenes), including MYB109 ortholog unigenes Cluster-2749.35807 (L_2fc = 2.5798) and Cluster-2749.70085 (L_2fc = 1.3722). In contrast, SET family had the largest number of downregulated genes (16 unigenes). Of the PKs families, TKL-Cr-3 family was uncovered to contain the largest number of upregulated genes. By comparison, CAMK_CDPK and Group-Cr-2 family contained the largest number of downregulated unigenes (Supplementary Table S5).

2.8. Real-Time Quantitative PCR Validation

To verify the RNA-seq results, an alternative strategy was selected for the upregulated unigenes. In total, five over-expressed unigenes were randomly selected for validation by qRT-PCR using the same RNA samples that were used for RNA-seq. Primers were designed to span exon-exon junctions (see Supplementary Table S6 and Figure S3). In most cases, the gene expression trends were similar between these two methods; the result is shown in Figure 3. The ortholog of cytosolic small heat shock protein encoding genes HSP17.6A, Cluster-2749.57700, which was detected by RNA-Seq as up-regulating genes in the salt treated samples (L_2fc = 9.76), was also detected to be significantly over-expressed by qRT-PCR method (Figure 3).

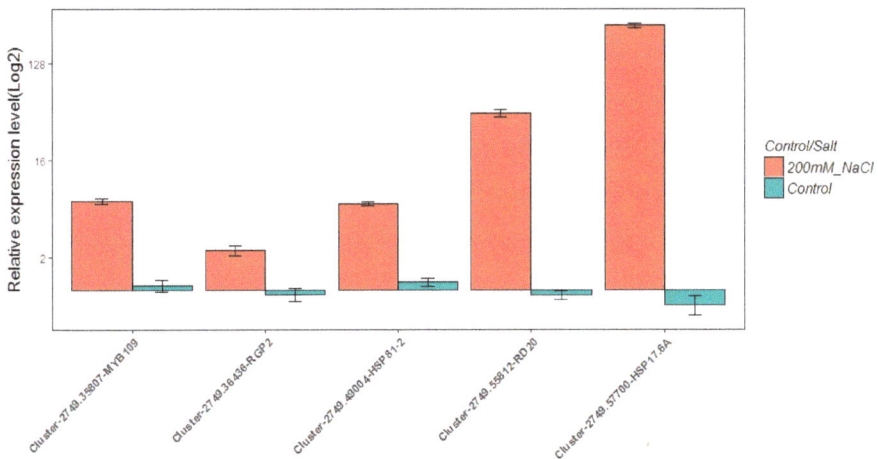

Figure 3. Real-time PCR verification of five up-regulated genes in *C. reinhardtii*. The red bars represent the qPCR results of samples under salt stress condition, while the corresponding blue bars represent the results of control samples. The individual black bars, representing the qPCR data, are the means ± SD of nine measurements (three technical replicates each for three biological samples).

3. Discussion

Salinity is one of the major environmental factors threatening crop productivity and plant growth worldwide [2,9,42]. Due to the complexity of abiotic stress-responding processes, although several hundreds of salt-responding genes have been reported in plants, understanding the genetic underpinnings of salt-responding traits in plantae remains a daunting challenge. The model alga *C. reinhardtii*, which contains one large cup-shaped chloroplast, has the ability to adapt rapidly to changing environmental conditions, such as high salinity, via the generation of novel traits [8,14,43,44]. Given previous results from analysis of salt stress in *C. reinhardtii* and other plants, we analyzed the Illumina RNA-seq data from this alga grown in BG-11 medium with the addition of 200 mM NaCl and analyzed in triplicate after 24 h of incubation [16,22]. In this study, a total of 5920 and

4715 unigenes were identified as up- and down-regulated genes in *C. reinhardtii* under salt stress by RNA-seq. Our study found some molecular cues for reducing the negative effects due to ionic/osmotic toxicity and photosynthesis impairment under saline conditions in *C. reinhardtii*.

Previous studies discovered that the cell density of *C. reinhardtii* cells obviously reduced when stressed by NaCl [8,14,16,21,22]. Neelam et al. demonstrated that at the morphological level, 150 or 200 mM NaCl salt stress led to palmelloid morphology, flagellar resorption, reduction in cell size, and slower growth rate in *C. reinhardtii* [21]. It should be noted that dead and dying cells have dys-regulated mRNA and contribute to transcript levels under saline stress. In our study, programmed cell death (PCD) in the *C. reinhardtii* cell was found with PCD-regulating proteins being significantly up-regulated, e.g., condensin complex subunit (Cluster-2749.11751: L_2fc = 9.368; Cluster-2749.12889: L_2fc = 7.766), sucrose-phosphatase 1 (Cluster-2749.35394: L_2fc = 4.304) and stress tolerance related fibrillin family member (Cluster-2749.70284: L_2fc = 1.920).

Saline stress leads to the overproduction of reactive oxygen species (ROS) in plants which are highly reactive and toxic and cause damage to lipids, carbohydrates, proteins and DNA which ultimately results in oxidative stress [8,9,14,45]. The accumulation of ROS also influences the expression of a number of genes and therefore controls many processes, such as growth, cell cycle, PCD, secondary stress responses and systemic signaling [8,9,14,45]. The excess Na^+ and oxidative stress in the intracellular or extracellular environment activates the acytoplasmic Ca^{2+} signal pathway for regulating an osmotic adjustment or homeostasis regulating of salt stress responses [24,29,39,46–51]. In our study, calcium-related pathway in the *C. reinhardtii* cell was found with several calcium ion binding proteins being significantly upregulated, e.g., peroxygenase 3 (Cluster-2749.55812: L_2fc = 10.431; Cluster-2749.59997: L_2fc = 7.680) and calreticulin (Cluster-2749.35394: L_2fc = 3.082).

The short-term (within 48 h) acclimation to salt stress in *C. reinhardtii* involves activation of phospholipid signaling, leading to the accumulation of phosphatidic acid (PA), which is a lipid second messenger in plant and animal systems [52–54]. In the case of *C. reinhardtii*, incubation in 150 mM NaCl leads to a three- to four-fold rise of PA levels within minutes [52,55]. Lysophosphatidic acid (LPA) has also been shown to accumulate in this alga under salt stress, with the dose-dependent response reaching a maximum at 300 mM NaCl [55,56]. In this study, soluble lysophosphatidic acid acyltransferase (Cluster-2749.8269: L_2fc = 9.126; Cluster-2749.9895: L_2fc = 8.720) was found to be significantly up-regulated in salt stress treated samples, which indicated the potential role of this gene in maintaining the lipid homeostasis by regulating PA under saline stress [55]. Further, analysis of glycerophospholipid metabolism pathways showed that the alga cells had significant up-regulation of FAD (flavin adenine dinucleotide)-dependent oxidoreductase family protein (Cluster-2749.52046; L_2fc = 2.695) that involves storing lipid catabolism and glycerol assimilation, and in glycerol-3-phosphate shuttle, which transports reduced power from cytosol to mitochondrion [8]. This suggests that the intracellular glycerol pool in *C. reinhardtii* cells likely increased as a response to salt stress, similar to what has been shown for the green alga *Dunaliella tertiolecta* [57,58].

Requirement of energy to maintain ion homeostasis is the major metabolic impact of salt stress. The reduction of oxidative stress and osmotic stress, and the up-regulation of heatshock proteins were speculated to aid protein renaturation and recover homeostasis [59–61]. In this study, the stress response is apparent in the *C. reinhardtii* cells with significant up-regulation of genes involved in oxidative/osmotic stress reduction process including glyceraldehyde-3-phosphate dehydrogenase C subunit 1 (Cluster-2749.27769: L_2fc = 1.930) and fumarase 1 (Cluster-2749.35832: L_2fc = 8.306). In bacterium *Escherichia coli*, trehalose is synthesized as a compatible solute and enables cells to exclude toxic cations and to acclimate to high concentrations of salt in the growth medium [62]. For maize, trehalose has helped to reduce the negative effects of saline stress as an osmoprotectant [63]. In our study, enzymes involved in trehalose synthesis significantly up-regulated, e.g., trehalose-6-phosphatase synthase S8 (Cluster-2749.17684: L_2fc = 6.453) and trehalose-6-phosphate synthase (Cluster-2749.61951: L_2fc = 1.123). These results indicated the potential underpinnings for these to maintain homeostasis in *C. reinhardtii* under saline conditions.

In plants, saline stress generally causes ion injury and osmotic stress, which interferes with numerous biochemical and physiological processes, including energy metabolism pathways such as photosynthesis [26,36,64,65] and photorespiration [8]. Previous pigment analyses have demonstrated that photosystem I-light harvesting complexes (LHCs) are damaged by ROS at high salt conditions, and PSII proteins involved in oxygen evolution are impaired [21,45]. In our study, impairment of photosynthesis in the *C. reinhardtii* cell population was found, with several photosystem I-light harvesting complex (LHC) proteins being significantly down-regulated (Figure 2B), e.g., photosystem I light harvesting complex gene LHCA2 (Cluster-2749.32743: L_2fc = −4.74; Cluster-2749.52511: L_2fc = −3.28), LHCA3 (Cluster-2749.43129: L_2fc = −6.583) and LHCA5 (Cluster-2749.40312: L2fc = −11.375; Cluster-2749.34085: L_2fc = −6.553). Further, we found most of the chloroplast encoded transcripts (e.g., PsaA, B, C, J, M) in photosystem I (PSI) were relatively unchanged in level while the nuclear genes (e.g., PsaD, E, G, F, H) down-regulated under saline conditions (Figure 2B). Existing studies have demonstrated the usage of acetate in the medium as alternative source of energy to compensate for the lowered efficiency in photosynthesis [66]. Consistent with this view, we found that acetyl-CoA synthetase (Cluster-2749.60516: L_2fc = 5.144; Cluster-2749.25511: L_2fc = 2.495), which combines acetate and CoA to form acetyl-CoA, was significantly up-regulated in the alga cells under saline conditions. In this study, a significant down-regulation was found in a key enzyme of the glyoxylate cycle—isocitrate lyase (ICL, [Cluster-2749.51492; L_2fc = −3.119]) [8,45,66,67]—which catalyzes the cleavage of isocitrate to succinate and glyoxylate. Together with malate synthase, ICL bypasses the two decarboxylation steps of the tricarboxylic acid cycle (TCA cycle) [8]. The spatio-temporal expression patterns of genes suggest that in alga cells acetyl-CoA is introduced into energy generation pathways for salt stress responses.

Glycolysis is considered to play an important role in plant development and adaptation to multiple abiotic stresses, such as cold, salt, and drought. It is the key respiratory pathway for generating ATP and carbohydrates metabolites [50,68–73]. In our work, salt stress significantly increased the expression of genes participating in the metabolism of main carbohydrates, such as starch, sucrose, soluble sugar and glucose (Figure 2A). For example, 31 genes of "glycolytic process" (GO:0006096) over-expressed during salt stress responding, including plastidic pyruvate kinase PKP-ALPHA (Cluster-2749.14688: L_2fc = 8.53) and PKP-BETA1 (Cluster-2749.26182: L_2fc = 3.68). This is consistent with Zhong et al. [68], who reported salt stress significantly increased the main carbohydrate contents of cucumber leaves [53]. Carbohydrates are involved not only in osmotic adjustment, but also can be used as protective agents for homeostasis regulating during salt stress tolerance [24,30,39,48,69,70,74–77]. Given that salt injury caused the destruction of photosynthesis, which might inhibit transport of carbohydrate and accumulate excess starch or sucrose, we speculate *C. reinhardtii* enhanced glycolysis metabolism to decrease carbohydrate accumulation in cells, which would promote the respiratory metabolism and mitochondrial electron transport, thus reducing the effects of ionic toxicity and osmotic stress caused by salt damage.

4. Materials and Methods

4.1. Chlamydomonas Material Preparation, Salt Stress Treatment and RNA Extraction

The *C. reinhardtii* strain GY-D55 wild type from LeadingTec (Shanghai, China) were grown in 150 mL of BG11 media, and placed on a shaking table with 120 rpm and maintained at light (16 h)/dark (8 h) at 23 °C, with an illumination of 100 μmol m^{-2}·s^{-1}. The density of cell cultures was determined by using the blood cell counting plate, with each value being the means of 6 repeats. Under this condition, *C. reinhardtii* cells were grown in BG11 for 14 d.

The methods published by Zhao [16] and Vega [22] were referenced for NaCl treatment in this study. A total of 50 mL medium with 800 mM NaCl was added to the 150 mL culture medium on a shaking table for finishing 200 mM NaCl treatment, the added NaCl was rapidly diluted, and then the pH value was adjusted to 7.0. A parallel set of cells that were unexposed to NaCl stress conditions and cultured in medium served as the experimental control. A total of 50 mL medium without NaCl was added into the control group. Each treatment had 3 repeats. For 24 h, 200 mM NaCl treatment significantly affected the cellular physiology of the alga, such as its photosynthetic and intracellular catalase activity; in this study, the culture time for *C. reinhardtii* under salt stress was 24 h.

After 24 h, 100 mL cell culture medium was extracted from the NaCl treated and control culture bottles, respectively. The collected cells were centrifuged at $3000 \times g$ for 5 min, and the collected cells were resuspended in 25 mL RNAlater (Ambion, Shanghai, China) solution for RNA extraction. The cells of each repeat were mixed and total RNAs were extracted separately using the TRIzol Reagent (Invitrogen, Carlsbad, CA, USA) following the manufacturer's procedures. RNA quality was assessed using the RNA Nano 6000 Assay Kit of the Agilent Bioanalyzer 2100 system (Agilent Technologies, Santa Clara, CA, USA) and the NanoDrop 2000 spectrophotometer (Thermo Scientific, Wilmington, NC, USA).

4.2. Illumina Library Construction and Sequencing

A total amount of 1.5 µg RNA per sample was used as input material for the RNA sample preparations. Sequencing libraries were generated using the NEBNext® Ultra™ RNA Library Prep Kit for Illumina® (NEB, San Diego, CA, USA) by following manufacturer's procedures, and index codes were added to attribute sequences to each sample. Briefly, mRNA was purified from total RNA using poly-T oligo-attached magnetic beads. The random hexamer primer and M-MuLV Reverse Transcriptase (RNase H⁻) were used to synthesize the first strand cDNA and the DNA Polymerase I and RNase H were used for second strand cDNA synthesis. Fragments of 150~200 bp cDNA were purified with the AMPure XP system (Beckman Coulter, Beverly, MA, USA). Then, 3 µL USER Enzyme (NEB, USA) was used with size-selected, adaptor-ligated cDNA at 37 °C for 15 min followed by 5 min at 95 °C. Then, PCR was performed with Phusion High-Fidelity DNA polymerase, Universal PCR primers and Index (X) Primer. Ten cycles were used for PCR enrichment. Finally, PCR products were purified (AMPure XP system) and library quality was assessed on the Agilent Bioanalyzer 2100 system. The clustering of the index-coded samples was performed on a cBot Cluster Generation System using TruSeq PE Cluster Kit v3-cBot-HS (Illumia, San Diego, CA, USA) according to the manufacturer's instructions. After cluster generation, the library preparations were sequenced on an Illumina HiSeq X platform (Illumina, San Diego, CA, USA), according to the manufacturer's procedures. All genetic data have been submitted to the NCBI Sequence Read Archive (SRA) database (https://www.ncbi.nlm.nih.gov/sra), SRA accession: PRJNA490089.

4.3. De Novo Transcriptome Assembling and Unigene Annotation

RNA sequencing and de novo transcriptome assembling were conducted to create reference sequence libraries for *C. reinhardtii*. The RNA sample of each repeat was sequenced separately. cDNA library construction and Illumina pair-end 150 pb sequencing (PE150) were performed at Novogene Co., Ltd. (Shanghai, China), according to instructions provided by Illumina Inc. Reads containing adapter, ploy-N and low-quality reads were removed from raw data for obtaining clean reads. The filtered high-quality reads were used for transcriptome assembling by the Trinity software with default parameters [78]. Clean datasets of 6 samples were pooled for de novo assembling and comprehensive sequence library construction. The Basic Local Alignment Search Tool (BLAST) searches of de novo assembled sequences against public databases (NR, NT, Swiss-Prot, Pfam, KOG/COG, Swiss-Prot, KEGG Ortholog database and Gene Ontology) with an E-value threshold of 10^{-10} were used for unigenes' annotation.

4.4. Calculation and Comparison of Unigene Expression

The independent transcripts libraries of 3 repeats under NaCl treatment conditions and 3 under control conditions were generated for *C. reinhardtii* by a PE150 sequencing analysis. The clean reads were aligned to the de novo assembled transcriptome and estimated by the RSEM [79] method. Gene expression levels were calculated by the fragment per kilobase of exon model per million mapped reads (FPKM) method. DESeq2 [80] was used to compare the expression levels between NaCl treated and control samples with an cutoff of adjusted *p*-value (padj) < 0.05 and | log2(foldchange) | > 1.

4.5. Gene Ontology (GO), Transcription Factors (TFs) and Protein Kinases (PKs) Prediction

The unigenes were transferred to the *C. reinhardtii* and *A. thaliana* gene IDs by using sequence similarity searching analysis against the genome of *C. reinhardtii* (ftp://ftp.psb.ugent.be/pub/plaza/plaza_public_dicots_04/Fasta/cds.all_transcripts.cre.fasta.gz) and *A. thaliana* (ftp://ftp.psb.ugent.be/pub/plaza/plaza_public_dicots_04/Fasta/cds.all_transcripts.ath.fasta.gz) with an E-value cutoff of 10^{-5}. The classifications of TFs and PKs of *C. reinhardtii* were downloaded from the iTAK database (http://bioinfo.bti.cornell.edu/cgi-bin/itak/index.cgi) [81]. The GO functional annotations file of *A. thaliana* was downloaded from Gene Ontology database (submitted 5 June 2018, http://geneontology.org/gene-associations/gene_association.tair.gz). The TFs and PKs of *C. reinhardtii* genes were transferred to their hit unigenes and the GO functional annotations of *A. thaliana* genes were assigned to their ortholog unigenes in *C. reinhardtii*.

4.6. GO, KEGG and MapMan Annotation and Enrichment

The GO enrichment analysis for DEGs of *C. reinhardtii* was performed by the topGO package of R. KEGG [82] is a database resource for understanding high-level functions and utilities of the biological system, such as the cell, the organism and the ecosystem, from molecular-level information, especially large-scale molecular datasets generated by genome sequencing and other high-throughput experimental technologies (http://www.genome.jp/kegg/). We used KOBAS [83] software to test the statistical enrichment of differential expression genes in KEGG pathways. MapMan (version 3.5.1 R2) [84] was also used to annotate the DEGs onto metabolic pathways. The DEGs of *C. reinhardtii* unigene IDs were transferred to the Arabidopsis Information Resource (TAIR) locus IDs during the MapMan analysis.

4.7. Real-Time Quantitative PCR (qRT-PCR) Verification

Real-time quantitative PCR (qRT-PCR) was performed to verify the expression patterns revealed by the RNA-seq analysis. The purified RNA of samples under salt stress and control conditions were treated with DNaseI and converted to cDNA using the PrimeScript RT Reagent Kit with gDNA Eraser (Takara, Dalian, China) according to the manufacturer's procedures. Five up-regulated unigenes in *C. reinhardtii* were selected for the qRT–PCR assay, including Cluster-2749.49004 (ortholog of HSP81-2), Cluster-2749.57700 (ortholog of HSP17.6A), Cluster-2749.55812 (ortholog of RD20), Cluster-2749.36436 (ortholog of RGP), and Cluster-2749.35807 (ortholog of MYB109). Gene-specific qRT–PCR primers (18–20 bp) (Table S6) were designed using Premier 5.0 software. qPCR was performed using SYBR Green qPCR Master Mix (DBI, Ludwigshafen, Germany) in ABI7500 Real-Time PCR System (ABI, Waltham, MA, USA). Three replicates were performed, and the amplicons were used for melting curve analysis to evaluate the amplification specificity. Relative gene expression was quantified using the $2^{-(\Delta\Delta Ct)}$ method [85]. Ortholog of the *A. thaliana* housekeeping GTP binding Elongation factor Tu family member AT5G60390 in *C. reinhardtii* (Cluster-2749.43263) was used to normalize the amount of template cDNA added in each reaction.

5. Conclusions

We performed a transcriptome analysis of short-term acclimation to salt stress (200 mM NaCl for 24 h) in *C. reinhardtii*. In total, 10,635 unigenes were identified as differentially expressed in *C. reinhardtii* under salt stress by RNA-seq, including 5920 that were up- and 4715 that were down-regulated. A series of molecular cues were screened by GO terms, MapMan and KEGG functional enrichment analyses, which were identified as potential mechanisms for salt stress responses. These mainly include maintaining the lipid homeostasis by regulating phosphatidic acid, acetate being used as an alternative source of energy for solving impairment of photosynthesis and enhancement of glycolysis metabolism to decrease the carbohydrate accumulation in cells. Our results may help understand the molecular and genetic underpinnings of salt responding traits in green alga *C. reinhardtii*.

Supplementary Materials: The following are available online at http://www.mdpi.com/1422-0067/19/11/3359/s1, Figure S1: The assembled transcriptome information of *C. reinhardtii*, Figure S2: The FPKM density distribution of *C. reinhardtii*, Table S1: Unigenes of *C. reinhardtii* annotated in *A. thaliana* genome by BLASTX analysis, Table S2: Information of the 5920 up- and 4715 down-regulated unigenes in *C. reinhardtii*, Table S3: The Gene Ontology (GO) enrichment results of the dys-regulated genes in *C. reinhardtii*, Table S4: The MapMan pathways enrichment results of the dys-regulated genes in *C. reinhardtii*, Table S5: Information of the differently expressed transcription factors (TFs) and protein kinases (PKs), Table S6: Information of the qRT–PCR primers.

Author Contributions: L.Z. and N.W. conceived and designed the study. L.Z., N.W., Z.Q., M.L., X.Z. performed the data collection and analysis. L.Z. and N.W. wrote the paper. L.Z., S.F. and X.Z. reviewed and edited the manuscript. All authors read and approved the manuscript.

Funding: This work was supported by National Natural Science Foundation of China (31800185) and A Project of Shandong Province Higher Educational Science and Technology Program (J18KA147).

Conflicts of Interest: The authors declare no conflict of interest.

Abbreviations

DEGs	differentially expressed genes
TFs	transcriptional factors
PKs	protein kinases
GO	gene ontology
FPKM	fragment per kilobase of exon model per million mapped reads
qRT-PCR	real-time quantitative PCR
BP	biological processes
PSI	photosystem I
PSII	photosystem II
PCD	programmed cell death
ROS	reactive oxygen species
PA	phosphatidic acid

References

1. Munns, R.; Tester, M. Mechanisms of salinity tolerance. *Annu. Rev. Plant Boil.* **2008**, *59*, 651–681. [CrossRef] [PubMed]
2. Song, J.; Wang, B.S. Using euhalophytes to understand salt tolerance and to develop saline agriculture: *Suaeda salsa* as a promising model. *Ann. Bot.* **2015**, *115*, 541–553. [CrossRef] [PubMed]
3. Epstein, E. Salt-tolerant crops: Origins, development, and prospects of the concept. *Plant Soil* **1985**, *89*, 187–198. [CrossRef]
4. Zhang, L.Y.; Zhang, X.J.; Fan, S.J. Meta-analysis of salt-related gene expression profiles identifies common signatures of salt stress responses in Arabidopsis. *Plant Syst. Evol.* **2017**, *303*, 757–774. [CrossRef]
5. Yuan, F.; Leng, B.Y.; Wang, B.S. Progress in Studying Salt Secretion from the Salt Glands in Recretohalophytes: How Do Plants Secrete Salt? *Front. Plant Sci.* **2016**, *7*, 977. [CrossRef] [PubMed]
6. Khona, D.K.; Shirolikar, S.M.; Gawde, K.K.; Hom, E.; Deodhar, M.A.; D'Souza, J.S. Characterization of salt stress-induced palmelloids in the green alga, *Chlamydomonas Reinhardtii*. *Algal Res.* **2016**, *16*, 434–448. [CrossRef]
7. Shen, X.Y.; Wang, Z.L.; Song, X.F.; Xu, J.J.; Jiang, C.Y.; Zhao, Y.X.; Ma, C.L.; Zhang, H. Transcriptomic profiling revealed an important role of cell wall remodeling and ethylene signaling pathway during salt acclimation in Arabidopsis. *Plant Mol. Boil.* **2014**, *86*, 303–317. [CrossRef] [PubMed]
8. Perrineau, M.M.; Zelzion, E.; Gross, J.; Price, D.C.; Boyd, J.; Bhattacharya, D. Evolution of salt tolerance in a laboratory reared population of *Chlamydomonas Reinhardtii*. *Environ. Microbiol.* **2014**, *16*, 1755–1766. [CrossRef] [PubMed]
9. Zhu, J.K. Plant salt tolerance. *Trends Plant Sci.* **2001**, *6*, 66–71. [CrossRef]

10. Shabala, S.; Munns, R.; Shabala, S. Salinity stress: Physiological constraints and adaptive mechanisms. In *Plant Stress Physiology*; CABI: Wallingford, UK, 2012.

11. Young, M.A.; Rancier, D.G.; Roy, J.L.; Lunn, S.R.; Armstrong, S.A.; Headley, J.V. Seeding conditions of the halophyte *Atriplex patula* for optimal growth on a salt impacted site. *Int. J. Phytoremediat.* **2011**, *13*, 674–680. [CrossRef]

12. Rao, A.R.; Sarada, R.; Ravishankar, G.A. Enhancement of carotenoids in green alga *Botyrocccus braunii* in various autotrophic media under stress conditions. *Int. J. Biomed. Pharm. Sci.* **2010**, *4*, 87–92.

13. Stanier, R.Y.; Kunisawa, R.; Mandel, M.; Cohenbazire, G. Purification and properties of unicellular blue-green algae (order Chroococcales). *Bacteriol. Rev.* **1971**, *35*, 171–205. [PubMed]

14. Liu, F.; Jin, Z.; Wang, Y.; Bi, Y.; Melton, R.J. Plastid Genome of Dictyopteris divaricata (Dictyotales, Phaeophyceae): Understanding the Evolution of Plastid Genomes in Brown Algae. *Mar. Biotechnol.* **2017**, *19*, 1–11. [CrossRef] [PubMed]

15. Sayed, O.H. Chlorophyll Fluorescence as a Tool in Cereal Crop Research. *Photosynthetica* **2003**, *41*, 321–330. [CrossRef]

16. Zuo, Z.; Chen, Z.; Zhu, Y.; Bai, Y.; Wang, Y. Effects of NaCl and Na_2CO_3 stresses on photosynthetic ability of *Chlamydomonas Reinhardtii*. *Biologia* **2014**, *69*, 1314–1322. [CrossRef]

17. Fedina, I.S.; Georgieva, K.; Grigorova, I. Response of Barley Seedlings to UV-B Radiation as Affected by Proline and NaCl. *J. Plant Physiol.* **2003**, *47*, 549–554. [CrossRef]

18. Khan, N.A. NaCl-Inhibited Chlorophyll Synthesis and Associated Changes in Ethylene Evolution and Antioxidative Enzyme Activities in Wheat. *Boil. Plant.* **2003**, *47*, 437–440. [CrossRef]

19. Demetriou, G.; Neonaki, C.; Navakoudis, E.; Kotzabasis, K. Salt stress impact on the molecular structure and function of the photosynthetic apparatus—The protective role of polyamines. *Biochim. Et Biophys. Acta* **2007**, *1767*, 272–280. [CrossRef] [PubMed]

20. Liu, X.D.; Shen, Y.G. Salt shock induces state II transition of the photosynthetic apparatus in dark-adapted *Dunaliella salina* cells. *Environ. Exp. Bot.* **2006**, *57*, 19–24. [CrossRef]

21. Neelam, S.; Subramanyam, R. Alteration of photochemistry and protein degradation of photosystem II from *Chlamydomonas reinhardtii* under high salt grown cells. *J. Photochem. Photobiol. B Boil.* **2013**, *124*, 63–70. [CrossRef] [PubMed]

22. Vega, J.M.; Garbayo, I.; Domínguez, M.J.; Vigara, J. Effect of abiotic stress on photosynthesis and respiration in: Induction of oxidative stress. *Enzym. Microb. Technol.* **2007**, *40*, 163–167. [CrossRef]

23. Atkinson, N.J.; Lilley, C.J.; Urwin, P.E. Identification of genes involved in the response of Arabidopsis to simultaneous biotic and abiotic stresses. *Plant Physiol.* **2013**, *162*, 2028–2041. [CrossRef] [PubMed]

24. Han, N.; Lan, W.J.; He, X.; Shao, Q.; Wang, B.S.; Zhao, X.J. Expression of a *Suaeda salsa* Vacuolar H^+/Ca^{2+} Transporter Gene in Arabidopsis Contributes to Physiological Changes in Salinity. *Plant Mol. Boil. Rep.* **2012**, *30*, 470–477. [CrossRef]

25. Qi, Y.C.; Liu, W.Q.; Qiu, L.Y.; Zhang, S.M.; Ma, L.; Zhang, H. Overexpression of glutathione S-transferase gene increases salt tolerance of arabidopsis. *Russ. J. Plant Physiol.* **2010**, *57*, 233–240. [CrossRef]

26. Zhang, S.R.; Song, J.; Wang, H.; Feng, G. Effect of salinity on seed germination, ion content and photosynthesis of cotyledons in halophytes or xerophyte growing in Central Asia. *J. Plant Ecol.* **2010**, *3*, 259–267. [CrossRef]

27. Lu, T.; Lu, G.; Fan, D.; Zhu, C.; Wei, L.; Qiang, Z.; Qi, F.; Yan, Z.; Guo, Y.; Li, W. Function annotation of the rice transcriptome at single-nucleotide resolution by RNA-seq. *Genome Res.* **2010**, *20*, 1238–1249. [CrossRef] [PubMed]

28. Liu, A.; Xiao, Z.; Li, M.W.; Wong, F.L.; Yung, W.S.; Ku, Y.S.; Wang, Q.; Wang, X.; Xie, M.; Yim, A.K. Transcriptomic reprogramming in soybean seedlings under salt stress. *Plant Cell Environ.* **2018**. [CrossRef] [PubMed]

29. Cui, F.; Sui, N.; Duan, G.; Liu, Y.; Han, Y.; Liu, S.; Wan, S.; Li, G. Identification of Metabolites and Transcripts Involved in Salt Stress and Recovery in Peanut. *Front. Plant Sci.* **2018**, *9*, 217. [CrossRef] [PubMed]

30. Guo, J.; Li, Y.; Han, G.; Song, J.; Wang, B. NaCl markedly improved the reproductive capacity of the euhalophyte *Suaeda Salsa*. *Funct. Plant Boil.* **2018**, *45*, 350. [CrossRef]

31. Cao, S.; Du, X.H.; Li, L.H.; Liu, Y.D.; Zhang, L.; Pan, X.; Li, Y.; Li, H.; Lu, H. Overexpression of *Populus tomentosa* cytosolic ascorbate peroxidase enhances abiotic stress tolerance in tobacco plants. *Russ. J. Plant Physiol.* **2017**, *64*, 224–234. [CrossRef]

32. Sui, N.; Tian, S.S.; Wang, W.Q.; Wang, M.J.; Fan, H. Overexpression of Glycerol-3-Phosphate Acyltransferase from *Suaeda salsa* Improves Salt Tolerance in Arabidopsis. *Front. Plant Sci.* **2017**, *8*, 1337. [CrossRef] [PubMed]
33. Wang, J.S.; Zhang, Q.; Cui, F.; Hou, L.; Zhao, S.Z.; Xia, H.; Qiu, J.J.; Li, T.T.; Zhang, Y.; Wang, X.J.; et al. Genome-Wide Analysis of Gene Expression Provides New Insights into Cold Responses in *Thellungiella Salsuginea*. *Front. Plant Sci.* **2017**, *8*, 713. [CrossRef] [PubMed]
34. Yuan, F.; Lyu, M.J.A.; Leng, B.Y.; Zhu, X.G.; Wang, B.S. The transcriptome of NaCl-treated *Limonium bicolor* leaves reveals the genes controlling salt secretion of salt gland. *Plant Mol. Boil.* **2016**, *91*, 241–256. [CrossRef] [PubMed]
35. Yuan, F.; Lyu, M.J.A.; Leng, B.Y.; Zheng, G.Y.; Feng, Z.T.; Li, P.H.; Zhu, X.G.; Wang, B.S. Comparative transcriptome analysis of developmental stages of the *Limonium bicolor* leaf generates insights into salt gland differentiation. *Plant Cell Environ.* **2015**, *38*, 1637–1657. [CrossRef] [PubMed]
36. Feng, Z.T.; Deng, Y.Q.; Fan, H.; Sun, Q.J.; Sui, N.; Wang, B.S. Effects of NaCl stress on the growth and photosynthetic characteristics of *Ulmus pumila* L. seedlings in sand culture. *Photosynthetica* **2014**, *52*, 313–320. [CrossRef]
37. Yuan, F.; Chen, M.; Yang, J.C.; Leng, B.Y.; Wang, B.S. A system for the transformation and regeneration of the recretohalophyte *Limonium bicolor*. *In Vitro Cell. Dev. Boil. Plant* **2014**, *50*, 610–617. [CrossRef]
38. Zhang, Q.; Zhao, C.Z.; Li, M.; Sun, W.; Liu, Y.; Xia, H.; Sun, M.N.; Li, A.Q.; Li, C.S.; Zhao, S.Z.; et al. Genome-wide identification of *Thellungiella salsuginea* microRNAs with putative roles in the salt stress response. *BMC Plant Boil.* **2013**, *13*, 180. [CrossRef] [PubMed]
39. Guo, Y.H.; Jia, W.J.; Song, J.; Wang, D.A.; Chen, M.; Wang, B.S. Thellungilla halophila is more adaptive to salinity than *Arabidopsis thaliana* at stages of seed germination and seedling establishment. *Acta Physiol. Plant.* **2012**, *34*, 1287–1294. [CrossRef]
40. Liu, J.; Zhang, F.; Zhou, J.J.; Chen, F.; Wang, B.S.; Xie, X.Z. Phytochrome B control of total leaf area and stomatal density affects drought tolerance in rice. *Plant Mol. Boil.* **2012**, *78*, 289–300. [CrossRef] [PubMed]
41. Zhou, J.C.; Fu, T.T.; Sui, N.; Guo, J.R.; Feng, G.; Fan, J.L.; Song, J. The role of salinity in seed maturation of the euhalophyte *Suaeda Salsa*. *Plant Biosyst.* **2016**, *150*, 83–90. [CrossRef]
42. Waśkiewicz, A.; Muzolf-Panek, M.; Goliński, P. *Phenolic Content Changes in Plants under Salt Stress*; Springer: New York, NY, USA, 2013; pp. 283–314.
43. Sudhir, P.; Murthy, S.D.S. Effects of salt stress on basic processes of photosynthesis. *Photosynthetica* **2004**, *42*, 481–486. [CrossRef]
44. Maršálek, B.; Zahradníčková, H.; Hronková, M. Extracellular Production of Abscisic Acid by Soil Algae under Salt, Acid or Drought Stress. *Z. Für Naturforschung C* **1992**, *47*, 701–704. [CrossRef]
45. Pineau, B.; Gérard-Hirne, C.; Selve, C. Carotenoid binding to photosystems I and II of Chlamydomonas reinhardtii, cells grown under weak light or exposed to intense light. *Plant Physiol. Bioch.* **2001**, *39*, 73–85. [CrossRef]
46. Ding, F.; Chen, M.; Sui, N.; Wang, B.S. Ca^{2+} significantly enhanced development and salt-secretion rate of salt glands of *Limonium bicolor* under NaCl treatment. *S. Afr. J. Bot.* **2010**, *76*, 95–101. [CrossRef]
47. Feng, Z.T.; Deng, Y.Q.; Zhang, S.C.; Liang, X.; Yuan, F.; Hao, J.L.; Zhang, J.C.; Sun, S.F.; Wang, B.S. K^+ accumulation in the cytoplasm and nucleus of the salt gland cells of *Limonium bicolor* accompanies increased rates of salt secretion under NaCl treatment using NanoSIMS. *Plant Sci.* **2015**, *238*, 286–296. [CrossRef] [PubMed]
48. Han, N.; Shao, Q.; Bao, H.Y.; Wang, B.S. Cloning and Characterization of a Ca^{2+}/H^+ Antiporter from Halophyte *Suaeda salsa* L. *Plant Mol. Boil. Rep.* **2011**, *29*, 449–457. [CrossRef]
49. Yang, S.; Li, L.; Zhang, J.; Geng, Y.; Guo, F.; Wang, J.; Meng, J.; Sui, N.; Wan, S.; Li, X. Transcriptome and Differential Expression Profiling Analysis of the Mechanism of $Ca^{(2+)}$ Regulation in Peanut (*Arachis hypogaea*) Pod Development. *Front. Plant Sci.* **2017**, *8*, 1609. [CrossRef] [PubMed]
50. Zheng, Y.; Liao, C.C.; Zhao, S.S.; Wang, C.W.; Guo, Y. The Glycosyltransferase QUA1 Regulates Chloroplast-Associated Calcium Signaling During Salt and Drought Stress in Arabidopsis. *Plant Cell Physiol.* **2017**, *58*, 329–341. [CrossRef] [PubMed]
51. Zhang, L.Y.; Zhang, Z.; Zhang, X.J.; Yao, Y.; Wang, R.; Duan, B.Y.; Fan, S.J. Comprehensive meta-analysis and co-expression network analysis identify candidate genes for salt stress response in Arabidopsis. *Plant Biosyst.* **2018**. [CrossRef]

52. Arisz, S.A.; Valianpour, F.; van Gennip, A.H.; Munnik, T. Substrate preference of stress-activated phospholipase D in Chlamydomonas and its contribution to PA formation. *Plant J. Cell Mol. Boil.* **2003**, *34*, 595–604. [CrossRef]

53. Zhou, J.J.; Liu, Q.Q.; Zhang, F.; Wang, Y.Y.; Zhang, S.Y.; Cheng, H.M.; Yan, L.H.; Li, L.; Chen, F.; Xie, X.Z. Overexpression of OsPIL15, a phytochromeinteracting factor- like protein gene, represses etiolated seedling growth in rice. *J. Integr. Plant Boil.* **2014**, *56*, 373–387. [CrossRef] [PubMed]

54. Sui, N.; Li, M.; Li, K.; Song, J.; Wang, B.S. Increase in unsaturated fatty acids in membrane lipids of *Suaeda salsa* L. enhances protection of photosystem II under high salinity. *Photosynthetica* **2010**, *48*, 623–629. [CrossRef]

55. Arisz, S.A.; Munnik, T. The salt stress-induced LPA response in Chlamydomonas is produced via PLA2 hydrolysis of DGK-generated phosphatidic acid. *J. Lipid Res.* **2011**, *52*, 2012–2020. [CrossRef] [PubMed]

56. Meijer, H.J.; Arisz, S.A.; Van Himbergen, J.A.; Musgrave, A.; Munnik, T. Hyperosmotic stress rapidly generates lyso-phosphatidic acid in Chlamydomonas. *Plant J.* **2010**, *25*, 541–548. [CrossRef]

57. Takagi, M.; Karseno; Yoshida, T. Effect of Salt Concentration on Intracellular Accumulation of Lipids and Triacylglyceride in Marine Microalgae Dunaliella Cells. *J. Biosci. Bioeng.* **2006**, *101*, 223–226. [CrossRef] [PubMed]

58. Goyal, A. Osmoregulation in Dunaliella, Part II: Photosynthesis and starch contribute carbon for glycerol synthesis during a salt stress in Dunaliella tertiolecta. *Plant Physiol. Biochem.* **2007**, *45*, 705–710. [CrossRef] [PubMed]

59. Yokthongwattana, C.; Mahong, B.; Roytrakul, S.; Phaonaklop, N.; Narangajavana, J.; Yokthongwattana, K. Proteomic analysis of salinity-stressed *Chlamydomonas reinhardtii* revealed differential suppression and induction of a large number of important housekeeping proteins. *Planta* **2012**, *235*, 649–659. [CrossRef] [PubMed]

60. Sun, Z.B.; Qi, X.Y.; Wang, Z.L.; Li, P.H.; Wu, C.X.; Zhang, H.; Zhao, Y.X. Overexpression of TsGOLS2, a galactinol synthase, in *Arabidopsis thaliana* enhances tolerance to high salinity and osmotic stresses. *Plant Physiol. Biochem.* **2013**, *69*, 82–89. [CrossRef] [PubMed]

61. Pang, C.H.; Li, K.; Wang, B.S. Overexpression of SsCHLAPXs confers protection against oxidative stress induced by high light in transgenic *Arabidopsis thaliana*. *Physiol. Plant.* **2011**, *143*, 355–366. [CrossRef] [PubMed]

62. Ferjani, A.; Mustardy, L.; Sulpice, R.; Marin, K.; Suzuki, I.; Hagemann, M.; Murata, N. Glucosylglycerol, a compatible solute, sustains cell division under salt stress. *Plant Physiol.* **2003**, *131*, 1628–1637. [CrossRef] [PubMed]

63. Zeid, I.M. Trehalose as osmoprotectant for maize under salinity-induced stress. *Res. J. Agric. Boil. Sci.* **2009**, *5*, 613–622.

64. Xu, J.J.; Li, Y.Y.; Ma, X.L.; Ding, J.F.; Wang, K.; Wang, S.S.; Tian, Y.; Zhang, H.; Zhu, X.G. Whole transcriptome analysis using next-generation sequencing of model species *Setaria viridis* to support C$_4$ photosynthesis research. *Plant Mol. Boil.* **2013**, *83*, 77–87. [CrossRef] [PubMed]

65. Sui, N.; Han, G.L. Salt-induced photoinhibition of PSII is alleviated in halophyte Thellungiella halophila by increases of unsaturated fatty acids in membrane lipids. *Acta Physiol. Plant.* **2014**, *36*, 983–992. [CrossRef]

66. Heifetz, P.B.; Boynton, J.E. Effects of Acetate on Facultative Autotrophy in *Chlamydomonas reinhardtii* Assessed by Photosynthetic Measurements and Stable Isotope Analyses. *Plant Physiol.* **2000**, *122*, 1439–1445. [CrossRef] [PubMed]

67. Soussi, M.; Ocaña, A.; Lluch, C. Effects of salt stress on growth, photosynthesis and nitrogen fixation in chick-pea (*Cicer arietinum* L.). *J. Exp. Bot.* **1998**, *49*, 1329–1337. [CrossRef]

68. Zhong, M.; Yuan, Y.; Shu, S.; Sun, J.; Guo, S.; Yuan, R.; Tang, Y. Effects of exogenous putrescine on glycolysis and Krebs cycle metabolism in cucumber leaves subjected to salt stress. *Plant Growth Regul.* **2016**, *79*, 319–330. [CrossRef]

69. Huang, J.; Li, Z.Y.; Biener, G.; Xiong, E.H.; Malik, S.; Eaton, N.; Zhao, C.Z.; Raicu, V.; Kong, H.Z.; Zhao, D.Z. Carbonic Anhydrases Function in Anther Cell Differentiation Downstream of the Receptor-Like Kinase EMS1. *Plant Cell* **2017**, *29*, 1335–1356. [CrossRef] [PubMed]

70. Wang, F.X.; Xu, Y.G.; Wang, S.; Shi, W.W.; Liu, R.R.; Feng, G.; Song, J. Salinity affects production and salt tolerance of dimorphic seeds of *Suaeda Salsa*. *Plant Physiol. Biochem.* **2015**, *95*, 41–48. [CrossRef] [PubMed]

71. Shao, Q.; Han, N.; Ding, T.L.; Zhou, F.; Wang, B.S. SsHKT1;1 is a potassium transporter of the C$_3$ halophyte *Suaeda salsa* that is involved in salt tolerance. *Funct. Plant Boil.* **2014**, *41*, 790–802. [CrossRef]

72. Li, K.; Pang, C.H.; Ding, F.; Sui, N.; Feng, Z.T.; Wang, B.S. Overexpression of *Suaeda salsa* stroma ascorbate peroxidase in Arabidopsis chloroplasts enhances salt tolerance of plants. *S. Afr. J. Bot.* **2012**, *78*, 235–245. [CrossRef]
73. Liu, S.S.; Wang, W.Q.; Li, M.; Wan, S.B.; Sui, N. Antioxidants and unsaturated fatty acids are involved in salt tolerance in peanut. *Acta Physiol. Plant.* **2017**, *39*, 207. [CrossRef]
74. Chen, M.; Song, J.; Wang, B.S. NaCl increases the activity of the plasma membrane H^+-ATPase in C_3 halophyte *Suaeda salsa* callus. *Acta Physiol. Plant.* **2010**, *32*, 27–36. [CrossRef]
75. Song, J.; Shi, G.W.; Gao, B.; Fan, H.; Wang, B.S. Waterlogging and salinity effects on two *Suaeda salsa* populations. *Physiol. Plant.* **2011**, *141*, 343–351. [CrossRef] [PubMed]
76. Meng, X.; Yang, D.Y.; Li, X.D.; Zhao, S.Y.; Sui, N.; Meng, Q.W. Physiological changes in fruit ripening caused by overexpression of tomato SlAN2, an R2R3-MYB factor. *Plant Physiol. Biochem.* **2015**, *89*, 24–30. [CrossRef] [PubMed]
77. Li, Y.Y.; Ma, X.L.; Zhao, J.L.; Xu, J.J.; Shi, J.F.; Zhu, X.G.; Zhao, Y.X.; Zhang, H. Developmental Genetic Mechanisms of C_4 Syndrome Based on Transcriptome Analysis of C_3 Cotyledons and C_4 Assimilating Shoots in *Haloxylon Ammodendron*. *PLoS ONE* **2015**, *10*, e0117175. [CrossRef] [PubMed]
78. Grabherr, M.G.; Haas, B.J.; Yassour, M.; Levin, J.Z.; Thompson, D.A.; Amit, I.; Adiconis, X.; Fan, L.; Raychowdhury, R.; Zeng, Q.; et al. Full-length transcriptome assembly from RNA-Seq data without a reference genome. *Nat. Biotechnol.* **2011**, *29*, 644–652. [CrossRef] [PubMed]
79. Li, B.; Dewey, C.N. RSEM: Accurate transcript quantification from RNA-Seq data with or without a reference genome. *BMC Bioinform.* **2011**, *12*, 323. [CrossRef] [PubMed]
80. Haas, B.J.; Papanicolaou, A.; Yassour, M.; Grabherr, M.; Blood, P.D.; Bowden, J.; Couger, M.B.; Eccles, D.; Li, B.; Lieber, M. De novo transcript sequence reconstruction from RNA-Seq: Reference generation and analysis with Trinity. *Nat. Protoc.* **2013**, *8*, 1494–1512. [CrossRef] [PubMed]
81. Zheng, Y.; Jiao, C.; Sun, H.; Rosli, H.G.; Pombo, M.A.; Zhang, P.; Banf, M.; Dai, X.; Martin, G.B.; Giovannoni, J.J.; et al. iTAK: A Program for Genome-wide Prediction and Classification of Plant Transcription Factors, Transcriptional Regulators, and Protein Kinases. *Mol. Plant* **2016**, *9*, 1667–1670. [CrossRef] [PubMed]
82. Okuda, S.; Yamada, T.; Hamajima, M.; Itoh, M.; Katayama, T.; Bork, P.; Goto, S.; Kanehisa, M. KEGG Atlas mapping for global analysis of metabolic pathways. *Nucleic Acids Res.* **2008**, *36*, W423–W426. [CrossRef] [PubMed]
83. Wu, J.; Mao, X.; Cai, T.; Luo, J.; Wei, L. KOBAS server: A web-based platform for automated annotation and pathway identification. *Nucleic Acids Res.* **2006**, *34*, W720–W724. [CrossRef] [PubMed]
84. Thimm, O.; Blasing, O.; Gibon, Y.; Nagel, A.; Meyer, S.; Kruger, P.; Selbig, J.; Muller, L.A.; Rhee, S.Y.; Stitt, M. MAPMAN: A user-driven tool to display genomics data sets onto diagrams of metabolic pathways and other biological processes. *Plant J. Cell Mol. Boil.* **2004**, *37*, 914–939. [CrossRef]
85. Livak, K.J.; Schmittgen, T.D. Analysis of relative gene expression data using real-time quantitative PCR and the 2(-Delta Delta C (T)) Method. *Methods* **2001**, *25*, 402–408. [CrossRef] [PubMed]

International Journal of
Molecular Sciences

MDPI

Article

Analysis of bZIP Transcription Factor Family and Their Expressions under Salt Stress in *Chlamydomonas reinhardtii*

Chunli Ji, Xue Mao, Jingyun Hao, Xiaodan Wang, Jinai Xue, Hongli Cui and Runzhi Li *

Institute of Molecular Agriculture and Bioenergy, Shanxi Agricultural University, Taigu 030801, China; jichunnli@sxau.edu.cn (C.J.); maoxue@sxau.edu.cn (X.M.); haojingyun@stu.sxau.edu.cn (J.H.); wangxiaodan@sxau.edu.cn (X.W.); xuejinai@sxau.edu.cn (J.X.); cuihongli@sxau.edu.cn (H.C.)
* Correspondence: lirunzhi@sxau.edu.cn; Tel.: +86-354-628-8344

Received: 20 July 2018; Accepted: 14 September 2018; Published: 17 September 2018

Abstract: The basic leucine-region zipper (bZIP) transcription factors (TFs) act as crucial regulators in various biological processes and stress responses in plants. Currently, bZIP family members and their functions remain elusive in the green unicellular algae *Chlamydomonas reinhardtii*, an important model organism for molecular investigation with genetic engineering aimed at increasing lipid yields for better biodiesel production. In this study, a total of 17 *C. reinhardtii* bZIP (CrebZIP) TFs containing typical bZIP structure were identified by a genome-wide analysis. Analysis of the CrebZIP protein physicochemical properties, phylogenetic tree, conserved domain, and secondary structure were conducted. *CrebZIP* gene structures and their chromosomal assignment were also analyzed. Physiological and photosynthetic characteristics of *C. reinhardtii* under salt stress were exhibited as lower cell growth and weaker photosynthesis, but increased lipid accumulation. Meanwhile, the expression profiles of six *CrebZIP* genes were induced to change significantly during salt stress, indicating that certain CrebZIPs may play important roles in mediating photosynthesis and lipid accumulation of microalgae in response to stresses. The present work provided a valuable foundation for functional dissection of CrebZIPs, benefiting the development of better strategies to engineer the regulatory network in microalgae for enhancing biofuel and biomass production.

Keywords: *Chlamydomonas reinhardtii*; bZIP transcription factors; salt stress; transcriptional regulation; photosynthesis; lipid accumulation

---

1. Introduction

Microalgae are considered to be one of the most promising feedstocks for renewable biofuel production. However, the shortage of inexpensive algal biomass currently hampers microalgae-based biofuel industrialization [1]. Microalgae accumulate high level of lipids, mainly in the form of triacylglycerol (TAG), when subjected to nutrient deprivation and other stresses [2–5]. In parallel, these adverse conditions also limit algal biomass accumulation. Consequently, genetic engineering to achieve an optimized balance between oil accumulation and biomass growth may represent an effective strategy for the improvement of microalgae biofuel yield. Therefore, it is necessary to comprehensively analyze the underlying molecular mechanisms that mediate stress-induced accumulation of oil in microalgae, particularly to identify the key transcription factors (TFs). The unicellular algae *Chlamydomonas reinhardtii* is the de facto model organism for research in microalgae. Various types of omics data for *C. reinhardtii* are available, including its full genome [6], the proteomics and metabolomics analysis, and the phenotype transition during N starvation [7–13]. These achievements provide the basis for further investigation into oil metabolism and regulation in *C. reinhardtii*, which would shed light on the development of rational strategies for sustainable production of microalgae biofuel.

Transcription factor (TF) encoding genes are considered as contributing to the diversity and evolution in plants. Identification of the transcriptional factors and their cognate transcriptional factor binding-sites is essential in manipulating the regulatory network for desired traits of the target molecules [14]. Moreover, the control of transcription initiation rates by transcription factors is an important means to modulate gene expression, and then regulate the organism growth and development [15]. The basic region-leucine zipper (bZIP) family is one of the most conserved and wildly distributed TFs present in multiple eukaryotes. To date, they have been extensively investigated in many plants including *Arabidopsis*, rice, tomato, maize, sorghum, carrot, and so forth [16–22]. The bZIP TFs have been found to mediate various biological processes, such as cell elongation [23], organ and tissue differentiation [24–26], energy metabolism [27], embryogenesis and seed maturation [28], and so forth. The bZIP TFs also participate in plant responses to biotic and abiotic stresses, including pathogen defense [29,30], hormone and sugar signaling [31,32], light response [33,34], salt and drought tolerance [20,35], and so forth. Typically, bZIP TFs contain a conserved 40–80 amino acid (aa) domain which has two structure motifs: A DNA-binding basic region and a leucine zipper dimerization domain [15]. The basic region composing of around 20 amino acid residues with an invariant N-X7-R/K-X9 motif is highly conserved, and the main function of this region is for nuclear localization and DNA binding. The leucine zipper containing a heptad repeat of leucine is less conserved, with the property for recognition and dimerization [21]. The diversified leucine zipper region is located exactly 9 aa downstream from the C-terminal of the basic region. Although bZIP family members were intensively reported to mediate diverse stress responses in higher plants, little attention has been paid to studying bZIP TFs and their downstream target genes on a genome-wide scale in microalgae.

A total of 147 putative TFs of 29 different protein families have been identified in *C. reinhardtii*, including 1 WRKY, 4 bHLH, 5 C2H2, 11 MYB, 2 MADS, 7 bZIP TFs, and so forth [36]. However, functions remain unclear for the majority of these TFs. The bZIP family is also one of the four largest TF families in oleaginous microalgae *Nannochloropsis* [14], showing that some bZIPs were putatively related with the transcriptional regulation of TAG biosynthesis pathways in *Nannochloropsis.* Therefore, the present study focused on the genome-wide identification of bZIP TFs in *C. reinhardtii* and their functional analysis, with an objective to elucidate the mechanism underlying the regulation of fatty acid and oil accumulation, and photosynthesis in microalgae, particularly under stresses.

In this study, the bZIP sequences of *C. reinhardtii* were intensively identified using a proteomic database, and a total of 17 CrebZIP TFs were obtained after removing the redundancy. Bioinformatics tools were employed to perform a detailed analysis of their genetic structure, chromosome distribution, classification, protein domain, and motifs, as well as evolutionary relationship. Furthermore, the physiological and photosynthetic characteristics of *C. reinhardtii* under salt stress were measured, including biomass concentration, lipid and pigment contents, as well as chlorophyll fluorescence variation. Finally, to infer the potential functions of these CrebZIPs, the expression profiles of *CrebZIP* genes under salt stress were quantitatively examined using quantitative real time (qRT)-PCR. Thus, these integrated data would provide new insights into comprehensive understanding of the stress-adaptive mechanisms and oil accumulation mediated by bZIP TFs in *C. reinhardtii* and other microalgae.

2. Results and Discussion

2.1. Identification of C. reinhardtii bZIP Family Members

To perform genome-wide identification of bZIP proteins in *C. reinhardtii*, BLAST and the Hidden Markov Model (HMM) profiles of the bZIP domain were used to screen the *C. reinhardtii* genome and proteome database, with bZIP sequences from *Arabidopsis* as the query. A total of 17 *CrebZIP* genes in

C. reinhardtii were identified and denominated as *CrebZIP1–CrebZIP17* based on their locations in the chromosome (Table 1).

The number of CrebZIPs obtained here was not consistent with previous reports. Corrêa et al. and Riano-Pachon et al. identified 7 putative CrebZIP TF coding sequences [15,36]. However, study on the evolution of bZIP family TFs among different plants by Que et al. detected 19 bZIP TFs in *C. reinhardtii*. They summarized that the number of bZIP TFs in algae (less than 20) and land plants (greater than 25) differed remarkably, and bZIP TFs might thus have expanded many times during plant evolution [22]. Such difference in bZIP numbers in *C. reinhardtii* might have resulted from different versions of the *C. reinhardtii* genome and protein database, and criteria used in those reports.

Conserved Domain Database (CDD) and Simple Modular Architecture Research Tool (SMART) analysis indicated that the 17 CrebZIP proteins all had typical bZIP conserved domains. Table 1 summarizes their physicochemical properties, including the protein length which ranged from 334 (CrebZIP13) to 2018 (CrebZIP1) amino acids, the corresponding molecular weight which varied from 3,4514.92 to 198,080.64 Da, and theoretical isoelectric point (pI) which varied from 4.96 (CrebZIP13) to 9.55 (CrebZIP12). The great difference in these properties may reflect their functional diversity in *C. reinhardtii*. The minus hydrophility of all CrebZIP proteins and their higher instability index (>40) showed that they were hydrophilic and unstable.

To get the protein structure information of these CrebZIP members, the secondary structure of the proteins was predicted by the PBIL LYON-GERLAND database. The secondary structure information is listed in Table 2, including α-helix, extended strand, and random coil. Of them, random coil accounted for a higher percentage (45.23–72.75%), while extended strand had the lowest proportion (0.42–8.73%). No β-bridge was detected in CrebZIPs.

Table 1. Physicochemical properties of *CrebZIP* gene coding proteins.

Gene Number	NCBI Accession Number	Phytozome Identifier	Chromosome Localization (bp)	Protein Length (aa)	Molecular Weight (Da)	Theoretical pI	Hydrophility	Instability Index
CrebZIP1	PNW88934.1	Cre01.g051174	Chr.1: 7078516–7088297	2018	198,080.64	6.00	−0.422	68.82
CrebZIP2	PNW83651.1	Cre05.g238250	Chr.5: 2933060–2936249	524	53,624.16	5.47	−0.045	58.98
CrebZIP3	PNW80451.1	Cre07.g318050	Chr.7: 782941–789359	802	82,929.15	6.16	−0.406	59.38
CrebZIP4	PNW80535.1	Cre07.g321550	Chr.7: 1255642–1260306	393	41,429.96	5.48	−0.510	44.72
CrebZIP5	PNW81157.1	Cre07.g344668	Chr.7: 4675427–4680202	750	72,176.41	6.18	−0.046	65.59
CrebZIP6	PNW79382.1	Cre09.g413050	Chr.9: 7331940–7342521	1053	106,414.13	6.42	−0.401	54.24
CrebZIP7	PNW77489.1	Cre10.g438850	Chr.10: 2769177–2772294	485	49,874.20	6.64	−0.678	58.82
CrebZIP8	PNW77864.1	Cre10.g454850	Chr.10: 4910181–4919296	1363	128,662.31	8.75	−0.249	64.61
CrebZIP9	PNW74780.1	Cre12.g510200	Chr.12: 2010086–2013670	353	35,933.29	6.36	−0.398	58.58
CrebZIP10	PNW74984.1	Cre12.g501600	Chr.12: 2939321–2944866	902	89,920.31	6.25	−0.274	56.84
CrebZIP11	PNW75863.1	Cre12.g557300	Chr.12: 7330581–7337724	1526	154,908.24	9.40	−0.719	65.07
CrebZIP12	PNW73681.1	Cre13.g568350	Chr.13: 981891–989596	1150	114,710.75	9.55	−0.563	68.90
CrebZIP13	XP_001693067.1	Cre13.g590350	Chr.13: 3885273–3888238	334	34,514.92	4.96	−0.172	58.83
CrebZIP14	PNW71231.1	Cre16.g692250	Chr.16: 591049–600037	1525	153,084.80	5.30	−0.558	72.50
CrebZIP15	PNW71414.1	Cre16.g653300	Chr.16: 1524635–1531081	867	90,393.04	6.11	−0.374	55.17
CrebZIP16	PNW72253.1	Cre16.g675700	Chr.16: 6123854–6131215	1172	115,285.37	6.48	−0.512	46.46
CrebZIP17	PNW71098.1	Cre17.g746547	Chr.17: 7009739–7013345	767	75,438.72	6.01	−0.379	55.72

Table 2. Secondary structure of CrebZIP proteins.

Gene Number	Alpha Helix (%)	Extended Strand (%)	Beta Bridge (%)	Random Coil (%)
CrebZIP1	24.58	2.68	0	72.75
CrebZIP2	53.24	1.53	0	45.23
CrebZIP3	36.41	8.73	0	54.86
CrebZIP4	45.29	1.53	0	53.18
CrebZIP5	41.60	0.80	0	57.60
CrebZIP6	43.49	2.94	0	53.56
CrebZIP7	45.80	0.42	0	53.78
CrebZIP8	30.01	3.67	0	66.32
CrebZIP9	36.54	8.22	0	55.24
CrebZIP10	38.03	3.77	0	58.20
CrebZIP11	26.34	4.59	0	69.07
CrebZIP12	39.57	3.22	0	57.22
CrebZIP13	50.30	3.59	0	46.11
CrebZIP14	39.08	1.70	0	59.21
CrebZIP15	46.25	1.85	0	51.90
CrebZIP16	33.19	1.88	0	64.93
CrebZIP17	27.38	3.00	0	69.62

2.2. Phylogenetic and Motif Analysis of CrebZIP Proteins

To explore the evolution and classification of CrebZIP TFs, we performed a phylogenetic analysis (Figure 1) of 17 CrebZIP and 11 AtbZIP protein sequences. AtbZIPs were selected according to the classification of *Arabidopsis* bZIP proteins. Ten groups of AtbZIP proteins named Group A, B, C, D, E, F, G, H, I, and S were defined according to the sequence similarity of the basic region and other conserved motifs [17]. Those AtbZIP proteins that did not fit into any group mentioned above were classified as Group U (unknown). One AtbZIP protein was selected from each group respectively, including AtbZIP12 (Group A), AtbZIP17 (Group B), AtbZIP9 (Group C), AtbZIP20 (Group D), AtbZIP34 (Group E), AtbZIP19 (Group F), AtbZIP16 (Group G), AtbZIP56 (Group H), AtbZIP18 (Group I), AtbZIP1 (Group S), and AtbZIP60 (Group U). As shown in Figure 1, most bZIP members from the same species tended to cluster together. Only three CrebZIPs (CrebZIP2, 7, and 15) were grouped together with three AtbZIPs (AtbZIP16, 17 and 20), respectively, forming three subfamilies. Analysis on sequence identity and the similarity of bZIP proteins between *C. reinhardtii* and *Arabidopsis* grouped in the same subfamilies, showed low levels of amino acid conservation between the two species. CrebZIP2 and AtbZIP16 had 10.8% identity and 14.4% similarity, while CrebZIP7 and AtbZIP17 exhibited 10.5% identity and 16.9% similarity. The third pair of CrebZIP15 and AtbZIP20 only had 6.6% identity and 11.6% similarity. The remaining 14 CrebZIPs were grouped into 8 subfamilies, with each containing two CrebZIP members except for CrebZIP1 and CrebZIP5, which were classified as two single-member subfamilies. It is possible that the 14 CrebZIPs may have independent ancestral origins different from the AtbZIPs, in consideration of the fact that *C. reinhardtii* is a lower plant, while *Arabidopsis* is a higher plant. To extend the bZIP analysis to larger lineages of green plants, bZIP members from two bryophytes including *Physcomitrella patens* and *Marchantia polymorpha* were also added into the phylogenetic analysis, as bryophytes are considered as one of the earliest diverging distant land-plant lineages. Table S1 shows 43 *PpbZIP* genes from *P. patens* and 14 *MpbZIP* genes from *M. polymorpha* that were identified. SMART analysis indicated that these bZIP proteins all had the typical bZIP conserved domains. The phylogenetic analysis of bZIP proteins from *C. reinhardtii*, *Arabidopsis*, *P. patens*, and *M. polymophra* indicated that most CrebZIP proteins were also not highly homologous with *P. patens* and *M. polymorpha* bZIPs (Figure S1). Two CrebZIPs (CrebZIP 2 and 6) and two MpbZIPs (MpbZIP4 and 12) clustered together, respectively. However, levels of identity and similarity were very low between the CrebZIPs and MpbZIPs grouped in the same subfamilies. The identity and similarity between CrebZIP2 and MpbZIP4 were 14.5% and 19.8%, respectively, while CrebZIP6 and MpbZIP12 only shared 8.6% identity and 13.4% similarity.

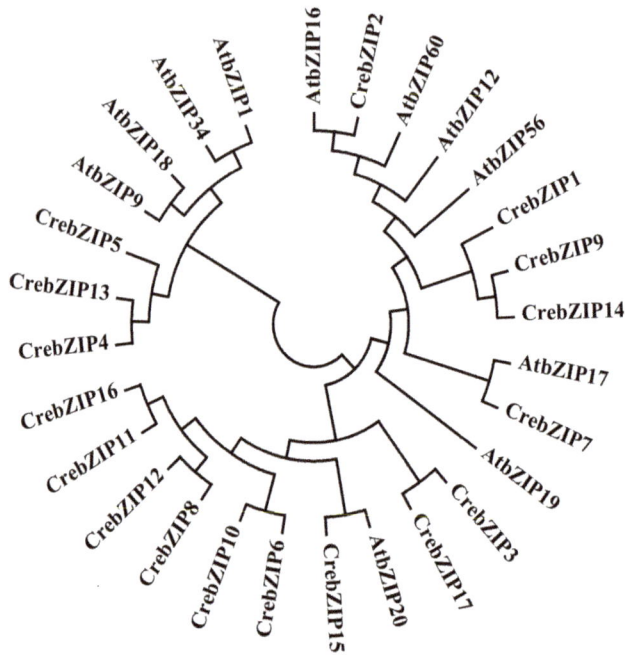

Figure 1. Phylogenetic analysis of *C. reinhardtii* and *Arabidopsis* basic leucine-region zipper (bZIP) proteins. ClustalW was employed to align the protein sequences of 17 CrebZIPs and 11 AtbZIPs representing subgroups A to I, S, and U in *Arabidopsis*. The phylogenetic tree was constructed using the neighbor-joining (NJ) method with MEGA7.0 software. The evolutionary distances were computed using the Poisson correction method with the number of amino acid substitutions per site as the unit. All positions containing gaps and missing data were excluded.

In view of the orthologous bZIP proteins playing a similar role, CrebZIP2 may function like AtbZIP16 (Group G), which mainly linked to light-regulated signal transduction and seed maturation. CrebZIP7 was aligned with AtbZIP17 (Group B), however no functional information was available for members of this group. CrebZIP15 was clustered together with AtbZIP20 from Group D. Members of this bZIP group mainly participated in defense against pathogens, and development [17].

To obtain insight into the divergence and function of CrebZIP TFs, the conserved motifs in the CrebZIPs were analyzed by MEME software. As depicted in Figure 2, all CrebZIP proteins contained the typical bZIP structure domain (motif 1). In addition, a glutamine (Q) enrichment region (motif 2) was detected in 10 CrebZIP proteins including CrebZIP1, 3, 6, 8, 10, 11, 12, 14, 15, and 16. In general, the basic region of bZIP protein has an invariant N-x7-R/K-x9 conserved motif residue rich in lysine (K) and arginine (R). The leucine zipper linked to the C-terminus of the basic region contains two sequential heptad repeat peptides, where a leucine (L) is located at the seventh of each peptide. In some cases, the leucine residue is replaced by isoleucine, valine, phenylalanine, or methionine. For CrebZIP proteins, the primary structure of the conserved domain was detected as N-x7-R/K-x9-L-x6-L-x6-L (motif 1), which is consistent with *Arabidopsis* bZIPs [17].

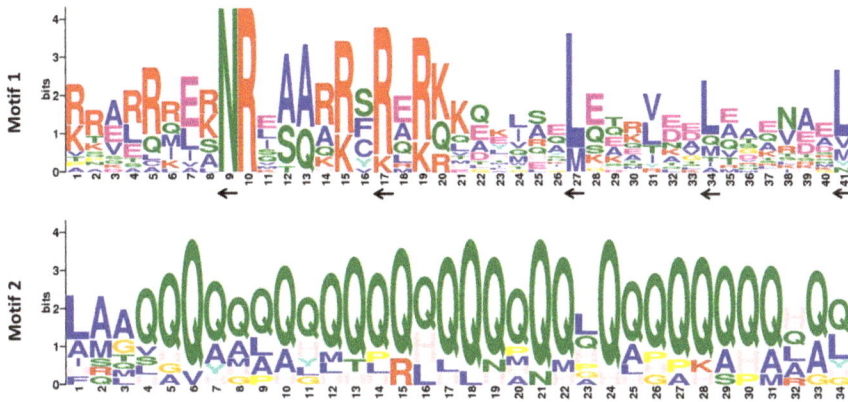

Figure 2. Analysis of the conserved domain in CrebZIP proteins. The conserved motifs in the CrebZIP proteins were identified with MEME software. Conserved sites in the key conserved domain (motif 1) are indicated by "←".

Previous studies showed that there were other special structural domains (e.g., proline-rich, glutamine-rich, and acidic domains) in plant bZIP proteins. These domains may have transcriptional activation function in regulating downstream target gene expressions [37]. For example, two glutamine-rich (~30% Gln) domains adjacent to the C terminal of a bZIP protein encoded by *PERIANTHIA (PAN)* in *Arabidopsis* were supposed to act as transcriptional activation domains [38]. A glutamine-rich region in the C-terminal halves of wheat bZIP family members HBP-1b (c38) and HBP-1b (c1), was reported to activate transcription of nuclear genes [39]. STGA1 (soybean TGA1), a member of the TGA (TGACG motif binding factor) subfamily of soybean bZIP TFs, contained a C-terminal glutamine-rich region as a putative transcription activation domain [40]. The conserved motifs shared by *Arabidopsis*, wheat, soybean, and other plants, suggested a similar function for the bZIP proteins. In this study, most *C. reinhardtii* bZIP protein sequences also contained a glutamine-rich region (motif 2), indicating that like higher plants, unicellular microalgae may retain structural domains of important functions during evolution, although the roles of these additional conserved motifs found in CrebZIP proteins are not yet clear.

2.3. Analysis of CrebZIP Gene Structure and Their Chromosomal Assignment

To further understand the evolutionary relationships among *CrebZIP* genes, GSDS (Gene Structure Display Server) was used to analyze their intron-exon structures. As shown in Figure 3, the number of exons varied from 3 to 16, demonstrating a great divergence among the 17 *CrebZIP* genes. The exon-intron structures of the genes were also highly different even in the same subfamily despite six exons were conserved in the subfamily composed of *CrebZIP8* and *CrebZIP12*. For instance, *CrebZIP6* and *CrebZIP10* grouped as a subfamily, with *CrebZIP6* having 16 exons and *CreZIP10* consisting of 5 exons. For the same subfamily of *CrebZIP9* and *CrebZIP14*, the former had 4 exons while the later contained 10 exons. Such variance in intron-exon structures was also found in *bZIP* genes of rice (*Oryza sativa*), soybean (*Glycine max*), and strawberry (*Fragaria ananassa*) plants. Among the *OsbZIP* genes having introns, the number of introns in open reading frames (ORF) varied from 1–12, 1–18, and 1–20 in rice [41], soybean [42], and strawberry plants [43], respectively. In agreement with these previous findings, this diversity in exon-intron organization indicated that both exon loss and gain occurred during the evolution of the *C. reinhardtii* bZIP gene family.

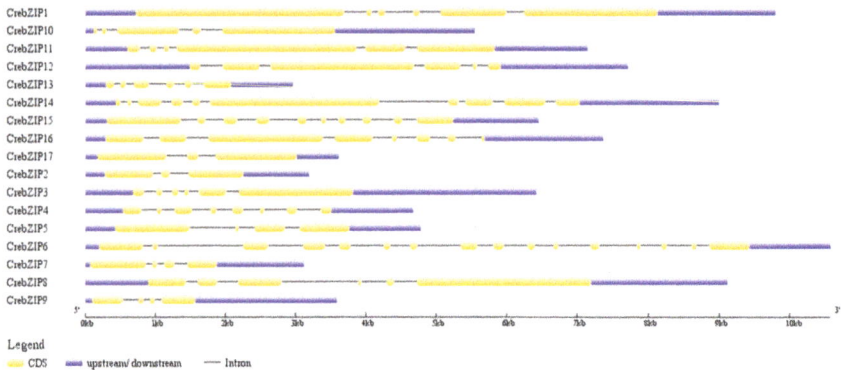

Figure 3. The Exon-intron organization of *CrebZIP* genes. The exons and introns are represented by yellow boxes and horizontal black lines, respectively. Untranslated regions are shown with blue boxes. The length of *CrebZIP* genes are indicated by the horizontal axis (kb).

Chromosome assignment of CrebZIP genes depicted by MapInspect software displayed 17 *CrebZIP* genes unevenly distributed on nine chromosomes of *C. reinhardtii*. Three *CrebZIP* genes were located on chromosomes 7, 12, and 16. Chromosomes 10 and 13 both had two *CrebZIP* genes. Only one *CrebZIP* gene was found on chromosomes 1, 5, 9, and 17 (Figure 4). In addition, several *CrebZIP* genes including *CrebZIP1, 2, 6, 13*, and *17*, were distributed near the ends of chromosomes.

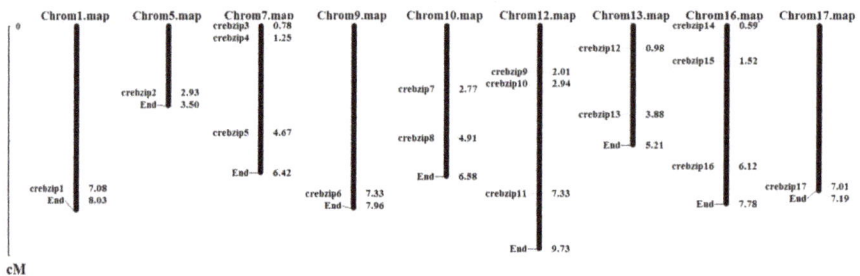

Figure 4. Distribution of *CrebZIP* gene family members on the *C. reinhardtii* chromosomes. The chromosome number is indicated at the top of each chromosome. Values next to the *bZIP* genes indicate the location on the chromosome.

2.4. Characterization of Cell Growth and Lipid Accumulation in C. reinhardtii Under Salt Stress

The effects of salt stress on *C. reinhardtii* growth and lipid accumulation are shown in Figure 5. Salt stress (150 mM NaCl treatment) significantly affected the cell growth, with OD values increasing slowly from 0.127 ± 0.002 to 0.276 ± 0.044 during 48 h cultivation, while the cell growth curve showed a rapid increase in the control with OD values from 0.127 ± 0.002 to 1.242 ± 0.052 (Figure 5a). In contrast, total lipid content in *C. reinhardtii* cells under salt stress significantly increased from 0.284 ± 0.029 to 0.437 ± 0.012 (Figure 5b), whereas lipid content in the control only increased at a small scale from 0.284 ± 0.029 to 0.348 ± 0.022, after 48 h cultivation. Consistent with these findings, Kato et al. observed that salinity stimulated lipid accumulation, but negatively affected biomass production in microalgae [44].

Oil accumulation in *C. reinhardtii* cells under salt stress was also examined by fluorescent microscopy using Nile Red staining (Figure 6). Notably, salt stress resulted in more oil droplets in *C. reinhardtii* cells, which was consistent with the total lipid content (Figure 5b) measured for the stressed cells. Similarly, other reports also showed that salinity stress induced enhancement of total lipid content and neutral lipid fractions within microalgae cells by affecting the fatty acid metabolism [45–47]. It is possible that in response to the decrease of cell membrane osmotic pressure and fluidity caused by salt stress, microalgae could employ an adaptive strategy to accumulate neutral lipid TAG, so as to maintain membrane integrity.

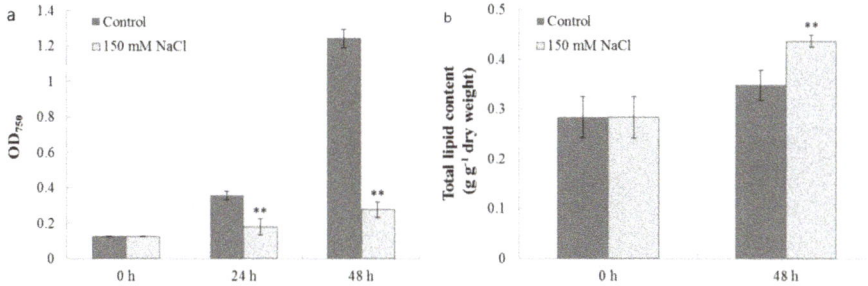

Figure 5. The effects of salt stress on (**a**) cell growth and (**b**) total lipid content of *C. reinhardtii*. Cell samples harvested after 0, 24, and 48 h cultivation from both the salt treatment and the control were used for measurements of cell growth and lipid content. Each value is the mean ± SD of three biological replicates. Asterisks indicate statistical significance (* 0.01 < p < 0.05, ** p < 0.01) between control and salt-treated cells according to the Tukey's test.

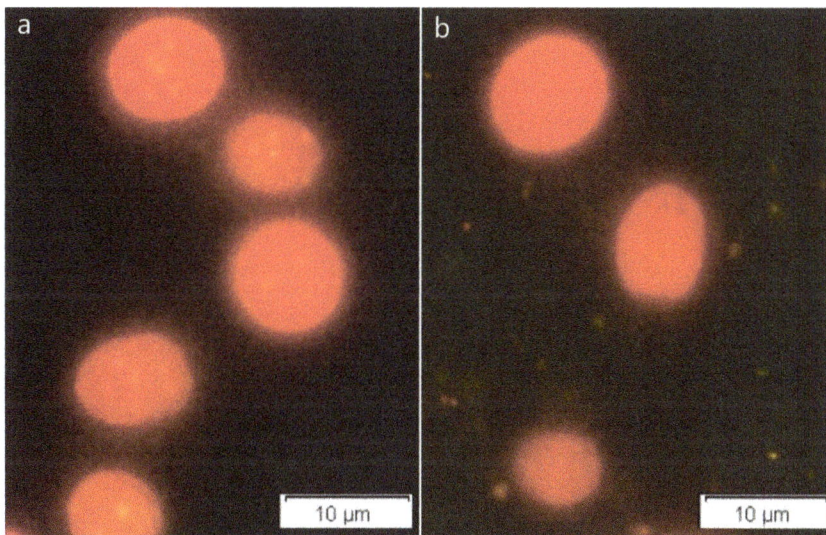

Figure 6. Nile Red staining lipid drops in *C. reinhardtii* under (**a**) salt stress and (**b**) the control treatments. The cell and lipid droplet colors were visualized with red and yellow, respectively, under fluorescent light.

2.5. Photosynthetic Properties of C. reinhardtii Cells under Salt Stress

To investigate the effects of salt stress on *C. reinhardtii* photosynthesis, the contents of three pigments, Chlorophyll a (Chla), Chlorophyll b (Chlb), and carotenoids (Car), were measured. As shown in Figure 7, salt stress also led to significant changes in pigment content of *C. reinhardtii*. Under salt stress, Chla and Chlb contents in *C. reinhardtii* both decreased from 6.147 ± 0.409 and 2.680 ± 0.161 to 4.903 ± 0.258 and 2.427 ± 0.0.055 mg 10^{-10} cells, respectively, during 48 h cultivation, showing significant difference from the control where Chla and Chlb contents increased during cultivation.

Chlorophyll is the main light-harvesting molecule for photosynthetic organisms. The reduction of chlorophyll content indicated weakened photosynthesis in *C. reinhardtii* under salt stress, possibly being the result of decreased synthesis or enhanced degradation of chlorophylls caused by the stress. It is known that salinity caused limitations in photosynthetic electron transport and then photosynthetic rate [48]. Moreover, salinity may stimulate chlorophyllase activity and accelerate chlorophyll degradation [49]. Unlike the chlorophyll case, the carotenoid content in *C. reinhardtii* grown under salinity increased from 2.240 ± 0.295 to 4.263 ± 0.180 mg 10^{-10} cells after 48 h cultivation, in comparison with the control where carotenoid content exhibited no obvious change during the cultivation. Higher carotenoid content under salt stress in *C. reinhardtii* was possibly due to positive adaptation to the stress, which was also observed in microalgae *Botryococcus braunii* by Rao et al. [50]. It was proved that carotenoids could protect the photosynthetic apparatus from photo-oxidative damage [51], providing the protection mechanism for microalgae cells against adverse stresses.

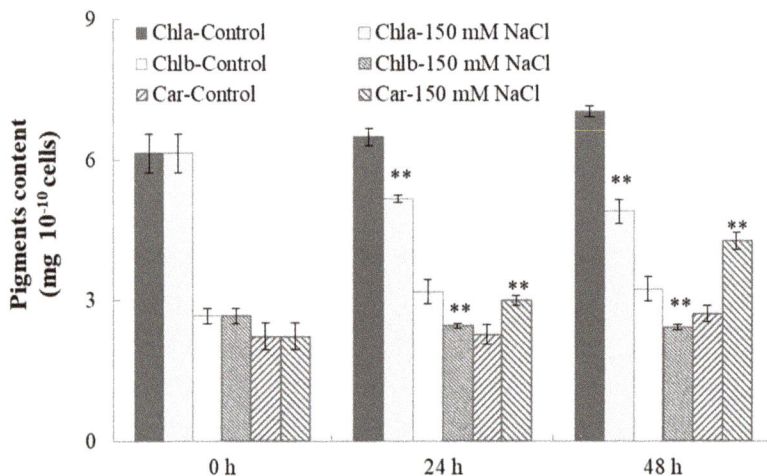

Figure 7. The effect of salt stress on pigment contents of *C. reinhardtii*. Cell samples harvested after 0, 24, and 48 h cultivation from both the salt treatment and the control were used to measure the contents of chlorophyll a, Chlorophyll b, and carotenoid. Each value is the mean $\pm$ SD of three biological replicates. Asterisks indicate statistical significance (* $0.01 < p < 0.05$, ** $p < 0.01$) between control and salt-treated cells according to the Tukey's test.

The photosynthetic performance in algae and plants was widely monitored by measuring chlorophyll fluorescence. For example, the F_v/F_m parameter was used for estimating the maximum quantum yield of PSII photochemistry. Non-photochemical quenching (NPQ) was employed to examine changes in the apparent rate constant for excitation decay by heat loss from PSII. NPQ was an essential part of the plant response to stress, as indicated by their slower growth [52]. Decreased F_v/F_m was often observed when plants were exposed to abiotic and biotic stresses [53]. The effect of salt stress on the chlorophyll fluorescence of *C. reinhardtii* is demonstrated in Figure 8. F_v/F_m values decreased from 0.695 ± 0.027 to 0.354 ± 0.038 in salt-treated cells during the 48 h cultivation, while it slightly increased from 0.695 ± 0.027 to 0.812 ± 0.022 in the control (Figure 8a). Meanwhile, NPQ values in cells detected under salt stress and in the control, both increased in this culture period. However, NPQ values under salinity increased more significantly (from 0.030 ± 0.002 to 0.199 ± 0.011) compared to the control (from 0.030 ± 0.002 to 0.061 ± 0.010) (Figure 8b). Analogously, Mou et al. [54] also observed lower F_v/F_m and associated induction of NPQ when *Chlamydomonas* sp. ICE-L was cultured under stress. In addition, they also summarized that the lower F_v/F_m indicated stress conditions, and the higher NPQ showed energy dissipation.

Taken together, salt stress affected various physiological and photosynthetic mechanisms associated with cell growth and development of *C. reinhardtii*. Microalgae cells altered their metabolism to adapt to the adverse environment by reducing biomass production and chlorophyll content, and simultaneously by increasing lipid and carotenoid contents. Moreover, salt stress impeded photosynthesis in *C. reinhardtii*, reflected by decreased F_v/F_m and increased NPQ. Salinity may induce excess production of reactive oxygen species (ROS) which caused cell damage. ROS also participated in regulating the expression of many genes and signal transduction pathways. Under stress, microalgae cells subsequently changed the physiological and photosynthetic status by: (1) Enhancing NPQ to reduce light energy absorption and to accelerate energy dissipation, thus mitigating the damage caused by excess excitation energy. (2) Accumulating lipid at the cost of reduced growth and carbohydrate storage as a response to the oxidative stress [55,56].

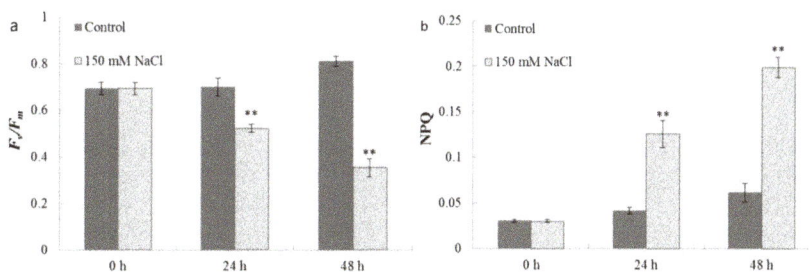

Figure 8. The effects of salt stress on (a) the maximum photo-chemical efficiency (F_v/F_m) and (b) non-photochemical quenching co-efficiency (NPQ) of *C. reinhardtii*. Cell samples harvested after 0, 24, and 48 h cultivation from both the salt treatment and the control, were used for measurements of F_v/F_m and NPQ. Each value is the mean ± SD of three biological replicates. Asterisks indicate statistical significance (* $0.01 < p < 0.05$, ** $p < 0.01$) between control and salt-treated cells according to the Tukey's test.

2.6. The Expression of CrebZIP Genes under Salt Stress

To investigate whether any CrebZIP functioned in regulating stress responses and oil accumulation in *C. reinhardtii*, expression analysis of all 17 *CrebZIP* genes following exposure to salt stress by quantitative real-time (qRT)-PCR were performed. Among the 17 *CrebZIP* genes detected, only six genes including *CrebZIP4, 5, 10, 11, 13*, and *16* showed significant expression changes during salt stress (Figure 9). The rest of 11 *CrebZIP* genes exhibited no obvious expression changes between the salt treatment and the control during 48 h cultivation. The *CrebZIP10, 11*, and *16* genes were up-regulated, whereas *CrebZIP4, 5*, and *13* genes were down-regulated under salt stress compared to the control (Figure 9), indicating that these CrebZIP TFs may be involved in the *C. reinhardtii* defense response to salt stress. Similarly, Zhu et al. also observed the obvious expression changes of several *bZIP* genes in tomato (*Solanum lycopersicum*) plants under salt and drought stresses. Furthermore, they concluded that SlbZIP1 mediates the stress tolerance in tomato plants through regulating an ABA-mediated pathway revealed by silencing of the gene and RNA-seq analysis [57]. Consequently, it was speculated that the six CrebZIP TFs may participate in regulation of photosynthesis and oil accumulation in *C. reinhardtii* under stress conditions, based on the association of *CrebZIP*'s expression and physiological phenotypes of the cells. Nevertheless, to verify whether the six CrebZIPs directly mediate the regulation of *C. reinhardtii* cell growth, photosynthesis, and lipid accumulation in response to stresses, mutants of gain-of-function and loss-of-function of these six CrebZIPs should be generated, and correspondingly, phenotypic analysis needs to be conducted on these mutants under stress in future study. The data from this study was the first evidence showing that CrebZIPs may mediate the regulation of stress responses, particularly the oil accumulation under salt stress, although a few CrebZIPs may be involved in regulating N-starving stress responses in *C. reinhardtii* [58].

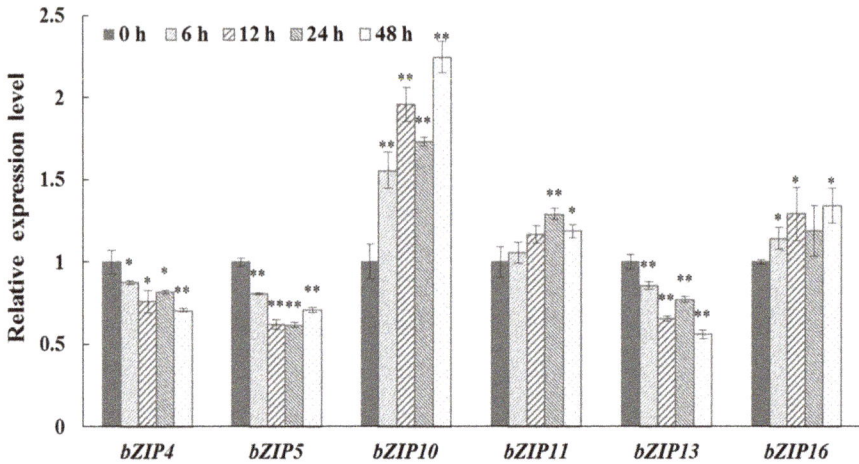

Figure 9. Expression profiles of six *CrebZIP* genes in *C. reinhardtii* under salt stress. Expression of *CrebZIP* genes was determined by qRT-PCR using total RNA from microalgae cells sampled at different time points of salt treatment. The α-tubulin gene was used as the internal reference gene. Each value is the mean ± SD of six biological replicates. Asterisks indicate statistical significance (* $0.01 < p < 0.05$, ** $p < 0.01$) between control and salt-treated cells according to the Tukey's test.

Although bZIP TFs in higher plants have been intensively explored, the study of bZIP proteins of microalgae have not, limiting the functional analysis of bZIP members in microalgae. The study of bZIP TFs in green plant evolution suggested that the ancestor of green plants possessed four bZIP genes functionally involved in oxidative stress, unfolded protein responses, and light-dependent regulations [15]. Takahashi et al. found that in algae *Vaucheria frigida* and *Fucus distichus*, each of two bZIP proteins, chromoproteins AUREO1 and AUREO2, contained one bZIP domain and one light–oxygen–voltage (LOV)-sensing domain, representing the blue light (BL) receptor. It was hypothesized that because bZIP proteins typically bind DNA by forming heterodimer, AUREO1 and AUREO2 may cooperatively regulate different kinds of BL responses by forming homo- and hetero-dimers [59]. Marie et al. reported that the two bZIP TFs AUREO1a and bZIP10 were likely to be involved in the blue light-dependent transcription of a cyclin gene that regulated the onset of the cell cycle in diatom (*Phaeodactylum tricornutum*) after a period of darkness [60]. Fischer et al. identified a singlet oxygen resistant 1 (SOR1), which was a putative bZIP protein in *C. reinhardtii*, that could stimulate the tolerance of *C. reinhardtii* to high O_2 formation by activating a reactive electrophile species (RES)-induced defense response, thereby enhancing the tolerance of this organism to photo-oxidative stress [61]. In addition to light-dependent regulation and oxidative stress resistance, bZIP TFs were also proved to participate in TAG accumulation in microalgae. Hu et al. revealed that bZIP family members were dominant (five such TFs) among 11 TFs that were potentially involved in the transcriptional regulation of TAG biosynthesis pathways in *Nannochloropsis* [14]. Consistent with these previous reports, the present study provides new data to show that some bZIP TFs may mediate the regulation of photosynthesis and lipid synthesis in microalgae, especially under stress conditions.

3. Materials and Methods

3.1. Microalgae Strains and Growth Conditions

The microalgae *C. reinhardtii* was purchased from the Freshwater Algae Culture Collection of the Institute of Hydrobiology, Chinese Academy of Science, China, and maintained in Tris-Acetate-Phosphate (TAP) medium [62]. The culture of *C. reinhardtii* was inoculated in 250 mL flasks under a continuous illumination of 100 μmol m^{-2} s^{-1} and a temperature of 25 ± 1 °C. A certain amount of NaCl was added into the TAP medium to a final concentration of 150 mmol L^{-1} in salt stress treatment, while algal cells were cultivated in normal TAP medium as the control.

3.2. Genome-Wide Identification of C. reinhardtii bZIP Gene Family

The bZIP protein sequence of *C. reinhardtii* was identified and downloaded through the following means: Using HMMMER V3.0 software with the Hidden Markov Model (HMM) PF00170 (from Pfam database, http://pfam.xfam.org/) of bZIP domain, from the Phytozome database (http://phytozome.jgi.doe.gov/pz/portal.html) and Plant Transcription Factor Database (PlnTFDB) with the keyword of bZIP. The screening results of the three means were merged, followed by manually removing the redundant or repetitive sequences.

The prediction of the bZIP structure domain was carried out within the amino acid sequence of the selected candidate bZIP protein family member in *C. reinhardtii* with the help of the Conserved Domain Database (CDD, http://www.ncbi.nlm.nih.gov/Structure/cdd/wrpsb.cgi) in the National Center for Biotechnology Information (NCBI, http://www.ncbi.nlm.nih.gov/), and Simple Modular Architecture Research Tool software (SMART, http://smart.embl-heidelberg.de). The candidate genes without bZIP structure domain were removed. The physicochemical properties of the bZIP protein of *C. reinhardtii*, including hydropathicity, molecular mass, instable index, and so forth, were predicted via ProtParam software from the Expasy database (http://web.expasy.org/protparam/).

3.3. Motif Recognition and Phylogenetic Analysis of C. reinhardtii bZIP Gene Family

Motifs of the selected genes were analyzed using the MEME suite (http://meme-suite.org/tools/meme). Multiple alignments of bZIP protein sequences were performed by ClustalW software and the phylogenetic tree was constructed with MEGA 7.0 software (https://www.megasoftware.net/home).

3.4. bZIP Protein Secondary Structure Prediction

The secondary structure of bZIP proteins in *C. reinhardtii* was predicted using the PBIL LYON-GERLAND database (https://npsa-prabi.ibcp.fr/cgi-bin/npsa_automat.pl?page=/NPSA/npsa_hnn.html).

3.5. bZIP Gene Structure and Chromosomal Assignment Analysis

The *bZIP* gene structure diagram was constructed by Gene Structure Display Server (GSDS) 2.0 software (http://gsds.cbi.pku.edu.cn/), and then quantitative analysis of introns and exons followed. The chromosome map of *CrebZIP* gene family members was depicted by MapInspect software according to the *CrebZIP* genes location on chromosomes of *C. reinhardtii*.

3.6. Measure of Microalgae Biomass Concentration

The biomass concentration of *C. reinhardtii* was indicated by optical cell density, which was measured with a UV-Visible spectrophotometer (UV-1200, Shanghai Jingke Instrument Co., Ltd., Shanghai, China) at 750 nm (OD_{750}). When necessary, the sample was diluted to give an absorbance in the range of 0.1–1.0. The experiments were conducted in triplicate.

3.7. Analysis of Total Lipid Content in Microalgae Cells

The total lipid content was determined by gravimetric analysis according to the method of Chen et al. [63]. The microalgae cells were collected by centrifugation and then lyophilized. About 50 mg of lyophilized microalgae sample was triturated in a mortar, and then the cell disruption was added into 7.5 mL chloroform/methanol (1:2, v/v) mixture. The mixture was placed at 37 °C overnight and then centrifuged to collect the supernatant. Residual biomass was extracted at least once more. All the supernatants were combined, and added into a chloroform and 1% sodium chloride solution to a final volume ratio of 1:1:0.9 (chloroform/methanol/water). The new mixture was centrifuged afterwards, and the subnatant was transferred to a pre-weighted vitreous vial. The sample solution was dried to constant weight at 60 °C under nitrogen flow. Finally, the total lipid content was obtained as a percentage of the dry weight (DW) of the microalgae. The experiments were conducted in triplicate.

3.8. Nile Red Staining and Microscopy for Assessment of Oil Accumulation in Microalgae Cells

Lipid droplets were visualized by fluorescent microscopy using Nile Red staining [64]. Microalgae cells under salt stress and control conditions were collected after 48 h cultivation by centrifugation. The cells pellets were washed with physiological saline solution three times. After the collected cells were re-suspended in the same solution, 10 μL Nile Red stain (0.1 g L^{-1} in acetone) and 200 μL dimethyl sulfoxide were added into an 800 μL microalgae suspension, with final concentration at about 1×10^6 cells mL^{-1}. The stained cells were incubated in the dark for 20 min at room temperature, and then immediately observed by fluorescent microscopy.

3.9. Determination of the Pigment Contents in Microalgae Cells

Contents of microalgae pigments, including chlorophyll (a and b) and carotenoid, were determined according to the methods described by Wellburn [65]. A certain amount of microalgae culture was centrifuged to collect cells. The cell pellets were mixed with methanol at 60 °C for 12 h in darkness, until the microalgae cells whitened completely. The mixture was centrifuged and supernatant extraction was collected. The optical densities of the extraction were measured with a spectrophotometer (UV-1601, Beijing Beifen-Ruili Aanlytical Instrument Co., Ltd., Beijing, China) at 666, 653, and 470 nm. The concentrations of chlorophyll a (Chla), chlorophyll b (Chlb), and carotenoid (Car) of the extraction were calculated as follows (mg L^{-1}):

$$Chla = 15.65 \times OD_{666} - 7.34 \times OD_{653} \tag{1}$$

$$Chlb = 27.05 \times OD_{653} - 11.21 \times OD_{666} \tag{2}$$

$$Car = (1000 \times OD_{470} - 2.86 \times Chla - 129.2 \times Chlb)/221 \tag{3}$$

The cell density of the microalgae culture was obtained using cytometry with a hemacytometer. The experiments were conducted in triplicate.

3.10. Chlorophyll Fluorescence Measurements

The chlorophyll fluorescence was measured using an Imaging-PAM (Pulse Amplitude Modulation) fluorescence monitor (Walz, Effeltrich, Germany) according to the method described by Mou et al. [54]. Samples were dark adapted for 15 min before the fluorescence measurement. Then the dark-level fluorescence yield (F_0), the maximum fluorescence yield (F_m), and the maximum light-adapted fluorescence yield (F'_m) were measured.

The maximum quantum yield of PSII was calculated as:

$$F_v/F_m = (F_m - F_0)/F_m \tag{4}$$

The non-photochemical quenching (NPQ) was calculated as:

$$NPQ = (F_m - F'_m)/F'_m \tag{5}$$

The experiments were conducted in triplicate.

3.11. RNA Isolation and Quantitative Real-Time (qRT)-PCR Analysis

CrebZIP genes expression under salt stress were analyzed by qRT-PCR. Total RNA of *C. reinhardtii* cells sampled after 0, 6, 12, 24, and 48 h cultivation under salt stress were extracted. RNA from each sample was used to synthesize the cDNAs with the cDNA Synthesis Kit (TaKaRa, Kusatsu, Japan). The primers for the 17 *CrebZIP* genes and one reference gene (α-tubulin) were designed using

Primer Premier 5.0 software. Sequences of all the primers used in this study are shown in Table 3. The qRT-PCR was performed in an ABI 7500 qRT-PCR system (Applied Biosystems, Foster City, CA, USA) with following reaction conditions: 95 °C for 3 min followed by 40 cycles of 95 °C for 15 s, 60 °C for 30 s, and 72 °C for 30 s. The experiments were repeated six times using independent RNA samples.

Table 3. Primer information of *CrebZIP* and reference genes for quantitative real-time (qRT)-PCR.

Gene Number	Forward Primer (5'-3')	Reverse Primer (5'-3')
CrebZIP1	CGGTCGATGACGCTAAGGC	TGGTCGGTGTCGGAGGAGT
CrebZIP2	GAGCGCAAGAAGCAGTACGTGACCT	GATGTTCCGCAGTGCCTCGTTCT
CrebZIP3	ATGCACCAAACGCCAAATCG	ATCTGCTGTAGGTCCGCCAGGGTA
CrebZIP4	GACAGCGGAGACTCAGACAT	CCCTTCTTACGCTGCCTGTA
CrebZIP5	CCGTCATCCATGCACCTACT	GCTGAAGATCAAGACCGCTG
CrebZIP6	CCCGTTTCCAGCCCATCAGAC	TGCAAGGCCGAGGGTGTTGATGAC
CrebZIP7	TGGGAGGCTCTGGCTTTGG	TGGCTGCTGCTGCTGCTGATG
CrebZIP8	GCAAAGGCAAGGGCAAGG	CGCAGGAGTTCTCAGCCGATT
CrebZIP9	GCGCTTCTTCCTCGCTACCA	CGCAAGCCCTTGTGCTGTTA
CrebZIP10	CCCTGACGTCCACCTTAACT	TCCCTCAGACAGCAACTCAG
CrebZIP11	ACGGACGTATGACATGAGCA	ACTGCTGGAACTGGTGGAA
CrebZIP12	CACAGCGACCGCAGCCATAA	GCCCAGCTTGTCCGAGAAGGA
CrebZIP13	GAGCTGCCTTCGACAAGC	CTGTGCTAGGCGATTCAGC
CrebZIP14	GATGGCGGGCTTATGTCGG	CAAGGCGTCCACGTTGTG
CrebZIP15	TGACTCATCCACGCACTTCCTC	TGATGTTGCACCAGCCCTGA
CrebZIP16	GGTTTCCTTCCCTACCCA	ACGGCACGGTTGTCAGCA
CrebZIP17	AAGCGCATTGTTGACGGAG	CGCAGCAGGTCTAATAAGTCG
Creα-tubulin	CTCGCTTCGCTTTGACGGTG	CGTGGTACGCCTTCTCGGC

4. Conclusions

The first genome-wide analysis of the bZIP TF family members in *C. reinhardtii* was performed in this study, with a total of 17 CrebZIP proteins identified. Detailed information was obtained for their evolutionary relationship, exon-intron organization and chromosome assignment, protein structural features, and conserved motifs. Moreover, expression profiling of *CrebZIP* genes by qRT-PCR indicated that six CrebZIPs might be involved in stress response and lipid accumulation in microalgae cells. Salt stress led to the reduction in biomass production, chlorophyll content, and F_v/F_m, but the enhancement in NPQ, carotenoid content, and oil accumulation in *C. reinhardtii*. Collectively, integration of findings in the present study provided new data to indicate that some CrebZIP TFs could play important roles in mediating regulation of cell growth, photosynthesis, and oil accumulation in microalgae, particularly under stress conditions. These *CrebZIP* genes could be utilized to further functionally characterize them, laying the foundation for elucidating their specific regulatory mechanisms and ultimately applying them in genetic improvement programs.

Supplementary Materials: Supplementary materials can be found at http://www.mdpi.com/1422-0067/19/9/2800/s1.

Author Contributions: R.L. designed the experiments and co-wrote the manuscript. C.J. performed the experiments and wrote the manuscript. C.J., X.M., J.H., X.W., J.X., and H.C. analyzed the data. All authors read and approved the final manuscript.

Funding: This work was financially supported by grants from the Coal-based Key Sci-Tech Project of Shanxi Province, China (Grant No. FT-2014-01), the Key Project of The Key Research and Development Program of Shanxi Province, China (Grant No. 201603D312005), and the Technology Innovation Fund of Shanxi Agricultural University, China (Grant No. 2016YJ01).

Conflicts of Interest: The authors declare no conflict of interest.

References

1. Wijffels, R.H.; Barbosa, M.J. An outlook on microalgal biofuels. *Science* **2010**, *329*, 796–799. [CrossRef] [PubMed]

2. Fan, J.H.; Cui, Y.B.; Wan, M.X.; Wang, W.L.; Li, Y.G. Lipid accumulation and biosynthesis genes response of the oleaginous *Chlorella pyrenoidosa* under three nutrition stressors. *Biotechnol. Biofuels* **2014**, *7*, 17. [CrossRef] [PubMed]
3. Wang, Z.T.; Ullrich, N.; Joo, S.; Waffenschmidt, S.; Goodenough, U. Algal lipid bodies: Stress induction, purification, and biochemical characterization in wild-type and starchless *Chlamydomonas reinhardtii*. *Eukaryot. Cell* **2009**, *8*, 1856–1868. [CrossRef] [PubMed]
4. Cakmak, T.; Angun, P.; Demiray, Y.E.; Ozkan, A.D.; Elibol, Z.; Tekinay, T. Differential effects of nitrogen and sulfur deprivation on growth and biodiesel feedstock production of *Chlamydomonas reinhardtii*. *Biotechnol. Bioeng.* **2012**, *109*, 1947–1957. [CrossRef] [PubMed]
5. He, Q.; Yang, H.; Wu, L.; Hu, C. Effect of light intensity on physiological changes, carbon allocation and neutral lipid accumulation in oleaginous microalgae. *Bioresour. Technol.* **2015**, *191*, 219–228. [CrossRef] [PubMed]
6. Merchant, S.S.; Prochnik, S.E.; Vallon, O.; Harris, E.H.; Karpowicz, S.J.; Witman, G.B.; Terry, A.; Salamov, A.; Fritz-Laylin, L.K.; Marechal-Drouard, L.; et al. The *Chlamydomonas* genome reveals the evolution of key animal and plant functions. *Science* **2007**, *318*, 245–250. [CrossRef] [PubMed]
7. Lv, H.; Qu, G.; Qi, X.; Lu, L.; Tian, C.; Ma, Y. Transcriptome analysis of *Chlamydomonas reinhardtii* during the process of lipid accumulation. *Genomics* **2013**, *101*, 229–237. [CrossRef] [PubMed]
8. Adrián, L.G.D.L.; Sascha, S.; Jacob, V.; Saheed, I.; Warren, C.; Bilgin, D.D.; Yohn, C.B.; Serdar, T.; Reiss, D.J.; Orellana, M.V. Transcriptional program for nitrogen starvation-induced lipid accumulation in *Chlamydomonas reinhardtii*. *Biotechnol. Biofuels* **2015**, *8*, 207. [CrossRef]
9. Rolland, N.; Atteia, A.; Decottignies, P.; Garin, J.; Hippler, M.; Kreimer, G.; Lemaire, S.D.; Mittag, M.; Wagner, V. *Chlamydomonas* proteomics. *Curr. Opin. Microbiol.* **2009**, *12*, 285–291. [CrossRef] [PubMed]
10. Wienkoop, S.; Weiss, J.; May, P.; Kempa, S.; Irgang, S.; Recuenco-Munoz, L.; Pietzke, M.; Schwemmer, T.; Rupprecht, J.; Egelhofer, V.; et al. Targeted proteomics for *Chlamydomonas reinhardtii* combined with rapid subcellular protein fractionation, metabolomics and metabolic flux analyses. *Mol. Biosyst.* **2010**, *6*, 1018–1031. [CrossRef] [PubMed]
11. Mastrobuoni, G.; Irgang, S.; Pietzke, M.; Aßmus, H.E.; Wenzel, M.; Schulze, W.X.; Kempa, S. Proteome dynamics and early salt stress response of the photosynthetic organism *Chlamydomonas reinhardtii*. *BMC Genom.* **2012**, *13*, 215. [CrossRef] [PubMed]
12. May, P.; Wienkoop, S.; Kempa, S.; Usadel, B.; Christian, N.; Rupprecht, J.; Weiss, J.; Recuenco-Munoz, L.; Ebenhöh, O.; Weckwerth, W.; et al. Metabolomics- and proteomics-assisted genome annotation and analysis of the draft metabolic network of *Chlamydomonas reinhardtii*. *Genetics* **2008**, *179*, 157–166. [CrossRef] [PubMed]
13. Valledor, L.; Furuhashi, T.; Recuencomuñoz, L.; Wienkoop, S.; Weckwerth, W. System-level network analysis of nitrogen starvation and recovery in *Chlamydomonas reinhardtii* reveals potential new targets for increased lipid accumulation. *Biotechnol. Biofuels* **2014**, *7*, 171. [CrossRef] [PubMed]
14. Hu, J.; Wang, D.; Jing, L.; Jing, G.; Kang, N.; Jian, X. Genome-wide identification of transcription factors and transcription-factor binding sites in oleaginous microalgae *Nannochloropsis*. *Sci. Rep.* **2014**, *4*, 5454. [CrossRef] [PubMed]
15. Corrêa, L.G.; Riaño-Pachón, D.M.; Schrago, C.G.; dos Santos, R.V.; Mueller-Roeber, B.; Vincentz, M. The role of bZIP transcription factors in green plant evolution: Adaptive features emerging from four founder genes. *PLoS ONE* **2008**, *3*, e2944. [CrossRef] [PubMed]
16. Riechmann, J.L.; Heard, J.; Martin, G.; Reuber, L.; Jiang, C.; Keddie, J.; Adam, L.; Pineda, O.; Ratcliffe, O.J.; Samaha, R.R. *Arabidopsis* transcription factors: Genome-wide comparative analysis among eukaryotes. *Science* **2000**, *290*, 2105–2110. [CrossRef] [PubMed]
17. Jakoby, M.; Weisshaar, B.; Drögelaser, W.; Vicentecarbajosa, J.; Tiedemann, J.; Kroj, T.; Parcy, F. bZIP transcription factors in *Arabidopsis*. *Trends Plant Sci.* **2002**, *7*, 106–111. [CrossRef]
18. Zou, M.; Guan, Y.; Ren, H.; Zhang, F.; Chen, F. A bZIP transcription factor, OsABI5, is involved in rice fertility and stress tolerance. *Plant Mol. Biol.* **2008**, *66*, 675–683. [CrossRef] [PubMed]
19. Yanez, M.; Caceres, S.; Orellana, S.; Bastias, A.; Verdugo, I.; Ruiz-Lara, S.; Casaretto, J.A. An abiotic stress-responsive bZIP transcription factor from wild and cultivated tomatoes regulates stress-related genes. *Plant Cell Rep.* **2009**, *28*, 1497–1507. [CrossRef] [PubMed]

20. Ying, S.; Zhang, D.F.; Fu, J.; Shi, Y.S.; Song, Y.C.; Wang, T.Y.; Li, Y. Cloning and characterization of a maize bZIP transcription factor, ZmbZIP72, confers drought and salt tolerance in transgenic *Arabidopsis*. *Planta* **2012**, *235*, 253–266. [CrossRef] [PubMed]

21. Wang, J.; Zhou, J.; Zhang, B.; Vanitha, J.; Ramachandran, S.; Jiang, S.Y. Genome-wide expansion and expression divergence of the basic leucine zipper transcription factors in higher plants with an emphasis on sorghum. *J. Integr. Plant Biol.* **2011**, *53*, 212–231. [CrossRef] [PubMed]

22. Que, F.; Wang, G.L.; Huang, Y.; Xu, Z.S.; Wang, F.; Xiong, A.S. Genomic identification of group A bZIP transcription factors and their responses to abiotic stress in carrot. *Genet. Mol. Res.* **2015**, *14*, 13274–13288. [CrossRef] [PubMed]

23. Fukazawa, J.; Sakai, T.; Ishida, S.; Yamaguchi, I.; Kamiya, Y.; Takahashi, Y. Repression of shoot growth, a bZIP transcriptional activator, regulates cell elongation by controlling the level of gibberellins. *Plant Cell* **2000**, *12*, 901–915. [CrossRef] [PubMed]

24. Abe, M.; Kobayashi, Y.; Yamamoto, S.; Daimon, Y.; Yamaguchi, A.; Ikeda, Y.; Ichinoki, H.; Notaguchi, M.; Goto, K.; Araki, T. FD, a bZIP protein mediating signals from the floral pathway integrator FT at the shoot apex. *Science* **2005**, *309*, 1052–1056. [CrossRef] [PubMed]

25. Silveira, A.B.; Gauer, L.; Tomaz, J.P.; Cardoso, P.R.; Carmelloguerreiro, S.; Vincentz, M. The *Arabidopsis* AtbZIP9 protein fused to the VP16 transcriptional activation domain alters leaf and vascular development. *Plant Sci.* **2007**, *172*, 1148–1156. [CrossRef]

26. Shen, H.; Cao, K.; Wang, X. A conserved proline residue in the leucine zipper region of AtbZIP34 and AtbZIP61 in *Arabidopsis thaliana* interferes with the formation of homodimer. *Biochem. Biophys. Res. Commun.* **2007**, *362*, 425–430. [CrossRef] [PubMed]

27. Baena-González, E.; Rolland, F.; Thevelein, J.M.; Sheen, J. A central integrator of transcription networks in plant stress and energy signalling. *Nature* **2007**, *448*, 938–942. [CrossRef] [PubMed]

28. Lara, P.; Oñatesánchez, L.; Abraham, Z.; Ferrándiz, C.; Díaz, I.; Carbonero, P.; Vicentecarbajosa, J. Synergistic activation of seed storage protein gene expression in *Arabidopsis* by ABI3 and two bZIPs related to OPAQUE2. *J. Biol. Chem.* **2003**, *278*, 21003–21011. [CrossRef] [PubMed]

29. Thurow, C.; Schiermeyer, A.S.; Butterbrodt, T.; Nickolov, K.; Gatz, C. Tobacco bZIP transcription factor TGA2.2 and related factor TGA2.1 have distinct roles in plant defense responses and plant development. *Plant J.* **2005**, *44*, 100–113. [CrossRef] [PubMed]

30. Kaminaka, H.; Näke, C.; Epple, P.; Dittgen, J.; Schütze, K.; Chaban, C.; Holt, B.F.; Merkle, T.; Schäfer, E.; Harter, K. bZIP10-LSD1 antagonism modulates basal defense and cell death in *Arabidopsis* following infection. *EMBO J.* **2006**, *25*, 4400–4411. [CrossRef] [PubMed]

31. Nieva, C.; Busk, P.K.; Domínguez-Puigjaner, E.; Lumbreras, V.; Testillano, P.S.; Risueño, M.C.; Pagès, M. Isolation and functional characterisation of two new bZIP maize regulators of the ABA responsive gene *rab28*. *Plant Mol. Biol.* **2005**, *58*, 899–914. [CrossRef] [PubMed]

32. Uno, Y.; Furihata, T.; Abe, H.; Yoshida, R.; Shinozaki, K.; Yamaguchi-Shinozaki, K. *Arabidopsis* basic leucine zipper transcription factors involved in an abscisic acid-dependent signal transduction pathway under drought and high-salinity conditions. *Proc. Natl. Acad. Sci. USA* **2000**, *97*, 11632–11637. [CrossRef] [PubMed]

33. Wellmer, F.; Kircher, S.; Rügner, A.; Frohnmeyer, H.; Schäfer, E.; Harter, K. Phosphorylation of the parsley bZIP transcription factor CRPF2 is regulated by light. *J. Biol. Chem.* **1999**, *274*, 29476–29482. [CrossRef] [PubMed]

34. Ulm, R.; Baumann, A.; Oravecz, A.; Máté, Z.; Adám, E.; Oakeley, E.J.; Schäfer, E.; Nagy, F. Genome-wide analysis of gene expression reveals function of the bZIP transcription factor HY5 in the UV-B response of *Arabidopsis*. *Proc. Natl. Acad. Sci. USA* **2004**, *101*, 1397–1402. [CrossRef] [PubMed]

35. Liu, C.; Mao, B.; Ou, S.; Wang, W.; Liu, L.; Wu, Y.; Chu, C.; Wang, X. OsbZIP71, a bZIP transcription factor, confers salinity and drought tolerance in rice. *Plant Mol. Biol.* **2014**, *84*, 19–36. [CrossRef] [PubMed]

36. Riano-Pachon, D.M.; Correa, L.G.; Trejos-Espinosa, R.; Mueller-Roeber, B. Green transcription factors: A *Chlamydomonas* overview. *Genetics* **2008**, *179*, 31–39. [CrossRef] [PubMed]

37. Liao, Y.; Zou, H.F.; Wei, W.; Hao, Y.J.; Tian, A.G.; Huang, J.; Liu, Y.F.; Zhang, J.S.; Chen, S.Y. Soybean GmbZIP44, GmbZIP62 and GmbZIP78 genes function as negative regulator of aba signaling and confer salt and freezing tolerance in transgenic *Arabidopsis*. *Planta* **2008**, *228*, 225–240. [CrossRef] [PubMed]

38. Chuang, C.F.; Running, M.P.; Williams, R.W.; Meyerowitz, E.M. The *PERIANTHIA* gene encodes a bZIP protein involved in the determination of floral organ number in *Arabidopsis thaliana*. *Genes Dev.* **1999**, *13*, 334–344. [CrossRef] [PubMed]

39. Mikami, K.; Sakamoto, A.; Iwabuchi, M. The HBP-1 family of wheat basic/leucine zipper proteins interacts with overlapping cis-acting hexamer motifs of plant histone genes. *J. Biol. Chem.* **1994**, *269*, 9974–9985. [PubMed]

40. Cheong, Y.H.; Park, J.M.; Yoo, C.M.; Bahk, J.D.; Cho, M.J.; Hong, J.C. Isolation and characterization of STGA1, a member of the TGA1 family of bZIP transcription factors from soybean. *Mol. Cells* **1994**, *4*, 405–412.

41. Nijhawan, A.; Jain, M.; Tyagi, A.K.; Khurana, J.P. Genomic survey and gene expression analysis of the basic leucine zipper transcription factor family in rice. *Plant Physiol.* **2008**, *146*, 333–350. [CrossRef] [PubMed]

42. Zhang, M.; Liu, Y.; Shi, H.; Guo, M.; Chai, M.; He, Q.; Yan, M.; Cao, D.; Zhao, L.; Cai, H. Evolutionary and expression analyses of soybean basic leucine zipper transcription factor family. *BMC Genom.* **2018**, *19*, 159. [CrossRef] [PubMed]

43. Wang, X.L.; Chen, X.; Yang, T.B.; Cheng, Q.; Cheng, Z.M. Genome-wide identification of bZIP family genes involved in drought and heat stresses in strawberry (*Fragaria vesca*). *Int. J. Genom.* **2017**, *2017*, 1–14. [CrossRef]

44. Kato, Y.; Ho, S.H.; Vavricka, C.J.; Chang, J.S.; Hasunuma, T.; Kondo, A. Evolutionary engineering of salt-resistant *Chlamydomonas* sp. strains reveals salinity stress-activated starch-to-lipid biosynthesis switching. *Bioresour. Technol.* **2017**, *245*, 1484–1490. [CrossRef] [PubMed]

45. Kalita, N.; Baruah, G.; Chandra, R.; Goswami, D.; Talukdar, J.; Kalita, M.C. *Ankistrodesmus falcatus*: A promising candidate for lipid production, its biochemical analysis and strategies to enhance lipid productivity. *J. Microbiol. Biotechnol. Res.* **2011**, *1*, 148–157.

46. Lu, N.; Wei, D.; Chen, F.; Yang, S.T. Lipidomic profiling and discovery of lipid biomarkers in snow alga *Chlamydomonas nivalis* under salt stress. *Eur. J. Lipid Sci. Technol.* **2012**, *114*, 253–265. [CrossRef]

47. Mohan, S.V.; Devi, M.P. Salinity stress induced lipid synthesis to harness biodiesel during dual mode cultivation of mixotrophic microalgae. *Bioresour. Technol.* **2014**, *165*, 288–294. [CrossRef] [PubMed]

48. Zhang, T.; Gong, H.; Wen, X.; Lu, C. Salt stress induces a decrease in excitation energy transfer from phycobilisomes to photosystem II but an increase to photosystem I in the cyanobacterium *Spirulina platensis*. *J. Plant Physiol.* **2010**, *167*, 951–958. [CrossRef] [PubMed]

49. Santos, C.V. Regulation of chlorophyll biosynthesis and degradation by salt stress in sunflower leaves. *Sci. Hortic.* **2004**, *103*, 93–99. [CrossRef]

50. Rao, A.R.; Dayananda, C.; Sarada, R.; Shamala, T.R.; Ravishankar, G.A. Effect of salinity on growth of green alga *Botryococcus braunii* and its constituents. *Bioresour. Technol.* **2007**, *98*, 560–564. [CrossRef] [PubMed]

51. Fedina, I.S.; Grigorova, I.D.; Georgieva, K.M. Response of barley seedlings to UV-B radiation as affected by NaCl. *J. Plant Physiol.* **2003**, *160*, 205–208. [CrossRef] [PubMed]

52. Finazzi, G.; Johnson, G.N.; Dall'Osto, L.; Zito, F.; Bonente, G.; Bassi, R.; Wollman, F.A. Nonphotochemical quenching of chlorophyll fluorescence in *Chlamydomonas reinhardtii*. *Biochemistry* **2006**, *45*, 1490–1498. [CrossRef] [PubMed]

53. Baker, N.R. Chlorophyll fluorescence: A probe of photosynthesis in vivo. *Annu. Rev. Plant Biol.* **2008**, *59*, 89–113. [CrossRef] [PubMed]

54. Mou, S.; Zhang, X.; Ye, N.; Dong, M.; Liang, C.; Qiang, L.; Miao, J.; Dong, X.; Zhou, Z. Cloning and expression analysis of two different *LhcSR* genes involved in stress adaptation in an Antarctic microalga, *Chlamydomonas* sp. ICE-L. *Extremophiles* **2012**, *16*, 193–203. [CrossRef] [PubMed]

55. Kan, G.; Shi, C.; Wang, X.; Xie, Q.; Wang, M.; Wang, X.; Miao, J. Acclimatory responses to high-salt stress in *Chlamydomonas* (Chlorophyta, Chlorophyceae) from Antarctica. *Acta Oceanol. Sin.* **2012**, *31*, 116–124. [CrossRef]

56. Wang, T.; Ge, H.; Liu, T.; Tian, X.; Wang, Z.; Guo, M.; Chu, J.; Zhuang, Y. Salt stress induced lipid accumulation in heterotrophic culture cells of *Chlorella protothecoides*: Mechanisms based on the multi-level analysis of oxidative response, key enzyme activity and biochemical alteration. *J. Biotechnol.* **2016**, *228*, 18–27. [CrossRef] [PubMed]

57. Zhu, M.; Meng, X.; Cai, J.; Li, G.; Dong, T.; Li, Z. Basic leucine zipper transcription factor SlbZIP1 mediates salt and drought stress tolerance in tomato. *BMC Plant Biol.* **2018**, *18*, 83. [CrossRef] [PubMed]

58. Miller, R.; Wu, G.; Deshpande, R.R.; Vieler, A.; Gärtner, K.; Li, X.; Moellering, E.R.; Zäuner, S.; Cornish, A.J.; Liu, B. Changes in transcript abundance in *Chlamydomonas reinhardtii* following nitrogen deprivation predict diversion of metabolism. *Plant Physiol.* **2010**, *154*, 1737–1752. [CrossRef] [PubMed]

59. Takahashi, F.; Yamagata, D.; Ishikawa, M.; Fukamatsu, Y.; Ogura, Y.; Kasahara, M.; Kiyosue, T.; Kikuyama, M.; Wada, M.; Kataoka, H. AUREOCHROME, a photoreceptor required for photomorphogenesis in stramenopiles. *Proc. Natl. Acad. Sci. USA* **2007**, *104*, 19625–19630. [CrossRef] [PubMed]

60. Huysman, M.J.; Fortunato, A.E.; Matthijs, M.; Costa, B.S.; Vanderhaeghen, R.; Van den Daele, H.; Sachse, M.; Inzé, D.; Bowler, C.; Kroth, P.G.; et al. AUREOCHROME1a-mediated induction of the diatom-specific cyclin dsCYC2 controls the onset of cell division in diatoms (*Phaeodactylum tricornutum*). *Plant Cell* **2013**, *25*, 215–228. [CrossRef] [PubMed]

61. Fischer, B.B.; Niyogi, K.K. *SINGLET OXYGEN RESISTANT 1* links reactive electrophile signaling to singlet oxygen acclimation in *Chlamydomonas reinhardtii*. *Proc. Natl. Acad. Sci. USA* **2012**, *109*, E1302–E1311. [CrossRef] [PubMed]

62. Harris, E. *The Chlamydomonas sourcebook: Introduction into Chlamydomonas and Its Laboratory Use*, 2nd ed.; Elsevier Science Publishing Co. Inc.: New York, NY, USA, 2008; ISBN 978-012-370-874-8.

63. Chen, L.; Liu, T.Z.; Zhang, W.; Chen, X.L.; Wang, J.F. Biodiesel production from algae oil high in free fatty acids by two-step catalytic conversion. *Bioresour. Technol.* **2012**, *111*, 208–214. [CrossRef] [PubMed]

64. Greenspan, P.; Mayer, E.P.; Fowler, S.D. Nile red: A selective fluorescent stain for intracellular lipid droplets. *J. Cell Biol.* **1985**, *100*, 965–973. [CrossRef] [PubMed]

65. Wellburn, A.R. The spectral determination of chlorophylls a and b, as well as total carotenoids, using various solvents with spectrophotometers of different resolution. *J. Plant Physiol.* **1994**, *144*, 307–313. [CrossRef]

International Journal of
Molecular Sciences

MDPI

Article

Comprehensive Analysis of Differentially Expressed Unigenes under NaCl Stress in Flax (*Linum usitatissimum* L.) Using RNA-Seq

Jianzhong Wu [1], Qian Zhao [2], Guangwen Wu [2], Hongmei Yuan [2], Yanhua Ma [1], Hong Lin [1], Liyan Pan [1], Suiyan Li [1] and Dequan Sun [1,*]

[1] Institute of Forage and Grassland Sciences, Heilongjiang Academy of Agricultural Sciences, Harbin 150086, China; wujianzhong176@163.com (J.W.); mayanhua@163.com (Y.M.); linhong@163.com (H.L.); panliyan@163.com (L.P.); lisuiyan@163.com (S.L.)
[2] Institute of Industrial Crop, Heilongjiang Academy of Agricultural Sciences, Harbin 150086, China; zhaoqian0401@sina.com (Q.Z.); wuguangwenflax@163.com (G.W.); yuanhm@163.com (H.Y.)
*   Correspondence: sundequan0451@163.com (D.S.); Tel.: +86-451-8666-8646

Received: 4 December 2018; Accepted: 11 January 2019; Published: 16 January 2019

Abstract: Flax (*Linum usitatissimum* L.) is an important industrial crop that is often cultivated on marginal lands, where salt stress negatively affects yield and quality. High-throughput RNA sequencing (RNA-seq) using the powerful Illumina platform was employed for transcript analysis and gene discovery to reveal flax response mechanisms to salt stress. After cDNA libraries were constructed from flax exposed to water (negative control) or salt (100 mM NaCl) for 12 h, 24 h or 48 h, transcription expression profiles and cDNA sequences representing expressed mRNA were obtained. A total of 431,808,502 clean reads were assembled to form 75,961 unigenes. After ruling out short-length and low-quality sequences, 33,774 differentially expressed unigenes (DEUs) were identified between salt-stressed and unstressed control (C) flax. Of these DEUs, 3669, 8882 and 21,223 unigenes were obtained from flax exposed to salt for 12 h (N1), 24 h (N2) and 48 h (N4), respectively. Gene function classification and pathway assignments of 2842 DEUs were obtained by comparing unigene sequences to information within public data repositories. qRT-PCR of selected DEUs was used to validate flax cDNA libraries generated for various durations of salt exposure. Based on transcriptome sequences, 1777 EST-SSRs were identified of which trinucleotide and dinucleotide repeat microsatellite motifs were most abundant. The flax DEUs and EST-SSRs identified here will serve as a powerful resource to better understand flax response mechanisms to salt exposure for development of more salt-tolerant varieties of flax.

Keywords: RNA-seq; DEUs; flax; NaCl stress; EST-SSR

1. Introduction

Worldwide, flax (*Linum usitatissimum* L.) is an economically important fiber crop, with the flax fiber industry rapidly expanding to meet increasing demand. However, in China flax competes with higher priority food crops, such as grains, that require ever increasing acreage to meet increasing food demands [1]. Consequently, flax cultivation there is confined to barren or even high-salinity plots, where flax varieties with high salt stress tolerance are urgently needed to increase yields of high quality fiber [2]. Unfortunately, the current lack of suitable salt tolerant varieties awaits development of salt-tolerant and high-yield flax germplasm resources. Toward this end, a current research focus is to identify flax genes involved in salt tolerance and salt stress responses [3]. To date, studies of flax salt stress responses have mainly focused on physiological and biochemical aspects instead of molecular response mechanisms. With the completion of the flax genome sequence [4] and continuous development of powerful molecular biological techniques, tools are now available to study regulatory responses to salt stress at the molecular level. Because the neutral salt NaCl is the main source of harmful salt in saline-alkali soils found on marginal lands, flax varieties that are specifically tolerant to NaCl are desired in China [5]. Only after a better understanding of complex salt tolerance mechanisms is mastered will great strides me made toward successful breeding of flax varieties with greater resistance to salt stress.

Research on mechanisms of salt tolerance in plants other than flax has developed rapidly and has revealed that specific molecular mechanisms of salt and alkali tolerance tend to be very complicated [6]. Plants employ adaptations and morphological changes to cope with various abiotic stresses through molecular, cellular, physiological and biochemical responses to stressors [7]. For example, plants regulate and balance osmotic pressure inside and outside of cells by accumulating metabolites to reduce or eliminate stress damage caused by water loss [8]. Nitric oxide (NO) was one of the factors which was operating the melatonin downstream to promote salinity tolerance in rapeseed based on the pharmacological, molecular and genetic data [9]. Arabidopsis EARLY FLOWERING3 (ELF3) enhances plants' resilience to salt stress [10]. Under salt stress, ELF3 suppresses GIGANTEA (GI) at the post-translational level and PHYTOCHROME INTERACTING FACTOR4 (PIF4) at the transcriptional level and PIF4 directly up-regulates the transcription of ORESARA1 and SAG29, which were the two genes that are positive regulators of salt stress response pathways.

Effects of salt stress on plant morphological development are mainly observed as reductions in seed germination, seedling growth and altered growth and development of plant tissues and organs [11–13]. Under salt stress, plant growth is generally inhibited by a water deficit manifesting as water exosmosis from cells. Exosmosis decreases plant growth rate significantly, causing wilting of plants and cell membrane damage that leads to plant cell death [14–16]. Meanwhile, Na$^+$ competition with various nutrients prevents plants from absorbing other key mineral elements, causing nutrient deficits such as K$^+$ deficiency, the most common NaCl-induced nutrient deficiency observed [17,18]. So far, a large number of studies on salt tolerance have been carried out in *Arabidopsis thaliana* [19,20], Oryza sativa [21,22] and other crops [23–28] using RNA-Seq technology. The integration of spatiotemporal expression patterns and response characteristics of different genes helps to identify a large number of differentially expressed genes (DEGs) and mechanisms related to salt stress.

In recent years, salt damage has seriously affected flax production in northeast China. Therefore, it is of great significance to understand salt stress response mechanisms and signal pathways, a challenge which currently has important economic, environmental and scientific urgency. Exploring the functional molecules of salt stress signals is fundamental to understanding the mechanism of salt-tolerant crops, so as to conduct genetic engineering and accelerate the breeding of salt-tolerant crops [29]. Ultimately, future agricultural breeding programs will benefit from enhancement of our understanding of resistance mechanisms to a variety of other stressors as well [30].

Transcriptome analysis, a recently developed tool, has greatly enhanced our understanding of plant stress resistance mechanisms [21,23,25–27,31]. However, this powerful technology has seldom been applied to the study of molecular mechanisms involved in flax tolerance, with only a few known resistance genes characterized to date. Here, we sequenced the flax transcriptome to identify differentially expressed unigenes (DEUs) for different NaCl stress exposure durations to better understand flax adaptive molecular responses mechanisms under NaCl stress. After flax transcriptome results were confirmed using qRT-PCR, large-scale analysis of EST-SSRs was conducted using public resources to understand the functions of genes involved in salt stress. The information obtained from this work lays the foundation for understanding molecular mechanisms that participate in the flax response to salt and other stressors, while also identifying relevant and useful genes and markers for future development of salt-tolerant flax varieties.

2. Results and Discussion

2.1. Response of Flax to NaCl Stress

Membrane systems are primary sites of salt stress injury, where such damage causes changes in or loss of plasma membrane semiperme ability, leading to increased electrolyte extravasation [8]. Because field experimentation is difficult to control and time-consuming, in its place laboratory studies have been conducted that measure plant leaf electrical conductivity to study salt stress. Based on previous preliminary conductivity results, three exposure times (N1 for 12 h, N2 for 24 h or N4 for 48 h) were selected to measure changes in gene expression profiles of flax exposed to 100 mM NaCl solution for each exposure duration in the laboratory.

2.2. Transcriptome Analysis

Illumina paired-end sequencing technology was employed to explore DEUs related to NaCl stress in flax using two biological replicates per time point. A total of 185,457,832 clean reads were generated after removing low quality regions and adapter sequences and sequences were mapped to the flax genome (ftp://climb.genomics.cn/pub/10.5524/100001_101000/100081/Flax.cds) using bowtie 2 (v2.1.0) (https://sourceforge.net/projects/bowtie-bio/files/bowtie2/2.1.0/) for short reads with default parameter settings (Table 1). Only 42.92% of reads could be mapped to the reference genome, possibly due to incomplete flax genome assembly. After RSEM (V1.2.4) (https://omictools.com/rsem-tool) [32] was used to evaluate the expression level of each gene, 304,809 of 422,316 unigenes remained with at least 1 fragment per kilobase of transcript per million mapped reads (FPKM) (Table 2). Sequence regions were assigned to exon, intron and intergenic types after comparing total mapped sequence reads with the reference genome. Since exon-type sequences accounted for more than 80% of genome-mapped sequences, as expected (Figure 1), these results confirmed a high degree of annotation accuracy. Moreover, comparison of FPKM box plots of gene expression levels of all genes for different experimental conditions demonstrated that sequence results were reliable, since each sample yielded equivalent reads and coverage depths between duplicates (Figure 2). Furthermore, comparison rates between sequence data and reference genes also demonstrated sequence reliability to some extent.

Table 1. Summary of data generated in the transcriptome sequencing of flax.

Sample *	Clean Reads	Total Mapped	Unique Mapped	Multi Mapped	Bases (Gb)	Q20 (%)	GC (%)
C1	60,125,334	25,807,774	9,037,386	16,770,388	7.48	99.36	48.35
C2	54,128,778	23,792,226	9,600,424	14,191,802	6.67	99.23	47.68
N11	62,422,020	26,632,880	9,637,906	16,994,974	7.77	99.33	47.67
N12	57,381,976	22,307,914	8,055,084	14,252,830	7.13	99.3	47.84
N21	60,066,230	25,957,230	9,698,022	16,259,208	7.48	99.37	47.87
N22	21,827,146	9,277,362	3,136,434	6,140,928	2.58	97.64	47.81
N41	56,385,726	25,227,542	9,202,726	16,024,816	7.01	99.35	47.96
N42	59,471,292	26,454,904	9,667,984	16,786,920	7.39	99.31	48.03

* Samples: Name of sequencing sample, C1/C2 for the two biological replicates of control, N11/N12, N21/N22 and N41/N42 represent for the two biological replicates of treatment with 100 mM NaCl for 12 h, 24 h and 48 h respectively; Clear reads: Number of clean reads participating in the comparison; Total mapped: Numbers of all reads compared to reference genes; Unique mapped: The number of reads compared to the unique location of the reference gene which was used for gene expression analysis; Multi mapped: Number of reads compared to multiple locations of reference genes; Bases (Gb): The total number of bases, Gb represent for billion base pairs; Q20: Percentage of sequencing error rate is smaller than 1% of base number; GC(%):The percentage of G + C of the total number bases.

Table 2. The number of genes in different expression levels.

FPKM Interval	0–0.1	0.1–1	1–3	3–15	15–60	>60
C1	1635 (3.03%)	9893 (18.31%)	11,852 (21.93%)	21,176 (39.19%)	7660 (14.17%)	1823 (3.37%)
C2	1805 (3.37%)	11,354 (21.21%)	11,228 (20.97%)	18,620 (34.78%)	8258 (15.42%)	2278 (4.25%)
N11	1805 (3.31%)	10,645 (19.52%)	11,210 (20.56%)	20,619 (37.82%)	8343 (15.30%)	1901 (3.49%)
N12	1784 (3.28%)	11,534 (21.18%)	11,789 (21.65%)	19,438 (35.70%)	7855 (14.43%)	2047 (3.76%)
N21	2226 (4.15%)	12,475 (23.27%)	11,564 (21.57%)	17,954 (33.49%)	7324 (13.66%)	2061 (3.84%)
N22	1015 (2.03%)	11,621 (23.24%)	11,373 (22.75%)	16,902 (33.80%)	7049 (14.10%)	2039 (4.08%)
N41	3221 (6.32%)	15,996 (31.38%)	10,872 (21.33%)	13,032 (25.57%)	5696 (11.17%)	2154 (4.23%)
N42	3804 (7.43%)	16,694 (32.61%)	10,687 (20.88%)	12,463 (24.35%)	5431 (10.61%)	2111 (4.12%)

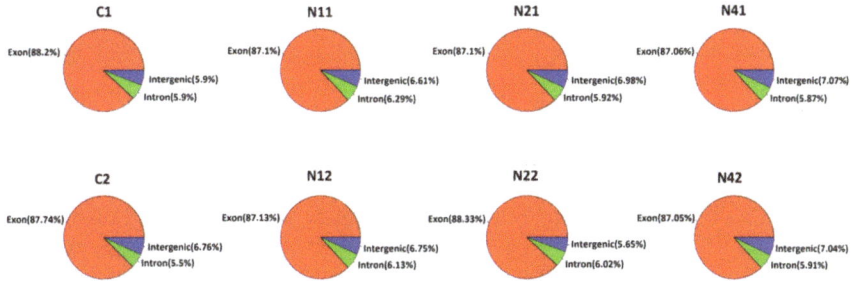

Figure 1. The comparison of clean reads with reference genome in different regions.

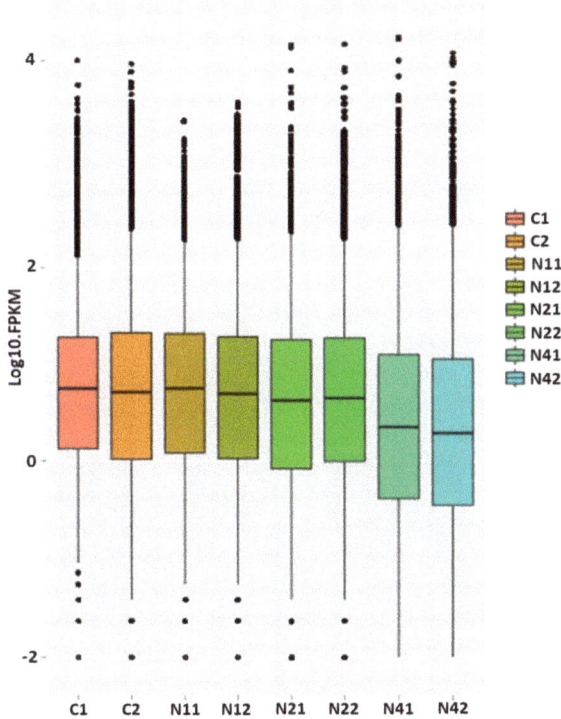

Figure 2. The box plot of FPKM in gene expression levels. Abscissa for sample name, ordinate for Log10.FPKM, box plot for each region for five statistics (From top to bottom: maximum, upper quartile, median, lower quartile and minimum), the outlier is shown in black dots.

2.3. Identification of Differentially Expressed Unigenes (DEUs)

Gene expression profiles in response to NaCl stress exposures of 12 h (N1), 24 h (N2) and 48 h (N4) were compared to the no-treatment control (water). DEUs were identified within the set of transcriptome sequences, with 33,774 significant DEUs identified (18,040 up-regulated and 15,734 down-regulated), with values of false discovery rate (FDR) $\leq$ 0.05 and 2$\times$ fold change significance cutoffs generated for various treatment time points.

At 12 h, 24 h and 48 h of stress exposure, 3669 DEUs (2219 up-regulated and 1450 down-regulated), 8882 DEUs (4865 up-regulated and 4017 down-regulated) and 21,223 DEUs (10,956 up-regulated and 10,267 down-regulated) were significantly differentially regulated in response to NaCl stress exposure, respectively (Figure 3). While different numbers of DEUs were observed for various pairings of stress exposure durations, 2581 DEUs (11.5%) were identified that were shared among all stress-exposed samples (Figure 4), in which 2576 co-expressed unigenes were detected (1322 up-regulated and 1254 down-regulated), with five unigenes not co-expressed (Supplementary Table S1). The proportion of DEUs common to paired exposure time points of 24 h/48 h (25.2%) was higher than corresponding proportions for 12 h/24 h (0.9%) and 12 h/48 h (1.9%). These results may be attributed to relatively greater effects of salt injury stress on flax from 24 h to 48 h than for other exposure windows, with involvement of a larger number of regulatory genes observed. This result is not surprising, since expression trends of common DEUs were not entirely consistent among different time periods. In general, DEUs analysis of flax under NaCl stress exposures should enhance our understanding of factors influencing gene expression during salt stress responses and provide clues to genes involved in salt tolerance.

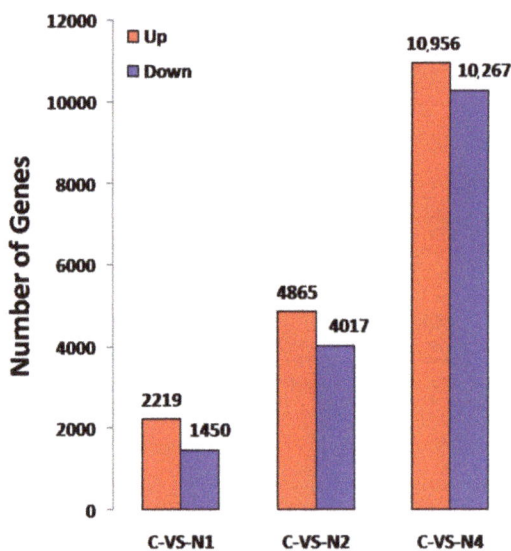

Figure 3. Comparison of up and down-regulation of DEGs. C-vs-N1, C-vs-N2 and C-vs-N4 representing the DEUs under the exposure time of 12 h, 24 h and 48 h in NaCl solution, respectively.

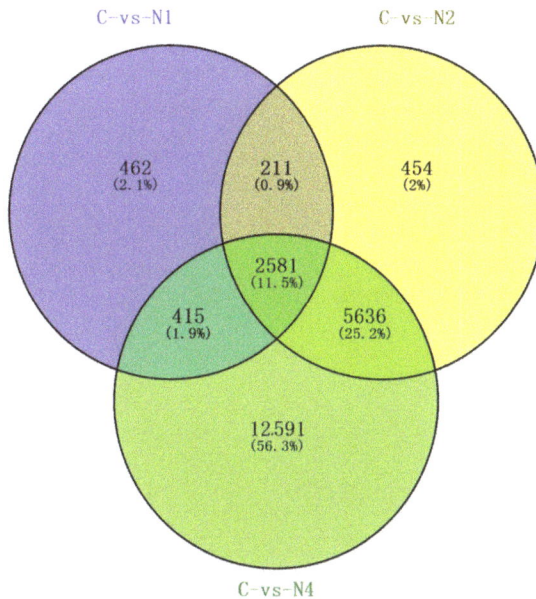

Figure 4. The Venn diagram of DEUs. Venn diagram representing the distribution of NaCl-responsive genes. The numbers in the Venn diagram indicated total numbers of regulated genes in the unique treatment.

2.4. Cluster Analysis of DEUs

Cluster analysis was used to determine expression patterns of DEUs under different experimental conditions. Generally, functions of unknown genes or unknown functions of known genes can be identified by clustering genes with the same or similar expression patterns into classes; genes with similar functions or genes that participate in the same metabolic processes or cell pathways tend to cluster together. DEU dynamic expression patterns for various NaCl stress exposure durations were identified in this study (Figure 5). After hierarchical clustering analysis of DEUs was conducted based on gene expression levels determined from FPKM values, DEUs within each single cluster were considered to be co-expressed genes. Color-coding of different cluster groupings highlights genes with similar expression patterns that shared similar functions or participated in the same biological processes. Moreover, duplicate biological samples for control or each NaCl stress treatment group were highly consistent, thus demonstrating reproducibility of RNA-seq results.

2.5. Functional Annotation

To understand DEU functions, we conducted Pfam, GO, KOG and KEGG enrichment analyses against the genetic background of the *Arabidopsis thaliana* genome (https://www.arabidopsis.org/). Of 2582 co-expressed unigenes, 2482 (96.13%) displayed significant similarity to known proteins (Supplementary Table S2), with 2104, 1473, 785 and 913 unigenes matched to homologous sequences using Pfam, GO, KOG and KEGG analyses, respectively. The most common enrichment analysis terms were related to pathway factors (Figure 6) such as plant hormone signal transduction, photosynthesis-antenna proteins and biosynthesis of amino acids as important in flax responses to NaCl exposure. Not surprisingly, here we identified a number of differentially expressed unigenes (DEUs), which were homologous to known stress regulating plant transcription factors, such as: bZIP (lus10033630), HD-ZIP (lus10025232), WRKY (Lus10020023, Lus10024380, Lus10022736, Lus10012870 and Lus10030517), NAC (Lus10042731, Lus10026617, Lus10036773, Lus10025118, Lus10041534,

Lus10003269 and Lus10042518), MYB (Lus10006740, Lus10019085 and Lus10006647), GRF (Lus10015651 and Lus10037668), GATA (Lus10031464, Lus10025829 and Lus10038273), ERF (Lus10005285, Lus10029987 and Lus10012226), CAMTA (Lus10024044) and B3 (Lus10018583 and Lus10039816). The basic leucine-region zipper (bZIP) transcription factors (TFs) act as crucial regulators in salt stress responses in plants [33]. In our study, Lus10033630 was homologous to AtbZIP34 which was required for Arabidopsis pollen wall patterning and the control of lipid metabolism and/or cellular transport in developing pollen.

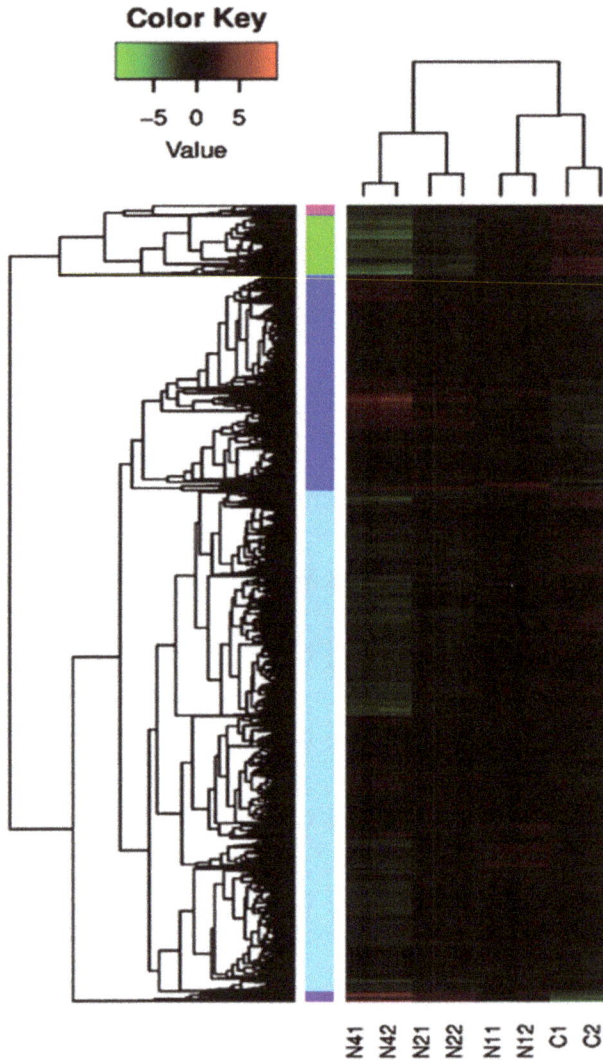

Figure 5. Cluster diagram of DEUs (FDR ≤ 0.05, fold change ≥ 2). The darker color represents the higher of the gene expression level. Each color block on the left represents a cluster of genes with similar expression levels. The Log2.FPKM value was used for clustering, with red for high expression gene and green for low expression gene. The color ranges from green to red, indicating higher gene expression.

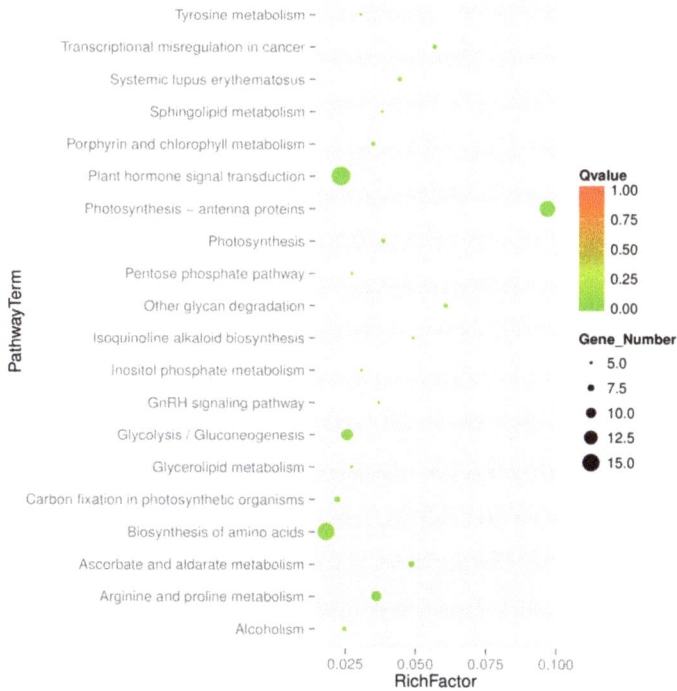

Figure 6. Gene enrichment of non-redundant unigenes. The longitudinal axis represents the different pathway and horizontal axis represents the Rich factor. The size of dots indicates the number of differentially expressed genes in this pathway and the color of the point corresponds to a different Qvalue range.

Interestingly, two salt-tolerant genes (lus10015754 and lus10000310) were obtained here to homologous with Arabidopsis Senescence-Associated Gene29 (SAG29), which was consistent with previous reports [10] and two salt-tolerant genes (lus10015285 and lus10025409) were homologous with SAG12 whose promoter control expression of isopentenyl transferase (ipt) gene in the decaying leaves of the lower part of the plant. Three genes (*Lus10040248*, *Lus10019786* and *Lus10037551*) belong to the Rho-like GTPases from plants (*ROP*) gene family, which might enhance salt tolerance by increasing root length, improving membrane injury and ion distribution [34]. Overall, all co-expressed unigenes could be aligned to the reference genome, suggesting that annotation and classification analyses performed here could be used to reliably predict flax gene functions.

2.6. RNA-Seq Expression Validation

To quantitatively assess the reliability of transcriptome data, six candidate DEUs were selected for analysis using real-time reverse transcription quantitative PCR (qRT-PCR) of biological duplicate samples. Consistent with RNA-sequencing analysis results, qRT-PCR showed significant log2-fold expression changes of DEUs among different salt-stress exposure treatments (Figure 7), thus demonstrating the reliability and accuracy of the transcriptome analysis of NaCl stress in flax conducted here.

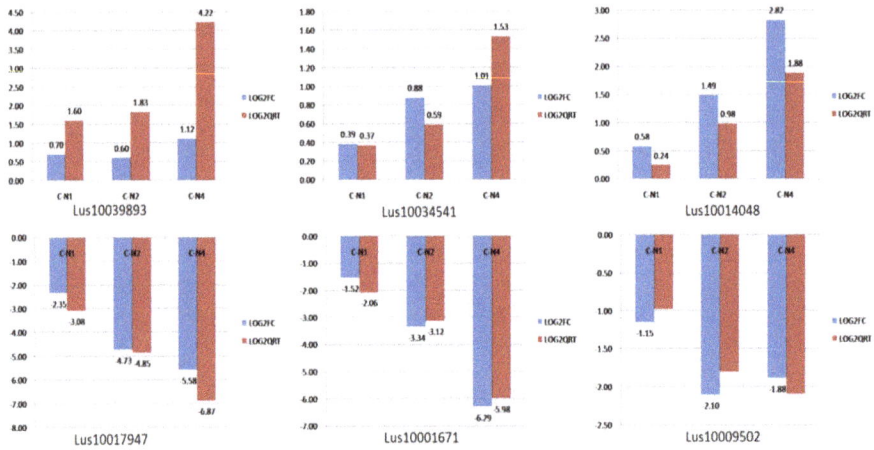

Figure 7. Validation of the RNA-seq data expression profile by qRT-PCR. The relative expression levels of 6 DEUs were calculated according to the $2^{-\Delta\Delta Ct}$ method using the *Actin* gene as an internal reference gene. The x-axis indicates the different exposure of 100 mM NaCl solution with 12 h (C-N1), 24 h (C-N2) and 48 h (C-N4). LOG2FC and LOG2QRT represent for the binary logarithm of fold changes of differentially expressed genes in RNA-seq and qRT-PCR, respectively.

2.7. Distribution Characteristics of EST-SSRs

A total of 1777 EST-SSRs with 2–6 bp repeat numbers were identified from sequence data (Table 3). Among SSR loci, trinucleotide microsatellites were the most abundant repeat motif (1002 SSRs, accounting for 56.39%), followed by dinucleotide microsatellites (623 SSRs, accounting for 35.06%), tetranucleotide microsatellites (68 SSRs, accounting for 3.83%), pentanucleotide microsatellites (58 SSRs, accounting for 3.26%) and hexanucleotide microsatellites (26 SSRs, accounting for 1.46%). A repeat iteration number of five (621) was the most common repeat iteration number observed, followed by six (502) and seven (258) repeat iteration numbers. AT/TA motifs were the dinucleotide microsatellite motifs most frequently observed of the six possible motifs (AC/TG, AG/TC, AT/TA, CT/GA, CG/GC and GT/CA), while CTT/GAA motifs were the most frequently represented trinucleotide microsatellites of the 30 possible motifs. The details of all EST-SSRs with the primer pairs are shown in Supplementary Table S3.

Table 3. Distribution of EST-SSR types.

Type	Repeat Number								Percentage (%)
	4	5	6	7	8	9	10	>10	Total	
Dinucleotide			238	148	82	57	38	60	623	35.06
Trinucleotide		558	250	107	49	36	2		1002	56.39
Quadnucleotide		53	12	3					68	3.83
Pentanucleotide	49	7	2						58	3.26
Hexanucleotide	23	3							26	1.46
Total	72	621	502	258	131	93	40	60	1777	

3. Materials and Methods

3.1. Material Planting and Processing

The fiber flax variety (Agatha), provided by Heilongjiang Academy of Agricultural Sciences in Harbin, China, provided material for high-throughput RNA-seq. Flaxseeds were placed in cups filled with sterilized vermiculite and maintained at 28 °C during the day and 22 °C at night in a growing room on a 16-h light/8-h dark cycle. Plants were irrigated every three days and the humidity in the growth room was maintained at 70%.

3.2. Stress Treatments and Sample Preparation

Three-week-old seedlings were rinsed to remove vermiculite and were placed into tanks filled with 100 mM NaCl solution or water for the control (CK). After seedlings were exposed to NaCl solution or water for 12 h, 24 h or 48 h, whole seedlings were harvested, frozen immediately in liquid nitrogen, then stored at −80 °C until RNA was prepared. Two biological replicates per treatment (each sample containing 10 plants each) were processed in parallel.

3.3. RNA Isolation and cDNA Library Construction

RNA isolation and cDNA library construction methods have been reported previously [35].

3.4. Illumina Sequencing, Assembly and Annotation

Transcriptome sequencing was performed using the Illumina HiSeq 2500 platform (https://www.illumina.com/systems/sequencing-platforms/hiseq-2500.html) to generate paired-end (PE) raw reads, each of ~100 bp in length. Clean reads were generated from raw reads by removal of adaptor sequences, ambiguous 'N' nucleotides (with a ratio of 'N' greater than 10%) and low quality sequences (with quality scores less than 5) and were assembled as described by using bowtie2 (2.1.0) software (http://bowtie-bio.sourceforge.net/bowtie2/index.shtml) against a reference genome (ftp://climb.genomics.cn/pub/10.5524/100001_101000/100081/). Only clean, high quality sequence data was used in subsequent analyses.

For homology-based annotation, the model plant genome of *Arabidopsis thaliana* (https://www.arabidopsis.org/) was selected for use as genetic background, with non-redundant sequences subjected to Pfam (http://pfam.xfam.org/), Gene Ontology (GO) (http://www.geneontology.org/), Eukaryotic Clusters of Orthologous Groups (KOG) (https://www.ncbi.nlm.nih.gov/COG/) and Kyoto Encyclopedia of Genes and Genomes (KEGG) (http://www.genome.jp/kegg/). For gene expression profiling analysis, functional assignments were mapped to GO terms [36]. Significantly enriched pathways were identified according to *p* values and enrichment factors [37].

3.5. Identification of DEUs and Cluster Analysis

DEUs were identified based on a negative binomial distribution using the edgeR package (https://www.rdocumentation.org/packages/edgeR/versions/3.14.0) [38]. Candidate genes exhibited false discovery rate (FDR) values ≤ 0.01, which were calculated from numbers of mapped reads. In addition, the fragments per kilobase of transcript per million mapped reads (FPKM) values of candidate genes were calculated using RSEM (http://deweylab.github.io/RSEM/; RNA-Seq by Expectation-Maximization) [32]. All expressed genes were divided into six categories: $0 < \text{FPKM} \leq 0.1$, $0.1 < \text{FPKM} \leq 1$, $1 < \text{FPKM} \leq 3$, $3 < \text{FPKM} \leq 15$, $15 < \text{FPKM} \leq 60$ and $\text{FPKM} > 60$. The formula for calculating expression level reflected by FPKM for each target gene is:

$$\text{FPKM} = 10C^9/NL$$

where C is the number of fragments of the target gene, N is the number of fragments for all genes and L is the base length of the target gene.

Finally, the fold change of FPKM for each sequence was calculated and genes with FPKM fold changes greater or equal to 2 were classified as DEUs. Common DEUs among different time points that exhibited FPKM fold change differences greater than 2 between biological replicates were eliminated. For each gene, normalized FPKM values for each transcript were clustered using the hclust function in R (http://www.r-project.org/) using a distance matrix representing FPKM level profiles of genes across the four sampled time points. The tree produced by the clustering process was split into two branches using the cutree function.

3.6. Real-Time qRT-PCR Validation

DEUs identified in transcriptome sequencing analysis were validated using qRT-PCR (quantitative RT-PCR) to further confirm RNA-Seq analysis gene expression results. Six genes (*Lus10017947*, *Lus10001671*, *Lus10009502*, *Lus10039893*, *Lus10034541* and *Lus10014048*) were selected for analysis of expression levels in flax treated with NaCl solution exposure times of 12 h, 24 h and 48 h using *L. ussitatissimum Act1* (GenBank accession no. AY857865) as the internal reference gene. Primer sets were designed from Illumina sequencing data using Primer Premier v6.24 (http://www. premierbiosoft.com/crm/jsp/com/pbi/crm/clientside/ProductList.jsp) (listed in Supplementary Table S4). Primers were synthesized by GENEWIZ, Inc. (Suzhou, China). Quantitative RT-PCR was performed with a SYBR® Premix Ex Taq™ II Kit (TaKaRa, Dalian, China) using an ABI 7500 Real-Time PCR System (Applied Biosystems, Foster City, CA, USA). Data for each experimental sample was corrected for sample loading differences using results of flax *Act1* qRT-PCR using the $2^{-\Delta\Delta Ct}$ method [39]. PCR amplification was performed using thermal cycling conditions of denaturation at 95 °C for 30 s followed by 40 cycles of amplification (95 °C for 5 s, 60 °C for 34 s). The control reaction (to normalize expression levels) and all samples were tested in triplicate.

3.7. Development of EST-SSR

All DEUs identified by transcriptome sequencing were analyzed to identify SSRs with dimer, trimer, tetramer, pentamer and hexamer motifs using SSRIT (http://archive.gramene.org/db/ markers/ssrtool). SSR primer pairs were designed with highest primer scores using Primer-BLAST (https://www.ncbi.nlm.nih.gov/tools/primer-blast/index.cgi?LINK_LOC=BlastHomeAd) with the following standard parameters: target amplicon length of 100–300 bp, annealing temperature variation from 55 to 65 °C and primer size of 18–28 bp.

4. Conclusions

As an important industrial crop, flax is currently cultivated on marginal lands in the northeast region of China. By analyzing the genome-wide transcriptome of flax during exposure to 100 mM NaCl solution, we identified 3669, 8882 and 21,223 potential salt stress-responsive DEUs after salt exposure for 12 h, 24 h and 48 h, respectively, compared with untreated control. All of these unigenes were identified as part of an extensive investigation of flax genes involved in the response to salt exposure to collectively provide high-resolution gene expression profiles for flax under control and salt stress conditions. Of the 2582 co-expressed unigenes, 2482 were annotated using at least one public database (GO, KOG, KO, Pfam, KEGG) and pathways linked to flax response to NaCl exposure were mainly associated with plant hormone signal transduction, photosynthesis-antenna proteins and biosynthesis of amino acids. Number of the genes are homologous to known stress regulators. Transcriptome sequencing results were verified by qRT-PCR. Ultimately, 1777 EST-SSRs based on transcriptome results were identified as important in flax response to NaCl exposure. The results described here thus lay the groundwork for further elucidation of molecular mechanisms of flax stress responses. Ultimately, this information will guide the development of flax varieties that are more tolerant to a wide range of environmental stresses.

Supplementary Materials: Supplementary materials can be found at http://www.mdpi.com/1422-0067/20/2/369/ s1.

Author Contributions: J.W. conceived and designed the study, interpreted the data, performed the experimental work and wrote the manuscript; Q.Z. drafted, wrote the manuscript, did the bioinformatics analysis and revised the manuscript; G.W. contributed to materials and revised the manuscript; H.Y., Y.M., H.L. and L.P. performed the experimental work and interpreted the data; S.L. and D.S. contributed to the experiments and discussion of results. All authors read and approved the final manuscript.

Funding: This research was funded by National Natural Science Foundation of China (grant number 31401451) and Heilongjiang Agricultural Science and Technology Innovation Project (grant number YYYF004).

Acknowledgments: The authors would also like to thank Tingbo Jiang for helpful critical comments on the manuscript.

Conflicts of Interest: The authors declare no conflict of interest.

References

1. Yu, Y.; Huang, W.G.; Chen, H.Y.; Wu, G.W.; Yuan, H.M.; Song, X.X.; Kang, Q.H.; Zhao, D.S.; Jiang, W.D.; Liu, Y.; et al. Identification of differentially expressed genes in flax (*Linum usitatissimum* L.) under saline-alkaline stress by digital gene expression. *Gene* **2014**, *549*, 113–122. [CrossRef] [PubMed]
2. El-Hariri, D.M.; Al-Kordy, M.A.; Hassanein, M.S.; Ahmed, M.A. Partition of photosynthates and energy production in different flax cultivars. *J. Nat. Fibers* **2005**. [CrossRef]
3. Pecenka, R.; Fürll, C.; Ola, D.C.; Budde, J.; Gusovius, H.J. Efficient use of agricultural land in production of energy: Natural insulation versus bio-energy. In Proceedings of the International Conference of Agricultural Engineering CIGR-AgEng, Agriculture and Engineering for a Healthier Life, Valencia, Spain, 8–12 July 2012.
4. Wang, Z.; Hobson, N.; Galindo, L.; Zhu, S.; Shi, D.; McDill, J.; Yang, L.; Hawkins, S.; Neutelings, G.; Datla, R.; et al. The genome of flax (*Linum usitatissimum*) assembled de novo from short shotgun sequence reads. *Plant J.* **2012**, *72*, 461–473. [CrossRef] [PubMed]
5. Marshall, G. Flax: Breeding and utilisation. *Plant Growth Regul.* **1991**, *10*, 171–172. [CrossRef]
6. Abdel, W.W.; Ahmed, S.A. Response surface methodology for production, characterization and application of solvent, salt and alkali-tolerant alkaline protease from isolated fungal strain, *Aspergillus niger*, WA 2017. *Int. J. Biol. Macromol.* **2018**, *115*, 447–458. [CrossRef]
7. Yamaguchi-Shinozaki, K.; Shinozaki, K. Transcriptional regulatory networks in cellular responses and tolerance to dehydration and cold stresses. *Annu. Rev. Plant Biol.* **2006**, *57*, 781–803. [CrossRef] [PubMed]
8. Zhu, J.K. Salt and Drought Stress Signal Transduction in Plants. *Annu. Rev. Plant Biol.* **2002**, *53*, 247–273. [CrossRef]
9. Zhao, G.; Zhao, Y.Y.; Yu, X.L.; Kiprotich, F.; Han, H.; Guan, R.Z.; Wang, R.; Shen, W.B. Nitric Oxide Is Required for Melatonin-Enhanced Tolerance against Salinity Stress in Rapeseed (*Brassica napus* L.) Seedlings. *Int. J. Mol. Sci.* **2018**, *19*, 1912. [CrossRef]
10. Sakuraba, Y.; Bulbul, S.; Piao, W.; Choi, G.; Paek, N.C. Arabidopsis EARLY FLOWERING3 increases salt tolerance by suppressing salt stress response pathways. *Plant J.* **2017**, *92*, 1106–1120. [CrossRef]
11. Li, Y. Effect of salt stress on seed germination and seedling growth of three salinity plants. *Pak. J. Biol. Sci.* **2008**, *11*, 1268–1272. [CrossRef]
12. Çavuşoğlu, K.; Kılıç, S.; Kabar, K. Some morphological and anatomical observations during alleviation of salinity (NaCl) stress on seed germination and seedling growth of barley by polyamines. *Acta Physiol. Plant.* **2007**, *29*, 551–557. [CrossRef]
13. Jiang, A.; Gan, L.; Tu, Y.; Ma, H.; Zhang, J.; Song, Z.; He, Y.C.; Cai, D.T.; Xue, X.D. The effect of genome duplication on seed germination and seedling growth of rice under salt stress. *Aust. J. Crop Sci.* **2013**, *7*, 1814–1821.
14. François, T.; Parent, B.; Caldeira, C.F.; Welcker, C. Genetic and physiological controls of growth under water deficit. *Plant Physiol.* **2014**, *164*, 1628–1635. [CrossRef]
15. Álvarez, S.; Rodríguez, P.; Broetto, F.; Sánchez-Blanco, M.J. Long term responses and adaptive strategies of pistacialentiscus under moderate and severe deficit irrigation and salinity: Osmotic and elastic adjustment, growth, ion uptake and photosynthetic activity. *Agric. Water Manag.* **2018**, *202*, 253–262. [CrossRef]
16. Li, H.; Chang, J.; Chen, H.; Wang, Z.; Gu, X.; Wei, C.; Zhang, Y.; Ma, J.; Yang, J.; Zhang, X. Exogenous melatonin confers salt stress tolerance to watermelon by improving photosynthesis and redox homeostasis. *Front. Plant Sci.* **2017**, *8*, 295. [CrossRef]
17. Nguyen, M.T.; Yang, L.E.; Fletcher, N.K.; Lee, D.H.; Kocinsky, H.; Bachmann, S.; Delpire, E.; McDonough, A. Effects of K$^+$-deficient diets with and without nacl supplementation on Na$^+$, K$^+$, and H$_2$O transporters' abundance along the nephron. *Am. J. Physiol. Renal Physiol.* **2012**, *303*, F92–F104. [CrossRef] [PubMed]

18. Rus, A.; Yokoi, S.; Sharkhuu, A.; Reddy, M.; Lee, B.H.; Matsumoto, T.K.; Koiwa, H.; Zhu, J.K.; Bressan, R.A.; Hasegawa, P.M. Athkt1 is a salt tolerance determinant that controls Na⁺ entry into plant roots. *Proc. Natl. Acad. Sci. USA* **2001**, *98*, 14150–14155. [CrossRef]
19. Atkinson, N.J.; Lilley, C.J.; Urwin, P.E. Identification of genes involved in the response of Arabidopsis to simultaneous biotic and abiotic stresses. *Plant Physiol.* **2013**, *162*, 2028–2041. [CrossRef]
20. Han, N.; Lan, W.J.; He, X.; Shao, Q.; Wang, B.S.; Zhao, X.J. Expression of a Suaeda salsa Vacuolar H+/Ca2+ Transporter Gene in Arabidopsis Contributes to Physiological Changes in Salinity. *Plant Mol. Boil. Rep.* **2012**, *30*, 470–477. [CrossRef]
21. Lu, T.; Lu, G.; Fan, D.; Zhu, C.; Wei, L.; Qiang, Z.; Qi, F.; Yan, Z.; Guo, Y.; Li, W. Function annotation of the rice transcriptome at single-nucleotide resolution by RNA-seq. *Genome Res.* **2010**, *20*, 1238–1249. [CrossRef]
22. Zhou, Y.; Yang, P.; Cui, F.; Zhang, F.; Luo, X.; Xie, J. Transcriptome Analysis of Salt Stress Responsiveness in the Seedlings of Dongxiang Wild Rice (Oryza rufipogon Griff.). *PLoS ONE* **2016**. [CrossRef] [PubMed]
23. Liu, A.; Xiao, Z.; Li, M.W.; Wong, F.L.; Yung, W.S.; Ku, Y.S.; Wang, Q.; Wang, X.; Xie, M.; Yim, A.K. Transcriptomic reprogramming in soybean seedlings under salt stress. *Plant Cell Environ.* **2018**, *12*, e0189159. [CrossRef]
24. Guo, J.; Li, Y.; Han, G.; Song, J.; Wang, B. NaCl markedly improved the reproductive capacity of the euhalophyte Suaeda Salsa. *Funct. Plant Boil.* **2018**, *45*, 350. [CrossRef]
25. Cui, F.; Sui, N.; Duan, G.; Liu, Y.; Han, Y.; Liu, S.; Wan, S.; Li, G. Identification of Metabolites and Transcripts Involved in Salt Stress and Recovery in Peanut. *Front. Plant Sci.* **2018**, *9*, 217. [CrossRef]
26. Yuan, F.; Lyu, M.J.A.; Leng, B.Y.; Zhu, X.G.; Wang, B.S. The transcriptome of NaCl-treated *Limonium bicolor* leaves reveals the genes controlling salt secretion of salt gland. *Plant Mol. Boil.* **2016**, *91*, 241–256. [CrossRef] [PubMed]
27. Wang, N.; Qian, Z.; Luo, M.; Fan, S.; Zhang, X.; Zhang, L. Identification of salt stress responding genes using transcriptome analysis in green alga chlamydomonas reinhardtii. *Int. J. Mol. Sci.* **2018**, *19*, 3359. [CrossRef]
28. Li, H.; Lin, J.; Yang, Q.S.; Li, X.G.; Chang, Y.H. Comprehensive analysis of differentially expressed genes under salt stress in pear (*Pyrus betulaefolia*) using RNA-seq. *Plant Growth Regul.* **2017**, *82*, 409–420. [CrossRef]
29. Xu, Y.Y.; Li, X.G.; Lin, J.; Wang, Z.H.; Yang, Q.S.; Chang, Y.H. Transcriptome sequencing and analysis of major genes involved in calcium signaling pathways in pear plants (*Pyrus calleryana* Decne.). *BMC Genom.* **2015**, *16*, 738. [CrossRef]
30. Hamed, K.B.; Ellouzi, H.; Talbi, O.Z.; Hessini, K.; Slama, I.; Ghnaya, T.; Bosch, S.M.; Savour, A.; Abdelly, C. Physiological response of halophytes to multiple stresses. *Funct. Plant Biol.* **2013**, *40*, 883–896. [CrossRef]
31. Dash, P.K.; Cao, Y.; Jailani, A.K.; Gupta, P.; Venglat, P.; Xiang, D.; Rai, R.; Sharma, R.; Thirunavukkarasu, N.; Abdin, M.Z.; et al. Genome-wide analysis of drought induced gene expression changes in flax (*Linum usitatissimum*). *Gm Crop. Food* **2014**, *5*, 106–119. [CrossRef]
32. Dewey, C.N.; Bo, L. RSEM: Accurate transcript quantification from RNA-Seq data with or without a reference genome. *BMC Bioinform.* **2011**, *12*, 323. [CrossRef]
33. Antónia, G.; David, R.; Matczuk, K.; Nikoleta, D.; David, C.; Twell, D.; Honys, D. AtbZIP34 is required for Arabidopsis pollen wall patterning and the control of several metabolic pathways in developing pollen. *Plant Mol. Biol.* **2009**, *70*, 581–601. [CrossRef]
34. Miao, H.X.; Sun, P.G.; Liu, J.H.; Wang, J.Y.; Xu, B.Y.; Jin, Z.Q. Overexpression of a Novel ROP Gene from the Banana (MaROP5g) Confers Increased Salt Stress Tolerance. *Int. J. Mol. Sci.* **2018**, *19*, 3108. [CrossRef] [PubMed]
35. Wu, J.; Zhao, Q.; Sun, D.; Wu, G.; Zhang, L.; Yuan, H.; Yu, Y.; Zhang, S.; Yang, X.; Li, Z.; et al. Transcriptome analysis of flax (*Linum usitatissimum* L.) undergoing osmotic stress. *Ind. Crop. Prod.* **2018**, *116*, 215–223. [CrossRef]
36. Harris, M.A.; Clark, J.; Ireland, A.; Lomax, J.; Ashburner, M.; Foulger, R.; Eilbeck, K.; Lewis, S.; Marshall, B.; Mungall, C.; et al. The Gene Ontology (GO) database and informatics resource. *Nucleic Acids Res.* **2014**, *32*, D258–D261. [CrossRef]
37. Kanehisa, M.; Goto, S.; Hattori, M.; Aoki-Kinoshita, K.F.; Itoh, M.; Kawashima, S.; Katayama, T.; Araki, M.; Hirakawa, M. From genomics to chemical genomics: New developments in KEGG. *Nucleic Acids Res.* **2006**, *34*, D354–D357. [CrossRef] [PubMed]

38. Robinson, M.D.; McCarthy, D.J.; Smyth, G.K. edgeR: A Bioconductor package for differential expression analysis of digital gene expression data. *Bioinformatics* **2010**, *26*, 139–140. [CrossRef] [PubMed]
39. Livak, K.J.; Schmittgen, T.D. Analysis of relative gene expression data using real-time quantitative PCR and the $2^{-\Delta\Delta CT}$ method. *Methods* **2001**, *25*, 402–408. [CrossRef]

International Journal of
Molecular Sciences

MDPI

Article

Transcriptome Sequence Analysis Elaborates a Complex Defensive Mechanism of Grapevine (*Vitis vinifera* L.) in Response to Salt Stress

Le Guan [1], Muhammad Salman Haider [1], Nadeem Khan [1], Maazullah Nasim [1], Songtao Jiu [2], Muhammad Fiaz [1], Xudong Zhu [1], Kekun Zhang [1] and Jinggui Fang [1,*]

[1]   College of Horticulture, Nanjing Agricultural University, Nanjing 210095, China; guanle@njau.edu.cn (L.G.); salman.hort1@gmail.com (M.S.H.); 2016104235@njau.edu.cn (N.K.); maazullah.nasim@gmail.com (M.N.); fiaz.m2002@gmail.com (M.F.); zhuxudong@njau.edu.cn (X.Z.); 2006204006@njau.edu.cn (K.Z.)
[2]   Department of Plant Science, School of Agriculture and Biology, Shanghai Jiao Tong University, Shanghai 200240, China; 2013104019@njau.edu.cn
*   Correspondence: fanggg@njau.edu.cn

Received: 28 October 2018; Accepted: 5 December 2018; Published: 12 December 2018

Abstract: Salinity is ubiquitous abiotic stress factor limiting viticulture productivity worldwide. However, the grapevine is vulnerable to salt stress, which severely affects growth and development of the vine. Hence, it is crucial to delve into the salt resistance mechanism and screen out salt-resistance prediction marker genes; we implicated RNA-sequence (RNA-seq) technology to compare the grapevine transcriptome profile to salt stress. Results showed 2472 differentially-expressed genes (DEGs) in total in salt-responsive grapevine leaves, including 1067 up-regulated and 1405 down-regulated DEGs. Gene Ontology (GO) and Kyoto Encyclopedia of Genes and Genomes (KEGG) annotations suggested that many DEGs were involved in various defense-related biological pathways, including ROS scavenging, ion transportation, heat shock proteins (HSPs), pathogenesis-related proteins (PRs) and hormone signaling. Furthermore, many DEGs were encoded transcription factors (TFs) and essential regulatory proteins involved in signal transduction by regulating the salt resistance-related genes in grapevine. The antioxidant enzyme analysis showed that salt stress significantly affected the superoxide dismutase (SOD), peroxidase (POD), catalase (CAT) and glutathione S-transferase (GST) activities in grapevine leaves. Moreover, the uptake and distribution of sodium (Na^+), potassium (K^+) and chlorine (Cl^-) in source and sink tissues of grapevine was significantly affected by salt stress. Finally, the qRT-PCR analysis of DE validated the data and findings were significantly consistent with RNA-seq data, which further assisted in the selection of salt stress-responsive candidate genes in grapevine. This study contributes in new perspicacity into the underlying molecular mechanism of grapevine salt stress-tolerance at the transcriptome level and explore new approaches to applying the gene information in genetic engineering and breeding purposes.

Keywords: grapevine; salt stress; ROS detoxification; phytohormone; transcription factors

1. Introduction

Grapevine (*Vitis vinifera*) is an economic fruit crop, primarily categorized into the table (fresh) and wine grapes [1]. Recent shifts in the environment have become the critical limiting factors for yield and grapevine products. Thus, it is indispensable to characterize the salt-tolerant grapevine varieties by screening salt resistance-related genes and genetically transform them to enable plants to withstand high salt concentrations. One-fifth of irrigated agricultural lands are affected by soil salinity, which leads to escalating the salt effects on plant growth investigations in the recent few years [2–4]. High soil salinity affects plants in multiple ways, such as inhibition of water uptake in the root zone, which makes it difficult for the plants to take up water; and results in dehydration of plant cells, leading to cell turgor and in response, plants have to increase osmotic pressure in their cells [5]. Also, due to the decrease in K^+/Na^+ value, the original balance of ions in plant cells might be interrupted, which has a toxic effect on enzymes, chlorophyll degradation and recurrent protein synthesis [6]. Simultaneously, salt induces cellular toxicity, which leads to undue reactive oxygen species (ROS) production and accumulation in different cellular compartments, resulting in lipid peroxidation (LPO) of biological membranes, ions leakage and DNA-strand cleavage [7].

Plants can evolve a complex defensive mechanism to counteract the salinity effects [8], which includes activation of numerous signaling sensors that conclusively excites various transcription factors (TFs) to induce stress-responsive genes, which enable plants to nurture and transcend the adverse conditions. In salinity, factors involved in signaling are: (i) discerning accretion or elimination of ions to stabilize the K^+/Na^+ balance and other ion levels via salt-inducible enzyme Na^+/H^+ antiporter (V-ATPase or PPase) and K^+ and Na^+ transporters (SOS family); (ii) biosynthesis of congenial solutes to adjust the vacuolar ionic balance and restore water in the biochemical reaction (Like polyols and mannitol); (iii) adjust the cell membrane structure; (iv) synthesis of multiple resistance-oriented proteins like ROS and pathogenesis-related proteins (PR family); and (v) induction of plant hormones (ABA, JA and IAA). These biological pathways improve the inclination of salt tolerance are likely to collaborate and may have the synergistic effect [6,9]. Besides, various transcription factors (TFs), such as HD-Zip, ERF, WRKY, bHLH are known to play a vital role in regulating salt resistance mechanism in plants [1,10].

Recently, next-generation sequencing (NGS) technology based high throughput RNA-seq technology has been extensively used to unveil and compare the transcriptome profile under abiotic stresses [1], which provides large-scale data to identify and characterize the DEGs. Previously, extensive studies have been carried out on antioxidant metabolism, ionomic uptake and transport, hormonal metabolism and stress signaling [11,12] but the underlying molecular mechanism of salt stress tolerance remain to be elucidated. Though several studies focusing on morphological variations, biochemical and physiological components are available in grapevine, however, there is no report on transcriptomic studies particularly molecular research associated with salt stress tolerance. Therefore, to comprehend the molecular mechanism of salt tolerance in grapevine, Illumina RNA-seq libraries were constructed from both control and salt-treated grapevine leaves. In addition, gene ontology (GO) enrichment analysis was also performed to investigate biochemical and physiological cues in response to salt stress. qRT-PCR analysis of critical salt stress-responsive genes was also carried out to validate RNA-seq results. The obtained information provides more profound insights into the grapevine molecular mechanism in improving breeding strategies for the development of transgenic plants, which can better resist the abiotic stress.

2. Results

2.1. Global Transcriptome Sequence Analysis

The transcriptomic sequencing of cDNA generated from both control and salt-treated grapevine leaf samples produced 21.2 and 21.4 million raw reads, respectively (Table S1). Following the filtering and trimming process, 20.2 and 20.6 million clean reads were retrieved from control and treatment group, respectively, corresponding to 8.16 Gb data, intimating the tag density from both control and salt-treatment, representing about 20 million reads, which is adequate for quantitative analysis of gene expression. For sequence alignments, SOAPaligner/soap2 software (http://soap.genomics.org.cn) was used as reference genome of grapevine (Version 1.0), suggesting total mapped reads as 67.4% matched complemented with both unique (57.42%) or multiple (9.96%) genomic positions (Table S1).

Transcriptome analysis can compare the number of DEGs and their expression pattern in different tissues. In our transcriptomic study, 21,746 transcripts were obtained from control and 21,541 transcripts from the treatment group. Among these expressed transcripts, 14,767 transcripts showed no significant changes in their expression level ($|\log_2 FC| < 1$), while 2472 transcripts were differentially expressed in the salt-treatment group ($|\log_2 FC| \geq 1$) at false discovery rate (FDR) <0.001), which includes 1067 (43.16%) up-regulated and 1405 (56.87%) down-regulated transcripts (Table S2). Moreover, 20 DEGs suggested their expression only in the control group and 27 DEGs were only expressed in the treatment group (Table S3).

2.2. GO and KEGG Analysis of DEGs in Response to Salt Stress

GO-based enrichment analysis functionally characterizes and annotated the 1, 591 (64.36% of 2, 472) DEGs into 45 functional groups, of which molecular function contains 15 groups, cellular component (15 groups) and biological process (9) (Figure 1 and Table S4) between control and salt-treated group. In molecular function (MF), "ATPase activity" (GO: 0042623) with 178 transcripts, followed by "phosphatase activity" (GO: 0008138) with 113 transcripts and least transcripts (4) were found in both "ABA binding (GO: 0010427)" and "Hsp90 protein binding (GO: 0010329)". In cellular component (CC), "photosynthetic membrane" possessed the highest number of transcripts (GO: 0034357, 106 transcripts), whereas, "thylakoid membrane" consisted of 97 transcripts (GO: 0042651). Furthermore, in biological process (BP), "response to oxidative stress" (GO: 0006979) harbored 164 transcripts, followed by "salinity response" (GO: 0009651) with 148, while "SOS response (GO: 0009432)", "stomatal closure (GO: 0090332)" and "cytochrome b6f complex (GO: 0010190)" with three transcripts each were the least group.

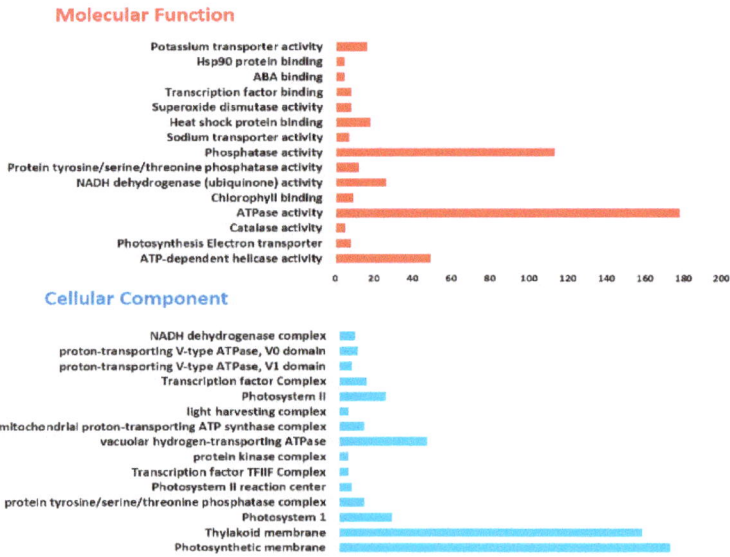

Figure 1. *Cont.*

Biological Process

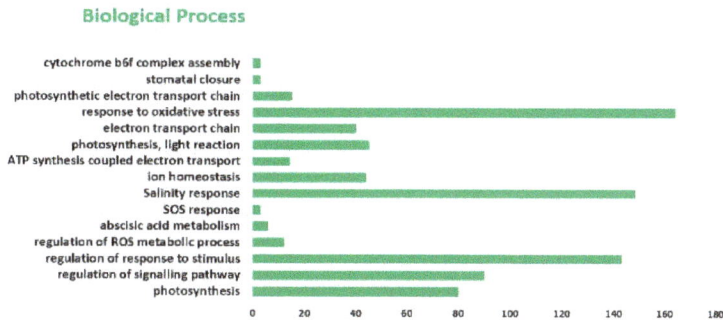

Figure 1. Gene ontology (Go) based annotations of 2472 DEGs. The main GO terms are categorized into "molecular function", "cellular component" and "biological process".

KEGG database simulates the functional annotation of the cells or the organism by sequence similarity and genome information. In this study, 453 (18.32% of 2472) transcripts were allocated to 30 pathways in KEGG database (Figure 2 and Table S5), while "Signal transduction" pathway with 79 transcripts was the most enriched pathway followed by "Folding, sorting and degradation" (65 transcripts) and "Carbohydrate metabolism" (63 transcripts).

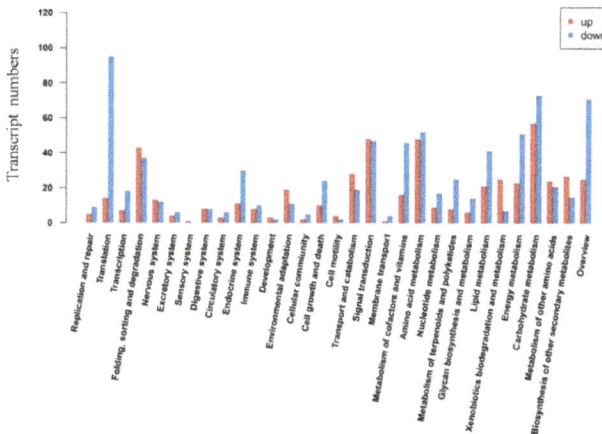

Figure 2. Kyoto Encyclopedia of Genetics and Genomics (KEGG) database analysis of DEGS (up and down-regulated) enriched in different biological pathways. The *X*-axis represents enriched pathways and *Y*-axis represents the total number of transcripts.

2.3. Photosynthetic Efficiency of Grapevine in Response to Salt Stress

To verify the extent of salt severity on grapevine physiology, photosynthetic efficiency and related parameters were estimated in the control and treatment group by using a portable Li-COR meter. Results suggested that net photosynthesis rate (A_N) was significantly reduced from 23.98 ± 1.33 (0 h) to 13.42 ± 1.31 (48 h) during the salt stress period. Likewise, an about 2-fold decrease in stomatal conductance significantly inhibited the net CO_2 assimilation rate (Ci; 35.78%) and transpiration rate (E; 51.33%) after 48 h of salt stress as compared to control plants (Figure 3). In the transcriptomic study, the DEGs encoding photosystem II CP47 (psbB) in PSII and photosystem I P700 (psaB) in PSI were down-regulated in salt-treated grapevine leaf samples as compared to control (Table S2), which is consistent with the physiological investigations of decreased net photosynthesis rate. Moreover, six DEGs encoding ATP-synthase and one DEG-related to the cytochrome b6-f complex were also down-regulated in grapevine leaf tissues after exposure to salt stress.

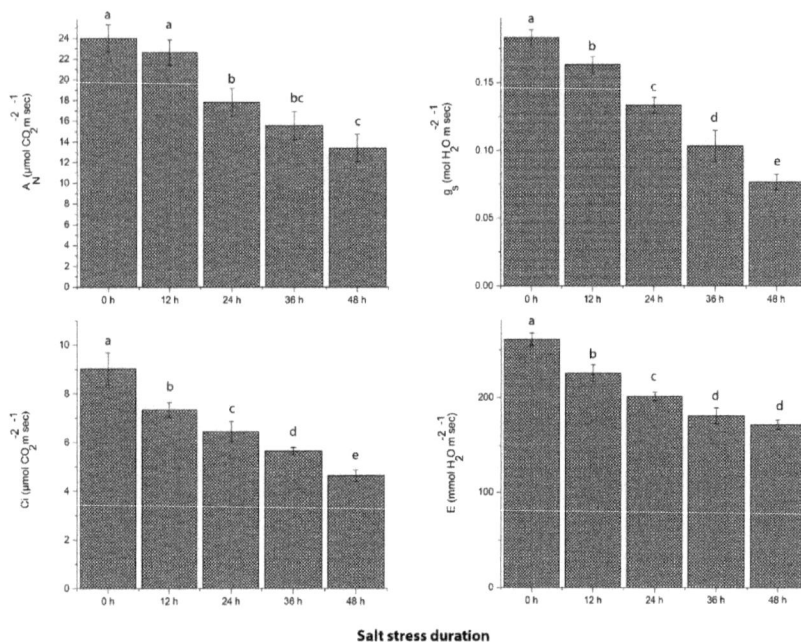

Figure 3. Estimation of photosynthetic efficiency, including net photosynthesis rate (A_N), stomatal conductance (g_S), transpiration rate (E) and net CO_2 assimilation rate (Ci) in grapevine leaves in response to salt stress as compared with control. Values represent mean $\pm$ SE ($n = 3$) and the significance level of 0.05 was used for different letters above bars.

2.4. Production and Scavenging of Reactive Oxygen Species (ROS) in Response to Salt Stress

ROS production is a universal plant response to almost all type of abiotic stresses. In response, plants accumulate various antioxidant enzymes (SOD, POD and CAT) that can quench free radicals, such as H_2O_2 and $O_2^{\bullet-}$ [12]. In this study, 44 DEGs were identified as enzymes in the ROS detoxification and scavenging system. These DEGs were functionally characterized into different ROS enzymes encoding Fe superoxide dismutase (Fe-SODs, 1 transcript), catalase (CAT, 2 transcripts), peroxidase (POD, 8 transcripts), glutathione S-transferase (GST, 16 transcripts), alternative oxidase (AOX, 1 transcript), glutathione-ascorbate (GSH-AsA) cycle (6 transcripts), the peroxiredoxin/thioredoxin (Prx/Trx, 9 transcripts) and polyphenol oxidase (PPO, 1 transcript) (Table 1 and Table S6).

Table 1. List of differentially-expressed genes related to redox metabolism and respiratory chain in grapevine perceived during salt stress.

Trait Name	Description	No. of Up-Regulated	No. of Down-Regulated	Sum
	Fe-SOD	0	1	1
ROS scavenging	POD	8	0	8
	CAT	2	0	2
	MDAR	1	0	1
GSH-AsA cycle	APx	1	0	1
	GR	0	2	2
	Grx	1	1	2
GPX pathway	GST	8	8	16
Prx/Trx	Trx	4	5	9
Cyanide-resistant respiration	AOX	0	1	1
Copper-containing enzymes	PPO	0	1	1

Fe-SOD: Fe superoxide dismutase; POD: peroxidase; CAT: catalase, APX: ascorbate peroxidase; MDAR: monodehydroascorbate reductase; GR: glutathione reductase; Grx: glutaredoxin; GST: glutathione S transferase; Trx: thioredoxin; AOX: alternative oxidase, PPO: polyphenol oxidase.

The metalloenzyme superoxide dismutase (SOD) provides primary defense line against ROS (superoxide radicals, $O_2^{\bullet-}$) and dismutates $O_2^{\bullet-}$ into O_2 and H_2O_2. SODs have three isozymes that are localized in different cellular compartments and vary in their functional properties, including copper-zinc (Cu/Zn-SOD), manganese (Mn-SOD) and iron (Fe-SOD). While only one Fe-SOD with slightly down-regulated expression level ($|\log_2 FC| > 1$) was found in this research, might be due to the severity of salt that suppressed the transcription of Fe-SOD gene in grapevine leaves. These findings are consistent with the previous reports [13–15] and were also confirmed by the SOD activity measurement, in which SOD activity was increased within 36h of salt stress but drastically decreased after 48 h (Figure 4a). In current findings, the activities of CAT and POD were progressively persuaded at 48 h of salt stress treatment (Figure 4b,c). Transcriptomics analysis showed that the DEGs encoding CAT and POD were up-regulated under salt stress, of which two POD transcripts, VIT_13s0067g02360 ($|\log_2 FC| = 3.68$) and VIT_08s0040g02200 ($|\log_2 FC| = 3.51$) were remarkably up-regulated in salt-treated group as compared to control, while remaining six POD and the two CAT transcripts were slightly up-regulated (their $|\log_2 FC|$ values were about 1), which is consistent with the physiological data of increased activities of antioxidant enzymes. These findings suggested a common response of antioxidant enzymes to detoxify ROS effects.

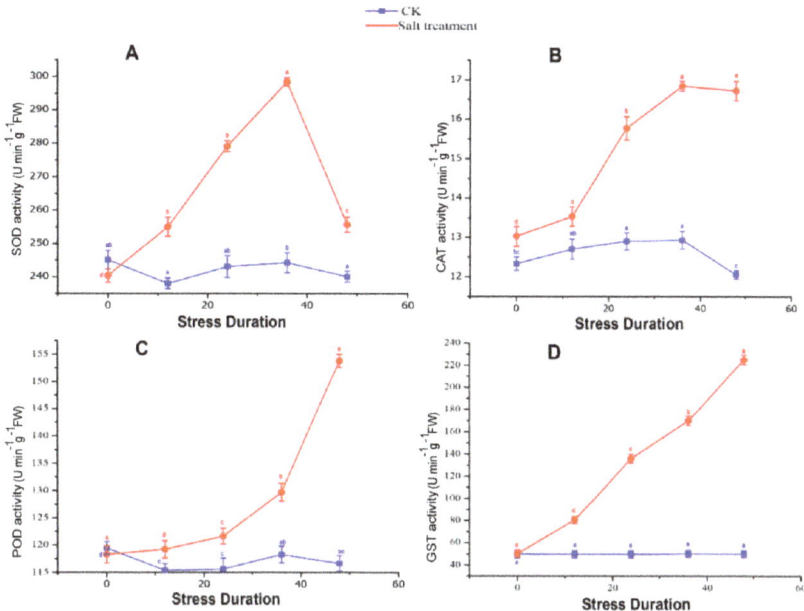

Figure 4. Changes in the enzyme activities of SOD (**A**), CAT (**B**), POD (**C**) and GST (**D**) in grapevine leaves grown for 48 h under control and salt stress. Values represent mean ± SE (*n* = 3) and the significance level of 0.05 was used for different letters above bars.

GSH and GST also play a crucial defense-related role against ROS caused by salt stress [16,17]. In this study, six transcripts involved in ascorbate-glutathione (AsA-GSH) cycle and 16 Glutathione S-transferase (GST) transcripts were detected, of which salt stress significantly induced two GST transcripts (VIT_05s0049g01070 and VIT_05s0049g01100), when compared with remaining four genes. GST activity was also significantly increased at 48 h of salt stress (Figure 3d), revealing its essential roles in the ROS scavenging process.

2.5. Heat Shock Protein (HSP) and Pathogenesis-Related Proteins (PR) in Response to Salt Stress

Heat shock proteins (HSPs) are the molecular chaperones that act as stress-responsive proteins, thus protecting plants from stress damage, which include mainly HSP100s, HSP90s, HSP70s, HSP60s (cpn60s) and small heat-shock proteins (sHSPs). Overall, 39 HSPs-related DEGs were further divided into high molecular weight HSPs (HMW HSPs; 4 transcripts), low molecular weight HSPs (LMW HSPs; 17 transcripts), heat stress transcription factors (6 transcripts) and other HSPs (12 transcripts) (Table 2 and Table S7). Three transcripts encoding HMW HSPs were down-regulated, while one transcript (VIT_18s0041g01230) was up-regulated. Similarly, 16 of the 17 LMW HSPs were up-regulated and some of them showed very high expression levels compared to the control, for example, VIT_16s0098g01060 ($|\log_2 FC| = 3.755743188$), VIT_13s0019g00860 ($|\log_2 FC| = 3.361853346$) and VIT_12s0035g01910 ($|\log_2 FC| = 3.135505213$), intimating that LMW HSPs play a more important role than the HMW HSPs in response to salt stress in grapevine. Six heat stress TFs (3 up-regulated and 3 down-regulated) showed very diverse transcription levels, suggesting their complex regulatory mechanism over HSPs. Also, 8 chaperone protein DnaJ (6 up-regulated and 2 down-regulated) transcripts identified in the current research, as DnaJ is a vital cofactor plays a central role in transducing stress-induced protein damage to induce heat shock gene transcription, while the up-regulation may suggest the extensive cellular protein damage by salt severity.

Plants can enhance tolerance mechanism against salt stress through over-expression of pathogenesis-related proteins. In the grapevine transcriptome, 37 DEGs encoding disease resistance proteins were identified and classified into 4 pathogenesis-related proteins (PR-1; all up-regulated), 2 chitinase (both down-regulated), 1 beta-1, 3-glucanase (down-regulated), 8 lipid transfer proteins (7 up-regulated, 1 down-regulated), 6 thaumatin-like proteins (1 up-regulated, 5 down-regulated), 1 germin protein (down-regulated), 13 disease resistance proteins (9 up-regulated, 4 down-regulated) and 2 snakin were perceived as up-regulated (Table 2 and Table S7).

Table 2. List of differentially-expressed genes related to heat-shock proteins (HSPs) and pathogens resistance (PRs) proteins in grapevine perceived during drought stress.

Trait Name	Description	No. of Up-Regulated	No. of Down-Regulated	Sum
Heat shock proteins	HMW HSPs	1	3	4
	LMW HSPs	16	1	17
	small HSPs	12	6	18
	other HSPs	7	5	12
	heat-stress transcription factors	3	3	6
PR-1	pathogenesis-related protein 1	4	0	4
PR-2	β-1,3-glucanase	0	1	1
PR-3,4,8,11	chitinase	0	2	2
PR-5	Thaumatin-like protein	1	5	6
PR-14	lipid transfer protein	7	1	8
PR-15	germin-like protein	0	1	1
	Disease resistance proteins	9	4	13
	snakin	2	0	2

HMW HSPs: High molecular weight heat shock proteins; LMW HSPs: Low molecular weight heat shock proteins.

2.6. Regulation of Hormonal Signaling in Response to Salt Stress

Hormones are pivotal to plants in stress adaptive signaling cascades and act as a central integrator to connect and reprogram different responses, such as photosynthesis and activities of ROS enzymes, protein structure and gene expression and accumulation of secondary metabolites [18–20]. In this experiment, various DEGs encoding hormone signaling was involved in abscisic acid (ABA), jasmonic acid (JA), auxin (IAA), gibberellin (GA), ethylene (ETH), brassinosteroid (BR) synthesis and signal transduction pathways (Table S8). Under salt stress, ABA is known to play a protective role in plants against LPO by assisting the accumulation of metabolites that act as osmolytes and also tends to close their stomata to reduce water loss by transpiration. Moreover, 8 transcripts encoding protein phosphatase 2C (PP2C) were down-regulated in the salt-treated grapevine leaves, while PP2C is deliberated as a negative regulator of the ABA signaling. Also, 2 ABA receptor PYL (1 up and 1 down-regulated) were also detected, which indicated that salt stress-induced not only the regulators but also the receptors in the ABA transduction pathway, by which ABA signaling pathway was enhanced quickly and then participated in the salt stress defense process.

Other plant hormones, like auxin and ethylene, also have important roles in plants to cope with salt stress. In this experiment, 23 auxin-related transcripts were detected, in which 2 auxin response factors (ARF) and 3 auxin-responsive proteins were down-regulated, while 12 auxin-induced proteins and 3 indole-3-acetic acid-induced proteins were up-regulated. Out of the 12 auxin-induced proteins, 4 transcripts (VIT_04s0023g00530, VIT_03s0038g01100, VIT_03s0038g01090 and VIT_04s0023g00520) were only expressed in the salt-treated samples, suggesting their close interaction with salt stress. In ethylene synthesis, 3 ACC oxidases (ACO) homologs were up-regulated, whereas 23 transcripts encoding ethylene-responsive TFs revealed variation in up-regulation (13 transcripts) and down-regulation (10 transcripts) in grapevine under salt stress.

2.7. Ion Transport Systems Mediating Na$^+$ Homeostasis in Response to Salt Stress

Salt tolerance mechanism works basically by reducing the undue accretion of Na$^+$ in the cytosol of the plant cell. The quantification of ionic concentrations suggested that Na$^+$ concentration increased significantly in leaf and root tissues (Figure 5). Leaves had higher Na$^+$ accumulation (5.51 ± 0.48), which was 40.47% more than Na$^+$ level of roots (3.28 ± 0.23). Moreover, Cl$^-$ concentration increased significantly at about 5-folds in leaves and 9-folds in roots as compared to their corresponding controls, respectively. In contrast, K$^+$ showed a decreasing trend in both tissues (leaf and root) after 48h of salt stress as compared with the control group (Figure 5). In the transcriptomic analysis, 14 DEGs were found to be involved in the ion transport systems, which include 2 vacuolar-type H$^+$-ATPase (V-type proton ATPase, 1 up and 1 down-regulated), 2 sodium/hydrogen exchanger (both down-regulated), 3 cyclic nucleotide-gated ion channel (CNGC, 2 up-regulated and one down-regulated), 2 potassium transporter (both down-regulated), 2 K$^+$ efflux antiporter (both down-regulated), 2 sodium-related cotransporter (sodium/pyruvate cotransporter, sodium/bile acid cotransporter (down-regulated) transcripts (Table S9). In the vacuolar membrane, V-ATPase is the central H$^+$ pump, which creates a transmembrane proton gradient and drives the Na$^+$/H$^+$ antiporter to transport the excessive Na$^+$ in the cytoplasm to vacuoles [21].

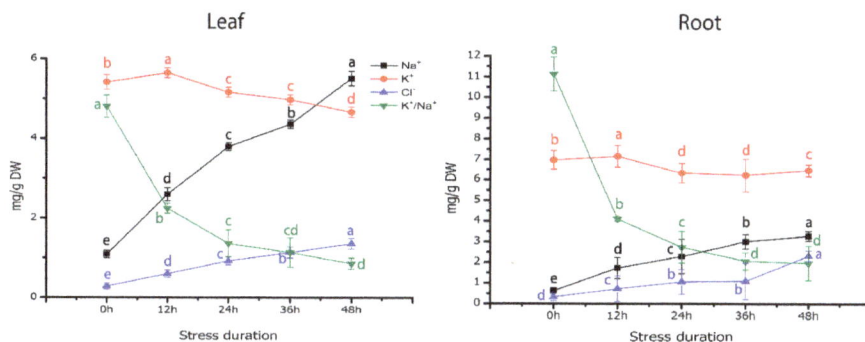

Figure 5. Ion concentrations of sodium (Na$^+$), potassium (K$^+$), chlorine (Cl$^-$) and K$^+$/Na$^+$ ratio in leaf and root samples of grapevine grown for 48 h of salt stress. Values represent means ± SE (*n* = 3) and the significance level of 0.05 was used for different letters above bars.

2.8. Transcription Factors in Response to Salt Stress

Transcription factors (TFs) are proteins that cooperate with other transcriptional regulators and bind cis-elements at the promoter region, thus up-regulate the downstream activities of many stress-related genes, results in inducing stress resistance in plants. Almost all the TFs identified in the present transcriptome data have already been reported to play a significant role to counter salt stress (Table S10). Results revealed five MYB transcripts (4 up-regulated, 1 down-regulated), 8 WRKY transcripts (all down-regulated), 1 C2H2 transcript (down-regulated), 4 DOF transcripts (3 up-regulated, 1 down-regulated), 6 HD-zip transcripts (all up-regulated), 5 bHLH transcripts (1 up-regulated and 4 down-regulated), 4 ZAT transcripts (1 up-regulated, 3 down-regulated), 6 NAC transcripts (1 up-regulated and 5 down-regulated), 3 PHD transcripts (all up-regulated) and 23 ERF TFs, intimating their critical roles in the grapevine resistance to salt stress (Figure 6).

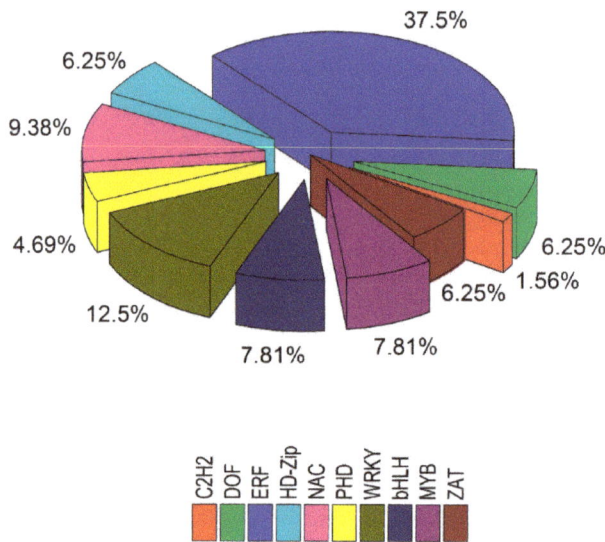

Figure 6. The sector diagram of major TFs identified and the total number of DEGs in grapevine leaf tissues after 48 h of salt stress compared with control.

2.9. qRT-PCR Validation of Illumina RNA-Seq Results

To validate the reliability of RNA-seq transcriptome, 16 DEGs were randomly selected to analyze the gene expression that was correlated with salt stress response and covering almost all the primary functions in various biological pathways, including transcription factors, metabolism, plant hormone signaling, disease resistance and ion transport (Table S11). The result suggested that expression of 16 DEGs treated with 0.8% soil salinity at the interval of 0, 12, 24, 36 and 48 h is inconsistent with the transcriptomic findings, validating the accuracy and reproducibility of the Illumina RNA-seq. Though, out of 16 DEGs, 12 DEGs showed recurrent expression pattern in response to salt stress, in which 8 genes were up-regulated (Figure 7a,c–e,g–j) and 4 genes were down-regulated (Figure 7k,n–p) with prolonged salt stress. Based on the expression patterns of these 12 genes, we selected them as candidate genes to further validate their expressional variations following the different concentrations of salt stress and recovery process.

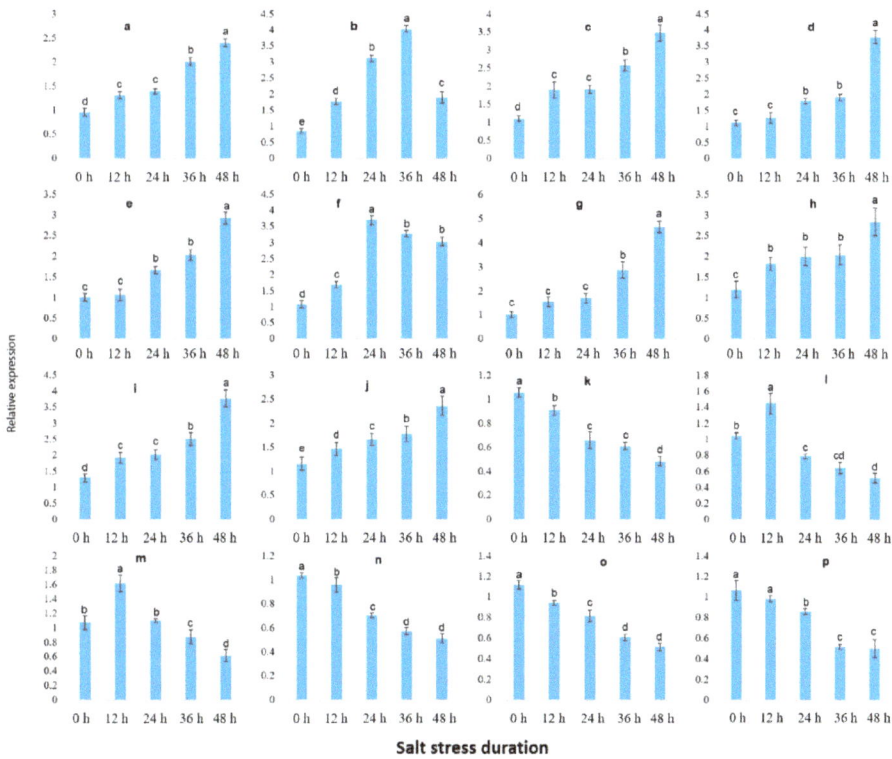

Figure 7. qRT-PCR validation of illumina Hiseq findings and screening of damage severity prediction marker genes. Values represent means ± SE (*n* = 3) and the significance level of 0.05 was used for different letters above bars. Genes have continual increasing or decreasing expression patterns were selected as candidate genes. **a**: VIT_05s0020g03740.t01, **b**: VIT_16s0050g02530.t01 **c**: VIT_19s0015g01070.t01, **d**: VIT_05s0049g00520.t01, **e**: VIT_13s0067g02360.t01, **f**: VIT_12s0035g01910.t01, **g**: VIT_04s0023g00530.t01, **h**: VIT_05s0049g01070.t01, **i**: VIT_06s0004g05670.t01, **j**: VIT_10s0003g01810.t01, **k**: VIT_00s0332g00110.t01, **l**: VIT_00s0201g00080.t01, **m**: VIT_07s0005g00160.t01, **n**: VIT_11s0052g01180.t01, **o**: VIT_14s0128g00020.t01, **p**: VIT_05s0062g00300.t01.

2.10. Salt Stress Recovery and the Selection and Validation of Marker Genes

Herein, 12 marker genes showing regular expression patterns, defined their potential as useful markers to determine the stress severity in grapevine plants. The growth status of grapevine plants was monitored, which indicated that salt severity turned grapevine leaves yellow and brownish blemishes were developed after a prolonged duration of salt stress and eventually die (Figures 8 and 9). At 1.5% salt concentration, grapevine plants can be recovered to normal growth conditions within 10 days of salt treatment by removing the salt stress, though few injured leaves could not survive even after the recovery, might be due to over-accumulation of salt. On the contrary, the plants died after prolonged salt stress duration (15 days), though there was no phenotypic evidence of death before going for recovery. Likewise, the critical time of recovery for 3.0% salt stress is 6 days. Similarly, qRT-PCR analysis of 12 candidate genes showed increased/ decreased expression level following the different doses of salt application. Furthermore, some genes showed unique expression pattern following 10 days of stress, such as VIT_05s0020g03740 showed an increasing trend within 9 days of salt stress but decreased significantly on the 10th day of salt stress (Figure 10a); whereas, the expression level of VIT_05s0049g00520, VIT_05s0049g01070 and VIT_06s0004g05670 was induced within 9 days after stress but sharply induced on the 10th day of salt stress (Figure 10d,h,i). Nevertheless, two transcripts (VIT_00s0332g00110 and VIT_05s0062g00300) showed a gradual decrease in their expression level till the 9th day but the sharp decrease was observed at the 10th day after NaCl application (Figure 10k,p).

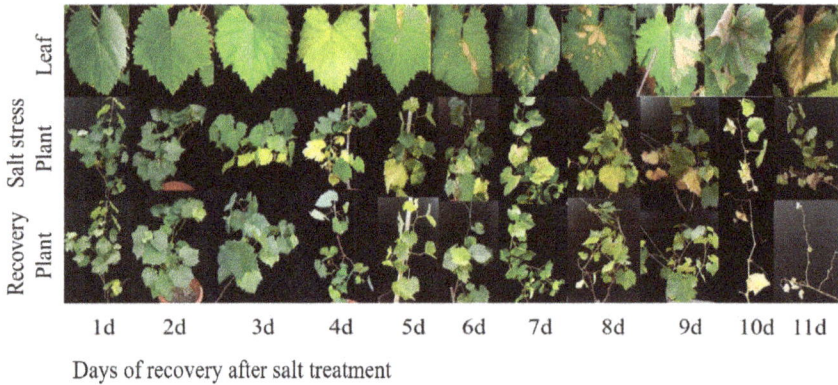

Figure 8. Grapevine growth status under salt stress and after removing salt stress. Grapevine plants were treated by 1.5% SS (salt stress) for 1, 2, 3,4, 5, 6, 7, 8, 9, 10 and 11 days (d), respectively and recovered by washing away the salt in the medium. Recovered plants were photographed 15 days after salt stress was removed.

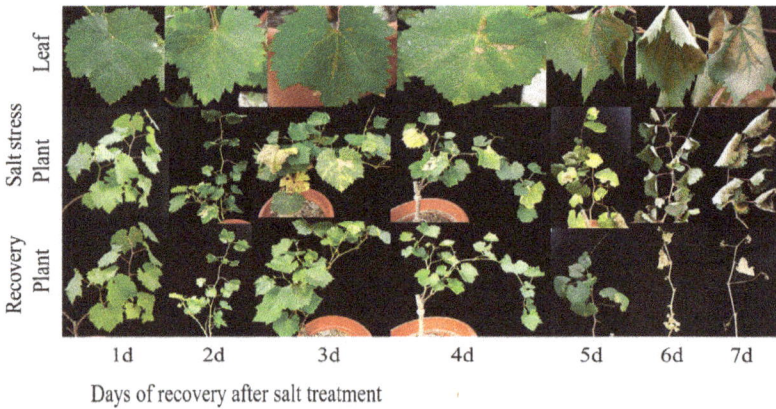

Figure 9. Grapevine growth status under salt stress and after removing salt stress. Grapevine plants were treated by 3.0% SS for 1, 2, 3, 4, 5, 6 and 7 days (d), respectively and recovered by washing away the salt in the medium. Recovered plants were photographed 15 days after salt stress was removed.

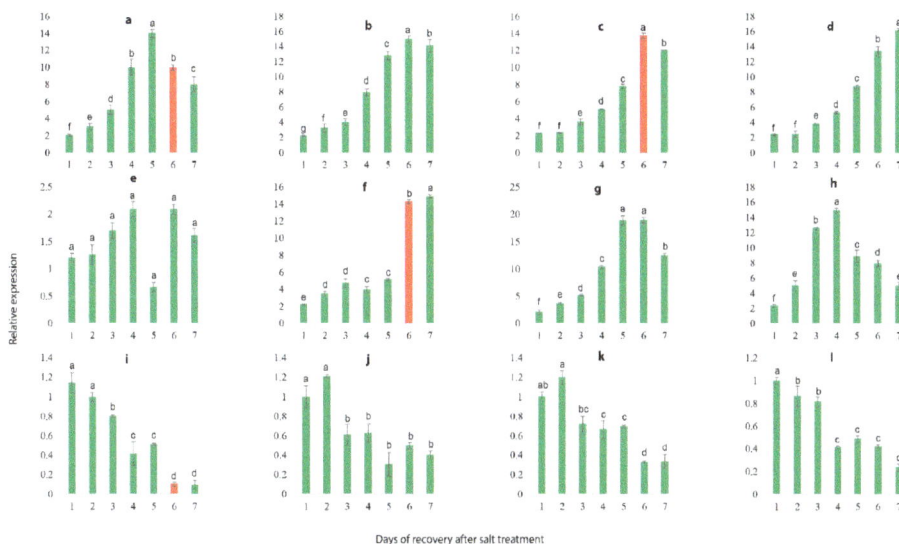

Figure 10. Expression patterns of the 12 candidate genes following the 7 days of 1.5% SS. Red bars indicate the sharp change in gene expression levels. Values represent mean ± SE (*n* = 3) and the significance level of 0.05 was used for different letters above bars. Genes with continual increasing or decreasing expression patterns were selected as candidate genes. **a**: VIT_05s0020g03740.t01, **b**: VIT_19s0015g01070.t01, **c**: VIT_05s0049g00520.t01, **d**: VIT_13s0067g02360.t01, **e**: VIT_04s0023g00530.t01, **f**: VIT_05s0049g01070.t01, **g**: VIT_06s0004g05670.t01, **h**: VIT_10s0003g01810.t01, **i**: VIT_07s0005g00160.t01, **j**: VIT_11s0052g01180.t01, **k**: VIT_14s0128g00020.t01, **l**: VIT_05s0062g00300.t01.

Interestingly, some of these genes showed a similar sharp expression level at the 6th day under 3.0% salt stress, such as transcript VIT_05s0020g03740 kept increasing until 5th day of stress but suddenly decreased on the 6th day of salt stress (Figure 11a), while the expression levels of VIT_05s0049g00520 and VIT_05s0049g01070 kept slow increasing trends up to 5 days of salt stress but showed a sharp increase on the 6th day (Figure 11d,h). Moreover, transcript VIT_00s0332g00110 showed a gradually decreasing trend up to 5 days of salt stress but significantly reduced on the 6th day of salt stress (Figure 11k). Based on above-mentioned findings, grapevine plants cannot be survived by curative processes after 10 days at 0.8% of NaCl and after 6 days at 1.5% of NaCl, which indicates that regardless of high or low concentrations of salt, these four genes with recurrent expression pattern could be used as potential markers to predict the severity imposed by salt stress.

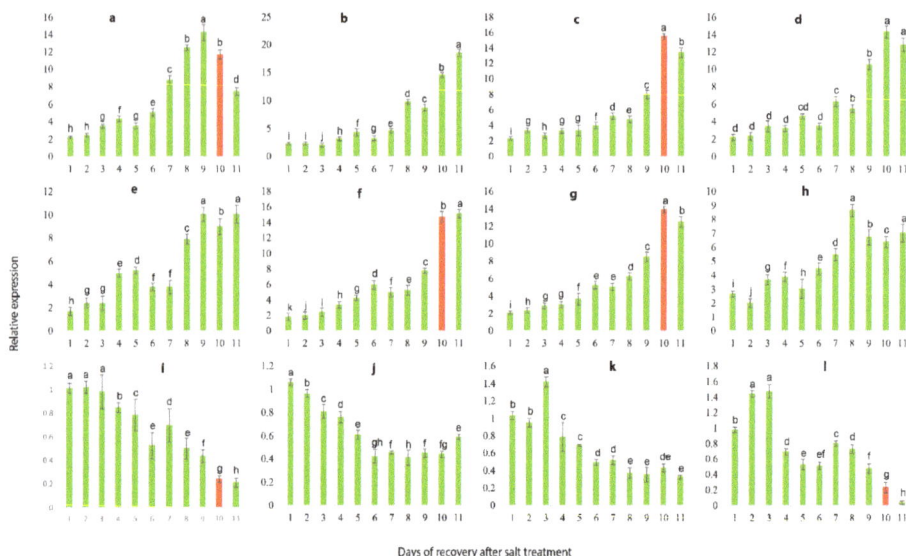

Days of recovery after salt treatment

Figure 11. Expression patterns of the 12 candidate genes following the 11 days of 3.0% SS. Red bars indicate the sharp change in gene expression levels. Values represent mean ± SE (n = 3) and the significance level of 0.05 was used for different letters above bars. Genes with continual increasing or decreasing expression patterns were selected as candidate genes. **a**: VIT_05s0020g03740.t01, **b**: VIT_19s0015g01070.t01, **c**: VIT_05s0049g00520.t01, **d**: VIT_13s0067g02360.t01, **e**: VIT_04s0023g00530.t01, **f**: VIT_05s0049g01070.t01, **g**: VIT_06s0004g05670.t01, **h**: VIT_10s0003g01810.t01, **i**: VIT_07s0005g00160.t01, **j**: VIT_11s0052g01180.t01, **k**: VIT_14s0128g00020.t01, **l**: VIT_05s0062g00300.t01.

3. Discussion

Salt stress is considered as most severe abiotic stress, which impairs all principal physiological functions, including photosynthesis, lipid metabolism and synthesis of proteins [22]. To confront the stress, plants are compelled to initiate protective responses, like restoring cellular ion concentrations and reducing the toxicity of ions like Na^+/H^+, K^+ and Cl^-. Moreover, the accretion of osmoprotectants and hydrophilic proteins, such as sugars, polyols, proline, glycine betaine (GB), amino acids (AA) and amines are crucial for governing the osmotic potential pressure. Also, the accumulation of ROS enzymes and antioxidants is vital to prevent tissue damage by eliminating the free radicals induced by salt stress [22,23].

Grapevine plants alter their physiology to combat salt stress severity. Current findings suggested that reduced stomatal conductance resulted in the inhibition of net photosynthesis rate and CO_2 exchange, which is considered as a primary response of grapevine to reduce transpiration rate to avoid salt accumulation in stomatal apertures of leaves. Zhang et al. [24] depicted that stress factors damage the photosynthetic pigments (chlorophylls and carotenoids) in both photosystems (PSI and PSII), which affect their light-absorbing efficiency, resulting in hindered photosynthetic capability. Our results are consistent with the findings of similar work on peach [25] and grapes [1], proposing that photosynthetic efficiency and CO_2 balance were affected by the reduced stomatal conductance. Moreover, the light-harvesting proteins (CP47) in PSII and chlorophyll binding proteins (P700) in PSI were down-regulated by the salt stress. Similar study intimated that salt stress induces ROS production, which damages the LHCs in PSI and impairs the PSII proteins involved in the evolution of oxygen [26]. Also, the modifications in leaf biochemistry decrease the synthesis of ATP amount, leading to regeneration of the RuBISCO, which results in down-regulation of photosynthetic metabolism [27], favor our findings of down-regulation of RuBISCO and ATP-related transcripts. In *Arabidopsis*, oxidative stress activates SnRK2 in ABA signaling, which regulates the stomatal conductance [27] and up-regulation of SnRK21 in our findings might be the reason for the inhibited photosynthetic activity of grapevine leaves.

Salt stress affects the large-scale metabolic activities that result in excessive ROS accumulation, which include singlet oxygen ($^1O_2$), superoxide radical ($O_2^{\bullet-}$), hydrogen peroxide (H_2O_2) and hydroxyl radical ($\bullet OH$), while similar results were observed in *Medicago truncatula* [28]. The ROS cytotoxicity activates the oxygen species, leading to disruption of optimum metabolic activities, which induce lipid peroxidation in plants [29,30]. Thus, the equilibrium between ROS production and quenching is critical under salt stress. Plants can unfold a complex antioxidative defense system to limit the oxidative damage, which mainly comprised of enzymatic antioxidant (SOD, CAT, POD and GST) and non-enzymatic antioxidants (AsA, GSH, proline and phenolic compounds) [31,32]. In the present study, the antioxidative defense system was activated in salt-treated leaves, although SOD-related transcripts were down-regulated, while CAT, POD and GST-related transcripts were significantly up-regulated. Similar research on *Pyrus pyrifolia* [33] and *Fagopyrum tataricum* [34] depicted that over-expression of GST transcripts significantly enhances salt stress tolerance. Moreover, the enhanced activities of ROS enzymes (CAT, POD and GST) are consistent with the transcriptomic data, which symbolize their vital functions in ROS detoxification. However, CAT activity increased significantly in our findings till 36 h of salt stress but drastically decrease at 48 h, which is in agreement with the down-regulation of SOD-related transcripts in grapevine under salt stress. Zhang et al. [35] reported that salt stress up-regulates the expression of CAT, POD and GST and increases the corresponding enzymes activities. Similarly, the complex accumulation pattern of antioxidant enzyme activities was observed in our findings, which is consistent with the findings of grapevine [36] and soybean [37] under salt stress.

HSPs are the molecular chaperones known to participate in the translocation and degradation of damaged proteins under abiotic stresses [38,39]. In the current study, HSP70 and HSP90 were down-regulated, while various heat stress TFs, small HSPs (sHSPs16–30 kDa) and other HSPs, like DnaJ, were up-regulated by salt stress. This irregular trend of HSPs may suggest that HSPs play an adaptive stress role by altering the growth and development of the plant. Several homologs of HSPs were also found to be activated in *Betula halophila* and *F. tataricum* under salt stress, intimating their regulatory role in various signaling-related pathways [34,40]. The disease resistance proteins can protect plants from pathogens by infection-induced responses of the immune system [41]. In our study, most pathogenesis-related proteins, non-specific lipid-transfer protein and disease resistance proteins were remarkably up-regulated, signifying that these genes not only function in disease resistance but also play essential roles in plant responses to salt stress [42].

Ionic compartmentalization and absorption are essential for growth under saline conditions because stress disrupts ion homeostasis [43]. Plant roots uptake Na^+ and other ions with water from the soil and translocate these ions to the leaves via transpiration stream. With the evaporation of water, a high level of salt gets accumulated in the apoplast and other cellular compartments. Ionic imbalance induces cellular toxicity via replacement of K^+ by Na^+ ions via interfering K^+ channels in the plasma membrane of the root [9,44], while all the potassium transporters were down-regulated in our findings and resulted in lower K^+ concentration in the salt-treated group as compared to control. The plant can resist the cytosolic salt accumulation in the vacuole and other cellular compartments to facilitate their metabolic functions [45,46]. This process involves the regulation of the expressions of some ionic channels and transporters-related genes, which enables the control of Na^+ transport within the plant [4,47]. In *Arabidopsis*, vacuolar $AtNa^+/H^+$ exchanger *SOS1* (Salt Overlay Sensitive) assists Na^+ extrusion from root cells [48,49] but Na^+/H^+ exchanger was down-regulated in our findings, which might be the reason of Na^+ accumulation in the grapevine roots. In addition, NHX1 was down-regulated in our results, while *AtNHX1* cloned plants resulted in high Na^+ in shoot tissues by altering the gene expression of Na^+ transporters [50]. The transcriptional activation of vacuolar-type ATPase (V-ATPase) in our findings suggested that it assists plants in reducing Na^+ accumulation by interacting H$^+$ pumps to counter salt stress [51]. The high-affinity K$^+$ transporters (HKTs) can mediate Na$^+$ transport and Na$^+$–K$^+$ symport, while the over-expressed *Arabidopsis AtHKT1* showed high Na$^+$ in leaves and reduced accumulation in roots [52], favor our findings of higher Na$^+$ accumulation in grapevine leaf tissues as compared to root. The inhibition of Na$^+$ influx is the correlative index of cyclic nucleotide-gated ion channels CNGCs that were up-regulated in this study, which is in good agreement with the findings reported in halophyte shrub (*Nitraria sibirica*) [53]. The genetic factors that control the accumulation and transport of Cl$^-$ from root to leaf tissues or enable plants to maintain low leaf Cl$^-$ level are the critical determinant of salt stress tolerance in plants [54], while higher accumulation of Cl$^-$ was observed in roots as compared to leaves in our findings, intimating that grapevine possesses the salt tolerance mechanism. However, in response to higher Cl$^-$ level, plants harness the Cl$^-$/H$^+$ transporters (CLCs) to maintain low Cl$^-$ accumulation, especially in aerial parts [55], whereas, no gene related to Cl$^-$ transport was found in our findings.

Phytohormones create a web of signals that are pivotal to plant growth, initiation of flowers, hypocotyl germination and abiotic stress response, which mainly include abscisic acid (ABA), auxin (AUX), jasmonic acid (JA), brassinosteroid (BR) and ethylene (ETH) [56,57]. High salt concentration triggers the ABA level in many plants, which is a well-known fact [25]. In our study, abscisic acid receptor PYL9 was up-regulated and its negative regulator PP2Cs were down-regulated, proposing that the ABA signaling pathway was activated in grapevine in response to salt stress. However, transcriptomic profiling of Jute (*Corchorus* spp.) revealed that DEGs encoding PYL were down-regulated under salt stress [58], which is contradicting with our findings as well as with the basic model of ABA signaling. Additionally, auxin stimulates cell elongation and cell division, also induces sugar and mineral accumulation at the site of application. Under salt stress, all ARFs and their repressors were down-regulated, whereas, most AUX/IAA proteins and IAA synthase were found up-regulated in our findings, suggesting that genes encoding IAA participate significantly in plant development in response to salt stress conditions, while similar results were reported in *V. vinifera* under oxidative stress [59]. JA generally reconciles specific signaling mechanisms involved in senescence, flowering and defense responses, while all the critical enzymes encoding JA were down-regulated, depicting that gene related to JA were suppressed by the salt severity in *Vitis vinifera*. Another study demonstrated that JA level was enhanced in salt-tolerant cultivars as compared to sensitive cultivars [60]. Salt stress inhibits the cell multiplication and expansion by suppressing the activities of growth-promoting hormones, including gibberellins and cytokinins [60], while these results are in favor of current findings.

TFs are regulatory proteins, demonstrated to be involved in regulating the stress-responsive gene expression in many plants responding to abiotic stress. Various MYB genes have been identified and known to induce plant responses to salt stress acclimation, such as *Arabidopsis* [61], rice [62] and wheat [63]. The over-expression of rice MYBs (*OsMYB48-1* and *OsMYB3R-2*) proposed alleviated tolerance to abiotic stresses, such as salt, cold and drought [64,65]. WRKY gene family is regarded as an essential TFs involved in salt stress response, such as *ZmWRKY33* in maize [66] and *GhWRKY39* in cotton [67], but, in our study, all the eight WRKY transcripts, including two WRKY33 were down-regulated under salt stress, which indicates the complexity of the WRKY regulatory mechanism and diverse nature in different stress conditions. Many other TFs with no direct response to salt stress but were triggered by other physiological changes like ROS and endogenous hormones. ERF family interact with ABA signaling pathway (dependent and/or independent) and respond to abiotic stresses [59,60]. In our study, 23 ERF transcripts were detected in the salt-treated grapevine leaves; meanwhile, ten ABA-related transcripts were also identified, intimating their essential roles in ABA-dependent ERF regulatory mechanism in grapevine. PHD finger proteins, especially PHD2, were reported to be involved in the salt stress response, which is known to induce by salt-induced oxidative stress [68]. All the three PHD transcripts detected in our transcriptome were up-regulated, which suggested the significance ROS synthesis caused by salt stress in the tested samples.

Specific genes regulate various plant traits and some of these genes expressed in unique patterns before the emergence of apparent traits, thus by detecting these unique expression signals, we can predict the occurrence of the corresponding phenotype. Hence, to measure the stress severity induced by NaCl stress, grapevine plants were subject to salt treatment with different doses of NaCl and time interval and were recovered by washing off the salt from roots. Results suggested that grapevine plants can be recovered within 10 days at 1.5% of NaCl dose and within 6 days under 3% salt stress. Similar expression patterns of genes (VIT_05s0049g00520, VIT_05s0049g01070, VIT_05s0020g03740 and VIT_00s0332g00110) observed in both NaCl treatments (1.5% and 3%) after 10 and 6 days, respectively, which makes them be the marker genes to estimate the salt severity. Selected genes were involved mainly in the maintenance of cellular structure and functions in plants. For instance, gene VIT_05s0020g03740 (non-specific lipid-transfer protein, LTP) and VIT_05s0049g00520 (proline-rich cell wall protein-like, PRPs) are known to play essential roles in maintaining the stability of the cell wall, membrane and osmotic pressure of the cell [69–71]. VIT_05s0049g01070 (glutathione S-transferase-like, GST) encoded as a critical protein, which has several physiological functions like ROS detoxification and protecting the DNA from damage [72]. Also, VIT_00s0332g00110 (Photosystem II reaction center protein) was involved in the most important physiological function (photosynthesis). Taken together, the transcriptional status of these marker genes reflects the vitality in grapevine plants. Moreover, a similar technique to predict marker genes can also be implicated on other crops where natural environmental disasters, such as temperature (low and high), water-logging and drought prevails occasionally. Taken together, grapevine possesses a complex regulatory mechanism of salt stress-tolerance, which mainly involves the regulation of key genes that are summarized in Figure 12.

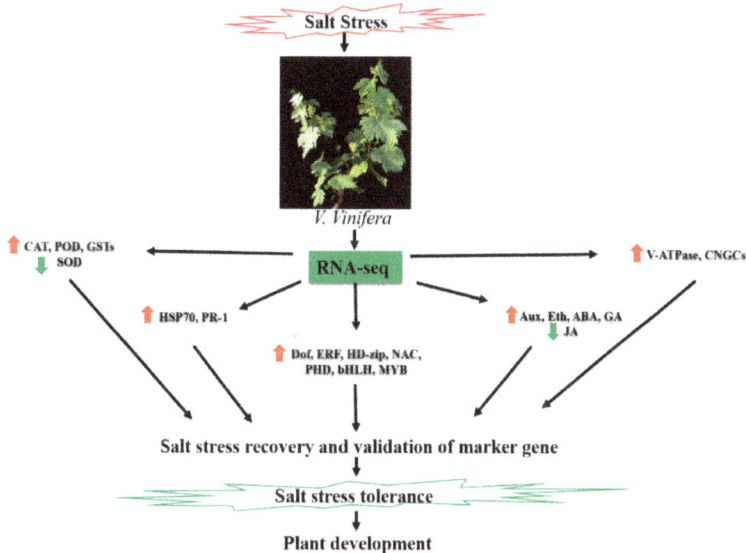

Figure 12. A schematic complex regulatory mechanism of salt stress tolerance in grapevine. Red arrows indicating up-regulated genes and green arrows indicating down-regulated genes. CAT, catalase; POD, peroxidase; GST, glutathione-s-transferase; SOD, superoxide dismutase; HSP70, heat shock protein 70 kDa; PR-1, Pathogenesis-related protein1; Dof, DNA-binding with one finger; ERF, ethylene responsive factor; HD-Zip, homeodomain-leucine zipper; NAC, NAC transcription factor; bHLH, basic helix-loop-helix; MYB, MYB transcription factor; Aux, auxin; Eth, ethylene; ABA, abscisic acid; GA, gibberellic acid; JA, jasmonic acid; V-ATPase, vacuolar-type ATPase; CNGCs, cyclic nucleotide-gated channels.

4. Materials and Methods

4.1. Plant Material and Salt Treatments

Two-year-old grapevine (Summer Black Cv.) pot grown plants obtained from Jiangsu Academy of Agriculture Sciences (JAAS), Nanjing, China and kept in greenhouse conditions (25 ± 5 °C), provided with 65% relative humidity (RH) and 16 h-light and 8 h-dark photoperiod at Nanjing Agricultural University, China. The grapevine plants were kept in a medium of soil-peat-sand at 3:1:1 (*v:v:v*) and used as experimental materials. Overall, ten grapevine plants were selected and categorized into salt-treated (5 plants) and control (5 plants) groups. NaCl (0.8%) was selected to induce salinity stress in grapevine plants. Fourth-unfolded leaf from both the NaCl-treated and control groups were collected at the interval of 0 (control), 12, 24, 36 and 48 h. Each sample has three replicates. Collected leaf samples were immediately frozen dried in liquid nitrogen and then stored at −80 °C until further analysis.

4.2. RNA Extraction, cDNA Library Construction and Illumina Deep Sequencing

Trizol reagent method was used to extract the total RNA from both salt-treated and control grapevine leaf samples (Invitrogen, Carlsbad, CA, USA). The RNA quantity was determined by using Micro-spectrophotometer (Nano-100, ALLSHENG, Hangzhou, China) and further mRNA purification and cDNA library construction were performed with the Ultra™ RNA Library Prep Kit for Illumina (MA, USA) by following the manufacturer's protocol. The final sampling collected after 48 h from salt treatment was sequenced against control (0 h) on an Illumina HiseqTM2500.

4.3. Mapping of reads, Gene Annotation and Analysis of Gene Expression Level

The raw sequence data were filtered by removing low-quality sequences and adapter reads by using HISAT [1]. After quality trimming, clean reads were mapped to the *V. vinifera* reference genome using Bowtie (1.1.2) by adapting standard mapping parameters [59]. In this data, >100 bp read length with <2 mismatches were mapped to reference genome. To calculate the gene expression and RPKM (and reads per kilobase per million, SAM tools and BamIndexStats.jar were used. Then, DEGseq2 was used to obtain differentially-expressed genes (DEGs) between Log_2 and the stationary phase [59]. The genes with FDR less than 0.001 and 2-fold change were pondered as DEGs.

4.4. Gene Ontology (GO) and Kyoto Encyclopedia of Genes and Genomics (KEGG)

For GO annotations, the DEGs were subjected GO database (http://www.geneontology.org/) by using program Blast2Go (http://www.blast2go.com/Ver.2.3.5). To classify genes or their products into terms (molecular function, biological process and cellular component) GO enrichment analysis by using GO-seq was used to under biological functions of DEGs [59]. For KEGG annotations, all the DEGs were mapped to the KEGG database (https://www.genome.jp/kegg/pathway.html) and looked for enriched pathways compared to the background genome [73].

4.5. Estimation of Photosynthesis Rate and Determination of Several Enzymatic and Ionic Concentrations

For photosynthesis rate (A_N), stomatal conductance rate (g_S), CO_2 exchange (Ci) and transpiration rate (E), 4th unfolded leaves were used from control and salt-treated grapevine plants between 9:00–11:00 AM, on full sunny day, using portable Li-COR (Li-6400XT, NE, USA) as briefly described by Haider et al. [1].

Leaf samples treated with 0.8% NaCl for 0, 12, 24, 36 and 48 h were used to determine the antioxidative enzymes activities, including SOD, CAT, POD and GST. The activity of SOD was measured using NBT at 560 nm; CAT activity was measured by monitoring disappearance of H_2O_2 at 240 nm, the POD was determined by guaiacol oxidation method following the method briefly explained by Haider et al. [74,75]. GST activity was determined using Glutathione S-transferase (GST) activity determination kit (Shanghai solarbio Bioscience & Technology Co., LTD, Shanghai, China) following the manufacturer's protocol.

For ionic concentrations, 0.5 g of leaf and root sample were first oven dried at 70 °C for 48 h and then ground to powder and digested in HNO3: HClO4 (2:1, *v:v*). The concentrations of selected ions (e.g., Na^+, K^+ and Cl^-) were determined using ICP-MS (Thermo Electron Corporation, MA, USA) as previously explained by Ma et al. [76]. The data were subjected to one-way analysis of variance (ANOVA) by using three replicates for each sample and expressed mean ± standard error (SE). Statistical analysis was carried out using Minitab (Ver 16) and SPSS (Ver 15.0) at $p < 0.05$ level of significance.

4.6. Quantitative Real-Time PCR (qRT-PCR) Analysis of DEGs and Validation of Illumina RNA-Seq Results

Sixteen genes selected from various pathways were used for the validation of the Illumina RNA-seq by qRT-PCR analysis. The primer pairs were designed using primer3 program (http://bioinfo.ut.ee/primer3-0.4.0/) and details of the primers are shown in supplementary Table S11. After extraction, total RNA was reverse-transcribed using the PrimeScript RT Reagent Kit with gDNA Eraser (Takara, Dalian, China). Each qPCR reaction contains 10 μL 2× SYBR Green Master Mix Reagent (Applied Biosystems, CA, USA), 2.0 μL cDNA sample and 400 nM of gene-specific primer in a final volume of 20 μL.qRT-PCR was carried out using an ABI PRISM 7500 real-time PCR system (Applied Biosystems, CA, USA). PCR conditions were 2 min at 95 °C, followed by 40 cycles of heating at 95 °C for 10 s and annealing at 60 °C for 40 s. A template-free control for each primer pair was set for each cycle. The All PCR reactions were normalized using the Ct value corresponding to the Grapevine actin gene (XM_010659103). Three biological replications were used and three measurements were performed on each replicate.

4.7. Salt Stress and Recovery Assay

To screen out the marker genes following the oxidative stress severity caused by salt, 15 grapevine plants were treated with two different acute salt concentrations (1.5% and 3.0%) and then plants recovered by washing off the NaCl solution.

Everyday 3 potted grapevine seedlings were recovered from NaCl stress by washing off the salts with the distilled water; this step was repeated till the salinity content from the medium was reduced to the average level (around 0.1%), 1/2 strength of Hoagland nutrient solution with standard NaCl content was watered again. The salt treated plants were sampled and photographed every day during the treatment and recovered plants. All qRT-PCR reactions for the selected marker genes were the same as previously mentioned.

5. Conclusions

A comparative transcriptome analysis was explored on two libraries constructed from salt-treated and control grapevine leaf samples. Results revealed that 2472 genes were differentially expressed and were significantly involved in antioxidant system, hormonal signaling, ion homeostasis and disease and pathogenesis-related pathways. Besides, many regulatory proteins encoding transcription factors were also identified that induce the function of other genes (e.g., *HSPs*) requisite for stress-adaptive responses and tolerance. The GO annotations assisted to screen out the series of molecular and physiological cues, which revealed their critical role in salt stress-tolerance mechanism. Moreover, salt stress significantly affected the photosynthetic efficiency and ions uptake and transport in *V. vinifera*. Though, antioxidant enzyme (CAT, POD and GST) activities were enriched to counter the lipid peroxidation. In this study, we have also screened out and validated the four candidate genes to predict salt severity in grapevine. Taken together, current study provided a deep overview of enriched genomic information along with physiological validation that will be useful for understanding the salt stress regulatory mechanism in grapevine.

Supplementary Materials: Supplementary materials can be found at http://www.mdpi.com/1422-0067/19/12/4019/s1.

Author Contributions: Conceived and designed the experiments: L.G., M.S.H., J.F. Perform the experiment: M.S.H., N.K., M.F., S.J., X.Z. Analyzed the data: L.G., K.Z., M.N., Manuscript writing: L.G., J.F. All the authors approved the final draft of the manuscript.

Funding: This work was supported by grants from the Jiangsu Agricultural Science and Technology Innovation Fund (CX (14) 2097), Natural Science Foundation of China (NSFC) (No. 31361140358), Natural Science Foundation of China (NSFC) (No. 31401846) and the Important National Science & Technology Specific Projects (No. 2012FY110100-3).

Conflicts of Interest: The authors declare that the research was conducted in the absence of any commercial or financial relationships that could be construed as a potential conflict of interest.

Abbreviations

CAT	Catalase
Cl^-	Chloride
DEGs	Differentially expressed genes
GO	Gene Ontology
GST	Glutathione S transferase
HSPs	Heat shock proteins
K^+	Potassium
KEGG	Kyoto Encyclopedia of Genes and Genomes
Na^+	Sodium
POD	Peroxidase
PRs	Pathogenesis-related proteins
qRT-PCR	Quantitative reverse transcriptome-PCR
ROS	Reactive oxygen species
RNA-seq	RNA-sequencing
SOD	Superoxide dismutase
SS	Salt stress
TFs	Transcription factors

References

1. Haider, M.S.; Kurjogi, M.M.; Khalil-Ur-Rehman, M.; Fiaz, M.; Pervaiz, T.; Jiu, S.; Haifeng, J.; Chen, W.; Fang, J. Grapevine immune signaling network in response to drought stress as revealed by transcriptomic analysis. *Plant Physiol. Biochem.* **2017**, *121*, 187–195. [CrossRef]
2. Allakhverdiev, S.I.; Sakamoto, A.; Nishiyama, Y.; Murata, N. Ionic and Osmotic Effects of Nacl-Induced Inactivation of Photosystems I and Ii in Synechococcus sp. *Plant Physiol.* **2000**, *123*, 1047–1056. [CrossRef]
3. Mahajan, S.; Tuteja, N. Cold salinity and drought stresses: An overview. *Arch. Biochem. Biophys.* **2005**, *444*, 139–158. [CrossRef]
4. Munns, R.; Tester, M. Mechanisms of salinity tolerance. *Annu. Rev. Plant Biol.* **2008**, *59*, 651. [CrossRef] [PubMed]
5. Chinnusamy, V.; Jagendorf, A.; Zhu, J.K. Understanding and Improving Salt Tolerance in Plants. *Crop Sci.* **2005**, *45*, 437–448. [CrossRef]
6. Tester, M.; Davenport, R. Na^+ Tolerance and Na^+ Transport in Higher Plants. *Ann. Bot.* **2003**, *91*, 503–527. [CrossRef] [PubMed]
7. Postnikova, O.A.; Shao, J.; Nemchinov, L.G. Analysis of the alfalfa root transcriptome in response to salinity stress. *Plant Cell Physiol.* **2013**, *54*, 1041–1055. [CrossRef]
8. Fan, X.D.; Wang, J.Q.; Yang, N.; Dong, Y.Y.; Liu, L.; Wang, F.W.; Wang, N.; Chen, H.; Liu, W.C.; Sun, Y.P. Gene expression profiling of soybean leaves and roots under salt, saline-alkali and drought stress by high-throughput Illumina sequencing. *Gene* **2013**, *512*, 392–402. [CrossRef]
9. Wang, B.; Lv, X.Q.; He, L.; Zhao, Q.; Xu, M.S.; Zhang, L.; Jia, Y.; Zhang, F.; Liu, F.L.; Liu, Q.L. Whole-transcriptome sequence analysis of verbena bonariensis in response to drought stress. *Int. J. Mol. Sci.* **2018**, *19*, 1751. [CrossRef] [PubMed]

10. Jaffar, M.A.; Song, A.; Faheem, M.; Chen, S.; Jiang, J.; Chen, L.; Fan, Q.; Chen, F. Involvement of cmwrky10 in drought tolerance of chrysanthemum through the aba-signaling pathway. *Int. J. Mol. Sci.* **2016**, *17*, 693. [CrossRef]

11. Wang, C.; Chen, H.F.; Hao, Q.N.; Shan, Z.H.; Zhou, R.; Zhi, H.J.; Zhou, X.A. Transcript profile of the response of two soybean genotypes to potassium deficiency. *PLoS ONE* **2012**, *7*, e39856. [CrossRef] [PubMed]

12. Wang, Y.; Tao, X.; Tang, X.M.; Xiao, L.; Sun, J.L.; Yan, X.F.; Li, D.; Deng, H.Y.; Ma, X.R. Comparative transcriptome analysis of tomato (Solanum lycopersicum) in response to exogenous abscisic acid. *BMC Genom.* **2013**, *14*, 841. [CrossRef] [PubMed]

13. Hernández, J.A.; Corpas, F.J.; Gómez, M.; Río, L.A.D.; Sevilla, F. Salt-induced oxidative stress mediated by activated oxygen species in pea leaf mitochondria. *Physiol. Plant.* **2010**, *89*, 103–110. [CrossRef]

14. Sreenivasulu, N.; Grimm, B.; Wobus, U.; Weschke, W. Differential response of antioxidant compounds to salinity stress in salt-tolerant and salt-sensitive seedlings of foxtail millet (Setaria italica). *Physiol. Plant.* **2010**, *109*, 435–442. [CrossRef]

15. Chaparzadeh, N.; D'Amico, M.L.; Khavari-Nejad, R.A.; Izzo, R.; Navari-Izzo, F. Antioxidative responses of Calendula officinalis under salinity conditions. *Plant Physiol. Biochem.* **2004**, *42*, 695–701. [CrossRef] [PubMed]

16. Marrs, K.A. The functions and regulation of glutathione-S-transferses in plants. *Annu. Rev. Plant Physiol. Plant Mol. Biol.* **1996**, *47*, 127–158. [CrossRef]

17. Edwards, R.; Dixon, D.P.; Walbot, V. Plant glutathione S-transferases: Enzymes with multiple functions in sickness and in health. *Trends Plant Sci.* **2000**, *5*, 193–198. [CrossRef]

18. Jakab, G.; Ton, J.; Flors, V.; Zimmerli, L.; Métraux, J.P.; Mauchmani, B. Enhancing Arabidopsis salt and drought stress tolerance by chemical priming for its abscisic acid responses. *Plant Physiol.* **2005**, *139*, 267–274. [CrossRef] [PubMed]

19. Kim, S.; Son, T.; Park, S.; Lee, I.; Lee, B.; Kim, H.; Lee, S. Influences of gibberellin and auxin on endogenous plant hormone and starch mobilization during rice seed germination under salt stress. *J. Environ. Biol.* **2006**, *27*, 181.

20. Peleg, Z.; Blumwald, E. Hormone balance and abiotic stress tolerance in crop plants. *Curr. Opin. Plant Biol.* **2011**, *14*, 290–295. [CrossRef] [PubMed]

21. Parida, A.K.; Das, A.B. Salt tolerance and salinity effects on plants: A review. *Ecotoxicol. Environ. Saf.* **2005**, *60*, 324–349. [CrossRef] [PubMed]

22. Carillo, P.; Annunziata, M.G.; Pontecorvo, G.; Fuggi, A.; Woodrow, P. Salinity stress and salt tolerance. In *Abiotic Stress in Plants-Mechanisms and Adaptations*; InTech: Rijeka, Croatia, 2011.

23. Horie, T.; Karahara, I.; Katsuhara, M. Salinity tolerance mechanisms in glycophytes. An overview with the central focus on rice plants. *Rice* **2012**, *5*, 11. [CrossRef] [PubMed]

24. Zhang, L.T.; Zhang, Z.S.; Gao, H.Y.; Xue, Z.C.; Yang, C.; Meng, X.L.; Meng, Q.W. Mitochondrial alternative oxidase pathway protects plants against photoinhibition by alleviating inhibition of the repair of photodamaged PSII through preventing formation of reactive oxygen species in Rumex, K.-1 leaves. *Physiol. Plant.* **2011**, *143*, 396–407. [CrossRef] [PubMed]

25. Haider, M.S.; Kurjogi, M.M.; Khalil-ur-Rehman, M.; Pervez, T.; Songtao, J.; Fiaz, M.; Jogaiah, S.; Wang, C.; Fang, J. Drought stress revealed physiological, biochemical and gene-expressional variations in "Yoshihime"peach (Prunus Persica, L.) cultivar. *J. Plant Interact.* **2018**, *13*, 83–90. [CrossRef]

26. Wang, N.; Qian, Z.; Luo, M.; Fan, S.; Zhang, X.; Zhang, L. Identification of Salt Stress Responding Genes Using Transcriptome Analysis in Green Alga Chlamydomonas reinhardtii. *Int. J. Mol. Sci.* **2018**, *19*, 3359. [CrossRef] [PubMed]

27. Hiroaki, F.; Verslues, P.E.; Jian-Kang, Z. Arabidopsis decuple mutant reveals the importance of SnRK2 kinases in osmotic stress responses in vivo. *Proc. Natl. Acad. Sci. USA* **2011**, *108*, 1717–1722.

28. Aydi, S.; Sassi, S.; Abdelly, C. Growth, nitrogen fixation and ion distribution in Medicago truncatula subjected to salt stress. *Plant Soil* **2008**, *312*, 59. [CrossRef]

29. Teakle, N.; Flowers, T.; Real, D.; Colmer, T. Lotus tenuis tolerates the interactive effects of salinity and waterlogging by "excluding"Na$^+$ and Cl$^-$ from the xylem. *J. Exp. Bot.* **2007**, *58*, 2169–2180. [CrossRef]

30. Tanveer, M.; Shabala, S. Targeting Redox Regulatory Mechanisms for Salinity Stress Tolerance in Crops. In *Salinity Responses and Tolerance in Plants*; Springer: Basel, Switzerland, 2018; Volume 1, pp. 213–234.

31. Mandhania, S.; Madan, S.; Sawhney, V. Antioxidant defense mechanism under salt stress in wheat seedlings. *Biol. Plant.* **2006**, *50*, 227–231. [CrossRef]

32. Abogadallah, G.M. Insights into the significance of antioxidative defense under salt stress. *Plant Signal. Behav.* **2010**, *5*, 369–374. [CrossRef]

33. Liu, D.; Liu, Y.; Rao, J.; Wang, G.; Li, H.; Ge, F.; Chen, C. Overexpression of the glutathione S-transferase gene from Pyrus pyrifolia fruit improves tolerance to abiotic stress in transgenic tobacco plants. *Mol. Biol.* **2013**, *47*, 515–523. [CrossRef]

34. Wu, Q.; Bai, X.; Zhao, W.; Xiang, D.; Wan, Y.; Yan, J.; Zou, L.; Zhao, G. De novo assembly and analysis of tartary buckwheat (fagopyrum tataricum Garetn.) transcriptome discloses key regulators involved in salt-stress response. *Genes* **2017**, *8*, 255. [CrossRef] [PubMed]

35. Yan, Z.; Zhou, L.; Yan, P.; Xiaojuan, W.; Dandan, P.; Yaping, L.; Xiaoshuang, H.; Xinquan, Z.; Xiao, M.; Linkai, H. Clones of FeSOD, MDHAR, DHAR Genes from White Clover and Gene Expression Analysis of ROS-Scavenging Enzymes during Abiotic Stress and Hormone Treatments. *Molecules* **2015**, *20*, 20939–20954.

36. Baneh, H.D.; Attari, H.; Hassani, A.; Abdollahi, R. Salinity effects on the physiological parameters and oxidative enzymatic activities of four Iranian grapevines (*Vitis vinifera* L.) cultivar. *Int. J. Agric. Crop Sci.* **2013**, *5*, 1022.

37. Weisany, W.; Sohrabi, Y.; Heidari, G.; Siosemardeh, A.; Ghassemi-Golezani, K. Changes in antioxidant enzymes activity and plant performance by salinity stress and zinc application in soybean (*Glycine max* L.). *Plant Omics* **2012**, *5*, 60.

38. Nollen, E.A.; Morimoto, R.I. Chaperoning signaling pathways. molecular chaperones as stress-sensingheat shock'proteins. *J. Cell Sci.* **2002**, *115*, 2809–2816.

39. Li, J.; He, Q.; Sun, H.; Liu, X. Acclimation-dependent expression of heat shock protein 70 in Pacific abalone (Haliotis discus hannai Ino) and its acute response to thermal exposure. *Chin. J. Oceanol. Limnol.* **2012**, *30*, 146–151. [CrossRef]

40. Shao, F.; Zhang, L.; Wilson, I.; Qiu, D. Transcriptomic Analysis of Betula halophila in Response to Salt Stress. *Int. J. Mol. Sci.* **2018**, *19*, 3412. [CrossRef]

41. Martin, G.B.; Bogdanove, A.J.; Sessa, G. Understanding the functions of plant disease resistance proteins. *Annu. Rev. Plant Biol.* **2003**, *54*, 23–61. [CrossRef]

42. Christensen, A.B.; Cho, B.H.; Næsby, M.; Gregersen, P.L.; Brandt, J.; Madriz-Ordeñana, K.; Collinge, D.B.; Thordal-Christensen, H. The molecular characterization of two barley proteins establishes the novel PR-17 family of pathogenesis-related proteins. *Mol. Plant Pathol.* **2002**, *3*, 135–144. [CrossRef]

43. Adams, P.; Thomas, J.C.; Vernon, D.M.; Bohnert, H.J.; Jensen, R.G. Distinct cellular and organismic responses to salt stress. *Plant Cell Physiol.* **1992**, *33*, 1215–1223.

44. Brini, F.; Masmoudi, K. Ion transporters and abiotic stress tolerance in plants. *ISRN Mol. Biol.* **2012**, *2012*, 927436. [CrossRef] [PubMed]

45. Reddy, M.; Sanish, S.; Iyengar, E. Compartmentation of ions and organic compounds in Salicornia brachiata Roxb. *Biol. Plant* **1993**, *35*, 547. [CrossRef]

46. Zhu, J.-K. Regulation of ion homeostasis under salt stress. *Curr. Opin. Plant Biol.* **2003**, *6*, 441–445. [CrossRef]

47. Rajendran, K.; Tester, M.; Roy, S.J. Quantifying the three main components of salinity tolerance in cereals. *Plant Cell Environ.* **2009**, *32*, 237–249. [CrossRef] [PubMed]

48. Huazhong, S.; Quintero, F.J.; Pardo, J.M.; Jian-Kang, Z. The putative plasma membrane Na(+)/H(+) antiporter SOS1 controls long-distance Na(+) transport in plants. *Plant Cell* **2002**, *14*, 465–477.

49. Quan-Sheng, Q.; Yan, G.; Dietrich, M.A.; Schumaker, K.S.; Jian-Kang, Z. Regulation of SOS1, a plasma membrane Na+/H+ exchanger in Arabidopsis thaliana, by SOS2 and SOS3. *Proc. Natl. Acad. Sci. USA* **2002**, *99*, 8436–8441.

50. Sottosanto, J.B.; Saranga, Y.; Blumwald, E. Impact of AtNHX1, a vacuolar Na + /H. + antiporter, upon gene expression during short- and long-term salt stress in Arabidopsis thaliana. *Bmc Plant Biol.* **2007**, *7*, 18. [CrossRef]

51. Zhu, J.K.; Shi, J.; Singh, U.; Wyatt, S.E.; Bressan, R.A.; Hasegawa, P.M.; Carpita, N.C. Enrichment of vitronectin-and fibronectin-like proteins in NaCl-adapted plant cells and evidence for their involvement in plasma membrane-cell wall adhesion. *Plant J.* **1993**, *3*, 637–646. [CrossRef]

52. Tomoaki, H.; Jo, M.; Masahiro, K.; Hua, Y.; Kinya, Y.; Rie, H.; Wai-Yin, C.; Ho-Yin, L.; Kazumi, H.; Mami, K. Enhanced salt tolerance mediated by AtHKT1 transporter-induced Na unloading from xylem vessels to xylem parenchyma cells. *Plant J. Cell Mol. Biol.* **2010**, *44*, 928–938.

53. Li, H.; Tang, X.; Zhu, J.; Yang, X.; Zhang, H. De Novo Transcriptome Characterization, Gene Expression Profiling and Ionic Responses of Nitraria sibirica Pall. under Salt Stress. *Forests* **2017**, *8*, 211. [CrossRef]

54. Henderson, S.W.; Baumann, U.; Blackmore, D.H.; Walker, A.R.; Walker, R.R.; Gilliham, M. Shoot chloride exclusion and salt tolerance in grapevine is associated with differential ion transporter expression in roots. *BMC Plant Biol.* **2014**, *14*, 1–18. [CrossRef] [PubMed]

55. Wei, L.I.; Wang, L.; Cao, J.; Bingjun, Y.U. Bioinformatics analysis of CLC homologous genes family in soybean genome. *J. Nanjing Agric. Univ.* **2014**, *37*, 35–43.

56. Magnan, F.; Ranty, B.; Charpenteau, M.; Sotta, B.; Galaud, J.P.; Aldon, D. Mutations in AtCML9, a calmodulin-like protein from Arabidopsis thaliana, alter plant responses to abiotic stress and abscisic acid. *Plant J.* **2008**, *56*, 575–589. [CrossRef] [PubMed]

57. Wang, H.; Liang, X.; Wan, Q.; Wang, X.; Bi, Y. Ethylene and nitric oxide are involved in maintaining ion homeostasis in *Arabidopsis callus* under salt stress. *Planta* **2009**, *230*, 293–307. [CrossRef] [PubMed]

58. Yang, Z.; Lu, R.; Dai, Z.; Yan, A.; Tang, Q.; Cheng, C.; Xu, Y.; Yang, W.; Su, J. Salt-Stress Response Mechanisms Using de Novo Transcriptome Sequencing of Salt-Tolerant and Sensitive *Corchorus* spp. Genotypes. *Genes* **2017**, *8*, 226. [CrossRef]

59. Haider, M.S.; Zhang, C.; Kurjogi, M.M.; Pervaiz, T.; Zheng, T.; Zhang, C.; Lide, C.; Shangguan, L.; Fang, J. Insights into grapevine defense response against drought as revealed by biochemical, physiological and RNA-Seq analysis. *Sci. Rep.* **2017**, *7*, 13134. [CrossRef]

60. Pedranzani, H.; Racagni, G.; Alemano, S.; Miersch, O.; Ramírez, I.; Peña-Cortés, H.; Taleisnik, E.; Machado-Domenech, E.; Abdala, G. Salt tolerant tomato plants show increased levels of jasmonic acid. *Plant Growth Regul.* **2003**, *41*, 149–158. [CrossRef]

61. Nagaoka, S.; Takano, T. Salt tolerance-related protein STO binds to a Myb transcription factor homologue and confers salt tolerance in Arabidopsis. *J. Exp. Bot.* **2003**, *54*, 2231–2237. [CrossRef]

62. Yang, A.; Dai, X.; Zhang, W.-H. A R2R3-type MYB gene, OsMYB2, is involved in salt, cold and dehydration tolerance in rice. *J. Exp. Bot.* **2012**, *63*, 2541–2556. [CrossRef]

63. Rahaie, M.; Xue, G.-P.; Naghavi, M.R.; Alizadeh, H.; Schenk, P.M. A MYB gene from wheat (*Triticum aestivum* L.) is up-regulated during salt and drought stresses and differentially regulated between salt-tolerant and sensitive genotypes. *Plant Cell Rep.* **2010**, *29*, 835–844. [CrossRef] [PubMed]

64. Dai, X.; Xu, Y.; Ma, Q.; Xu, W.; Wang, T.; Xue, Y.; Chong, K. Overexpression of an R1R2R3 MYB gene, OsMYB3R-2, increases tolerance to freezing, drought and salt stress in transgenic Arabidopsis. *Plant Physiol.* **2007**, *143*, 1739–1751. [CrossRef] [PubMed]

65. Xiong, H.; Li, J.; Liu, P.; Duan, J.; Zhao, Y.; Guo, X.; Li, Y.; Zhang, H.; Ali, J.; Li, Z. Overexpression of OsMYB48-1, a novel MYB-related transcription factor, enhances drought and salinity tolerance in rice. *PLoS ONE* **2014**, *9*, e92913. [CrossRef] [PubMed]

66. Li, H.; Gao, Y.; Xu, H.; Dai, Y.; Deng, D.; Chen, J. ZmWRKY33: A WRKY maize transcription factor conferring enhanced salt stress tolerances in Arabidopsis. *Plant Growth Regul.* **2013**, *70*, 207–216. [CrossRef]

67. Shi, W.; Liu, D.; Hao, L.; Wu, C.A.; Guo, X.; Li, H. GhWRKY39, a member of the WRKY transcription factor family in cotton, has a positive role in disease resistance and salt stress tolerance. *Plant Cell Tissue Organ Cult.* **2014**, *118*, 17–32. [CrossRef]

68. Wei, W.; Huang, J.; Hao, Y.J.; Zou, H.F.; Wang, H.W.; Zhao, J.Y.; Liu, X.Y.; Zhang, W.K.; Ma, B.; Zhang, J.S. Soybean GmPHD-Type Transcription Regulators Improve Stress Tolerance in Transgenic Arabidopsis Plants. *PLoS ONE* **2009**, *4*, e7209. [CrossRef] [PubMed]

69. Singh, N.; Bressan, R.A.; Carpita, N.C. Cell Walls of Tobacco Cells and Changes in Composition Associated with Reduced Growth upon Adaptation to Water and Saline Stress. *Plant Physiol.* **1989**, *91*, 48–53.

70. Cameron, K.D.; Teece, M.A.; Smart, L.B. Increased accumulation of cuticular wax and expression of lipid transfer protein in response to periodic drying events in leaves of tree tobacco. *Plant Physiol.* **2006**, *140*, 176–183.

71. Gothandam, K.M.; Nalini, E.; Karthikeyan, S.; Jeongsheop, S. OsPRP3, a flower specific proline-rich protein of rice, determines extracellular matrix structure of floral organs and its overexpression confers cold-tolerance. *Plant Mol. Biol.* **2010**, *72*, 125–135. [CrossRef] [PubMed]

72. Kampkötter, A.; Volkmann, T.E.; de Castro, S.H.; Leiers, B.; Klotz, L.O.; Johnson, T.E.; Link, C.D.; Henkle-Dührsen, K. Functional analysis of the glutathione S-transferase 3 from Onchocerca volvulus (Ov-GST-3): A parasite GST confers increased resistance to oxidative stress in Caenorhabditis elegans. *J. Mol. Biol.* **2003**, *325*, 25–37. [CrossRef]

73. Pervaiz, T.; Haifeng, J.; Salman, H.M.; Cheng, Z.; Cui, M.; Wang, M.; Cui, L.; Wang, X.; Fang, J. Transcriptomic Analysis of Grapevine (cv. Summer Black) Leaf, Using the Illumina Platform. *PLoS ONE* **2016**, *11*, e0147369.

74. Haider, M.S.; Khan, I.A.; Naqvi, S.A.; Jaskani, M.J.; Khan, R.W.; Nafees, M.; Pasha, I. Fruit developmental stages effects on biochemical attributes in date palm. *Pak. J. Agric. Sci.* **2013**, *50*, 577–583.

75. Haider, M.S.; Khan, I.A.; Jaskani, M.J.; Naqvi, S.A.; Khan, M.M. Biochemical attributes of dates at three maturation stages. *Emir. J. Food Agric.* **2014**, *11*, 953–962. [CrossRef]

76. Ma, Y.; Wang, J.; Zhong, Y.; Geng, F.; Cramer, G.R.; Cheng, Z.M. Subfunctionalization of cation/proton antiporter 1 genes in grapevine in response to salt stress in different organs. *Hortic. Res.* **2015**, *2*, 15031. [CrossRef] [PubMed]

International Journal of
Molecular Sciences

MDPI

Article

A Genome-Wide Association Study Reveals Candidate Genes Related to Salt Tolerance in Rice (*Oryza sativa*) at the Germination Stage

Jie Yu [1,2,†], Weiguo Zhao [1,3,†], Wei Tong [1,2], Qiang He [1,4], Min-Young Yoon [1,5], Feng-Peng Li [1,6], Buung Choi [1,7], Eun-Beom Heo [1,8], Kyu-Won Kim [9,*] and Yong-Jin Park [1,9,*]

[1] Department of Plant Resources, College of Industrial Sciences, Kongju National University, Yesan 32439, Korea; agnesyu121@ahau.edu.cn (J.Y.); wgzsri@126.com (W.Z.); wtong@ahau.edu.cn (W.T.); qiangh06@gmail.com (Q.H.); myyoon0721@gmail.com (M.-Y.Y.); lifengpeng2013@gmail.com (F.-P.L.); pckorea1587@gmail.com (B.C.); hueunbum@gmail.com (E.-B.H.)
[2] State Key Laboratory of Tea Plant Biology and Utilization, Anhui Agricultural University, Hefei 230036, China
[3] School of Biotechnology, Jiangsu University of Science and Technology, Sibaidu, Zhenjiang, Jiangsu 212018, China
[4] National Key Facility for Crop Resources and Genetic Improvement, Institute of Crop Science, Chinese Academy of Agricultural Sciences, Beijing 100081, China
[5] Leader of Eco. Energy & Bio (LEEBCOR), 190-26 Hwangyeonggongwon-ro, Asan-si, Chungcheongnam-do 31529, Korea
[6] Suzhou GENEWIZ Biotechnology Co. LTD, C3 218 Xinghu Road Suzhou Industrial Park, Suzhou 215123, China
[7] Chemical Safety Division, National Institute of Agricultural Sciences (NIAS), Wanju 55365, Korea
[8] Breeding & Research Institute, Koregon Co. LTD, Anseong Center 60-34, Gokcheon-gil, Bogae-Myeon, Anseong-Si, Gyeonggi-Do 17509, Korea
[9] Center of Crop Breeding on Omics and Artificial Intelligence, Kongju National University, Yesan 32439, Korea
* Correspondence: sh.kyuwon@gmail.com (K.-W.K.); yjpark@kongju.ac.kr (Y.-J.P.); Tel.: +82-41-330-1200 (K.-W.K. & Y.-J.P.)
† These authors contributed equally to this work.

Received: 21 September 2018; Accepted: 9 October 2018; Published: 12 October 2018

Abstract: Salt toxicity is the major factor limiting crop productivity in saline soils. In this paper, 295 accessions including a heuristic core set (137 accessions) and 158 bred varieties were re-sequenced and ~1.65 million SNPs/indels were used to perform a genome-wide association study (GWAS) of salt-tolerance-related phenotypes in rice during the germination stage. A total of 12 associated peaks distributed on seven chromosomes using a compressed mixed linear model were detected. Determined by linkage disequilibrium (LD) blocks analysis, we finally obtained a total of 79 candidate genes. By detecting the highly associated variations located inside the genic region that overlapped with the results of LD block analysis, we characterized 17 genes that may contribute to salt tolerance during the seed germination stage. At the same time, we conducted a haplotype analysis of the genes with functional variations together with phenotypic correlation and orthologous sequence analyses. Among these genes, *OsMADS31*, which is a MADS-box family transcription factor, had a down-regulated expression under the salt condition and it was predicted to be involved in the salt tolerance at the rice germination stage. Our study revealed some novel candidate genes and their substantial natural variations in the rice genome at the germination stage. The GWAS in rice at the germination stage would provide important resources for molecular breeding and functional analysis of the salt tolerance during rice germination.

Keywords: rice; genome-wide association study; salt stress; germination; natural variation

1. Introduction

Salt stress undesirably affects plant growth during all developmental stages. Therefore, it is a major threat to crop productivity [1] and this situation has lasted in some parts of the world for over 3000 years with growth every year. As a monocotyledonous model plants, rice feeds more than one half of the world's population [2]. However, rice is also sensitive to salt stress and is currently listed as the most salt-sensitive cereal crop, which results in most cultivated varieties having a salinity threshold of 3 dSm^{-1} [3].

Seed germination is usually a very important stage in the seedling stable stand establishment and determines the success of crop production [4]. The effects of salt stress on seed germination are extremely complex involving various physical and biochemical cues. Generally, salt stress is negatively correlated with seed germination and seedling growth [5] in most plants such as *Oryza sativa* [6], *Zea mays* [7], *Helianthus annuus* [8], and *Brassica* spp. [9]. The impact of salt stress on seed germination is attributed to seed water uptake and ion toxic effect. During the seed germination, salinity alters the imbibition of water by reducing the osmotic potential of the germination medium [10], damages the ultrastructure of cells, tissue, and organs [11], changes the activity of enzymes [12], disturbs hormonal balance [13], alters protein metabolism [14], and reduces the use of seed reserves [15]. However, various environmental (external) and plant physiological (internal) factors affect seed germination under saline conditions including temperature, light, water and gasses, seed age, seed dormancy, nature of seed coat, seed morphology, and seedling vigor [16].

Recently, improving rice salt tolerance during the germination stage became more important because salinity may rapidly reduce the germination rate and percentage, which, in turn, may lead to a reduction of crop yields [17]. Many efforts have been made to improve seed germination and seedling vigor by optimizing the non-genetic factors [6]. However, success in improving salt tolerance in rice is made by identifying a major quantitative trait locus (QTL), which contributes to salt tolerance in rice. QTL analysis of seed germination have been reported in rice [18], soybean [19], wheat [20], *Arabidopsis* [21], and *Brassica rapa* [22]. However, it is difficult to develop rice elite varieties with a high level of salt tolerance due to a lack of understanding the mechanisms of salt tolerance during the seed germination stage. Moreover, QTLs conferring salt stress tolerance in rice were identified mainly at the seedling stage [23], but there are few reports on rice seed germination [18,24,25]. Wang et al. [18] detected 16 QTLs for the imbibition rate and germination percentage under control and salt stress with the recombinant inbred line (RIL) population derived from IR26/Jiucaiqing. Abe et al. [26] identified a candidate gene, *OsGA20ox1*, for a major QTL controlling seedling vigor in rice. Zheng et al. [27] identified 11 QTLs for salt tolerance at the germination and early seedling stage in *japonica* rice.

Genome-wide association study (GWAS) is an efficient method for detecting valuable natural variations in trait-associated loci as well as allelic variations in candidate genes underlying quantitative and complex traits [28,29]. Instead of SSR (simple sequence repeat) markers, which are commonly used for association mapping [30,31], SNPs (single-nucleotide polymorphisms) have become more popular for GWASs with the rapid development of NGS (next-generation sequencing) and the existence of high-density SNP markers by re-sequencing [32]. In rice, there are some successful reports to dissect genetic architecture of complex traits through GWAS [28,29,32,33]. However, limited studies have been carried out in rice to identify genes/QTLs for salt tolerance using GWAS. Kumar et al. [34] identified total 64 SNPs significantly associated with Na$^+$/K$^+$ ratio and other traits for reproductive stage salinity tolerance using GWAS. Yu et al. [35] identified 93 candidate genes significantly associated with salt tolerance at the rice seedling stage. Shi et al. [36] identified 22 significant salt tolerance associated SNPs based on the stress-susceptibility indices (SSIs) of vigor index (VI) and the mean germination time (MGT). Naveed et al. [37] identified 20 QTN for salinity tolerance at the germination and seedling stages in rice. Seed germination plays an important role in the cycle of plant growth. To our knowledge, there is little research until now on the identification of genes particularly for the germination stage salinity tolerance using GWAS in rice. Here, we applied GWAS mapping using ~1.65 million SNPs/indels covering all 12 rice chromosomes in a diverse rice collection to identify candidate genes and natural variation that may contribute to salt tolerance during the rice germination stage with the aim to guide breeding of salt-tolerant rice varieties.

2. Results

2.1. Phenotypic Screening and Evaluation

Individual value plots for GP with 0, 200 and 300 mM NaCl from a screening experiment using 12 randomly selected samples are shown in Figure 1a. According to the screening result, 200 mM NaCl fully exhibited their phenotypic variance, which resulted in the most diverse phenotypic distribution and facilitated discrimination of accessions with different salt tolerance levels. Thus, the treatment of 200 mM NaCl was chosen as the target salinity level for determining the salt tolerance of all accessions.

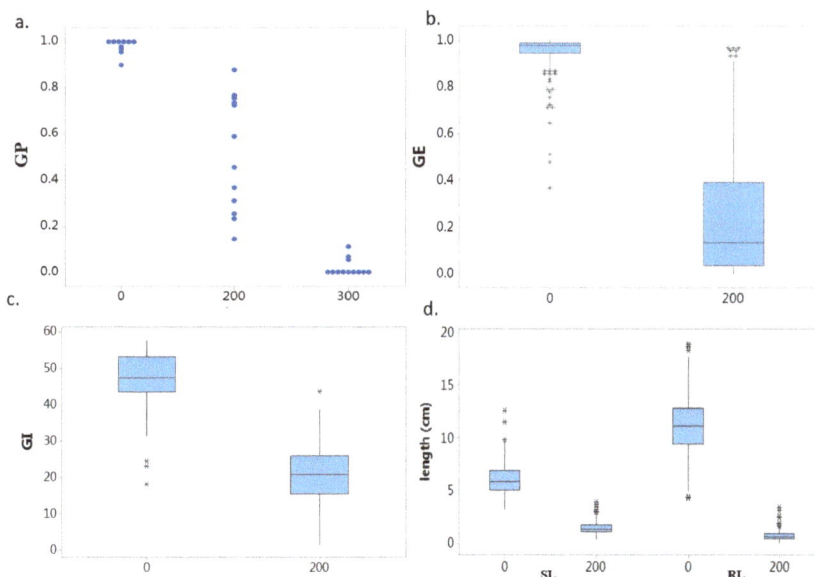

Figure 1. Determination of the optimum NaCl concentration and main phenotypes under salt stress and control conditions. (**a**) Individual value plot of germination percentage in the presence of 0, 200, and 300 mM NaCl (each dot represents an individual). (**b–d**) Box plots for phenotypic values in the presence of 0 and 200 mM NaCl (the asterisks are extreme outliers). GP: germination percentage, GE: germination energy, GI: Germination index, SL: shoot length, RL: root length.

The following traits: GP, GE, GI, SL, and RL were examined under 0 and 200 mM NaCl salt stress during the rice germination stage. Descriptive statistics of the phenotypes related to salt tolerance during the germination stage of the current collection were presented in Table 1. Box plots of phenotypes including GE, GI, SL, and RL in the presence of 0 and 200 mM NaCl are shown in Figure 1b–d. The data suggested that seed germination traits were negatively influenced by salt stress. Salt stress inhibits shoot and root elongation dramatically but GE and RL were more affected than SL [18]. These findings indicated that most germination parameters under salt stress exhibited lower performance than under control conditions, which may restrict plant growth.

The correlation coefficients of phenotypes under control and salt-stress conditions were also evaluated (Table 2). RL was significantly and positively correlated with SL only under control conditions. Excepting RL and SL was also significant positively correlated with GI. GP, GE, and GI were all significantly and positively correlated with each other under control conditions while all phenotypes were significantly and positively correlated with each other in salt stress conditions.

These results suggested that all phenotypes evaluated in this study could be used for GWAS and some overlapped results could be found among the phenotypes.

Table 1. Descriptive statistics for the traits in the control and salt-treated (200 mM NaCl) rice accessions.

Trait	Salinity Level (NaCl/mM)	Mean ± SD [a]	Range	Median	IQR [b]
GP	0	0.97 ± 0.06	0.47–1.00	0.99	0.97–1.00
	200	0.87 ± 0.18	0.14–1.00	0.94	0.83–0.98
GE	0	0.95 ± 0.08	0.37–1.00	0.98	0.94–0.99
	200	0.25 ± 0.29	0–0.97	0.13	0.03–0.39
GI	0	47.66 ± 6.71	18.33–57.87	47.55	43.40–53.44
	200	20.66 ± 8.53	1.51–43.72	21.05	15.48–26.03
SL	0	6.07 ± 1.38	3.25–12.62	5.82	5.08–6.90
	200	1.45 ± 0.63	0.38–3.87	1.29	1.03–1.76
RL	0	11.28 ± 2.89	4.28–18.89	11.11	9.46–12.81
	200	0.69 ± 0.49	0.013–3.44	0.55	0.39–0.87

[a] Standard deviation. [b] Interquartile range. GP: Germination percentage. GE: Germination energy. GI: Germination index. SL: shoot length. RL: root length.

Table 2. Pearson correlation coefficients among traits under control and salt stress (200 mM NaCl) conditions.

	Trait	GP	GE	GI	SL	RL
	GP					
	GE	0.946 ***				
Control	GI	0.624 ***	0.710 ***			
	SL	−0.001 ns	0.037 ns	0.168 **		
	RL	0.073 ns	0.087 ns	0.110 ns	0.254 ***	
	GP					
	GE	0.394 ***				
200 mM	GI	0.758 ***	0.849 ***			
	SL	0.439 ***	0.738 ***	0.725 ***		
	RL	0.357 ***	0.618 ***	0.594 ***	0.712 ***	

GP: germination percentage. GE: germination energy. GI: germination index. SL: shoot length. RL: root length.
*, **, ***, ns: significant at the 0.05, 0.01, and 0.001 probability level and not significant, respectively.

2.2. Principal Components Analysis (PCA)

PCA was performed with the 1.65 million high-quality SNPs/indels to mine the population structure in all rice accessions. Two components were suggested by the scree plot (Supplementary file 2: Figure S1a). Clear subpopulation structures were observed based on the first two PCs (PC1 and PC2), which resulted in two subpopulations, *indica* and *japonica*, with the admixture accessions located between the two groups (Figure 2a).

For the PCA using the phenotypes, we examined correlations between subspecies in salt tolerance levels using four main phenotypes that drove the differences among accessions. We used TASSEL to perform a PCA of R-GE, R-GI, R-RL, and R-SL in the rice collection. Most of the phenotypic variation (>91%) in the collection was explained by the first two PCs (Supplementary file 1: Figure S1b). Thus, we generated a PCA plot using PC1 and PC2. However, rice accessions in our study were not clustered into clearly defined groups (such as *indica* or *japonica*) based on the above four phenotypes (Figure 2b). This indicates that salt tolerance levels in rice (*O. sativa*) are not strongly correlated with the *indica* or *japonica* subgroups.

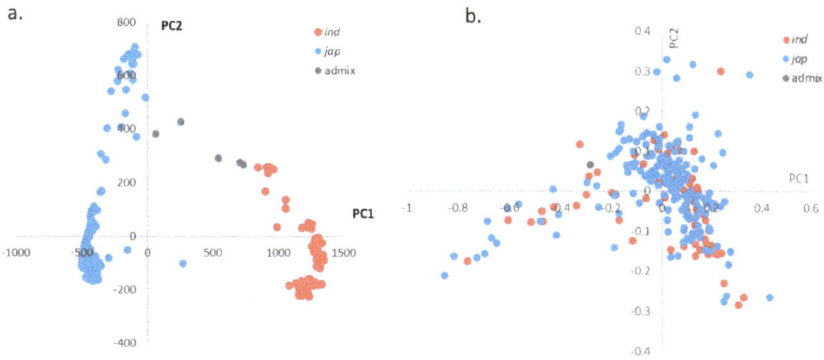

Figure 2. Principal components analysis (PCA) using genotype and phenotype data. (**a**). For genotype data, 295 accessions were divided into ind (*indica*) and jap (*japonica*) based on PC1 and PC2 along with an admixture group. (**b**). For phenotype data, no clear grouping was observed.

2.3. GWAS and Candidate Gene Identification

To generate the genotype dataset for GWAS, more than ~1.65 million SNPs/indels were identified across the accessions and subjected to GWAS applied with the CMLM [38]. The GWAS results were shown on Figure 3 and Supplementary file 1: Figure S2. We took associations held by the peaks with $-\log10 (p)$ value > 5 and adjusted p-value (FDR, false discovery rate) < 0.05 for further analysis since the cutoff of $-\log10 (p)$ value was five when the FDR $\leq$ 0.05. Under the salt stress condition, the significant signals were detected for GE, RL, SL, R/S, and R-R/S. In total, 10 SNPs were found significant for these traits and only one SNP (chr12_1628276) were found common for RL, R/S, and R-R/S traits. Excluding the common one, only 3, 2, 2, 1, and 1 SNPs were found uniquely associated for R-R/S, GE, RL, SL, and R/S, respectively. In addition, we also found two SNPs (chr02_1090174 and chr05_20164893) were the two strongest, significantly associated for GE in all observed traits ($p < 10^{-9}$).

We further conducted a genome-wide LD analysis of the candidate peak regions and determined LD blocks harboring significant SNPs/indels that characterized in the last step as regions containing putative candidate genes. LD block analysis was detected in a 400 kb range centered on the highest $-\log10 (p)$ value (Figure 3c,d). Annotation of SNPs/indels from the 200 kb up-stream and down-stream ranges, together with the LD block analysis, resulted in the identification of 79 genes included in these peaks and some candidate genes have been reported previously to contribute to salt tolerance (Supplementary file 2: Table S2). Among the known genes, seven were associated with chromosome 1 (all in RL), 22 with chromosome 2 (7 in GE, 6 in R/S and 9 in R-R/S), 10 with chromosome 3 with R-R/S, 15 with chromosome 4 (3 in SL and 12 in RL), 6 with chromosome 5 (all in GE), 6 with chromosome 11 (all in R-R/S), and 13 with chromosome 12 (the peak region was identical in RL, R/S, and R-R/S). In the LD block analysis, most highly associated SNPs/indels were located in small or large LD blocks, which indicated that they were in significant linkage disequilibrium. Thus, these candidate genes may contribute to salt stress independently or co-operatively with other variations in other genes harboring these SNPs/indels. Simultaneously, we screened candidate genes containing many highly associated SNPs/indels in the genic region as well as some highly associated signals not located in known genes but suggesting that these unknown genes may also be related to salt tolerance (Supplementary file 4: Table S3). Some of those SNPs/indels were located in the coding region of the unknown genes rather than in the surrounding 200 kb regions. These genes could also be important determinants of salt tolerance in rice.

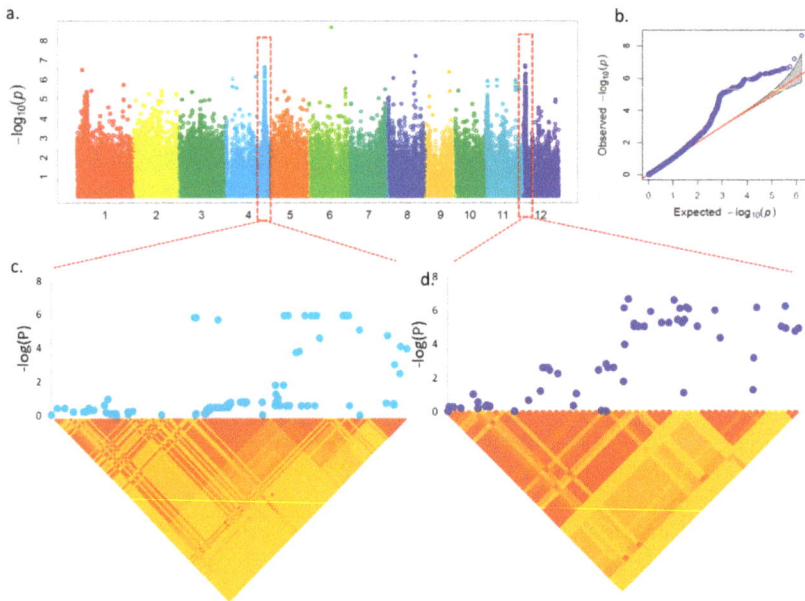

Figure 3. Genome-wide association mapping and LD block analysis for root length (RL) under salt stress (200 mM) conditions. (**a**) Manhattan plot from association mapping using the CMLM. (**b**) QQ plot of expected and observed P values. (**c**) The peak region on chromosome 4 along with the LD blocks. (**d**) The peak region on chromosome 12 along with the LD blocks. In (**c**,**d**) pair-wise LD between SNPs is indicated as *D′* values: red indicates a value of 1 and yellow indicates 0. The LD region was 200 kb upstream and downstream of the top −log (*p*) value in the peak range.

2.4. Natural Variations in Candidate Genes and Sequence Analysis

Based on the associated peaks identified in the GWAS and by determining the LD blocks test, we identified many candidate genes associated with salt tolerance during the rice germination stage. To mine functional and novel candidate genes, we investigated 17 final candidate genes (Table 3) that contained highly associated SNPs/indels within the coding region. Many of these SNPs/indels have been reported to play a role in the salt stress in rice such as *OsAGO2* (Os04g0615700) [39], *OsZIFL13* (Os12g0133300) [40], and *OsHAK11* (Os04g0613900) [41], which are related to salt stress in rice. These genes are involved in the salt tolerance in rice by different pathways [42].

Natural variations of these 17 genes were mined and then functional variations were screened after checking the positions of the variations in genes and the corresponding amino acid change. Among the genes, *OsMADS31*, which is involved in floral organ specification and implicated in plant growth and development, was identified and predicted to be involved in salt tolerance [43]. As shown in Figure 4, one natural SNP substitution (T/A) was detected and caused an F/L amino acid change, which is presented by type 1 (reference sequence) and type 2 (variation) (Figure 4a). Furthermore, we generated a haplotype network of the whole collection, which was dominated by two common haplotypes including primarily the *japonica* type (type 1) and the *indica* type (type 2), respectively (Figure 4e). A phenotypic difference was observed in type 1 with 236 accessions and an average RL of 0.6293 and type 2 with 58 accessions and an RL of 0.9327 (Figure 4b). We conducted further orthologue alignment of *OsMADS31* in several rice groups and other species (Figure 4c). Type 2 (candidate SNP) showed an F/L amino acid change compared to other rice groups and species (type 1, *Oryza brachyantha, Oryza rufipogon, Oryza punctate, Hordeum vulgare, Triticum aestivum, Aegilops tauschii*, and *Triticum urartu*). However, type 2 shared this F/L with three other rice species (*Oryza glaberrima, Oryza barthii* and *Oryza glumaepatula*). *Oryza glaberrima* and *Oryza barthii* are African rice and its wild type have higher salt tolerance than *Oryza sativa* species. *Oryza glumaepatula* is a wild rice found in South America usually in deep and sometimes flowing water, which may also have salt tolerance characteristics based on the presence of related genes. Four salt tolerant accessions with type 2 haplotype and four salt sensitive accessions without the haplotype were used for the real-time expression analysis. Generally, the relative RNA expression level of Os*MADS31* was higher in type 1 than in type 2, which indicates that the gene expression is down-regulated in salt conditions when compared to the control (Figure 4d).

Table 3. Candidate genes with highly associated signals in the coding region that overlapped with the GWAS and LD analysis.

Chr_Pos [a]	Trait	*p*-Value	FDR [b]	Gene ID	Description
chr02_19605493	R/S	1.72×10^{-7}	0.00779	Os02g0532500	Germin family protein, Germin-like protein 2-4
				Os02g0532900	Glycoside hydrolase family 17 protein
				Os02g0533300	Carbonic anhydrase, CAH1-like domain, containing protein
				Os02g0533800	Similar to ATPase inhibitor
chr04_31168058	RL	2.42×10^{-7}	0.00859	Os04g0612900	Vacuolar ATPase assembly integral membrane protein VMA21-like domain-containing protein
				Os04g0613900	Similar to Potassium transporter 18, OsHAK11
				Os04g0614000	Similar to Peroxisomal 2,4-dienoyl-CoA reductase
				Os04g0614100	MADS-box domain-containing protein, OsMADS31
				Os04g0614600	Similar to Viroid RNA-binding protein, aminotransferase
				Os04g0614500	Pyridoxal phosphate-dependent transferase, major region, subdomain 1 domain-containing protein
				Os04g0615100	Similar to Lecithine cholesterol acyltransferase-like protein
				Os04g0615700	Protein argonaute 2, OsAGO2
chr12_1628276	RL, R/S, R-R/S	2.02×10^{-7}	0.00859	Os12g0133100	Major facilitator superfamily protein, OsZIFL12
				Os12g0133300	zinc-induced facilitator-like 13, OsZIFL13
				Os12g0133400	4′-phosphopantetheinyl transferase domain-containing protein
				Os12g0133700	Stress-activated protein kinase pathway-regulating phosphatase 1
				Os12g0133800	Similar to Auxin efflux carrier protein, OsPIN1d

[a] The position was based on the annotation data on Os-Nipponbare-Reference-IRGSP-1.0 (RAP-DB, http://rapdb. dna.affrc.go.jp/). [b] FDR: False discovery rate. FDR Adjusted *p* values were calculated by GAPIT applying the Benjamini-Hochberg (1995) FDR-controlling procedure. RL: root length. R/S: root/shoot ratio. R-R/S: relative root/shoot ratio.

We also found several other functional SNPs/indels in the 17 candidate genes that were correlated with a phenotypic difference (Supplementary file 5: Table S4). These candidate genes may be related to rice salt tolerance, according to both previous reports and the natural variation mining in the current study. Novel polymorphisms of those genes may also contribute to salt tolerance that make the rice resistant to salinity.

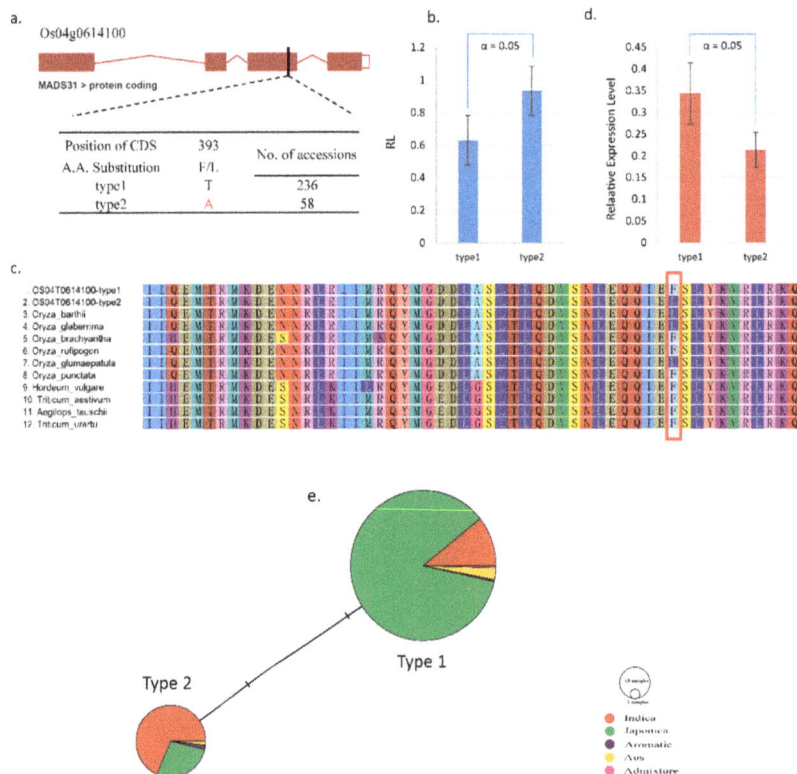

Figure 4. Haplotyping and sequence analysis of Os04g0614100, which was correlated with a phenotypic difference. (**a**) One functional SNP of the candidate gene in the CDS region. Type 1 is the reference and type 2 is the SNP. (**b**) The phenotypic difference based of the functional SNP. (**c**) Amino acid sequence alignment using several orthologues in various rice subgroups and species. Red box indicates the target amino acid change caused by the functional SNP. (**d**) RNA expression levels in rice accessions with type 1 and type 2. (**e**) Haplotype network analysis. Circle size is proportional to the number of samples within a given haplotype. Lines between haplotypes represent mutational steps between alleles. Colors denote rice designation.

3. Discussion

3.1. Salt Tolerance at Rice Germination Stage

The seed germination is one of the most critical steps in the life cycle of a crop. Seed germination begins with water uptake while salinity prevents water imbibition, which inhibits seed germination [15]. Experiments have shown that increased salinity delays the initiation of germination, which leads to a reduced germination percentage. However, salt tolerance during the early growth stages is not always correlated during subsequent growth stages [44,45]. The seeds of crops in different genotypes may germinate adequately under salt stress. However, the seedling may not become fully established later. We observed differential inhibition of the root length and shoot length in our study, which suggested that salinity can influence the germination quality of the seed.

By using the optimized salinity (200 mM NaCl) for discriminating accessions with different salt tolerance levels, we characterized the salt-tolerance-related phenotypes in a collection comprising 295 rice varieties. Phenotypic differences between the control and 200 mM NaCl salinity conditions suggested that rice growth during the seed germination stage can be markedly inhibited by salt stress, which results in very low germination energy and index (GE and GI) as well as reduced root and shoot lengths. This may suppress rice seed germination especially in some direct-sowing areas and decreases plant density and yield markedly. Therefore, development of rice varieties with salt-tolerant seeds would prevent salinity-mediated plant and yield loss during the early growth stage.

3.2. Salt Tolerance Is Not Strongly Correlated with Rice Subgroups

According to Lee et al. [46], the salt tolerance of *indica* rice was higher than that of *japonica* rice at the seedling stage, which was determined by measuring shoot Na^+ and K^+ absorption. However, as revealed in a recent study of the salt tolerance of 115 *O. sativa* and *O. glaberrima* accessions, salt tolerance was not strongly correlated with *O. sativa* cultivar groups [35]. Most of the *japonica* types were salt sensitive, but accessions from the *indica* group and *O. glaberrima* showed a wide range of sensitivities [47]. In our study, we performed a PCA of all rice germplasm using both genotype and phenotype data. Inconsistent with the genotypic PCA, which separated the collection into clear groups, phenotypic PCA using germination-related phenotypes showed no clear grouping (Figure 2b). This indicated that salt tolerance levels during the seed germination stage are not well correlated with the rice (*O. sativa*) subgroup.

3.3. GWAS and Candidate Gene Identification

In some direct-sowing areas, salt tolerance in rice during the seed germination stage is particularly important. To improve rice productivity in such areas, novel genes and alleles associated with complex quantitative salinity tolerance traits must be identified in diverse rice accessions and salt-tolerant varieties bred. An alternate and complementary approach is GWAS, which takes advantage of historical recombination events and, thus, enables a high-resolution genome wide mapping for the identification of target genomic regions in response to complex quantitative traits in rice [29]. In this study, we used a core set of rice collections and multiple bred varieties to investigate candidate loci and genes that regulate important phenotypes under salt stress in rice at the germination stage. Twelve GWAS peaks representing new QTLs on chromosomes 1, 2, 3, 4, 5, 11, and 12 during the rice germination stage were identified. The current association mapping can serve as source of novel salt tolerance genes and alleles. Thus, we found abundant candidate regions with high association peaks in five traits and were distributed on seven chromosomes. Now many QTL analysis of rice salt tolerance have been reported, but it is difficult to directly compare the chromosomal location of marker–trait associations detected in this study with previously reported QTLs because different materials at different stages, descriptive traits, and molecular maps have been used. Wang et al. [18] detected 16 QTLs for the imbibition rate and the germination percentage. Kumar et al. [34] identified 64 SNPs (loci) significantly associated with salt stress-related traits by GWAS. Leon et al. [48] identified 85 additive QTLs for seedling salinity tolerance by GBS. Yu et al. [35] identified 25 SNPs (loci) significantly associated with salt stress-related traits by GWAS. Shi et al. [36] identified 22 SNPs based on SSIs of VI and MGT by GWAS. In this study, we also found that some SNPs associated with salt-tolerance traits overlapped or located in similar or proximal regions such as a significant SNP (chr04_31168058) near *qRTL4.10* identified by Leon et al. [48] and SNPs (chr04_31164404) identified by Yu et al. (2017). This SNP is also located near the SNPs (chr04_34164920 and chr04_ 34292214) identified by Kumar et al. [34] associated with Na^+/K^+ ratio. Additionally, two QTLs for salt tolerance and potassium concentration were mapped just prior to this region, respectively, by Lin et al. [49] and Cai and Morishima [50]. The above results also indicated that chromosome 4 including many candidate genes in this region was found to be important for salt tolerance.

So far, about 70 salt tolerance QTLs had been located in rice using biparental mapping populations, but fine mapping and narrowing down reports are limited [34]. Driven by LD blocks to define the genomic regions for searching candidate genes has advantages over the fixed-window approach in which a certain distance from a significant SNP is considered to be the region containing candidate genes [51] by eliminating falsely included or excluded genes [52]. The wide candidate regions ranged from <1 kb to >1 Mb depending on the chromosomal position, which suggests that the resolution of the association mapping is highly dependent on the LD of the neighboring regions of the significant SNPs [34]. Since some of the LD blocks harboring significant SNPs did not contain an annotated gene, this method might have produced some false negatives or the identified region may have contained important DNA-binding or gene regulation sites, in which case, the causal gene was not detected in the LD block [53]. In this work, we used a 400 kb range of the strongest signal to locate the candidate genes, which is in line with previous studies [54]. From an LD block analysis, we obtained 79 candidate genes that had significant SNP/indel associations in LD block regions. Therefore, these regions and candidate genes have a statistically and genetically supported background and, therefore, may be important for the salt tolerance of rice during the germination stage. Apart from SNPs that had an association with previous known QTLs for salinity tolerance, there were a few SNPs, which hit specific genes that were known or functionally characterized for salt stress. Among 79 candidate genes including seven protein kinases (PK) (1 Serine/threonine protein kinase and 1 *OsCDPK26*), six ion exchanger and transporter related genes, five transcription factors (TFs), two electron carrier (peroxidase, Os01g0172600, and oxidoreductase, Os05g0411200), and two major facilitator superfamily proteins (Os12g0133100, *OsZIFL12*; Os12g0133300, *OsZIFL13*). In addition, we also found one stress-associated protein 18 (SAPs) (*OsSAP18*, Os02g0121600), one vacuolar ATPase assembly integral membrane protein (Os04g0612900), two argonaute family proteins (AGOs) (*OsAGO2*, Os04g0615700, and *OsAGO3*, Os04g0615800), one chloroplast precursor (*Ferritin1*, Os11g0106700), one calmodulin-like protein 3 (*OsCML3*, Os12g0132300), one Auxin efflux carrier protein (*OsPIN1d*, Os12g0133800), one Glycoside hydrolase (Os02g0532900), and one Glycosyltransferase (*ALG3*, Os01g0172000). The above results indicated that the candidate genes may play an essential role in salt tolerance mechanisms [34], which also indicated that salt tolerance genes are involved in ion pumps, calcium, the salt overly sensitive (SOS) pathway, mitogen-activated protein kinases (MAPK), glycine betaine, proline, and the reactive oxygen species pathways in a high salinity environment [42].

Furthermore, 17 candidate genes with high $-\log10$ (p) value-associated signals inside the coding region were also mined and may play an important role in salt tolerance. *OsHAKs* are candidates for high-affinity K^+ uptake transporters in the rice root. The transcription of *OsHAK11* (Os04g0613900) is significantly induced by salt stress and K^+ starvation, respectively [55]. *AGOs* (Os04g0615700) play important roles in the regulation of development and stress responses, antiviral immune response, transposons, and the regulation of chromatin structure and can affect the growth and development as well as the response to abiotic and biotic stress [56]. *OsPIN* (Os12g0133800), which encodes a member of the auxin efflux carrier proteins, is involved in the root elongation growth and lateral root formation patterns via the regulation of auxin distribution in rice [57]. The Germin family protein (Os02g0532500) had been revealed to be connected with a plant cell defense and diseases and to be highly resistant to sodium dodecyl sulfate (SDS) and proteases and important for early plant development and germination in plants [58]. *SAP* (Os12g0133700) is the A20/AN1 zinc-finger containing proteins, which can regulate the stress signaling in plants [59]. The Zinc-induced facilitator-like (ZIFL) family genes (Os12g0133100, *OsZIFL12*, Os12g0133300, *OsZIFL13*) are up-regulated under stress conditions [40].

Based on these regions and candidate genes, it may be possible to mine the natural variations of rice in response to salt stress in some tolerant accessions and apply those alleles to sensitive accessions via breeding methods. To the best of our knowledge, this is the first large-scale GWAS focusing on salt stress during the rice germination stage. These candidate regions and genes will facilitate the development of salt-tolerant rice varieties.

3.4. Novel Natural Variations of Candidate Genes

Investigation of new natural variations in focal traits can extend the tolerant varieties' functional alleles to other non-tolerant varieties. Breeding methods can then be used to transfer them to elite lines to produce tolerant varieties. Using the results of the GWAS and LD analysis, the haplotypes of candidate genes can be targeted and the functional alleles involved in responses can be identified.

New alleles in rice have been reported [60,61] and provide insight for researchers and breeders into the underlying mechanisms, which facilitates the breeding of improved varieties. According to Arora et al. [42], *OsMADS31* expression was low and not markedly affected by salt and cold stress. However, the expression was relatively down-regulated in seedlings under a salt condition. Nevertheless, *OsMADS31* expression was higher in seeds than during the panicle stage. In this study, we found that *OsMADS31* was associated with salt tolerance in rice at the seed germination stage with a down-regulated expression in the salt condition (Figure 4d). The contribution of MADS-box genes to flower organ specification is well developed in eudicots, but not very well in rice. Therefore, the roles of MADS genes and other candidate genes identified here using GWAS at the seed germination stage in response to salt stress should be investigated further. Moreover, by adapting functional studies (such as those performed using TALEN and CRISPR/Cas 9), the functions of genes and gene variations can be determined. Natural variations that have functional signals could be a good starting point for the exploration of gene-based assays of phenotypically different individuals such as salt tolerant vs. sensitive, drought resistant vs. susceptible, and more.

Overall, we investigated the genetics architecture of natural variation in rice salt-tolerance-related traits at the germination stage by GWAS mapping in 295 rice accessions. A total of 79 candidate genes were determined by LD blocks analysis. In addition, by detecting the highly associated variations located inside the genic region that overlapped with the results of LD block analysis, we finally characterized 17 genes that may contribute to salt tolerance during the seed germination stage. The salt tolerance related novel candidate genes would provide important resources for molecular breeding and functional analysis of the salt tolerance during the rice germination.

4. Materials and Methods

4.1. Materials

A core set of 137 rice accessions and 158 bred varieties from the National Gene Bank of the Rural Development Administration (RDA-Genebank, Korea) [62,63] was re-sequenced in the current study (Supplementary file 1: Table S1). We conducted a field experiment during the rice-growing season at the Kongju National University experimental farm and young leaves from a single plant were collected and immediately kept at −80 °C prior to genomic DNA extraction using the DNeasy Plant Mini Kit (Qiagen, Hilden, Germany). Qualified DNA was sent for the whole genome re-sequencing.

4.2. Whole Genome Re-Sequencing and Variation Detection

The genomes of all 295 rice accessions were sequenced with an average coverage of approximately 7.8× on an Illumina HiSeq 2000 or 2500 Sequencing Systems Platform (Illumina Inc., San Diego, CA, USA). Raw reads were aligned against the rice reference genome (IRGSP 1.0) [64] for genotypes calling and only SNPs/indels without the missing value and a minor allele frequency (MAF) > 0.05 and containing genotype calls for all 295 accessions that were used. Lastly, ~1.65 million high-quality SNPs/Indels were obtained and used for the further GWAS [65].

4.3. Evaluation of Salt Stress and Phenotyping

We first carried out the pre-screening experiment using 12 randomly selected samples to determine the optimum level of NaCl concentration for the evaluation of salt stress during the germination stage. Seed germination were initially screened by germinating 30 seeds per genotype in petri dishes with two layers of filter papers soaked in two different NaCl concentrations: 200 and 300 mM NaCl. The germination percentage was recorded daily for 10 days. At the concentration of 300 mM NaCl, seeds hardly germinated and the seedlings did not grow out enough to be able to measure root and shoot length. Therefore, in this study, we used 0 mM NaCl (non-stress) and 200 mM NaCl (salt stress) for phenotyping all 295 accessions.

The following experiments were performed in petri dishes containing two-layered filter paper. Thirty seeds were first washed in water, then sterilized in 1% sodium hypochlorite solution for 10 min, and washed three times in deionized distilled water. Thereafter, seeds of each accession were soaked in petri dishes and then incubated at 30 °C with 40% relative humidity. Petri dishes were randomized in an incubator and three replicates of each accession under control and salt conditions (200 mM) were adopted. The solution was replaced every two days to maintain the NaCl concentration and the distilled water volume, respectively. The daily germination seed was measured and filter papers were replaced as necessary. Plumule emergence was taken as an index of germination. The length reached about 2 mm. At the end of day 10, we measured the root length (RL) and shoot length (SL) of the seedlings and the R/S (root/shoot ratio) was also calculated. Based on these experiments, several germination stage-related phenotypes (list below) were calculated and subjected to a GWAS. The mean value of the three biological replicates was calculated and used in the further analysis.

Germination Percentage (GP)

GP was recorded daily for 10 days and was calculated using the formula below.
GP = Number of germinated seeds at 10 days/Total number of seeds tested × 100%

Germination energy (GE)

GE was recorded daily for four days and was calculated using the formula below.
GE = Number of germinated seeds at four days/Total number of seeds tested × 100%

Germination index (GI)

GI was calculated using the formula below.
GI=Σ(Gt/t), where Gt is the number of seeds that germinated on day t (Alvarado et al. 1987, and Ruan et al. 2002).

Relative germination energy (R-GE)

R-GE was calculated by using the formula below.
R-GE = $GE_{200}/GE_{control}$.

Relative germination index (R-GI)

R-GI was calculated by using the formula below.
R-GI = $GI_{200}/GI_{control}$.

Relative root length (R-RL)

R-RLwas calculated by using the formula below R-RL = $RL_{200}/RL_{control}$.

Relative shoot length (R-SL)

R-SLwas calculated by using the formula below R-SL = $SL_{200}/SL_{control}$.

Relative R/S (R-R/S)

R-R/S was calculated by using the formula below R-R/S = $R/S_{200}/R/S_{control}$.

4.4. Principal Components and GWAS Analysis

Principal components analysis (PCA) of the genotype with ~1.65 million SNPs/indels and four main salt-tolerance-related phenotypes: R-GE, R-GI, R-RL, and R-SL was conducted using GAPIT and Trait Analysis by Association, Evolution and Linkage (TASSEL) 5.0 [66]. Principal component analyses (PCA) in genotypic and phenotypic were also performed using GAPIT and TASSEL 5 [66].

GWAS was performed in the GAPIT package (Genome Association and Prediction Integrated Tool) in which an advanced kinship clustering algorithm was implemented [38]. Only SNPs with adjusted p-values < 0.05 were considered significantly associated. Gene loci containing the SNPs with significantly associated peaks in the Manhattan plot of the GWAS result were considered to be candidate genes related to salt tolerance.

4.5. Linkage Disequilibrium (LD) Block, Haplotype Analysis, and Expression Analysis

LD analysis. LD analysis was calculated using TASSEL 5 [66] based on the high-quality variations (with neither missing genotype calls over all accessions nor MAF < 0.05) in a 400 kb range determined by the most closely associated SNP/indel. An LD block was recognized when the top 95% confidence intervals of the D' value exceeded 0.98 and the lower bounds exceeded 0.70 [67]. Loci with significant variations harbored by LD blocks were then defined as the candidate genes.

Haplotype analysis. With about 7.3× depth of genome coverage, we constructed the haplotyping of the identified candidate genes. Nucleotide polymorphisms on the target genes were captured according to the rice reference genome (IRGSP 1.0). The orthologous genes of the target candidate genes in several other plants were provided by Ensembl Plants (http://plants.ensembl.org). Alignments of orthologous gene sequences were conducted using Geneious (http://www.geneious.com) [68]. In addition, the TCS [69] haplotype network was conducted by PopART v 1.7 [70].

Gene expression analysis by qRT-PCR. Germinated seeds with shoot and root after 10 days in control (H_2O) and salt (200 mM NaCl) conditions were collected and used for expression analysis. Total RNA was prepared using an RNA extraction kit (Qiagen, Hilden, Germany). cDNA was synthesized according to the manufacturer's instructions using the PrimeScript™ RT reagent Kit (TaKaRa, Shiga, Japan). Real-time PCR was carried out using the SYBR Green method with the primers of Os*MADS31* (MADS-F: TGGCTTCACTGACTCTGCAA, MADS-R: TACATACCCGGCTGTGCATC). Relative expression levels were calculated using the $2^{-\Delta\Delta CT}$ method [71] with *Ubiquitin 5* (*UBQ 5*) as the internal control [72] under three replicated tests.

Supplementary Materials: The following are available online at http://www.mdpi.com/1422-0067/19/10/3145/s1.

Author Contributions: Conceptualization, Y.-J.P. Formal analysis, Q.H. Funding acquisition, Y.-J.P. Investigation, M.-Y.Y. and F.L. Methodology, M.-Y.Y. and B.C. Project administration, K.-W.K. and Y.-J.P. Software, W.T. and E.-B.H. Writing–original draft, J.Y. and W.Z.

Funding: National Research Foundation of Korea (NRF), Rural Development Administration, Republic of Korea (RDA).

Acknowledgments: This work was supported by the National Research Foundation of Korea (NRF) grant funded by the Korea government (MSIT) (NRF-2017R1A2B3011208). This work was carried out with the support of "Cooperative Research Program for Agriculture Science and Technology Development (Project No. PJ013405)" Rural Development Administration, Republic of Korea.

Conflicts of Interest: The authors declare no conflict of interest. The funders had no role in the design of the study, in the collection, analyses, or interpretation of data, in the writing of the manuscript, or in the decision to publish the results.

Abbreviations

GWAS	genome-wide association study
LD	linkage disequilibrium
NGS	next-generation sequencing
SNP	single-nucleotide polymorphism
INDEL	insertion and deletion
MAF	minor allele frequency
RL	root length
SL	shoot length
RS	root/shoot ratio
GP	Germination percentage
GE	Germination energy
GI	Germination index
CMLM	compressed mixed linear model
PCA	Principal component analysis

References

1. Sakadevan, K.; Nguyen, M.L. Extent, impact, and response to soil and water salinity in arid and semiarid regions. *Adv. Agron.* **2010**, *109*, 55.
2. Mather, K.A.; Caicedo, A.L.; Polato, N.R.; Olsen, K.M.; McCouch, S.; Purugganan, M.D. The extent of linkage disequilibrium in rice (*Oryza sativa* L.). *Genetics* **2007**, *177*, 2223–2232. [CrossRef] [PubMed]
3. Wang, Z.; Chen, Z.; Cheng, J.; Lai, Y.; Wang, J.; Bao, Y.; Huang, J.; Zhang, H. QTL analysis of Na^+ and K^+ concentrations in roots and shoots under different levels of NaCl stress in rice (*Oryza sativa* L.). *PLoS ONE* **2012**, *7*, e51202. [CrossRef] [PubMed]
4. Almansouri, M.; Kinet, J.M.; Lutts, S. Effect of salt and osmotic stresses on germination in durum wheat (*Triticum durum* Desf.). *Plant and Soil* **2001**, *231*, 243–254. [CrossRef]
5. Rehman, S.; Harris, P.J.; Bourne, W.F.; Wilkin, J. The relationship between ions, vigour and salinity tolerance of acacia seeds. *Plant Soil* **2000**, *220*, 229–233. [CrossRef]
6. Xu, S.; Hu, B.; He, Z.; Ma, F.; Feng, J.; Shen, W.; Yang, J. Enhancement of salinity tolerance during rice seed germination by presoaking with hemoglobin. *Int. J. Mol. Sci.* **2011**, *12*, 2488–2501. [CrossRef] [PubMed]
7. Khodarahmpour, Z.; Ifar, M.; Motamedi, M. Effects of NaCl salinity on maize (*Zea mays* L.) at germination and early seedling stage. *Afr. J. Biotechnol.* **2014**, *11*, 298–304. [CrossRef]
8. Mutlu, F.; Bozcuk, S. Salinity-induced changes of free and bound polyamine levels in sunflower (*Helianthus annuus* L.) roots differing in salt tolerance. *Pak. J. Bot.* **2007**, *39*, 1097–1102.
9. Ulfat, M.; Athar, H.U.R.; Ashraf, M.; Akram, N.A.; Jamil, A. Appraisal of physiological and biochemical selection criteria for evaluation of salt tolerance in canola (*Brassica napus* L.). *Pak. J. Bot.* **2007**, *39*, 1593–1608.
10. Khan, M.A.; Weber, D.J. *Ecophysiology of High Salinity Tolerant Plants*; Springer Science & Business Media: New York, NY, USA, 2006.
11. Koyro, H.W. Ultrastructural Effects of Salinity in Higher Plants. In *Salinity, Environment-Plants-Molecules*; Springer: New York, NY, USA, 2002; ISBN 978-1-4020-0492-6.
12. Gomes-Filho, E.; Lima, C.R.F.M.; Costa, J.H.; da Silva, A.C.M.; Lima, M.d.G.S.; de Lacerda, C.F.; Prisco, J.T. Cowpea ribonuclease, properties and effect of NaCl-salinity on its activation during seed germination and seedling establishment. *Plant Cell Rep.* **2008**, *27*, 147–157. [CrossRef] [PubMed]
13. Khan, M.A.; Rizvi, Y. Effect of salinity, temperature, and growth regulators on the germination and early seedling growth of *Atriplex griffithii* var. stocksii. *Can. J. Bot.* **1994**, *72*, 475–479. [CrossRef]
14. Yupsanis, T.; Moustakas, M.; Eleftheriou, P.; Damianidou, K. Protein phosphorylation-dephosphorylation in alfalfa seeds germinating under salt stress. *J. Plant Physiol.* **1994**, *143*, 234–240. [CrossRef]
15. Othman, Y.; Al-Karaki, G.; Al-Tawaha, A.; Al-Horani, A. Variation in germination and ion uptake in barley genotypes under salinity conditions. *World J. Agric. Sci.* **2006**, *2*, 11–15.
16. Wahid, A.; Rasul, E.; Rao, A.U.R. *Germination of Seeds and Propagules under Salt Stress*; Handbook of Plant and Crop Stress: Boca Raton, FL, USA, 1999; Volume 2, pp. 153–167.
17. Foolad, M.; Hyman, J.; Lin, G. Relationships between cold-and salt-tolerance during seed germination in tomato, Analysis of response and correlated response to selection. *Plant Breed.* **1999**, *118*, 49–52. [CrossRef]
18. Wang, Z.F.; Wang, J.F.; Bao, Y.M.; Wu, Y.Y.; Zhang, H.S. Quantitative trait loci controlling rice seed germination under salt stress. *Euphytica* **2011**, *178*, 297–307. [CrossRef]
19. Csanádi, G.; Vollmann, J.; Stift, G.; Lelley, T. Seed quality QTLs identified in a molecular map of early maturing soybean. *Theor. Appl. Genet.* **2001**, *103*, 912–919. [CrossRef]
20. Bai, C.; Liang, Y.; Hawkesford, M.J. Identification of QTLs associated with seedling root traits and their correlation with plant height in wheat. *J. Exp. Bot.* **2013**, *64*, 1745–1753. [CrossRef] [PubMed]
21. DeRose-Wilson, L.; Gaut, B.S. Mapping salinity tolerance during *Arabidopsis thaliana* germination and seedling growth. *PLoS ONE* **2011**, *6*, e22832. [CrossRef] [PubMed]
22. Basnet, R.K.; Duwal, A.; Tiwari, D.N.; Xiao, D.; Monakhos, S.; Bucher, J.; Visser, R.G.F.; Groot, S.P.C.; Bonnema, G.; Maliepaard, C. Quantitative trait locus analysis of seed germination and seedling vigor in *Brassica rapa* reveals QTL hotspots and epistatic interactions. *Front. Plant Sci.* **2015**, *6*, 1032. [CrossRef] [PubMed]
23. Hoang, T.M.L.; Tran, T.N.; Nguyen, T.K.T.; Williams, B.; Wurm, P.; Bellairs, S.; Mundree, S. Improvement of salinity stress tolerance in rice, challenges and opportunities. *Agronomy* **2016**, *6*, 54. [CrossRef]
24. Wang, Z.; Wang, J.; Bao, Y.; Wang, F.; Zhang, H. Quantitative trait loci analysis for rice seed vigor during the germination stage. *Zhejiang Univ. Sci. B (Biomed. & Biotechnol.)* **2010**, *11*, 958–964.

25. Cheng, J.; He, Y.; Yang, B.; Lai, Y.; Wang, Z.; Zhang, H. Association mapping of seed germination and seedling growth at three conditions in *indica* rice (*Oryza sativa* L.). *Euphytica* **2015**, *206*, 103–115. [CrossRef]

26. Abe, A.; Takagi, H.; Fujibe, T.; Aya, K.; Kojima, M.; Sakakibara, H. OsGA20ox1.; a candidate gene for a major QTL controlling seedling vigor in rice. *Theor. Appl. Genet.* **2012**, *125*, 647–657. [CrossRef] [PubMed]

27. Zheng, H.; Liu, B.; Zhao, H.; Wang, J.; Liu, H.; Sun, J.; Xing, J.; Zou, D. Identification of QTLs for salt tolerance at the germination and early seedling stage using linkage and association analysis in *japonica* rice. *Chin. J. Rice Sci.* **2014**, *28*, 358–366.

28. Huang, X.; Wei, X.; Sang, T.; Zhao, Q.; Feng, Q.; Zhao, Y.; Li, C.; Zhu, C.; Lu, T.; Zhang, Z.; et al. Genome-wide association studies of 14 agronomic traits in rice landraces. *Nat. Genet.* **2010**, *42*, 961–967. [CrossRef] [PubMed]

29. Zhao, K.; Tung, C.W.; Eizenga, G.C.; Wright, M.H.; Ali, M.L.; Price, A.H.; Norton, G.J.; Islam, M.R.; Reynolds, A.; Mezey, J.; et al. Genome-wide association mapping reveals a rich genetic architecture of complex traits in *Oryza sativa*. *Nat. Commun.* **2011**, *2*, 467. [CrossRef] [PubMed]

30. Zhao, W.; Park, E.J.; Chung, J.W.; Park, Y.J.; Chung, I.M.; Ahn, J.K.; Kim, G.H. Association analysis of the amino acid contents in rice. *J. Integr. Plant Biol.* **2009**, *51*, 1126–1137. [CrossRef] [PubMed]

31. Li, G.; Na, Y.W.; Kwon, S.W.; Park, Y.J. Association analysis of seed longevity in rice under conventional and high-temperature germination conditions. *Plant Syst. Evol.* **2014**, *300*, 389–402. [CrossRef]

32. Huang, X.; Zhao, Y.; Wei, X.; Li, C.; Wang, A.; Zhao, Q.; Li, W.; Guo, Y.; Deng, L.; Zhu, C.; et al. Genome-wide association study of flowering time and grain yield traits in a worldwide collection of rice germplasm. *Nat. Genet.* **2012**, *44*, 32–39. [CrossRef] [PubMed]

33. Ma, X.; Feng, F.; Wei, H.; Mei, H.; Xu, K.; Chen, S.; Li, T.; Liang, X.; Liu, H.; Luo, L. Genome-wide association study for plant height and grain yield in rice under contrasting moisture regimes. *Front. Plant Sci.* **2016**, *7*, 1801. [CrossRef] [PubMed]

34. Kumar, V.; Singh, A.; Mithra, S.V.; Krishnamurthy, S.L.; Parida, S.K.; Jain, S.; Tiwari, K.K.; Kumar, P.; Rao, A.R.; Sharma, S.K.; et al. Genome wide association mapping of salinity tolerance in rice (*Oryza sativa*). *DNA Res.* **2015**, *22*, 133–145. [CrossRef] [PubMed]

35. Yu, J.; Zhao, W.; He, Q.; Kim, T.S.; Park, Y.J. Genome-wide association study and gene set analysis for understanding candidate genes involved in salt tolerance at the rice seedling stage. *Mol. Genet. Genomics* **2017**, *292*, 1391–1403. [CrossRef] [PubMed]

36. Shi, Y.; Gao, L.; Wu, Z.; Zhang, X.; Wang, M.; Zhang, C.; Zhang, F.; Zhou, Y.; Li, Z. Genome-wide association study of salt tolerance at the seed germination stage in rice. *BMC Plant Biol.* **2017**, *17*, 92. [CrossRef] [PubMed]

37. Naveed, S.A.; Zhang, F.; Zhang, J.; Zheng, T.Q.; Meng, L.J.; Pang, Y.L.; Xu, J.L.; Li, Z.K. Identification of QTN and candidate genes for salinity tolerance at the germination and seedling stages in rice by genome-wide association analyses. *Sci. Rep.* **2018**, *8*, 6505. [CrossRef] [PubMed]

38. Lipka, A.E.; Tian, F.; Wang, Q.; Peiffer, J.; Li, M.; Bradbury, P.J.; Gore, M.A.; Buckler, E.S.; Zhang, Z. GAPIT, genome association and prediction integrated tool. *Bioinformatics* **2012**, *28*, 2397–2399. [CrossRef] [PubMed]

39. Kapoor, M.; Arora, R.; Lama, T.; Nijhawan, A.; Khurana, J.P.; Tyagi, A.K.; Kapoor, S. Genome-wide identification, organization and phylogenetic analysis of Dicer-like, argonaute and RNA-dependent RNA polymerase gene families and their expression analysis during reproductive development and stress in rice. *BMC Genom.* **2008**, *9*, 451. [CrossRef] [PubMed]

40. Ricachenevsky, F.K.; Sperotto, R.A.; Menguer, P.K.; Sperb, E.R.; Lopes, K.L.; Fett, J.P. ZINC-INDUCED FACILITATOR-LIKE family in plants, lineage-specific expansion in monocotyledons and conserved genomic and expression features among rice (*Oryza sativa*) paralogs. *BMC Plant Biol.* **2011**, *11*, 20. [CrossRef] [PubMed]

41. Okada, T.; Nakayama, H.; Shinmyo, A.; Yoshida, K. Expression of OsHAK genes encoding potassium ion transporters in rice. *Plant Biotechnol.* **2008**, *25*, 241–245. [CrossRef]

42. Wang, J.; Chen, L.; Wang, Y.; Zhang, J.; Liang, Y.; Xu, D. A computational systems biology study for understanding salt tolerance mechanism in rice. *PLoS ONE* **2013**, *8*, e64929. [CrossRef] [PubMed]

43. Arora, R.; Agarwal, P.; Ray, S.; Singh, A.K.; Singh, V.P.; Tyagi, A.K.; Kapoor, S. MADS-box gene family in rice, genome-wide identification, organization and expression profiling during reproductive development and stress. *BMC Genom.* **2007**, *8*, 242. [CrossRef] [PubMed]

44. Zeng, L.; Shannon, M.; Grieve, C. Evaluation of salt tolerance in rice genotypes by multiple agronomic parameters. *Euphytica* **2002**, *127*, 235–245. [CrossRef]

45. Ferdose, J.; Kawasaki, M.; Taniguchi, M.; Miyake, H. Differential sensitivity of rice cultivars to salinity and its relation to ion accumulation and root tip structure. *Plant Prod. Sci.* **2009**, *12*, 453–461. [CrossRef]

46. Lee, K.S.; Choi, W.Y.; Ko, J.C.; Kim, T.S.; Gregorio, G.B. Salinity tolerance of *japonica* and *indica* rice (*Oryza sativa* L.) at the seedling stage. *Planta.* **2003**, *216*, 1043–1046. [CrossRef] [PubMed]

47. Platten, J.D.; Egdane, J.A.; Ismail, A.M. Salinity tolerance.; Na⁺ exclusion and allele mining of HKT1;5 in *Oryza sativa* and *O. glaberrima*, many sources.; many genes.; one mechanism? *BMC Plant Biol.* **2013**, *13*, 32. [CrossRef] [PubMed]

48. Leon, T.B.; Steven, L.; Prasanta, K.S. Molecular dissection of seedling salinity tolerance in rice (*Oryza sativa* L.) using a high-density GBS-based SNP linkage map. *Rice* **2016**, *9*, 52. [CrossRef] [PubMed]

49. Lin, H.X.; Yanagihara, S.; Zhuang, J.Y.; Senboku, T.; Zheng, K.L.; Yashima, S. Identification of QTLs for salt tolerance in rice via molecular markers. *Chin. J. Rice Sci.* **1998**, *12*, 72–78.

50. Cai, H.W.; Morishima, H. QTL clusters reflect character associations in wild and cultivated rice. *Theor. Appl. Genet.* **2002**, *104*, 1217–1228. [PubMed]

51. Courtois, B.; Audebert, A.; Dardou, A.; Roques, S.; Ghneim-Herrera, T.; Droc, G.; Frouin, J.; Rouan, L.; Goze, E.; Kilian, A.; et al. Genome-wide association mapping of root traits in a *japonica* rice panel. *PLoS ONE* **2013**, *8*, e78037. [CrossRef] [PubMed]

52. Chen, C.; DeClerck, G.; Tian, F.; Spooner, W.; McCouch, S.; Buckler, E. PICARA.; an analytical pipeline providing probabilistic inference about a priori candidates genes underlying genome-wide association QTL in plants. *PLoS ONE* **2012**, *7*, e46596. [CrossRef] [PubMed]

53. Sur, I.; Tuupanen, S.; Whitington, T.; Aaltonen, L.A.; Taipale, J. Lessons from functional analysis of genome-wide association studies. *Cancer Res.* **2013**, *73*, 4180–4184. [CrossRef] [PubMed]

54. Xu, X.; Liu, X.; Ge, S.; Jensen, J.D.; Hu, F.; Li, X.; Dong, Y.; Gutenkunst, R.N.; Fang, L.; Huang, L.; et al. Resequencing 50 accessions of cultivated and wild rice yields markers for identifying agronomically important genes. *Nat. Biotechnol.* **2012**, *30*, 105–111. [CrossRef] [PubMed]

55. Reddy, I.N.B.L.; Kim, B.K.; Yoon, I.S.; Kim, K.H.; Kwon, T.R. Salt tolerance in rice, focus on mechanisms and approaches. *Rice Sci.* **2017**, *24*, 123–144. [CrossRef]

56. Yang, Y.; Zhong, J.; Ouyang, Y.D.; Yao, J. The integrative expression and co-expression analysis of the AGO gene family in rice. *Gene* **2013**, *528*, 221–235. [CrossRef] [PubMed]

57. Inahashi, H.; Shelley, I.J.; Yamauchi, T.; Nishiuchi, S.; Takahashi-Nosaka, M.; Matsunami, M.; Ogawa, A.; Noda, Y.; Inukai, Y. OsPIN2, which encodes a member of the auxin efflux carrier proteins, is involved in root elongation growth and lateral root formation patterns via the regulation of auxin distribution in rice. *Physiol. Plantarum* **2018**. [CrossRef] [PubMed]

58. Rebecca, M.D.; Patrick, A.R.; Patricia, M.M.; Jan, E.L. Germins, a diverse protein family important for crop improvement. *Plant Sci.* **2009**, *177*, 499–510.

59. Kothari, K.S.; Dansana, P.K.; Giri, J.; Tyagi, A.K. Rice stress associated protein 1 (OsSAP1) interacts with aminotransferase (OsAMTR1) and pathogenesis-related 1a protein (OsSCP) and regulates abiotic stress responses. *Front. Plant Sci.* **2016**, *7*, 1057. [CrossRef] [PubMed]

60. Konishi, S.; Izawa, T.; Lin, SY.; Ebana, K.; Fukuta, Y.; Sasaki, T.; Yano, M. An SNP caused loss of seed shattering during rice domestication. *Science* **2006**, *312*, 1392–1396. [CrossRef] [PubMed]

61. Hu, B.; Wang, W.; Ou, S.; Tang, J.; Li, H.; Che, R.; Zhang, Z.; Chai, X.; Wang, H.; Wang, Y. Variation in NRT1.1B contributes to nitrate-use divergence between rice subspecies. *Nat. Genet.* **2015**, *47*, 834–838. [CrossRef] [PubMed]

62. Kim, K.W.; Chung, H.K.; Cho, G.T.; Ma, K.H.; Chandrabalan, D.; Gwag, J.G.; Kim, T.S.; Cho, E.G.; Park, Y.J. PowerCore, a program applying the advanced M strategy with a heuristic search for establishing core sets. *Bioinformatics* **2007**, *23*, 2155–2162. [CrossRef] [PubMed]

63. Zhao, W.; Cho, G.T.; Ma, K.H.; Chung, J.W.; Gwag, J.G.; Park, Y.J. Development of an allele-mining set in rice using a heuristic algorithm and SSR genotype data with least redundancy for the post-genomic era. *Mol. Breed.* **2010**, *26*, 639–651. [CrossRef]

64. Kawahara, Y.; de la Bastide, M.; Hamilton, J.P.; Kanamori, H.; McCombie, W.R.; Ouyang, S.; Schwartz, D.C.; Tanaka, T.; Wu, J.; Zhou, S.; et al. Improvement of the *Oryza sativa* Nipponbare reference genome using next generation sequence and optical map data. *Rice* **2013**, *6*, 4. [CrossRef] [PubMed]

65. Kim, T.S.; He, Q.; Kim, K.W.; Yoon, M.Y.; Ra, W.H.; Li, F.P.; Tong, W.; Yu, J.; Oo, W.H.; Choi, B.; et al. Genome-wide resequencing of KRICE_CORE reveals their potential for future breeding.; as well as functional and evolutionary studies in the post-genomic era. *BMC Genom.* **2016**, *17*, 408. [CrossRef] [PubMed]

66. Bradbury, P.J.; Zhang, Z.; Kroon, D.E.; Casstevens, T.M.; Ramdoss, Y.; Buckler, E.S. TASSEL, software for association mapping of complex traits in diverse samples. *Bioinformatics* **2007**, *23*, 2633–2635. [CrossRef] [PubMed]

67. Gabriel, S.B.; Schaffner, S.F.; Nguyen, H.; Moore, J.M.; Roy, J.; Blumenstiel, B.; Higgins, J.; DeFelice, M.; Lochner, A.; Faggart, M. The structure of haplotype blocks in the human genome. *Science* **2002**, *296*, 2225–2229. [CrossRef] [PubMed]

68. Kearse, M.; Moir, R.; Wilson, A.; Stones-Havas, S.; Cheung, M.; Sturrock, S.; Buxton, S.; Cooper, A.; Markowitz, S.; Duran, C. Geneious basic, an integrated and extendable desktop software platform for the organization and analysis of sequence data. *Bioinformatics* **2012**, *28*, 1647–1649. [CrossRef] [PubMed]

69. Clement, M.; Posada, D.; Crandall, K.A. TCS, a computer program to estimate gene genealogies. *Mol. Ecol.* **2000**, *9*, 1657–1659. [CrossRef] [PubMed]

70. Leigh, J.W.; Bryant, D. PopART, Full-feature software for haplotype network construction. *Methods Ecol. Evol.* **2015**, *6*, 1110–1116. [CrossRef]

71. Livak, K.J.; Schmittgen, T.D. Analysis of relative gene expression data using real-time quantitative PCR and the $2^{-\Delta\Delta CT}$ method. *Methods* **2001**, *25*, 402–408. [CrossRef] [PubMed]

72. Jain, M.; Nijhawan, A.; Tyagi, A.K.; Khurana, J.P. Validation of housekeeping genes as internal control for studying gene expression in rice by quantitative real-time PCR. *Biochem. Biophys. Res. Commun.* **2006**, *345*, 646–651. [CrossRef] [PubMed]

International Journal of
Molecular Sciences

MDPI

Article

Association Analysis of Salt Tolerance in Asiatic cotton (*Gossypium arboretum*) with SNP Markers

Tussipkan Dilnur [†], Zhen Peng [†], Zhaoe Pan, Koffi Kibalou Palanga, Yinhua Jia, Wenfang Gong and Xiongming Du *

State Key Laboratory of Cotton Biology, Institute of Cotton Research, Chinese Academy of Agricultural Sciences, Anyang 455000, China; tdilnur@mail.ru (T.D.); cripengzhen09@126.com (Z.P.); panzhaoe@163.com (Z.P.); palangaeddieh@yahoo.fr (K.K.P.); jiayinhua_0@sina.com (Y.J.); gwf018@126.com (W.G.)
* Correspondence: dujeffrey8848@hotmail.com; Tel.: +86-0372-256-2252
† These authors contributed equally to this work.

Received: 25 March 2019; Accepted: 30 April 2019; Published: 1 May 2019

Abstract: Salinity is not only a major environmental factor which limits plant growth and productivity, but it has also become a worldwide problem. However, little is known about the genetic basis underlying salt tolerance in cotton. This study was carried out to identify marker-trait association signals of seven salt-tolerance-related traits and one salt tolerance index using association analysis for 215 accessions of Asiatic cotton. According to a comprehensive index of salt tolerance (CIST), 215 accessions were mainly categorized into four groups, and 11 accessions with high salinity tolerance were selected for breeding. Genome-wide association studies (GWAS) revealed nine SNP rich regions significantly associated with relative fresh weight (RFW), relative stem length (RSL), relative water content (RWC) and CIST. The nine SNP rich regions analysis revealed 143 polymorphisms that distributed 40 candidate genes and significantly associated with salt tolerance. Notably, two SNP rich regions on chromosome 7 were found to be significantly associated with two salinity related traits, RFW and RSL, by the threshold of $-\log_{10}P \geq 6.0$, and two candidate genes (Cotton_A_37775 and Cotton_A_35901) related to two key SNPs (Ca7_33607751 and Ca7_77004962) were possibly associated with salt tolerance in *G. arboreum*. These can provide fundamental information which will be useful for future molecular breeding of cotton, in order to release novel salt tolerant cultivars.

Keywords: *Gossypium arboretum*; salt tolerance; single nucleotide polymorphisms; association mapping

1. Introduction

Soil salinity accumulation has become a serious environmental problem [1] that could negatively affect plant growth, geographical distribution, and agricultural products [2,3]. Salinization consists of the accumulation of water-soluble salts in the soil, including ions of potassium (K^+), magnesium (Mg^{2+}), calcium (Ca^{2+}), chloride (Cl^-), sulfate (SO_4^{2-}), carbonate (CO_3^{2-}), bicarbonate (HCO_3^-) and sodium (Na^+). The causes of land salinization can be divided into two categories: 1) primary (natural) and 2) secondary (anthropogenic) [4]. The primary reason includes arid climates, high underground water levels, seawater infiltration, and so on [5]. The secondary reason is irrigation practices. Soil salinization is reducing the area that can be used for agriculture by 1%–2% every year, hitting hardest in the arid and semi-arid regions. Therefore, the development of salt-tolerant crops is a pressing scientific goal [6], but the ability of plants to deal with these adverse factors is different [7]. Cotton is one of the advantageous salt-tolerant crop with a threshold salinity level of 7.7 dS·m^{-1} [8]. However, high salt concentrations can still hinder growth during the germination and seedling stages, which are the two most susceptible stages of plants [2,9].

The mechanisms involved in the response to salinity in cotton have been well described by Peng et al. [3,10]. Under salinity stress conditions, soluble salts are accumulated in the root zone of plants, then causing osmotic and ionic stress and mineral perturbations [3,11], leading to dramatic reductions in crop quality and yield [5]. However, the genetic control of salt tolerance is only partially understood, because of the diversity of the regulation mechanisms, and the complexity of the genetic architecture of salt tolerance [9]. As the fundamental aim of genetics is to connect genotype to phenotype, the identification and characterization of genes associated with agronomical important traits is essential for both understanding the genetic basis of phenotypic variation and efficient crop improvement. Modern molecular biology techniques and new statistical methods have opened new horizons for the cotton breeders; thus, linkage mapping and association mapping are the two important methods employed for QTL analysis. Molecular marker-quantitative trait association is one of the powerful approaches for exploring the molecular basis of phenotypic variations in plant [12], and could be used to increase the efficiency of a breeding program, especially for salinity tolerance [13,14]. The present studies of genetic map construction are mainly reflected in three different DNA based molecular markers such as simple sequence repeats (SSR) [15], single-nucleotide polymorphism (SNP) markers [16,17] and Intron length polymorphisms (ILD) markers [18].

Single Nucleotide Polymorphism is often abbreviated to SNP; it describes a variation in a single nucleotidae that occurs at a specific position in the genome [19]. Genome-wide association studies based on linkage disequilibrium (LD) is an effective strategy tool to study phenotype-genotype association. Compared with traditional QTL mapping, GWAS can use SNPs obtained by genome re-sequencing as molecular markers to dissect complex traits [20]. One SNP occurs every 100–300 bp in any genome; therefore SNPs markers have higher polymorphism than SSRs and other molecular markers [19]. GWAS has been successfully applied in rice, Arabidopsis, maize, wheat, barley and other crops to identify characteristic-related SNP markers of their important trait [21–24]. In cotton, various genetic maps based on SSR and SNP markers have been constructed using bi-parental mapping populations and natural population of *Gossypium hirsutum*; however, there are fewer studies, and no causal genes responsible for the salt tolerance traits from *Gossypium arboretum* have been identified [25–28].

Asiatic cotton (*Gossypium arboretum*) was introduced into China from ancient India, Burma or Vietnam over 2000 years ago [29]. Du et al (2018) reported that the natural population of Gossypium. arboretum was classified into three main groups represented South China, Yangtze River region, and Yellow River region groups respectively that exhibited strong geographical distribution [30]. A draft genome of cotton diploid *Gossypium arboretum* (the size is 1.7 Gb,$2n = 2\times = 26$) was recently reported by Li et al. (2015) [31]. The genetic basis of Asiatic cotton will provide a fundamental resource for genetics research of the important agronomic traits for cotton breeding. Therefore, the present study was performed in consideration of the following objectives: (i) to screen salinity tolerance at germination stage; (ii) to analyze the marker-trait associations by using SNP markers; (iii) to identify the causal genes which are responsible for the salt tolerance traits from *G. arboreum*.

2. Results

2.1. Phenotypic Diversity of G. arboretum Population

Seven salt tolerance related traits, including GR, FW, SL, WC, ChlC, EC, and MDA, were measured for all 215 *G. arboretum* accessions under 0 mM (C) and 150 mM (S) NaCl treatment (Figure 1 and Table S1). ANOVA analysis of seven salt-tolerance-related traits as measured for genetic diversity shows significant difference among the accessions ($P < 0.0001$) (Table 1). Correlation of GR with FW and SL was highly significant ($P < 0.001$), while GR with ChlC was also significant ($P < 0.01$). Correlation results of FW with SL, ChlC and MDA were also highly significant with $P < 0.001$. Correlation between SL and ChlC was highly significant ($P < 0.001$). A positive correlation was found between ChlC and EC ($P < 0.01$), while a negative correlation was found between ChlC and MDA ($P < 0.05$). The correlation between related EC and MDA was significant at $P < 0.05$. Interestingly, correlation analysis revealed that WC had significant correlation ($P < 0.05$) with MDA (Table 2). According to the CIST, 215 accessions were mainly categorized into four clusters. Cluster 1 contained 12 accessions that were highly sensitive to salt treatment (<0.6), Cluster 2 contained 26 accessions that were moderate tolerant to salt treatment (0.6–1.5), Cluster 3 included 153 accessions that were tolerant (1.5–2.5), and Cluster 4 had 24 accessions that were highly tolerant to salt treatment (>2.5) (Figure 1, Figure S2 and Table S1). Based on this result, the high tolerant accessions (24) were selected (Table S1). Basing on the comprehensive index of salt tolerance, we finally selected 11 high tolerant accessions (top 5%) for breeding using, including GuangXiZuoXianZhongMian, LiaoYang-1, ZhaoXianHongJieMian, PingLeXiaoHua, KaiYuanTuMian, YuXi33, ChangShuXiaoBaiZi, PingGuoJiuPingZhongMian, FuChuanJiangTangZhongMian, TangShanBaiZiZhongMian, and ShiJiaZhuangJianMian (Table S1).

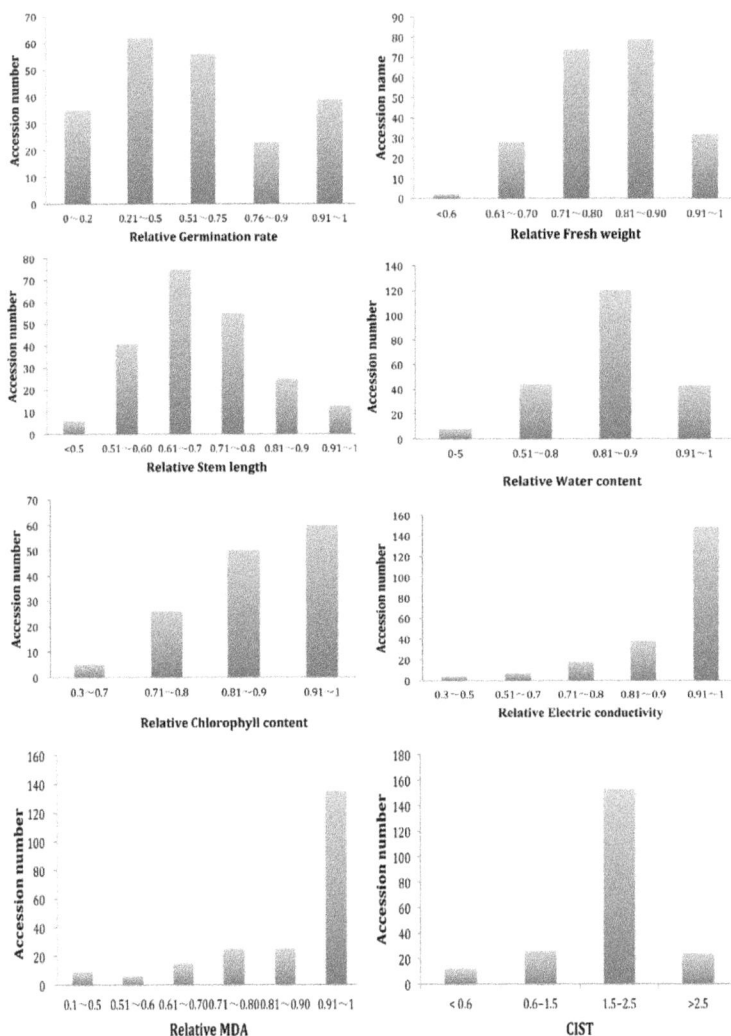

Figure 1. Relative value frequency distribution diagram of seven salt tolerance traits and one salt tolerance index of 215 *G. arboreum* accessions.

Table 1. Analysis of the traits related salt treatment in *G. arboretum* accessions.

Traits [1]	Mean	SD	Min	Max	CV	Mean Square	F	P > F
GR	19.936	10.77	0	40	54.02	338.55	8.16	<0.0001
FW	0.371	0.089	0.02	0.77	24.2	0.128	34.16	<0.0001
SL	4.969	1.481	0.40	10.2	29.84	40.77	49.76	<0.0001
WC	1.112	3.481	0.36	96.97	312.9	31.83	4.01	<0.0001
ChlC	39.47	9.96	3.50	50.2	25.53	6763.5	8.31	<0.0001
EC	31.47	42.54	0.116	844.4	135.19	541.2	6.49	<0.0001
MDA	0.007	0.0046	0	0.04	59.8	0.000095	14.36	<0.0001

[1] *GR* germination rate; *FW* fresh weight; *SL* stem length; *WC* water content; *ChlC* relative chlorophyll content; *EC* electric conduct; *MDA* methylene dioxyamphetamine.

Table 2. Analysis of traits related salt treatment of *G. arboretum* accessions (Pearson correlation coefficient).

Trait [1]	GR	FW	SL	WC	ChlC	EC	MDA
GR	1	0.22 ***	0.295 ***	−0.0178	0.112 **	0.0841	0.012
FW		1	0.575 ***	−0.043	0.070 ***	0.0613	0.135 ***
SL			1	−0.017	0.115 ***	−0.051	0.030
WC				1	0.048	0.002	0.059 *
ChlC					1	0.1 **	−0.073 *
EC						1	0.079 *
MDA							1

[1] For trait abb. Look at Table 1. * Significant at *P* < 0.05; ** Significant at *P* < 0.01; *** Significant at *P* < 0.001 for the correlation coefficient.

2.2. Association Mapping of Salt Tolerance Related Traits Using SNP Markers

The total of 1,568,133 high-quality SNPs (MAF > 0.05, missing rate < 40%) in 215 *G. arboreum* accessions were used for GWAS of the salt traits. The SNP markers associated with the seven salt-tolerance-related traits and one salt tolerance index were identified based on the threshold value, $\log_{10}(P) \geq 4.0$, using the MLMM model in the EMMAX software (Table S2 and Figure S3). The threshold of $-\log_{10}P \geq 4.0$ was also derived from the quantile–quantile (QQ) plots, For RGR, $-\log_{10}(P)$ values and QQ plots suggested relatively weak genetic association (Figure 2a). Most of the upward deviation from the linear line occurred at around $-\log_{10}(P) = 4.0$, which presumably indicates true positives (Figure 2b–h). For RGR, $-\log_{10}(P)$ values and QQ plots suggested relatively weak genetic association (Figure 2a). By applying the threshold of $-\log_{10}(P) \geq 4.0$, the 2062 SNP markers covered all 13 chromosomes and 100 SNP markers that were unknown location (Table 3). Chromosome 3 had the maximum number of SNPs (332 SNPs), and Chromosome 12 had the minimum (57) number of SNPs. Among the nucleotide polymorphisms, 1708 SNPs were interginic, 96 SNPs were intronic, 68 SNPs were exonic, 112 SNPs were upstream, 69 SNPs were downstream and 9 SNPs were upstream and downstream (Table S3).

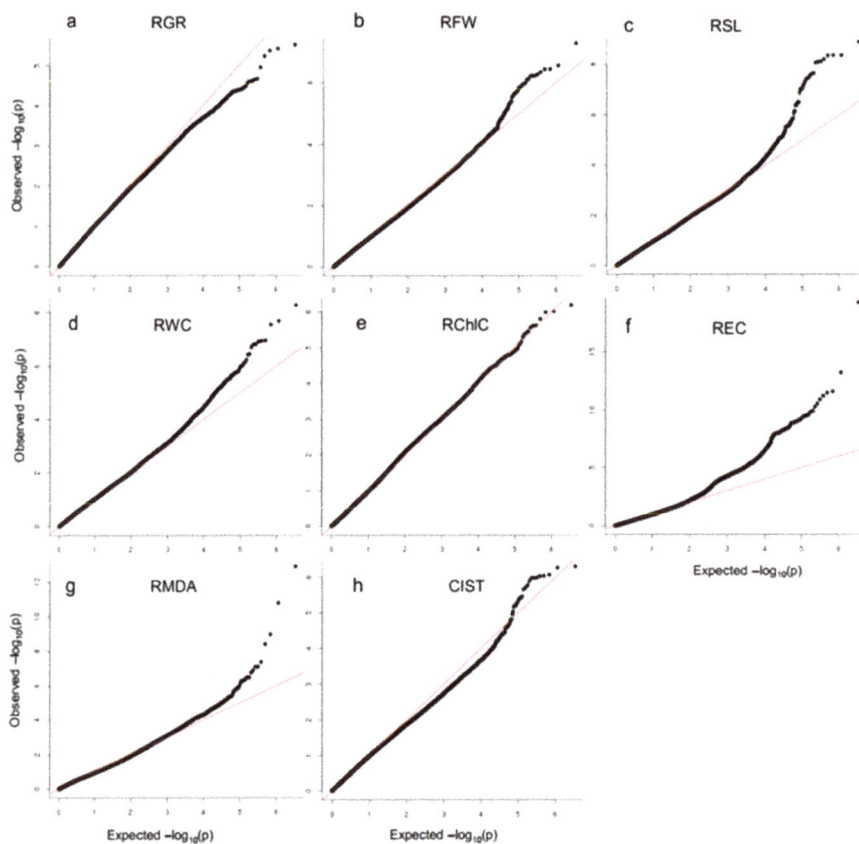

Figure 2. Quantile-quantile plots of versus expected $-\log_{10}P$ values of GWAS result. The red dashed line in each plot represents an idealized case where theoretical test statistics quantile match simulated test statistic quantile. (**a**) Relative germination rate (RGR). (**b**) Relative fresh weight (RFW). (**c**) Relative stem length (RSL). (**d**) Relative water content (RWC). (**e**) Relative chlorophyll content (RChlC). (**f**) Relative electric conductivity (REC). (**g**) Relative MDA (RMDA). (**h**) Comprehensive index of salt tolerance (CIST).

Among these 2062 marker-trait associations, 61 markers were associated with RGR, 187 markers were associated with RFW, 255 markers were associated with RSL, 370 markers were associated with RWC, 190 markers were associated with RChlC, 583 markers were associated with REC, 335 markers were associated with RMDA and 81 markers were associated with CIST (Table 3, Table S2) The most SNPs related positive salt tolerance indicators (RFW, RSL and RWC) were on chromosome 7. The most SNPs related negative salt tolerance indicators (REC, RMDA) were on chromosome 3 (Table 3).

Table 3. Associated SNPs of different salinity traits distribution on chromosome [1].

Chromosome	Total	RGR	RFW	RSL	RWC	RChlC	REC	RMDA	CIST
Chr-1	163	1	26	22	12	15	14	70	3
Chr-2	152	5	10	10	39	3	12	49	24
Chr-3	332	1	3	8	7	11	278	24	0
Chr-4	201	3	12	26	27	51	59	19	4
Chr-5	155	2	7	17	21	4	94	8	2
Chr-6	112	10	16	11	22	2	21	23	7
Chr-7	295	2	67	108	67	6	24	14	7
Chr-8	99	15	10	8	27	4	9	24	2
Chr-9	104	4	2	8	9	47	7	24	3
Chr-10	80	1	7	10	34	4	11	7	6
Chr-11	112	5	5	8	17	20	24	18	15
Chr-12	57	8	4	2	31	3	5	3	1
Chr-13	100	1	4	10	29	16	6	32	2
Chr-UN	100	3	14	7	28	4	19	20	5
Total	2062	61	187	255	370	190	583	335	81

[1] the SNP exceeding a significant threshold($-\log_{10}(P) \geq 4.0$); 2 For trait abb. Look at Table 1; CIST comprehensive index of salt tolerance.

2.3. SNP Rich Regions Associated with RFW and RSL

In the following, we focused on nine SNP rich regions associated RFW, RSL, RChlC, RWC and CIST for which MLMM analysis yield more significant associations considering $-\log_{10}P \geq 4.0$ values and the position of strong peaks in the Manhattan plots (Table S4).

Two SNP rich regions on chromosome 7 were found to associate with two biomass-related traits such as RFW and RSL (Figure 3a,b) when setting the threshold of $-\log_{10}P \geq 6.0$ on the Manhattan plots. The first candidate region (Group 1) starting at 33,513,007 bp and ending at 33,616,148 bp (103,141 bp), on chromosome 7, which contained 6 polymorphism SNPs (Figure 3a,b) and located within five genes, Cotton_A_37774, Cotton_A_37775, Cotton_A_37776, Cotton_A_37777 and Cotton_A_40811. All six SNPs were related to RSL, and three SNPs were related to related fresh weight (RFW) (Tables S4 and S6). Two intronic SNPs (Ca7_33606785 and Ca7_33607751) related with both of RSL and RFW), including key SNP Ca7_33607751 ($-\log_{10}(P)$ = 7.14, possession 33693051 bp) in this region, were in gene of Cotton_A_37775 (Table S5 Group 1). Four intergenic SNPs were related to related fresh weight (RFW) (Table S5 Group 1). Cotton_A_37775 was annotated as heat shock protein, a homolog of Arabidopsis thaliana AT5G52640. Haplotype analysis showed a low level of linkage disequilibrium (LD) (lowest $r2$ = 30, highest $r2$ = 100) between the associated SNPs in Group 1 (Figure 3c,d). There were three genotypes for the key SNP Ca7_33607751. Genotypes CC, TT and CT contained 7, 135, and 6 accessions respectively, whereas the accessions carrying CC genotype showed the highest RFW and RSL; the accessions carrying CT genotype showed medium-higher RWF (0.909) and RSL (0.775); And the accessions carrying TT genotype showed the lowest RFW (0.81) and RSL (0.698) (Figure 3e,f).

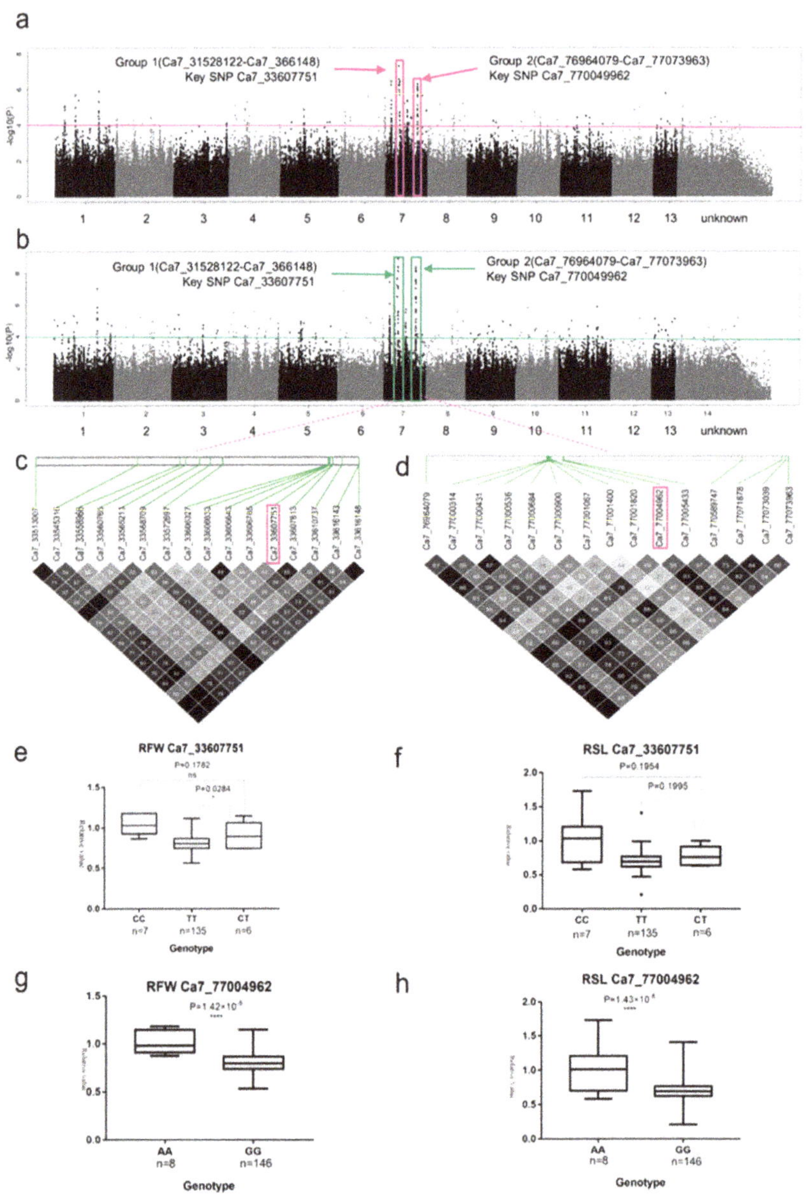

Figure 3. GWAS results for RFW and RSL, and analysis of the peaks on chromosome 7 (Group 1 and Group 2). (**a**) Manhattan plot for RFW. The horizontal line represents the significant threshold ($-\log_{10}P$ = 4). The pink color surrounds represent SNP rich regions (Group 1 and Group 2). (**b**) Manhattan plot for RSL. The horizontal line represents the significant threshold ($-\log_{10}P$ = 4). The pink color surrounds represent SNP rich regions (Group 1 and Group 2). (**c**) LD surrounding the peak of Group 1. (**d**) LD surrounding the peak of Group 2. (**e**) Phenotypic differences for RFW based on the key SNP (Ca7_33607751) of Group 1. (**f**) Phenotypic differences for RSL based on the key SNP (Ca7_33607751) of Group 1. (**g**) Phenotypic differences for RFW based on the key SNP (Ca7_77004962) of Group 2. (**h**) Phenotypic differences for RSL based on the key SNP (Ca7_77004962) of Group 2.

We then focused on the second highest peak (Group 2) on chromosome 7, which were common related to two biomass-related traits such as RFW and RSL. Group 2 was estimated to be 76,964,079 –77,073,963 bp (109,884 bp), and to contain 9 polymorphisms, which were located within 6 genes (Figure 3a,b and Tables S4 and S6). Among the 9 polymorphisms, 2 SNPs (Ca7_77000431 and Ca7_77004962) were related with both traits such as RFW and RSL, 7 SNPs were related only RSL. Most of these SNPs (6 of 9), including key SNP Ca7_77004962 ($-\log_{10}(P)$ = 8.36) were located between two genes *Cotton_A_35901* and *Cotton_A_35900* (Table S6). Haplotype analysis showed a low level of LD (lowest r^2 = 15, highest r^2 = 100) between the associated SNPs in Group 2 (Figure 3d). There were two genotypes for the key SNP Ca7_77004962. Genotypes AA and GG contained 8 and 146 accessions, respectively, whereas the accessions carrying AA genotype showed the highest RFW (1) and RSL (1); the accessions carrying GG genotype showed lower RWF (0.808) and RSL (0.697) (Figure 3g,h).

2.4. SNP Rich Regions Associated with RChlC

Similarly, we analyzed four SNP rich regions on chromosome 4, 9 and 11, which closely associated with RChlC (Figure 4a). The candidate region on chromosome 4 (Group 3) was predicted to map from 32,124,549 bp to 32,259,755 bp (135,206 bp), and contained 14 polymorphisms, which were located within 6 genes (Figure 4a, Table S4 and Table S5 Group 3). Eight SNPs were distributed near the gene of *Cotton_A_26219* (Table S5). We found that two haplotypes in three SNPs (Ca4_32185154, Ca4_32191704 and Ca4_32197265) by using high pairwise LD correlation ($r^2 \geq 96$) (Figure 4b,f). These three SNPs were near *Cotton_A_26219*. Haplotype A and Haplotype B contained 27 and 157 accessions respectively (Figure 4f). The accessions carrying haplotype A showed lower RChlC (0.80) than haplotype B (0.97) (Figure 4i).

We estimated the candidate region on chromosome 9 (Group 4) to be 56,869,491–56,879,931 bp (10,440 bp) and assigned 14 polymorphisms, which were located within 3 genes (Figure 4a, Table S4 and Table S5 Group 4). Most of these polymorphisms (8 of 14), including the key SNP Ca9_56878752 (exonic) were in *Cotton_A_10864*. Five coding region SNPs in *Cotton_A_10865*, which is annotated as F-Box protein and leucine-rich repeat protein 14 (Table S5 Group 4). We found that two haplotypes in seven SNPs (Ca9_56875436, Ca9_56875607, Ca9_56876686, Ca9_56876694, Ca9_56877663 in Cotton_A_10865, and Ca9_56878677, Ca9_56878752 in *Cotton_A_10864*) by using high pairwise LD correlation ($r^2 > 80$) (Figure 4c,g). Haplotype A and Haplotype B contained 9 and 84 accessions, respectively (Figure 4g). The accessions carrying haplotype A showed higher RChlC (1) than haplotype B (0.93) (Figure 4j).

The candidate region on chromosome 9 (Group 5) was predicted to map from 92,702,808 bp to 92,735,184 bp (32,376 bp), and contained 19 polymorphisms (Figure 4a, Table S4 and Table S5 Group 5). All 19 significant SNP markers (key SNP Ca9_92711930, $-\log_{10}P$ = 5.80) was located between two pathogenesis-related thaumatin superfamily protein genes: *Cotton_A_15275* (average distance 13,817 bp) and *Cotton_A_15276* (average distance 31,662 bp) (Table S5 Group 5). We found that two haplotypes in thirteen SNPs (from Ca9_92710379 to Ca9_92717732) by using high pairwise LD correlation ($r^2 \geq 91$) (Figure 4d,h). Haplotype A and Haplotype B contained 12 and 97 accessions, respectively (Figure 4h). The accessions carrying haplotype A showed higher RChlC (1) than haplotype B (0.92) (Figure 4k).

The candidate region on chromosome 11 (Group 6) was predicted to map from 47,006,642 bp to 47,011,718 bp (5076 bp), and contained 11 polymorphisms (Figure 5a, Table S4 and Table S5 Group 6). All 11 significant SNP markers (key SNP Ca11_47011718, $-\log_{10}P \geq 5.61$) was located between two genes: *Cotton_A_28249* (average distance from SNPs 7649 bp) and *Cotton_A_28248* (average distance 5083 bp from SNPs). There were three genotypes for the key Ca11_47011718 (Table S5 Group 6). Genotypes AA, GG and AG contained 190, 13, and 2 accessions respectively, whereas the accessions carrying AA (0.933) genotype showed lower RChlC; And the accessions carrying GG (1) and AG (1) genotype showed the highest RChlC (Figure 5i).

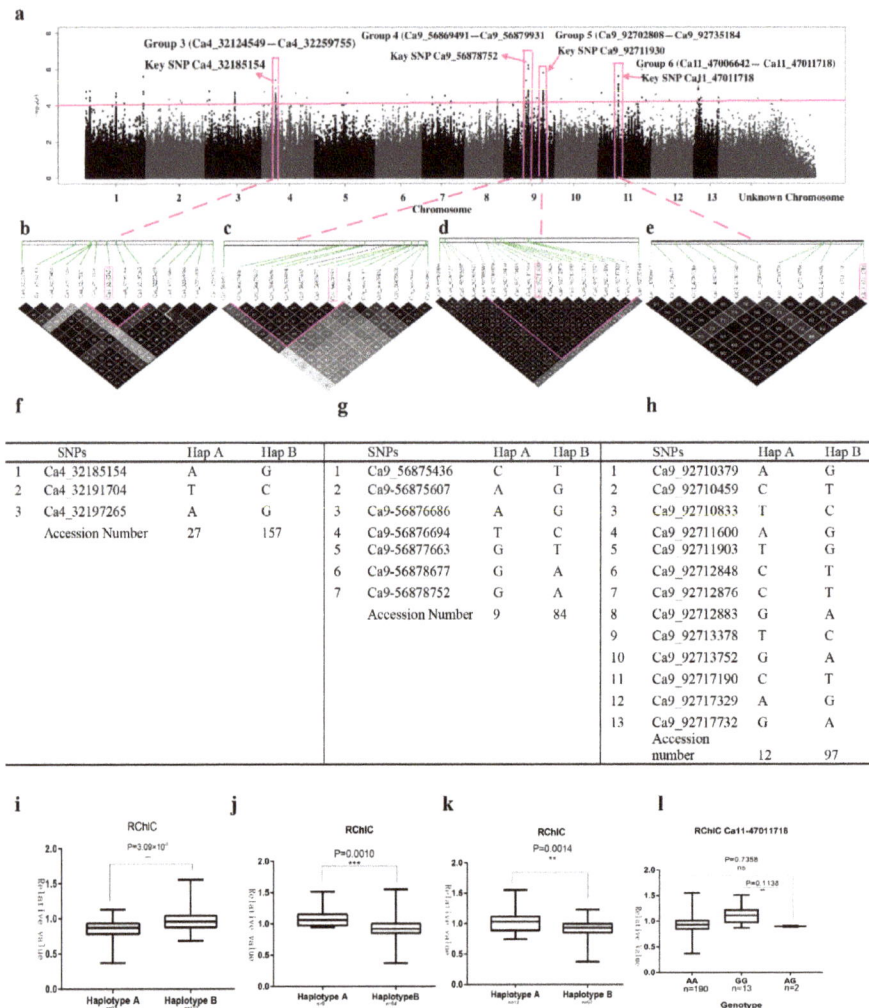

Figure 4. GWAS results for RChlC and analysis of the peaks on chromosome 4, 9 and 11. (**a**) Manhattan plot for RChlC. The horizontal line represents the significant threshold ($-\log_{10}P = 4$). The pink color surrounds represent SNP rich regions (Group 3, Group 4, Group 5 and Goup 6). (**b**) LD surrounding the peak of Group 3. (**c**) LD surrounding the peak of Group 4. (**d**) LD surrounding the peak of Group 5. (**e**) LD surrounding the peak of Group 6. (**f**) Haplotypes in Group 3. (**g**) Haplotypes in Group 4 (**h**) Haplotypes in Group 5. (**i**) Phenotypic differences of RChlC between two haplotypes in Group 3. (**j**) Phenotypic differences of RChlC between two haplotypesin Group 4. (**k**) Phenotypic differences of RChlC between two haplotypesin Group 5. (**l**) Phenotypic differences for RChlC based on the key SNP (Ca11_47011718) of Group 6.

2.5. SNP Rich Region Associated with RWC

The candidate region on chromosome 7 (Group 7) was predicted to map from 39,823,916 bp to 39,843,478 bp (19,562 bp), and contained 29 polymorphisms in four genes (*Cotton_A_05854*, *Cotton_A_05853*, *Cotton_A_05852* and *Cotton_A_05852*) (Figure 5a, Table S4 and Table S5 Group 7). Among these four genes, most important gene is *Cotton_A_05853*, because, 15 of these SNPs (11 intronic and 3 intergenic), including intronic key SNP Ca7_39832729 ($-\log_{10}(P) > 5.35$) were located in *Cotton_A_05853* (*AT3G12800*), annotated in The Arabidopsis Information Resource (TAIR) as short-chain dehydrogenase-reductase B and oxidation-reduction process (Table S5 Group 7). We found that two haplotypes of eight SNPs in *Cotton_A_05853* (from Ca7_39832407 to Ca7_39832920) by using high pairwise LD correlation ($r^2 \geq 90$) (Figure 5b). Haplotype A and Haplotype B contained 42 and 6 accessions, respectively (Figure 5c). The accessions carrying haplotype A showed higher RWC (0.86) than haplotype B (0.50) (Figure 5d).

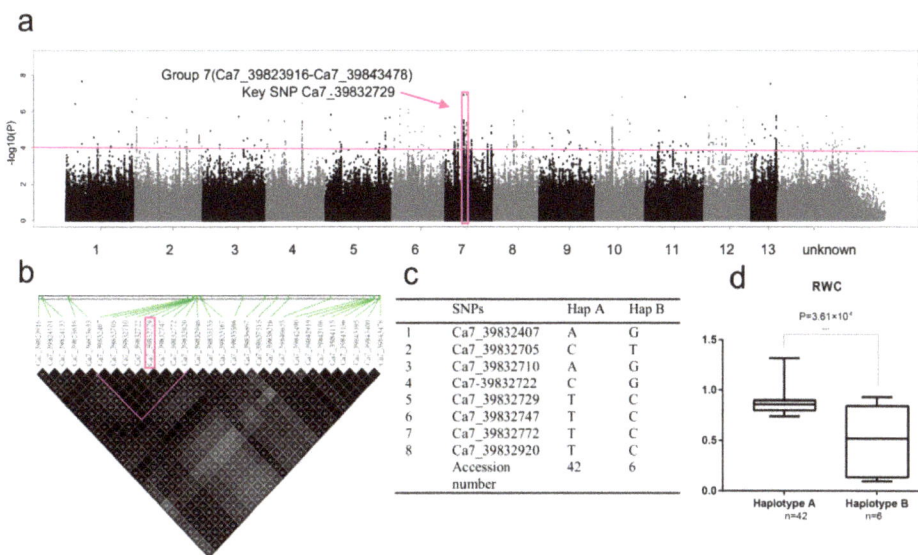

Figure 5. GWAS results for RWC and analysis of the peak on chromosome 7. (**a**) Manhattan plot for RWC. The horizontal line represents the significant threshold ($-\log_{10}P = 4$). The pink color surround represents SNP rich region (Group 7). (**b**) LD surrounding the peak of Group 7. (**c**) Haplotypes in Group 7. (**d**) Phenotypic differences of RWC between two haplotypes.

2.6. SNP Rich Regions Associated with CIST

The candidate region on chromosome 2 (Group 8) was predicted to map from 86,947,129 to 87,017,559 bp (11,697 bp) and to contain 16 polymorphisms in two genes (*Cotton_A_22673*, *Cotton_A_22672*) (Figure 6a, Table S4 and Table S5 Group 8). Fifteen significant intergenic SNPs, including key SNP Ca2_86954790 ($-\log_{10}(P) \geq 6.32$) were located near the gene of *Cotton_A_22673* (average distance 63,317 bp). Only one significant non-synonymous SNP is in *Cotton_A_22672*, which without annotation (Table S5 Group 8). We found that two haplotypes of six SNPs near to *Cotton_A_22673* (from Ca2_86947129 to Ca2_86956247) by using high pairwise LD correlation ($r^2 \geq 93$) (Figure 6b). Haplotype A and Haplotype B contained 27 and 157 accessions, respectively (Figure 6d). The accessions carrying haplotype A showed lower CIST (1.042) than haplotype B (1.935) (Figure 6f).

The candidate region on chromosome 11 (Group 9) was predicted to map from 39,493,066 to 39,504,763 bp (11,697 bp) and to contain 9 polymorphisms in two genes (*Cotton_A_21725* and *Cotton_A_21726*) (Figure 6a, Table S4 and Table S5 Group 9). *Cotton_A_21726* contained eight significant intronic SNPs, including key SNP Ca11_39504708 ($-\log_{10}P = 5.98$), annotated Glycosyl hydrolase family 10 proteins. Only one significant downstream SNP was in *Cotton_A_21725*, annotated as DNA/RNA polymerases superfamily protein (Table S5 Group 9). We found that three haplotypes in all nine SNPs in this region by using very high pairwise LD correlation ($r^2 = 100$) (Figure 6c,e). Haplotype A, Haplotype B and Haplotype C contained 35, 5 and 36 accessions, respectively (Figure 6e).

The accessions carrying haplotype A (1.973) and C (2.056) showed similar CIST, but the accessions carrying haplotype B showed the lowest CIST (−0.1398) (Figure 6g).

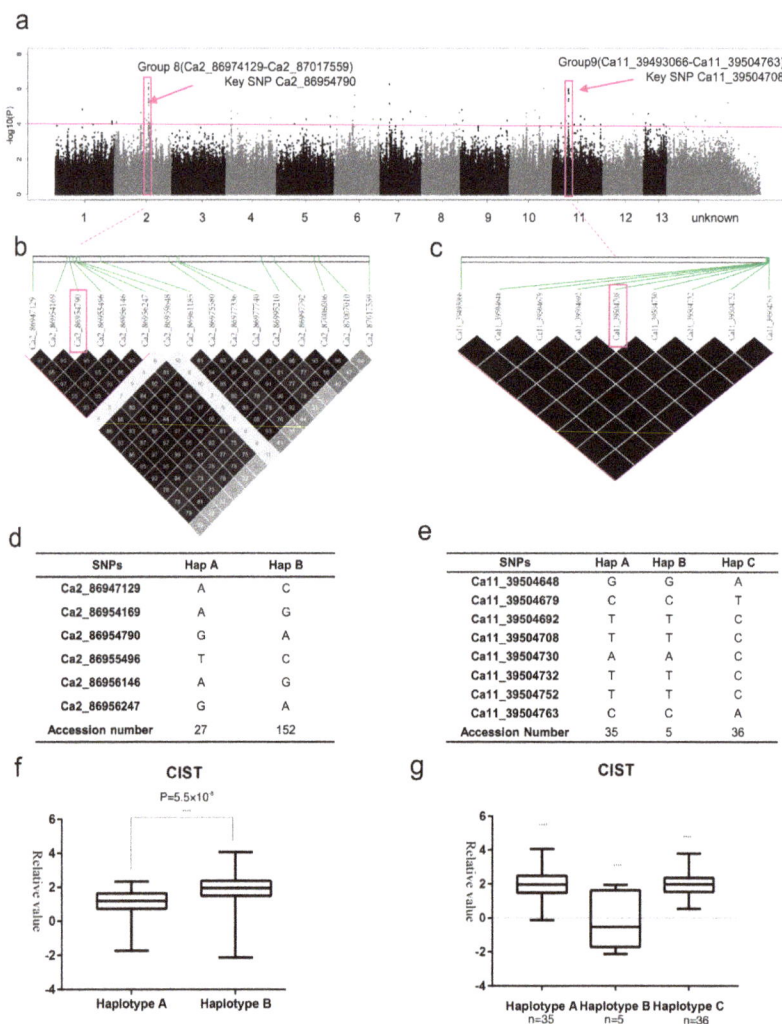

Figure 6. GWAS results for CIST and analysis of the peaks on chromosome 2 and 11. (**a**) Manhattan plot for CIST. The horizontal line represents the significant threshold ($-\log_{10}P = 4$). The pink color surrounds represent SNP rich regions (Group 8 and Group 9). (**b**) LD surrounding the peak of Group 8. (**c**) LD surrounding the peak of Group 9. (**d**) Haplotypes in Group 8. (**e**) Haplotypes in Group 9. (**f**) Phenotypic differences of CIST between two haplotypes in Group 8. (**g**) Phenotypic differences of CIST among three haplotypes in Group 9.

3. Discussion

3.1. Genetic Variation in Salt Tolerance Related Traits of G. arboretum Accessions

G. arboretum is considered superior to upland cotton varieties based on the following traits: precociousness, wide adaptability, drought tolerance and disease resistance from fusarium wilt, insects and bell mild disease [32]. However, it has not been well characterized at the molecular level. Thus, our current study mainly focused on identification and screening of salt tolerant germplasm during seedling stage, to find genetic relationships and to study marker-trait associations SNP markers. The *G. arboretum* accessions, considered as an invaluable gene pool for cotton improvement, were used in this study. Understanding of the genetic diversity of *G. arboretum* can facilitate the efficient use of these resources in the development of superior cotton cultivars with favorable agronomic traits.

Abiotic stress leads to a series of morphological, physiological, biochemical and molecular changes that have adverse effects on the plant growth, development and productivity. In fact, the salinity is a major abiotic stress that limits cotton growth and development at the germination and seedling stages [27]. In this study, the suitable optimum NaCl concentration was determined by monitoring germination rates of accessions. Shixiya No. 1, Sichuansuining and Chaoxianjinhuaxiaozi (included in the 215 accessions) under different NaCl concentrations (0, 50, 100, 150, 200, 250, and 300 mM). Under 200, 250, and 300 mM NaCl treatments, most seeds were unable to geminate. In addition, most of the seeds germinated under 100 and 150 mM NaCl treatments (Figure S1). Based on these results and the reports of Chen et al. (2008), 150 mM NaCl concentration was considered suitable salt treatment for *G. arboretum* [29]. Under 150 mM NaCl treatment, significant differences among different cotton accessions were observed during germination and seeding stages. The assessment of diversity in a species is important in plant breeding programs, and for effective conservation, management and utilization of genetic resources of the species [30]. The salt tolerance related traits (GR, FW, SL, WC, ChlC, EC, MDA) were comparable to that of the salt tolerance study and reported previous works [33]. Notably, plant germination rate, stem length, fresh weight are the major components of plant yield and were used as selection criteria in breeding. The ANOVA for the important salt tolerance parameters revealed significant differences ($P < 0.0001$) among the genotypes, implying that sufficient phenotypic polymorphism existed between individual *G. arboretum* accessions in this study. The correlation analysis is important to identify the mutual associations among the traits [34]. There was efficiency correlation between seven salt-tolerance-related traits; it is useful for multiple trait selection at one time for development of improved cotton varieties. The classification information derived from these studies may be used to facilitate the development of salt tolerant cotton accessions that could give economic yield in salinity prone areas.

3.2. Association Mapping of Salt Tolerance Traits Using SNP Markers

Genome-wide association study (GWAS) can effectively associate genotypes with phenotypes in natural populations and simultaneously detect many natural allelic variations and candidate genes in a single study, in contrast to QTL linkage mapping [35]. To cope with environmental stress, plants activate a large set of genes leading to the accumulation of specific stress-associated proteins. In this study, we first performed a genome-wide association analysis of salt tolerance related traits with 215 of natural accessions of *G. arboretum*. This study uncovered 2062 loci ($-\log_{10}P \geq 4.0$) associated salt tolerance traits and identified a set of candidate genes that could be exploited to alter salt tolerance development to improve *G. arboretum* accessions. The analysis of genomic distribution of SNPs in this study revealed that the most of SNPs related positive salt tolerance indicators (RFW, RSL and RWC) were on chromosome 7 and the most of SNPs related negative salt tolerance indicators (REC, RMDA) were on chromosome 3.

The nine SNP rich regions analysis revealed 143 polymorphisms that distributed 40 candidate genes and significantly associated RFW, RSL, RChlC, RWC and CIST. We found that twelve and eight SNPs in Group 1 and Group 2 were associated with two biomass traits such as RFW and RSL, because there is highly significant correlation ($P < 0.001$) between FW and SL (Table S5 Group 1 and Group 2).

In the first SNP rich region (Group 1), the most plausible candidate indented in the peak on chromosome 7 was *Cotton_A_37775* that was involved in Heat shock protein (Hsps), which plays a crucial role in salt stress response. *Cotton_A_37775* contains 1,951 amino acid and shares 79% identify at the amino acid level with a homolog of *Arabidopsis thaliana AT5G52640*, which annotated Heat shock protein (Hsp90.1). The 90 kDa heat shock protein (Hsp90) is a widespread family of molecular chaperones found in prokaryotes and all eukaryotes [36]. Hsp90 chaperone machinery is a key regulator of proteostasis under both physiological and stress conditions in eukaryotic cells. A large number of co-chaperones interact with HSP90 and regulate the ATPase-associated conformational changes of the HSP90 dimer that occur during the processing of clients [37]. The basic functions of HSP90s are of assisting protein folding, protein degradation and protein trafficking, and they also play an important role in signal transduction networks, cell cycle control and morphological evolution. Although HSP90s are constitutively expressed in most organisms, their expression is up-regulated in response to stresses such as cold, heat, salt stress, heavy metals, phytohormones, light and dark transitions [38]. The most prominent response of plants under high temperature stress is the rapid production of heat shock proteins (HSPs). Ding (2006) found that heat shock treatment in early growing period benefited the abundance of HSPs in cotton leaves in high temperature season, and then increased the ability thermo-tolerance [39]. In the plant *Arabidopsis thaliana* (A. thaliana), HSP90 homologs are encoded by seven different genetic loci. Of these, one is expressed in the endoplasmic reticulum (*HSP90.7*), one in the mitochondrion (*HSP90.6*), one in the chloroplast (*HSP90.5*), and four in the cytosol. The gene encoding one cytosolic protein (*HSP90.1/At5g52640*) is highly stress-inducible, whereas the other three (*HSP90.2/At5g56030, HSP90.3/At5g56010*, and *HSP90.4/At5g56000*) are constitutively expressed and are the products of very recent duplication events [40]. The homolog *At5g52640* is found up-regulated in response to viruses stresses by playing a role in cell migration [41].

In the second SNP rich region (Group 2), we identified a gene, *Cotton_A_35901* (301 amino acid), encoding a SNARE-like superfamily protein homologue, which has not been previously reported in cotton. Complexes of SNARE proteins mediate intracellular membrane fusion between vesicles and organelles to facilitate transport cargo proteins in plant cells [36]

In the third SNP rich region (Group 3), we found that two haplotypes in three SNPs (Ca4_32185154, Ca4_32191704 and Ca4_32197265) by using high pairwise LD correlation ($r^2 \geq 96$). These three intergenic haplotype SNPs located between two genes *Cotton_A_26219* and *Cotton_A_26218*. *Cotton_A_26219* have no annotation. Anther gene *Cotton_A_26218* is a homologue of *A.thaliana AT1G18800*, which encoding nucleosome assembly protein (nrp1-1 nrp2-1). Juan (2012) show that the nucleosome assembly protein (NAP1) family histone chaperones are required for somatic homologous recombination (HR) in *A. thaliana* [42]. HR is essential for maintaining genome integrity and variability. To orchestrate HR in the context of chromatin is a challenge, both in terms of DNA accessibility and restoration of chromatin organization after DNA repair. Histone chaperones function in nucleosome assembly/disassembly and could play a role in HR.

Depletion of either the NAP1 group or NAP1-RELATED PROTEIN (NRP) group proteins caused a reduction in HR in plants under normal growth conditions as well as under a wide range of genotoxic or abiotic stresses. In *Arabidopsis thaliana*, *AT1G18800* is required for maintaining cell proliferation and cellular organization in root tips [43].

The analysis in the fourth SNP rich region (Group 4) revealed two haplotypes in seven SNPs distributed in two candidate genes, Cotton_A_10865 and Cotton_A_10864 on Chr9, were associated with RChlC. Cotton_A_10865 is homologous to *AT4G08980* (F-BOX WITH WD-40), encodes an F-box gene that is a novel negative regulator of AGO1 protein levels and may play a role in abscisic aci (ABA) signaling and/or response. ABA signaling also plays a major role in mediating physiological responses to environmental stresses such as salt, osmotic, and cold stress. The accumulation of ABA in response to water or salt stress is a cell signaling process, encompassing initial stress signal perception, cellular signal transduction and regulation of expression of genes encoding key enzymes in ABA biosynthesis and catabolism [44]. In addition, there are a lot of studies to prove ABA response to salt stress in different crops such as *A. thaliana* [44–47], rice [48], wheat [49], corn [50]. The homologue of *Cotton_A_10864* is *AT5G65270* (RAB GTPASE HOMOLOG A4A) in *A. thaliana*. Several genes in the Rab GTPase family have been shown to be responsive to abiotic stress, including response of *SsRab2* to water stress in *Sporobolus stapfianus* [51], *MfARL1* to salt stress in *A. thaliana* [47], *OsRab7* to cold stress in rice [52] and *AtRabG3e* to salt/osmotic stress in *A. thaliana* [53].

In the fifth SNP rich region (Group 5), the two genes (*Cotton_A_15275* and *Cotton_A_15276*) in Group 5 were pathogenesis-related thaumatin superfamily protein. Pathogenesis-related (PR) proteins play an important role in plants as a protein-based defensive system against abiotic and biotic stress, particularly pathogen infections. It is also named thaumatin-like proteins (TLPs), because it has sequence similarity with thaumatin [54,55].

In the six SNP rich region (Group 6), we found two genes *Cotton_A_28248* and *Cotton_A_28249*. *Cotton_A_28248* has no annotation. *Cotton_A_28249* is homologous to *AT4G32050* and encodes neurochondrin family protein, which is an atypical RIIα-specific A-kinase anchoring protein. In the seventh SNP rich region (Group 7), we found two haplotypes in seven SNPs distributed in one gene, *Cotton_A_05853*, has no annotation.

In the eighth SNP rich region (Group 8), we found two candidate genes *Cotton_A_22673* and *Cotton_A_22672*. *Cotton_A_22673* is annotated carbon-sulfur lyases, homologous to *A.thaliana AT5G09970*. *AT5G09970* locate in mitochondrion and encodes a DYW-class PPR protein required for RNA editing at four sites in mitochondria of *A. thaliana*.

In the ninth SNP rich region (Group 9), we found two candidate genes *Cotton_A_21725* and *Cotton_A_21726*. *Cotton_A_21725* encodes DNA/RNA polymerases superfamily protein. Several genes are induced under the influence of various abiotic stresses. Among these are DNA repair genes, which are induced in response to the DNA damage. Since the stresses affect the cellular gene expression machinery, it is possible that molecules involved in nucleic acid metabolism including helicases are likely to be affected. The light-driven shifts in redox-potential can also initiate the helicase gene expression. Helicases are ubiquitous enzymes that catalyze the unwinding of energetically stable duplex DNA (DNA helicases) or duplex RNA secondary structures (RNA helicases). Most helicases are members of DEAD-box protein superfamily and play essential roles in basic cellular processes such as DNA replication, repair, recombination, transcription, ribosome biogenesis and translation initiation. Therefore, helicases might be playing an important role in regulating plant growth and development under stress conditions by regulating some stress-induced pathways [56]. *Cotton_A_21726* encodes glycoside hydrolasefamily 10. GHs (glycosyl hydrolases) enzymes that catalyze the hydrolysis of glycosidic bonds between sugars and other moieties, can be classified into more than 100 families [55]. We could not find any relationship between glycoside hydrolase family 10-prtain and salt tolerance. But the glycoside hydrolase family 5 gene is expressed in rice leaves and seedling shoots, whereas its expression is induced by stress-related hormones, submergence and salt in whole seedlings [57].

The identified genetic variation and candidate genes deepen our understanding of the molecular mechanisms underlying salt tolerance traits. The genes discussed in this study may be considered as candidate genes for salt tolerance in cotton.

4. Materials and Methods

4.1. Plant Materials and Sample Preparation

The genetic materials used in the present study includes 215 accessions of *G. arboretum*, among them 209 accessions belong to China, 3 accessions from the United States, and 3 accessions from India, Pakistan and Japan. The germplasm was assembled from Germplasm bank of Institute of Cotton Research of Chinese Academy of Agricultural Sciences (CAAS), Anyang, China. The detailed list of accessions along with their origin is described in Table S1.

The phenotypic analysis and genetic experiment for these selected accessions were performed in the laboratory of Cotton Research Institute of CAAS, Anyang, China. Seedlings were grown in a phytotron incubating chamber under 14 h/10 h (light/dark) cycle, 28/14 °C (day/night), 450 $\mu mol \cdot m^{-2} \cdot s$ light intensity and a relative humidity of 60–80% conditions [34]. For each genotype, 200 hand-selected seeds for each variety were sterilized with 15% hydrogen peroxide for 4 h, and subsequently rinsed with sterile distilled water at least 4 times, followed by seed submersion in sterile water for 12 h at room temperature. For identification of the seed germination rates (GR), 120 healthy seeds from each accession were selected and placed in germination boxes (200 × 150 mm diameter), containing two sheets of filter paper and soaked with 20 mL of the NaCl solutions (0 or 150 mM) respectively. For identification of other traits, such as fresh weight (FW), stem length (SL), water content (WC), chlorophyll content (ChlC), electric conductivity (EC) and methylene dioxyamphetamine (MDA), twenty-five healthy seeds each were planted in the 300 mL volumetric flask (Figure S1), containing 140 g sand (sterilized in autoclave at 160 °C for 2 h), 40 mL sterilized water (for control 0 mM) and 40 mL NaCl solutions (for 150 mM).

4.2. Trait Evaluation

After seven days of the germination, the 215 *G. arboretum* accession seedlings grown in the soil were evaluated for seven salt-tolerance-related traits, such as GR, FW, SL, WC, ChlC, EC and MDA. The seeds were considered to have germinated when the radicle and plumule length was equivalent to the seed length or half of the seed length. The germination rate was calculated GR (%) = (number of germinated seeds/total seed number used in the test) × 100. Fifteen plants from each accession were harvested and cleaned by sterilized water and immediately used for determination of FW, SL and ChlC.

For leave water content (WC) estimation, detached leaves were floated on deionized water for 8 h at 4 °C and turgid weights (TW) were determined. Later, dry weight (DW) of leaves was measured after oven-dried at 105 °C for 10 min, and then 80 °C for 24 h respectively. WC was calculated WC (%) = (FW − DW)/(TW − DW) × 100 [33].

For EC measurement, 0.5 g of leaves were rinsed with double distilled water (ddH$_2$O) and put in volumetric flask containing 40 mL of ddH$_2$O. Afterwards, the test flasks were incubated at room temperature for 4 h. The electrical conductivity of the solution (C1) was measured using a conductivity meter EM38 (ICT, Australian). The test flasks were boiled for 10min and then cooled at room temperature to measure the electrical conductivity (C2). The REC was calculated using the formula C1/C2 × 100%. [33]

For MDA measurement, 0.5 g of leaves was rinsed with double distilled water (ddH$_2$O) and then crushed in Thiobarbituric acid extract solution (5 mL, 0.5%). Absorbances were monitored at three different wavelengths i.e., 450 nm, 532 nm and 600 nm. MDA was calculated according to Le et al. (2000) method. MDA (X) = [6.452*(OD532 − OD600 − 0.559*OD450] *Vt/(Fw × 1000) [33]. Where X is MDA in μmol/g, Vt is total volume of extraction solution in mL and FW is fresh weighting.

All phenotypic data and physiological indicators were performed at least three biological replicates. The relative value of GR phenotypic data was used for further GWAS. The formula is RGR = GR$_{150}$/GR$_{control}$, i.e., the same as that for other traits, RFW, RSL, RWC, RChlC, REC and RMDA.

4.3. DNA Extraction

Young leaves were collected from five plants of each accession and stored at −80 °C. The DNA was isolated from the frozen leave using CTAB method [34] with some modifications. DNA concentration was quantified using a NanoDrop2000 instrument (Thermo Scientific, USA), and normalized to 50 ng/mL.

4.4. Phenotypic Diversity

Analysis of variance (ANOVA) and phenotypic correlations between different salt related physiological traits were performed using statistical software package SAS 9.21. Relative value for each trait and CIST were calculated according to these formulae:

Relative value = value under stress treatment (S)/value under control treatment (C);

CIST = positive index (RGR + RSL + RFW + RChlC + RWC)/negative index (REC + RMDA).

Int. J. Mol. Sci. **2019**, *20*, 2168

4.5. Genome-Wide Association Analysis

The SNP markers associated with the seven traits and one salt tolerance index were identified using the MLMM model in the EMMAX software [58]. Key SNP analysis was identified by IGV_2.3.83_4 (Integrative genomic viewer, version 2.3.83_4). Lastly, ~1.57 million high-quality SNPs (MAF > 0.05, missing rate < 40%) were obtained and used for the further GWAS. The genome-wide significance thresholds of all tested traits were evaluated using $-\log 10(P) \geq 4.0$.

4.6. Statistical Analysis

For phenotypic experiment, each data represented the mean of three or more biological replicate treatments, with each treatment consisting of at least five plants. The statistical analyses were performed using Tukey's two-way analysis of variance (ANOVA) in IBM SPSS Statistics v19.0 (SPSS Inc., Chicago, IL, USA). *P*-values < 0.001 were considered statistically significant. Pearson correlation also was used SPSS V19.0. The different star between two traits represent significance at *P*-value < 0.05, 0.01, 0.001.

5. Conclusions

We firstly investigated the genetic architecture of natural variation in Asian cotton salt-tolerance-related traits at the seedlings stage by GWAS mapping in 215 accessions. The SNP markers and candidate genes in this study may be used as references for other association mapping studies of salt tolerance. The salt tolerance related novel candidate genes will provide an important resource for molecular breeding and functional analysis of the salt tolerance during the cotton germination.

Supplementary Materials: Supplementary Materials can be found at http://www.mdpi.com/1422-0067/20/9/2168/s1. Supplementary Table S1. The source, Pedigree information and salt tolerance classification according to the comprehensive index of salt tolerance (CIST) of 215 *G. arboretum* accessions after 7 days of seed growth. Supplementary Table S2. The SNPs showing–$\log_{10}P$ value ≥ 4.0 related with seven phenotypic traits and comprehensive index of salt tolerance (CIST) in 215 *G. arboreum* accessions. Supplementary Table S3. Basic information about genic and intergenic SNP markers. Supplementary Table S4. The basic information of nine SNP rich regions analyzed in this study. Supplementary Table S5. The SNPs in the nine SNP rich regions analyzed in this study. Supplementary Table S6. SNPs and associated genes of two SNP rich regions on chromosome considering $-\log_{10}(P) \geq 6.0$.

Author Contributions: X.D., T.D. and Z.P. (Zhen Peng) initiated the research; D.X. and T.D. designed the experiments. T.D. performed the experiments, T.D., W.G., and Z.P. (Zhaoe Pan) collected the data from the greenhouse, T.D., X.D. and Y.J. performed the analysis, T.D. drafted the manuscript, X.D., Z.P. (Zhen Peng) and K.K.P. contributed to the final editing of manuscript. All authors contributed in the interpretation of results and approved the final manuscript.

Funding: This research was funded by the National Science and Technology Support Program of China (2013BAD01B03).

Acknowledgments: We are grateful to the National mid-term GenBank for cotton in Institute of Cotton Research of Chinese Academy of Agricultural Sciences (ICR, CAAS) for providing the 215 *G. arboretum* accessions used in this study.

Conflicts of Interest: The authors declare no conflict of interest.

Abbreviations

RGR	Relative germination rate
RFW	Relative fresh weight
RSL	Relative stem length
RWC	Relative water content
RChlC	Relative chlorophyll content;
REC	Relative electric conduct;
RMDA	Relative methylene dioxyamphetamine
CIST	Comprehensive index of salt tolerance
MMLM	Multi-Locus Mixed Model
QTL	Quantitative trait loci;
SSR	Simple sequence repeats

SNP	Single-nucleotide polymorphism
TW	Turgid weights;
ANOVA	Analysis of variance;
QQ	Quantile–quantile plots
LD	Linkage disequilibrium;
GWAS	Genome-wide association study;
HR	Homologous recombination
ABA	Abscisic acid;
GHs	Glycosyl hydrolases

References

1. Wang, Y.; Deng, C.; Liu, Y.; Niu, Z.; Li, Y. Identifying change in spatial accumulation of soil salinity in an inland river watershed, China. *Sci. Total Environ.* **2018**, *621*, 177–185. [CrossRef] [PubMed]
2. Nematzadeh, G.A. Salt-related Genes Expression in Salt-Tolerant and Salt-Sensitive Cultivars of Cotton (*Gossypium* sp. L.) under NaCl Stress. *J. Plant Mol. Breed.* **2018**. [CrossRef]
3. Peng, Z.; He, S.; Sun, J.; Pan, Z.; Gong, W.; Lu, Y.; Du, X. Na+ compartmentalization related to salinity stress tolerance in upland cotton (*Gossypium hirsutum*) seedlings. *Sci. Rep.* **2016**, *6*, 34548. [CrossRef] [PubMed]
4. Paul, D.; Lade, H. Plant-growth-promoting rhizobacteria to improve crop growth in saline soils: A review. *Agron. Sustain. Dev.* **2014**, *34*, 737–752. [CrossRef]
5. Gao, W.; Xu, F.C.; Guo, D.D.; Zhao, J.R.; Liu, J.; Guo, Y.W.; Singh, P.K.; Ma, X.N.; Long, L.; Botella, J.R. Calcium-dependent protein kinases in cotton: Insights into early plant responses to salt stress. *BMC Plant Biol.* **2018**, *18*, 15. [CrossRef] [PubMed]
6. Munns, R.; Tester, M. Mechanisms of Salinity Tolerance. *Annu. Rev. Plant Biol.* **2008**, *59*, 651–681. [CrossRef] [PubMed]
7. Gray, S.B.; Brady, S.M. Plant Developmental Responses to Climate Change. *Dev. Biol.* **2016**, *419*, 64–77. [CrossRef]
8. Ahmad, S.; Ashraf, M.; Khan, N. Genetic basis of salt-tolerance in cotton (*Gossypium hirsutum* L.). *Sci. Technol. Dev.* **2004**, *23*, 45–50.
9. Frouin, J.; Languillaume, A.; Mas, J.; Mieulet, D.; Boisnard, A.; Labeyrie, A.; Bettembourg, M.; Bureau, C.; Lorenzini, E.; Portefaix, M. Tolerance to mild salinity stress in japonica rice: A genome-wide association mapping study highlights calcium signaling and metabolism genes. *PLoS ONE* **2018**, *13*, e0190964. [CrossRef] [PubMed]
10. Peng, Z.; He, S.; Gong, W.; Sun, J.; Pan, Z.; Xu, F.; Lu, Y.; Du, X. Comprehensive analysis of differentially expressed genes and transcriptional regulation induced by salt stress in two contrasting cotton genotypes. *BMC Genom.* **2014**, *15*, 760. [CrossRef]
11. Parida, A.K.; Das, A.B. Salt tolerant and salinity effects on plants. *Ecotoxicol. Environ. Saf.* **2005**, *60*, 324–349. [CrossRef] [PubMed]
12. Yang, X.; Gao, S.; Xu, S.; Zhang, Z.; Prasanna, B.M.; Li, L.; Li, J.; Yan, J. Characterization of a global germplasm collection and its potential utilization for analysis of complex quantitative traits in maize. *Mol. Breed.* **2011**, *28*, 511–526. [CrossRef]
13. Abbasi, Z.; Majidi, M.M.; Arzani, A.; Rajabi, A.; Mashayekhi, P.; Bocianowski, J. Association of SSR markers and morpho-physiological traits associated with salinity tolerance in sugar beet (*Beta vulgaris* L.). *Euphytica* **2015**, *205*, 785–797. [CrossRef]
14. Kantartzi, S.; Stewart, J.M. Association analysis of fibre traits in *Gossypium arboreum* accessions. *Plant Breed.* **2008**, *127*, 173–179. [CrossRef]
15. Du, L.; Cai, C.; Wu, S.; Zhang, F.; Hou, S.; Guo, W. Evaluation and exploration of favorable QTL alleles for salt stress related traits in cotton cultivars (*G. hirsutum* L.). *PLoS ONE* **2016**, *11*, e0151076. [CrossRef]
16. Cai, C.; Zhu, G.; Zhang, T.; Guo, W. High-density 80 K SNP array is a powerful tool for genotyping *G. hirsutum* accessions and genome analysis. *BMC Genom.* **2017**, *18*, 654. [CrossRef]
17. Sun, Z.; Li, H.; Zhang, Y.; Li, Z.; Ke, H.; Wu, L.; Zhang, G.; Wang, X.; Ma, Z. Identification of SNPs and Candidate Genes Associated with Salt Tolerance at the Seedling Stage in Cotton (*Gossypium hirsutum* L.). *Front. Plant Sci.* **2018**, *9*, 1011. [CrossRef] [PubMed]

18. Cai, C.; Wu, S.; Niu, E.; Cheng, C.; Guo, W. Identification of genes related to salt stress tolerance using intron-length polymorphic markers, association mapping and virus-induced gene silencing in cotton. *Sci. Rep.* **2017**, *7*, 528. [CrossRef]

19. Wang, S.; Chen, J.; Zhang, W.; Hu, Y.; Chang, L.; Fang, L.; Wang, Q.; Lv, F.; Wu, H.; Si, Z. Sequence-based ultra-dense genetic and physical maps reveal structural variations of allopolyploid cotton genomes. *Genome Biol.* **2015**, *16*, 108. [CrossRef] [PubMed]

20. Tan, Z.; Zhang, Z.; Sun, X.; Li, Q.; Sun, Y.; Yang, P.; Wang, W.; Liu, X.; Chen, C.; Liu, D. Genetic Map Construction and Fiber Quality QTL Mapping Using the CottonSNP80K Array in Upland Cotton. *Front. Plant Sci.* **2018**, *9*, 225. [CrossRef]

21. Atwell, S.; Huang, Y.S.; Vilhjálmsson, B.J.; Willems, G.; Horton, M.; Li, Y.; Meng, D.; Platt, A.; Tarone, A.M.; Hu, T.T. Genome-wide association study of 107 phenotypes in Arabidopsis thaliana inbred lines. *Nature* **2010**, *465*, 627. [CrossRef] [PubMed]

22. Xue, Y.; Warburton, M.L.; Sawkins, M.; Zhang, X.; Setter, T.; Xu, Y.; Grudloyma, P.; Gethi, J.; Ribaut, J.-M.; Li, W. Genome-wide association analysis for nine agronomic traits in maize under well-watered and water-stressed conditions. *Theor. Appl. Genet.* **2013**, *126*, 2587–2596. [CrossRef]

23. Chen, G.; Zhang, H.; Deng, Z.; Wu, R.; Li, D.; Wang, M.; Tian, J. Genome-wide association study for kernel weight-related traits using SNPs in a Chinese winter wheat population. *Euphytica* **2016**, *212*, 173–185. [CrossRef]

24. McCouch, S.R.; Wright, M.H.; Tung, C.-W.; Maron, L.G.; McNally, K.L.; Fitzgerald, M.; Singh, N.; DeClerck, G.; Agosto-Perez, F.; Korniliev, P. Open access resources for genome-wide association mapping in rice. *Nat. Commun.* **2016**, *7*, 10532. [CrossRef]

25. Abdelraheem, A.; Fang, D.D.; Zhang, J. Quantitative trait locus mapping of drought and salt tolerance in an introgressed recombinant inbred line population of Upland cotton under the greenhouse and field conditions. *Euphytica* **2018**, *214*, 8. [CrossRef]

26. Zhao, Y.L.; Wang, H.M.; Shao, B.X. SSR-based association mapping of salt tolerance in cotton (Gossypium hirsutum L.). *Genet. Mol. Res.* **2016**, *15*, gmr-15027370. [CrossRef] [PubMed]

27. Saeed, M.; Guo, W.Z.; Zhang, T.Z. Association mapping for salinity tolerance in cotton (*Gossypium hirsutum* L.) germplasm from US and diverse regions of China. *Aust. J. Crop Sci.* **2014**, *8*, 338–346.

28. Jia, Y.H.; Sun, J.L.; Wang, X.W.; Zhou, Z.L.; Pan, Z.E.; He, S.P.; Pang, B.Y. Molecular Diversity and Association Analysis of Drought and Salt Tolerance in *Gossypium hirsutum* L. Germplasm. *J. Integr. Agric.* **2014**, *13*, 1845–1853. [CrossRef]

29. Liu, F.; Zhou, Z.L.; Wang, C.Y.; Wang, Y.H.; Cai, X.Y.; Wang, X.X.; Zhang, Z.S.; Wang, K.B. Genetic diversity and relationship analysis of *Gossypium arboreum* accessions. *Genet. Mol. Res.* **2015**, *14*, gmr-14522. [CrossRef] [PubMed]

30. Du, X.; Huang, G.; He, S.; Yang, Z.; Sun, G.; Ma, X.; Li, N.; Zhang, X.; Sun, J.; Liu, M. Resequencing of 243 diploid cotton accessions based on an updated A genome identifies the genetic basis of key agronomic traits. *Nat. Genet.* **2018**, *50*, 796. [CrossRef] [PubMed]

31. Li, F.; Fan, G.; Wang, K.; Sun, F.; Yuan, Y.; Song, G.; Li, Q.; Ma, Z.; Lu, C.; Zou, C. Genome sequence of the cultivated cotton Gossypium arboreum. *Nat. Genet.* **2014**, *46*, 567. [CrossRef]

32. Mehetre, S.S.; Aher, A.R.; Gawande, V.L.; Patil, V.R.; Mokate, A.S. Induced polyploidy in Gossypium: A tool to overcome interspecific incompatibility of cultivated tetraploid and diploid cottons. *Curr. Sci.* **2003**, *84*, 1510–1512.

33. Peng, Z.; He, S.; Sun, J.; Xu, F.; Jia, Y.; Pan, Z.; Wang, L. An Efficient Approach to Identify Salt Tolerance of Upland Cotton at Seedling Stage. *Acta Agron. Sin.* **2014**, *40*, 476–486. [CrossRef]

34. Zhao, Y.; Wang, H.; Wei, C.; Li, Y. Genetic Structure, Linkage Disequilibrium and Association Mapping of Verticillium Wilt Resistance in Elite Cotton (*Gossypium hirsutum* L.) Germplasm Population. *PLoS ONE* **2014**, *9*, e86308. [CrossRef]

35. Sun, Z.; Wang, X.; Liu, Z.; Gu, Q.; Zhang, Y.; Li, Z.; Ke, H.; Yang, J.; Wu, J.; Wu, L. Genome-wide association study discovered genetic variation and candidate genes of fibre quality traits in *Gossypium hirsutum* L. *Plant Biotechnol. J.* **2017**, *15*, 982–996. [CrossRef] [PubMed]

36. Song, H.M.; Zhao, R.M.; Fan, P.X.; Wang, X.C.; Chen, X.Y.; Li, Y.X. Overexpression of *AtHsp90.2*, *AtHsp90.5* and *AtHsp90.7* in Arabidopsis thaliana enhances plant sensitivity to salt and drought stresses. *Planta* **2009**, *229*, 955–964. [CrossRef] [PubMed]

37. Schopf, F.H.; Biebl, M.M.; Buchner, J. The HSP90 chaperone machinery. *Nat. Rev. Mol. Cell Biol.* **2017**, *18*, 345. [CrossRef] [PubMed]

38. Zhou, X.H.; Li, X.S.; Wang, P.; Yan, B.L.; Teng, Y.J.; Yi, L.F. Molecular cloning and expression analysis of *HSP90* gene from Porphyra yezoensis Ueda (*Bangiales, Rhodophyta*). *J. Fish. China* **2010**, *34*, 1844–1852.

39. Ding, Z.; Yang, G.; Wu, J. Effect of Heat Shock at Germinating Period on Growing Developmental of Cotton. *J. Wuhan Bot. Res.* **2006**, *24*, 579–582.

40. Sangster, T.A.; Bahrami, A.; Wilczek, A.; Watanabe, E.; Schellenberg, K.; Mclellan, C.; Kelley, A.; Kong, S.W.; Queitsch, C.; Lindquist, S. Phenotypic Diversity and Altered Environmental Plasticity in *Arabidopsis thaliana* with Reduced Hsp90 Levels. *PLoS ONE* **2007**, *2*, e648. [CrossRef] [PubMed]

41. Busch, W.; Wunderlich, H.F. Identification of novel heat shock factor-dependent genes and biochemical pathways in *Arabidopsis thaliana*. *Plant J. Cell Mol. Biol.* **2010**, *41*, 1–14. [CrossRef]

42. Juan, G.; Yan, Z.; Wangbin, Z.; Jean, M.; Aiwu, D.; Wen-Hui, S. NAP1 family histone chaperones are required for somatic homologous recombination in Arabidopsis. *Plant Cell* **2012**, *24*, 1437–1447.

43. Zhu, Y.; Dong, A.; Meyer, D.; Pichon, O.; Renou, J.P.; Cao, K.; Shen, W.H. Arabidopsis NRP1 and NRP2 encode histone chaperones and are required for maintaining postembryonic root growth. *Plant Cell* **2006**, *18*, 2879–2892. [CrossRef]

44. Waśkiewicz, A.; Beszterda, M.; Goliński, P. ABA: Role in Plant Signaling Under Salt Stress. *Salt Stress Plants* **2013**, 175–196. [CrossRef]

45. Asaoka, R.; Uemura, T.; Nishida, S.; Fujiwara, T.; Ueda, T.; Nakano, A. New insights into the role of Arabidopsis RABA1 GTPases in salinity stress tolerance. *Plant Signal. Behav.* **2013**, *8*, e25377. [CrossRef]

46. Park, M.Y.; Chung, M.S.; Koh, H.S.; Lee, D.J.; Ahn, S.J.; Kim, C.S. Isolation and functional characterization of the Arabidopsis salt-tolerance 32 (*AtSAT32*) gene associated with salt tolerance and ABA signaling. *Physiol. Plant.* **2010**, *135*, 426–435. [CrossRef]

47. Wang, T.Z.; Xia, X.Z.; Zhao, M.G.; Tian, Q.Y.; Zhang, W.H. Expression of a Medicago falcata small GTPase gene, *MfARL1* enhanced tolerance to salt stress in Arabidopsis thaliana. *Plant Physiol. Biochem.* **2013**, *63*, 227–235. [CrossRef]

48. Sripinyowanich, S.; Klomsakul, P.; Boonburapong, B.; Bangyeekhun, T.; Asami, T.; Gu, H.; Buaboocha, T.; Chadchawan, S. Exogenous ABA induces salt tolerance in indica rice (*Oryza sativa* L.): The role of OsP5CS1 and OsP5CR gene expression during salt stress. *Environ. Exp. Bot.* **2013**, *86*, 94–105. [CrossRef]

49. Noaman, M.M.; Dvorak, J.; Dong, J.M. Genes inducing salt tolerance in wheat, Lophopyrum elongatum and amphiploid and their responses to ABA under salt stress. In *Prospects for Saline Agriculture*; Springer: Dordrecht, The Netherlands, 2002; pp. 139–144.

50. Zhao, K.F.; Fan, H.; Harris, P. Effect of exogenous aba on the salt tolerance of corn seedlings under salt stress. *Acta Bot. Sin.* **1995**, *37*, 295–300.

51. O'Mahony, P.J.; Oliver, M.J. Characterization of a desiccation-responsive small GTP-binding protein (*Rab2*) from the desiccation-tolerant grass Sporobolus stapfianus. *Plant Mol. Biol.* **1999**, *39*, 809–821. [CrossRef]

52. Nahm, M.Y.; Kim, S.W.; Yun, D.; Lee, S.Y.; Cho, M.J.; Bahk, J.D. Molecular and biochemical analyses of *OsRab7*, a rice Rab7 homolog. *Plant Cell Physiol.* **2003**, *44*, 1341–1349. [CrossRef] [PubMed]

53. Mazel, A.; Leshem, Y.; Tiwari, B.S.; Levine, A. Induction of salt and osmotic stress tolerance by overexpression of an intracellular vesicle trafficking protein AtRab7 (*AtRabG3e*). *Plant Physiol.* **2004**, *134*, 118–128. [CrossRef] [PubMed]

54. Wang, X.J.; Tang, C.L.; Lin, D.; Cai, G.L.; Liu, X.Y.; Bo, L.; Han, Q.M.; Buchenauer, H.; Wei, G.R.; Han, D.J. Characterization of a pathogenesis-related thaumatin-like protein gene TaPR5 from wheat induced by stripe rust fungus. *Physiol. Plant.* **2010**, *139*, 27–38. [CrossRef]

55. Liu, J.J.; Sturrock, R.; Ekramoddoullah AK, M. The superfamily of thaumatin-like proteins: Its origin, evolution, and expression towards biological function. *Plant Cell Rep.* **2010**, *29*, 419–436. [CrossRef]

56. Vashisht, A.A.; Tuteja, N. Stress responsive DEAD-box helicases: A new pathway to engineer plant stress tolerance. *J. Photochem. Photobiol. B Biol.* **2006**, *84*, 150–160. [CrossRef]

57. Opassiri, R.; Pomthong, B.; Akiyama, T.; Nakphaichit, M.; Onkoksoong, T.; Cairns, M.K.; Cairns, J.R.K. A stress-induced rice (*Oryza sativa* L.) β-glucosidase represents a new subfamily of glycosyl hydrolase family 5 containing a fascin-like domain. *Biochem. J.* **2007**, *408*, 241–249. [CrossRef] [PubMed]

58. Kang, H.M.; Sul, J.H.; Service, S.K.; Zaitlen, N.A.; Kong, S.-Y.; Freimer, N.B.; Sabatti, C.; Eskin, E. Variance component model to account for sample structure in genome-wide association studies. *Nat. Genet.* **2010**, *42*, 348. [CrossRef] [PubMed]

International Journal of
Molecular Sciences

MDPI

Article

Overexpression of the Stress-Inducible *SsMAX2* Promotes Drought and Salt Resistance via the Regulation of Redox Homeostasis in *Arabidopsis*

Qiaojian Wang [1,2,†], Jun Ni [2,*,†], Faheem Shah [2], Wenbo Liu [2], Dongdong Wang [1,2], Yuanyuan Yao [2], Hao Hu [2], Shengwei Huang [2], Jinyan Hou [2], Songling Fu [1,*] and Lifang Wu [2,*]

[1] College of Forestry and Landscape Architecture, Anhui Agricultural University, Hefei 230000, Anhui, China; wangqj521@126.com (Q.W.); 15755059531@163.com (D.W.)
[2] Key Laboratory of High Magnetic Field and Ion Beam Physical Biology, Hefei Institutes of Physical Science, Chinese Academy of Sciences, Hefei 230000, Anhui, China; faheemhorticulturist@gmail.com (F.S.); liuwenbo9261@sina.com (W.L.); 17355356851@163.com (Y.Y.); huhaoasd@mail.ustc.edu.cn (H.H.); swhuang@ipp.ac.cn (S.H.); jyhou@ipp.ac.cn (J.H.)
* Correspondence: nijun@ipp.ac.cn (J.N.); fusongl001@outlook.com (S.F.); lfwu@ipp.ac.cn (L.W.); Tel.: +86-551-6559-5672 (J.N. & F.S.); +86-551-6559-1413 (L.W.)
† These authors contributed equally to this work.

Received: 8 January 2019; Accepted: 12 February 2019; Published: 15 February 2019

Abstract: Recent studies have demonstrated that strigolactones (SLs) also participate in the regulation of stress adaptation; however, the regulatory mechanism remains elusive. In this study, the homolog of *More Axillary Branches 2*, which encodes a key component in SL signaling, in the perennial oil plant *Sapium sebiferum* was identified and functionally characterized in *Arabidopsis*. The results showed that the expression of *SsMAX2* in *S. sebiferum* seedlings was stress-responsive, and *SsMAX2* overexpression (OE) in *Arabidopsis* significantly promoted resistance to drought, osmotic, and salt stresses. Moreover, *SsMAX2* OE lines exhibited decreased chlorophyll degradation, increased soluble sugar and proline accumulation, and lower water loss ratio in response to the stresses. Importantly, anthocyanin biosynthesis and the activities of several antioxidant enzymes, such as superoxide dismutase (SOD), peroxidase (POD), and ascorbate peroxidase (APX), were enhanced in the *SsMAX2* OE lines, which further led to a significant reduction in hydrogen peroxide levels. Additionally, the *SsMAX2* OE lines exhibited higher expression level of several abscisic acid (ABA) biosynthesis genes, suggesting potential interactions between SL and ABA in the regulation of stress adaptation. Overall, we provide physiological and biochemical evidence demonstrating the pivotal role of *SsMAX2* in the regulation of osmotic, drought, and salt stress resistance and show that *MAX2* can be a genetic target to improve stress tolerance.

Keywords: *SsMAX2*; *Sapium sebiferum*; drought, osmotic stress; salt stress; redox homeostasis; strigolactones; ABA

1. Introduction

Abiotic stresses, such as drought, salt, cold, and flooding, significantly affect vegetative and reproductive growth and cause devastating yield losses each year. Plants have developed different coping mechanisms to deal with these stresses, mainly through the regulation of phytohormonal networks and dynamic changes of intracellular chemicals [1,2]. Phytohormones play a central role in the regulation of both vegetative and reproductive growth as well as adaptation to adverse growth conditions [3]. Hormones, such as abscisic acid (ABA), cytokinin, auxin, and salicylic acid (SA), have been proposed to be directly involved in the regulation of stress tolerance [4–7]. Strigolactones (SLs), which are a group of terpenoid compounds, play a key role in the regulation of shoot branching and the symbiosis with fungi in interactions with other hormones [8–11]. Recently, it was revealed that SLs also regulate plant adaptations to abiotic stresses [12,13]. In *Arabidopsis*, exogenous application of GR24, a SL analog, significantly improved salt and drought resistance, while the mutation of SL signaling gene *MAX2* made it sensitive to abiotic stresses [13,14]. However, the regulatory mechanism of SLs in stress tolerance still remains largely elusive.

In the perennial woody plant, the biological and molecular functions of strigolactones in the regulation of plant growth and stress adaptation have barely been studied. In the bioenergy plant *Jatropha curcas*, SLs antagonistically regulate the axillary bud outgrowth in interactions with cytokinin and gibberellin [15,16]. Several recent reports have also demonstrated that manipulation of the expression of SL biosynthesis genes can lead to significant change of the shoot branching phenotype in *Populus* and *Malus* [17,18], indicating that SLs have significant functions on the morphogenesis of woody plants. Plant growth and seed yield of woody plants are also threatened by abiotic stresses, such as salt, drought, and cold. Based on recent discoveries on the role of SLs in stress regulation in *Arabidopsis*, a functional study of the key genes in SL biosynthesis or signaling in woody plants would provide potential targets for genetic modifications to generate new cultivars with higher tolerance to abiotic stresses.

Sapium sebiferum, the seeds of which contain high level of fatty acids, has been considered as one of the most promising bioenergy plants. The oil from its seed coat and kernel can be manufactured into resources for lubricants, candles, cosmetics, and biodiesels [19,20]. It is widely distributed in most areas of China and even in the marginal land. However, while the plant adapts well to flooding and cold conditions, it is more sensitive to drought and salt stresses. The selection of high-yield cultivars with high resistance to drought and salt stresses is therefore the foremost goal in the molecular breeding of *Sapium sebiferum*. In this study, we identified that the expression of the SL signaling component *MAX2* was strongly responsive to abiotic stresses in *S. sebiferum* seedlings. Then, the biological functions of *SsMAX2* in the regulation of drought, osmotic, and salt tolerance were evaluated in *Arabidopsis*. The regulatory mechanism of *SsMAX2* to stress tolerance was further investigated at the physiological, molecular, and biochemical levels. This study not only reveals a pivotal role of *SsMAX2* in the regulation of drought, osmotic, and salt stress resistance but also provides evidence that *SsMAX2* can be a useful target for genetic engineering to produce stress-resistant plants.

2. Results

2.1. Gene Cloning of SsMAX2 from Sapium sebiferum Seedlings and the Gene Expression Profile in Response to Abiotic Stresses

The MAX2 homolog with 76% sequence similarity with AtMAX2 was identified from the *S. sebiferum* transcriptome database (Figure 1A). Then, the phylogenetic analysis of the MAX2 sequences from more than 20 plant species was carried out (Table S1). The results showed that SsMAX2 had the highest sequence identity with several perennial woody plants, such as *Jatropha curcas*, *Ricinus communis*, and Hevea brasiliensis, which also belong to the Euphorbiaceae family (Figure 1A).

MAX2 was the key component for SL signaling. Mutation of *MAX2* led to a significant increase in axillary branching and decrease in hypocotyl elongation. Our results are in accord with previous findings that constitutive expression of *SsMAX2* in *Arabidopsis* inhibits shoot branching and hypocotyl elongation, while *max2 Arabidopsis* mutant exhibits elongated hypocotyl growth (Figure 1B,C) and increased shoot branching (Figure 1D,E), demonstrating that *SsMAX2* has conserved functions with its homologs from *Arabidopsis*, rice, and pea.

To investigate whether *SsMAX2* is involved in the regulation of abiotic regulation, we first characterized the time-course expression profile of *SsMAX2* of *S. sebiferum* seedlings in response to osmotic and salt stresses. The results showed that osmotic treatment, which was mimicked by mannitol, significantly increased *SsMAX2* expression at 3 h after treatment (Figure 2A), whereas salt treatment induced a significant increase in *SsMAX2* expression at 12 h (Figure 2B). This demonstrated that *SsMAX2* is a stress-responsive gene, which might be involved in the regulation of adaptation to abiotic stresses.

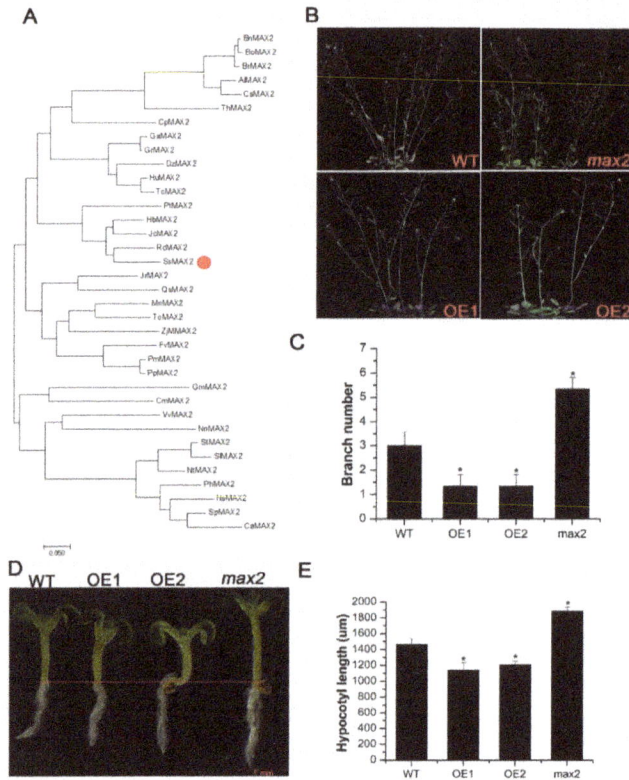

Figure 1. Phylogenetic analysis and functional characterization of *SsMAX2* in *Arabidopsis*. (**A**) Phylogenetic analysis of *SsMAX2* with its homologs from other species; (**B,C**) Rosette branching of 30-day-old plants of wild-type (WT), *SsMAX2* overexpression1 (OE1), OE2, and *max2*; (**D,E**) Hypocotyl length of 5-day-old seedlings of WT, *SsMAX2* OE1, OE2, and *max2* in half Murashige and Skoog (MS) medium. Data are presented as means ± SD of 20 replicates. Significant differences were determined by Student's *t*-test. Significance level: * $p < 0.05$.

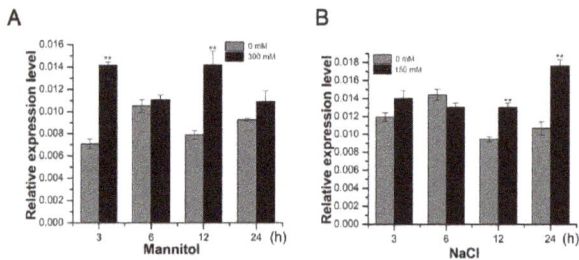

Figure 2. Expression profile of *SsMAX2* of 4-week-old *S. sebiferum* seedlings in response to (**A**) osmotic and (**B**) salt stresses. 300 mM mannitol and 150 mM NaCl were applied to 20-day-old *S. sebiferum* seedlings. *SsACT* was used as the internal control. Data are presented as means ± SD of three biological replicates. Significant differences were determined by Student's *t*-test. Significance level: ** $p < 0.01$.

2.2. SsMAX2 Conferred Drought and Osmotic Stress Tolerance in Arabidopsis

As the expression of *SsMAX2* in *S. sebiferum* seedlings was significantly induced by osmotic stress (Figure 2), we further investigated whether the constitutive expression of *SsMAX2* in *Arabidopsis* could confer drought and osmotic stress resistance. Results from the petri experiment showed that the *SsMAX2* OE lines exhibited significantly higher adaptation to osmotic stress, which was mimicked by the mannitol treatment (Figure 3A). Moreover, after withholding water for 11 days, all *Arabidopsis* lines exhibited significant dehydration, especially the wild-type (WT) and *max2* mutant seedlings. Seven days after re-watering, almost half of the *SsMAX2* OE lines survived (Figure 3B,C). Furthermore, chlorophyll fluorescence parameters, such as maximum photochemical efficiency of PSII (Fv/Fm), were investigated. The results showed that the *SsMAX2* OE lines exhibited much higher ratio of Fv/Fm under drought stress compared with the WT and *max2* mutant seedlings. Additionally, the water loss ratio in the leaf, which is an important characteristic of drought adaptation in plants, was much lower in the *SsMAX2* OE lines. These results suggest that *SsMAX2* positively regulates drought and osmotic stress adaptation in *Arabidopsis*. Interestingly, significant anthocyanin increase in the leaves of the *SsMAX2* OE lines was detected after drought treatment compared with the WT and *max2* mutant seedlings (Figure 4A). As previously reported, anthocyanin plays a key role in the regulation of the endogenous reactive oxygen species (ROS) level in response to abiotic stresses [21]. In this study, the leaves of two *SsMAX2* OE lines exhibited triple anthocyanin content compared to the WT and *max2* mutant seedlings (Figure 4B). Accordingly, the expression of the anthocyanin biosynthesis genes *chalcone synthase* (*CHS*), *chalcone isomerase* (*CHI*), *flavanone 3-hydroxylase* (*F3H*), *flavanone 3′-hydroxylase* (*F3′H*), *dihydroflavonol reductase* (*DFR*), and *anthocyanin synthase* (*ANS*) was more significantly upregulated in the *SsMAX2* OE lines and downregulated in the *max2* mutant in response to drought stress (Figure 4C), suggesting SL may also regulate anthocyanin biosynthesis. These results demonstrate that overexpression (OE) of *SsMAX2* confer drought and osmotic stress tolerance, and the significant upregulation of anthocyanin accumulation in the *SsMAX2* OE lines may contribute to drought and osmotic stress resistance in the *SsMAX2* OE lines.

2.3. SsMAX2 Conferred Salt Tolerance in Arabidopsis

We further investigated the salt responses of different *Arabidopsis* lines (two *SsMAX2* OE lines, wild-type, and *max2* mutant). The results showed that seedlings of different lines exhibited no significant growth variations under normal conditions (Figure 5A). However, the WT and *max2* plants showed significant blushing phenotype after seven days of growth in half MS medium containing 100 mM NaCl, while the *SsMAX2* OE lines exhibited significantly higher tolerance to salt stress, even in the 150 mM NaCl medium (Figure 5A). The salt stress experiment was also conducted on different *Arabidopsis* lines growing in the soil. The results were in accord with that of the petri experiment, with the *SsMAX2* OE lines exhibiting robust salt tolerance (Figure 5B). The survival rate of the *SsMAX2* OE lines could reach as high as 65% at 7 d after 150 mM treatment (Figure 5C). It is worth noting here that the chlorophyll content of the *max2* mutant was significantly lower than that of *SsMAX2* OE and WT plants (Figure 5D). Stress can induce senescence and cause significant chlorophyll degradation in the leaf [22]. The results showed that the decrease in stress-induced chlorophyll in the leaves of the *SsMAX2* OE lines was significantly lower than that of the *max2* mutant and wild-type plants (Figure 5D). This suggests that *MAX2* may positively regulate chlorophyll synthesis and that *SsMAX2* overexpression in *Arabidopsis* can retard leaf senescence induced by salt stress.

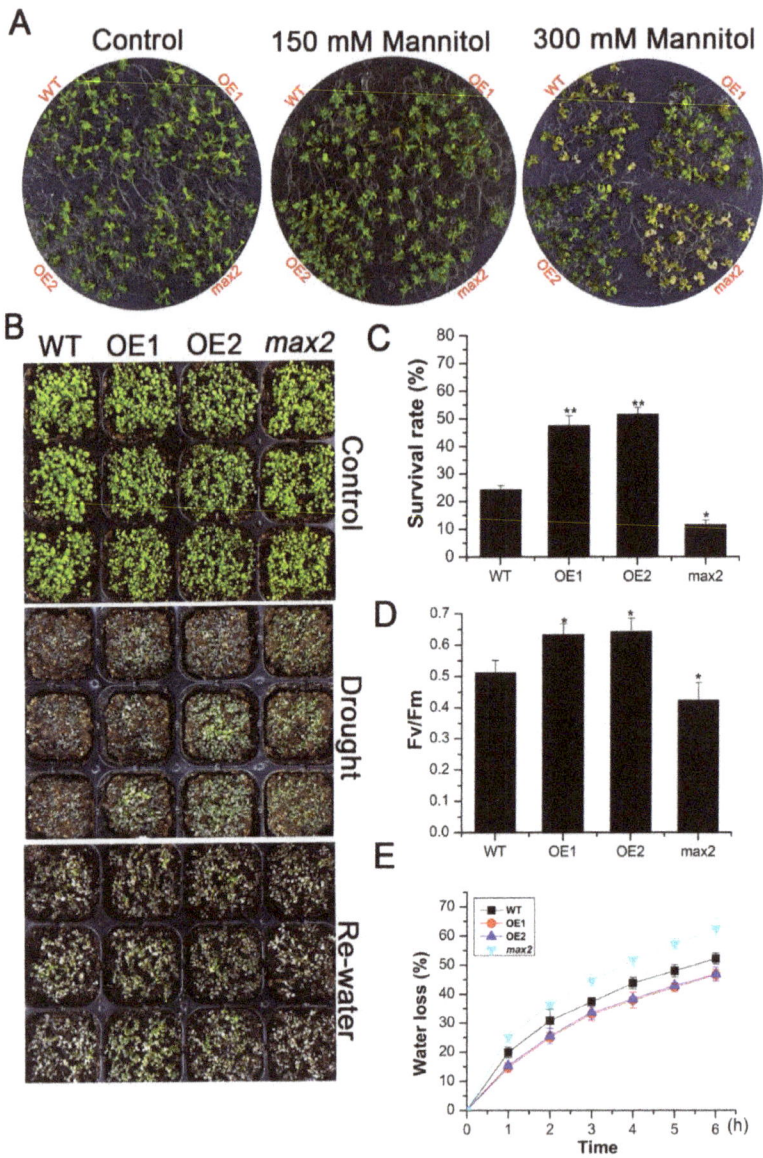

Figure 3. Constitutive expression of *SsMAX2* promoted osmotic and drought stress resistance in *Arabidopsis*. (**A**) Phenotype of *SsMAX2* OE lines, *max2*, and wild-type (WT) in half MS medium containing 150 and 300 mM mannitol; (**B**) Phenotype of different *Arabidopsis* lines after withholding water and rewatering treatment; (**C**) Survival rate of seedlings after drought treatment; (**D**) Maximum photochemical efficiency of PSII (Fv/Fm) of different lines under drought stress; (**E**) Water loss rate. Data are presented as means ± SD of three biological replicates. Significant differences were determined by Student's *t*-test. Significance level: * $p < 0.05$, ** $p < 0.01$.

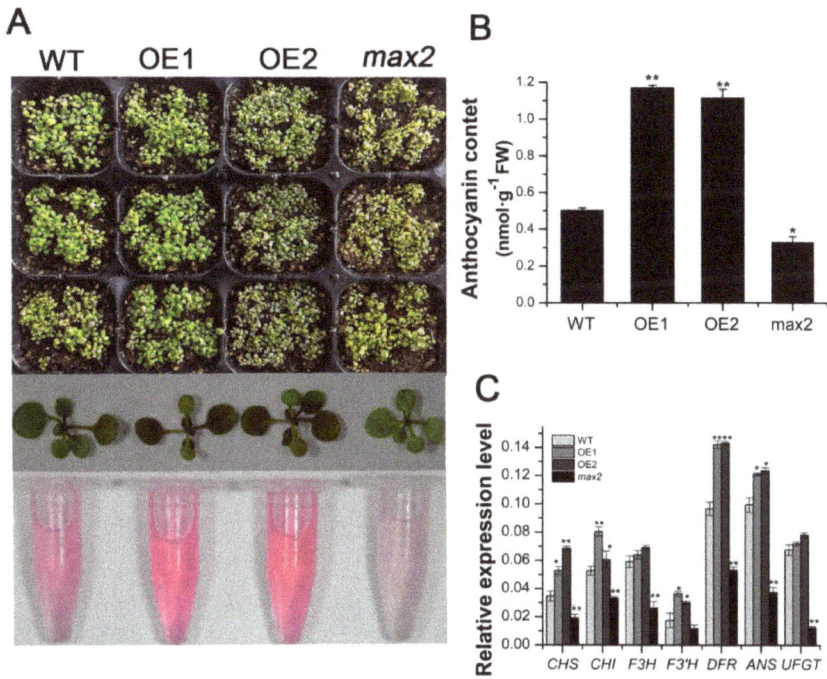

Figure 4. *SsMAX2* promoted anthocyanin accumulation in *Arabidopsis* leaves under drought stress. (**A**) Drought-induced anthocyanin accumulation in the leaves; (**B**) Anthocyanin level of different lines; (**C**) Relative expression of anthocyanin biosynthesis genes in the leaves of different lines in response to drought stress. Data are presented as means ± SD of three biological replicates. Significant differences between WT and the other groups were determined by Student's *t*-test. Significance level: * $p < 0.05$, ** $p < 0.01$.

2.4. SsMAX2 Promoted Seed Germination under Both Salt and Osmotic Stresses

We further investigated whether the *SsMAX2* OE lines could also improve stress resistance during the seed germination stage. We evaluated the effects of different concentrations of mannitol and NaCl on the seed germination of different *Arabidopsis* lines. The results showed that, under normal conditions, the germination rate of *SsMAX2* OE1 lines, *max2*, and WT showed no significant variations (Figure 6A,B). However, the seed germination of WT and *max2* mutant was more likely to be inhibited, whereas *SsMAX2* OE lines exhibited much higher germination ratio, with increasing concentrations of both mannitol and NaCl (Figure 6A,B). The results also showed that, even under 200 mM mannitol and 150 mM NaCl, the seed germination of *SsMAX2* OE lines was still over 50% (Figure 6A,B). Furthermore, the time-course assay of the seed germination showed that the seed germination of the *max2* mutant was significantly delayed compared with that of the *SsMAX2* OE lines under both salt and drought stress (Figure 6C–E). These results suggest that *SsMAX2* confer significant salt and osmotic stress resistance during seed germination.

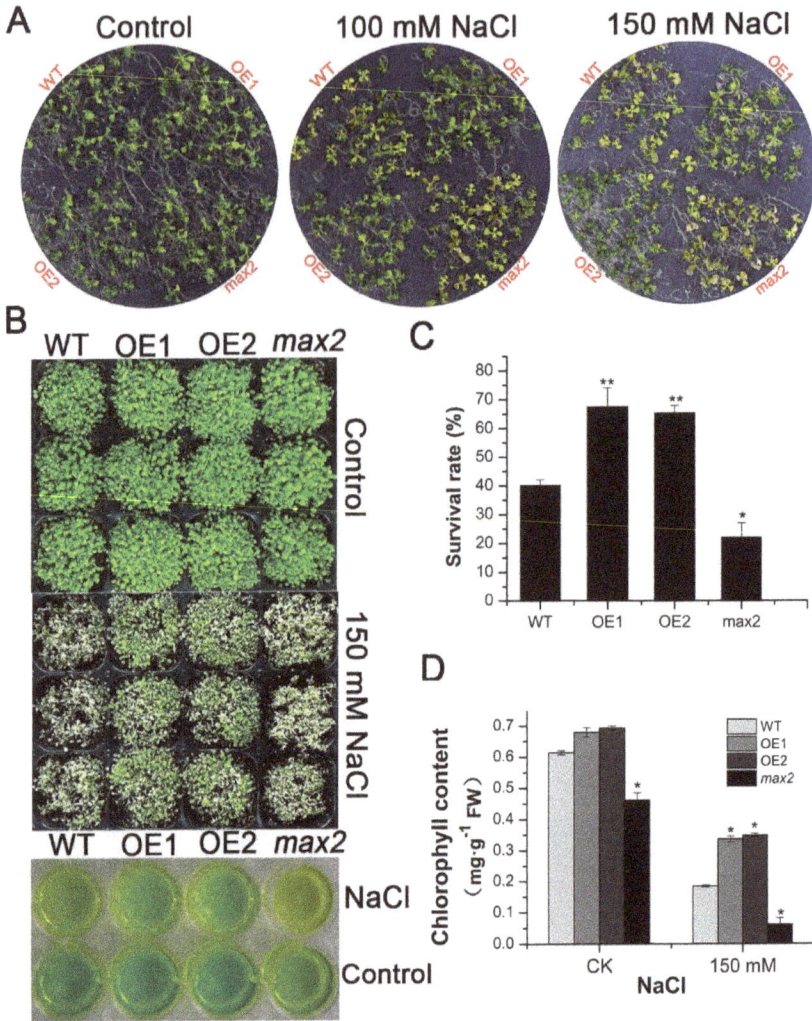

Figure 5. *SsMAX2* conferred salt resistance in *Arabidopsis*. (**A**) Growth phenotype of 5-day-old *SsMAX2* OE lines, *max2*, and WT seedlings under salt stress; (**B,C**) The growth and survival rate 7 days after 150 mM NaCl treatment on 15-day-old seedlings; (**D**) Chlorophyll content of the seedling leaves before or 7 days after salt stress treatment. Data are presented as means ± SD of three biological replicates. Significant differences between WT and the other groups were determined by Student's *t*-test. Significance level: * $p < 0.05$, ** $p < 0.01$.

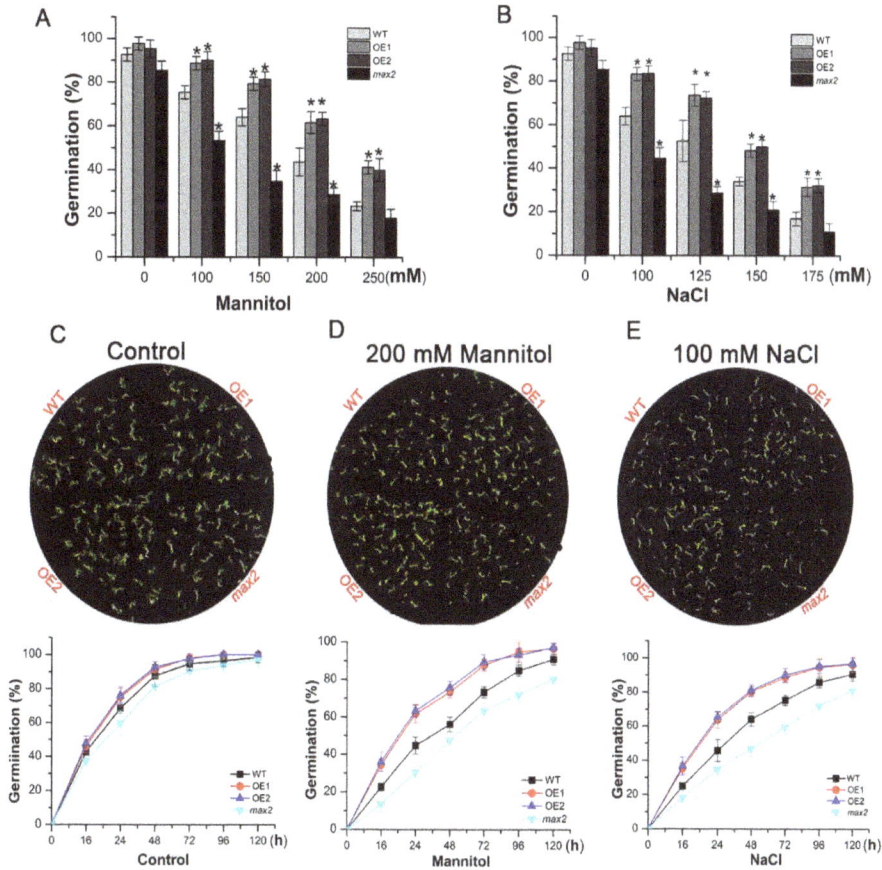

Figure 6. *SsMAX2* OE lines exhibited higher salt and osmotic stress tolerance during the seed germination stage. (**A,B**) Seed germination rate of the *SsMAX2* OE lines, *max2*, and WT under various concentrations of NaCl and mannitol treatment; (**C–E**) Time-course assay of the seed germination of different lines under 200 mM mannitol and 100 mM NaCl treatment. Data are presented as means ± SD of three biological replicates. Significant differences between WT and the other groups were determined by Student's *t*-test. Significance level: * $p < 0.05$.

2.5. SsMAX2 Regulated the Hydrogen Peroxide, Malondialdehyde (MDA), Proline, and Soluble Sugar Accumulation in the Seedlings in Response to the Stresses

The significant increase in endogenous peroxide or superoxide chemical levels induced by the abiotic stresses is responsible for the initiation of leaf senescence and death [23]. Malondialdehyde (MDA) is an important marker for lipid peroxidation due to overproduction of ROS in the cell [24]. Here, the results showed that both osmotic and salt stress could cause a significant increase in hydrogen peroxide and MDA in all lines, while *SsMAX2* OE lines had a significant lower level of both hydrogen peroxide and MDA (Figure 7A–D), suggesting a tightly regulated ROS and MDA homeostasis in the *SsMAX2* OE lines. This was also in accord with the physiological results, which showed that *SsMAX2* OE lines had better resistance and delayed leaf senescence to the stresses.

Proline and soluble sugars play an important role in maintaining osmotic homeostasis in plant cells [23]. Our results showed that the *SsMAX2* OE lines accumulated higher proline and soluble sugars in leaves than WT and *max2* mutant (Figure 7E,F). As both drought and salt stress can break the osmosis homeostasis in the plant, the enhanced accumulation of proline and soluble sugars can significantly prevent water loss from leaves under osmotic stresses. The results show that decreased water loss in the *SsMAX2* OE lines contribute to drought and salt stress resistance.

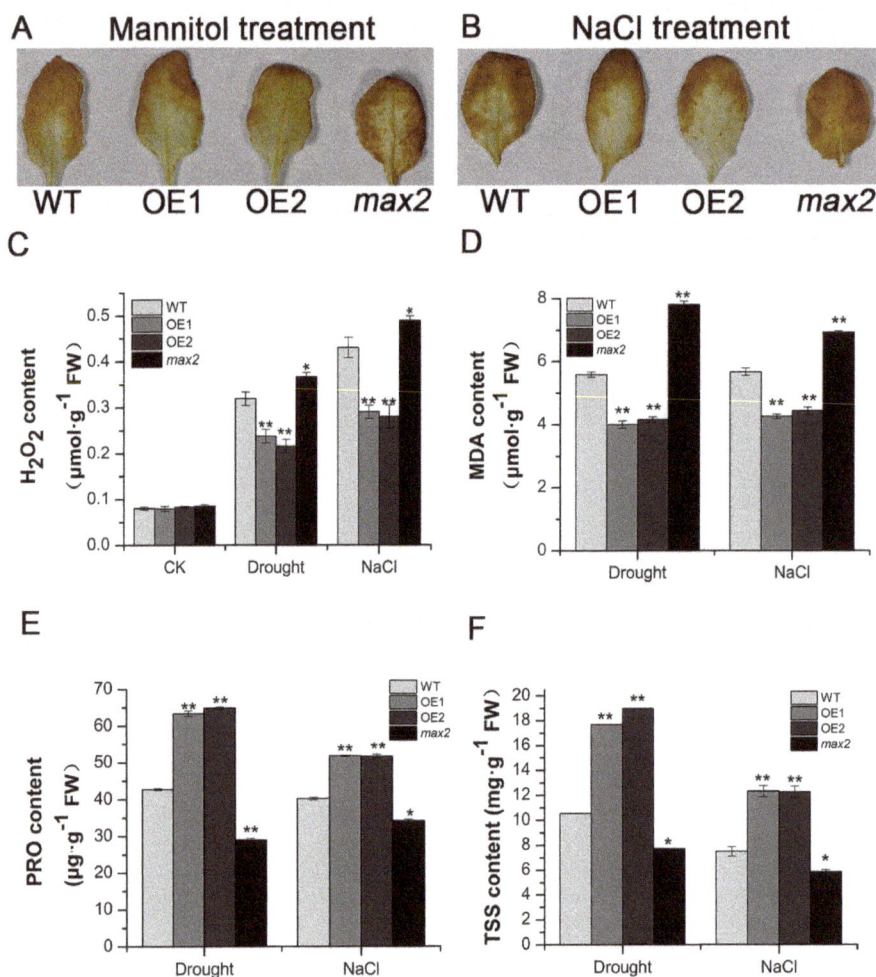

Figure 7. *SsMAX2* OE lines exhibited lower hydrogen peroxide and malondialdehyde (MDA) levels and increased proline and soluble sugar levels in response to osmotic and salt stresses. (**A,B**) DAB staining of the leaves of *SsMAX2* OE lines, *max2*, and WT after mannitol and NaCl treatment. The (**C**) hydrogen peroxide, (**D**) MDA, (**E**) proline, and (**F**) soluble sugar level of 15-day-old seedlings of the *SsMAX2* OE lines, *max2*, and WT were determined 5 days after withholding water or 150 mM NaCl treatment. Data are presented as means ± SD of three biological replicates. Significant differences between WT and the other groups were determined by Student's *t*-test. Significance level: * $p < 0.05$, ** $p < 0.01$.

2.6. SsMAX2 Increased the Enzyme Activity of Superoxide Dismutase (SOD), Peroxidase (POD), and Ascorbate Peroxidase (APX)

As the hydrogen level in the *SsMAX2* OE lines was significantly lower than that in WT and *max2* mutant, we further investigated whether the key enzymes involved in the regulation of ROS degradation were also affected in response to drought and salt treatment. POD, SOD, and CAT are the main oxidative enzymes involved in the regulation of ROS homeostasis in the cell [25,26]. The results showed that, under both drought and salt stress, the activity of POD, SOD, and CAT of the *SsMAX2* OE lines was significantly higher than that of *max2* and WT (Figure 8), whereas the max2 mutant exhibited the lowest activity of the antioxidative enzymes, further demonstrating a tightly controlled ROS scavenge ability controlled by SL signaling. These results suggest that *MAX2* may be involved in the regulation of plant ROS homeostasis via controlling the activities of oxidative enzymes.

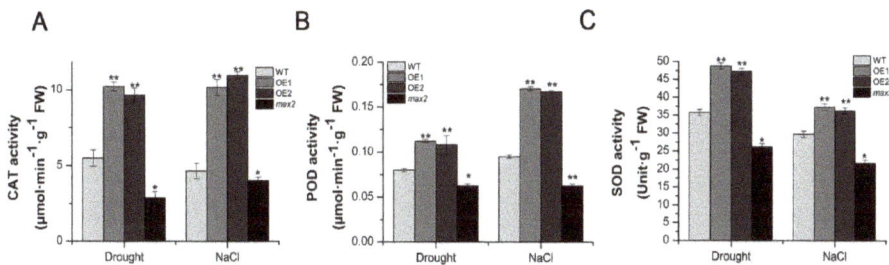

Figure 8. *SsMAX2* increased the activity of the antioxidant enzymes. The enzyme activities of (**A**) CAT, (**B**) POD, and (**C**) SOD of the extracts of 15-day-old seedlings were separately determined 5 days after withholding water or 150 mM NaCl treatment. Data are presented as means ± SD of three biological replicates. Significant differences between WT and the other groups were determined by Student's *t*-test. Significance level: * $p < 0.05$, ** $p < 0.01$.

2.7. Diverse Regulation of the Abscisic Acid (ABA) Biosynthesis Genes in SsMAX2 OE Lines and max2 in Response to Drought and Salt Stress

ABA is the key phytohormone that directly regulates abiotic stresses. Thus, in this study, we further investigated whether the expression of ABA biosynthesis genes (*CYP707-A1, -A2, -A3, NCED3,* and *OAA3*) was diversely regulated in response to drought and salt stress. After salt treatment, the significant upregulation of *NCED3, OAA3,* and *CYO707A1* in the *SsMAX2* OE lines could be detected at 6 h after treatment compared with the WT and *max2* mutant (Figure 9). The expression of *CYP707A3* and *NCED3* was relatively higher in the *SsMAX2* OE lines (Figure 9C,D). Specifically, the basic expression of *CYP707A2* in the *max2* mutant was higher than the WT and *SsMAX2* OE lines (Figure 9B), which is in accord with previously published results in *Arabidopsis* [27]. These results suggest a potential interaction between SL and ABA in the regulation of abiotic stress adaptation.

3. Discussion

Sapium sebiferum, which is one of the most important commercial woody plants in China, has received considerable attention due to its high oil content in the seed coat and kernel, excellent sightseeing value as a landscape plant, and its high adaptation to the adverse marginal land. Recent studies on the plant have mainly focused on flower sex determination, seed yield, oil extraction and production, and herb values [28–32]. However, only few reports have demonstrated its antistress abilities. Abiotic stresses, such as drought and salt, can significantly reduce the yield output in *S. sebiferum* [33], which significantly limits its industrial potential. As the transgenic approaches have been widely used and proven to be very effective in the regulation of abiotic stress tolerance in many species [34–36], the generation of high-stress-resistant cultivars via genetic modifications is the foremost mission in *S. sebiferum* breeding.

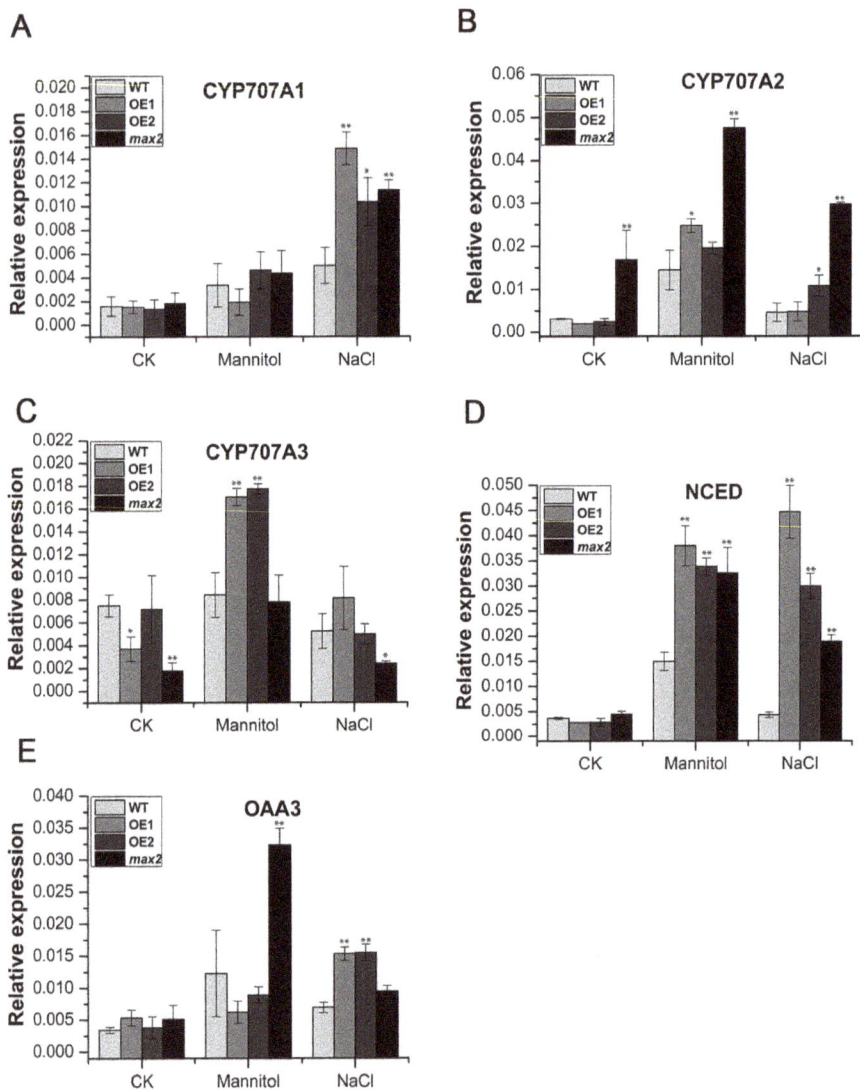

Figure 9. Expression of key ABA biosynthesis genes of the *SsMAX2* OE lines, *max2*, and WT was diversely regulated in response to salt and osmotic treatment. The gene expression of (**A**) *CYP707A1*, (**B**) *CYP707A2*, (**C**) *CYP707A3*, (**D**) *NCED3*, and (**E**) *OAA3* was determined at 6 h after mannitol and NaCl treatment by qPCR. *SsACT* was used as the internal control. Data are presented as means ± SD of three biological replicates. Student's *t*-test was used to determine the significant differences between WT and other *Arabidopsis* lines. Significance level: * $p < 0.05$, ** $p < 0.01$.

Previous studies have demonstrated that SLs are involved in the regulation of shoot branching, senescence, and photomorphogenesis [37,38]. Some recent researches have revealed the pivotal role of SLs in the regulation of stress adaptation [39]. Thus, the identification and characterization of SL biosynthesis and signaling genes in woody plants are important for generating stress-tolerant cultivators. *More Axillary Branches 2 (MAX2)*, which encodes a F-box E3 ligase in *Arabidopsis*, is the key component involved in SL signal transduction [40]. *MAX2* plays a key role in the regulation of shoot branching, photomorphogenesis, and stress adaptation [13,27,41,42]. In this work, we identified and functionally characterized the *MAX2* homolog (*SsMAX2*) from the oil plant *S. sebiferum*. Constitutive expression of *SsMAX2* in *Arabidopsis* inhibited the shoot branching and the hypocotyl elongation (Figure 1), further confirming the biological functions of *SsMAX2* as *MAX2* homolog in *Arabidopsis*. In *S. sebiferum* seedlings, it was interesting to find that the expression of *MAX2* was significantly upregulated in response to drought and salt treatment (Figure 2). Many abiotic stress-inducible genes, such as *NAC5* [43], *XTH3* [44], and *UGT87A2* [45], are important in controlling the adaptation to stresses. Thus, the significant upregulation of *SsMAX2* indicates that it may be correlated with stress adaptation. In comparison with the *max2 Arabidopsis* mutant, our results further demonstrated that *SsMAX2* OE lines exhibited significant drought and salt tolerance (Figures 3 and 5). The seed germination in the *SsMAX2* OE lines also had higher drought and salt tolerance (Figure 6), whereas the *max2* mutant was more sensitive to the stresses, as previously described [14,42]. These results demonstrate that *MAX2* participates in the regulation of stress adaptation. However, the regulatory mechanism of how *MAX2* controls the increased tolerance to abiotic stresses remains elusive.

Plants have developed many mechanisms to cope with biotic and abiotic stresses, such as accumulation of secondary metabolites, activated oxidative enzyme, or nonenzyme systems [36,46,47]. Anthocyanins, which consist of a group of phenolic compounds, act as important antioxidants in plants suffering from abiotic stresses [48]. Increased anthocyanin production has been shown to significantly enhance tolerance to abiotic stresses in *Arabidopsis*, grapevine, and bamboo [49–51]. Our results also showed that, under drought stress, significant accumulation of anthocyanins in the leaves of the *SsMAX2* OE lines was detected, while its content was relatively lower in the *max2* mutant and WT plants (Figure 4). The qPCR results further showed that the expression of the key genes in the anthocyanin biosynthesis pathway was significantly induced in the *SsMAX2* OE lines (Figure 4C). These results suggest that SLs may be involved in the regulation of anthocyanin biosynthesis in *Arabidopsis*, which is important for adaptation to salt stress. It is worth noting here that the *SsMAX2* OE lines had higher chlorophyll content in the leaves, while the *max2* mutant exhibited much lower chlorophyll content than WT plants under normal growth conditions (Figure 5). Furthermore, drought-stress-induced chlorophyll degradation in the leaves was much lower than that in WT and *max2* plants (Figure 5D), suggesting *MAX2* may positively regulate chlorophyll biosynthesis or accumulation, the level of which can be an important indicator of adaptation to abiotic stresses [52]. Many researchers have suggested that the accumulation of soluble sugars and amino acids is key for osmotic stress resistance [53,54] due to their direct role in the regulation of water uptake and loss. Our results also showed that the level of both proline and soluble sugars was significantly higher in the *SsMAX2* OE lines after drought and salt treatment in comparison with the WT and *max2* seedlings (Figure 7E,F), suggesting a pivotal role of *MAX2* in the regulation of cellular metabolite homeostasis.

The generation of oxidative chemicals induced by abiotic stresses in cells is the main cause of cell apoptosis and death [46]. Plants have developed a tightly regulated mechanism to maintain endogenous oxidative chemicals at a certain level. Many reports have demonstrated that exogenous application of ROS cleavage chemicals (e.g., melatonin) or overexpression of oxidative chemical cleavage enzymes (e.g., SOD, POD, and APX), can significantly promote tolerance to abiotic stresses due to the efficient cleavage of the stress-induced oxidative chemical level [55,56]. In this study, we presumed that *MAX2* could be involved in the regulation of oxidative chemical levels in plants. The results showed that both drought and salt treatment significantly induced hydrogen peroxide accumulation in *Arabidopsis* seedlings. However, the level of this was significantly lower in the *SsMAX2* OE lines compared with that of the *max2* mutant and wild-type (Figure 7A,B). Accordingly, the activity analysis of key oxidative enzymes, such as CAT, POD, and SOD, also proved that *SsMAX2* OE plants had higher capability in the cleavage of hydrogen peroxide induced by salt and drought stress, whereas *max2* mutant exhibited much lower enzyme activity (Figure 8). These results suggest that the SL signaling may be directly involved in the regulation of redox homeostasis, although the molecular mechanism still needs further investigation.

ABA is the key phytohormone that positively regulates abiotic stress adaptation. Many reports have demonstrated that ABA accumulation can happen immediately when plants are subjected to drought, salt, cadmium, or cold stresses [57]. Exogenous application with ABA or overexpression of ABA biosynthesis genes can significantly promote abiotic stress resistance in many species [58]. A recent study also demonstrated that ABA and SL coordinately regulated salt stress tolerance in *Sesbania cannabina* [59]. In this study, the *SsMAX2* OE lines exhibited higher expression level of key ABA biosynthesis-related genes, such as *CYP707A1*, *CYP707A3*, *NCED*, and *OAA3*, compared with the *max2* mutant or WT plants (Figure 9), further indicating that *SsMAX2*-induced stress tolerance may be partially ABA-dependent.

In this study, we isolated and functionally characterized the *MAX2* homolog in the oil plant *Sapium sebiferum*. We not only investigated the gene function in controlling shoot branching and hypocotyl elongation but, most importantly, we characterized the novel function of *SsMAX2* in the regulation of drought and salt adaptation. We showed that *MAX2* potentially controls chlorophyll biosynthesis and degradation, anthocyanin biosynthesis, soluble sugars, and proline accumulation (Figure 10). The physiological and biochemical results demonstrate that *SsMAX2* plays a pivotal role in the regulation of redox homeostasis via the regulation of antioxidative enzymes (Figure 10). The results also suggest that there may be potential interactions between SL and ABA in the regulation of abiotic stress adaptation (Figure 10). Further research will be focused on the identification of the molecular network of SL in the regulation of stress adaptation.

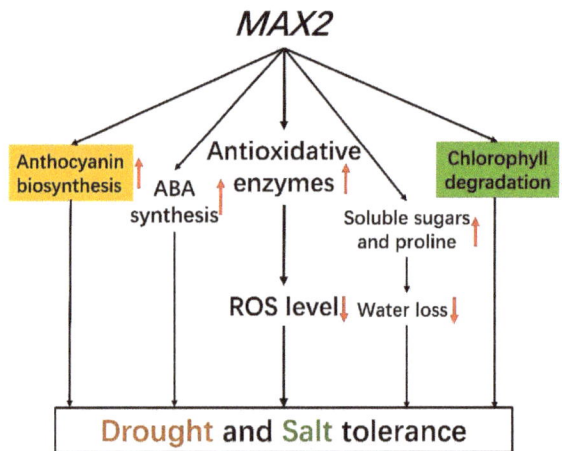

Figure 10. Model of *SsMAX2* in the regulation of drought and salt adaptation. The arrows indicate the downregulated or upregulated activities.

4. Materials and Methods

4.1. Plant Materials and Growth Conditions

The Col-0 ecotype *Arabidopsis* and *max2* mutant (CS9565) were ordered from the Arabidopsis Biological Resource Center (ABRC). *Arabidopsis* seeds were sterilized with 75% ethanol solution for one minute, followed by sterilization with 8% sodium hypochlorite for 15 min, and then germinated in half MS medium. *Arabidopsis* seedlings were grown in the growth chamber (22 °C; 16-h light/8-h dark photoperiod; 120 mol·m^{-2}·s^{-1} radiation strength; 75% humidity). All plants were fertilized with half Hoagland solution every other week.

Sapium sebiferum seeds were germinated in the peat soil as previously described [60]. Two-week-old *S. sebiferum* seedlings were transplanted into the garden pot (10 cm × 10 cm) and grown in the chamber as described above.

4.2. Gene Cloning, Vector Construction, and Arabidopsis Transformation

The protein sequence of AtMAX2 (AT2G42620.1) was used to blast against the local *S. sebiferum* transcriptome database using the NCBI local blast package: BLAST 2.7.1. The coding sequence (CDS) and amino acid sequence information is listed in Table S2. The complete CDS of *SsMAX2* was cloned into the pOCA30 expression vector and transformed into the *Agrobacterium* EHA105. The *Arabidopsis* transformation was carried out following a previously described method [61]. The candidate transgenic *Arabidopsis* plants were firstly screened on half MS containing 40 mg/L kanamycin and then further confirmed by RT-PCR. Two out of over 10 independent homozygous transgenic lines were used for further experiments.

4.3. Drought and Salt Treatment

In the petri experiment, different concentrations of mannitol and NaCl were separately used for drought and salt stress treatment. In the soil experiment, plants of different lines were grown in moss peat soil. To induce drought conditions, water was withheld for a number of days, as indicated in the figures. The plant growth was monitored after three days after rewatering. For salt stress, the pods of *Arabidopsis* were directly submerged in different concentrations of sodium chloride until the soil was completely saturated. Then, the plants were put under normal growth conditions.

4.4. RNA Extraction and Quantitative Real-Time PCR (qPCR)

Total RNA was extracted from *Arabidopsis* and *S. sebiferum* seedlings according to the instructions of HP Total Plant RNA kit (Omega, Shanghai, China). RNA concentration and integrity were further analyzed by a micro-analyzer and gel electrophoresis. For cDNA synthesis, 1–1.5 g of total RNA was used using the TransScript II One-Step gDNA removal and cDNA Synthesis SuperMix (Transgen, Beijing, China). qPCR was performed using the Premix Ex Taq^{TM} II (Transgen, Beijing, China) on the LightCycler 96 System (Roche, Basel, Switzerland). The qPCR program was set as follows: preheating, 95 °C, 10 min; amplification (45 cycles), 95 °C, 10 s, 60 °C, 20 s, and 72 °C, 20 s; melting curve: 95 °C, 2 min, 60 °C, 30 s, then continuously increased to 95 °C. The calculation of the relative gene expression was based on the $2^{-\Delta\Delta Cq}$ method as described previously [62]. The detailed primer information for each gene is listed in Table S3.

4.5. Total Chlorophyll and Anthocyanin Determination

The total chlorophyll content was analyzed based on a previously described method [63]. The leaf tissue was homogenized in liquid nitrogen and subsequently extracted in 80% acetone containing 1 M KOH overnight. After centrifugation at $12,000 \times g$ for 10 min, the supernatant was used for chlorophyll determination using a Scandrop spectrophotometer (Analytikjena, Jena, Germany). The anthocyanin was determined as previously described [64] with minor modifications. Briefly, the *Arabidopsis* leaves were ground into fine powders in liquid nitrogen. Then, the powders were transferred to methanol containing 1% HCl at 4 °C in the dark for 24 h. The aqueous phase was then used for anthocyanin determination in the spectrophotometer with the following formula: OD = $(A_{530} - A_{620}) - 0.1 (A_{650} - A_{620})$.

4.6. Determination of the Water Loss Rate

Approximately 0.5 g fresh leaves of 15-day-old *Arabidopsis* plants were collected and weighed immediately. The leaves were kept in a petri dish in open air. The water loss rate was calculated every hour based on the change in leaf weight, as previously described [65].

4.7. Diaminobenzidine (DAB) Staining of Hydrogen Peroxide in the Leaves

Diaminobenzidine (DAB) staining was used for in situ detection of hydrogen peroxide in *Arabidopsis* leaves as previously described [56]. The detached *Arabidopsis* leaves of different *Arabidopsis* lines were submerged in the DAB solution (1 mg·mL^{-1}, pH 3.8) overnight at room temperature.

The leaves were submerged in ethanol until the chlorophyll was washed off. Then, the leaves were used for hydrogen peroxide detection.

4.8. Determination of Hydrogen Peroxide, MDA, Proline and Total Soluble Sugar Level, and Antioxidant Enzyme Activity

The plant samples were firstly ground into powder in liquid nitrogen and then suspended in ice-cold phosphate buffer (0.1 M, pH = 7). The sample was vortexed at maximum speed for 1 min and centrifuged at 12,000 rpm for 15 min. The supernatants were then used for further determination of the hydrogen peroxide level and antioxidant enzyme activity. The hydrogen peroxide level of different samples was determined using the Hydrogen Peroxide Assay Kit (Jiancheng Bioengineering Institute, Nanjing, China). The absorbance at 405 nm was determined by a Scandrop spectrophotometer (Analytikjena, Germany). The hydrogen peroxide level was calculated based on a previously described formula [56]. The MDA, proline, and total soluble sugar level were separately determined using the MDA Assay Kit, the Proline Assay Kit, and the Plant Soluble Sugar Content Test Kit (Jiancheng Bioengineering Institute, Nanjing, China), as previously described [24].

The antioxidant enzyme activities were also determined using the spectrophotometric method. SOD, POD, and CAT activities were separately determined using the Total Superoxide Dismutase (T-SOD) Assay Kit, Peroxidase Assay Kit, and Catalase (CAT) Assay Kit (Jiancheng Bioengineering Institute, Nanjing, China), as previously described [56].

4.9. Chlorophyll Fluorescence Measurement

The photosynthesis rate of the *Arabidopsis* plants of different lines after drought treatment was analyzed using the Portable Photosynthesis Rate Detector AGHJ-PPF (Anhui Institute of Optics and Fine Mechanics, Chinese Academy of Sciences) after a 30 min dark adaptation, as previously described [66]. Parameters including the maximum photochemical efficiency of PSII (Fv/Fm), minimal fluorescence (F0), maximal fluorescence (Fm), and PSII (as shown in Figure S1) were calculated according to a previous method [66].

4.10. Phylogenetic Analysis

The homologs of MAX2 of the other species were obtained by blasting against nucleic acid sequences in GenBank. Phylogenetic analysis was carried out using MEGA software Version 7.0 [67]. One thousand bootstrap replicates were performed for the phylogenetic tree construction.

4.11. Statistical Analysis

Multiple comparisons between different samples were carried out using Statistical Product and Service Solutions (SPSS, Chicago, IL, USA) with one-way ANOVA, followed by the Tukey's test ($p < 0.05$). Student's *t*-test was used to analyze the significant difference between the indicated groups and control.

Supplementary Materials: Supplementary materials can be found at http://www.mdpi.com/1422-0067/20/4/837/s1. Table S1: Information of the MAX2 homologs from other species. Table S2: CDS and amino acid sequence information of SsMAX2. Table S3: Information of all the primers used in this study. Figure S1: Parameters of the photosynthesis rate analysis by the Portable Photosynthesis Rate Detector AGHJ-PPF.

Author Contributions: Conceptualization, J.N.; data curation, Q.W., J.N., F.S., W.L., D.W., Y.Y., H.H., S.H., J.H., S.F., and L.W.; investigation, Q.W., J.N., and F.S.; supervision, L.W.; validation, Q.W.; writing—original draft, J.N.; writing—review & editing, J.N.

Funding: This work was funded by the Anhui Natural Science Foundation (1708085QC70), the National Natural Science Foundation of China (31500531), the Grant of the President Foundation of Hefei Institutes of Physical Science of Chinese Academy of Sciences (YZJJ201502&YZJJ201619), and the Science and Technology Service program of Chinese Academy of Sciences (KFJ-STS-ZDTP-002).

Acknowledgments: We thank Kaiqin Ye for critically reading the manuscript.

Int. J. Mol. Sci. **2019**, *20*, 837

Conflicts of Interest: The authors declare no conflict of interest.

Abbreviations

ABA	Abscisic acid
MDA	Malondialdehyde
MS	Murashige and Skoog
OE	Overexpression
PEG	polyethylene glycol
qPCR	Quantitative real-time PCR
ROS	Reactive oxygen species
rpm	Round per minute
RT-PCR	Reverse transcription PCR
SL	Strigolactone
WT	Wild-type

References

1. Zvi, P.; Eduardo, B. Hormone balance and abiotic stress tolerance in crop plants. *Curr. Opin. Plant Biol.* **2011**, *14*, 290–295.
2. Raja, V.; Majeed, U.; Kang, H.; Andrabi, K.I.; John, R. Abiotic stress: Interplay between ROS, hormones and MAPKs. *Environ. Exp. Bot.* **2017**, *137*, 142–157. [CrossRef]
3. Munne-Bosch, S.; Muller, M. Hormonal cross-talk in plant development and stress responses. *Front. Plant Sci.* **2013**, *4*, 529. [CrossRef] [PubMed]
4. Jones, A.M. A new look at stress: Abscisic acid patterns and dynamics at high-resolution. *New Phytol.* **2016**, *210*, 38–44. [CrossRef]
5. Zwack, P.J.; Rashotte, A.M. Interactions between cytokinin signalling and abiotic stress responses. *J. Exp. Bot.* **2015**, *66*, 4863–4871. [CrossRef]
6. Kang, G.Z.; Li, G.Z.; Guo, T.C. Molecular mechanism of salicylic acid-induced abiotic stress tolerance in higher plants. *Acta Physiol. Plant.* **2014**, *36*, 2287–2297. [CrossRef]
7. Korver, R.A.; Koevoets, I.T.; Testerink, C. Out of shape during stress: A key role for auxin. *Trends Plant Sci.* **2018**, *23*, 783–793. [CrossRef]
8. Van Zeijl, A.; Liu, W.; Xiao, T.T.; Kohlen, W.; Yang, W.C.; Bisseling, T.; Geurts, R. The strigolactone biosynthesis gene *DWARF27* is co-opted in rhizobium symbiosis. *BMC Plant Biol.* **2015**, *15*, 260. [CrossRef]
9. Gomez-Roldan, V.; Fermas, S.; Brewer, P.B.; Puech-Pages, V.; Dun, E.A.; Pillot, J.P.; Letisse, F.; Matusova, R.; Danoun, S.; Portais, J.C.; et al. Strigolactone inhibition of shoot branching. *Nature* **2008**, *455*, 189–194. [CrossRef]
10. Yamaguchi, S.; Kyozuka, J. Branching hormone is busy both underground and overground. *Plant Cell Physiol.* **2010**, *51*, 1091–1094. [CrossRef]
11. Foo, E. Auxin influences strigolactones in pea mycorrhizal symbiosis. *J. Plant Physiol.* **2013**, *170*, 523–528. [CrossRef] [PubMed]
12. Cardinale, F.; Krukowski, P.K.; Schubert, A.; Visentin, I. Strigolactones: Mediators of osmotic stress responses with a potential for agrochemical manipulation of crop resilience. *J. Exp. Bot.* **2018**, *69*, 2291–2303. [CrossRef] [PubMed]
13. Ha, C.V.; Leyva-Gonzalez, M.A.; Osakabe, Y.; Tran, U.T.; Nishiyama, R.; Watanabe, Y.; Tanaka, M.; Seki, M.; Yamaguchi, S.; Dong, N.V.; et al. Positive regulatory role of strigolactone in plant responses to drought and salt stress. *Proc. Natl. Acad. Sci. USA* **2014**, *111*, 851–856. [CrossRef] [PubMed]
14. Bu, Q.; Lv, T.; Shen, H.; Luong, P.; Wang, J.; Wang, Z.; Huang, Z.; Xiao, L.; Engineer, C.; Kim, T.H. Regulation of drought tolerance by the F-box protein MAX2 in *Arabidopsis*. *Plant Physiol.* **2014**, *164*, 424–439. [CrossRef] [PubMed]
15. Ni, J.; Gao, C.C.; Chen, M.S.; Pan, B.Z.; Ye, K.Q.; Xu, Z.F. Gibberellin promotes shoot branching in the perennial woody plant *Jatropha curcas*. *Plant Cell Physiol.* **2015**, *56*, 1655–1666. [CrossRef] [PubMed]

16. Ni, J.; Zhao, M.L.; Chen, M.S.; Pan, B.Z.; Tao, Y.B.; Xu, Z.F. Comparative transcriptome analysis of axillary buds in response to the shoot branching regulators gibberellin A3 and 6-benzyladenine in *Jatropha curcas*. *Sci. Rep.* **2017**, *7*, 11417. [CrossRef] [PubMed]

17. Muhr, M.; Prufer, N.; Paulat, M.; Teichmann, T. Knockdown of strigolactone biosynthesis genes in *Populus* affects *BRANCHED1* expression and shoot architecture. *New Phytol.* **2016**, *212*, 613–626. [CrossRef]

18. Foster, T.M.; Ledger, S.E.; Janssen, B.J.; Luo, Z.W.; Drummond, R.S.M.; Tomes, S.; Karunairetnam, S.; Waite, C.N.; Funnell, K.A.; van Hooijdonk, B.; et al. Expression of *MdCCD7* in the scion determines the extent of sylleptic branching and the primary shoot growth rate of apple trees. *J. Exp. Bot.* **2018**, *69*, 2379–2390. [CrossRef]

19. Wang, R.; Hanna, M.A.; Zhou, W.W.; Bhadury, P.S.; Chen, Q.; Song, B.A.; Yang, S. Production and selected fuel properties of biodiesel from promising non-edible oils: *Euphorbia lathyris* L., *Sapium sebiferum* L. and *Jatropha curcas* L. *Bioresour. Technol.* **2011**, *102*, 1194–1199. [CrossRef]

20. Xu, J.S.; Chikashige, T.; Meguro, S.; Kawachi, S. Effective utilization of stillingia or Chinese tallow-tree (*Sapium sebiferum*) fruits. *Mok. Gakk.* **1991**, *37*, 494–498.

21. Xu, Z.; Mahmood, K.; Rothstein, S.J. ROS induces anthocyanin production via late biosynthetic genes and anthocyanin deficiency confers the hypersensitivity to ROS-generating stresses in *Arabidopsis*. *Plant Cell Physiol.* **2017**, *58*, 1364–1377. [CrossRef] [PubMed]

22. Aarti, P.D.; Tanaka, R.; Tanaka, A. Effects of oxidative stress on chlorophyll biosynthesis in cucumber (*Cucumis sativus*) cotyledons. *Physiol. Plant.* **2010**, *128*, 186–197. [CrossRef]

23. Liu, C.; Xu, Y.; Feng, Y.; Long, D.; Cao, B.; Xiang, Z.; Zhao, A. Ectopic expression of mulberry G-Proteins alters drought and salt stress tolerance in tobacco. *Int. J. Mol. Sci.* **2018**, *20*, 89. [CrossRef] [PubMed]

24. Wang, X.; Gao, F.; Bing, J.; Sun, W.; Feng, X.; Ma, X.; Zhou, Y.; Zhang, G. Overexpression of the *Jojoba aquaporin* gene, *ScPIP1*, enhances drought and salt tolerance in transgenic *Arabidopsis*. *Int. J. Mol. Sci.* **2019**, *20*, 153. [CrossRef] [PubMed]

25. Wang, J.; Chen, G.; Zhang, C. The effects of water stress on soluble protein content, the activity of SOD, POD and CAT of two ecotypes of reeds (*Phragmites communis*). *Acta Bot. Boreal.-Occident. Sin.* **2002**, *22*, 561–565.

26. Wang, C.-T.; Ru, J.-N.; Liu, Y.-W.; Li, M.; Zhao, D.; Yang, J.-F.; Fu, J.D.; Xu, Z.-S. Maize *WRKY* transcription factor *ZmWRKY106* confers drought and heat tolerance in transgenic plants. *Int. J. Mol. Sci.* **2018**, *19*, 3046. [CrossRef] [PubMed]

27. Shen, H.; Zhu, L.; Bu, Q.Y.; Huq, E. MAX2 affects multiple hormones to promote photomorphogenesis. *Mol. Plant* **2012**, *5*, 750–762. [CrossRef]

28. Ni, J.; Shah, F.A.; Liu, W.; Wang, Q.; Wang, D.; Zhao, W.; Lu, W.; Huang, S.; Fu, S.; Wu, L. Comparative transcriptome analysis reveals the regulatory networks of cytokinin in promoting the floral feminization in the oil plant *Sapium sebiferum*. *BMC Plant Biol.* **2018**, *18*, 96. [CrossRef]

29. Wang, Y.Q.; Peng, D.; Zhang, L.; Tan, X.F.; Yuan, D.Y.; Liu, X.M.; Zhou, B. Overexpression of *SsDGAT2* from *Sapium sebiferum* (L.) roxb increases seed oleic acid level in *Arabidopsis*. *Plant Mol. Biol. Rep.* **2016**, *34*, 638–648.

30. Fu, R.; Zhang, Y.; Guo, Y.; Chen, F. Chemical composition, antioxidant and antimicrobial activity of Chinese tallow tree leaves. *Ind. Crop Prod.* **2015**, *76*, 374–377. [CrossRef]

31. Divi, U.K.; Zhou, X.R.; Wang, P.H.; Butlin, J.; Zhang, D.M.; Liu, Q.; Vanhercke, T.; Petrie, J.R.; Talbot, M.; White, R.G.; et al. Deep sequencing of the fruit transcriptome and lipid accumulation in a non-seed tissue of Chinese tallow, a potential biofuel crop. *Plant Cell Physiol.* **2016**, *57*, 125–137. [CrossRef] [PubMed]

32. Wang, X.; Luo, X.Y. Study on herbicidal activities of different organs of *Sapium sebiferum*. *Weed Sci.* **2011**, *2011*, 4.

33. Zhu, W.; Li, X. Stress resistance of *Sapium sebiferum* and its forestation at wind gap. *Prot. For. Sci. Technol.* **2017**, *2017*, 10.

34. El-Esawi, M.A.; Alayafi, A.A. Overexpression of rice *Rab7* gene improves drought and heat tolerance and increases grain yield in rice (*Oryza sativa* L.). *Genes* **2019**, *10*, 56. [CrossRef] [PubMed]

35. Polle, A.; Chen, S.L.; Eckert, C.; Harfouche, A. Engineering drought resistance in forest trees. *Front. Plant Sci.* **2019**, *9*, 18. [CrossRef]

36. Zwanenburg, B.; Blanco-Ania, D. Strigolactones: New plant hormones in the spotlight. *J. Exp. Bot.* **2018**, *69*, 2205–2218. [CrossRef] [PubMed]

37. Waters, M.T.; Gutjahr, C.; Bennett, T.; Nelson, D.C. Strigolactone signaling and evolution. *Annu. Rev. Plant Biol.* **2017**, *68*, 291–322. [CrossRef] [PubMed]

38. Mostofa, M.G.; Li, W.; Nguyen, K.H.; Fujita, M.; Lam-Son Phan, T. Strigolactones in plant adaptation to abiotic stresses: An emerging avenue of plant research. *Plant Cell Environ.* **2018**, *41*, 2227–2243. [CrossRef]

39. Stirnberg, P.; Furner, I.J.; Ottoline Leyser, H.M. MAX2 participates in an SCF complex which acts locally at the node to suppress shoot branching. *Plant J.* **2010**, *50*, 80–94. [CrossRef]

40. Stirnberg, P.; Van, D.S.K.; Leyser, H.M. MAX1 and MAX2 control shoot lateral branching in *Arabidopsis*. *Development* **2002**, *129*, 1131–1141.

41. An, J.-P.; Li, R.; Qu, F.-J.; You, C.-X.; Wang, X.-F.; Hao, Y.-J. Apple F-Box protein MdMAX2 regulates plant photomorphogenesis and stress response. *Front. Plant Sci.* **2016**, *7*, 1685. [CrossRef] [PubMed]

42. Takasaki, H.; Maruyama, K.; Kidokoro, S.; Ito, Y.; Fujita, Y.; Shinozaki, K.; Yamaguchsi-Shinozaki, K.; Nakashima, K. The abiotic stress-responsive NAC-type transcription factor OsNAC5 regulates stress-inducible genes and stress tolerance in rice. *Mol. Genet. Genom.* **2010**, *284*, 173–183. [CrossRef] [PubMed]

43. Cho, S.K.; Kim, J.E.; Park, J.A.; Eom, T.J.; Kim, W.T. Constitutive expression of abiotic stress-inducible hot pepper *CaXTH3*, which encodes a xyloglucan endotransglucosylase /hydrolase homolog, improves drought and salt tolerance in transgenic *Arabidopsis* plants. *FEBS Lett.* **2006**, *580*, 3136–3144. [CrossRef] [PubMed]

44. Li, P.; Li, Y.J.; Wang, B.; Yu, H.M.; Li, Q.; Hou, B.K. The *Arabidopsis UGT87A2*, a stress-inducible family 1 glycosyltransferase, is involved in the plant adaptation to abiotic stresses. *Physiol. Plant.* **2016**, *159*, 416–432. [CrossRef] [PubMed]

45. Keunen, E.; Remans, T.; Bohler, S.; Vangronsveld, J.; Cuypers, A. Metal-induced oxidative stress and plant mitochondria. *Int. J. Mol. Sci.* **2011**, *12*, 6894–6918. [CrossRef]

46. Roy, S.J.; Tucker, E.J.; Tester, M. Genetic analysis of abiotic stress tolerance in crops. *Curr. Opin. Plant Biol.* **2011**, *14*, 232–239. [CrossRef] [PubMed]

47. Nguyen, H.-C.; Lin, K.-H.; Ho, S.-L.; Chiang, C.-M.; Yang, C.-M. Enhancing the abiotic stress tolerance of plants: From chemical treatment to biotechnological approaches. *Physiol. Plant.* **2018**, *164*, 452–466. [CrossRef] [PubMed]

48. Eryılmaz, F. The relationships between salt stress and anthocyanin content in higher plants. *Biotechnol. Biotechnol. Equip.* **2006**, *20*, 47–52. [CrossRef]

49. Naing, A.H.; Il Park, K.; Ai, T.N.; Chung, M.Y.; Han, J.S.; Kang, Y.W.; Lim, K.B.; Kim, C.K. Overexpression of snapdragon *Delila* (*Del*) gene in tobacco enhances anthocyanin accumulation and abiotic stress tolerance. *BMC Plant Biol.* **2017**, *17*, 65. [CrossRef] [PubMed]

50. Lotkowska, M.E.; Tohge, T.; Fernie, A.R.; Xue, G.P.; Balazadeh, S.; Muellerroeber, B. The *Arabidopsis* transcription factor *MYB112* promotes anthocyanin formation during salinity and under high light stress. *Plant Physiol.* **2015**, *169*, 1862–1880. [CrossRef]

51. Castellarin, S.D.; Pfeiffer, A.; Sivilotti, P.; Degan, M.; Peterlunger, E.; Di Gaspero, G. Transcriptional regulation of anthocyanin biosynthesis in ripening fruits of grapevine under seasonal water deficit. *Plant Cell Environ.* **2007**, *30*, 1381–1399. [CrossRef] [PubMed]

52. Na, Y.W.; Jeong, H.J.; Lee, S.Y.; Choi, H.G.; Kim, S.H.; Rho, I.R. Chlorophyll fluorescence as a diagnostic tool for abiotic stress tolerance in wild and cultivated strawberry species. *Hort. Environ. Biotech.* **2014**, *55*, 280–286. [CrossRef]

53. Nuccio, M.L.; Rhodest, D.; McNeil, S.D.; Hanson, A.D. Metabolic engineering of plants for osmotic stress resistance. *Curr. Opin. Plant Biol.* **1999**, *2*, 128–134. [CrossRef]

54. Wani, S.H.; Gosal, S.S. Genetic engineering for osmotic stress tolerance in plants—Role of proline. *J. Genet. Evol.* **2011**, *3*, 14–25.

55. Shi, H.; Wang, X.; Tan, D.X.; Reiter, R.J.; Chan, Z. Comparative physiological and proteomic analyses reveal the actions of melatonin in the reduction of oxidative stress in Bermuda grass (*Cynodon dactylon* (L). Pers.). *J. Pineal Res.* **2015**, *59*, 120–131. [CrossRef] [PubMed]

56. Ni, J.; Wang, Q.; Shah, F.A.; Liu, W.; Wang, D.; Huang, S.; Fu, S.; Wu, L. Exogenous melatonin confers cadmium tolerance by counterbalancing the hydrogen peroxide homeostasis in wheat seedlings. *Molecules* **2018**, *23*, 799. [CrossRef] [PubMed]

57. Seiler, C.; Rajesh, K.; Reddy, P.S.; Strickert, M.; Rolletschek, H.; Scholz, U.; Wobus, U.; Sreenivasulu, N. ABA biosynthesis and degradation contributing to ABA homeostasis during barley seed development under control and terminal drought-stress conditions. *J. Exp. Bot.* **2011**, *62*, 2615–2632. [CrossRef]

58. Vishwakarma, K.; Upadhyay, N.; Kumar, N.; Yadav, G.; Singh, J.; Mishra, R.K.; Kumar, V.; Verma, R.; Upadhyay, R.G.; Pandey, M. Abscisic acid signaling and abiotic stress tolerance in plants: A review on current knowledge and future prospects. *Front. Plant Sci.* **2017**, *8*, 161. [CrossRef] [PubMed]
59. Ren, C.G.; Kong, C.C.; Xie, Z.H. Role of abscisic acid in strigolactone-induced salt stress tolerance in arbuscular mycorrhizal *Sesbania cannabina* seedlings. *BMC Plant Biol.* **2018**, *18*, 74. [CrossRef] [PubMed]
60. Shah, F.A.; Ni, J.; Chen, J.; Wang, Q.; Liu, W.; Chen, X.; Tang, C.; Fu, S.; Wu, L. Proanthocyanidins in seed coat tegmen and endospermic cap inhibit seed germination in *Sapium sebiferum*. *Peer J.* **2018**, *6*, 10. [CrossRef] [PubMed]
61. Zhang, X.; Henriques, R.; Lin, S.S.; Niu, Q.W.; Chua, N.H. Agrobacterium-mediated transformation of *Arabidopsis thaliana* using the floral dip method. *Nat. Protoc.* **2006**, *1*, 641–646. [CrossRef] [PubMed]
62. Livak, K.J.; Schmittgen, T.D. Analysis of relative gene expression data using real-time quantitative PCR and the $2^{-\Delta\Delta Ct}$ method. *Methods* **2001**, *25*, 402–408. [CrossRef] [PubMed]
63. Adriana, P.; Gaby, T.; Sylvain, A.; Iwona, A.; Simone, M.; Thomas, M.; Karl-Hans, O.; Bernhard, K.U.; Ji-Young, Y.; Liljegren, S.J. Chlorophyll breakdown in senescent *Arabidopsis* leaves. Characterization of chlorophyll catabolites and of chlorophyll catabolic enzymes involved in the degreening reaction. *Plant Physiol.* **2005**, *139*, 52–63.
64. Cinzia, S.; Alessandra, P.; Elena, L.; Amedeo, A.; Pierdomenico, P. Sucrose-specific induction of the anthocyanin biosynthetic pathway in *Arabidopsis*. *Plant Physiol.* **2006**, *140*, 637–646.
65. Zhang, K.W.; Xia, X.Y.; Zhang, Y.Y.; Gan, S.S. An ABA-regulated and Golgi-localized protein phosphatase controls water loss during leaf senescence in *Arabidopsis*. *Plant J.* **2012**, *69*, 667–678. [CrossRef] [PubMed]
66. Yin, G.F.; Zhao, N.J.; Shi, C.Y.; Chen, S.; Qin, Z.S.; Zhang, X.L.; Yan, R.F.; Gan, T.T.; Liu, J.G.; Liu, W.Q. Phytoplankton photosynthetic rate measurement using tunable pulsed light induced fluorescence kinetics. *Opt. Express* **2018**, *26*, A293–A300. [CrossRef] [PubMed]
67. Kumar, S.; Stecher, G.; Tamura, K. MEGA7: Molecular evolutionary genetics analysis version 7.0 for bigger datasets. *Mol. Biol. Evol.* **2016**, *33*, 1870–1874. [CrossRef]

International Journal of
Molecular Sciences

MDPI

Article

Rice Overexpressing *OsNUC1-S* Reveals Differential Gene Expression Leading to Yield Loss Reduction after Salt Stress at the Booting Stage

Chuthamas Boonchai [1], Thanikarn Udomchalothorn [1,2], Siriporn Sripinyowanich [3], Luca Comai [4], Teerapong Buaboocha [5,6] and Supachitra Chadchawan [1,6,*]

[1] Center of Excellence in Environment and Plant Physiology, Department of Botany, Faculty of Science, Chulalongkorn University, Bangkok 10330, Thailand; Gwang_desu@hotmail.com (C.B.); Thanikarn@sut.ac.th (T.U.)

[2] Surawiwat School, Suranaree University of Technology, Nakhon Ratchasima 30000, Thailand

[3] Faculty of Liberal Arts and Science, Kasetsart University, Kamphaeng Saen Campus, Nakhon Pathom 73140, Thailand; wanich_s@hotmail.com

[4] Department of Plant Biology and Genome Center, University of California Davis, Davis, CA 95616, USA; lcomai@ucdavis.edu

[5] Department of Biochemistry, Faculty of Science, Chulalongkorn University, Bangkok 10330, Thailand; Teerapong.B@chula.ac.th

[6] Omics Science Center, Faculty of Science, Chulalongkorn University, Bangkok 10330, Thailand

[*] Correspondence: Supachitra.C@chula.ac.th or s_chadchawan@hotmail.com; Tel.: +66-2-2185485

Received: 17 November 2018; Accepted: 27 November 2018; Published: 7 December 2018

Abstract: Rice nucleolin (OsNUC1), consisting of two isoforms, OsNUC1-L and OsNUC1-S, is a multifunctional protein involved in salt-stress tolerance. Here, *OsNUC1-S*'s function was investigated using transgenic rice lines overexpressing *OsNUC1-S*. Under non-stress conditions, the transgenic lines showed a lower yield, but higher net photosynthesis rates, stomatal conductance, and transpiration rates than wild type only in the second leaves, while in the flag leaves, these parameters were similar among the lines. However, under salt-stress conditions at the booting stage, the higher yields in transgenic lines were detected. Moreover, the gas exchange parameters of the transgenic lines were higher in both flag and second leaves, suggesting a role for *OsNUC1-S* overexpression in photosynthesis adaptation under salt-stress conditions. Moreover, the overexpression lines could maintain light-saturation points under salt-stress conditions, while a decrease in the light-saturation point owing to salt stress was found in wild type. Based on a transcriptome comparison between wild type and a transgenic line, after 3 and 9 days of salt stress, the significantly differentially expressed genes were enriched in the metabolic process of nucleic acid and macromolecule, photosynthesis, water transport, and cellular homeostasis processes, leading to the better performance of photosynthetic processes under salt-stress conditions at the booting stage.

Keywords: RNA binding protein; nucleolin; salt stress; photosynthesis; light saturation point; booting stage; transcriptome

---

1. Introduction

Rice (*Oryza sativa* L.) is a staple food and a main source of energy for humans, especially in Asia. There are several biotic or abiotic stresses that limit rice growth and yield, such as soil salinity, drought, and soil nutrition [1]. Salinity is a severe abiotic stress worldwide that directly contributes to the economic outcome of agriculturists. It negatively affects plants at both physiological and cellular levels. The plant water absorption is disrupted, leading to reductions in plant growth and development [2]. Moreover, ion toxicity also causes changes in plant metabolism, including the photosynthesis processes and energy production [3]. It directly affects photosynthetic components, including chlorophyll *a*, chlorophyll *b*, and carotenoids, because salt stress increases enzyme activities involved in chlorophyll degradation, which leads to a decrease in chlorophyll levels [4]. Moreover, it can also induce reactive oxygen species (ROS) production, which can trigger protein and lipid damage [5]. The level of plant injury depends on species, developmental stage, age, and also the severity of the salinity.

Rice is strongly affected by salt stress at both the seedling and reproductive stages [6]. These can remarkably affect rice plants that are grown in the paddy field of rain-fed areas containing rock salt underneath, like the northeastern region of Thailand [7], because rice is normally germinated at the beginning of the rainy season, when the salinity is low due to a certain amount of rain. However, when the land is drying from the evaporation due to the lack of rain, the salinity can migrate from the rock salt underneath through the surface [7], causing salt stress at the certain developmental stage of rice. If this occurs at the reproductive stage, it becomes the major effect for plant yield reduction [8].

Nucleolin, a multifunctional protein, is localized in various cellular locations, including the nucleus. It consists of three domains, the N-terminus, containing several acidic stretches, the central region, containing an RNA-binding domain, and the C-terminus, containing a glycine/arginine-rich domain [9]. Because of differences in the numbers of each motif, different functions were found in different species. For example, the *NUC-L1* gene in *Arabidopsis thaliana* is related to its growth and development [10], whereas in pea, this gene is regulated by light [11].

In rice, two forms of Nucleolin1 were found, a longer (*OsNUC1-L*, GenBank Accession No. AK103446) and a shorter (*OsNUC1-S*, GenBank Accession No. AK063918) form. *OsNUC1-S* lacks an N-terminal region, but the cDNA sequences of the other regions are the same as the longer form. In 2013, Sripinyowanich and colleagues studied *OsNUC1-S* overexpression in *Arabidopsis* and found that this form promoted salt tolerance by increasing the number of lateral roots and enhancing root growth. In addition, the overexpression of this gene in rice leads to shoot fresh weight increasing when seedlings are grown under salt-stress conditions [12]. In the reproductive stage, the function of this gene is still unknown. Recently, the overexpression of the *OsNUC1-L* form resulted in an increase in salt tolerance in both rice and *Arabidopsis* by enhancing photosynthesis [13].

Here, we investigate the effects of *OsNUC1-S* overexpression on photosynthetic responses under normal and salt-stress conditions in transgenic rice at the reproductive stage and use a transcriptome analysis to explore the gene expression levels affected by *OsNUC1-S* overexpression. The experiments were performed in flag leaves and second leaves as both leaves have been reported to have a major role in generating carbohydrate resource from the photosynthesis process for seed production [14].

2. Results

2.1. Overexpression of OSNUC1-S Affects Rice Yield

The transgenic rice lines, TOSL1, TOSL2, and TOSL3, with *OsNUC1-S* overexpression were used in these experiments. *OsNUC1* expression was compared among wild type (WT) and the transgenic lines, when grown in the control and salt stress condition at reproductive stage. Significantly higher *OsNUC1* gene expression could be detected in transgenic lines when compared to WT, especially in the salt stress condition (Figure 1).

To investigate if the overexpression of *OsNUC1-S* affected rice productivity, the tiller number per plant, panicle number per plant, panicle length, fertility rate (%), and seed number per plant of the transgenic rice, TOSL1, TOSL2, TOSL3, and WT, were evaluated as shown in Table 1. In the normal grown condition (control), over-expression of *OsNUC1-S* caused the reduction in % fertility, leading to the reduction in seed number per plant. However, under salt stress, the transgenic lines tended to have the higher tiller number per plant, panicle number per plant, panicle length, % fertility, and seed number per plant, except the TOSL2 line that showed similar % fertility and seed number per plant to WT. The reduction percentage of seeds/plants in WT was 64%, while the transgenic lines had the seeds/plant reduction of 17%, 47%, and 27%. These data suggested that there should be some changes in metabolisms in the transgenic lines due to *OsNUC1-S* over-expression.

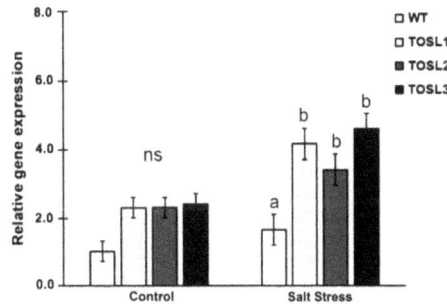

Figure 1. *OsNUC1* gene expression in wild type (WT) and *OsNUC1-S* over-expression lines, TOSL1, TOSL2, and TOSL3, in the control and salt stress condition. Analysis of variance was performed and means were compared with Tukey's range test analysis. The data were presented as the mean $\pm$ SE and a different letter above the bar showed the significant difference in means ($p < 0.05$). ns represents no statistically difference among means.

Table 1. Effects of salt stress at reproductive stage on tiller number per plant, panicle number per plant, panicle length, % fertility, and seed number per plant.

Reproductive Characters	Control *				Salt Stress *			
	WT	TOSL1	TOSL2	TOSL3	WT	TOSL1	TOSL2	TOSL3
Tiller number/pl	5.75 $\pm$ 0.55 ab	5.25 $\pm$ 0.55 abc	7.00 $\pm$ 0.55 a	5.50 $\pm$ 0.55 abc	3.00 $\pm$ 0.55 c	3.75 $\pm$ 0.55 bc	4.50 $\pm$ 0.55 abc	4.00 $\pm$ 0.55 bc
Panicle number/pl	5.25 $\pm$ 0.33 a	3.50 $\pm$ 0.33 b	3.50 $\pm$ 0.33 b	3.00 $\pm$ 0.33 bc	1.75 $\pm$ 0.33 c	2.50 $\pm$ 0.33 bc	2.75 $\pm$ 0.33 bc	2.00 $\pm$ 0.33 bc
Panicle length	10.15 $\pm$ 0.71	9.28 $\pm$ 0.71	8.87 $\pm$ 0.71	8.87 $\pm$ 0.71	7.38 $\pm$ 0.71	8.27 $\pm$ 0.71	9.00 $\pm$ 0.71	9.73 $\pm$ 0.71
% Fertility	65.36 $\pm$ 9.52 a	45.18 $\pm$ 9.52 ab	48.39 $\pm$ 9.52 ab	44.34 $\pm$ 9.52 ab	15.27 $\pm$ 9.52 b	47.03 $\pm$ 9.52 ab	14.17 $\pm$ 9.52 b	36.84 $\pm$ 9.52 ab
Seeds per plant	193.50 $\pm$ 11.15 a	108.75 $\pm$ 11.15 bc	130.00 $\pm$ 11.15 b	115.75 $\pm$ 11.15 bc	70.25 $\pm$ 11.15 c	89.75 $\pm$ 11.15 bc	68.50 $\pm$ 11.15 c	84.50 $\pm$ 11.15 bc

* The experiment was performed with random complete block design in four replicates, each of which consisted of two individuals. Analysis of variance was performed and means were compared with Tukey's range test analysis. The data were presented as the mean $\pm$ SE and a different letter above the bar showed the significant difference in means ($p < 0.05$).

2.2. Overexpression of OSNUC1-S Increased the Photosynthetic Rate, Stomatal Conductance, and Transpiration Rate under Salt-Stress Conditions

Salt stress can cause a decrease in productivity in rice, and the major organs generating the carbohydrate for grain filling are flag leaves and second leaves. Therefore, we investigated the photosynthetic activities in both of these leaves in WT and *OsNUC1-S* over-expressing lines under control and salt stress conditions. Under the control condition, only second leaves of transgenic rice overexpressing *OSNUC1-S* showed higher net photosynthetic rates (P_N) and stomatal conductance levels (g_s), while the flag leaves had similar levels, except TOSL2, which had a lower P_N than the other lines. These results may reflect positional effects of the transformation. After 9 days of salt stress, the overexpression of *OsNUC1-S* increased the P_N and g_s of all the transgenic lines (Figure 2). The P_N values in flag leaves of transgenic plants were approximately two-fold those of the WT (Figure 2A), while in second leaves, the P_N values of transgenic plants increased up to 2.5-fold those of WT (Figure 2B). Similar effects were also found for the g_s values of both flag and second leaves

(Figure 2C,D). There was no effect on the intercellular CO_2 concentration (Ci) (Figure 2E,F), but *OsNUC1-S*'s overexpression resulted in an increased transpiration rate of the second leaves under control conditions and in the flag leaves under salt-stress conditions (Figure 2G,H).

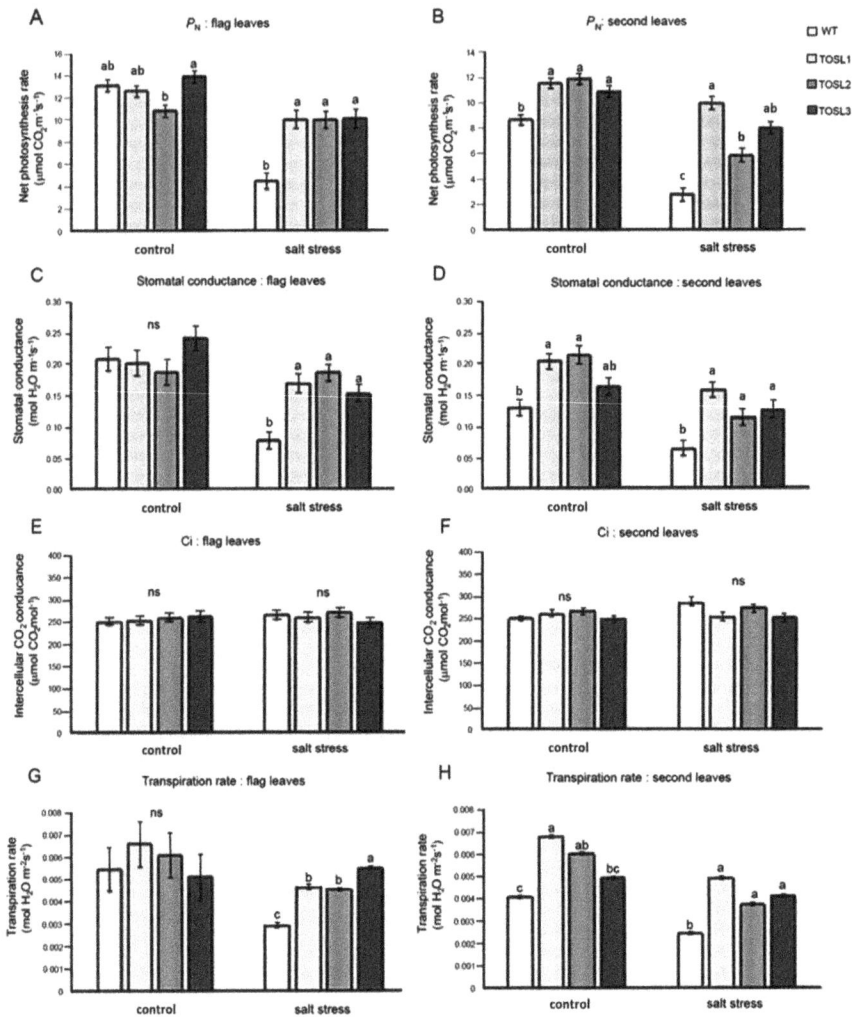

Figure 2. The net photosynthetic rate (P_N) (**A,B**), stomatal conductance (g_s) (**C,D**), intercellular CO_2 concentration (Ci) (**E,F**), and transpiration rate (*E*) (**G,H**) of flag leaves (**A,C,E,G**) and second leaves (**B,D,F,H**), when wild type (WT) and the *OsNUC1–S* overexpressing lines, TOSL1, TOSL2, and TOSL3, were grown under control or salt-stress conditions. The data were presented as the mean ± SE and a different letter above the bar showed the significant difference in means ($p < 0.05$) based on Tukey's range test analysis. ns represents no statistically difference among means.

2.3. Overexpression of OSNUC1-S Affected Both Light-Response Curves and CO_2-Response Curves of Flag and Second Leaves under Salt-Stress Conditions

We investigated the light-response curves of these lines. The light-response curves of the flag leaves in all the lines were similar when the light intensity varied from 100 to 2000 $\mu mol \cdot m^{-2} \cdot s^{-1}$, except TOSL2, which had a lower P_N than other lines. The light-saturation points were approximately

1000 $\mu mol \cdot m^{-2} \cdot s^{-1}$ in all lines (Figure 3A). The salt stress caused decreases in the P_N values of all the lines, but the transgenic lines had significantly higher P_N values when the light intensity was greater than 200 $\mu mol \cdot m^{-2} \cdot s^{-1}$. The light-saturation point of the WT decreased to 600 $\mu mol \cdot m^{-2} \cdot s^{-1}$, while the transgenic lines had light-saturation points similar to those of plants grown under control conditions (1000 $\mu mol \cdot m^{-2} \cdot s^{-1}$) (Figure 3B).

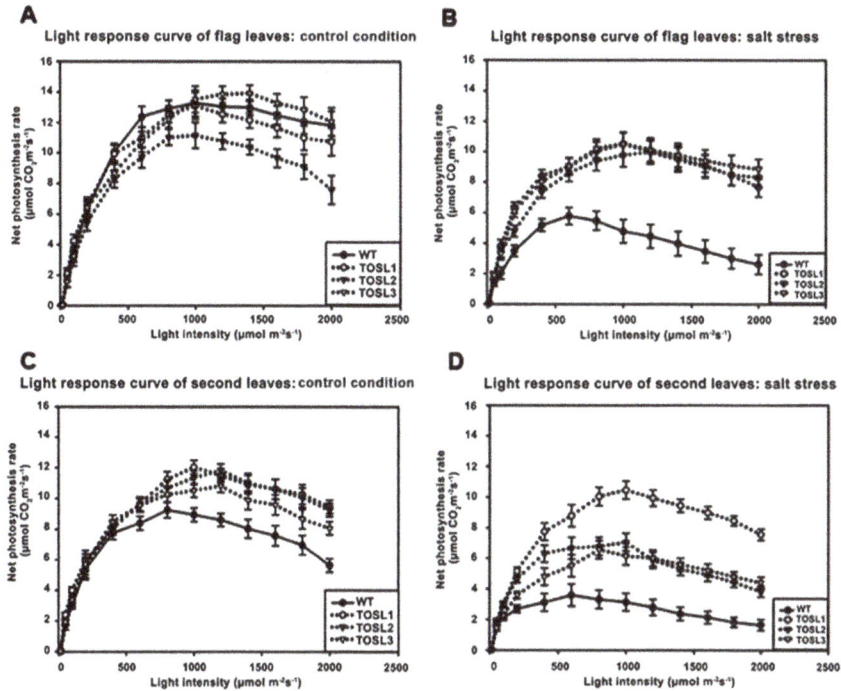

Figure 3. Light-response curves of flag leaves (**A,B**) and second leaves (**C,D**), when wild type and transgenic lines, TOSL1, TOSL2, and TOSL3, were grown under control (**A,C**) and salt-stress (**B,D**) conditions.

The second leaves of the WT had lower light-saturation points than those of the transgenic lines when grown under control conditions. The P_N values of the WT's second leaves started to decline when the light intensity was greater than 800 $\mu mol \cdot m^{-2} \cdot s^{-1}$; however, for the transgenic lines, the light-saturation point was 1000 $\mu mol \cdot m^{-2} \cdot s^{-1}$ (Figure 3C).

Salt stress caused a decrease in the light-saturation point in WT second leaves, but it did not affect the light-saturation points of the transgenic lines' second leaves. In WT second leaves, the light-saturation point declined to 600 $\mu mol \cdot m^{-2} \cdot s^{-1}$, and a more than two-fold reduction in the P_N was found. A similar reduction in the P_N was found in the transgenic lines, except TOSL1, which maintained a P_N in its second leaves that was similar to the P_N under control growth conditions. However, all the transgenic lines' second leaves had light-saturation points of 1000 $\mu mol \cdot m^{-2} \cdot s^{-1}$ (Figure 3D).

The CO_2-response curves of flag and second leaves of all the lines were also investigated as the CO_2 concentration increased from 200 to 1000 $\mu mol \cdot mol^{-1}$. Under control conditions, the flag leaves of all the lines showed similar CO_2-response curves. The CO_2-saturation point was ~800 $\mu mol \cdot mol^{-1}$ (Figure 4A). Under salt-stress conditions, significantly higher P_N values of the flag leaves were found in all the transgenic lines when the CO_2 concentration was greater than 200 $\mu mol \cdot mol^{-1}$. However, salt stress had no effect on the CO_2-saturation points of flag leaves in any line (Figure 4B).

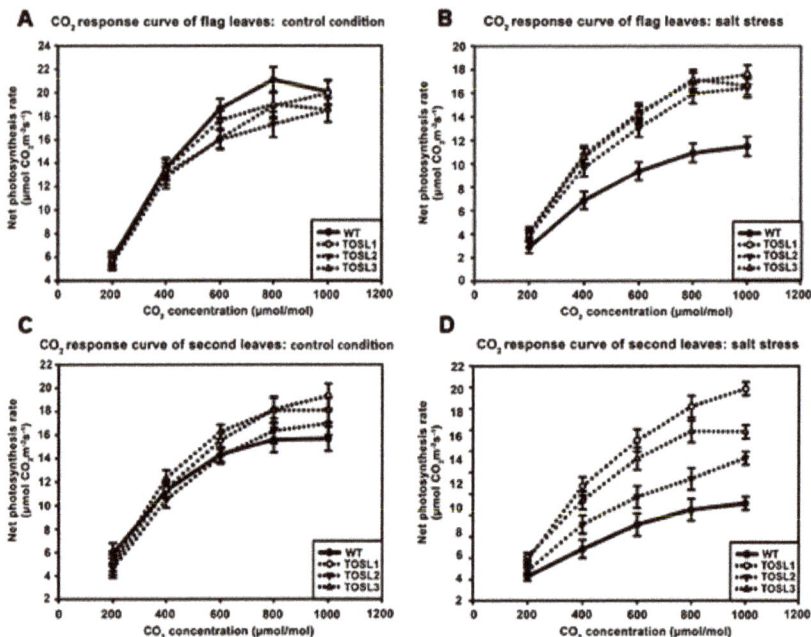

Figure 4. CO_2-response curves of flag leaves (**A,B**) and second leaves (**C,D**), when wild type and transgenic lines, TOSL1, TOSL2, and TOSL3, were grown under control (**A,C**) and salt-stress (**B,D**) conditions.

In the second leaves, the CO_2-response curves and CO_2-saturation points of all the lines were similar and consistent with the flag leaf's response to the lower P_N in plants grown under control conditions (Figure 4C). Salt stress caused decreases in the P_N values of all the lines, but it did not affect the CO_2 saturation point of WT second leaves. On the contrary, TOSL1 and TOSL2's second leaves had increased CO_2 saturation points to over 1000 mmol·mol^{-1}, while TOSL3 showed the same CO_2 saturation point as WT second leaves, when grown under salt-stress conditions (Figure 4D).

2.4. Salt Stress Affects Photosystem II (PSII) Photochemistry Efficiency and Photosynthetic Pigment Contents in Flag and Second Leaves

To investigate the effect of salt stress on the efficiency of PSII photochemistry, the F_v/F_m ratio was investigated. The salt-stress level did not significantly affect the PSII efficiency in the flag leaves of any lines (Figure 5A). However, in the second leaves, a significant reduction in the PSII efficiency (F_v/F_m) was found in WT after 9 days of salt treatment, while in all the transgenic lines, a reduced effect was found (Figure 5B). Thus, second leaves were more susceptible to salt stress than flag leaves, and *OsNUC1-S*'s overexpression contributed to the PSII photochemistry efficiency under salt-stress conditions.

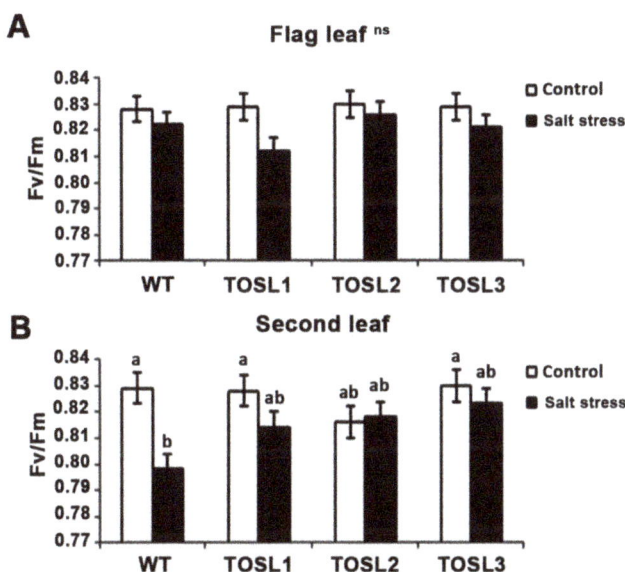

Figure 5. The F_v/F_m ratios of flag leaves (**A**) and second leaves (**B**) under control conditions and after being treated with 150 mM NaCl solution for 9 days. Three independent *OsNUC1-S* transgenic rice lines with difference transgene expression levels and WT plants were used. The data were presented as the mean ± SE and a different letter above the bar showed the significant difference in means ($p < 0.05$) based on Tukey's range test analysis. ns represents no statistically difference among means.

OsNUC1-S's overexpression tended to increase both the chlorophyll and carotenoid contents in flag leaves, but not in the second leaves, under optimal growth conditions (Figure 6A–F). Salt stress caused decreases in all of the pigments in the flag leaves, but significantly higher carotenoid contents were found in both flag and second leaves in the transgenic lines when compared with WT (Figure 6E,F).

2.5. OsNUC1-S's Overexpression Increased Carbohydrate Metabolism and Sugar Transport in Flag Leaves of Rice Grown Under Control Conditions

Because of the effects of *OsNUC1-S*'s overexpression on the photosynthetic characteristics, the transcriptome approach was used to investigate changes at the transcript level. TOSL3 was chosen as the representative for investigations of the flag leaf transcriptome. Based on a transcriptome comparison of DEGs (differentially expressed genes) between WT and TOSL3, when grown under control conditions, the DEGs were highly enriched in the cellular macromolecule metabolic processes, including the macromolecule biosynthetic process. Genes involved in transmembrane transport, regulation of cellular process, and the pigment metabolic process were also found.

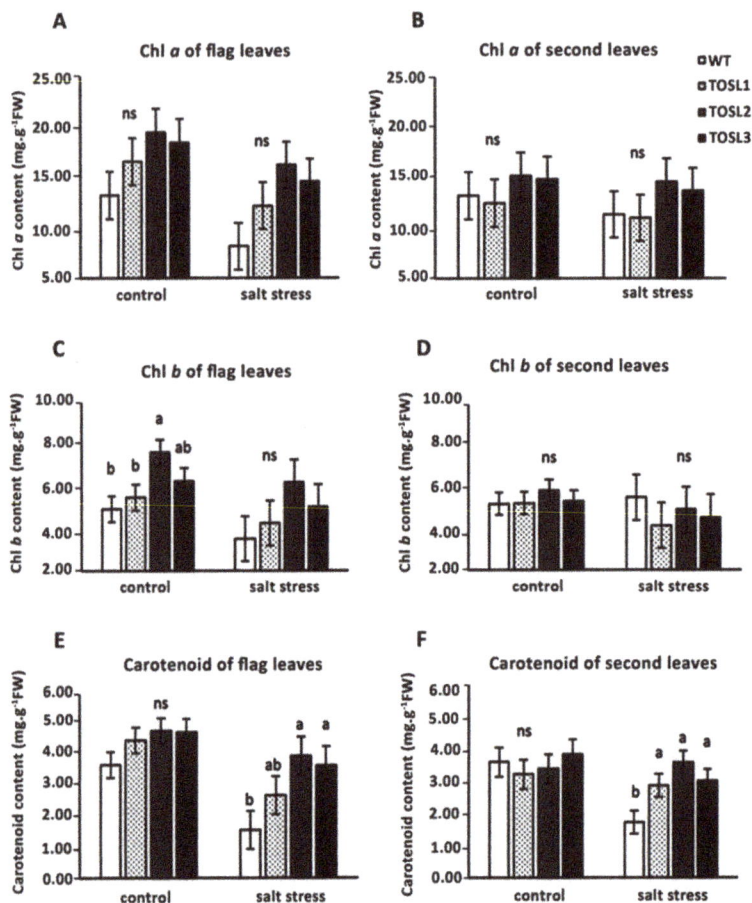

Figure 6. Photosynthetic pigments in flag leaves (**A,C,E**) and second leaves (**B,D,F**) 9 days after the salt-stress treatment at the booting stage. The data were presented as the mean ± SE and a different letter above the bar showed the significant difference in means ($p < 0.05$) based on Tukey's range test analysis. ns represents no statistically significant difference among means.

For the cellular component enrichment, a plasma membrane and chloroplast envelope were reported. This supported *OsNUC1-S*'s role in the enhancement of macromolecule production for grain filling in the flag leaves. Interestingly, the molecular function enrichment was found for only linoleate 13S-lipoxygenase activity. This enzyme is involved in plant growth and development [15], and also in the wounding response through jasmonic acid (JA) signaling [16,17] (Figure 7A).

Six days later, the DEGs resulting from *OsNUC1-S* expression changed. For the biological process, the genes functioning in carbohydrate transport were enriched, supporting the role of flag leaves in seed development. The consistent enrichment of molecular function in substrate-specific transmembrane transport was found (Figure 7B).

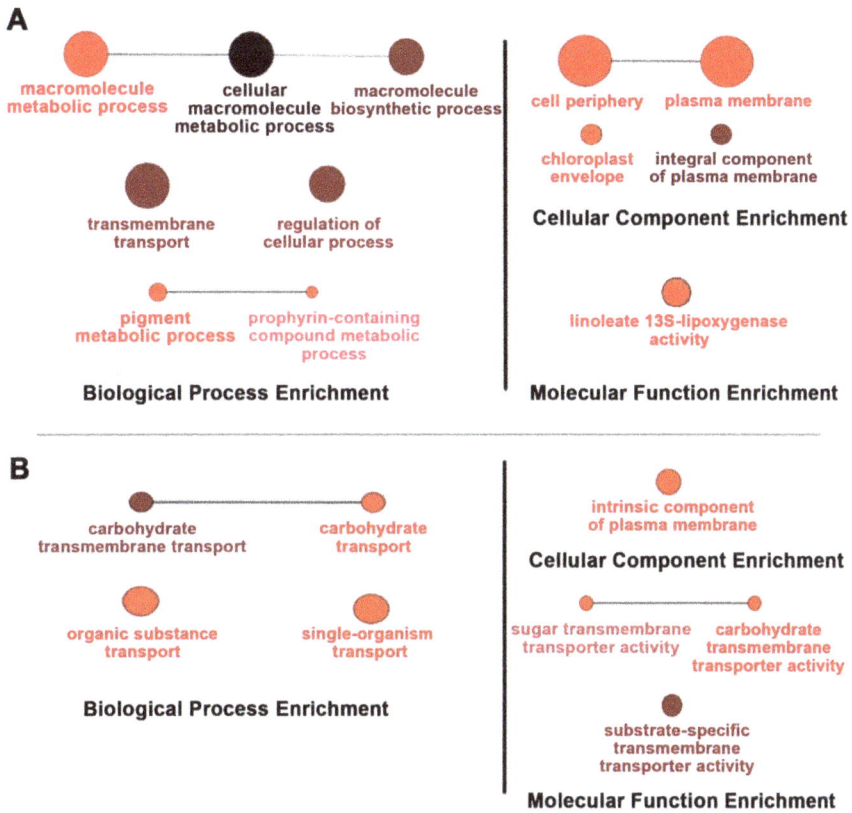

Figure 7. Gene enrichment analysis of the differentially expressed genes in the TOSL3 line, which overexpresses *OsNUC1-S*. The tissues for the transcriptome analysis were collected 3 days (**A**) and 9 days (**B**) after the first day of the booting stage. The darker colors represent the higher significance and the larger size of node represents the higher number of genes in the group.

2.6. The Overexpression of OsNUC1-S Increased the Expression of Genes Involved in Water Transport and Cellular Homeostasis, Nucleic Acid and Macromolecule Metabolic Processes, and Photosynthetic Processes

Under salt-stress conditions, the effects of *OsNUC1-S* overexpression were different from the effects found in normally grown plants. After 3 days of salt stress, the flag leaves of the transgenic lines were enriched with transcripts of genes involved in water transport and cellular homeostasis, nucleic and macromolecule metabolic processes, and photosynthetic processes. This was also consistent with the cellular enrichment in chloroplasts and nuclei. For the KEGG pathways, enrichment occurred in photosynthesis, carbon fixation, porphyrin and chlorophyll metabolism, and carotenoid biosynthesis, all of which involved activities in chloroplasts. Moreover, genes in glyoxylate and dicarboxylate metabolism, and terpenoid backbone biosynthesis, were also enriched (Figure 8).

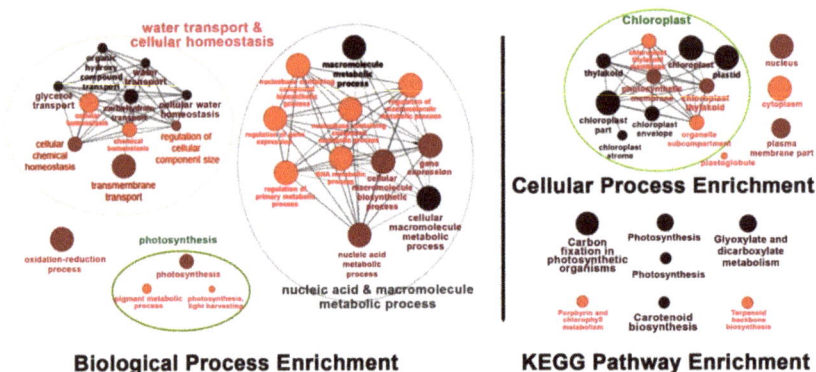

Figure 8. The enrichment for the biological process, cellular process, and Kyoto Encyclopedia of Genes and Genomes (KEGG) pathways of the differentially expressed genes in the flag leaves of transgenic rice overexpressing *OsNUC1-S* after 3 days of salt stress. The darker colors represent the higher significance and the larger size of node represents the higher number of genes in the group.

When the plants were under a prolonged stress for 9 days, similar processes, cellular compartments, and metabolic pathways were found to be affected, except that pigment and porphyrin-containing compound metabolic processes were detected, suggesting more pigment synthesis-related processes occurred, while the enrichment in light harvesting was not found. A new set of enriched genes found at this time point was in carboxylic acid metabolic processes and responses to oxidative stress. Chloroplasts were the main organelles that were enriched with the transcripts of genes affected by *OsNUC1-S* expression. The enrichment of the genes in the nucleus and plasma membrane also had a similar pattern to that found in flag leaves after 3 days of salt stress. The pathway enrichment after salt stress for 9 days was similar to that found after 3 days of stress, with more pathways in amino sugar and nucleotide sugar metabolism, pyruvate metabolism, ascorbate and aldarate metabolism, and pentose phosphate pathways (Figure 9). The list of the differentially expressed genes between wild type and transgenic rice with *OsNUC1-S* overexpression after salt stress treatment for 3 and 9 days is shown in Supplementary Table S1.

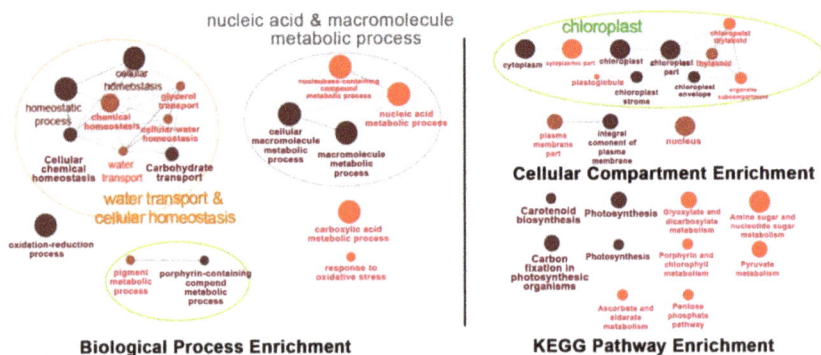

Figure 9. The enrichment for biological process, cellular process, and KEGG pathways of the differentially expressed genes in the flag leaves of transgenic rice overexpressing *OsNUC1-S* after 9 days of salt stress. The darker colors represent the higher significance and the larger size of the node represents the higher number of genes in the group.

To validate the RNA-Seq data, the up-regulated genes after 3 days of salt stress, *LOC_Os01g58470* (*CEST*) (Figure 10A), *LOC_Os03g39610* (*CAB*) (Figure 10B), and *LOC_Os04g33830* (*PSAO*) (Figure 10C), were chosen as representatives for quantitative RT-PCR (qRT-PCR). The expression in the transgenic line, TOSL3 was about two- to three-fold higher than WT under salt stress condition, based on RNA-Seq analysis (Supplementary Table S1). The comparable expression was also detected by qRT-PCR as shown in Figure 10A–C. Based on RNA-Seq analysis, *LOC_Os01g64960* (*PsbS1*) (Figure 10D) showed the highest fold change when compared to WT after salt stress for 9 days. The qRT-PCR of this gene expression revealed about five-fold higher expression than WT after 9 days (Figure 10D). The increase in the expression of these genes was correlated with the RNA-Seq data.

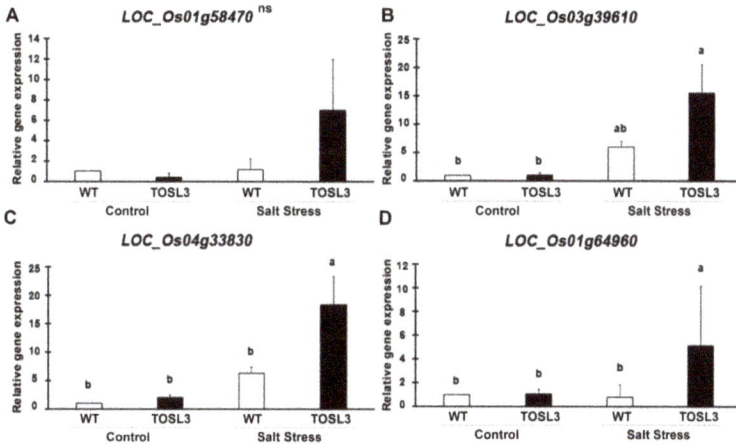

Figure 10. Gene expression analysis of *LOC_Os01g58470* (**A**), *LOC_Os03g39610* (**B**), *LOC_Os04g33830* (**C**), and *LOC_Os01g64960* (**D**) in WT and TOSL3 plants under control and salt stress conditions. Bar represents standard deviation of three biological replicates. The measurement was performed with three technical replicates. The data were presented as the mean $\pm$ SE and a different letter above the bar showed the significant difference in means ($p < 0.05$) based on Tukey's range test analysis. ns represents no statistically significant difference among means.

3. Discussion

OsNUC1-S, one of the mRNA splice forms, lacks an N-terminal region that contains several acidic stretches and a nuclear localization signal. However, based on the localization study [10], the protein localizes to both the cytoplasm and nucleus. The overexpression of *OsNUC1-S* did not change the photosynthetic activity levels in flag leaves when plants were grown under control conditions (Figure 2), but it resulted in reduced rice seed numbers and fertility rates (Table 1). The transcriptomic analysis found increased levels of carbohydrate metabolism and sugar transmembrane transport (Figure 7). This was in agreement with the role of flag leaves as the energy source for rice grain development [18]. Interestingly, TOSL3 was enriched in linoleate 13S-lipoxygenase activity (Figure 7). Lipoxygenase is the enzyme involved in JA biosynthesis and in responding to wounding and stress [17]. The exogenous application of JA induces flag leaf senescence by regulating chlorophyll degradation, membrane deterioration, and SAG (senescence associated gene) expression levels [19]. Moreover, the overexpression of the *TIFY* gene could enhance the rice grain yield, possibly owing to a reduction in JA sensitivity [20]. Therefore, the increase in lipoxygenase activity in the *OsNUC1-S* overexpression line could increase the JA level, causing decreases in the rice grain yield and fertility rate (Table 1). Both flag and second leaves contribute to the grain-filling process [14], but flag leaves provide more than 50% of the assimilates for grain filling [21]. In a comparison between WT and transgenic lines, a greater P_N was found only in the second leaves of the transgenic lines when grown under salt-stress conditions. This may be related to the maximization of the photosynthetic capacity of the flag leaves.

Under salt stress, a threefold reduction of P_N was detected in WT, while only a 10–15% reduction was found in the flag leaves of transgenic lines (Figure 2A). This suggested a role for *OsNUC1-S* in photosynthetic enhancement. The increase in photosynthetic activity was supported by the enrichment of genes involved in the photosynthetic processes, as well as in water transport and cellular homeostasis activities (Figure 8). The enrichment in nucleic acid and macromolecule metabolic processes supports a role for nucleolin that involves RNA modifications. Thus, OsNUC1-S may have a specific target for its activity, which results in greater changes in the expression levels of some genes compared with others when *OsNUC1-S* is overexpressed. The enrichment in carbon fixation, as well as porphyrin and chlorophyll metabolism, suggests that *OsNUC1-S* enhanced both the light reaction and the carbon fixation process. The increased light-saturation point (Figure 3) and carbon fixation activity (Figure 4) were consistent with the transcriptome data. Interestingly, when plants were impacted by salt, the function of *OsNUC1-S* changed. Both the transcriptome and physiological data indicated that this gene promoted salt tolerance through enhanced photosynthesis. Genes encoding proteins in the photosynthetic processes were strongly expressed in transgenic lines compared with WT during salt stress (Supplementary Table S1, Figure 10). Most of the up-regulated genes were located in PSI and PSII of the light reaction. Moreover, some genes in the Calvin cycle also increased, possibly causing the higher P_N values in transgenic plants. The enriched genes encoding chlorophyll A-B-binding protein (Supplementary Table S1), which functions in light harvesting, were also up-regulated in transgenic plants. This supported the light-response curve results in which a higher light-saturation point occurred in transgenic plants under salt stress.

The second leaves in WT were more susceptible to salt stress than flag leaves. The former had a significant decrease in F_v/F_m after 9 days of salt stress, while salt stress had no effect on flag leaves. Na^+ can be transferred from root to shoot, and in rice, *OsHKT1;5* excludes Na^+ from the phloem to prevent Na^+ transfer to the younger leaf blades [22]. Therefore, the second leaves were affected by salt stress to a larger extent than the flag leaves. This may be the mechanism that prevents Na^+ toxicity from reaching the flag leaf, which has the main role in carbon fixation for grain development.

4. Materials and Methods

4.1. Plant Material and Salt Treatment

The experiment was conducted in a planting house for transgenic plants in the Botany Department, Faculty of Science, Chulalongkorn University in Thailand. Three independent transgenic rice lines expressing *OsNUC1-S* driven by the Ubiquitin promoter have been produced by Sripinyowanich and colleagues [12] in a "Nipponbare" rice genetic background, and the homozygous T_3 generations of these transgenic lines were used in this experiment. "Nipponbare" rice seeds were obtained from the National Laboratory for Protein Engineering and Plant Genetic Engineering, Peking-Yale Joint Research Center for Plant Molecular Genetics and AgroBiotechnology, Peking University, People's Republic of China. Either wild type (WT) or transgenic seeds were germinated for 7 days. Then, seedlings were transferred to pots containing 5 kg of soil. When plants were at the booting stage, 150 mM NaCl solution was added to cause salt stress at 8–10 ds·m^{-1}. All photosynthetic parameters were measured after 9 days salt treatment using a Gas Analysis System (LI-COR, LI-6400, USA) with a 1200 μmol·m^{-2}·s^{-1} light intensity and 380 μmol mol^{-1} carbon dioxide (CO_2) concentration.

For chlorophyll fluorescence parameters, either the flag leaf or second leaf was used to determine maximum quantum yield of photosystem II (PSII) photochemistry (F_v/F_m) using Pocket PEA (Hansatech Instruments, Ltd., Norfolk, UK). In total, 30 min was used for dark adaptation, and then a 1-s saturation flash was used to measure the potential maximum photochemical efficiency of PSII.

4.2. RNA-seq and Data Analysis

For the transcriptomic analysis, flag leaves of WT and TOSL3, the transgenic line with *OsNUC1-S* overexpression, were collected 3 and 9 days after the first day of the booting stage. Total RNA was extracted using Invitrogen's Concert™ Plant RNA Reagent and then treated with DNaseI (NEB). A Dynabeads mRNA purification kit (Invitrogen, Carlsbad, CA, USA) and a KAPA Stranded mRNA-Seq Kit were used for mRNA isolation and cDNA libraries' preparation, respectively. Fragment sizes of ~300 bp were selected and connected with adaptors. Then, the fragments were enriched by PCR for 12 cycles. All libraries were sequenced using the Genome Analyzer (Illumina HiSeq4000, San Diego, CA, USA). Adaptors were subsequently removed from all short-sequence reads before grouping, following the protocol of Missirian et al. [23]. All the sequences were aligned and mapped to the rice genome, and then, the differentially expressed genes (DEGs) were identified using the DESeq program [24]. The genes showing the differential expression were selected to validate with quantitative RT-PCR. *LOC_Os01g58470*, *LOC_Os01g64960*, *LOC_Os03g39610*, and *LOC_Os04g33830* predicted to localize in plastids were selected. The primers for detection of the gene expression are shown in Table 2. *OsEF-1α* gene expression was used as an internal control. For q-RT-PCR, briefly, 1 μL of RNA extraction from either WT's or transgenic rice's flag leaves was used to synthesize the cDNA with an AccuPower® RT PreMix (BIONEER, Oakland, CA, USA). Gene expression of the target genes was detected by quantitative PCR using Luna® Universal qPCR Master MiX (BioLabs, San Diego, CA, USA). The thermal cycle was performed at 95 °C for 60 s, then 40 cycles of 95 °C for 15 s, 58.5 °C for 30 s, followed by 95 °C for 30 s, and the extension was done at 70 °C for 5 s. The relative expression of interesting genes was calculated using the Pfaffl method [25] with the formula:

$$\text{Ratio} = (E_{target})^{\Delta CPtarget(control-sample)} / (E_{ref})^{\Delta CPref(control-sample)}$$

Table 2. Primers for gene expression detection.

Gene	Primer	Sequence (5′–3′)	Tm (°C)
OsEF-1α	EF-1α-F	ATGGTTGTGGAGACCTTC	53.7
	EF-1α-R	ATGGTTGTGGAGACCTTC	58.2
LOC_Os01g58470	Os01g58470-F	AGGCATTGATCCTGAGACAG	54.3
	Os01g58470-R	AGAGCAGAATATCCCACTGC	54.4
LOC_Os01g64960	Os01g64960-F	GCATCGCCTTCTCCATCA	57.1
	Os01g64960-R	GAAGACGACGTTGAAGAGGA	57.3
LOC_Os03g39610	Os03g39610-F	GGAGGCGGTGTGGTTCAAGG	61.0
	Os03g39610-R	GCGGTAGCCCTCGACGAATC	60.3
LOC_Os04g3383	Os04g33830-F	CCGTTCTGGCTGTGGTT	55.4
	Os04g33830-R	CGTCCGTACAGTCAAGCTAA	54.3

4.3. Gene Ontology (GO) Term Analysis

Genes that had a *p*-value < 0.5 and log$_2$fold change less than −1 or more than 1 were classified into GO terms using the ClueGO tool [26]. They were analyzed into three GO terms, cellular compartment, biological process, and molecular function, and subjected to a KEGG [27] pathway analysis.

4.4. Pigment Extraction and Quantification

Chlorophyll *a*, chlorophyll *b*, and total carotenoid contents were studied according to the method of Lichtenthaler [28]. Briefly, 50 mg fresh weight of either flag leaves or second leaves were extracted with 10 mL of 80% acetone and then incubated in the dark for 24 h at room temperature. Pigment extracts were measured at 470, 646.8, and 663.2 nm using a spectrophotometer (Agilent Technology, Santa Clara, CA, USA).

4.5. Yield Collection

After a 9 days salt-stress treatment, plants were recovered by the addition of water until soil salinity was below 2 ds·m^{-1}. Seeds were harvested when they were fully developed and then desiccated. The numbers of panicles per plant and seeds per panicle, as well as the fertility rates (%), were determined.

4.6. Statistical Analysis

A randomized complete block design with four replications was used for the experimental plots. Analysis of variance was performed to detect the differences among means and a Tukey's range test was used to detect significant differences between each mean at *p*-value < 0.05 using SPSS 21.0 statistical software. All the results are presented as mean ± standard error of the mean.

5. Conclusions

Based on the experimental results, a role of *OsNUC1-S* in the salt tolerance of rice during the reproductive stage was suggested to be due to the enhancement of photosynthetic processes in both flag and second leaves through the modification of gene expression levels in water transport, photosynthesis, cellular homeostasis, and carotenoid biosynthesis. These could help maintain the grain yield after a salt stress (Figure 11).

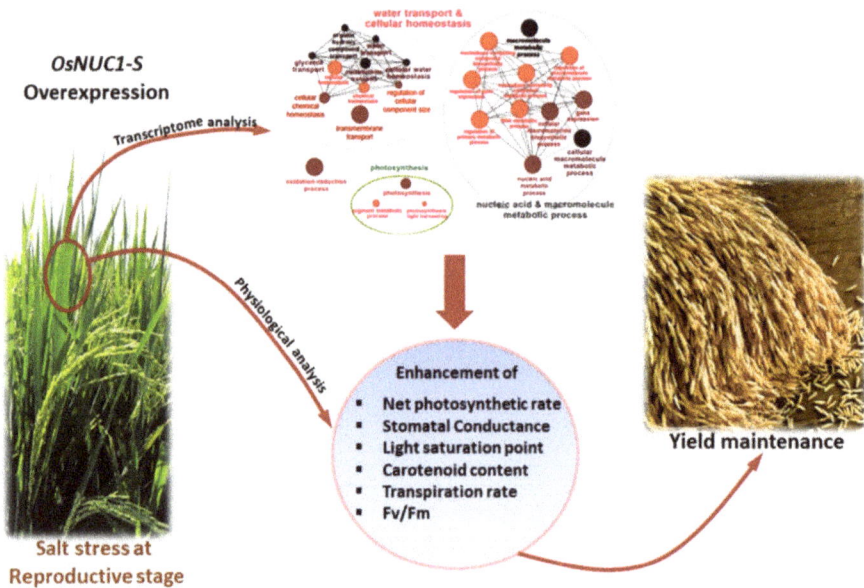

Figure 11. The scheme summarized the role of *OsNUC1-S* overexpression contributing to salt tolerance in rice.

Supplementary Materials: The following are available online at http://www.mdpi.com/1422-0067/19/12/3936/s1.

Author Contributions: Conceptualization, S.C. and L.C.; methodology, S.C., T.B. and L.C.; software, L.C. and T.U.; validation, C.B. and S.S.; formal analysis, C.B. and S.C.; investigation, C.B.; resources, S.S.; data curation, S.C. and T.B.; writing—original draft preparation, C.B.; writing—review and editing, S.C., T.B. and L.C.; supervision, S.C., L.C. and T.B.; project administration, S.C.; funding acquisition, S.C.

Funding: This research was supported by the Ratchadaphiseksomphot Endowment Fund (2015-16), Chulalongkorn University (grant numbers RES560530061-FW, CU-58-012-FW) and Sci-SuperIV fund from Faculty of Science, Chulalongkorn University. C.B. was supported by the Overseas Research Experience Scholarship (ORES) for Graduate Students from the Graduate School and the Faculty of Science, Chulalongkorn University and Science Achievement Scholarship of Thailand.

Acknowledgments: We thank Lesley Benyon from Edanz Group (www.edanzediting.com/ac) for editing a draft of this manuscript.

Conflicts of Interest: The authors declare no conflict of interest.

Abbreviations

Ci	intercellular CO_2 concentration
F_v/F_m	intercellular CO_2 concentration
gs	stomatal conductance
OsNUC1	rice *NUCLEOLIN1*
P_N	net photosynthesis rate
WT	wild type

References

1. Fageria, N. Yield physiology of rice. *J. Plant Nutr.* **2007**, *30*, 843–879. [CrossRef]
2. Hakim, M.; Juraimi, A.; Begum, M.; Hanafi, M.; Ismail, M.R.; Selamat, A. Effect of salt stress on germination and early seedling growth of rice (*Oryza sativa* L.). *Afr. J. Biotechnol.* **2010**, *9*, 1911–1918.
3. Bethke, P.C.; Drew, M.C. Stomatal and nonstomatal components to inhibition of photosynthesis in leaves of *Capsicum annuum* during progressive exposure to NaCl salinity. *Plant Physiol.* **1992**, *99*, 219–226. [CrossRef] [PubMed]
4. Reddy, M.; Vora, A. Changes in pigment composition, Hill reaction activity and saccharides metabolism in Bajra (*Pennisetum typhoides* S & H) leaves under NaCl salinity. *Photosynthetica* **1986**, *20*, 50–55.
5. Petrov, V.; Hille, J.; Mueller-Roeber, B.; Gechev, T.S. ROS-mediated abiotic stress-induced programmed cell death in plants. *Front. Plant Sci.* **2015**, *66*, 69. [CrossRef] [PubMed]
6. Lutts, S.; Kinet, J.; Bouharmont, J. Changes in plant response to NaCl during development of rice (*Oryza sativa* L.) varieties differing in salinity resistance. *J. Exp. Bot.* **1995**, *46*, 1843–1852. [CrossRef]
7. Touch, S.; Pipatpongsa, T.; Takeda, T.; Takemura, J. The relationships between electrical conductivity of soil and reflectance of canopy, grain, and leaf of rice in northeastern Thailand. *Int. J. Remote Sens.* **2015**, *36*, 1136–1166.
8. Zhang, J.; Lin, Y.J.; Zhu, L.F.; Yu, S.M.; Sanjoy, K.K.; Jin, Q.Y. Effects of 1-methylcyclopropene on function of flag leaf and development of superior and inferior spikelets in rice cultivars differing in panicle types. *Field Crops Res.* **2015**, *177*, 64–74. [CrossRef]
9. Tajrishi, M.M.; Tuteja, R.; Tuteja, N. Nucleolin: The most abundant multifunctional phosphoprotein of nucleolus. *Commun. Integr. Biol.* **2011**, *4*, 267–275. [CrossRef]
10. Petricka, J.J.; Nelson, T.M. Arabidopsis nucleolin affects plant development and patterning. *Plant Physiol.* **2007**, *144*, 173–186. [CrossRef]
11. Reichler, S.A.; Balk, J.; Brown, M.E.; Woodruff, K.; Clark, G.B.; Roux, S.J. Light differentially regulates cell division and the mRNA abundance of pea nucleolin during de-etiolation. *Plant Physiol.* **2001**, *125*, 339–350. [CrossRef] [PubMed]
12. Sripinyowanich, S.; Chamnanmanoontham, N.; Udomchalothorn, T.; Maneeprasopsuk, S.; Santawee, P.; Buaboocha, T.; Qu, L.-J.; Gu, H.; Chadchawan, S. Overexpression of a partial fragment of the salt-responsive gene *OsNUC1* enhances salt adaptation in transgenic *Arabidopsis thaliana* and rice (*Oryza sativa* L.) during salt stress. *Plant Sci.* **2013**, *213*, 67–78. [CrossRef] [PubMed]
13. Udomchalothorn, T.; Plaimas, K.; Sripinyowanich, S.; Boonchai, C.; Kojonna, T.; Chutimanukul, P.; Comai, L.; Buaboocha, T.; Chadchawan, S. *OsNucleolin1-L* expression in Arabidopsis enhances photosynthesis via transcriptome modification under salt stress conditions. *Plant Cell Physiol.* **2017**, *58*, 717–734. [CrossRef] [PubMed]

14. Lee, S.; Jeong, H.; Lee, S.; Lee, J.; Kim, S.-J.; Park, J.-W.; Woo, H.R.; Lim, P.O.; An, G.; Nam, H.G. Molecular bases for differential aging programs between flag and second leaves during grain-filling in rice. *Sci. Rep.* **2017**, *7*, 8792. [CrossRef]

15. Junghans, T.G.; De Almeida-Oliveira, M.G.; Moreira, M.A. Lipoxygenase activities during development of root and nodule of soybean. *Pesqui. Agropecu. Bras.* **2004**, *39*, 625–630. [CrossRef]

16. Lenglet, A.; Jaślan, D.; Toyota, M.; Mueller, M.; Müller, T.; Schönknecht, G.; Marten, I.; Gilroy, S.; Hedrich, R.; Farmer, E.E. Control of basal jasmonate signalling and defence through modulation of intracellular cation flux capacity. *New Phytol.* **2017**, *216*, 1161–1169. [CrossRef]

17. Brodhun, F.; Cristobal-Sarramian, A.; Zabel, S.; Newie, J.; Hamberg, M.; Feussner, I. An iron 13S-lipoxygenase with an α-linolenic acid specific hydroperoxidase activity from *Fusarium oxysporum*. *PLoS ONE* **2013**, *8*, e64919. [CrossRef]

18. Kholupenko, I.; Voronkova, N.; Burundukova, O.; Zhemchugova, V. Demand for assimilates determines the productivity of intensive and extensive rice crops in Primorskii krai. *Russ. J. Plant Physiol.* **2003**, *50*, 112–118. [CrossRef]

19. Liu, L.; Li, H.; Zeng, H.; Cai, Q.; Zhou, X.; Yin, C. Exogenous jasmonic acid and cytokinin antagonistically regulate rice flag leaf senescence by mediating chlorophyll degradation, membrane deterioration, and senescence-associated genes expression. *J. Plant Growth Regul.* **2016**, *35*, 366–376. [CrossRef]

20. Hakata, M.; Muramatsu, M.; Nakamura, H.; Hara, N.; Kishimoto, M.; Iida-Okada, K.; Kajikawa, M.; Imai-Toki, N.; Toki, S.; Nagamura, Y. Overexpression of *TIFY* genes promotes plant growth in rice through jasmonate signaling. *Biosci. Biotechnol. Biochem.* **2017**, *81*, 906–913. [CrossRef]

21. Li, Z.; Pinson, S.R.; Stansel, J.W.; Paterson, A.H. Genetic dissection of the source-sink relationship affecting fecundity and yield in rice shape (*Oryza sativa* L.). *Mol. Breed.* **1998**, *4*, 419–426. [CrossRef]

22. Kobayashi, N.I.; Yamaji, N.; Yamamoto, H.; Okubo, K.; Ueno, H.; Costa, A.; Tanoi, K.; Matsumura, H.; Fujii-Kashino, M.; Horiuchi, T.; et al. OsHKT1;5 mediates Na$^+$ exclusion in the vasculature to protect leaf blades and reproductive tissues from salt toxicity in rice. *Plant J.* **2017**, *91*, 657–670. [CrossRef] [PubMed]

23. Missirian, V.; Henry, I.; Comai, L.; Filkov, V. POPE: Pipeline of parentally-biased expression. In Proceedings of the ISBRA: 2012 Bioinformatics Research and Applications, Dallas, TX, USA, 21–23 May 2012; pp. 177–188.

24. Anders, S.; Huber, W. Differential expression analysis for sequence count data. *Genome Biol.* **2010**, *11*, R106. [CrossRef] [PubMed]

25. Pfaffl, M.W. A new mathematical model for relative quantification in real-time RT-PCR. *Nucleic Acids Res.* **2001**, *29*, e45. [CrossRef] [PubMed]

26. Bindea, G.; Mlecnik, B.; Hackl, H.; Charoentong, P.; Tosolini, M.; Kirilovsky, A.; Fridman, W.-H.; Pagès, F.; Trajanoski, Z.; Galon, J. ClueGO: A Cytoscape plug-in to decipher functionally grouped gene ontology and pathway annotation networks. *Bioinformatics* **2009**, *25*, 1091–1093. [CrossRef] [PubMed]

27. Kanehisa, M.; Goto, S. KEGG: Kyoto encyclopedia of genes and genomes. *Nucleic Acids Res.* **2000**, *28*, 27–30. [CrossRef] [PubMed]

28. Lichtenthaler, H.K. Chlorophylls and carotenoids: Pigments of photosynthetic biomembranes. *Methods Enzymol.* **1987**, *148*, 350–380.

International Journal of
Molecular Sciences

MDPI

Article

Overexpression of a Novel *ROP* Gene from the Banana (*MaROP5g*) Confers Increased Salt Stress Tolerance

Hongxia Miao [1,†], Peiguang Sun [2,†], Juhua Liu [1], Jingyi Wang [1], Biyu Xu [1,*] and Zhiqiang Jin [1,2,*]

[1]   Key Laboratory of Biology and Genetic Resources of Tropical Crops, Ministry of Agriculture,
     Institute of Tropical Bioscience and Biotechnology, Chinese Academy of Tropical Agricultural Sciences,
     Xueyuan Road 4, Haikou 571101, China; miaohongxia@itbb.org.cn (H.M.); liujuhua@itbb.org.cn (J.L.);
     wangjingyi@itbb.org.cn (J.W.)
[2]   Key Laboratory of Genetic Improvement of Bananas, Hainan Province, Haikou Experimental Station,
     Chinese Academy of Tropical Agricultural Sciences, Xueyuan Road 4, Haikou 570102, China;
     sunpeiguang@catas.cn
*   Correspondence: biyuxu@126.com (B.X.); jinzhiqiang@itbb.org.cn (Z.J.); Tel.: +86-898-6696-0172 (B.X.)
†   These authors contributed equally to this work.

Received: 25 August 2018; Accepted: 1 October 2018; Published: 11 October 2018

Abstract: Rho-like GTPases from plants (ROPs) are plant-specific molecular switches that are crucial for plant survival when subjected to abiotic stress. We identified and characterized 17 novel ROP proteins from *Musa acuminata* (MaROPs) using genomic techniques. The identified MaROPs fell into three of the four previously described ROP groups (Groups II–IV), with MaROPs in each group having similar genetic structures and conserved motifs. Our transcriptomic analysis showed that the two banana genotypes tested, Fen Jiao and BaXi Jiao, had similar responses to abiotic stress: Six genes (*MaROP-3b, -5a, -5c, -5f, -5g,* and *-6*) were highly expressed in response to cold, salt, and drought stress conditions in both genotypes. Of these, *MaROP5g* was most highly expressed in response to salt stress. Co-localization experiments showed that the MaROP5g protein was localized at the plasma membrane. When subjected to salt stress, transgenic *Arabidopsis thaliana* overexpressing *MaROP5g* had longer primary roots and increased survival rates compared to wild-type *A. thaliana*. The increased salt tolerance conferred by *MaROP5g* might be related to reduced membrane injury and the increased cytosolic K^+/Na^+ ratio and Ca^{2+} concentration in the transgenic plants as compared to wild-type. The increased expression of salt overly sensitive (SOS)-pathway genes and calcium-signaling pathway genes in *MaROP5g*-overexpressing *A. thaliana* reflected the enhanced tolerance to salt stress by the transgenic lines in comparison to wild-type. Collectively, our results suggested that abiotic stress tolerance in banana plants might be regulated by multiple *MaROPs*, and that *MaROP5g* might enhance salt tolerance by increasing root length, improving membrane injury and ion distribution.

Keywords: banana (*Musa acuminata* L.); ROP; genome-wide identification; abiotic stress; salt stress; *MaROP5g*

1. Introduction

Small GTPases (GTP)-binding proteins, present in a wide variety of eukaryotes, are the central regulators of numerous signal transduction processes [1,2]. These proteins are structurally classified into at least five families, including Rat sarcoma (RAS), Ras homolog (RHO), Rat brain (RAB), RAS-related nuclear (RAN), and adenosine diphosphate (ADP) ribosylation factor (ARF) [1–3]. In all reported eukaryotes, RAS and RHO families are signaling switches, whereas these proteins in other families are primarily involved in the regulation of vesicle and large molecule movement [1,2]. However, higher plants have a unique RHO subfamily of small GTP-binding proteins known as ROPs

(Rho-like GTPases from plants) [4,5]. These proteins are also known as RAS-related C3 botulinum toxin substrates (RACs), due to their sequence similarity to Rac GTPases [6].

ROPs are plant-specific molecular switches that regulate intracellular signaling pathways by cycling between an active form and an inactive, guanosine diphosphate (GDP)-bound form. Biological activities associated with ROPs are diverse, which include polar growth, development, environmental stress responses, and host-pathogen interactions [7–13]. Since the first *ROP* gene was isolated from peas [14], multiple *ROPs*, with typical RhoGEF domains and molecular masses between 21 and 24 kDa, have been described in numerous plant species: 11 in *Arabidopsis thaliana* [15], 9 in *Zea mays* [16], 11 in *Brassica napus* [17], 7 in *Vitis vinifera* [9], 7 in *Oryza sativa* [18], 7 in *Medicago truncatula* [7], 6 in *Nicotiana tabacum* [7], 5 in *Hevea brasiliensis* [19], and 9 in *Solanum lycopersicum* [20]. ROPs can be classified into four groups (I–IV) based on their molecular structure and motif conservation [5,7,16].

ROP expression levels and biological functions are affected by various abiotic stressors. When exposed to cold, the transcriptional expression of apple *ROP* increases, leading to a decrease in the concentrations of ethylene and reactive oxygen species in the fruits [21]. In *A. thaliana*, *ROP11* expression affected seed germination, seedling growth, stomatal closure, abscisic acid (ABA)-mediated responses, and drought stress responses [22]. The overexpression of *ROP1* in tobacco increased salt sensitivity in response to salt stress by increasing H_2O_2 production [23]. The Na^+/K^+ ratio in transgenic *A. thaliana* expressing the *Medicago falcate* small GTPase gene (*MfARL1*) was lower than that in wild-type (WT) *A. thaliana*; the transgenic plants consequently had an increased tolerance for salt stress [24]. Knock out of the *A. thaliana* ROP effector (RIC1) increased the survival rate of plants under salt stress by improving the reassembly of depolymerized microtubules [25]. Taken together, these studies have revealed the many important roles of ROPs in the regulation of plant response to abiotic stresses.

The banana (*Musa acuminata*) is one of the most intensively produced and globally important fruit crops [26]. As a large monocotyledonous herbaceous annual, banana plants are frequently harmed or destroyed by various abiotic stress conditions during growth and development [27]. In particular, saline soil is a major abiotic stressor which limits banana cultivation worldwide [28,29]. Genome-wide identification of genes involved in the resistance of banana plants to cold, drought, and salt stress increases our knowledge of plant mechanisms for environmental stress tolerance, while the functional identification of relevant genes acts as a framework for future genetic studies focused on increasing the resistance of the banana plant to these stressors [30–33]. However, genome-wide investigations of the *ROP* gene family, and thus an integrated assessment of the potential functions of this important molecular switch, are still lacking in banana.

To address this information gap, we identified *ROPs* genome-wide in *M. acuminata*, known as *MaROPs*, and analyzed their phylogenetic relationships, gene structures, protein motifs, and expression changes in response to a number of abiotic stressors, including cold, drought, and salt. More importantly, we noted that the expression of the *MaROP5g* gene was associated with salt stress in banana. The overexpression of *MaROP5g* in *A. thaliana* conferred increased salt tolerance by lengthening roots, improving recovery of membrane injury and ion distributions. This comprehensive study of *MaROPs* in *M. acuminata* enhances our understanding of *ROPs* in response to abiotic stress conditions in banana plants, and provides a foundation for future studies aiming to improve the abiotic stress resistance of crop plants, especially with respect to salt stress.

2. Results

2.1. Identification and Phylogenetic Analysis of Banana MaROP Genes

We used the basic local alignment search tool (BLAST) and the hidden Markov models (HMM) to identify MaROPs with typical RhoGEF domains (PF00621) in the *M. acuminata* genome, using the sequences of AtROPs and OsROPs as queries [34,35]. We identified 17 MaROPs in the *M. acuminata* genome, and designated these MaROP-2a, -2b, -2c, -3a, -3b, -4, -5a, -5b, -5c, -5d, -5e, -5f, -5h, -5g,

-6, -7a, and -7b, following the nomenclature of their respective orthologous proteins in *O. sativa*. The 17 predicted MaROP proteins ranged from 195 amino acid residues (MaROP5d) to 214 amino acid residues (MaROP4), with relative molecular masses between 21.297 kDa (MaROP5d) and 23.784 kDa (MaROP3a), and isoelectric points between 8.61 and 9.43 (Table S1).

To investigate the evolutionary relationships among ROP proteins, we constructed a maximum-likelihood (ML) phylogenetic tree based on our multiple sequence alignment of 17 ROP proteins from *M. acuminata*, 11 from *A. thaliana*, and 7 from *O. sativa*. The MaROP proteins fell into 3 distinct groups (Figure 1): Group II contained 6 MaROPs (MaROP-2a, -2b, -2c, -3a, -3b, and -4), 3 AtROPs (AtROP-9, -10, and -11), and 4 OsROPs (OsAPL-1, -2, -3, and -4); Group III contained MaROP-7a, -7b, AtROP7, and OsROP7; and Group IV contained 9 MaROPs (MaROP-5a, -5b, -5c, -5d, -5e, -5f, -5g, -5h, and -6), 6 AtROPs (AtROP-1, -2, -3, -4, -5, and -6), and 3 OsROPs (OsROP-2, -5, and -6). Group I containing AtROP8 served as an outgroup to the phylogenetic analysis.

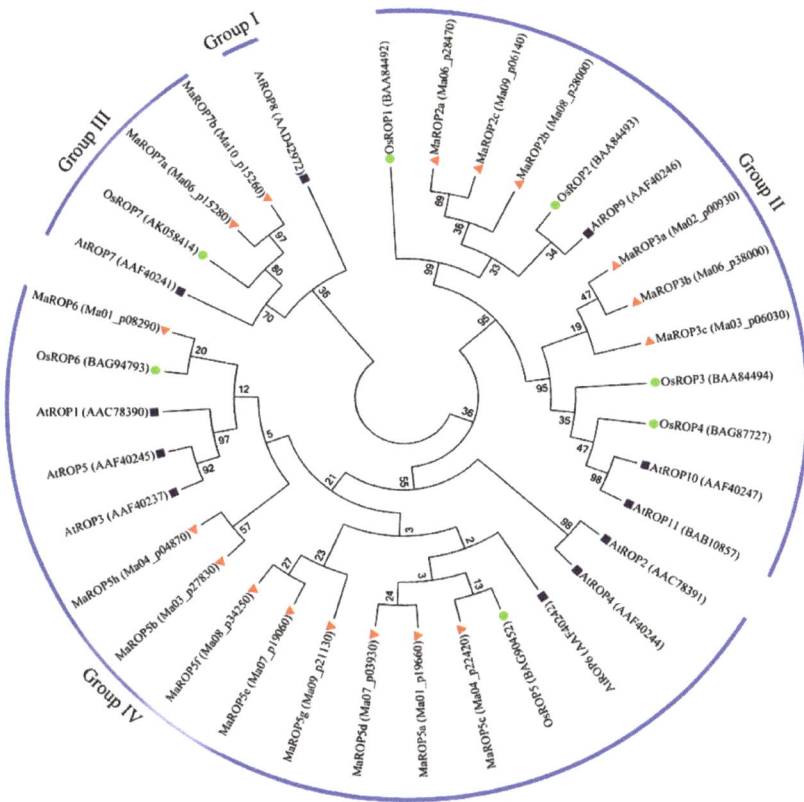

Figure 1. Phylogenetic analysis of Rho-like GTPases from plants (ROPs) from *A. thaliana*, rice, and bananas. The maximum-likelihood phylogenetic tree was drawn with MEGA5.2, using 1000 bootstraps. Four subgroups were identified (Groups I–IV). Circles, squares, and triangles represent ROP proteins from rice, *A. thaliana*, and bananas, respectively.

2.2. Gene Structure and Conserved Motif Analysis of Banana MaROP Genes

Evolutionary analysis supported the classification of the 17 *MaROP* genes into three distinct groups (Groups II–IV), which is consistent with their exon–intron structural divergence within families (Figure 2). Our analysis of the exon–intron structure using the Gene Structure Display Server showed that the *MaROP* genes contained 8 exons in Group II, 7 exons in Group III, and 6–7 exons in Group IV, suggesting the conservation of the exon–intron structures of *MaROPs* within the same group.

To explore MaROPs structural diversity and potential functionality, we analyzed the conserved motifs of the identified MaROPs and predicted their functional annotations. We identified 10 conserved motifs across the 17 MaROP proteins with Multiple Em (Motif Elicitation), which are annotated with InterPro (Figure 2; Table S2). Motifs 1–4 were annotated as a RhoGEF domain (PF00621), the characteristic domain of the ROP protein family. Motifs 1–5 were found across all of the 17 MaROPs, while motif 6 was only present in MaROP-2a, -2b, and -2c; motifs 6–8 were found in MaROP-3a and -3b; and motifs 9 and 10 were found in MaROP4. It is probable that this pattern of conservation and variation across the motifs reflected the evolutionary relatedness and functional divergence of the 17 MaROPs.

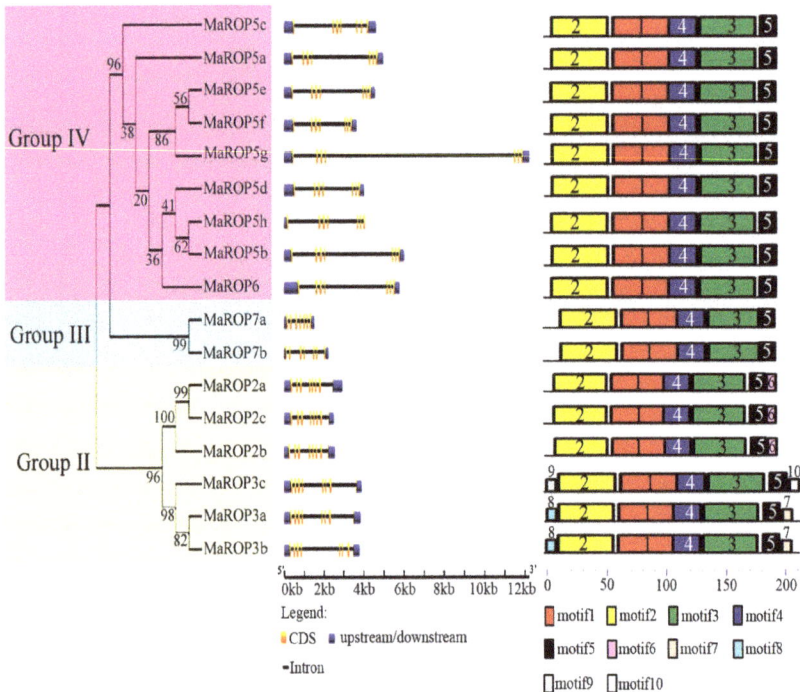

Figure 2. Phylogenetic, gene structure, and motif analyses of banana *Musa acuminata* ROP (MaROP) proteins. MaROPs were classified into Groups II–IV based on their phylogenetic relationships. Exon–intron structure analyses were performed with the Gene Structure Display Server (GSDS). Blue boxes indicate upstream/downstream; yellow boxes indicate exons; black lines indicate introns. All of the proteins were identified using the Multiple EM for Motif Elicitation (MEME) database, using the complete predicted amino acid sequences of each MaROP. Motifs 1–4 were annotated as a RhoGEF domain.

2.3. Expression Analysis of MaROP Genes in Response to Cold, Salt, and Osmotic Stresses

To investigate the response of the *MaROP* genes in response to different abiotic stressors, we analyzed the *MaROP* expression in banana leaves following exposure to cold, salt, and osmotic stress conditions (Figure 3A; Table S3). Compared to the control, significant differences in the expression of 14 *MaROP* genes (82%) were detected following the exposure to abiotic stress treatments (Figure 3A; Table S3). In BaXi Jiao (BX), the expression levels of *MaROP-3b, -5a, -5c, -5f, -5g,* and *-6* were significantly upregulated, as indicated by the fragments per kilobase of exon per million fragments mapped (FPKM) value, which is higher than 2.0 by all three of the abiotic stressors. *MaROP5d* was upregulated by cold treatment only (FPKM > 2.0), and *MaROP2c* was downregulated by osmotic treatment only (FPKM < 0.5). In Fen Jiao (FJ), *MaROP-3b, -5a, -5c, -5f, -5g, -6* were significantly expressed (FPKM > 8.9) in the presence of abiotic stressors. Under osmotic treatment, *MaROP6* was upregulated (FPKM > 2.3) in FJ, but maintained a low level of expression in BX (FPKM < 0.66). In addition, *MaROP-3a* and *-5h* are significantly downregulated in the presence of stress in BX and FJ, compared to control. *MaROP5g* had a higher level of expression (FPKM > 24) in response to salt stress than any of the other *ROPs* across both the BX and FJ genotypes, implying *MaROP5g* might play an important role in the regulation of salt stress tolerance in banana plants.

2.4. Validation of Differential Expression of Six MaROP Genes by Quantitative Real-Time Polymerase Chain Reaction (qRT-PCR) Analysis

Our RNA-seq analysis indicated that *MaROP-3b, -5a, -5c, -5f, -5g,* and *-6* showed significant expressions by abiotic stressors. Such a feature of these six genes was further verified by quantitative real-time polymerase chain reaction (qRT-PCR) analysis. After normalization, all the examined *MaROPs*, with the exception of *MaROP3b* in FJ-salt and *MaROP5c* in FJ-salt, were well-correlated and generally consistent with our RNA-seq analyses (r = 0.8789–0.9992; Figure 3B; Table S4), indicating the reliability of our transcriptomic results in both banana varieties. Of particular interest was *MaROP5g*, which showed high expression following salt stress treatment compared to other *MaROP* genes, as determined by both RNA-seq and qRT-PCR experiments.

2.5. Full-Length cDNA, Subcellular Localization, and Expression Pattern of MaROP5g under Salt Stress

Based on the results of the RNA-seq and qRT-PCR analyses, we used PCR to amplify the full-length cDNA of *MaROP5g* from banana roots. The full-length *MaROP5g*cDNA had a 591 bp open reading frame (ORF), encoding 196 amino acids. The predicted MaROP5g protein had a typical RhoGEF domain and several additional characteristics of the ROP protein family (Figure S1; Table S2).

We measured the transcriptional response of *MaROP5g* in BX and FJ plant roots to salt stress. Compared to 0 h (no stress condition), the roots of the BX plants became black following 6 h of salt stress treatment (Figure 4A). However, there was no discernible phenotypic change in the roots of FJ plants under the same treatment (Figure 4C). The expression of *MaROP5g* quickly increased from 0 h, reached the maximum level at 4 h, and then gradually decreased at 6 h (Figure 4B). The expression pattern of *MaROP5g* was similar between FJ and BX (Figure 4D), but *MaROP5g* showed lower expression in FJ compared to BX under salt stress. These results suggested that the regulation of *MaROP5g* expression by salt treatment was genotype-dependent, as the roots of the two tested banana genotypes may have variable sensitivity to salt stress treatments. BX showed more sensitivity than FJ under salt stress treatment.

To localize the MaROP5g protein in the cell, we introduced the *MaROP5g* ORF into a pCAMBIA1304-GFP vector upstream of the *GFP* gene to create a MaROP5g-GFP fusion construct, which was used to transform *A. thaliana*. Co-localization experiments showed that the MaROP5g-GFP (green fluorescent protein) fusion protein was localized to the FM4-64-labeled plasma membrane in *A. thaliana* root tips (Figure 4E).

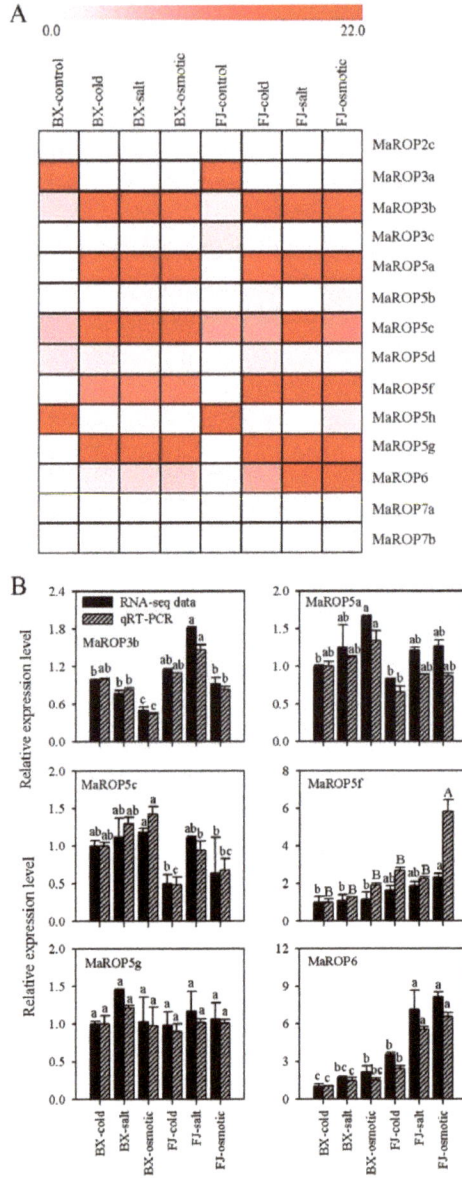

Figure 3. Differential expression of Banana *MaROPs* in response to cold, salt, and osmotic stresses in BaXi Jiao (BX; *M. acuminata* cv. Cavendish; AAA group) and Fen Jiao (FJ; *M. acuminata*; group AAB) banana varieties, as determined with transcriptomic analysis and qRT-PCR. (**A**) Heat map clusters were created based on the fragments per kilobase of exon per million fragments mapped (FPKM) value of the *MaROPs*. Magnitude of differences in gene expression is indicated with a one-color scheme. (**B**) Data are presented as means ± standard deviations of *n* = 3 biological replicates. Different lowercase letters above bars indicate significant differences at *p* < 0.05, and different uppercase letters above bars indicate extremely significant differences at *p* < 0.01, using Duncan's multiple range tests.

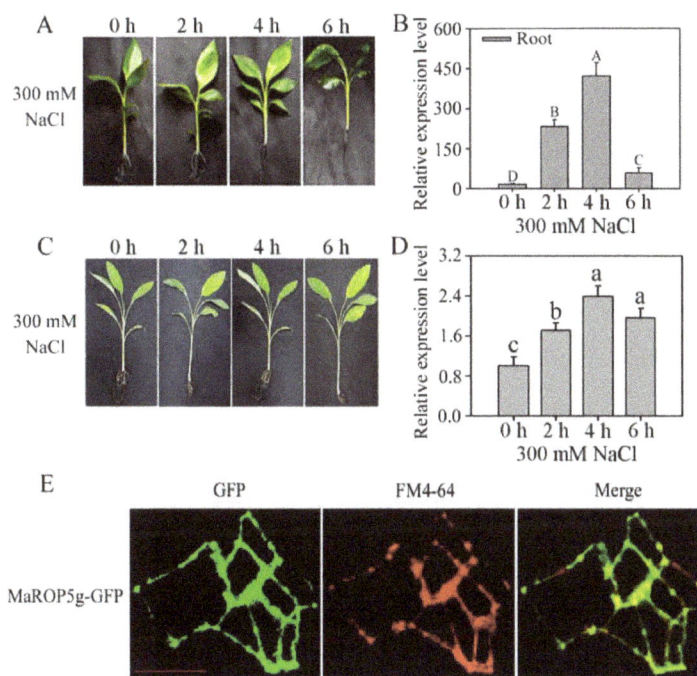

Figure 4. *MaROP5g* expression analyses in both banana varieties roots after different periods of exposure to salt stress and subcellular localization. (**A**) Phenotypes of BX roots exposed to salt stress; (**B**) expression of *MaROP5g* in BX roots exposed to salt stress; (**C**) phenotypes of FJ roots exposed to salt stress; (**D**) expression of *MaROP5g* in FJ roots exposed to salt stress. Data are presented as means ± standard deviations of *n* = 3 biological replicates. Different lowercase letters above bars indicate significant differences at *p* < 0.05, and different uppercase letters above bars indicate extremely significant differences at *p* < 0.01, using Duncan's multiple range tests. (**E**) MaROP5g subcellular localization. GFP fluorescence is green, and FM4-64 is red. Merge was created by merging the GFP and FM4-64 fluorescent images. Scale bars = 10 μm.

2.6. MaROP5g Overexpression Enhances Tolerance to Salt Stress

To investigate the role of *MaROP5g* in response to salt stress, *MaROP5g* was introduced into the pCAMBIA1304 vector under the control of the 35S promoter. After a floral-dip transformation of *A. thaliana*, we analyzed three transgenic lines with single-copy transgene (R3, R8, and R42) from the T3 generation, selected through genomic DNA Southern blot analysis (Figure S2A). The expression level of *MaROP5g* in three transgenic lines was 89–109 folds compared to WT and empty vector (VC), as revealed by qRT-PCR analysis (Figure S2B).

Under no-salt (control) and high-salt conditions, the seed germination rate and root growth were greater in the transgenic seedlings as compared to those in the WT seedlings. After salt treatments ranging from 100 to 200 mM, the transgenic seedlings had grown longer primary roots, as compared to WT seedlings (Figure 5A–C). Furthermore, when adult *A. thaliana* growing in soil were treated with 350 mM salt daily for 15 days, the transgenic lines grew better (Figure 5D,E) and were more likely to survive (Figure 5F), as compared to those of WT. Thus, transgenic *A. thaliana* lines overexpressing *MaROP5g* were more tolerant to salt stress than those of WT.

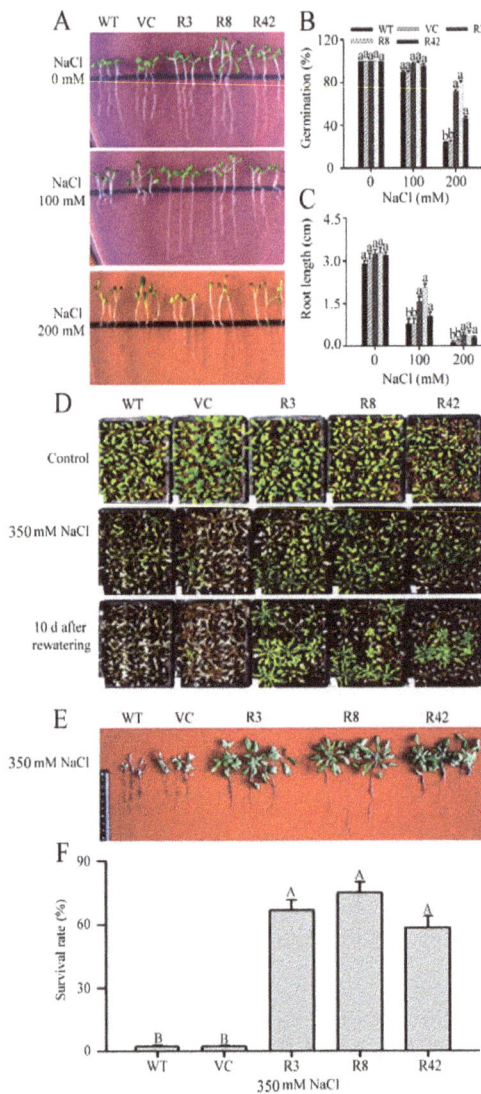

Figure 5. Comparison of wild-type and *A. thaliana* overexpressing *MaROP5g* under standard growing conditions (control) and different salt stress treatments. (**A**) Phenotypes of WT and transgenic lines under control and salt conditions; (**B**) Germination of WT and transgenic lines under control and salt conditions; (**C**) Root length of WT and transgenic lines under control and salt conditions; (**D,E**) Phenotypes of WT and transgenic mature plants under control or salt conditions or after rewatering; (**F**) Survival rates of WT and transgenic mature plants under control or salt conditions. WT: wild-type. VC: vector. R3, R8, R42: *MaROP5g* transgenic plants. ANOVA was used to compare the significance of differences, using Dunnett's tests in the comparison between WT and each overexpression lines. Data are presented as means ± standard deviations of $n = 3$ biological replicates. Different lowercase letters above bars indicate significant differences at $p < 0.05$, different uppercase letters above bars indicate extremely significant differences at $p < 0.01$.

2.7. MaROP5g Overexpression Reduced Malonaldehyde (MDA) Content and Ion Leakage (IL), Increased Ca^{2+} and K$^+$/Na$^+$ Ratio under Salt Stress

Malonaldehyde (MDA) is usually employed as an index of oxidative damages in plants [36]. Ion leakage (IL) is an important indicator of membrane injury [36]. To investigate whether *MaROP5g* expression influences MDA and IL content, we measured MDA and IL content in the shoots and roots of transgenic lines and WT plants, following high-salt treatments vs. no-salt control (Figure 6). Following high-salt treatment, MDA content was lower in the shoots and roots of the transgenic plants as compared to those of WT (Figure 6A–G). Similarly, IL value was lower in the shoots and roots of transgenic plants as compared to those of WT under high salt treatment (Figure 6B–H). Taken together, these results suggest that the overexpression of *MaROP5g* in transgenic *A. thaliana* plants may have prevented or reduced membrane injury under salt stress as compared to WT.

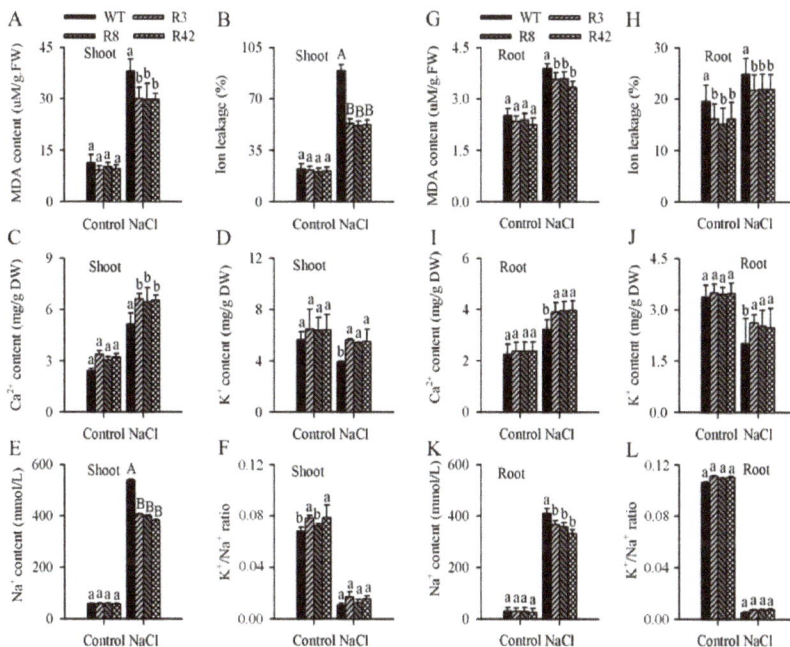

Figure 6. Physiological analyses and ion concentration in roots from wild-type and *A. thaliana* overexpressing *MaROP5g*. (**A–F**) Malonaldehyde content, ion leakage, Ca^{2+} concentration, K$^+$ concentration, Na$^+$ ion concentration, and K$^+$/Na$^+$ ratio of wild-type and transgenic shoots under normal conditions and salt treatment. (**G–L**) Malonaldehyde content, ion leakage, Ca^{2+} concentration, K$^+$ concentration, Na$^+$ ion concentration, and K$^+$/Na$^+$ ratio of wild-type and transgenic roots under normal conditions and salt treatment. WT: Wild-type. R3, R8, R42: *MaROP5g* transgenic plants. ANOVA was used to compare the significance of differences, using Student's *t* tests in the comparison between control and NaCl treatment. Data are presented as means ± standard deviations of *n* = 3 biological replicates. Different lowercase letters above bars indicate significant differences at *p* < 0.05, and different uppercase letters above bars indicate extremely significant differences at *p* < 0.01.

Under highly saline conditions, plant cells survive by retaining a high cytosolic Ca^{2+} concentration and a high K$^+$/Na$^+$ ratio [37,38]. Under high-salt treatment, the concentrations of Ca^{2+} and K$^+$ in the shoots or roots of transgenic *A. thaliana* plants were greater (Figure 6C,D), while the Na$^+$ concentration was lower in the shoots or roots of transgenic *A. thaliana* plants as compared to those of WT (Figure 6E). Therefore, the shoots or roots of the transgenic lines maintained higher K$^+$/Na$^+$ ratios than did those of the WT plants during salt treatment (Figure 6F). These results suggested that the overexpression of *MaROP5g* in plants subjected to salt stress increased cellular Ca^{2+} and K^+ accumulation, decreased cellular Na^+ accumulation, and improved the K^+/Na^+ ratio.

2.8. MaROP5g Overexpression Increased the Expression of Salt Overly Sensitive (SOS)-Pathway and Ca^{2+}-Sensing Genes

To gain an in-depth understanding of *MaROP5g* function in response to salt stress, we measured the expression of three Salt Overly Sensitive (SOS)-pathway genes (namely *SOS1*, *SOS2*, and *SOS3*) and several genes encoding calcium-signaling pathway proteins, including calcineurin B-like (CBL) proteins, CBL-interacting protein kinases (CIPKs), and calcium-dependent protein kinases (CDPKs) [39], in both WT and *MaROP5g*-overexpressing *A. thaliana* (Figure 7). Under standard growth conditions, we observed no significant differences in the transcription levels of the tested genes in the transgenic lines as compared to those in WT plants. However, under salt stress, the gene expression levels of *SOS1*, *SOS2*, *SOS3*, *CBL*, *CIPK*, and *CDPK* were higher in the transgenic lines as compared to those in WT. This indicated that *MaROP5g* overexpression in response to salt stress led to the up-regulation of both SOS-pathway genes and calcium-signaling pathway genes.

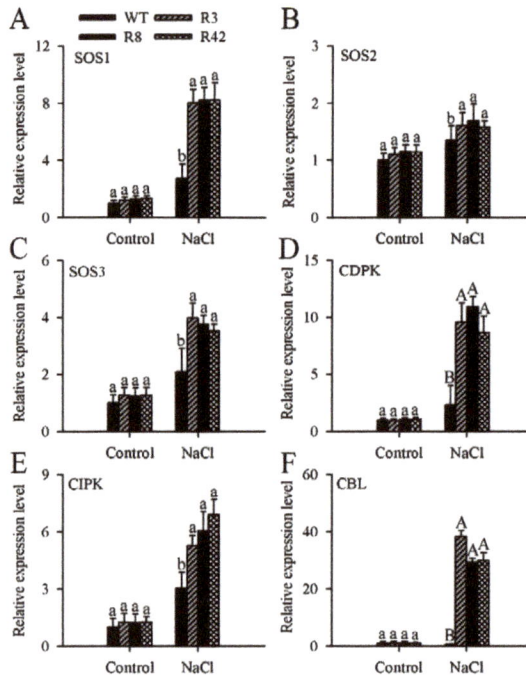

Figure 7. Expression of Salt Overly Sensitive (SOS)- and calcium-signaling genes in wild-type and *A. thaliana* overexpressing *MaROP5g*.WT: Wild-type. R3, R8, R42: *MaROP5g* transgenic plants. (**A–F**) Expression patterns of *SOS1*, *SOS2*, *SOS3*, *CDPK*, *CIPK*, and *CBL* genes in wild-type and transgenic roots under normal conditions and salt treatment. ANOVA was used to compare the significance of differences, using Student's *t* tests in the comparison between control and NaCl treatment. Data are presented as means ± standard deviations of $n = 3$ biological replicates. Different lowercase letters above bars indicate significant differences at $p < 0.05$, and different uppercase letters above bars indicate extremely significant differences at $p < 0.01$.

3. Discussion

Despite its economic and social importance, research on banana plants has generally been slower relative to many other crops, especially with respect to the abiotic stress responses [31,32]. ROP is an important molecular switch involved in plant signal transduction processes, which has been suggested to play crucial roles in the regulation of the environmental stress responses in numerous plant species [8,11,21,40]. We have identified 17 *MaROPs* by searching the *M. acuminata* genome, which were classified into three groups (II–IV), following the nomenclature derived for *OsROPs* [18]. Of the 17 *MaROPs*, none was categorized into Group I, congruent with *ROPs* in other higher plants, such as *O. sativa* [18], *Z. mays* [16], *Medicago truncatula* [7], and *N. tabacum* [7]. The recovered phylogenetic relationships were further supported by our analyses of gene structure and conserved motifs. The *MaROP* genomic sequences in Groups II–IV were found to contain 6–8 exons and 6–7 introns. Similar structural features have been observed in *ROPs* of other plant species, including *N. tabacum* [23], *A. thaliana* [15], and *M. truncatula* [7]. Moreover, all of the identified MaROPs had the typical RhoGEF domain (PF00621), and MaROP proteins within each group shared similarly conserved motifs (Figure 2), which is consistent with the observations in *M. truncatula* [7].

Bananas are extremely sensitive to abiotic stress and can suffer severe losses in yield and quality when exposed to cold, salt, or drought conditions [36]. We found that 82% of the 17 *MaROPs* showed transcriptional changes following cold, salt, and osmotic stress treatments (Figure 3). Interestingly, except the high expression genes (*MaROP-3b, -5a, -5c, -5f, -5g,* and *-6*) under three stress treatments, *MaROP-3a* and *-5h* are significantly downregulated in the presence of stress in BX and FJ compared to control. To our knowledge, this is the first report showing that banana *MaROPs* exhibit extensive and diverse responses to abiotic stressors. The induction of *ROP* expression by cold, salt, and drought has previously been reported in other plants, such as *Malus× domestica* Borkh [21], *A. thaliana* [22], and *N. tabacum* [23].

It is particularly important to note that the expression of *MaROP5g* among the 17 *MaROP* genes was most highly induced following salt stress treatment across both banana genotypes tested (Figures 3 and 4A–D), which may imply its positive role in mediating banana's response to salt stress. Further, although *MaROP5g* expression in roots of both BX and FJ genotypes can be induced by salt stress treatment, BX showed more sensitivity than FJ under salt stress treatment. This may suggest that BX, with its genome constitution as AAA, is more sensitive to salt treatment in comparison to the B-genome-containing genotype FJ. Such an observation is consistent with previous studies that FJ, with AAB genotypes, exhibited higher tolerance to abiotic stresses relative to BX [27].The MaROP5g protein was located on the plasma membrane (Figure 4E), consistent with *A. thaliana* ROP2 [41] and rice OsRac5 [18]. To better understand the function of *MaROP5g* during salt stress, we generated a number of *MaROP5g*-overexpressing transgenic *A. thaliana* lines. Under salt stress, the transgenic seedlings and adult plants exhibited a higher survival rate and increased root length as compared to WT (Figure 5), suggesting that *MaROP5g* overexpression might contribute to the maintenance of a healthy growth status, through the improvement of root development and distributions [24,25], and hence enhance salt stress tolerance.

As cell membranes are one of the primary targets of various environmental stresses, MDA is commonly used as an index of oxidative damages [36], and IL is used as an important indicator of membrane injury in plant research [36]. MDA content and IL were measured to assess the role of *MaROP5g* overexpression in reducing membrane injury under salt conditions. *MaROP5g* overexpression resulted in decreased IL and MDA content relative to WT, indicating that *MaROP5g*-overexpressing plants may experience less membrane injury and maintain a healthy physiological status under salt conditions.

In plants, high K^+ and low Na^+ concentrations are beneficial for the maintenance of physiological processes under salt stress [42]. In recent years, a high cytosolic K^+/Na^+ ratio has become an accepted marker of salinity tolerance [38]. Previous studies have reported that the expression of *MfARL1* resulted in a reduced Na^+/K^+ ratio in transgenic *A. thaliana* as compared to WT, due to a lower accumulation of Na+ [24]. Under salt stress, *MaROP5g* overexpression decreased the accumulation of cellular Na+, increased the Ca^{2+} concentration, and improved the K+/Na+ ratio in transgenic *A. thaliana* as compared to WT plants (Figure 6). Therefore, the increased salt stress tolerance conferred by *MaROP5g* overexpression may be due not only to the decreased Na+ accumulation, but also to the increased Ca^{2+} concentration in transgenic lines as compared to WT.

Many different ion transporters and channel proteins, such as SOS, CDPK, CBL, and CIPK, play crucial roles in maintaining ion homeostasis during salt stress [39,42,43]. For example, under salt stress, the SOS1 and SOS2 proteins regulate Na+/K+ homeostasis in *A. thaliana*; once the calcium binding protein SOS3 senses an increase in cytosolic calcium concentration, the SOS3–SOS2 protein kinase complex activates the SOS1 ion transporter [42,43]. In addition, calcium (Ca^{2+}), as a second messenger, plays an important role in salt stress processes [39,44]. Ca^{2+} increase can be decoded and recognized by Ca^{2+} sensors, including CBLs, CIPKs, and CDPKs [44]. CBLs recognize the increase in cytosolic Ca^{2+} concentration triggered by Na+ accumulation [39]. CIPKs and CDPKs may unify and coordinate ionic homeostasis at the cellular and organismal level [44]. We examined the expression of these SOS- and calcium-signaling pathway genes in the *MaROP5g*-overexpressing transgenic *A. thaliana* seedlings in relation to WT seedlings. Following salt treatment, SOS- and calcium-signaling pathway genes were a significant expression in the transgenic seedlings as compared to WT seedlings. This suggested that the *MaROP5g*-overexpressing transgenic plants were more responsive to SOS- and calcium-signaling compared to WT plants, implying that *MaROP5g*-overexpressing plants had improved Na+ and Ca^{2+} ionic homeostasis under salt stress conditions.

4. Experimental Section

4.1. Plant Materials

BaXi Jiao (BX; *M. acuminata* cv. Cavendish; AAA group) is a triploid banana cultivar that is of high yield and high quality and can be stored for an extended period of time [31,32]. Fen Jiao (FJ; *M. acuminata*; group AAB), another triploid banana cultivar, has good flavor, rapid ripening, and a high tolerance for abiotic stress [31,32]. Both of these banana cultivars were planted and maintained at the banana plantation of the Chinese Academy of Tropical Agricultural Sciences (Danzhou, Hainan, China; 19°11′–19°52′ N, 108°56′–109°46′ E). All of the banana plants were grown in 70% relative humidity at 28 °C, with 200 µmol·m^{-2}·s^{-1} light in cycles of 16 h light/8 h dark (Sylvania GRO LUX fluorescent lamps; Utrecht, The Netherlands).

For the salt and drought-simulation experiments, five-leaf stage banana plants of both cultivars were irrigated with 300 mmol·L^{-1} NaCl or 200 mmol·L^{-1} mannitol, respectively, for 7 days, as previously described by Hu et al. [27]. For the cold experiments, five-leaf stage banana plants of both cultivars were subjected to 4 °C for 22 h, as previously described by Hu et al. [27]. For the control experiments, five-leaf stage banana plants of both cultivars were irrigated with equal volume water with the stress groups at 28 °C. After the completion of each treatment, we harvested and immediately froze in liquid nitrogen the leaves and root systems of each plant, which were stored at −80 °C. Twelve five-leaf stage banana plants and three biological replicates were performed for each treatment.

4.2. Identification and Phylogeny of the MaROP Gene Family

The banana ROP protein sequences were downloaded from the DH-Pahang genome database (*M. acuminata*, A-genome, 2*n* = 22) (available online: http://banana-genome.cirad.fr) [33]. ROP amino acid sequences from *A. thaliana* (AtROPs) and *O. sativa* (OsROPs) were downloaded from the TAIR (The Arabidopsis Information Resource) (available online: http://www.arabidopsis.org) and RGAP (Rice Genome Annotation Project) (available online: http://rice.plantbiology.msu.edu) databases, respectively. HMMER (available online: http://hmmer.org) was used to predict conserved RhoGEF domains (PF00621; available online: http://pfam.sanger.ac.uk) in the ROP proteins [34]. The basic local alignment search tool (BLAST) (available online: http://www.ncbi.nlm.nih.gov/BLAST/) was used to identify putative MaROPs, based on the sequences of the AtROPs and OsROPs [35]. The conserved domains of the putative MaROPs were identified with the Conserved Domain Database (available online: http://www.ncbi.nlm.nih.gov/cdd) and validated with PFAM (available online: http://pfam. sanger.ac.uk) [45–47]. Identity numbers of all of the putative MaROPs that we have identified was presented in Table S1. All of the MaROP, AtROP, and OsROP sequences were aligned with Multiple Sequence Alignment (MUSCLE), and a bootstrapped ML phylogenetic tree (1000 replicates) was constructed in MEGA 5.2 (available online: http://www.megasoftware.net/) using this alignment [48].

4.3. Protein Properties and Gene Structure

We predicted the molecular masses and isoelectric points of the putative MaROP proteins with the Expert Protein Analysis System database (available online: http://expasy.org/) [49]. We constructed a bootstrapped ML phylogenetic tree (1000 replicates) in MEGA 5.2 software by aligning all MaROP sequences with MUSCLE [47]. MaROP protein motifs were identified with Multiple Em for Motif Elicitation (available online: http://meme-suite.org) and annotated using InterProScan (available online: http://www.ebi.ac.uk/Tools/pfa/iprscan) [50,51]. Structural features of the *MaROP* genes were identified with Gene Structure Display Server (available online: http://gsds.cbi.pku.edu.cn) by comparing the nucleotide sequences to predicted coding regions for all *MaROPs* [52]. *MaROP* promoter sequences were obtained from the banana genome database (available online: http://banana-genome. cirad.fr) [33]. Based on fragments 2000 bp upstream of each *MaROP*, a transcription start site was predicted with the Berkeley *Drosophila* Genome Project database (available online: http://www.fruitfly. org/seq_tools/promoter.html) and the *cis*-acting elements were predicted with PlantCARE (available online: http://bioinformatics.psb.ugent.be/webtools/plantcare/html) [53,54].

4.4. Transcriptomic Analysis

We isolated total RNA from the leaf tissues of the banana seedlings subjected to each of the three treatments (salt, osmosis, and cold) and control (no stress conditions), which were constructed into respective cDNA libraries [31,32]. Deep paired-end sequencing was performed with an Illumina GAII according to manufacturer's instructions. There are two replicates for each sample. The sequencing depth was 5.34X on average. Adaper sequences in the raw reads were removed using FASTX-tookit (Illlumina, San Diego, CA, USA). Using Tophat v.2.0.10, clean reads were mapped to the DH-Pahang genome [33]. The transcriptome assemblies were performed by Cufflinks [27]. Gene expression levels were calculated as fragments per kilobase of exon per million fragments mapped (FPKM). DEGseq was used to identify differently expressed genes [55]. Under three stress conditions, the expression level of each gene was compared with the control. A heat map was created based on the FPKM value of the *MaROPs*, compared to the control by MeV 4.9.0 software (available online: https://sourceforge.net/projects/mev-tm4/).

4.5. QRT-PCR Analysis

The gene expression of *MaROPs* in response to cold, salt, and osmotic (drought) stress was measured with qPCR, using a SYBR Premix ExTaq kit (TaKaRa, Shiga, Japan) on a Stratagene Mx3000P detection system (Stratagene, San Diego, CA, USA). The primer pairs with high specificity and efficiency were selected based on their melting curve and on agarose gel electrophoresis (Table S5). The amplification efficiencies of the primer pairs chosen ranged from 0.9 to 1.1. *MaActin* (EF672732) and *MaUBQ2* (HQ853254) were used as the internal controls. The expression levels of *MaROP* relative to *MaActin* and *MaUBQ2* were calculated with the $2^{-\Delta\Delta CT}$ method [56]. Three biological replicates for each sample were performed.

4.6. Full-Length cDNA of MaROP5g and Gene Expression During Salt Treatment

Based on our RNA-seq results, we selected the *ROP* gene *MaROP5g* for further analysis. The entire coding region of *MaROP5g* was amplified with PCR, using single-stranded cDNA obtained from the roots of banana plants subjected to salt stress as the source template, using a specific primer pair (5′-gcaccatggagatgagcgcgtcgaggt-3′ and 5′-gcgactagtcaatatggagcaacctttc-3′). The resulting *MaROP5g* fragment was verified with DNAMAN (available online: http://www.lynnon.com/) and compared to the genome database of DH-Pahang using BLAST [33,35].

Twelve ex vitro banana plants with uniform growth at the five-leaf stage were selected and divided into four groups for salt treatments, which were irrigated with half-strength Hoagland solution, supplemented with 300 mM NaCl for 0, 2, 4, or 6 h (*n* = 4 per time period), as previously described by Xu et al. [36]. Samples were frozen individually in liquid nitrogen and stored at −70 °C. Compared to the expression of 0 h in BX, the relative expression level of *MaROP5g* at 2, 4, and 6 h under salt stress in BX was calculated. Compared to the expression of 0 h in FJ, the relative expression level of *MaROP5g* at 2, 4, and 6 h under salt stress in FJ banana plants was calculated.

4.7. Subcellular Localization of MaROP5g

The ORF of *MaROP5g* was digested with the restriction enzymes *Nco* I and *Spe* I and inserted into a pCAMBIA1304-GFP expression vector to generate a MaROP5g-GFP fusion protein, under the control of the cauliflower mosaic virus (CaMV) 35S promoter. The recombinant pCAMBIA1304-MaROP5g-GFP plasmid was transferred to *Agrobacterium tumefaciens* strain LBA4404, and used to transform *A. thaliana* through a floral-dip method [57]. Root tips (3–5 mm) of *A. thaliana* seedlings (5-day old) with a stable expression of MaROP5g-GFP were incubated in 1 mL 1/2 Murashige and Skoog (MS) medium containing 10 μg FM4-64 (Invitrogen, Carlsbad, CA, USA) for 5 min at 25 °C, according to the Riqal et al. [58] methods. The GFP (488 nm emission filter) and FM4-64 (543 nm emission filter) signals were visualized using confocal laser scanning microscopy (CLSM; Nikon, A1, Tokyo, Japan). According to the Protein Subcellular Localization Prediction Tool (PSORT) software (available online: www.genscript.com/psort.html) prediction, the XXRR-like motif in the N-terminus of MaROP5g protein was identified as a membrane retention.

4.8. Plant Transformation and Generation of Transgenic Plants

The ORF of *MaROP5g* was digested with the restriction enzymes *Nco* I and *Spe* I, and inserted into a pCAMBIA1304 vector. The recombinant pCAMBIA1304-MaROP5g plasmid was transferred to *A. tumefaciens* strain LBA4404 [57]. Transgenic *A. thaliana* plants were generated using the floral dip-mediated infiltration method [57]. Seeds from T_0 transgenic plants were plated in kanamycin selection medium (50 mg·L^{-1}). The homozygous T_3 lines were used for further functional investigation of *MaROP5g*.

4.9. Southern Blot Analyses

Genomic DNA isolated from the T_3 generation kanamycin-resistant transgenic lines was digested with the *EcoR* I restriction enzyme. A 436 bp region of *MaROP5g* was amplified by PCR, using a pair of specific oligo primers (5′-gtggtggatggtaacacagtta-3′ and 5′-aacctttctgttgcttttttttc-3′). Based on this sequence, we prepared a hybridization probe for use with DIG-dUTP (Roche Applied Science, Mannheim, Germany), following the manufacturer's instructions. After hybridization, the HyBond N$^+$ membrane (Amersham) was washed and exposed to X-ray film (Kodak BioMax MS, Kodak Eastman, Rochester, NY, USA), following the method described by Miao et al. [59].

4.10. Salt Stress Treatments in WT and Transgenic Plants

The seeds of both transgenic and WT *A. thaliana* (Columbia ecotype; control) were first vernalized for 2 days at 4 °C in the dark, and surface sterilization in 75% ethanol for 10 min, prior to germination on half-strength MS medium or directly in soil. These *A. thaliana* plants were maintained at 22 °C with 70% humidity and a 16 h light/8 h dark cycle (Sylvania GRO LUX fluorescent lamps; Utrecht, The Netherlands). To analyze *A. thaliana* phenotypes in early seedlings under normal conditions, four day-old seedlings were transferred to 1/2 MS medium for 15 days, photos were taken, and root lengths were measured. To test salt stress tolerance in early seedlings, four day-old seedlings were transferred to either 1/2 MS or 1/2 MS supplemented with 100–200 mM NaCl for 15 days, after which photos were taken and the root length was measured. To test salt stress tolerance in adult plants, 4-week-old *A. thaliana* plants were irrigated with 350 mM NaCl for 15 days, and then photos were taken and survival rates were assessed (leaves fall and roots rot were identified as death). To measure the expression of SOS- and calcium-signal pathway genes in the WT and transgenic lines, 15-day-old seedlings were transferred to 1/2 MS supplemented with 350 mM NaCl for up to 10 h. Whole leaves were used to quantify relative gene expression, using qRT-PCR (see Table S5 for primer sequences).

4.11. Measurement of IL and MDA Content

Four-week-old *A. thaliana* plants were irrigated with 350 mM NaCl for 15 days and leaf samples were collected to examine IL and MDA. IL was detected according to the method described by Xu et al. [36]. Leaf samples were cut into strips and incubated in 10 mL of distilled water at 25 °C for 8 h. The initial conductivity (C1) was determined with a conductivity meter (DDBJ-350). The samples were then boiled for 10 min to yield complete IL. After cooling down, the electrolyte conductivity (C2) was measured. IL was calculated according to the equation: IL (%) = C1/C2:100. MDA content was measured according to the thiobarbituric acid colorimetric method, as described by Xu et al. [36].

4.12. Ca^{2+}, Na^+, and K^+ Concentrations

We irrigated 4-week-old WT and transgenic plants with 350 mM NaCl for 15 days. We then collected the plant roots and washed them with ultrapure water. Plant roots were heated to 105 °C for 10 min and then dried at 80 °C for 48 h. We dissolved 50 mg of each dried sample in 6 mL nitric acid and 2 mL H_2O_2 (30%), and then heated the solution to 180 °C for 15 min. The digested samples were diluted to a total volume of 50 mL with ultrapure water, transferred to clean tubes, and analyzed with atomic absorption spectroscopy (Analyst400, Perkin Elmer, Waltham, MA, USA).

4.13. Statistical Analysis

Three biological replicates for each sample were performed. Statistical analyses were performed using SPSS 19.0 (Chicago, IL, USA). We used analyses of variance (ANOVAs) to compare the significance of differences based on Dunnett's tests or Student's *t* tests. Specifically, Dunnett's tests were used to compare between WT and each overexpression line, while Student's *t* tests were used to compare between control and NaCl treatment. $p < 0.05$ was considered a significant level and $p < 0.01$ an extremely significant level.

5. Conclusions

In this study, for the first time for banana, we identified 17 *MaROP* genes in the *M. acuminata* genome, and classified these genes into three groups (II–IV) based on phylogeny, gene structure, and conserved protein motifs. The expression patterns of the *MaROP* genes in response to abiotic stress as reported here may shed light on the possible involvement of these genes in the regulation of abiotic stress signaling pathways. Of particular interest was *MaROP5g*, the overexpression of which increased the plant's tolerance for salt stress not only by maintaining a healthy growth status, but also by reducing membrane injury and improving ion distribution. Our results lay a foundation for genetic improvements in the banana plant, increasing resistance to various abiotic stressors, particularly salt. It is necessary to point out that these conclusions were drawn from the heterologous expression of banana *MaROP5g* in *A. thaliana* as a model plant system, which may or may not be valid in other plant systems. Further studies are required to characterize the function of *MaROP5g* in banana.

Supplementary Materials: Supplementary materials can be found at http://www.mdpi.com/1422-0067/19/10/3108/s1.

Author Contributions: Conceived and designed the experiments: B.X. and Z.J. Performed the experiments: H.M. and P.S. Analyzed the data: H.M. and P.S. Contributed reagents/materials/analysis tools: H.M., P.S., J.L., J.W., B.X., and Z.J. Wrote the paper: H.M. and B.X.

Acknowledgments: We thank Qing Liu (Commonwealth Scientific and Industrial Research Organization Agriculture and Food, Australia) for technical assistance. This work was supported by the Modern Agro-industry Technology Research System (No. CARS-31), the National Natural Science Foundation of China (NSFC, No. 31401843), the Central Public-interest Scientific Institution Basal Research Fund for Innovative Research Team Program of CATAS (No. 1630052017010), and the Central Public-interest Scientific Institution Basal Research Fund for Chinese Academy of Tropical Agricultural Sciences (No. 1630052016006).

Conflicts of Interest: The authors declare no conflict of interest.

References

1. Takai, Y.; Sasaki, T.; Matozaki, T. Small GTP-binding proteins. *Physiol. Rev.* **2001**, *81*, 153–208. [CrossRef] [PubMed]
2. Ono, E.; Wong, H.L.; Kawasaki, T.; Hasegawa, M.; Kodama, O.; Shimamoto, K. Essential role of the small GTPase Rac in disease resistance of rice. *Proc. Natl. Acad. Sci. USA* **2001**, *98*, 759–764. [CrossRef] [PubMed]
3. Rocha, N.; Payne, F.; Huang-Doran, I.; Sleigh, A.; Fawcett, K.; Adams, C.; Stears, A.; Saudek, V.; O'Rahilly, S.; Barroso, I.; Semple, R.K. The metabolic syndrome-associated small G protein ARL15 plays a role in adipocyte differentiation and adiponectin secretion. *Sci. Rep.* **2017**, *71*, 17593. [CrossRef] [PubMed]
4. Eliáš, M.; Klimeš, V. Rho GTPases: Deciphering the evolutionary history of a complex protein family. *Methods Mol. Biol.* **2012**, *827*, 13–34. [PubMed]
5. Feiquelman, G.; Fu, Y.; Yalovsky, S. RopGTPases structure-function and signaling pathways. *Plant Physiol.* **2018**, *176*, 57–59. [CrossRef] [PubMed]
6. Winge, P.; Brembu, T.; Bones, A.M. Cloning and characterization of rac-like cDNAs from *Arabidopsis thaliana*. *Plant Mol. Biol.* **1997**, *35*, 483–495. [CrossRef] [PubMed]
7. Liu, W.; Chen, A.M.; Luo, L.; Sun, J.; Cao, L.P.; Yu, G.Q.; Zhu, J.B.; Wang, Y.Z. Characterization and expression analysis of *Medicago truncatula* ROP GTPase family during the early stage of symbiosis. *J. Integr. Plant Biol.* **2010**, *52*, 639–652. [CrossRef] [PubMed]
8. Zheng, Z.L.; Yang, Z. The rop GTPase switch turns on polar growth in pollen. *Trends Plant Sci.* **2000**, *5*, 298–303. [CrossRef]
9. Abbal, P.; Pradal, M.; Sauvage, F.X.; Chatelet, P.; Paillard, S.; Canaguier, A.; Adam-Blondon, A.F.; Tesniere, C. Molecular characterization and expression analysis of the Rop GTPase family in *Vitis vinifera. J. Exp. Bot.* **2007**, *58*, 2641–2652. [CrossRef] [PubMed]
10. Zhang, Y.; McCormick, S. The regulation of vesicle trafficking by small GTPases and phospholipids during pollen tube growth. *Sex Plant Reprod.* **2010**, *23*, 87–93. [CrossRef] [PubMed]
11. Huang, J.B.; Liu, H.; Chen, M.; Li, X.; Wang, M.; Yang, Y.; Wang, C.; Huang, J.; Liu, G.; Liu, Y.; et al. Rop3 GTPase contributes to polar auxin transport and auxin responses and is important for embryogenesis and seedling growth in Arabidopsis. *Plant Cell* **2014**, *26*, 3501–3518. [CrossRef] [PubMed]
12. Poraty-Gavra, L.; Zimmermann, P.; Haigis, S.; Bednarek, P.; Hazak, O.; Stelmakh, O.R.; Sadot, E.; Schulze-Lefert, P.; Gruissem, W.; Yalovsky, S. The Arabidopsis Rho of plants GTPase AtROP6 functions in developmental and pathogen response pathways. *Plant Physiol.* **2013**, *161*, 1172–1188. [CrossRef] [PubMed]
13. Ma, Q.H.; Zhu, H.H.; Han, J.Q. Wheat ROP proteins modulate defense response through lignin metabolism. *Plant Sci.* **2017**, *262*, 32–38. [CrossRef] [PubMed]
14. Yang, Z.; Watson, J.C. Molecular cloning and characterization of rho, a ras-related small GTP-binding protein from the garden pea. *Proc. Natl. Acad. Sci. USA* **1993**, *90*, 8732–8736. [CrossRef] [PubMed]
15. Li, H.; Shen, J.J.; Zheng, Z.L.; Lin, Y.K.; Yang, Z.B. The Rop GTPase switch controls multiple developmental processes in Arabidopsis. *Plant Physiol.* **2001**, *126*, 670–684. [CrossRef] [PubMed]
16. Christensen, T.M.; Vejlupkova, Z.; Sharma, Y.K.; Arthur, K.M.; Spatafora, J.W.; Albright, C.A.; Meeley, R.B.; Duvick, J.P.; Quatrano, R.S.; Fowler, J.E. Conserved subgroups and developmental regulation in the monocot *rop* gene family. *Plant Physiol.* **2003**, *133*, 1791–1808. [CrossRef] [PubMed]
17. Chan, J.; Pauls, P.K. Brassica napus Rop GTPases and their expression in microspore cultures. *Planta* **2007**, *225*, 469–484. [CrossRef] [PubMed]

18. Chen, L.; Shiotani, K.; Togashi, T.; Miki, D.; Aoyama, M.; Wong, H.L.; Kawasaki, T.; Shimamoto, K. Analysis of the Rac/Rop small GTPase family in rice: Expression, subcellular localization and role in disease resistance. *Plant Cell Physiol.* **2010**, *51*, 585–595. [CrossRef] [PubMed]

19. Qin, Y.X.; Huang, Y.C.; Fang, Y.J.; Qi, J.Y.; Tang, C.R. Molecular characterization and expression analysis of the small GTPase ROP members expressed in laticifers of the rubber tree (*Hevea brasiliensis*). *Plant Physiol. Biochem.* **2014**, *74*, 193–204. [CrossRef] [PubMed]

20. Liang, Q.X.; Cao, G.Q.; Zhao, S.P.; Huang, Q.C.; Ying, F.Q.; Chen, W. Analysis of ROP signaling in the leaf epidermis of mutant tomato with low-energy ion beam. *Genet. Mol. Res.* **2015**, *14*, 3807–3816. [CrossRef] [PubMed]

21. Zermiani, M.; Zonin, E.; Nonis, A.; Begheldo, M.; Ceccato, L.; Vezzaro, A.; Baldan, B.; Trentin, A.; Masi, A.; Fadanelli, L.; et al. Ethylene negatively regulates transcript abundance of ROP-GAP rheostat-encoding genes and affects apoplastic reactive oxygen species homeostasis in epicarps of cold stored apple fruits. *J. Exp. Bot.* **2015**, *66*, 7255–7270. [CrossRef] [PubMed]

22. Li, Z.; Kang, J.; Sui, N.; Liu, D. ROP11 GTPase is a negative regulator of multiple ABA responses in Arabidopsis. *J. Integr. Plant Biol.* **2012**, *54*, 169–179. [CrossRef] [PubMed]

23. Cao, Y.; Li, Z.; Chen, T.; Zhang, Z.; Zhang, J.; Chen, S. Overexpression of a tobacco small G protein gene NtRop1 causes salt sensitivity and hydrogen peroxide production in transgenic plants. *Sci. China C Life Sci.* **2008**, *51*, 383–390. [CrossRef] [PubMed]

24. Wang, T.Z.; Xia, X.Z.; Zhao, M.G.; Tian, Q.Y.; Zhang, W.H. Expression of a Medicago falcata small GTPase gene, MfARL1 enhanced tolerance to salt stress in *Arabidopsis thaliana*. *Plant Physiol. Biochem.* **2013**, *63*, 227–235. [CrossRef] [PubMed]

25. Li, C.; Lu, H.; Li, W.; Yuan, M.; Fu, Y. A ROP2-RIC1 pathway fine-tunes microtubule reorganization for salt tolerance in Arabidopsis. *Plant Cell Environ.* **2017**, *40*, 1127–1142. [CrossRef] [PubMed]

26. Paul, J.Y.; Khanna, H.; Kleidon, J.; Hoang, P.; Geijskes, J.; Daniells, J.; Zaplin, E.; Rosenberg, Y.; James, A.; Mlalazi, B.; et al. Golden bananas in the field: Elevated fruit pro-vitamin A from the expression of a single banana transgene. *Plant Biotechnol. J.* **2017**, *15*, 520–532. [CrossRef] [PubMed]

27. Hu, W.; Wang, L.; Tie, W.; Yan, Y.; Ding, Z.; Liu, J.; Li, M.; Peng, M.; Xu, B.; Jin, Z. Genome-wide analyses of the bZIP family reveal their involvement in the development, ripening and abiotic stress response in banana. *Sci. Rep.* **2016**, *6*, 30203. [CrossRef] [PubMed]

28. Sreedharan, S.; Shekhawat, U.K.; Ganapathi, T.R. Constitutive and stress-inducible overexpression of a native aquaporin gene (MusaPIP2;6) in transgenic banana plants signals its pivotal role in salt tolerance. *Plant Mol. Biol.* **2015**, *88*, 41–52. [CrossRef] [PubMed]

29. Lee, W.S.; Gudimella, R.; Wong, G.R.; Tammi, M.T.; Khalid, N.; Harikrishna, J.A. Transcripts and MicroRNAs responding to salt stress in *Musa acuminata* Colla (AAA Group) cv. Berangan roots. *PLoS ONE* **2015**, *10*, e0127526. [CrossRef] [PubMed]

30. Hu, W.; Yan, Y.; Shi, H.; Miao, H.; Tie, W.; Ding, Z.; Wu, C.; Liu, Y.; Wang, J.; Xu, B.; Jin, Z. The core regulatory network of the abscisic acid pathway in banana: Genome-wide identification and expression analyses during development, ripening, and abiotic stress. *BMC Plant Biol.* **2017**, *17*, 145. [CrossRef] [PubMed]

31. Miao, H.X.; Sun, P.G.; Liu, Q.; Miao, Y.L.; Liu, J.H.; Zhang, K.X.; Hu, W.; Zhang, J.B.; Wang, J.Y.; Wang, Z.; et al. Genome-wide analyses of SWEET family proteins reveal involvement in fruit development and abiotic/biotic stress responses in banana. *Sci. Rep.* **2017**, *7*, 3536. [CrossRef] [PubMed]

32. Miao, H.X.; Sun, P.G.; Liu, Q.; Liu, J.H.; Xu, B.Y.; Jin, Z.Q. The AGPase family proteins in banana: Genome-wide identification, phylogeny, and expression analyses reveal their involvement in the development, ripening, and abiotic/biotic stress responses. *Int. J. Mol. Sci.* **2017**, *18*, 1581. [CrossRef] [PubMed]

33. D'Hont, A.; Denoeud, F.; Aury, J.M.; Baurens, F.C.; Carreel, F.; Garsmeur, O.; Noel, B.; Bocs, S.; Droc, G.; Rouard, M.; et al. The banana (*Musa acuminata*) genome and the evolution of monocotyledonous plants. *Nature* **2012**, *488*, 213–217. [CrossRef] [PubMed]

34. Eddy, S.R. A new generation of homology search tools based on probabilistic inference. *Genome Inform.* **2009**, *23*, 205–211. [PubMed]

35. Altschul, S.F.; Gish, W.; Miller, W.; Myers, E.W.; Lipman, D.J. Basic local alignment search tool. *J. Mol. Biol.* **1990**, *215*, 403–410. [CrossRef]

36. Xu, Y.; Hu, W.; Liu, J.H.; Zhang, J.B.; Jia, C.H.; Miao, H.X.; Xu, B.Y.; Jin, Z.Q. A banana aquaporin gene, MaPIP1; 1, is involved in tolerance to drought and salt stresses. *BMC Plant Biol.* **2014**, *14*, 59. [CrossRef] [PubMed]

37. Han, S.; Wang, C.W.; Wang, W.L.; Jiang, J. Mitogen-activated protein kinase 6 controls root growth in Arabidopsis by modulating Ca2+-based Na+ flux in root cell under salt stress. *J. Plant Physiol.* **2014**, *171*, 26–34. [CrossRef] [PubMed]

38. Ruiz-Lozano, J.M.; Porcel, R.; Azcón, C.; Aroca, R. Regulation by arbuscular mycorrhizae of the integrated physiological response to salinity in plants: New challenges in physiological and molecular studies. *J. Exp. Bot.* **2012**, *63*, 4033–4044. [CrossRef] [PubMed]

39. Miranda, R.S.; Alvarez-Pizarro, J.C.; Costa, J.H.; Paula, S.O.; Pirsco, J.T.; Gomes-Filho, E. Putative role of glutamine in the activation of CBL/CIPK signaling pathways during salt stress in sorghum. *Plant Signal. Behav.* **2017**, *12*, e1361075. [CrossRef] [PubMed]

40. Hoefle, C.; Huesmann, C.; Schultheiss, H.; Bornke, F.; Hensel, G.; Kumlehn, J.; Huckelhoven, R. A barley ROP GTPase ACTIVATING PROTEIN associates with microtubules and regulates entry of the barley powdery mildew fungus into leaf epidermal cells. *Plant Cell* **2011**, *23*, 2422–2439. [CrossRef] [PubMed]

41. Jeon, B.W.; Hwang, J.U.; Hwang, Y.; Song, W.Y.; Fu, Y.; Gu, Y.; Bao, F.; Cho, D.; Kwak, J.M.; Yang, Z.; et al. The Arabidopsis small G protein ROP2 is activated by light in guard cells and inhibits light-induced stomatal opening. *Plant Cell* **2008**, *20*, 75–87. [CrossRef] [PubMed]

42. Shi, H.; Ishitani, M.; Kim, C.; Zhu, J.K. The *Arabidopsis thaliana* salt tolerance gene *SOS1* encodes a putative Na⁺/H⁺ antiporter. *Proc. Natl. Acad. Sci. USA* **2000**, *97*, 6896–6901. [CrossRef] [PubMed]

43. Liu, J.; Ishitani, M.; Halfter, U.; Kim, C.S.; Zhu, J.K. The *Arabidopsis thaliana* SOS2 gene encodes a protein kinase that is required for salt tolerance. *Proc. Natl. Acad. Sci. USA* **2000**, *97*, 3730–3734. [CrossRef] [PubMed]

44. Köster, P.; Wallrad, L.; Edel, K.H.; Faisal, M.; Alatar, A.A.; Kudla, J. The battle of two ions: Ca²⁺ signaling against Na⁺ stress. *Plant Biol.* **2018**, *7*. [CrossRef] [PubMed]

45. Marchler-Bauer, A.; Bo, Y.; Han, L.; He, J.; Lanczycki, C.J.; Lu, S.; Lu, S.; Chitsaz, F.; Derbyshire, M.K.; Geer, R.C.; et al. CDD/SPARCLE: Functional classification of proteins via subfamily domain architectures. *Nucleic Acids Res.* **2017**, *45*, 200–203. [CrossRef] [PubMed]

46. Finn, R.D.; Coggill, P.; Eberhardt, R.Y.; Eddy, S.R.; Mistry, J.; Mitchell, A.L.; Potter, S.C.; Punta, M.; Qureshi, M.; Sangrador-Vegas, A.; et al. The Pfam protein families database: Towards a more sustainable future. *Nucleic Acids Res.* **2016**, *44*, 279–285. [CrossRef] [PubMed]

47. Larkin, M.A.; Blackshields, G.; Brown, N.P.; Chenna, R.; Mcgettigan, P.A.; Mcwilliam, H.; Valentin, F.; Wallace, I.M.; Wilm, A.; Lopez, R.; et al. Clustal W and Clustal X version 2.0. *Bioinformatics* **2007**, *23*, 2947–2948. [CrossRef] [PubMed]

48. Tamura, K.; Peterson, D.; Peterson, N.; Stecher, G.; Nei, M.; Kumar, S. MEGA5: Molecular evolutionary genetic analysis using maximum likelihood, evolutionary distance, and maximum parsimony methods. *Mol. Biol. Evol.* **2011**, *28*, 2731–2739. [CrossRef] [PubMed]

49. Gasteiger, E.; Gattiker, A.; Hoogland, C.; Ivanyi, I.; Appel, R.D.; Bairoch, A. ExPASy: The proteomics server for in-depth protein knowledge and analysis. *Nucleic Acids Res.* **2003**, *31*, 3784–3788. [CrossRef] [PubMed]

50. Bailey, T.L.; Boden, M.; Buske, F.A.; Frith, M.; Grant, C.E.; Clementi, L.; Ren, J.; Li, W.W.; Noble, W.S. MEME SUITE: Tools for motif discovery and searching. *Nucleic Acids Res.* **2009**, *37*, 202–208. [CrossRef] [PubMed]

51. Jones, P.; Binns, D.; Chang, H.Y.; Fraser, M.; Li, W.; McAnulla, C.; McWilliam, H.; Maslen, J.; Mitchell, A.; Nuka, G.; et al. InterProScan 5: Genome-scale protein function classification. *Bioinformatics* **2014**, *30*, 1236–1240. [CrossRef] [PubMed]

52. Hu, B.; Jin, J.; Guo, A.Y.; Zhang, H.; Luo, J.; Gao, G. GSDS 2.0: An upgraded gene feature visualization server. *Bioinformatics* **2015**, *31*, 1296–1297. [CrossRef] [PubMed]

53. Celniker, S.E.; Wheeler, D.A.; Kronmiller, B.; Carlson, J.W.; Haipern, A.; Patel, S.; Adams, M.; Champe, M.; Dugan, S.P.; Frise, E.; et al. Finishing a whole-genome shotgun: Release 3 of the *Drosophila melanogaster* euchromatic genome sequence. *Genome Biol.* **2002**, *3*, RESEARCH0079. [CrossRef] [PubMed]

54. Lescot, M.; Déhais, P.; Thijs, G.; Marchal, K.; Moreau, Y.; Peer, Y.V.; Rouzé, P.; Rombauts, S. PlantCARE, a database of plant *cis*-acting regulatory elements and a portal to tools for in silico analysis of promoter sequences. *Nucleic Acids Res.* **2002**, *30*, 325–327. [CrossRef] [PubMed]

55. Wang, L.; Feng, Z.; Wang, X.; Wang, X.; Zhang, X. DEGseq: An R package for identifying differentially expressed genes from RNA-seq data. *Bioinformatics* **2010**, *26*, 136–138. [CrossRef] [PubMed]

56. Livak, K.J.; Schmittgen, T.D. Analysis of relative gene expression data using real-time quantitative PCR and the $2^{-\Delta\Delta CT}$ Method. *Methods* **2001**, *25*, 402–408. [CrossRef] [PubMed]
57. Clough, S.J.; Bent, A.F. Floral dip: A simplified method for *Agrobacterium*-mediated transformation of *Arabidopsis thaliana*. *Plant J.* **1998**, *16*, 735–743. [CrossRef] [PubMed]
58. Riqal, A.; Doyle, S.M.; Robert, S. Live cell imaging of FM4-64, a tool for tracing the endocytic pathways in Arabidopsis root cells. *Methods Mol. Biol.* **2015**, *1242*, 93–103.
59. Miao, H.X.; Qin, Y.H.; Teixeira da Silva, J.A.; Ye, Z.X.; Hu, G.B. Cloning and expression analysis of *S-RNase* homologous gene in *Citrus reticulata* Blanco cv. Wuzishatangju. *Plant Sci.* **2011**, *180*, 358–367. [CrossRef] [PubMed]

International Journal of
Molecular Sciences

MDPI

Article

Vascular Plant One-Zinc-Finger (VOZ) Transcription Factors Are Positive Regulators of Salt Tolerance in Arabidopsis

Kasavajhala V. S. K. Prasad [1], Denghui Xing [1,2] and Anireddy S. N. Reddy [1,*]

[1]   Department of Biology and Cell and Molecular Biology Program, Colorado State University,
      Fort Collins, CO 80523, USA; kpsatya@mail.colostate.edu (K.V.S.K.P.); david.xing@mso.umt.edu (D.X.)
[2]   Genomics Core Lab, Division of Biological Sciences, University of Montana, Missoula, MT 59812, USA
*   Correspondence: reddy@colostate.edu; Tel.: +970-491-5773; Fax: +970-491-0649

Received: 16 September 2018; Accepted: 20 November 2018; Published: 23 November 2018

Abstract: Soil salinity, a significant problem in agriculture, severely limits the productivity of crop plants. Plants respond to and cope with salt stress by reprogramming gene expression via multiple signaling pathways that converge on transcription factors. To develop strategies to generate salt-tolerant crops, it is necessary to identify transcription factors that modulate salt stress responses in plants. In this study, we investigated the role of VOZ (VASCULAR PLANT ONE-ZINC FINGER PROTEIN) transcription factors (VOZs) in salt stress response. Transcriptome analysis in WT (wild-type), *voz1-1*, *voz2-1* double mutant and a *VOZ2* complemented line revealed that many stress-responsive genes are regulated by VOZs. Enrichment analysis for gene ontology terms in misregulated genes in *voz* double mutant confirmed previously identified roles of VOZs and suggested a new role for them in salt stress. To confirm VOZs role in salt stress, we analyzed seed germination and seedling growth of WT, *voz1*, *voz2-1*, *voz2-2* single mutants, *voz1-1 voz2-1* double mutant and a complemented line under different concentrations of NaCl. Only the double mutant exhibited hypersensitivity to salt stress as compared to WT, single mutants, and a complemented line. Expression analysis showed that hypersensitivity of the double mutant was accompanied by reduced expression of salt-inducible genes. These results suggest that VOZ transcription factors act as positive regulators of several salt-responsive genes and that the two VOZs are functionally redundant in salt stress.

Keywords: *Arabidopsis thaliana*; VOZ; transcription factor; salt stress; transcriptional activator

1. Introduction

Throughout their life span, plants are constantly subjected to diverse abiotic and biotic stresses, which severely inhibit plant growth and development, and cause huge losses in crop yields [1,2]. As sessile organisms, plants respond to these stresses rapidly by altering their gene expression patterns, which ultimately change biochemical and physiological processes that enable them to survive under stress conditions [2–4]. Plant transcription factors (TFs) play a key role in reprogramming gene expression in response to stresses [3–6]. Many of these transcription factors act by regulating the expression of down-stream genes that are important for stress tolerance [2–4,7]. VOZ (VASCULAR PLANT ONE-ZINC FINGER PROTEIN) is a plant-specific TF family with two members, *VOZ1* and *VOZ2*. Previous studies have shown that *VOZ1* is specifically expressed in the phloem tissue while *VOZ2* is highly expressed in the roots [8]. Yasui et al. [9] have shown their localization in vascular bundles and predominant subcellular presence in the cytoplasm, while they function in the nucleus. VOZ TFs (for brevity we refers to them as VOZs) were identified as proteins that bind to a *cis*-element *GCGTNx7ACGC* in the promoter of *AVP1* (V-PPAse) [8]. VOZs have two conserved domains (viz., A and B) and share about 53% similarity. VOZ2 regulates the expression of the target genes by binding to *cis*-elements via Domain B as a dimer. The B domain has a zinc finger motif and a basic region [8].

VOZs were also classified into NAC (for NAM (no apical meristem)) subgroup VIII-2 as they share homology with NAC proteins in the C-terminal basic region [10]. VOZs regulate flowering through their interaction with *PHYB* and promote the expression of *FLC* and *FT* [9,11]. More recent genetic, biochemical and cell biological studies have shown that VOZs interact with and modulate the function of CONSTANS (CO) in promoting flowering [12]. VOZs also play a key role in plant responses to abiotic (cold, drought and heat) and biotic (pathogens) stresses. VOZs function as a positive regulator of plant responses against bacterial and fungal pathogens, and as a negative regulator of two abiotic stresses-old and drought [13,14]. The expression levels of both *VOZ1* and *VOZ2* were also altered in response to biotic and abiotic stresses in an opposite manner [13]. Overexpression of the VOZ2 conferred biotic stress tolerance, however it showed sensitivity to freezing and drought stress in Arabidopsis [13]. Recent reports indicate that VOZ1 and VOZ2 act as transcriptional repressors for *DREB2C* and *DREB2A* respectively, which mediate heat stress response in plants [15,16]. Despite a few reports describing the role of VOZs in some abiotic stresses [17–20], the full scope of VOZs' function in other stresses and their potential target genes are not well understood. In the current study, analysis transcriptomes from WT, *voz1-1 voz2-1* double mutant and a complemented line suggested a new role for VOZs in salt stress. Analysis of seed germination and seedling growth of WT, *voz1*, *voz2-1*, *voz2-2* single mutants, *voz1-1 voz2-1* double mutant and a complemented line under different concentrations of NaCl revealed that only the double mutant is hypersensitive to salt stress. Through analysis of the upstream regions of genes regulated by VOZs for canonical and non-canonical binding sites, we have identified potential new targets of VOZ transcription factors. Furthermore, expression of salt-induced genes is impaired in the VOZ double mutant. Collectively, these results suggest that VOZs act as positive regulators of salt stress response and that the two VOZs are functionally redundant in salt stress.

2. Results

Arabidopsis VOZ TF family contains two members—*VOZ1* and *VOZ2*. Both these TFs were reported to be involved in flowering and response to abiotic (cold, drought and heat) and biotic stresses [9,11–16]. Here we performed RNA-Seq analysis of gene expression with RNA from wild type (WT), a *voz* double knockout (DKO—*voz1-1 voz2-1*) mutant and a complemented line (COMP2-4). Double mutant (DKO) lines exhibited suppressed growth and leaf vein clearing in older leaves in comparison to WT and COMP2-4 line (Figure 1a). Expression of *VOZ2* in *voz1-1 voz2-1* (COMP2-4) rescued the DKO phenotype (Figure 1a). Hence, we have chosen 30-day-old plants to identify new potential targets of the VOZs by comparing the transcriptomes of WT, DKO and complemented line (COMP2-4). Prior to RNA-Seq and phenotypic analysis, the genotypes of all the lines were verified by genomic PCR (Figure 1b and Figure S1a–c) with gene-specific and T-DNA or transposon-specific primers. RT-PCR analysis with *VOZ1* and *VOZ2* specific primers confirmed the absence of transcripts in *DKO* line (Figure 1c).

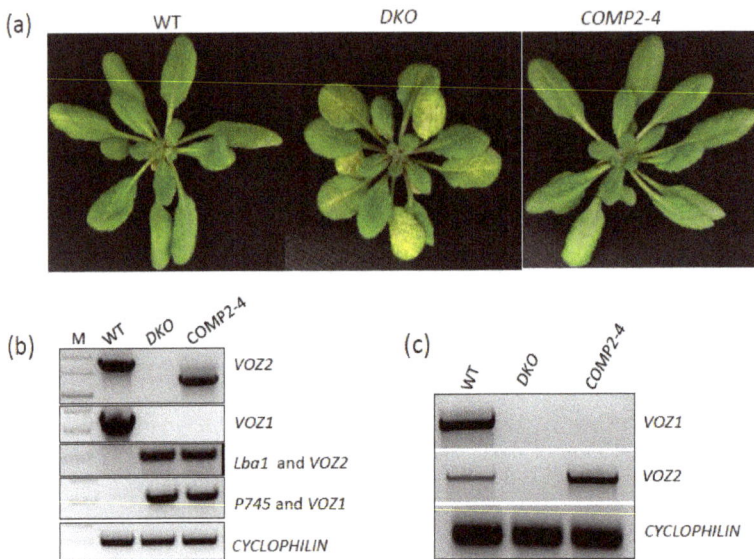

Figure 1. Validation of genotypes used for RNA-Seq. (**a**) Top panel: Phenotype of 30-day-old plants of wild-type (WT), double knockout (*DKO*) mutant (*voz1-1 voz2-1*) and *DKO* complemented line (COMP2-4) grown at 21 °C under day neutral conditions at 60% humidity. (**b**) Genomic PCR of three genotypes used for RNA-Seq. Top panel (PCR with *VOZ2*-specific primers); second panel (PCR with *VOZ1*-specific primers); third panel (PCR with T-DNA specific Lba1 and *VOZ2*-specific reverse primer); fourth panel (PCR with Tn insertion specific primer P745 and VOZ1-specific forward primer); bottom panel (PCR with *CYCLOPHILIN*-specific primers). In all cases expected size PCR product was obtained. (**c**) Analysis of expression of *VOZ1* (top panel), *VOZ2* (middle panel) and *CYCLOPHILIN* (bottom panel) using sqRT-PCR in 30-day-old seedlings of WT, *DKO* mutant (*voz1-1 voz2-1*) and *DKO* complemented line (COMP2-4).

2.1. Loss of Function of VOZs Resulted in Misregulation of Genes

For RNA-Seq, two biological replicates of WT, *DKO* and COMP2-4 were used. About 37 to 75 million short reads per sample were obtained (Table S1). The reads were mapped to the Arabidopsis genome (TAIR 10) and ~90% of these were uniquely mapped (Table S1). The expression levels of individual transcripts were determined by the number of reads per kilobase per million (RPKM). The expression patterns of the genes were well correlated among the replicates. However, the expression patterns were poorly correlated between WT and *DKO*, as indicated by an R^2 value of 0.77, suggesting a significant effect of VOZs on gene expression (Figure S2). The Cufflinks package was used to identify differentially expressed (DE) genes by comparing the transcriptome of WT with *DKO*. In the *DKO*, 112 genes were misregulated (significance adj. $p \le 0.05$ and fold change >log2) as compared to WT (Additional File 1 sheet1). Further, expression levels of the majority of these were either partially or fully restored in the *VOZ2* complemented line (COMP2-4) (Figure 2a–c; Additional File 1 sheet2), suggesting that DE genes are either direct or indirect targets of VOZs and loss of these TFs caused significant effects on expression of many genes. The majority of the DE genes (101) were up-regulated, while only 11 were down-regulated in the *DKO* mutant. About 83% of up-regulated genes are partially complemented while ~27% of them are fully complemented by overexpression of *VOZ2*. In the case of down-regulated DE genes, ~27% of genes were fully complemented and 72% exhibited partial complementation (Figure 2). The misregulation of a number of DE genes was verified using RT-qPCR (Figure 3a,b), corroborating the RNA-Seq data.

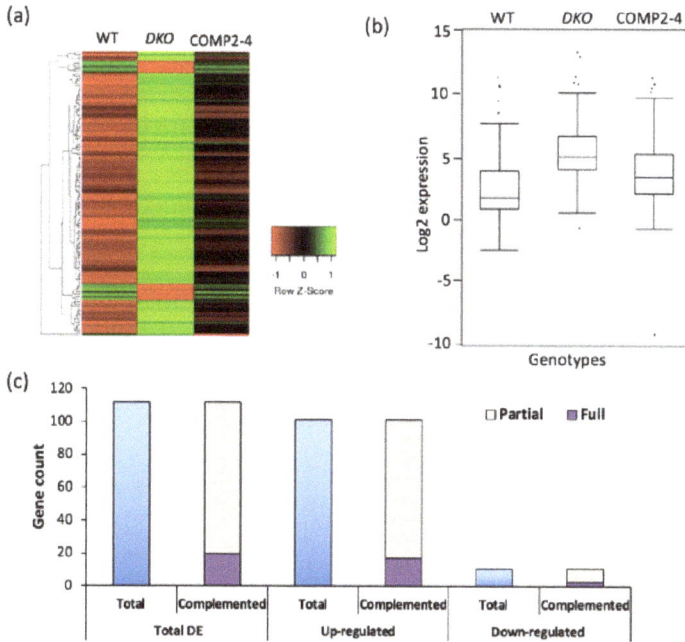

Figure 2. Analysis of differentially expressed genes. (**a**) Heatmap representation of differentially expressed genes in *WT*, *DKO* and *COMP2-4* (COMP) plants. Expression values were used to generate the heatmap using the Heatmapper. Columns represent samples and rows represent genes. Color scale indicates the gene expression level. Green indicates high expression and red Indicates low expression. (**b**) Box-and-whisker plots showing expression of differentially expressed (DE) genes in different genotypes. (**c**) Gene counts of total, up and down-regulated DE genes that are either fully or partially complemented in *COMP2-4* line.

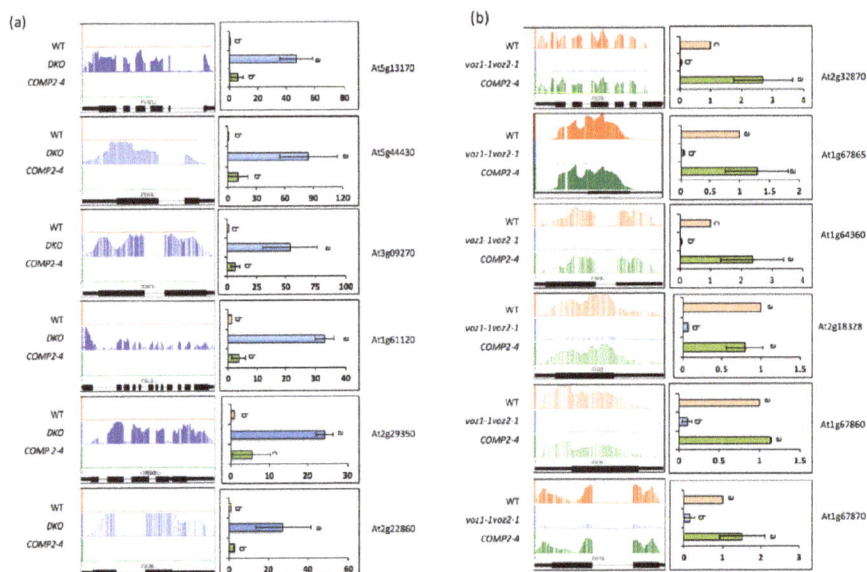

Figure 3. Validation of up- and down-regulated genes in *DKO*. (**a**) RT-qPCR validation of randomly selected up-regulated genes. (**b**) RT-qPCR of randomly selected down-regulated genes. Left panels in (**a,b**) show relative sequence read abundance (Integrated Genome Browser view) as histograms in WT, *DKO (voz1-1 voz2-1)* and COMP2-4 lines. The Y-axis indicates read depth with the same scale for all three lines. The gene structure is shown below the read depth profile. The lines represent introns and the boxes represent exons. The thinner boxes represent 5′ and 3′ UTRs. Right panels in (**a,b**) show fold change in expression level relative to WT. WT values were considered as 1. Student's *t*-test was performed and significant differences (*p* < 0.05) among samples are labeled with different letters. The error bars represent SD. The genes that were randomly picked include At1g61120 (terpene synthase 4), At1g64360 (enescence-associated and QQS-related), At1g67860 (hypothetical protein), At1g67865 (hypothetical protein), At1g67870 (hypothetical protein), At2g18328 (RAD-like4), At2g22860 (phytosulfokine 2 Precursor), At2g29350 (senescence-associated gene 13), At3g09270 (glutathione S-transferase TAU8), At2g32870 (TRAF-like protein), At5g13170 (senescence-associated gene29), and At5g44430 (plant defensing 1.2C).

2.2. VOZs Regulate Expression of Several Transcription Factors

It is possible that the effect of VOZs on the expression of its DE genes is mediated via its regulation of other TFs, hence we analyzed the DE genes for enrichment of TFs. Arabidopsis has over 1700 genes encoding TFs that are grouped into 58 families. Among DE genes, we observed eight TFs belonging to four families (Figure 4a, Additional File 2 and Figure S3). Of these families, bHLH (Basic-Helix-Loop-Helix) (*p* ≤ 0.03), MYB (v-myb avian myeloblastosis viral oncogene homolog)-related (*p* ≤ 0.00009) and NAC (*p* ≤ 0.00002) are highly enriched (Figure 4a). The number of TFs in each family and the direction of their expression in the *DKO* are presented in Figure S3 and Additional File 2. Interestingly, members of bHLH, NAC and C_2H_2 TF families are up-regulated whereas the members of MYB-related families are down-regulated in *DKO*. Significantly, five out of eight members of TF families showed expression levels similar to that of WT in the complemented line, indicating that VOZs regulates the expression of these transcription factors.

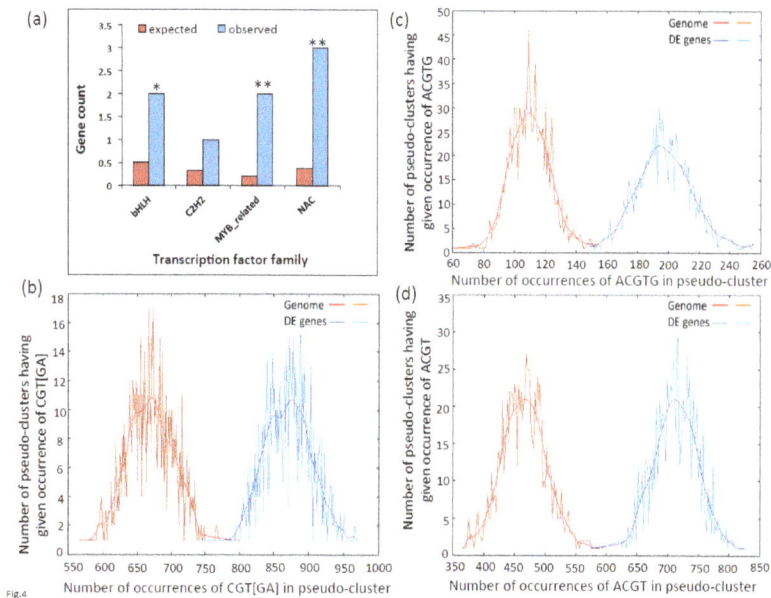

Fig.4

Figure 4. Enrichment of transcription factor families and VOZ binding sites in the promoters of DE genes. (**a**) DE genes are enriched for specific TF families. Observed: Number of genes associated with particular TF family in DE genes. Expected: Number of genes expected in each individual TF family in the genome. Asterisks on the bar represent significant overrepresentation of TFs with a (* $p \leq 0.05$) and (** $p \leq 0.0001$), respectively. (**b–d**) POBO analysis of NAC consensus sequence (*CGT[GA]*), G-box core sequence (*ACGTG*) and LS-7 *cis*-element (*ACGT*), respectively, in the −1000 bp upstream of TSS. One thousand pseudoclusters were generated from top 112 DE genes and genome background. The jagged lines show the motif frequencies from which the best-fit curve is derived. *CGT[GA], ACGTG* and *ACGT* elements are significantly overrepresented (two-tailed $p < 0.0001$) in the upstream sequences of DE genes.

2.3. Promoters of Differentially Expressed Genes are Enriched for G-Box, NAC and LS-7 Elements

As VOZs share significant sequence similarity with NAC subgroup VIII-2 TFs (particularly in the C-terminal region) and also reported to bind to the palindromic NAC binding sequence (palNAC-BS), we analyzed the promoters of DE genes (−1000 bp upstream of TSS) for the enrichment for G-box core sequence (*ACGTG*), NAC-consensus sequence (*CGT[GA]*) and TGA TFs recognition LS-7 element (*ACGT*), which are thought to bind VOZs, using POBO analysis. This analysis revealed a significant enrichment of *cis*-elements ($p < 0.0001$) of *ACGTG*, *CGT[GA]* and *ACGT* in the promoter regions of DE genes (Figure 4b–d). Ninety percent of the DE genes contain *CGT[GA]* (1 to 12), 58% have *ACGTG* (1 to 8) and 84% have *ACGT* (1 to 12) binding sites in their upstream (−500 bp) region (Table S2). Furthermore, POBO analysis of up- and down-regulated DE genes separately also exhibited significant ($p \leq 0.0001$) enrichment of these binding sites (Figure 5). These results indicate VOZs might regulate the expression of some of these DE genes directly through these elements.

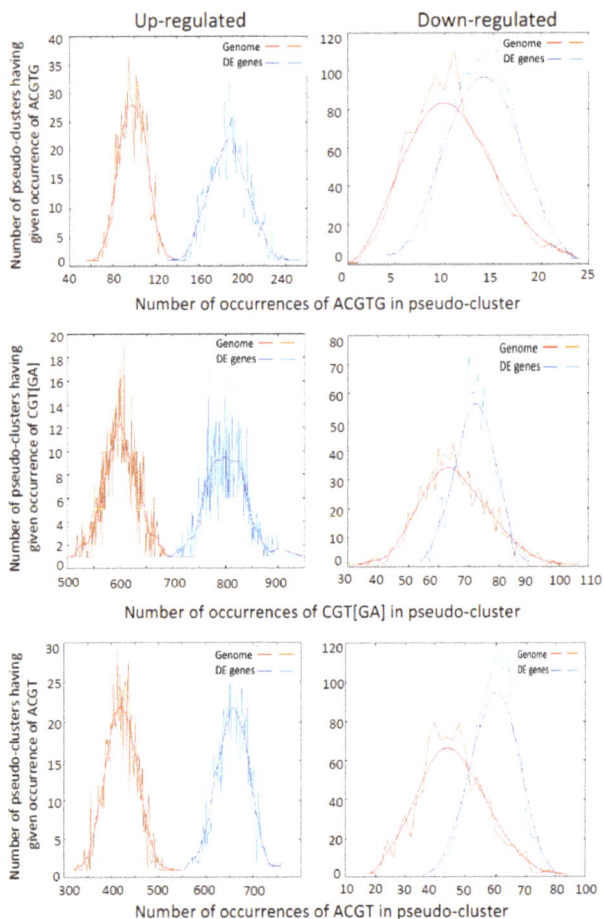

Figure 5. VOZ-binding sites in the promoters of up- and down-regulated DE genes. POBO analysis of VOZs binding motif, G-box core sequence (ACGTG) (top panels), NAC consensus sequence (CGT[GA]) (middle panels), and LS-7 *cis*-element (ACGT) (bottom panels) in the −1000 bp upstream of TSS. A total of 1000 pseudoclusters were generated from 101 up-regulated (left panels) and 11 down-regulated genes (right panels) and genome background. The jagged lines show the motif frequencies from which best-fitted curve is derived. VOZs binding sites are significantly (two-tailed $p < 0.0001$) over-represented in the upstream sequences of both up- and down-regulated genes.

2.4. GO Term Enrichment Analysis for Biological Processes in Differentially Expressed Genes

Previously, VOZ proteins have been shown to play an important role in flowering, plant immunity, cold, heat, and drought stresses [9,11–17]. We performed gene ontology (GO) analysis not only to verify if DE genes function in previously reported processes but also to gain insight into other functional roles of VOZs. A singular GO term enrichment analysis for biological processes was performed with GeneCodis separately for up-regulated and down-regulated DE genes. No significant enrichment of any GO term for down-regulated DE genes was found. However, a total of 38 GO terms for biological processes were enriched in up-regulated DE genes (Figure 6a and Additional File 3). Consistent with previously reported functions of VOZs, GO terms associated with the processes involving plant response to pathogen/pests and water deprivation, osmotic stress and oxidative stress were enriched. GO terms that are of special interest are "response to salt stress" and "hyperosmotic salinity response" for the following reasons: (i) both are among the top 10 most enriched GO terms; (ii) these two GO terms together have 10 genes (second most of all GO categories); (iii) role of VOZ proteins in salt stress is not known; and (iv) expression of the majority of these genes is altered in opposite direction in the complemented line (Table S3).

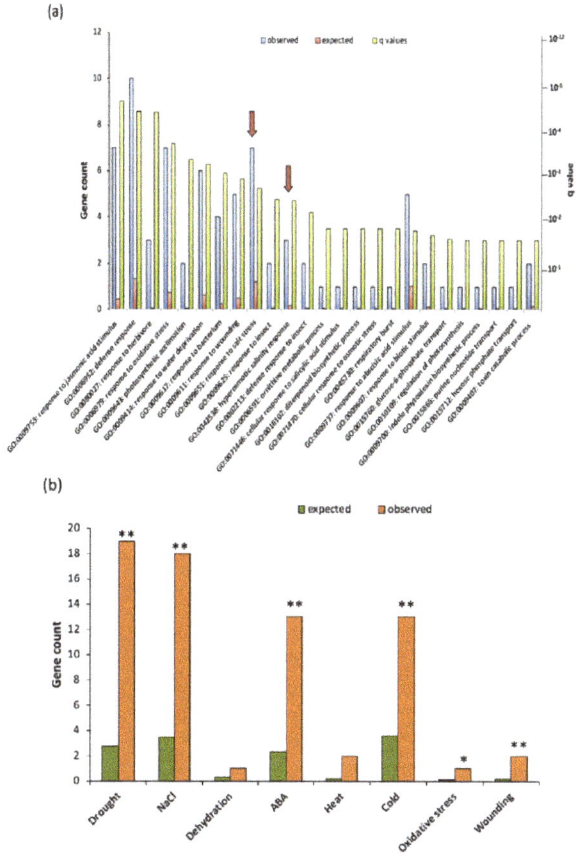

Figure 6. Gene ontology (GO) enrichment analysis of DE genes. (**a**) GO term enrichment analysis for biological processes of up-regulated genes. For each GO term, the expected and observed gene numbers along with the statistical significance (*p*-value) for the enrichment is presented. Observed: Number of DE genes associated with a GO term for biological processes. Expected: Number of genes expected for each GO term in the genome. "Response to salt stress" and "Hyperosmotic salinity response" GO terms are indicated with an arrow. (**b**) A significant number of DE genes are associated with abiotic stress response in comparison with genome background with a ** $p \le 0.0001$ and * $p \le 0.05$.

2.5. VOZs Regulate Expression of Many Abiotic Stress-Responsive Genes

Since G-box, NAC and TGA *cis*-elements occurred significantly in the promoter region of DE genes (Figures 4 and 5), we performed enrichment analysis to determine the number of DE genes associated with different stresses. For this analysis, we compared the DE gene list with all the listed abiotic stress genes at http://caps.ncbs.res.in/stifdb/browse.html#se. This analysis also indicated a substantial enrichment ($p \le 0.001$) of different abiotic stress-responsive genes with a significant number of genes associated with drought and cold stress, which is consistent with the reported functions of VOZs (Figure 6b and Additional File 4) of Nakai et al. [14]. However, VOZ's function in salt stress is not known. Enrichment analysis indicated that about 16% of DE genes (18 genes) in *DKO* are associated with salt stress (Additional File 4). Furthermore, in the COMP2-4 line, the expression of salt stress-responsive genes found in both these analyses was partially or fully restored to WT levels (Figure S5 and Table S3a,b).

2.6. VOZs Regulate Salt Stress Tolerance

2.6.1. VOZ Double Mutant Exhibits Hypersensitivity to Salt Stress

Previous studies have shown that VOZs play an important role in drought, cold and heat [13,15,16], but their regulatory role in salt stress is not known. Since salt stress-responsive genes are enriched in DE genes, we investigated the role of VOZs in salt stress tolerance. Wild type, double mutant (DKO), COMP 2-4 and single mutants of VOZs (*voz1-1, voz2-1, voz2-2*) were tested for salt tolerance. Seed germination rate, seedling growth and root length were scored by growing them on different concentrations (0, 100 or 150 mM) of NaCl. In general, seed germination rate was significantly affected under salt stress in a NaCl concentration-dependent manner in all genotypes. However, irrespective of NaCl concentration, seeds of DKO genotype exhibited delayed germination in comparison to WT, COMP2-4 and single *voz* mutants (Figure 7a). Further, seedling growth (fresh weight) was also significantly affected in a NaCl concentration-dependent manner in all the genotypes. Similar to the rate of seed germination, the growth of DKO seedlings was severely suppressed when compared to that of WT, COMP2-4 and single mutants (Figure 7b,c). A suppression of the primary root length in a NaCl concentration-dependent manner was also observed (Figure 7b). Particularly at 100 and 150 mM NaCl, a difference in root growth was evident among the genotypes (Figure 7b—left bottom panel and 7c). A significant suppression in the primary root growth was observed in DKO lines as compared to WT, COMP2-4 and single mutant lines, indicating the increased sensitivity of DKO mutant to salt stress (Figure 7b,c). Even at 150 mM NaCl, WT, COMP2-4 and single mutants were found to be relatively more tolerant to salt stress, indicating DKO seedlings exhibit hypersensitivity to salt stress.

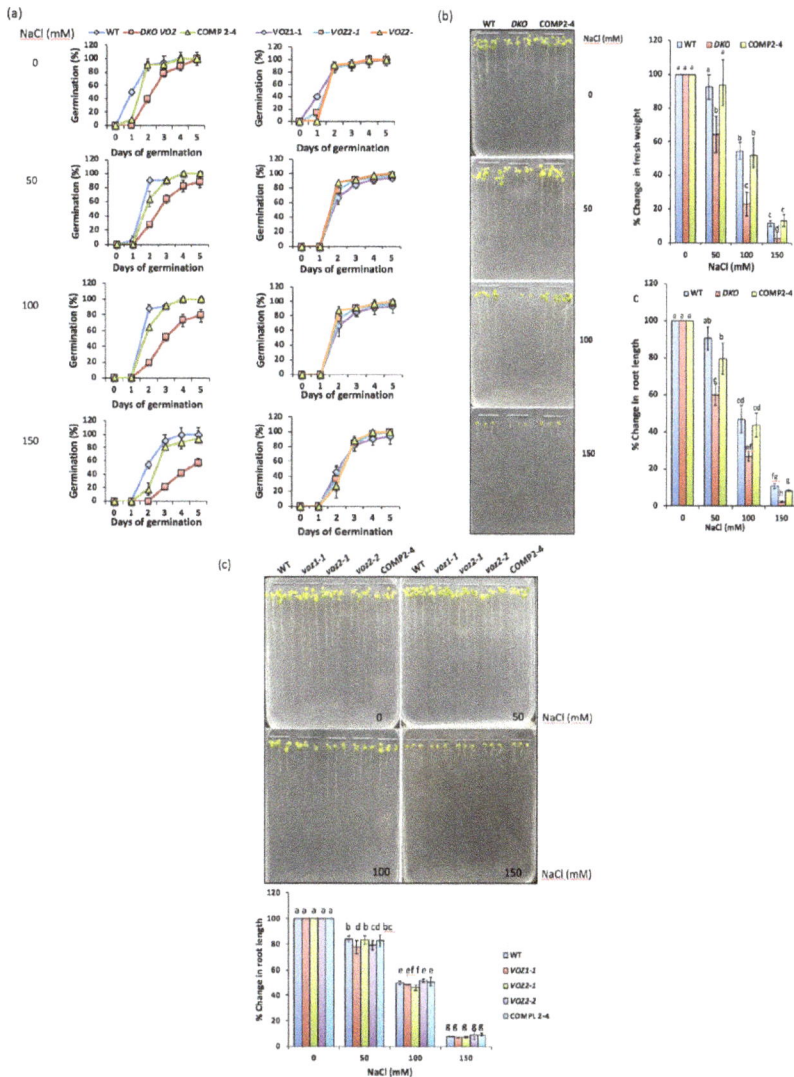

Figure 7. Germination and seedling growth of WT, mutants and complemented line in the presence of NaCl. (**a**) *VOZ* Double mutant (DKO) exhibits delayed germination under salt stress. The time course of seed germination of WT, DKO, COMP2-4 (left panels), *voz1-1*, *voz2-1* and *voz2-2* (right panels) in the presence of 0, 50, 100 and 150 mM NaCl. Each value shown here is mean of three biological replicates with *n* = 10. The error bars represent SD. (**b**) *VOZ* Double mutant (DKO) is hypersensitive to salt stress. Left panel: Growth of seedlings of WT, DKO and COMP2-4 on MS (Murashige and Skoog medium) plates containing different concentrations of NaCl. Seeds were plated on 1/2 strength MS medium supplemented with 0, 50, 100 and 150 mM of NaCl and were allowed to germinate and grow for two weeks. The photographs were taken after two weeks. Right panel, top: Seedling fresh weight. Right panel, bottom: Seedling root length at different concentrations of NaCl was measured for all genotypes and plotted as % relative to growth on normal (0 mM) MS medium. Three biological replicates were used. Eight to ten seedlings for each genotype per treatment for each biological replicate were included.

Student's *t*-test was performed and significant differences ($p \leq 0.05$) among samples are labeled with different letters. The error bars represent SD. (c) Single mutants of VOZs are not hypersensitive to salt stress. Top: Growth of seedlings of WT, COMP2-4, *voz1-1*, *voz2-1* and *voz2-2* on MS plates containing different concentrations of salt. Seeds were plated on half-strength MS medium supplemented with 0, 50, 100 or 150 mM of NaCl and were allowed to germinate and grow for two weeks. The photographs were taken after two weeks. Bottom: Seedling root length at different concentrations of NaCl was measured for all genotypes and plotted as % relative to growth on normal (0 mM) MS medium. Three biological replicates were used. For each genotype, eight to ten seedlings per treatment and for each biological replicate were used. Student's *t*-test was performed and significant differences ($p \leq 0.05$) among samples are labeled with different letters. The error bars represent SD.

2.6.2. VOZs Activate the Expression of Salt-Responsive Genes

Prior to analyzing the transcript levels of different salt-responsive genes, we first quantified *VOZ1* and *VOZ2* transcripts in WT seedlings grown on medium supplemented with 0, 50, 100 and 150 mM NaCl. Alterations in transcript levels of *VOZ1* and *VOZ2* under salt stress were observed. The expression of both *VOZ1* and *VOZ2* was significantly enhanced with increasing concentration of NaCl (Figure 8). For example, at 100 mM NaCl, ~2.5- and 2.0-fold increases in transcript levels of *VOZ1* and *VOZ2*, respectively, were observed. However, 150 mM NaCl reduced the salt-induced elevation of transcript levels of *VOZs*, probably due to severe growth inhibition at this concentration. To gain further insights into the role of VOZs in salt stress, the expression level of salt-responsive genes under both "enrichment of salt-responsive genes" and the GO category of "response to salt stimulus" was compared in WT, *DKO* and COMP2-4 RNA-Seq data. The majority of salt-responsive genes were represented in *DKO* and COMP2-4 datasets and their expression profiles were opposite to each other (Figure S5 and Table S3a,b). Motif analysis of upstream (-1000 bp) regions of these genes indicated significant enrichment ($p < 0.0001$) for VOZ binding sites, viz., G-box (*ACGTG*) and NAC bind sites (*CGT[GA]*) and LS-7 (*ACGT*) (Figure S6a,b (lower panels) and Table S3). Arabidopsis genes (At1g16850 (transmembrane protein), At5g59310 (LTP4), At2g37760 (AKR4c8), At5g59820 (ZAT12), At4g23600 (COR13), At5g24770 (VSP2), At1g10585 (bHLH DNA-binding superfamily protein), At2g43510 (ATTI1) and At4g37990 (ATCAD8)), which are closest to genes involved in salt tolerance in other plant species [18–28] and also contain either of VOZ-binding motifs in their promoter (Table S3a,b), were selected as representatives to analyze their expression under control and salt stress conditions. The expression pattern of these nine genes was analyzed by RT-qPCR. Under control conditions, the expression levels of all nine genes were significantly higher in *DKO* as compared to WT or COMP2-4 in 30-day-old plants (Additional File 1). However, in 15-day-old seedlings there was no up-regulation of salt-responsive genes in the *DKO* in untreated seedlings, suggesting differential regulation of these genes by VOZs depending on the developmental stage of the plants. Upon exposure to salt (100 mM), expression of these genes was highly induced in the WT seedlings (Figure 9). However, loss of both VOZs (*DKO*) caused a significant reduction in salt induction of these genes, suggesting that VOZs are essential for the increased expression of these salt-responsive genes under salt stress. The expression levels of three genes, viz., At1g16850, At5g59310 and At3g04720, were partially restored to WT level under salt stress in the COMP2-4 line (Figure S4). Similar to our results, expression of several cold-responsive genes that are highly expressed in the DKO were not restored in the *VOZ2* complemented line [13,14].

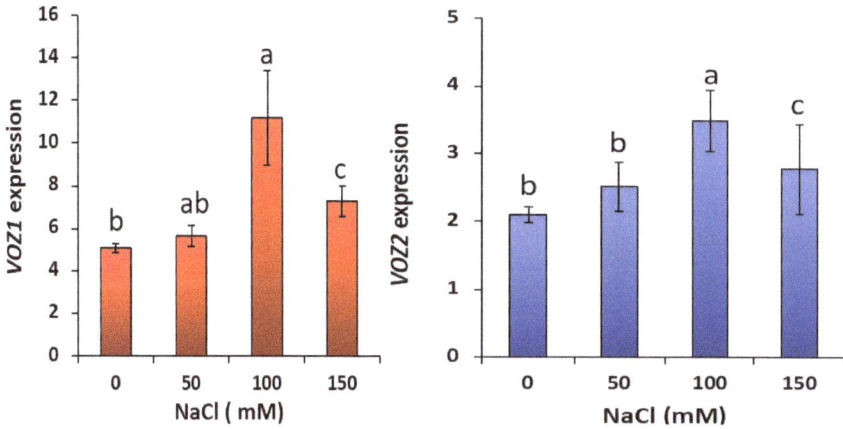

Figure 8. Expression of VOZs in response to salt stress. Expression of *VOZ1* (top panel) and *VOZ2* (bottom panel) in 10-day-old seedlings of WT seedlings grown on 1/2 MS medium supplemented with 0, 50, 100 or 150 mM NaCl was determined by RT-qPCR. The expression of *VOZs* was normalized with *ACTIN2*.

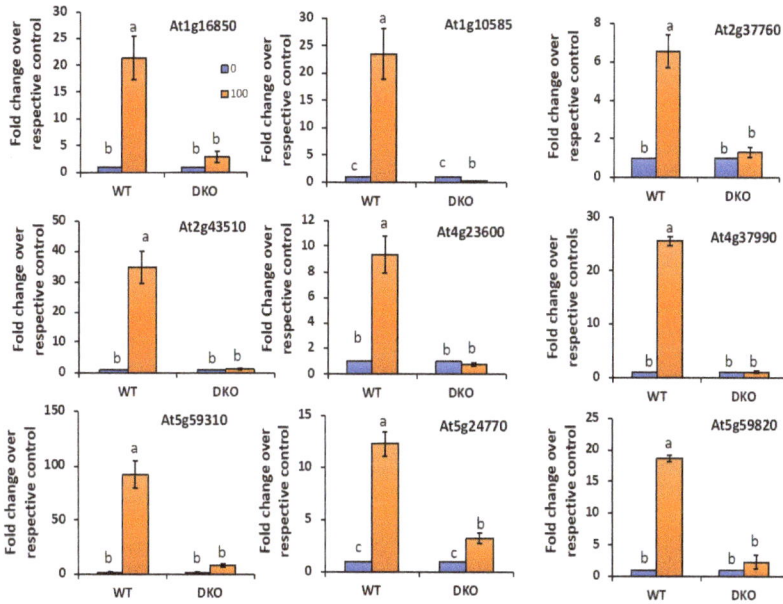

Figure 9. VOZs positively regulate the expression of salt-responsive genes. Expression of salt-responsive genes in 10-day-old seedlings of WT and DKO lines on 1/2 MS medium supplemented with 0 or 100 mM NaCl was determined by RT-qPCR. The expression level of salt-responsive genes was normalized with *ACTIN2*. Fold change in expression level relative to their respective controls (0 mM) is presented. 0 mM values were considered as 1. Three biological replicates were used. Student's *t*-test was performed and significant differences ($p \leq 0.05$) among samples are labeled with different letters. The error bars represent SD.

The majority of the salt-responsive genes contain *cis*-elements in their promoter regions to which known TFs bind. These include *CACGTG, CACG[G/A]C, CATGTG, VCGCGB* and *MCGTGT* that bind G_box bHLH, N_box_bHLH, Nac_box_NAC, and *CAMTA* TFs, respectively. To understand the regulation of these salt-responsive DE genes by VOZs, POBO analysis was carried out for the enrichment of these *cis*-elements in the upstream regions of salt-responsive genes. A significant enrichment ($p < 0.0001$) for *RSRE*s (*VCGCGB* and *MCGTGT*) was observed in the upstream region (-1000 bp) of the DE genes (Figure S7). Further, enrichment for *CACGTG*, but not *CACGGC* and *CATGTG*, was found in the promoter regions of the salt stress-responsive genes (Figure S6a,b—top and middle panels). Significantly, enrichment of VOZ binding consensus motifs (*CGT[GA]*, *ACGTG* and *ACGT*) was also observed (Figure S6a,b—bottom panels). Only two genes (viz. At2g37760 and At5g59820) have a canonical binding site (*ACGT$_{GATTCAC}$ACGC*) for VOZs in their promoter regions. These results suggest that the regulation of salt-responsive genes by VOZs is accomplished via certain *cis*-elements (*CGT[GA]*, *ACGTG* and *ACGT*) within the consensus motifs of VOZs.

3. Discussion

Previous studies with *voz* double mutant reported several developmental defects, such as smaller plants, impaired root growth, delayed flowering, round lamina of juvenile leaves, reduced trichome number on abaxial side, and siliques with aborted seeds [11,14]. Expression of either *VOZ1* or *VOZ2* under their native promoter or expression of *VOZ2* under *CaMV35S* promoter in *DKO* completely rescued the phenotype [11,13]. In this study, we observed another phenotype. Thirty-day-old seedlings of *DKO* plants grown under day neutral conditions at 21 °C exhibited vein-clearing phenotype in older leaves (Figure 1a). This vein-clearing phenotype became more apparent with the age of the plants. Single mutants of *VOZ*s did not exhibit this developmental phenotype. Over-expression of *VOZ2* in *DKO* rescued the phenotype (Figure 1A), suggesting that VOZ1 and VOZ2 are functionally redundant in vein-clearing phenotype.

3.1. VOZs are Involved in Regulation of Many Stress-Responsive Genes

Our global transcriptome analysis using RNA-Seq revealed that a significant number of genes that are involved in diverse stress responses are regulated either directly or indirectly by VOZs (Figures 2a and 6, Additional File 1, sheet 1,2). Previously Nakai et al. [14] and more recently Kumar et al. [12] compared the expression of genes in WT and *DKO* using microarrays. Our study significantly differs from these in several ways. Here, we used next-generation sequencing that significantly increases the depth of transcriptome analysis and avoids some problems associated with microarrays. More importantly, the use of a complemented line in which double mutant phenotypes are rescued allowed us to identify the genes that are regulated specifically by VOZs (Figure 1a,b and Additional File 1 sheet 2). Despite using RNAseq, our study revealed a smaller number of DE genes as compared to the previous studies [12,14]. This difference in the number of DE genes and limited overlap between DE genes among different studies could be due to the difference in the age and developmental stage of plants used (30-day-plants in this study vs. 14-day-old-seedlings in previous studies) and/or due to the differences in methodologies (microarrays vs RNA-Seq). In fact, developmental regulation of expression levels of TFs has been previously reported [6,29,30]. Nevertheless, our study identified a new set of 94 genes that are regulated by VOZs (Additional File 1) when compared with Nakai et al. [14] and Kumar et al. [12]. Interestingly, GO enrichment of biological process using DE genes in *DKO* from Nakai et al. [14], we found enrichment of only two biological processes, viz., farmesyl diphosphate metabolic process and sequiterpenoid biosynthetic process. This is in contrast to this study wherein we observed enrichment for 38 GO terms that are consistent with reported functions of VOZs (Figure 6a, Additional File 3). In addition, GO enrichment analysis suggested a new role for VOZs in salt stress, which was confirmed experimentally. Reproducibility among replicates, full or partial restoration of expression of ~98% of DE genes in COMP2-4 to WT level (Figure 2b) and RT-qPCR validation of expression of a number (22 genes) of randomly selected DE genes (Figure 3, Figure S5) indicates that the identified DE genes in this study are bona fide direct or indirect targets of VOZs. Enrichment of DE genes in multiple abiotic stress-responses indicates that VOZs play a major role in crosstalk between multiple stress signal transduction pathways (Figure 6b and Additional File 4). GO analysis of the DE genes indicated high enrichment of GO terms associated with diverse processes that are critical for plant responses to biotic stresses, such as bacteria and fungi, and abiotic stresses including drought, cold, salt and oxidative stress (Figure 6a). Enrichment of genes involved in response to hormones such as abscisic acid (ABA) and jasmonic acid (JA) was also observed (Figure S8 and Additional File 5). Together, these results suggest that VOZs could be integrators of a variety of stress responses. Consistent with these results, VOZs were reported to play an important role in multiple stress responses [13,14].

3.2. Genes with Binding Motifs for VOZs are Both Up- and Down-Regulated

Electrophoretic mobility shift assays showed that a VOZ protein binds to $GCGTN_{x7}ACGC$ sequence in vitro in the V-PPase gene (*AVP1*) promoter [8]. The palindromic sequence-binding site was considered as canonical VOZ TF binding site. We analyzed it to see if the promoter region (−2000 bp upstream to TSS) of DE genes is enriched for this motif, but found no significant enrichment. Only one (DE gene AT5G16360) has two VOZ binding sites (−1590, −1604 bp) in its upstream region. We followed our analysis by screening the promoter regions of DE genes with two suboptimal ($GCGTN_{x7}ACGT$ and $GCGTN_{x8}ACGC$) and other motifs ($GCGTN_{x7}AAGC$, $GCTTN_{x7}ACGC$, $ACGTN_{x7}ACGC$) that are reported to bind VOZs [8,12]. This analysis resulted in the identification of another DE gene (AT2G14247) containing $GCGTN_{x7}ACGT$ motif (−336 bp) and none containing $GCGTN_{x8}ACGC$ in their promoter regions. Recently, Kumar et al. [12] using systematic evolution of ligands by exponential enrichment (SELEX) assay and electrophoretic mobility shift assay (EMSA) not only confirmed $GCGT_{GTGATAC}ACGC$ as VOZ2 binding site but also revealed additional binding sites. Based on this study, we scanned the upstream region of all the DE genes and detected six additional genes (At1g23150, At2g17840, At2g22860, At2g37760, At4g00780, At5g59820) that have these new VOZ-binding sites. In addition to these *cis*-elements, other studies also identified binding of VOZs to alternate palindromic NAC-binding sequences (palNAC-BS) that are similar to other NAC proteins that are responsive to abiotic stress [10,14]. Analysis of DE genes showed that >90% contain *CGT[GA]*, 58% contain *ACGTG* and 83% contain *ACGT* elements and these motifs are significantly enriched in their promoter regions (Figure 4, Table S2). Both up- and down-regulated genes showed enrichment for VOZ-binding sites (Figure 5). Significant enrichment of VOZ-binding motifs in DE genes indicates that VOZs likely regulate the expression of those genes directly by binding to these *cis*-elements.

3.3. VOZs Likely Regulate Expression of Some Genes Indirectly

In the promoter regions of some DE genes we did not find any of the VOZ-binding sites and these genes are likely regulated indirectly by other TFs. We found enrichment of four TF families (bHLH, C2H2, NAC and MYB-related) in DE genes (Figure 4) and TFs in three of these families were up-regulated (Figure S3). Many members of these TF families have multiple binding sites for VOZs in their promoter regions and exhibited expression levels similar to WT in the complemented line (Table S2; Additional File 1). For example, members of bHLH, NAC and C2H2 are up-regulated in *DKO* while they are down-regulated in *COMP2-4* line. In contrast, members of a MYB-related family were down-regulated in *DKO* and up-regulated in *COMP2-4*. It is possible that these TFs may regulate the expression of DE genes that do not contain canonical VOZ binding motif [12]. Together, these data indicate a complex network of regulation of expression of TFs by VOZs.

Recent studies identified *RSRE* element *VCGCGB* as the core element that is enriched in a majority of early-activated genes under stresses [31]. As this element is identical to the binding site of TFs signal responsive/calmodulin-binding transcription activators (SRS/CAMTAs) (*VCGCGB*), many studies showed SRs, in general, and SR1/CAMTA3, in particular, in regulation of multiple biotic and abiotic stress responses [5,6,32–34]. Significantly, SRs regulate genes involved in abiotic stress response, particularly through *MCGTGT* element. Our analysis of the promoter region of all the DE genes indicated that a significant enrichment of the *VCGCGB* and *MCGTGT* elements, suggesting that VOZs might regulate the abiotic stress responses through SRs (Figure S7). Further, the fact that the majority of these genes are misregulated in *DKO* and are implicated in various stress signaling pathways, also suggests an important role for SRs in VOZ-mediated regulation. In support of this, it has been shown that VOZs and CAMTAs interact with the *AVP1* promoter and regulate its expression [8,35]. POBO analysis indicated the enrichment of *RSRE* motif in the promoter regions of the up-regulated DE genes (Figure S7). As shown in Figure 6, significant enrichment of GO term for "responses to salt stress" and "water deprivation" was observed only in the up-regulated DE genes (Additional File 3). Further, enrichment for *VCGCGB* and *MCGTGT* in DE genes and enrichment for GO terms "water deprivation" and "cellular response to osmotic" suggest that VOZs could be regulating drought response genes through utilization of *MCGTGT* and *VCGCGB* by SRs (Figure 6, Figure S7, and Additional File 3). In fact, a significant alteration in cold and drought-responsive genes expression was observed in *DKO* line even under non-stress conditions [14]. One possibility is that VOZs form heterodimers with CAMTAs/SRs or other TFs in regulating some genes. The fact that VOZs and CAMTAs bind to the *AVP1* promoter lends supports to this. However, thus far direct interaction of VOZs and CAMTAs has not been reported. Recently, it has been shown that VOZs interacts with CONSTANS, another TF, in regulating flowering [12].

3.4. VOZ Confers Salt Tolerance by Activating the Expression of Salt-Responsive Genes

Double mutant line (*voz1-1 voz2-1*) were more sensitive to salt stress in terms of seed germination rate, seedling growth and root growth when compared with the WT, COMP2-4 and single mutant lines. Thus, our results suggest that (a) *VOZ1* and *VOZ2* have redundant functions in salt tolerance and (b) *VOZs* act as positive regulators of plants response to salt stress. This positive regulation of salt stress by VOZs is similar to that observed under biotic stress and differs from that of the cold and drought stress response, where it functions as a negative regulator [14]. Previously, Nakai et al. [13,14] have shown that *DKO* was significantly tolerant to cold and drought whereas it is sensitive to bacterial and fungal pathogens. They further reported that the over-expression of *VOZ2* confers tolerance to freezing and drought but curtails tolerance to biotic stresses. Taken together these results suggest that VOZs have opposing functions under salt stress as compared to cold and drought stresses. To further understand the regulation (direct versus indirect) by VOZs, salt-responsive genes were identified and subjected to POBO analysis for enrichment of *RSRE* (*VCGCGB*), *NAC* (*CGT[GA]*), *G-box* (*ACGTG*) and *ACGT* in their upstream region. This analysis revealed significant enrichment for *RSRE* (*VCGCGB*), *CGT[GA]*), *ACGTG*, *ACGT* and *MCGTGT* (Figure 5 and Figure S7). Hence, it is possible that some of these genes could be direct targets of VOZs i.e., they bind to these motifs to regulate expression. Alternatively, other TFs such as CAMTAs/SRs could also participate along with VOZs in this regulation as discussed above.

In summary, our results showed that a large number of genes associated with biotic and abiotic stress responses are regulated by VOZs. Most of these genes are likely direct targets as they contain one or more type of VOZ-binding sites in their promoter region. Analysis of DE genes suggested a new role for VOZs in salt stress. We experimentally showed that VOZs function as positive regulators of salt tolerance. The model in Figure 10 summarizes the role of VOZs in salt stress response. Plants in response to salt stress activate expression of VOZs. This activation of VOZs, in turn, regulates the expression level of salt-responsive genes either directly or indirectly thereby conferring salt tolerant phenotype. The absence of VOZs in *DKO* significantly curtails salt-induced activation of the salt-responsive genes leading to hypersensitive phenotype.

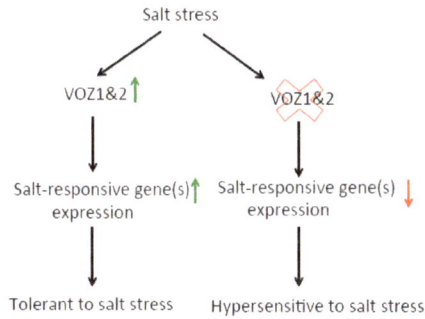

Figure 10. A proposed model for the role of VOZs in salt stress response (see text for details). Green and red arrows indicate the increased and decreased expression levels, respectively.

4. Materials and Methods

4.1. Plant Materials and Growth Conditions

All experiments were performed with *Arabidopsis thaliana* Columbia-0 ecotype. Seeds of single (*voz1-1, voz2-1*) and double mutants (*DKO; voz1-1 voz2-2*) of *VOZ1* and *VOZ2* used in this study were characterized previously [11]. The complemented line (COMP-4) was generated by transforming *DKO* with *VOZ2* cDNA under *CaMV35S* promoter. Plants were grown in soil in a growth chamber at 21 °C, 60% relative humidity under 12/12 h light/dark conditions.

4.2. Salt Stress Treatment

To study the effect of salt stress on seed germination and seedling growth, surface sterilized seeds of wild type (WT) and mutants were sown on half-strength MS medium (containing 0.5 mg/L MES and 1% sucrose) pH 5.7 and supplemented with 0, 50, 100 or 150 mM NaCl. The surface-sterilized seeds of all the lines were stratified at 4 °C in dark for 5 d prior to sowing on plates. The plates with seeds were allowed to germinate under long day condition (16 h light/8 h dark) at 22 °C. Germination rate and fresh weight of seedling and root growth was determined by recording the number of seeds that exhibited emergence of radicle, weight of the seedlings and length of the roots after two weeks of growth, respectively. All experiments were performed three times with a minimum of three replicates.

4.3. RNA-Seq

Total RNA from WT, *DKO* and COMP2-4 genotypes was isolated using miRNAeasy kit (Qiagen, Germantown, MD, USA#217004). Traces of genomic DNA were removed using on column DNAse digestion. RNA-Seq was performed essentially as described previously in Prasad et al. [6].

4.4. Mapping of the Reads and Identification of Differentially Expressed (DE) Genes

The reads were aligned to the TAIR 10 version of the Arabidopsis genome, and DE gene list was generated using the criteria as described earlier [6]. VENNY (http://bioinfogp.cnb.csic.es/tools/venny/) a web-based tool was used for identification of common genes in one or more datasets. Heatmap of DE genes was generated using log2 transformed expression values of each gene using Heatmapper [36]. Box-and-whisker plot of DE genes was generated using the log2 transformed expression values in WT, *DKO* and *COMP2-4* with JMP Pro, version 13, statistical software (SAS, Cary, NC, USA).

4.5. Bioinformatics Analysis of DE Genes for VOZs Binding Motifs

Identification of DE genes containing VOZ binding motifs $GCGTN_{x7}ACGC$, *ACGTG*, *CGT[GA]* and *ACGT* in their promoter was carried out using "Patmatch" (Version 1.1) tool (www.arabidopsis.org).

With this tool, we identified motifs on both strands of upstream sequences (−500 bp) preceding the TSS in TAIR10 database. Both up- and down-regulated DE genes were included as input for scoring both type and number of VOZs binding motifs.

4.6. GO Term Enrichment Analysis

GO term enrichment analysis was performed using GeneCodis [37]. Single enrichment analysis with TAIR GO annotations was performed using the hypergeometric test with Benjamin-Hochberg false discovery rate (FDR) correction with a significance of $p \leq 0.05$. The DE genes that are up- or down-regulated were analyzed separately.

4.7. Identification of TFs, Abiotic Stress and Hormone-Responsive Genes in DE List

To identify various TFs in the DE genes, a list of all TFs was obtained from Plant TF Database (version 3.0) (http://planttfdb.cbi.pku.edu.cn) [38] and all DE genes were queried against the total TF list. TAIR 10 ID of all TF genes was used as input for identifying the DE genes encoding the TFs and classifying them based on the similarity with Total TF family list. The TFs and the genes responsive to various abiotic stress conditions were obtained from STIFB (Stress Responsive TF Database) (http://caps.ncbs.res.in/stifdb2/). Promoters of the genes that contained *cis*-element for binding of the TFs that are involved in abiotic stress response were retrieved for the analysis. DE genes were queried against the list of the genes for a specific abiotic stress. Further, on the basis of overlap of locus ID (TAIR ID) between the lists of genes, they were further categorized into different subsets. Similarly, plant hormone biosynthesis and signaling genes in DE list were identified by comparing the DE genes with that of genes list of each individual hormone available from the Arabidopsis Hormone Database 2.0 (http://ahd.cbi.pku.edu.cn).

4.8. Promoter Analysis for Enrichment of Cis-Elements

To identify the *cis*-elements in promoters, either 500 or 1000 bp sequence upstream of the transcription start site was extracted from TAIR using an online tool for bulk sequence retrieval. For the estimation of the enrichment for particular *cis*-elements, promoter sequences (−500 or −1000 bp) were used as input for POBO analysis [39]. The upstream sequences of −500 or −1000 bp of the genes in the data set were used as an input into the web portal and analyzed for *cis*-element/motif against *Arabidopsis thaliana* background (clean). The following parameters were used for this analysis: number of promoters in cluster is equivalent to number of input sequences; number of pseudoclusters to generate =1000. For statistical significance, a linked GraphPad application calculates a two-tailed *p*-value using generated *t*-value and degrees of freedom for determination of the statistical differences between the input sequences and the background.

4.9. Validation of DE Genes Using RT-qPCR Analysis

Primers for validation of DE genes using Real-time qPCR (RT-qPCR) were designed using Primer Quest web tool (http://www.idtdna.com/Primerquest/Home/Index) from IDT, Coralville, IA, USA (Additional File 6). DE genes were randomly selected and analyzed for their expression levels using RT-qPCR. cDNA from 30-day-old plants was prepared and expression of each gene in all genotypes was estimated essentially as described [6]. For each genotype, cDNA from two independent biological replicates was used. Three technical replicates were used for each sample. *ROC5 (CYCLOPHILIN)* was used as a reference gene. Fold change in expression was calculated and plotted with respect to WT. The expression level in WT for each gene is considered as 1.

4.10. RT-qPCR Analysis of Salt-Responsive Genes

Ten-day-old control and salt-treated seedlings of different genotypes were used for extraction of total RNA. A quantity of 1 µg of RNA was used for the preparation of cDNA using Superscript III reverse

Int. J. Mol. Sci. **2018**, *19*, 3731

transcriptase system as described in Prasad et al. [6]. The cDNA was diluted 6 times and 1.5 µL per reaction was used as a template. Expression analysis was performed using RT-qPCR as described above. The data obtained were normalized with *ACTIN2* and fold change in the expression level was calculated relative to their respective control, i.e., 0 mM NaCl. The expression level in control was considered as 1. A minimum of 3 technical replicates and 3 biological replicates were used for each experiment.

Supplementary Materials: Supplementary information is available online at http://www.mdpi.com/1422-0067/19/12/3731/s1.

Author Contributions: A.S.N.R. conceived and directed the project. K.V.S.K.P. and A.S.N.R. designed the experiments. K.V.S.K.P. performed RNA-Seq, all bioinformatics analysis with the DE gene list and all experiments pertinent to salt stress and RT-qPCR analysis of gene expression. DX analyzed RNA-Seq data and generated DE genes list. K.V.S.K.P. and A.S.N.R. wrote the manuscript. DX read and commented on the manuscript.

Funding: This work was supported by a grant from the National Science Foundation (MCB#5333470) to A.S.N.R.

Conflicts of Interest: The authors declare no conflicts interests. The founding sponsors had no role in the design of the study, in the collection, analysis or interpretation of the data, in writing of the manuscript, and in the decision to publish the results.

Abbreviation: *DKO (voz1-1 voz2-1)*, *COMP2-4 (CaMV35S: VOZ2: OCS)*; DE (Differentially expressed genes).

References

1. Cheeseman, J.M. The evolution of halophytes, glycophytes and crops, and its implications for food security under saline conditions. *New Phytol.* **2015**, *206*, 557–570. [CrossRef] [PubMed]
2. Zhu, J.K. Abiotic stress signaling and responses in plants. *Cell* **2016**, *167*, 313–324. [CrossRef] [PubMed]
3. Reddy, A.S.; Ali, G.S.; Celesnik, H.; Day, I.S. Coping with stresses: Roles of calcium- and calcium/calmodulin-regulated gene expression. *Plant Cell* **2011**, *23*, 2010–2032. [CrossRef] [PubMed]
4. Ohama, N.; Sato, H.; Shinozaki, K.; Yamaguchi-Shinozaki, K. Transcriptional Regulatory Network of Plant Heat Stress Response. *Trends Plant Sci.* **2017**, *22*, 53–65. [CrossRef] [PubMed]
5. Kim, Y.S.; An, C.; Park, S.; Gilmour, S.J.; Wang, L.; Renna, L.; Brandizzi, F.; Grumet, R.; Thomashow, M.F. CAMTA-Mediated Regulation of Salicylic Acid Immunity Pathway Genes in Arabidopsis Exposed to Low Temperature and Pathogen Infection. *Plant Cell* **2017**, *29*, 2465–2477. [CrossRef] [PubMed]
6. Prasad, K.V.S.K.; Abdel-Hameed, A.A.E.; Xing, D.; Reddy, A.S.N. Global gene expression analysis using RNA-seq uncovered a new role for SR1/CAMTA3 transcription factor in salt stress. *Sci. Rep.* **2016**, *6*, 27021. [CrossRef] [PubMed]
7. Khan, S.A.; Li, M.Z.; Wang, S.M.; Yin, H.J. Revisiting the Role of Plant Transcription Factors in the Battle against Abiotic Stress. *Int. J. Mol. Sci.* **2018**, *19*, 1634. [CrossRef] [PubMed]
8. Mitsuda, N.; Hisabori, T.; Takeyasu, K.; Sato, M.H. VOZ; isolation and characterization of novel vascular plant transcription factors with a one-zinc finger from Arabidopsis thaliana. *Plant Cell Physiol.* **2004**, *45*, 845–854. [CrossRef] [PubMed]
9. Yasui, Y.; Mukougawa, K.; Uemoto, M.; Yokofuji, A.; Suzuri, R.; Nishitani, A.; Kohchi, T. The phytochrome-interacting vascular plant one-zinc finger1 and VOZ2 redundantly regulate flowering in Arabidopsis. *Plant Cell* **2012**, *24*, 3248–3263. [CrossRef] [PubMed]
10. Jensen, M.K.; Kjaersgaard, T.; Nielsen, M.M.; Galberg, P.; Petersen, K.; O'Shea, C.; Skriver, K. The Arabidopsis thaliana NAC transcription factor family: Structure-function relationships and determinants of ANAC019 stress signalling. *Biochem. J.* **2010**, *426*, 183–196. [CrossRef] [PubMed]
11. Celesnik, H.; Ali, G.S.; Robison, F.M.; Reddy, A.S. Arabidopsis thaliana VOZ (Vascular plant One-Zinc finger) transcription factors are required for proper regulation of flowering time. *Biol. Open* **2013**, *2*, 424–431. [CrossRef] [PubMed]
12. Kumar, S.; Choudhary, P.; Gupta, M.; Nath, U. VASCULAR PLANT ONE-ZINC FINGER1 (VOZ1) and VOZ2 Interact with CONSTANS and Promote Photoperiodic Flowering Transition. *Plant Physiol.* **2018**, *176*, 2917–2930. [CrossRef] [PubMed]
13. Nakai, Y.; Fujiwara, S.; Kubo, Y.; Sato, M.H. Overexpression of VOZ2 confers biotic stress tolerance but decreases abiotic stress resistance in Arabidopsis. *Plant Signal. Behav.* **2013**, *8*, e23358. [CrossRef] [PubMed]

14. Nakai, Y.; Nakahira, Y.; Sumida, H.; Takebayashi, K.; Nagasawa, Y.; Yamasaki, K.; Akiyama, M.; Ohme-Takagi, M.; Fujiwara, S.; Shiina, T.; et al. Vascular plant one-zinc-finger protein 1/2 transcription factors regulate abiotic and biotic stress responses in Arabidopsis. *Plant J.* **2013**, *73*, 761–775. [CrossRef] [PubMed]

15. Koguchi, M.; Yamasaki, K.; Hirano, T.; Sato, M.H. Vascular plant one-zinc-finger protein 2 is localized both to the nucleus and stress granules under heat stress in Arabidopsis. *Plant Signal. Behav.* **2017**, *12*, e1295907. [CrossRef] [PubMed]

16. Song, C.; Lee, J.; Kim, T.; Hong, J.C.; Lim, C.O. VOZ1, a transcriptional repressor of DREB2C, mediates heat stress responses in Arabidopsis. *Planta* **2018**, *247*, 1439–1448. [CrossRef] [PubMed]

17. Yasui, Y.; Kohchi, T. VASCULAR PLANT ONE-ZINC FINGER1 and VOZ2 repress the FLOWERING LOCUS C clade members to control flowering time in Arabidopsis. *Biosci. Biotechnol. Biochem.* **2014**, *78*, 1850–1855. [CrossRef] [PubMed]

18. Oberschall, A.; Deák, M.; Török, K.; Sass, L.; Vass, I.; Kovács, I.; Fehér, A.; Dudits, D.; Horváth, G.V. A novel aldose/aldehyde reductase protects transgenic plants against lipid peroxidation under chemical and drought stresses. *Plant J.* **2000**, *24*, 437–446. [CrossRef] [PubMed]

19. Simpson, P.J.; Tantitadapitak, C.; Reed, A.M.; Mather, O.C.; Bunce, C.M.; White, S.A.; Ride, J.P. Characterization of Two Novel Aldo-Keto Reductases from Arabidopsis: Expression Patterns, Broad Substrate Specificity, and an Open Active-Site Structure Suggest a Role in Toxicant Metabolism Following Stress. *J. Mol. Biol.* **2009**, *392*, 465–480. [CrossRef] [PubMed]

20. Zhou, J.; Li, F.; Wang, J.L.; Ma, Y.; Chong, K.; Xu, Y.Y. Basic helix-loop-helix transcription factor from wild rice (OrbHLH2) improves tolerance to salt- and osmotic stress in Arabidopsis. *J. Plant Physiol.* **2009**, *166*, 1296–1306. [CrossRef] [PubMed]

21. Safi, H.; Saibi, W.; Alaoui, M.M.; Hmyene, A.; Masmoudi, K.; Hanin, M.; Brini, F. A wheat lipid transfer protein (TdLTP4) promotes tolerance to abiotic and biotic stress in Arabidopsis thaliana. *Plant Physiol. Biochem.* **2015**, *89*, 64–75. [CrossRef] [PubMed]

22. Kim, I.J.L.; Yun, B.W.; Jamil, M. GA Mediated OsZAT-12 Expression Improves Salt Resistance of Rice. *Int. J. Agric. Biol.* **2016**, *18*, 330–336.

23. Le, C.T.; Brumbarova, T.; Ivanov, R.; Stoof, C.; Weber, E.; Mohrbacher, J.; Fink-Straube, C.; Bauer, P. ZINC FINGER OF ARABIDOPSIS THALIANA12 (ZAT12) Interacts with FER-LIKE IRON DEFICIENCY-INDUCED TRANSCRIPTION FACTOR (FIT) Linking Iron Deficiency and Oxidative Stress Responses. *Plant Physiol.* **2016**, *170*, 540–557. [CrossRef] [PubMed]

24. Vogel, J.T.; Zarka, D.G.; Van Buskirk, H.A.; Fowler, S.G.; Thomashow, M.F. Roles of the CBF2 and ZAT12 transcription factors in configuring the low temperature transcriptome of Arabidopsis. *Plant J.* **2005**, *41*, 195–211. [CrossRef] [PubMed]

25. Gong, Z.; Koiwa, H.; Cushman, M.A.; Ray, A.; Bufford, D.; Kore-eda, S.; Matsumoto, T.K.; Zhu, J.; Cushman, J.C.; Bressan, R.A.; et al. Genes that are uniquely stress regulated in salt overly sensitive (sos) mutants. *Plant Physiol.* **2001**, *126*, 363–375. [CrossRef] [PubMed]

26. Shan, L.; Li, C.L.; Chen, F.; Zhao, S.Y.; Xia, G.M. A Bowman-Birk type protease inhibitor is involved in the tolerance to salt stress in wheat. *Plant Cell Environ.* **2008**, *31*, 1128–1137. [CrossRef] [PubMed]

27. Li, R.; Wang, W.J.; Wang, W.G.; Li, F.S.; Wang, Q.W.; Xu, Y.; Wang, S.H. Overexpression of a cysteine proteinase inhibitor gene from Jatropha curcas confers enhanced tolerance to salinity stress. *Electron. J. Biotechnol.* **2015**, *18*, 368–375. [CrossRef]

28. Srinivasan, T.; Kumar, K.R.; Kirti, P.B. Constitutive expression of a trypsin protease inhibitor confers multiple stress tolerance in transgenic tobacco. *Plant Cell Physiol.* **2009**, *50*, 541–553. [CrossRef] [PubMed]

29. Reddy, A.S.; Reddy, V.S.; Golovkin, M. A calmodulin binding protein from Arabidopsis is induced by ethylene and contains a DNA-binding motif. *Biochem. Biophys. Res. Commun.* **2000**, *279*, 762–769. [CrossRef] [PubMed]

30. Yang, T.B.; Poovaiah, B.W. A calmodulin-binding/CGCG box DNA-binding protein family involved in multiple signaling pathways in plants. *J. Boil. Chem.* **2002**, *277*, 45049–45058. [CrossRef] [PubMed]

31. Walley, J.W.; Coughlan, S.; Hudson, M.E.; Covington, M.F.; Kaspi, R.; Banu, G.; Harmer, S.L.; Dehesh, K. Mechanical stress induces biotic and abiotic stress responses via a novel cis-element. *PLoS Genet.* **2007**, *3*, e172. [CrossRef] [PubMed]

32. Du, L.; Ali, G.S.; Simons, K.A.; Hou, J.; Yang, T.; Reddy, A.S.; Poovaiah, B.W. Ca(2+)/calmodulin regulates salicylic-acid-mediated plant immunity. *Nature* **2009**, *457*, 1154–1158. [CrossRef] [PubMed]

33. Laluk, K.; Prasad, K.V.; Savchenko, T.; Celesnik, H.; Dehesh, K.; Levy, M.; Mitchell-Olds, T.; Reddy, A.S. The calmodulin-binding transcription factor SIGNAL RESPONSIVE1 is a novel regulator of glucosinolate metabolism and herbivory tolerance in Arabidopsis. *Plant Cell Physiol.* **2012**, *53*, 2008–2015. [CrossRef] [PubMed]

34. Galon, Y.; Nave, R.; Boyce, J.M.; Nachmias, D.; Knight, M.R.; Fromm, H. Calmodulin-binding transcription activator (CAMTA) 3 mediates biotic defense responses in Arabidopsis. *FEBS Lett.* **2008**, *582*, 943–948. [CrossRef] [PubMed]

35. Mitsuda, N.; Isono, T.; Sato, M.H. Arabidopsis CAMTA family proteins enhance V-PPase expression in pollen. *Plant Cell Physiol.* **2003**, *44*, 975–981. [CrossRef] [PubMed]

36. Babicki, S.; Arndt, D.; Marcu, A.; Liang, Y.; Grant, J.R.; Maciejewski, A.; Wishart, D.S. Heatmapper: Web-enabled heat mapping for all. *Nucleic Acids Res.* **2016**, *44*, W147–W153. [CrossRef] [PubMed]

37. Tabas-Madrid, D.; Nogales-Cadenas, R.; Pascual-Montano, A. GeneCodis3: A non-redundant and modular enrichment analysis tool for functional genomics. *Nucleic Acids Res.* **2012**, *40*, W478–W483. [CrossRef] [PubMed]

38. Jin, J.; Zhang, H.; Kong, L.; Gao, G.; Luo, J. PlantTFDB 3.0: A portal for the functional and evolutionary study of plant transcription factors. *Nucleic Acids Res.* **2014**, *42*, D1182–D1187. [CrossRef] [PubMed]

39. Kankainen, M.; Holm, L. POBO, transcription factor binding site verification with bootstrapping. *Nucleic Acids Res.* **2004**, *32*, W222–W229. [CrossRef] [PubMed]

International Journal of
Molecular Sciences

MDPI

Article

The Salt-Stress Response of the Transgenic Plum Line J8-1 and Its Interaction with the Salicylic Acid Biosynthetic Pathway from Mandelonitrile

Agustina Bernal-Vicente [1], Daniel Cantabella [1,2], Cesar Petri [3], José Antonio Hernández [1,*] and Pedro Diaz-Vivancos [1,4]

[1]   Biotechnology of Fruit Trees Group, Department Plant Breeding, CEBAS-CSIC,
      Campus Universitario de Espinardo, 25, 30100 Murcia, Spain; tina.cartagena@hotmail.com (A.B.-V.);
      daniel.cantabella@irta.cat (D.C.); pdv1@um.es (P.D.-V.)
[2]   IRTA, XaRTA-Postharvest, Edifici Fruitcentre, Parc Científic i Tecnològic Agroalimentari de Lleida,
      25003 Lleida, Catalonia, Spain
[3]   Departamento de Producción Vegetal, Universidad Politécnica de Cartagena, Paseo Alfonso XIII, 48, 30203
      Cartagena, Spain; cesar.petri@upct.es
[4]   Department of Plant Biology, Faculty of Biology, University of Murcia, Campus de Espinardo,
      E-30100 Murcia, Spain
*   Correspondence: jahernan@cebas.csic.es; Tel.: +34-968-396200

Received: 18 October 2018; Accepted: 6 November 2018; Published: 8 November 2018

Abstract: Salinity is considered as one of the most important abiotic challenges that affect crop productivity. Plant hormones, including salicylic acid (SA), are key factors in the defence signalling output triggered during plant responses against environmental stresses. We have previously reported in peach a new SA biosynthetic pathway from mandelonitrile (MD), the molecule at the hub of the cyanogenic glucoside turnover in *Prunus* sp. In this work, we have studied whether this new SA biosynthetic pathway is also present in plum and the possible role this pathway plays in plant plasticity under salinity, focusing on the transgenic plum line J8-1, which displays stress tolerance via an enhanced antioxidant capacity. The SA biosynthesis from MD in non-transgenic and J8-1 micropropagated plum shoots was studied by metabolomics. Then the response of J8-1 to salt stress in presence of MD or Phe (MD precursor) was assayed by measuring: chlorophyll content and fluorescence parameters, stress related hormones, levels of non-enzymatic antioxidants, the expression of two genes coding redox-related proteins, and the content of soluble nutrients. The results from in vitro assays suggest that the SA synthesis from the MD pathway demonstrated in peach is not clearly present in plum, at least under the tested conditions. Nevertheless, in J8-1 NaCl-stressed seedlings, an increase in SA was recorded as a result of the MD treatment, suggesting that MD could be involved in the SA biosynthesis under NaCl stress conditions in plum plants. We have also shown that the plum line J8-1 was tolerant to NaCl under greenhouse conditions, and this response was quite similar in MD-treated plants. Nevertheless, the MD treatment produced an increase in SA, jasmonic acid (JA) and reduced ascorbate (ASC) contents, as well as in the coefficient of non-photochemical quenching (qN) and the gene expression of *Non-Expressor of Pathogenesis-Related 1* (*NPR1*) and *thioredoxin H* (*TrxH*) under salinity conditions. This response suggested a crosstalk between different signalling pathways (NPR1/Trx and SA/JA) leading to salinity tolerance in the transgenic plum line J8-1.

Keywords: chlorophyll fluorescence; J8-1 plum line; mandelonitrile; *Prunus domestica*; redox signalling; salicylic acid; salt-stress; soluble nutrients

1. Introduction

Salinity or salt stress significantly affects crop productivity, and it is considered as one of the most important abiotic challenges that plant scientists must confront today. Due to the use of saline waters for irrigation, the percentage of land affected by salinity is continuously growing worldwide. When plants are submitted to salt stress conditions, physiological, biochemical, and nutritional disorders occur, limiting plant growth and development and, ultimately, productivity. These deleterious effects are due to the accumulation of toxic ions (Na^+ and Cl^-), leading to especially Ca^{2+} and K^+ deficiency among other nutrient imbalances, and the reduced water uptake produced by osmotic stress [1,2]. In addition, salinity also induced an oxidative stress mediated by reactive oxygen species (ROS) at the subcellular level [2].

Furthermore, it is well known that plant hormones are key factors in the defence signalling output triggered during both abiotic and biotic environmental stress conditions. Among these hormones, SA has attracted much attention, although other plant hormones, such as abscisic acid (ABA) and jasmonic acid (JA), have also been suggested as modulators of plant defence responses.

Considering the important roles of SA during plant responses against stress conditions, SA is of potential agro-economic interest as a modulator of plant plasticity. Although the regulation of SA biosynthesis and the SA-mediated stress tolerance mechanism have not been fully characterised [3], researchers have found that the exogenous application of SA or analogues induce tolerance to several stress conditions [4]. In the same line, we have previously reported in peach (*P. persica* L.) plants that mandelonitrile (MD) is also involved in SA biosynthesis and improves plant performance under biotic and abiotic stress conditions [5]. In *Prunus*, MD is at the hub of cyanogenic glycoside (CNglcs) synthesis and turnover [6]. CNglcs are specialized secondary metabolites that have been linked to plant plasticity improvement against environmental stress conditions. However, CNglcs turnover is highly species dependent [6]. As a result, further studies must be performed to elucidate whether this new SA biosynthetic pathway from MD is also present in other *Prunus* species, and to determine its possible role in plant plasticity under stress conditions. Accordingly, other authors have suggested that SA biosynthesis varies depending on different factors, including the plant species and the environmental conditions [7–9].

One common consequence of exposure to stress conditions is the establishment of oxidative signalling that triggers transduction cascades controlling plant development and defence [10]. The major low-molecular-weight antioxidants ascorbate (ASC) and glutathione (GSH) determine the specificity of this oxidative signalling. Thus, ASC and GSH have been shown to be multifunctional metabolites that are important in redox homeostasis and signalling as well as in developmental and defence reactions [11]. The *NON-EXPRESSOR OF PR-PROTEINS1* (*NPR1*) transcription factor, which is activated by SA, is one of the few known redox-regulated signalling proteins in plants, highlighting the crosstalk between the antioxidant metabolism and plant hormones during environmental stress responses. On the other hand, the roles of thioredoxins (Trx) in redox signalling as regulators of scavenging mechanisms and as components of signalling pathways are well established [12]. It has been suggested that SA signalling activates Trx-h5, leading to NPR1 reduction and releasing active monomers that are translocated from the cytosol into the nucleus; this, in turn, activates the expression of defence genes [13,14].

In the present manuscript, we have analysed whether the SA biosynthetic pathway from MD, previously observed in peach [5], is also present in plum (*P. domestica* cv. *Claudia verde*) plants in the presence or absence of NaCl. Moreover, in order to gain deeper knowledge of the SA-mediated defence network in *Prunus*, we have used transgenic plum plants over-expressing four copies of the cytosolic ascorbate peroxidase gene. These transgenic plants with enhanced antioxidant capacity, named line J8-1, have shown higher regeneration efficiency and enhanced vigour as well as tolerance to salt stress under in vitro conditions [15,16]. Moreover, line J8-1 has displayed enhanced tolerance to water stress under greenhouse conditions [17]. Thus, line J8-1 could be an excellent model to study the crosstalk among stress tolerance, oxidative stress and SA in plum plants. Taking into account all the mentioned above, we analysed the effect of MD and Phe (MD precursor) treatments on plant performance (chlorophyll content, chlorophyll fluorescence parameters and leaf and root water contents), on the content of stress-related hormones, on the redox state and the expression of two redox-related genes, and on the soluble leaf and root nutrient content in the transgenic J8-1 line under control and salt stress conditions.

2. Results

The following experiments were designed in order to elucidate whether MD could be a precursor of SA biosynthesis in plum plants, as occurred in peach [5]. Moreover, to further study the crosstalk among stress tolerance, oxidative stress, and SA under salinity conditions, the effect of MD and Phe (MD precursor) has been investigated in the transgenic plum line J8-1 submitted to NaCl.

2.1. Metabolomic Analysis of SA Biosynthesis in Plum Plants

We have previously described in peach that the cyanogenic glycoside (CNglcs) pathway is involved in SA biosynthesis, suggesting the existence of a third SA biosynthetic pathway, being MD the intermediary molecule between both pathways [5]. Taking into account the fact that the CNglcs pathway is highly dependent on the plant species [6], here we have studied whether this SA biosynthetic pathway, from MD, is also functional in plum plants under control and salinity conditions.

When micropropagated non-transgenic plum (cv. *Claudia verde*) shoots were fed with [^{13}C]Phe or with [^{13}C]MD, in the absence of NaCl, increased levels of Phe, MD and benzoic acid were recorded, whereas amygdalin only increased in Phe-treated shoots (Figure 1). However, contrary to that which we observed in peach [5], none of the treatments produced a significant rise in SA content (Figure 1). In the presence of NaCl, benzoic acid (BA) content only increased in MD-treated micropropagated shoots, whereas only the Phe treatment produced an accumulation of MD and SA (Figure 1).

The SA biosynthesis from the CNglcs pathway was also studied in micropropagated shoots from the transgenic plum line J8-1. In the absence of stress, the [^{13}C]MD treatment decreased MD and BA levels, while [^{13}C]Phe-fed micropropagated J8-1 shoots displayed increased amounts of Phe and amygdalin and lesser amounts of BA (Figure 2). Similar to results in cv. *Claudia verde* and contrary to that which occurred in peach plants [5], neither [^{13}C]MD nor [^{13}C]Phe increased the SA content under in vitro conditions. Salt stress induced a significant decrease in Phe, MD, and amygdalin in both control and treated (MD or Phe) J8-1 shoots. However, both treatments ameliorated the decrease in benzoic acid observed in control shoots (Figure 2). Regarding SA levels, no statistically significant differences were observed in NaCl-submitted shoots (Figure 2).

Under our experimental conditions, we were able to detect [^{13}C]-Phe, -MD and -SA, but no [^{13}C]-benzoic acid was observed in either plum plant, cv. *Claudia verde* or the J8-1 line. Regarding the percentage of [^{13}C]-labelled compounds, similar values were recorded in both plum plants, and no significant differences were observed among the different treatments and conditions. We observed less than 10% of [^{13}C]Phe, and [^{13}C]MD but [^{13}C]SA values ranged between 20% and 25% of the total amount detected (Supplemental Figure S1). It is noteworthy to mention that, although differences were not statistically significant, the highest levels of [^{13}C]MD and [^{13}C]SA were observed in [^{13}C]MD-fed micropropagated shoots. Moreover, under salinity conditions, no [^{13}C]Phe was detected, probably because its rapid turnover under stress conditions (Supplemental Figure S1).

These results suggest that the SA synthesis from the MD pathway demonstrated in peach is not clearly present in plum, at least under in vitro conditions. For this reason, further experiments in order to investigate the crosstalk among stress tolerance, oxidative stress and SA under salinity were performed on the transgenic line J8-1, displaying an enhanced antioxidant capacity.

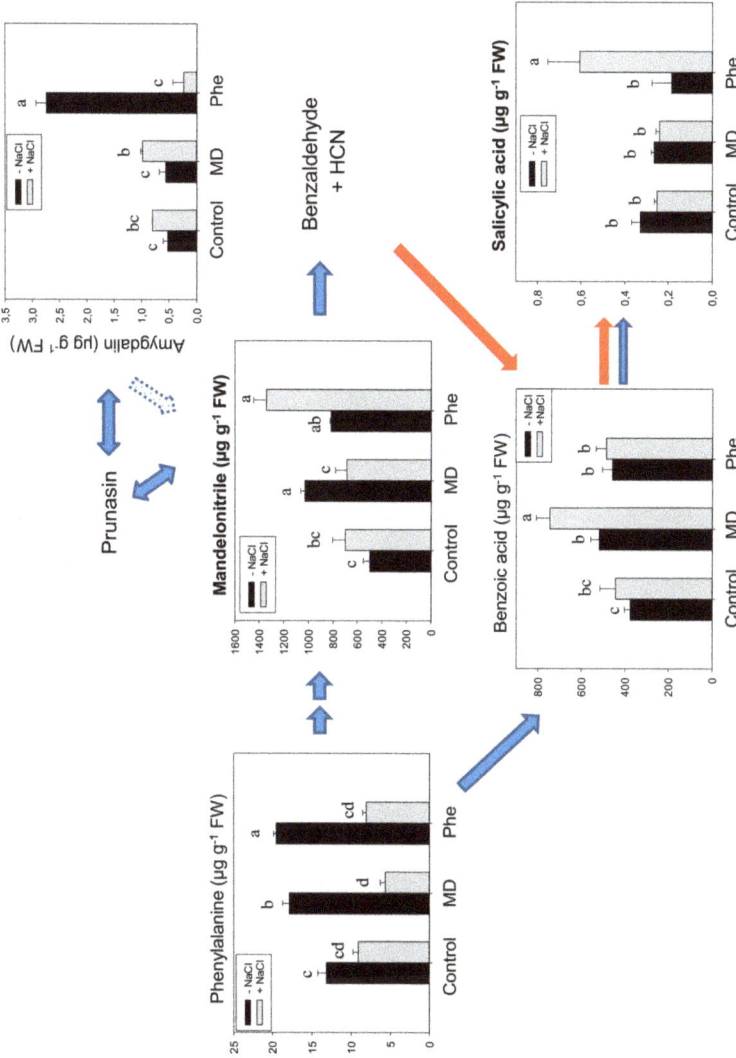

Figure 1. Salicylic acid (SA) biosynthetic and cyanogenic glucoside (CNglcs) pathways in salt-stressed (100 mM NaCl) plum cv. *Claudia verde* shoots micropropagated in the presence or absence of [^{13}C]MD or [^{13}C]Phe. Total levels (µM g^{-1} FW) of amygdalin, benzoic acid, mandelonitrile, phenylalanine, and salicylic acid are shown. Data represent the mean ± SE of at least 12 repetitions of each treatment. Different letters indicate significant differences in each graph according to Duncan's test ($p \leq 0.05$). Blue arrows indicate the previously described SA biosynthesis in higher plants [3] (dot arrow, putative), whereas red arrows show the recently described pathway [5].

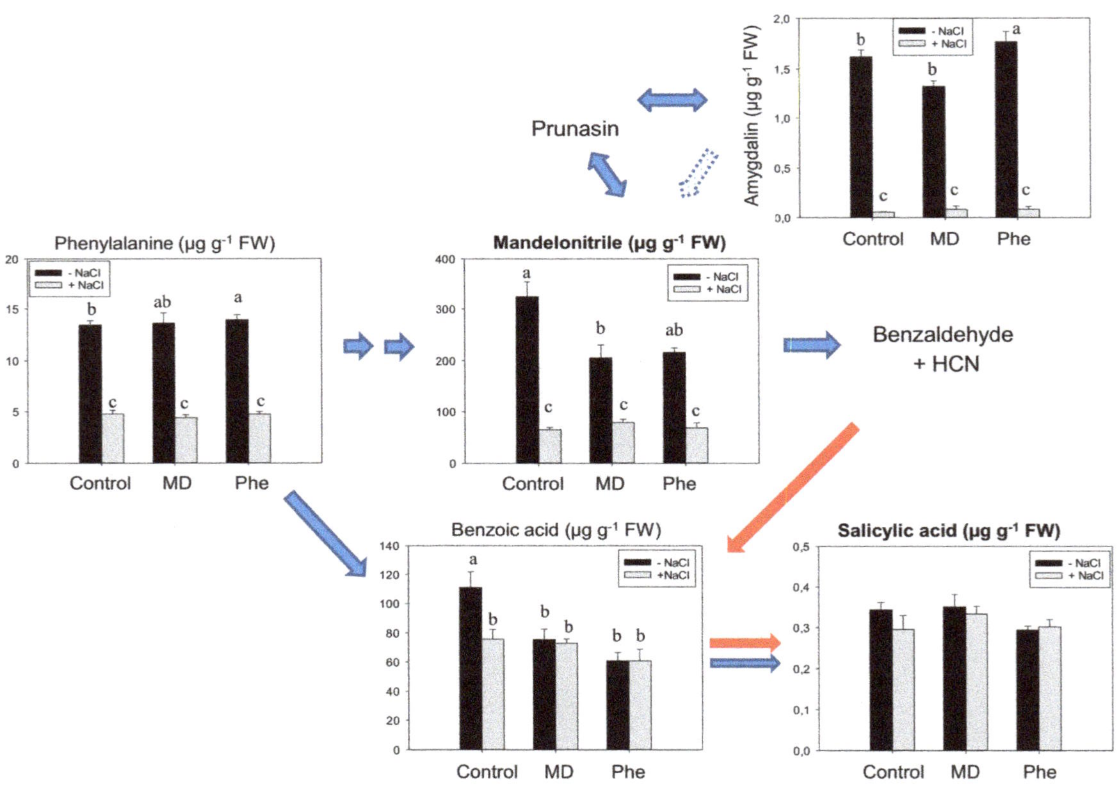

Figure 2. Salicylic acid (SA) biosynthetic and cyanogenic glucoside (CNglcs) pathways in salt-stressed (100 mM NaCl) transgenic J8-1 plum shoots micropropagated in the presence or absence of [^{13}C]MD or [^{13}C]Phe. Total levels (µM g^{-1} FW) of amygdalin, benzoic acid, mandelonitrile, phenylalanine, and salicylic acid are shown. Data represent the mean ± SE of at least 12 repetitions of each treatment. Different letters indicate significant differences in each graph according to Duncan's test ($p \leq 0.05$). Blue arrows indicate the previously described SA biosynthesis in higher plants [3] (dot arrow, putative), whereas red arrows show the recently described pathway [5].

2.2. Effect on Stress-Related Hormones: SA, ABA and JA

It is known that cross-talk among different hormonal signals is involved in different physiological responses as well as in response to environmental challenges. In that sense, the content of SA and other well-known stress-related hormones like ABA and JA was determined in leaves from J8-1 seedlings.

In the absence of NaCl, similar to that observed in micropropagated shoots, neither MD nor Phe affected SA levels (Figure 3). Under NaCl stress, however, a significant increase in SA concentration was observed, especially in the presence of MD. In fact, MD-treated J8-1 seedlings showed a 2.3- and 1.7-fold SA increase compared to control and Phe-treated plants, respectively (Figure 3).

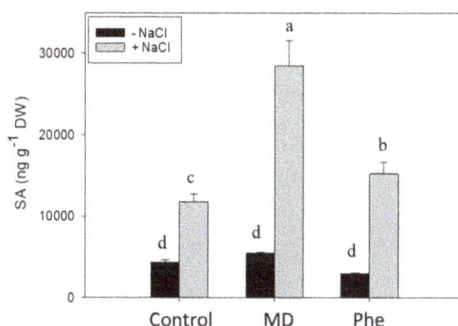

Figure 3. Total SA level (ng g-1 DW) in the leaves of J8-1 seedlings grown in the presence or absence of MD or Phe and submitted to salt stress (6 g/L NaCl). Data represent the mean ± SE of at least four repetitions of each treatment. Different letters indicate significant differences according to Duncan's test ($p \leq 0.05$).

In addition, we also analysed the levels of other hormones related to stress such as ABA and JA. In the absence of NaCl, both treatments increased ABA levels, with the Phe-treated J8-1 seedlings showing the highest levels (Figure 4A). The presence of NaCl produced a change in this response. In that regards, control plants showed a 1.7-fold increase in ABA levels, whereas in Phe-treated plants, ABA levels declined up to 1.8-fold. Regarding MD-treated J8-1 seedlings, a small but significant decrease in ABA was recorded, in relation to the levels observed in the absence of NaCl (Figure 4A). With regard to the JA concentration, only the MD treatment in the presence of NaCl produced statistically significant changes, increasing considerably the JA levels (Figure 4B).

2.3. Plant Growth, Chlorophyll Contents, and Chlorophyll Fluorescence

In previous works, we reported the tolerance of the transgenic plum line J8-1 to salinity (up to 150 mM) under in vitro conditions [16] and to water-stress under ex vitro conditions (up to 15 days of water deprivation) [17]. The NaCl-tolerance was also confirmed under ex vitro conditions as observed by the effect of MD and Phe (MD precursor) treatments on plant performance (chlorophyll content, chlorophyll fluorescence parameters and leaf and root water contents) under salinity stress conditions. Accordingly, NaCl treatment (6 g/L) did not have a significant effect on plant growth (Supplemental Figure S2) or on the leaf water content either in the absence or presence of MD or Phe treatments (Supplemental Figure S3). On the other hand, salinity increased the root water content in control and MD-treated seedlings (Supplemental Figure S3).

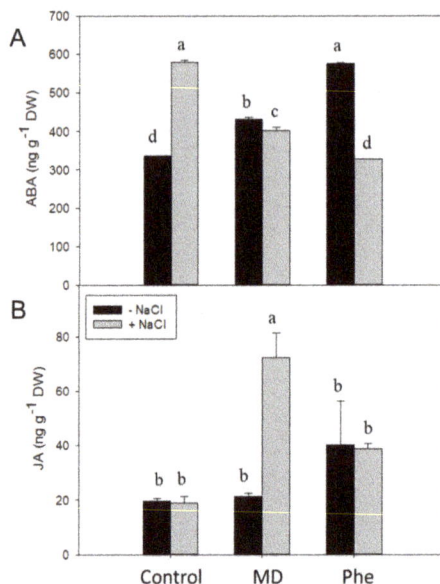

Figure 4. Effect on the stress-related hormones ABA and JA. Total ABA (**A**) and JA (**B**) levels (ng g^{-1} DW) in the leaves of J8-1 seedlings grown in the presence or absence of MD or Phe submitted to salt stress (6 g/L NaCl). Data represent the mean ± SE of at least four repetitions of each treatment. Different letters indicate significant differences according to Duncan's test ($p \leq 0.05$).

We also analysed the effect of NaCl in the presence or absence of MD and Phe treatments on the chlorophyll content in leaves from J8-1 seedlings. In the absence of NaCl, MD treatment increased the Chla content, whereas Phe produced a rise in Chla and Chlb. Under salinity conditions, an increase in Chla and Chlb was observed in non-treated plants as well as in the presence of MD. However, a decrease in Chla was produced in Phe-treated plants (Figure 5).

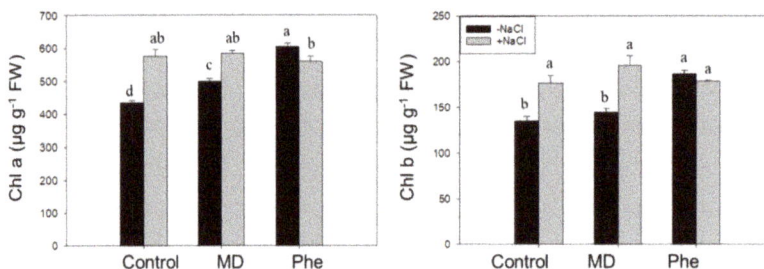

Figure 5. Effect of NaCl (6 g/L) on Chla and Chlb content in control, MD, and Phe treated J8-1 plum seedlings. Data represents the mean ± SE of at least four repetitions. Different letters indicate statistical significance according to Duncan's test ($p < 0.05$).

In addition to chlorophyll content determination, different photochemical [Y(II) and qP] and non-photochemical [(Y(NPQ) and qN] quenching chlorophyll fluorescence related parameters were also analysed. Under control conditions, both treatments increased qP, whereas a decrease in qN occurred in Phe-treated plants. Under NaCl stress, non-treated plants showed an increase in qP and qN (Figure 6). The MD treatment decreased the photochemical quenching parameters, but an increase in the non-photochemical quenching parameters took place. Regarding the Phe treatment, a decrease in Y(II) was observed, but qP did not show statistically significant changes, whereas, similar to the MD treatment, a significant increase in the non-photochemical quenching parameters occurred (Figure 6).

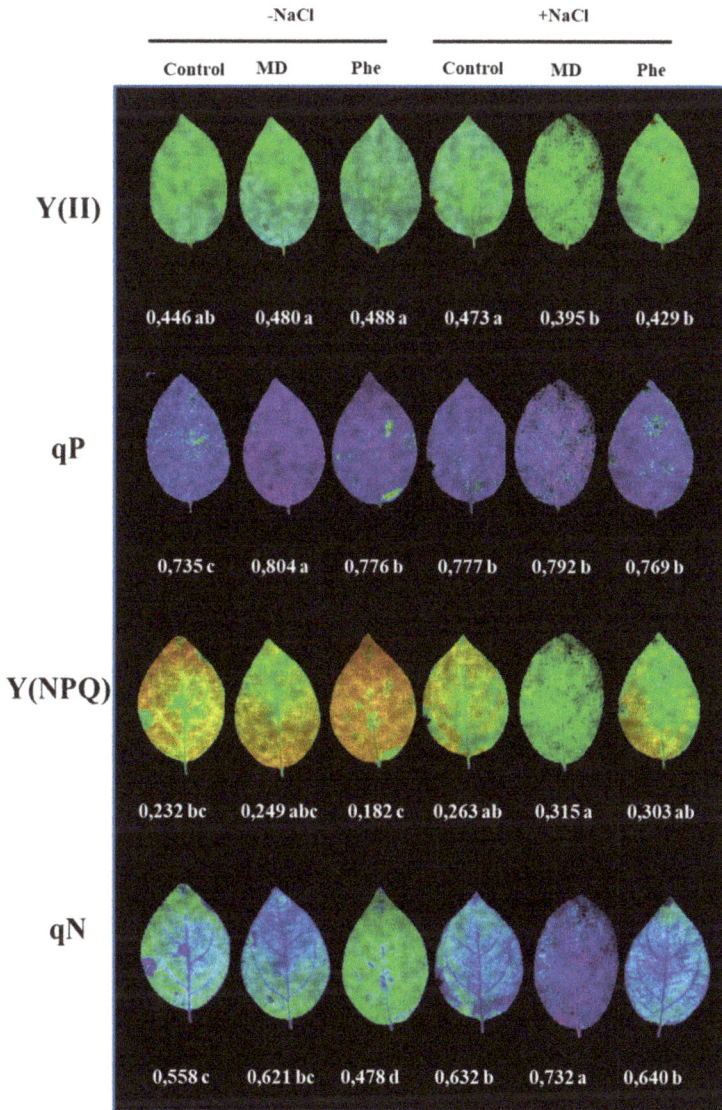

Figure 6. The effect of salt stress (6 g/L NaCl) on the chlorophyll fluorescence parameters in J8-1 seedling leaves. Representative images of the quantum yield of photochemical energy conversion in PS II [Y(II)], the photochemical quenching (qP) and the quantum yield of regulated non-photochemical energy loss in PS II and its coefficient [y(NPQ) and qN] are shown. Zero represents the lowest value and 1 the maximum value for each parameter. The averages of the values of the different parameters analysed are displayed below each image. Data represent the mean ± SE of at least six repetitions. Different letters indicate statistical significance according to Duncan's test ($p < 0.05$).

2.4. Redox State and the Gene Expression of Redox-Related Genes

It is well known that the stress hormone SA can interact with the antioxidant metabolism modulating cellular redox homeostasis. For this raison, we determined the redox state in micropropagated J8-1 shoots and in leaves from J8-1 seedlings by analysing the ascorbate and glutathione levels in the absence and in the presence of NaCl.

Under control conditions, micropropagated shoots did not show significant changes in ascorbate or glutathione levels (Tables 1 and 2). In the presence of NaCl, MD treatment decreased the total (TASC) and reduced ascorbate (ASC) levels, whereas Phe increased the TASC content. As a result, a decrease in the redox state of ascorbate (ASC/TASC) occurred in Phe-treated plants (Table 1). Regarding glutathione levels, an increase in its reduced form (GSH), as well as in the total glutathione (TGSH) level, was only observed in MD-treated micropropagated shoots under salt stress. However, no changes in the redox state of glutathione (GSH/TGSH) were observed in any treatments (Table 2).

Table 1. Effect of salt stress (100 mM NaCl), in the presence or absence of MD and Phe treatments, on total ascorbate (TASC) and reduced ascorbate (ASC) content in micropropagated J8-1 shoots. Data represent the mean ± SE of at least four repetitions. Different letters in the same column indicate significant differences according to Duncan's test ($p \leq 0.05$).

	Treatment	TASC (μmol g^{-1} FW)	ASC (μmol g^{-1} FW)	Ascorbate Redox State
−NaCl	Control	1.2 ± 0.16 c	0.9 ± 0.08 ab	0.74 ± 0.03 a
	MD	1.3 ± 0.03 bc	0.9 ± 0.04 ab	0.73 ± 0.01 a
	Phe	1.7 ± 0.03 c	1.2 ± 0.05 a	0.69 ± 0.02 a
+NaCl	Control	1.7 ± 0.09 ab	1.2 ± 0.03 a	0.69 ± 0.05 a
	MD	1.2 ± 0.06	0.8 ± 0.04 b	0.72 ± 0.00 a
	Phe	ab 2.1 ± 0.26 a	1.2 ± 0.16 a	0.57 ± 0.02 b

Table 2. Effect of salt stress (100 mM NaCl), in the presence or absence of MD and Phe treatments, on total glutathione (TGSH) and reduced glutathione (GSH) content in micropropagated J8-1 shoots. Data represent the mean ± SE of at least four repetitions. Different letters in the same column indicate significant differences according to Duncan's test ($p \leq 0.05$).

	Treatment	TGSH (nmol g^{-1} FW)	GSH (nmol g^{-1} FW)	Glutathione Redox State
−NaCl	Control	91.1 ± 4.82 c	86.7 ± 4.25 c	0.95 ± 0.01 ab
	MD	90.1 ± 1.94 c	86.1 ± 1.47 c	0.96 ± 0.00 a
	Phe	104.9 ± 7.60 c	98.3 ± 6.99 bc	0.93 ± 0.00 ab
+NaCl	Control	104.9 ± 6.66 bc	96.9 ± 6.62 bc	0.92 ± 0.00 b
	MD	140.5 ± 6.40 a	131.7 ± 6.38 a	0.94 ± 0.00 ab
	Phe	120.5 ± 6.55 ab	112.0 ± 5.63 ab	0.93 ± 0.01 ab

The response was rather different in J8-1 seedlings. In this case, under our experimental conditions, we were not able to detect oxidised ascorbate, so only ASC content is shown (Table 3). In absence of NaCl, and similar to that observed under in vitro conditions, no significant differences were apparent in the ASC and GSH levels. When plants were subjected to saline stress, MD treatment increased ASC but decreased GSH. However, MD and Phe treatments induced an accumulation of oxidised glutathione (GSSG), leading to a decrease in the redox state of glutathione in both cases (Table 3).

Table 3. Effect of salt stress (100 mM NaCl), in the presence or absence of MD and Phe treatments, on ascorbate (ASC) and glutathione (GSH, reduced; GSSG, oxidized) contents in the leaves of J8-1 seedlings. Data represent the mean ± SE of at least four repetitions. Different letters in the same column indicate significant differences according to Duncan's test ($p \leq 0.05$).

	Treatment	ASC (μmol g^{-1} FW)	GSH (nmol g^{-1} FW)	GSSG (nmol g^{-1} FW)	Glutathione Redox State
−NaCl	Control	4.6 ± 0.7 b	109.5 ± 1.9 a	11.4 ± 1.1 b	0.91 ± 0.01 a
	MD	3.7 ± 0.2 bc	99.6 ± 5.2 a	37.5 ± 1.4 a	0.72 ± 0.02 c
	Phe	4.5 ± 0.6 b	114.5 ± 5.4 a	30.1 ± 2.8 a	0.79 ± 0.02 b
+NaCl	Control	3.4 ± 0.4 bc	103.5 ± 7.5 a	13.5 ± 1.0 b	0.88 ± 0.01 a
	MD	7.1 ± 1.5 a	79.3 ± 8.0 b	36.3 ± 3.3 a	0.68 ± 0.02 c
	Phe	1.8 ± 0.2 c	114.2 ± 6.9 a	30.2 ± 2.1 a	0.79 ± 0.01 b

We also studied the effect of MD and Phe treatments on the *Non-Expressor of Pathogenesis-Related 1* (*NPR1*) and *thioredoxin H* (*TrxH*) gene expression levels in NaCl-stressed J8-1 micropropagated plum shoots and in leaves from J8-1 plum seedlings. In the absence of NaCl, micropropagated shoots treated with Phe showed reduced *NPR1* expression but induced *TrxH* expression (Figure 7A,B). In the presence of NaCl, both treatments increased the expression of the studied redox-related genes. The induction was especially striking for the effect of Phe on *NPR1* expression (nearly a five-fold increase) and for the increase in *TrxH* expression (13-fold) observed as a result of MD treatment (Figure 7A,B).

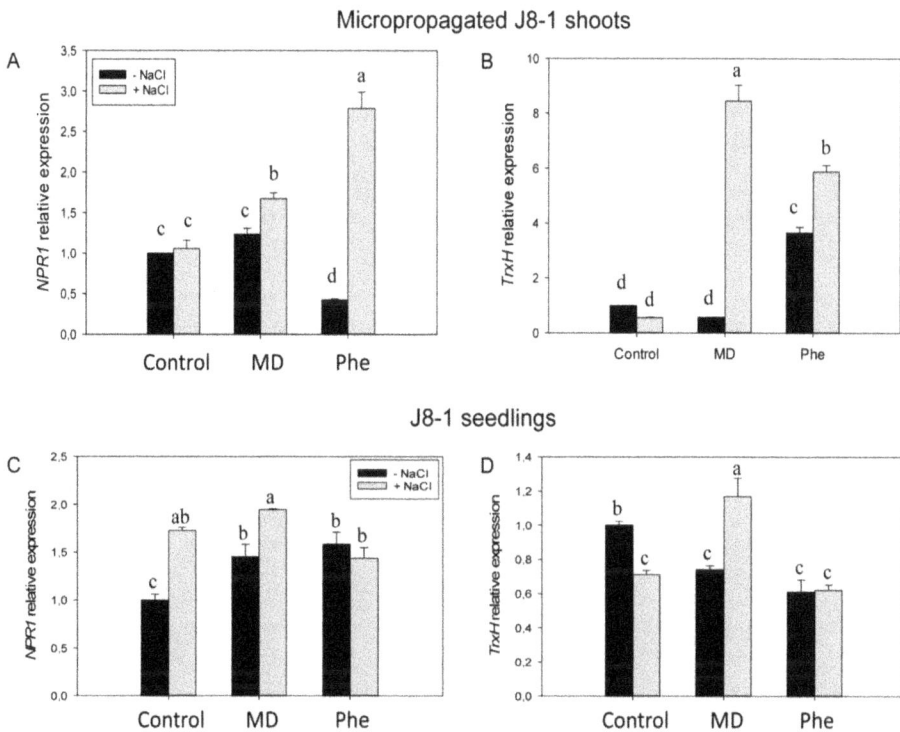

Figure 7. Gene expression of *NPR1* and *TrxH* in micropropagated J8-1 shoots (**A,B**) and in the leaves of J8-1 seedlings (**C,D**) grown in the presence or absence of MD or Phe and submitted to salt stress. Data represent the mean ± SE of at least five repetitions of each treatment. Different letters indicate significant differences in each graph according to Duncan's test ($p \leq 0.05$).

The effect of the treatments on the expression of both redox-related genes in the transgenic plum seedlings was somewhat different. In the absence of NaCl, both MD and Phe treatments induced *NPR1* expression but reduced *TrxH* expression in a similar manner (Figure 7C,D). Under salinity conditions, control and MD-treated seedlings increased *NPR1* expression, whereas *TrxH* expression was repressed in control seedlings but was again induced by MD (Figure 7C,D).

2.5. Effect of MD and Phe on Soluble Leaf and Root Nutrient Content under Salt Stress Conditions

Salt stress produced ion toxicity associated with excess Cl^- and Na^+ uptake, leading to Ca^{2+} and K^+ deficiency and other nutrient imbalances [2]. Therefore, the effect of MD and Phe treatments on soluble K^+, Ca^{2+}, Na^+ and Cl^- levels was analysed in leaves and roots from transgenic plum seedlings grown in the presence and absence of NaCl. Under control conditions, MD and Phe increased leaf K^+ content but decreased leaf Ca^{2+} content. No effects of MD or Phe treatments on Na^+ and Cl^- levels were observed (Figure 8). In NaCl-stressed seedlings, an increase in all the analysed nutrients occurred in the leaves from non-treated plants. Similar results were observed in the MD and Phe treatments, in which the leaves of salt-stressed seedlings also displayed increased Ca^{2+}, Na^+, and Cl^- levels, although the K^+ level slightly decreased due to MD and was not affected by Phe. It is important to note the low leaf Na^+ levels found in the transgenic plum seedlings (Figure 8).

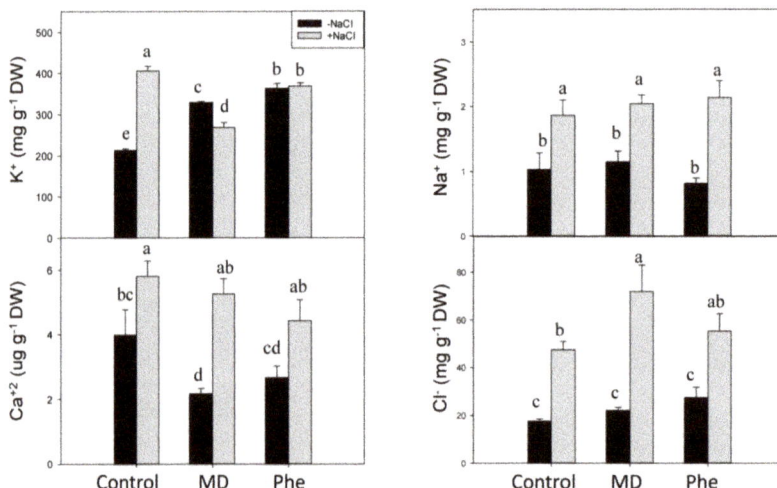

Figure 8. Effect of salt stress (6 g/L NaCl) on soluble K^+, Ca^{2+}, Na^+, and Cl^- contents in leaves from control, MD- and Phe-treated J8-1 seedlings. Data represent the mean ± SE of at least four repetitions. Different letters indicate statistical significance according to Duncan's test ($p < 0.05$).

In the absence of NaCl stress, the only significant change observed in roots was an increase in soluble K^+ in Phe-treated plants (Figure 9). In the presence of NaCl, an accumulation of the phytotoxic ions Na^+ and Cl^- was observed in control and MD- and Phe-treated seedlings. However, the Na^+ accumulation in roots was lower in MD-treated plants than in the other treatments. The K^+ content increased in non-treated NaCl-stressed roots, whereas no significant differences were observed in Ca^{2+} levels in any case (Figure 9).

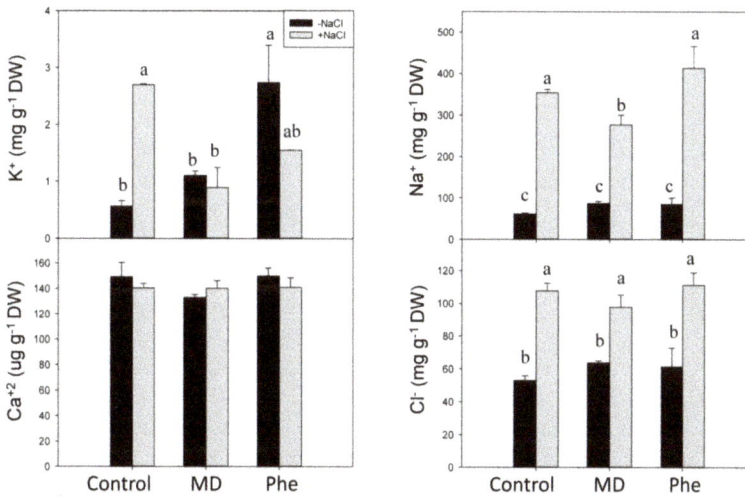

Figure 9. Effect of salt stress (6 g/L NaCl) on soluble K^+, Ca^{2+}, Na^+, and Cl^- contents in roots from control, MD- and Phe-treated J8-1 seedlings. Data represent the mean $\pm$ SE of at least four repetitions. Different letters indicate statistical significance according to Duncan's test ($p < 0.05$).

3. Discussion

3.1. Involvement of MD on SA Biosynthesis in Plum

In a previous work, we reported that the transgenic plum line J8-1 was tolerant up to 150 mM NaCl under in vitro conditions. This response correlated with high ascorbate peroxidase (APX) activity and gene expression and glutathione and ascorbate contents [16]. We also demonstrated that APX overexpression in line J8-1 can play a major role in the response of J8-1 seedlings to drought conditions by inducing changes at the physiological, biochemical, proteomic, and genetic levels [17]. In our opinion, it is of interest to characterise the response of this transgenic line to NaCl stress under ex vitro conditions. In addition, we recently reported that MD is the intermediary molecule between a suggested new SA biosynthetic pathway and CNglcs turnover in peach plants [5]. All these findings led us to investigate not only the response of the J8-1 line to salinity, but also whether the new SA pathway described in peach plants [5] is also present in this line and the possible role of this pathway on plant performance.

In micropropagated peach shoots fed with [13C]MD, nearly 20% of the total SA quantified appeared as [13C]SA, demonstrating that MD can be an intermediary molecule in this novel pathway controlling amygdalin and SA biosynthesis [5] However, when micropropagated plum "*Claudia Verde*" shoots were fed with [13C]MD or [13C]Phe, no increases in SA were detected, although significant increases in MD, Phe and benzoic acid (SA-precursor) were observed. We also assayed this possibility using micropropagated J8-1 shoots. However, the results concerning the involvement of MD as a putative intermediary of SA biosynthesis in this plum line were negative. These results led to the hypothesis that the new SA synthesis pathway from MD, previously demonstrated in peach, seemed not to be operative in plum under in vitro conditions. Nevertheless, we observed that NaCl stress affected the CNglcs pathway, mainly at the MD and amygdalin levels. Due to the fact that amygdalin is derived from MD by the addition of two glucose molecules, it is logical to assume that the amygdalin content decreases. The glucose molecules could be used for osmotic adjustment or to obtain energy for different metabolic processes.

As a conclusion, it seems that MD is not an intermediary for SA biosynthesis in micropropagated plums as it was found to be in micropropagated peach shoots, unless the synthesized SA appears as conjugated form. However, an increase in SA was produced in J8-1 plum seedlings when NaCl stress was imposed, especially in MD-treated plants, suggesting that MD could be involved in the SA biosynthesis in plum grown in the presence of NaCl under greenhouse conditions. These findings are different to the results found in peach, where the contribution of this pathway to the SA pool does not seem to be relevant under salt stress or *Plum pox virus*-infection conditions [18].

3.2. Plant Performance of Plum under NaCl Stress

As expected, line J8-1 was also tolerant to high NaCl levels under greenhouse conditions, as observed by the lack of negative effects on plant growth (measured as shoot biomass fresh weight) or in the leaf and root water levels, both in control and MD- and Phe-treated plants. In addition, chlorophyll levels can be seen as a biochemical marker of salt tolerance in plants; the maintenance or the increase in Chl content under NaCl stress can be considered as a protection mechanism for the photosynthesis process. In this regards, both non-treated and MD-treated plants showed increases in Chla and Chlb levels. Moreover, J8-1 plants increased/and or maintained the non-photochemical quenching parameters under NaCl stress, especially in the presence of MD. The maintenance of the non-photochemical quenching parameters under stress situations has been associated with a capacity to dissipate light energy safely, and it can be seen as an adaptive mechanism to protect the chloroplasts under NaCl conditions, avoiding the over-generation of ROS, as described for other plant species [2,19–21].

Thus, according to plant growth, leaf and root water content, chlorophyll contents and chlorophyll fluorescence data, the transgenic plum line J8-1 can be considered as salt-tolerant.

3.3. Stress-Related Hormones and NaCl Response

The effect of MD and Phe treatments on the stress-related hormones ABA and JA in line J8-1 was quite different to that observed in peach plants. In peach, in the absence of NaCl, the treatments had no effect on ABA and reduced the JA levels. Under saline conditions, MD decreased ABA and JA concentrations, whereas Phe produced a decrease in JA [18]. The transgenic line J8-1 showed contrasting results. In this case, the differences could be due to the use of different plant species showing a different NaCl tolerance and the different NaCl levels used: peach plants were subjected to 2 g/L NaCl, whereas plum plants were treated with 6 g/L NaCl.

Some studies have suggested a positive interaction between SA, JA, ethylene and ABA signalling pathways, improving the response of plants to environmental stresses [22]. JA and SA can regulate plant responses to abiotic stresses. Accordingly, the exogenous application of both JA and SA has been found to enhance salt-tolerance in some plant species by increasing their antioxidative capacity [23, 24]. In addition, an increase in the SA/JA ratio has been suggested as a marker of salt stress [25]. In the current study, an increased SA/JA ratio due to salinity was observed in control and MD- and Phe-treated seedlings, the increase being much greater in MD-treated plants. This response was mainly due to the sharp increase in SA levels in MD-treated J8-1 seedlings under salinity conditions. In peach plants, the MD treatment slightly decreased the SA/JA ratio, and no effect of NaCl on plant development was observed [18]. In a study of the salt-tolerant sweet-potato genotype ND98, the JA concentrations in the leaves and roots increased after 12 h of saline treatment (200 mM NaCl), and this response correlated with a regulated stomatal closure [26].

ABA is a well-known regulator of stomatal regulation and, hence, of the abiotic stress response. In the current study, Phe treatment decreased the ABA content, whereas a small decrease occurred in MD-treated plants and an increase was observed in untreated seedlings under salinity conditions. Moreover, SA is also involved in stomatal regulation, and SA treatment has been found to decrease the stomatal aperture in Arabidopsis [27]. Accordingly, the SA increase, and the scanty effect in ABA observed in MD-treated plants in the current study suggests more efficient stomatal regulation under saline conditions.

Considered together, all of this data suggests that MD could have a positive effect on the J8-1 response to salinity through an increase in SA and JA and tight control of the ABA levels.

3.4. NaCl Effects on Redox State and Ion Homeostasis

In micropropagated J8-1 shoots, GSH levels increased under salt stress, especially in the MD treatment. Furthermore, this increase correlated with the induction of *NPR1* and *TrxH*, suggesting a role for GSH in the stress-induced expression of these redox-related genes. It seems that GSH could play a role in *NPR1* induction under in vitro conditions. In micropropagated peach shoots, treatment with the artificial precursor of cysteine, L-2-oxothiazolidine-4-carboxylic acid (OTC), produced an increase in GSH and in the GSH/GSSG ratio as well as in the *NPR1* expression both in healthy and in *Plum pox virus*-infected shoots [28]. On the other hand, in J8-1 seedlings, the increase in *NPR1* and *TrxH* expression due to the MD treatment correlated with a decrease in the GSH redox state. As mentioned, under stress conditions, the MD treatment produced a sharp increase in SA content, and researchers have shown that SA-induced changes in glutathione lead to a more oxidised environment that modulates the plant defence responses [29–31]. Similarly, MD treatment also leads to a more oxidised environment in peach plants via changes in non-enzymatic and enzymatic antioxidant levels that could be responsible for the modification of the function of redox-regulated proteins such as NPR1 [5]. In absence of stress this MD-induced oxidised environment was due to a decrease in ASC content and GSH redox state in both peach [5] and J8-1 seedlings, whereas in J8-1 seedlings submitted to salinity the decrease in the GSH redox state was accompanied by an increase in ASC. This increase in ASC could be also related to the salt stress tolerance displayed by J8-1 seedlings [4,11,16].

Thioredoxins (Trx) are ubiquitous disulfide reductases that regulate the redox status of target proteins and seem to be involved in the protection of plant cells in stress situations that induce oxidative stress [12]. Trx can prevent the oxidative damage of important macromolecules, thus protecting plants against the stress-induced lipid peroxidation of membranes or repairing oxidised proteins [12]. Proteomic tools have made it possible to identify many potential targets of Trx, including many proteins related with important cellular processes [32]. One of the proteins regulated by thioredoxins is NPR1. Cytosolic Trxs catalyze the redox changes in NPR1 from oligomeric to monomeric forms, with SA inducing TRX-5h to catalyse NPR1 monomer release and to prevent re-oligomerization [13]. Different studies have reported the induction of TrxHs by abiotic or biotic stresses [33], suggesting that these proteins can act as antioxidants in vivo [34].

In the current study, the induction of *TrxH* gene expression by MD was more evident under in vitro conditions, where the absence of the root system leads to more severe symptoms under saline conditions. In plum seedlings (ex vitro conditions), the induction of *TrxH* was lower (only a 1.6-fold increase), which was similar to that observed in the induction of *Trxh1* in rice plants treated with 100 mM NaCl [35]. In addition, the effect of salinity on *TrxH* expression was very similar to that observed in peach [18]: in the absence of chemical treatments, salinity reduced the expression of *TrxH*, whereas induction occurred in the presence of MD, and no changes were produced in the presence of Phe.

Elevated SA levels may mediate adaptive responses against salt stress through NPR1-dependent and NPR1-independent pathways. Salt stress (100 mM NaCl) was found to have a strong effect on plant growth in the Arabidopsis *npr1-5* mutant, which lacks the NPR1-dependent SA signalling pathway. However, the effect of NaCl stress on the plant growth of the Arabidopsis *nudt7* mutant, which constitutively expressed NPR1-dependent and NPR1-independent SA signalling, was more attenuated [36]. In addition, the *npr1-5* mutant was unable to control the Na$^+$ influx and prevent K$^+$ loss in shoots and roots, in contrast to the results observed in the *nudt7* mutant. These authors concluded that the constitutive expression of NPR1-dependent SA signalling enhanced salt tolerance by controlling Na$^+$ entry into roots and shoots as well as minimising K$^+$ loss during NaCl challenges, which is an important component of salt and oxidative stress tolerance in Arabidopsis [36]. This information agrees with our results, since under saline conditions, MD-treated plants showed elevated SA levels as well as *NPR1* and *TrxH* expression and less Na$^+$ accumulation in roots than the other treatments. This effect of SA on Na$^+$ levels has also been observed in pea plants [37]. These authors reported that SA treatment reduced Na$^+$ accumulation in pea roots in the presence on 70 mM NaCl.

In the salt-tolerant sweet potato genotype ND98, the JA content increased in the leaves and roots after 12 h of saline treatment (200 mM NaCl), and this response correlated with regulated stomatal closure, reduced Na$^+$ accumulation and increased K$^+$ concentrations. Furthermore, this genotype showed a more balanced ion homeostasis than the salt-sensitive genotype [26]. In the current study, this response was also partially observed in MD-treated plants under saline stress. In this case, the increase in JA produced by MD treatment in the presence of NaCl correlated with a lower Na$^+$ accumulation in roots compared with the other treatments.

In MD- and Phe-treated peach seedlings, an accumulation of saline ions in roots was recorded, suggesting that both treatments could trigger different mechanisms leading to the development of adaptive responses against salinity [38]. These results contrast to those obtained in MD-treated J8-1 seedlings submitted to salinity conditions, which showed a strong increase in leaf soluble Ca^{2+}, which correlated with increased SA and JA contents. However, the soluble Ca^{2+} accumulation in leaves was also observed in control and Phe-treated plants that showed increased SA but no change in JA levels. It is important to note that the Ca^{2+} levels observed in leaves from MD- and Phe-treated J8-1 seedlings were lower than in control plants, and a similar response was observed in peach seedlings, suggesting that Ca^{2+} ions could be chelated by organic molecules like MD and Phe [38]. In pea plants, treatment with exogenous SA (50–100 µM) induced an increase of Ca^{2+} in shoots but not in roots, although under NaCl stress (70 mM), the presence of SA did not prevent a NaCl-induced decrease in Ca^{2+} levels [37]. It has been suggested that SA-induced Ca^{2+} contents can also lead to stomatal closure [39]. In J8-1 plants, increases in leaf SA levels under salinity conditions seem to be related to increases in Ca^{2+} in leaves that could lead to tight stomatal control, thus providing protection to membranes under stress conditions.

4. Material and Methods

4.1. Plant Material

The assays were performed on micropropagated plum [*Prunus domestica* cv. *Claudia verde* and transgenic line J8-1 [16,17]] shoots and J8-1 seedlings, which were submitted to NaCl stress in the presence or absence of MD and Phe (MD precursor) treatments.

The micropropagated plum shoots were subcultured at four-week intervals for micropropagation and samples were taken at the end of the second subculture in the presence of MD and Phe treatments. In the micropropagated shoots, salt stress was imposed by adding 100 mM NaCl to the micropropagation media in the presence or absence of 200 µM [^{13}C]MD or [^{13}C]Phe (Campro Scientific GmbH, Germany), as described in Diaz-Vivancos et al. (2017) [5]. Seedlings were obtained from rooted and acclimatized to ex vitro conditions J8-1 plantlets. Under greenhouse conditions, J8-1 seedlings were grown in 2 L pots during two months. Then seedlings were submitted to an artificial rest period (eight weeks) in a cold chamber to ensure uniformity and fast growth. After the rest period, seedlings were irrigated once a week with 6 g/L NaCl in the presence or absence of 1 mM MD or Phe for seven weeks. Samples were taken at the end of this period. For all the conditions, 12 seedlings were assayed, and another 12 plants were kept as control.

4.2. Metabolomic Analysis

Micropropagated shoots leaf samples (0.5 g FW) were extracted in 50% methanol (1/3 *w/v*) and then filtered in PTFE 0.45 µm filters (Agilent Technologies, Palo Alto, CA, USA).The levels of Phe, MD, amygdalin, benzoic acid and SA were determined in micropropagated shoots at the Metabolomics Platform at CEBAS-CSIC (Murcia, Spain) using an Agilent 1290 Infinity UPLC system coupled to a 6550 Accurate-Mass quadrupole TOF mass spectrometer (Agilent Technologies, Palo Alto, CA, USA) [5]. The hormone levels (ABA, JA and SA) in the dry leaves of J8-1 seedlings treated with MD or Phe (0.2 g DW) were determined using a UHPLC-mass spectrometer (Q-Exactive, ThermoFisher Scientific, Barcelona, Spain) at the Plant Hormone Quantification Platform at IBMCP (Valencia, Spain).

4.3. Chlorophyll Determination and Chlorophyll Fluorescence

Approximately 0.2 g of the leaves from the J8-1 seedlings submitted to NaCl stress in the presence or absence of MD and Phe were incubated in 50 mL of 80% acetone (v/v) for 72 h under darkness. The chlorophyll a (Chla) and chlorophyll b (Chlb) content was analysed by measuring the absorbance at 663 and 645 nm [40].

Chlorophyll fluorescence parameters were measured in detached leaves from J8-1 seedlings submitted to NaCl stress in the presence or absence of MD and Phe treatments using a chlorophyll fluorimeter (IMAGIM-PAM M-series, Heinz Walz, Effeltrich, Germany). After a dark incubation period (15 min), the leaves' minimum and maximum fluorescence yields were monitored. Kinetic analyses were carried out as previously described [21], and the effective PSII quantum yield [Y(II)], the coefficients of photochemical quenching (qP) and non-photochemical quenching (qN), and the quantum yield of regulated energy dissipation [(Y(NPQ)] were recorded.

4.4. Ascorbate and Glutathione Analysis

Micropropagated J8-1 line shoot and seedling leaf samples were snap-frozen in liquid nitrogen and stored at −80 °C until use. The frozen samples were homogenised (1/3 w/v) with 1 M $HClO_4$ containing 1 mM polyvinylpolypyrrolidone and 1 mM EDTA. Homogenates were centrifuged at 12,000× g for 10 min, and the supernatant was neutralised with 5 M K_2CO_3 to pH 5.5–6. The homogenate was centrifuged at 12,000× g for 1 min to remove $KClO_4$. The supernatant obtained was used for ascorbate and glutathione determination as previously described [41,42].

4.5. Gene Expression

We studied the expression levels of the redox-regulated genes *NPR1* (*Non-Expressor of Pathogenesis-Related Gene 1*) and *TrxH* (*thioredoxin H*). Briefly, micropropagated shoots and leaf samples from line J8-1 were snap-frozen in liquid nitrogen and stored at −80 °C until use. RNA was extracted using the Power Plant RNA Isolation kit (Mo Bio), according to the manufacturer's instructions. The primer sequences were as follows: *NPR1* (forward 5′-tgcacgagctcctttagtca-′3; reverse 5′-cggcttactgcgatcctaag-′3); *TrxH* (forward 5′-tggcggagttggctaagaag-′3; 5′-ttcttggcacccacaacctt-′3); *β-actin* (forward 5′tgcctgccatgtatgttgccatcc′3; reverse 5′aacagcaaggtcagacgaaggat′3).

The expression levels of *NPR1*, *TrxH*, and the *β-actin* gene, used for normalisation, were determined as described in [15] by real-time RT-PCR using the GeneAmp 7500 sequence detection system (Applied Biosystems, Foster City, CA, USA). Relative quantification of gene expression was calculated by the Delta-Delta Ct method.

4.6. Determination of Soluble K^+, Ca^{2+}, Na^+, and Cl^- Content

The effect of NaCl stress in the presence and absence of MD and Phe on soluble K^+, Ca^{2+}, Na^+, and Cl^- content was determined in leaves and roots of J8-1 seedlings grown under greenhouse conditions. First, leaf and root samples (at least five replicates per treatment) were oven-dried at 65°C and ground to a fine powder. Then, approximately 0.1 g was extracted with milliQ water (1/10 w/v) at 50 °C for 3 h and shake-incubated for 24 h at 30 °C.

The concentrations of the soluble nutrients analysed were determined by ion-selective electrodes (IonMeter, Nsensors ©) that were previously calibrated with standard solutions of NaCl (for Na^+ and Cl^-), $CaCl_2$ (for Ca^{+2}), and KCl (for K^+).

Int. J. Mol. Sci. **2018**, *19*, 3519

4.7. Statistical Analysis

The data were analysed by one-way or two-way ANOVA using SPSS 22 software (Chicago, IL, USA). Means were separated with the Duncan's Multiple Range Test ($p < 0.05$).

5. Conclusions

As a general conclusion, in this work we have demonstrated that the plum line J8-1 is tolerant to NaCl in terms of plant growth and plant performance (chlorophyll content and chlorophyll fluorescence parameters, shoot biomass and leaf and root water contents) under the tested conditions. In the presence of NaCl, the MD treatment produced the highest SA and JA increases, but it also induced the expression of *NPR1* and *TrxH* transcripts. These results, similar to those reported by other authors, suggest that the *NPR1/TrxH* interaction, along with SA and JA accumulation, may play an important role in the tolerant response of the J8-1 plum line to salt stress. The biosynthetic pathways of SA, JA and ABA take place in the chloroplast [43], and this organelle is rapidly affected by salt stress [2]. Therefore, a connection of the SA, JA, and ABA pathways and qN with the expression of *NPR1* and *TrxH*, mediated by the redox state of the chloroplast can be suggested.

Finally, the results led us to think that the new SA synthesis pathway demonstrated in peach seemed not to be operative in plum under in vitro conditions. However, MD could be involved in the SA biosynthesis under NaCl stress conditions in plum plants under greenhouse conditions. In the transgenic plum line J8-1 a crosstalk between different signalling pathways (NPR1/Trx and SA/JA) leading to salinity tolerance is suggested.

Supplementary Materials: Supplementary materials can be found at http://www.mdpi.com/1422-0067/19/11/3519/s1.

Author Contributions: A.B.V. and P.D.V performed and analyzed the metabolomics, plant growth, chlorophyll content and ascorbate and glutathione determination assays. D.C. performed and analyzed the gene expression and nutrient content experiments. C.P. performed the micropropagation and MD and Phe treatments of micropropagated shoots. J.A.H. performed and analyzed the chlorophyll fluorescence assay. P.D.V. and J.A.H. designed the experiments, performed data curation and write, review and edit the manuscript. Supervision, project administration and funding acquisition by P.D.V. and J.A.H. All authors discussed and commented on the content of the paper.

Funding: This work was supported by the Spanish Ministry of Economy and Competitiveness (Projects AGL2014-52563-R and INIA-RTA2013-00026-C03-00).

Acknowledgments: PDV and CP thank CSIC and UPCT, respectively, as well as the Spanish Ministry of Economy and Competitiveness for their 'Ramon and Cajal' research contract, co-financed by FEDER funds. This work was supported by the Spanish Ministry of Economy and Competitiveness (Projects AGL2014-52563-R and INIA-RTA2013-00026-C03-00).

References

1. Hossain, M.S.; Dietz, K.J. Tuning of Redox Regulatory Mechanisms, Reactive Oxygen Species and Redox Homeostasis under Salinity Stress. *Front. Plant Sci.* **2016**, *7*, 548. [CrossRef] [PubMed]
2. Acosta-Motos, J.; Ortuño, M.; Bernal-Vicente, A.; Diaz-Vivancos, P.; Sanchez-Blanco, M.; Hernandez, J. Plant Responses to Salt Stress: Adaptive Mechanisms. *Agronomy* **2017**, *7*, 18. [CrossRef]
3. Miura, K.; Tada, Y. Regulation of water, salinity, and cold stress responses by salicylic acid. *Front. Plant Sci.* **2014**, *5*, 4. [CrossRef] [PubMed]
4. Khan, M.I.; Fatma, M.; Per, T.S.; Anjum, N.A.; Khan, N.A. Salicylic acid-induced abiotic stress tolerance and underlying mechanisms in plants. *Front. Plant Sci.* **2015**, *6*, 462. [CrossRef] [PubMed]
5. Diaz-Vivancos, P.; Bernal-Vicente, A.; Cantabella, D.; Petri, C.; Hernández, J.A. Metabolomic and Biochemical Approaches Link Salicylic Acid Biosynthesis to Cyanogenesis in Peach Plants. *Plant Cell Physiol.* **2017**, *58*, 2057–2066. [CrossRef] [PubMed]
6. Gleadow, R.M.; Møller, B.L. Cyanogenic Glycosides: Synthesis, Physiology, and Phenotypic Plasticity. *Annu. Rev. Plant Biol.* **2014**, *65*, 155–185. [CrossRef] [PubMed]

7. Catinot, J.; Buchala, A.; Abou-Mansour, E.; Metraux, J.P. Salicylic acid production in response to biotic and abiotic stress depends on isochorismate in *Nicotiana benthamiana*. *FEBS Lett.* **2008**, *582*, 473–478. [CrossRef] [PubMed]

8. Ogawa, D.; Nakajima, N.; Seo, S.; Mitsuhara, I.; Kamada, H.; Ohashi, Y. The phenylalanine pathway is the main route of salicylic acid biosynthesis in *Tobacco mosaic virus* infected tobacco leaves. *Plant Biotechnol.* **2006**, *23*, 395–398. [CrossRef]

9. Liu, X.; Rockett, K.S.; Kørner, C.J.; Pajerowska-Mukhtar, K.M. Salicylic acid signalling: New insights and prospects at a quarter-century milestone. *Essays Biochem.* **2015**, *58*, 101–113. [CrossRef] [PubMed]

10. Foyer, C.H.; Noctor, G. Oxidant and antioxidant signalling in plants: A re-evaluation of the concept of oxidative stress in a physiological context. *Plant Cell Environ.* **2005**, *28*, 1056–1071. [CrossRef]

11. Foyer, C.H.; Noctor, G. Ascorbate and glutathione: The heart of the redox hub. *Plant Physiol.* **2011**, *155*, 2–18. [CrossRef] [PubMed]

12. Vieira Dos Santos, C.; Rey, P. Plant thioredoxins are key actors in the oxidative stress response. *Trends Plant Sci.* **2006**, *11*, 329–334. [CrossRef] [PubMed]

13. Tada, Y.; Spoel, S.H.; Pajerowska-Mukhtar, K.; Mou, Z.; Song, J.; Wang, C.; Zuo, J.; Dong, X. Plant Immunity Requires Conformational Charges of NPR1 via S-Nitrosylation and Thioredoxins. *Science* **2008**, *321*, 952–956. [CrossRef] [PubMed]

14. Brosché, M.; Kangasjärvi, J. Low antioxidant concentrations impact on multiple signalling pathways in *Arabidopsis thaliana* partly through NPR1. *J. Exp. Bot.* **2012**, *63*, 1849–1861. [CrossRef] [PubMed]

15. Faize, M.; Faize, L.; Petri, C.; Barba-Espin, G.; Diaz-Vivancos, P.; Clemente-Moreno, M.J.; Koussa, T.; Rifai, L.A.; Burgos, L.; Hernandez, J.A. Cu/Zn superoxide dismutase and ascorbate peroxidase enhance in vitro shoot multiplication in transgenic plum. *J. Plant Physiol.* **2013**, *170*, 625–632. [CrossRef] [PubMed]

16. Diaz-Vivancos, P.; Faize, M.; Barba-Espin, G.; Faize, L.; Petri, C.; Hernández, J.A.; Burgos, L. Ectopic expression of cytosolic superoxide dismutase and ascorbate peroxidase leads to salt stress tolerance in transgenic plums. *Plant Biotechnol. J.* **2013**, *11*, 976–985. [CrossRef] [PubMed]

17. Diaz-Vivancos, P.; Faize, L.; Nicolas, E.; Clemente-Moreno, M.J.; Bru-Martinez, R.; Burgos, L.; Hernandez, J.A. Transformation of plum plants with a cytosolic ascorbate peroxidase transgene leads to enhanced water stress tolerance. *Ann. Bot.* **2016**, *117*, 1121–1131. [CrossRef] [PubMed]

18. Bernal-Vicente, A.; Petri, C.; Hernández, J.A.; Diaz-Vivancos, P. The effect of abiotic and biotic stress on the salicylic acid biosynthetic pathway from mandelonitrile in peach. *J. Plant Physiol.* (under review).

19. Acosta-Motos, J.-R.; Diaz-Vivancos, P.; Alvarez, S.; Fernandez-Garcia, N.; Jesus Sanchez-Blanco, M.; Antonio Hernandez, J. Physiological and biochemical mechanisms of the ornamental *Eugenia myrtifolia* L. plants for coping with NaCl stress and recovery. *Planta* **2015**, *242*, 829–846. [CrossRef] [PubMed]

20. Acosta-Motos, J.R.; Diaz-Vivancos, P.; Alvarez, S.; Fernandez-Garcia, N.; Jesus Sanchez-Blanco, M.; Hernandez, J.A. NaCl-induced physiological and biochemical adaptative mechanisms in the ornamental *Myrtus communis* L. plants. *J. Plant Physiol.* **2015**, *183*, 41–51. [CrossRef] [PubMed]

21. Cantabella, D.; Piqueras, A.; Acosta-Motos, J.R.; Bernal-Vicente, A.; Hernandez, J.A.; Diaz-Vivancos, P. Salt-tolerance mechanisms induced in *Stevia rebaudiana* Bertoni: Effects on mineral nutrition, antioxidative metabolism and steviol glycoside content. *Plant Physiol. Biochem.* **2017**, *115*, 484–496. [CrossRef] [PubMed]

22. Boatwright, J.L.; Pajerowska-Mukhtar, K. Salicylic acid: An old hormone up to new tricks. *Mol. Plant Pathol.* **2013**, *14*, 623–634. [CrossRef] [PubMed]

23. Qiu, Z.; Guo, J.; Zhu, A.; Zhang, L.; Zhang, M. Exogenous jasmonic acid can enhance tolerance of wheat seedlings to salt stress. *Ecotoxicol. Environ. Saf.* **2014**, *104* (Suppl. C), 202–208. [CrossRef] [PubMed]

24. He, Y.; Zhu, Z.J. Exogenous salicylic acid alleviates NaCl toxicity and increases antioxidative enzyme activity in *Lycopersicon esculentum*. *Biol. Plant.* **2008**, *52*, 792. [CrossRef]

25. Acosta-Motos, J.R.; Ortuño, M.F.; Álvarez, S.; López-Climent, M.F.; Gómez-Cadenas, A.; Sánchez-Blanco, M.J. Changes in growth, physiological parameters and the hormonal status of *Myrtus communis* L. plants irrigated with water with different chemical compositions. *J. Plant Physiol.* **2016**, *191* (Suppl. C), 12–21. [CrossRef] [PubMed]

26. Zhang, H.; Zhang, Q.; Zhai, H.; Li, Y.; Wang, X.; Liu, Q.; He, S. Transcript profile analysis reveals important roles of jasmonic acid signalling pathway in the response of sweet potato to salt stress. *Sci. Rep.* **2017**, *7*, 40819. [CrossRef] [PubMed]

27. Khokon, A.R.; Okuma, E.; Hossain, M.A.; Munemasa, S.; Uraji, M.; Nakamura, Y.; Mori, I.C.; Murata, Y. Involvement of extracellular oxidative burst in salicylic acid-induced stomatal closure in *Arabidopsis*. *Plant Cell Environ.* **2011**, *34*, 434–443. [CrossRef] [PubMed]

28. Clemente-Moreno, M.J.; Diaz-Vivancos, P.; Piqueras, A.; Antonio Hernandez, J. Plant growth stimulation in *Prunus* species plantlets by BTH or OTC treatments under in vitro conditions. *J. Plant Physiol.* **2012**, *169*, 1074–1083. [CrossRef] [PubMed]

29. Yang, Y.; Qi, M.; Mei, C. Endogenous salicylic acid protects rice plants from oxidative damage caused by aging as well as biotic and abiotic stress. *Plant J.* **2004**, *40*, 909–919. [CrossRef] [PubMed]

30. Herrera-Vasquez, A.; Salinas, P.; Holuigue, L. Salicylic acid and reactive oxygen species interplay in the transcriptional control of defense genes expression. *Front. Plant Sci.* **2015**, *6*, 171. [CrossRef] [PubMed]

31. Vlot, A.C.; Dempsey, D.A.; Klessig, D.F. Salicylic acid, a multifaceted hormone to combat disease. *Annu. Rev. Phytopathol.* **2009**, *47*, 177–206. [CrossRef] [PubMed]

32. Gelhaye, E.; Rouhier, N.; Jacquot, J.P. The thioredoxin *h* system of higher plants. *Plant Physiol. Biochem.* **2004**, *42*, 265–271. [CrossRef] [PubMed]

33. Tsukamoto, S.; Morita, S.; Hirano, E.; Yokoi, H.; Masumura, T.; Tanaka, K. A Novel cis-Element That Is Responsive to Oxidative Stress Regulates Three Antioxidant Defense Genes in Rice. *Plant Physiol.* **2005**, *137*, 317–327. [CrossRef] [PubMed]

34. Issakidis-Bourguet, E.; Mouaheb, N.; Meyer, Y.; Miginiac-Maslow, M. Heterologous complementation of yeast reveals a new putative function for chloroplast m-type thioredoxin. *Plant J.* **2001**, *25*, 127–135. [CrossRef] [PubMed]

35. Zhang, C.-J.; Zhao, B.-C.; Ge, W.-N.; Zhang, Y.-F.; Song, Y.; Sun, D.-Y.; Guo, Y. An Apoplastic H-Type Thioredoxin Is Involved in the Stress Response through Regulation of the Apoplastic Reactive Oxygen Species in Rice. *Plant Physiol.* **2011**, *157*, 1884–1899. [CrossRef] [PubMed]

36. Jayakannan, M.; Bose, J.; Babourina, O.; Rengel, Z.; Shabala, S. Salicylic acid in plant salinity stress signalling and tolerance. *Plant Growth Regul.* **2015**, *76*, 25–40. [CrossRef]

37. Barba-Espin, G.; Clemente-Moreno, M.J.; Alvarez, S.; Garcia-Legaz, M.F.; Hernandez, J.A.; Diaz-Vivancos, P. Salicylic acid negatively affects the response to salt stress in pea plants. *Plant Biol.* **2011**, *13*, 909–917. [CrossRef] [PubMed]

38. Bernal-Vicente, A.; Cantabella, D.; Hernández, J.A.; Diaz-Vivancos, P. The effect of mandelonitrile, a recently described salicylic acid precursor, on peach plant response against abiotic and biotic stresses. *Plant Biol.* **2018**, *20*, 986–994. [CrossRef] [PubMed]

39. Liu, X.; Zhang, S.; Lou, C. Involvement of nitric oxide in the signal transduction of salicylic acid regulating stomatal movement. *Chin. Sci. Bull.* **2003**, *48*, 449–452. [CrossRef]

40. Arnon, D.I. Copper enzymes in isolated chloroplasts. *Polyphenoloxidase in Beta vulgaris*. *Plant Physiol.* **1949**, *24*, 1–15. [PubMed]

41. Vivancos, P.D.; Dong, Y.P.; Ziegler, K.; Markovic, J.; Pallardo, F.V.; Pellny, T.K.; Verrier, P.J.; Foyer, C.H. Recruitment of glutathione into the nucleus during cell proliferation adjusts whole-cell redox homeostasis in *Arabidopsis thaliana* and lowers the oxidative defence shield. *Plant J.* **2010**, *64*, 825–838. [CrossRef] [PubMed]

42. Pellny, T.K.; Locato, V.; Vivancos, P.D.; Markovic, J.; De Gara, L.; Pallardo, F.V.; Foyer, C.H. Pyridine Nucleotide Cycling and Control of Intracellular Redox State in Relation to Poly (ADP-Ribose) Polymerase Activity and Nuclear Localization of Glutathione during Exponential Growth of *Arabidopsis* Cells in Culture. *Mol. Plant* **2009**, *2*, 442–456. [CrossRef] [PubMed]

43. Czarnocka, W.; Karpiński, S. Friend or foe? Reactive oxygen species production, scavenging and signaling in plant response to environmental stresses. *Free Radic. Biol. Med.* **2018**, *122*, 4–20. [CrossRef] [PubMed]

International Journal of
Molecular Sciences

MDPI

Article

Overexpression of Transglutaminase from Cucumber in Tobacco Increases Salt Tolerance through Regulation of Photosynthesis

Min Zhong [1], Yu Wang [1], Yuemei Zhang [1], Sheng Shu [1], Jin Sun [1,2] and Shirong Guo [1,2,*]

[1]   Key Laboratory of Southern Vegetable Crop Genetic Improvement, Ministry of Agriculture,
      College of Horticulture, Nanjing Agricultural University, Nanjing 210095, China;
      2016204040@njau.edu.cn (M.Z.); ywang@njau.edu.cn (Y.W.); zym941128@163.com (Y.Z.);
      shusheng@njau.edu.cn (S.S.); jinsun@njau.edu.cn (J.S.)
[2]   Suqian Academy of Protected Horticulture, Nanjing Agricultural University, Suqian 223800, China
[*]   Correspondence: srguo@njau.edu.cn; Tel.: +86-25-8439-5267

Received: 20 January 2019; Accepted: 1 February 2019; Published: 19 February 2019

Abstract: Transglutaminase (TGase) is a regulator of posttranslational modification of protein that provides physiological protection against diverse environmental stresses in plants. Nonetheless, the mechanisms of TGase-mediated salt tolerance remain largely unknown. Here, we found that the transcription of cucumber *TGase* (*CsTGase*) was induced in response to light and during leaf development, and the CsTGase protein was expressed in the chloroplast and the cell wall. The overexpression of the *CsTGase* gene effectively ameliorated salt-induced photoinhibition in tobacco plants, increased the levels of chloroplast polyamines (PAs) and enhanced the abundance of D1 and D2 proteins. TGase also induced the expression of photosynthesis related genes and remodeling of thylakoids under normal conditions. However, salt stress treatment reduced the photosynthesis rate, PSII and PSI related genes expression, D1 and D2 proteins in wild-type (WT) plants, while these effects were alleviated in *CsTGase* overexpression plants. Taken together, our results indicate that TGase-dependent PA signaling protects the proteins of thylakoids, which plays a critical role in plant response to salt stress. Thus, overexpression of TGase may be an effective strategy for enhancing resistance to salt stress of salt-sensitive crops in agricultural production.

Keywords: TGase; photosynthesis; salt stress; polyamines; cucumber

---

1. Introduction

Photosynthesis is the basic manufacturing process in plants; it can increase carbon gains and improve crop yield and quality [1]. Salinity, along with other environmental stresses such as drought and chilling, induces inhibition of photosynthetic activity by disruption of the chloroplast structure and reduction in CO_2 assimilation [2]. The effects of environmental stresses on photosynthesis in cucumber have been extensively studied. We have described several defense systems that protect the photosynthetic apparatus by exogenous polyamines (PAs) application. For example, exogenous spermidine (Spd) delays chlorophyll degradation under heat stress, and the regulation of fatty acids and accumulation of PAs in thylakoid membranes are induced by exogenous putrescine (Put) under salt stress [3,4]. PAs are low molecular weight aliphatic amines; the biochemical properties of PAs are quite simple, but their regulation of processes is strikingly complex and wide processes [5]. The majority of PAs in higher plants are Spd, spermine (Spm), and their precursor Put, which derive from arginine in chloroplasts [6,7]. On the other hand, as the polycationic nature of PAs at physiological pH, PAs could be free molecules conjugated with organic acids or bound to negatively charged macromolecules such as proteins, nucleic acids, and chromatin through transglutaminases (TGases) enzymatic activity, stabilizing their structures [8,9]. These interactions are essential for the effects of PAs on plant cell growth and developmental processes and plant response to various stresses.

Moreover, PAs, as organic cations and permeant buffers, have been reported to protect the photosynthetic apparatus by regulating the size of the antenna proteins of light harvesting chlorophyll a/b protein complexes (LHCII) and the larger subunit of ribulose bisphosphate carboxylase-oxygenase during stresses, such as UV-B radiation and salt stress [10,11]. PAs are also synthesized and oxidized in chloroplasts, while the addition of PAs inhibits the destruction of thylakoids and prevents the loss of pigment during salt stress [12,13]. The levels of endogenous PAs are also related to chlorophyll biosynthesis and the rate of photosynthesis during stresses [14]. PA accumulation in the lumen promotes an increase in ATP and the electric field in *vivo* and in *vitro* [15]. In addition, endogenous PAs might be involved in the assembly of photosynthetic membrane complexes such as thylakoid membranes [16].

TGases are crucial factors of the thylakoid system but are often rather ignored. TGases catalyze proteins by establishing ε-(γ-glutamyl) links, then regulate proteins post-translational modification and the covalent binding of PAs to protein substrates [17,18]. TGases are widely distributed in microorganisms, animals, and plants. However, research on these enzymes in plants is more rarely reported than in animal systems, in which was detected for the first time in *Arabidopsis thaliana* the presence of only one gene, *AtPng1p*, which encodes a putative *N*-glycanase containing the Cys-His-Asp triad of the TGase catalytic domain and was expressed ubiquitously [19]. Although TGases are found in several organs in lower and higher plants, they are activated in a Ca^{2+} dependent manner and are involved in fertilization, abiotic and biotic stresses, senescence, and programmed cell death, under different light environment conditions, including natural habitats, but the function of TGases in chloroplasts has received the most attention [18,20]. The activity of TGases has been shown to be light sensitive, and some proteins of the photosystems (LHCII, CP29, CP26, and CP24) have been shown to be endogenous substrates of TGase in chloroplasts [21]. Meanwhile, TGase not only localized in the chloroplast grana and close to LHCII but also localized in the walls of the bulliform cells of leaves. Its activity was light dependent and its abundance depended on the degree of grana development [22]. Moreover, the expression and activity of TGase was involved in length of light exposure in maize [23].

In earlier works, we described the effects of salt and heat stresses on changes of free PA contents in leaves of cucumber and tomato [24,25]. Specifically, the content of PAs decreased under stress conditions, resulting in severe damage to photosynthetic organs such as chloroplasts, which were severely deformed into irregular shapes, and increased starch granules. TGase was induced by salt stress and involved in the protection of the photosynthetic apparatus [26]. In this article, we reveal that TGase has a positive role in PAs accumulation and induce the transcript of photosynthetic genes in chloroplast, which may play crucial roles in the regulation of photosynthetic organ stability. To our knowledge, this is the first report to show that the TGase positively regulates plant's photosynthetic through accumulation of PAs to enhance salt tolerance.

2. Results

2.1. Expression Profile Analysis of TGase

To elucidate the molecular function of TGase, we analyzed the gene expression of *TGase* in cucumber. First, to investigate the effect of light on the expression of *TGase* in cucumber plants, we monitored the transcriptional levels of *TGase* in light-treated cucumber seedlings. The transcript level of *TGase* was gradually induced by light and reached the highest level at 16 h (Figure 1A). These results indicate that light plays a vital role in regulating *TGase* expression. To further explore the expression of *TGase* at different developmental stages, we investigated the transcript levels of *TGase* in leaves ranging from 1 to 8 weeks old by quantitative real-time PCR (qPCR). The expression level of *TGase* increased with the growth and development of the leaves (Figure 1B). *TGase* transcript levels in 4 and 8 week-old plants was significantly higher than those in 1-week-old plants (Figure 1B). Furthermore, *TGase* transcript levels increased during leaf development from young to mature leaves (Figure 1C). Some reports showed that TGase was widely present in plant tissue [17]. We extracted RNA from roots, stems, leaves, flowers, and fruits, and then analyzed the transcript levels of *TGase* in these tissues via qPCR. Our results also showed that *TGase* was present in all investigated tissues and was highly expressed in leaves and flowers, but minimally expressed in roots and stems (Figure 1D).

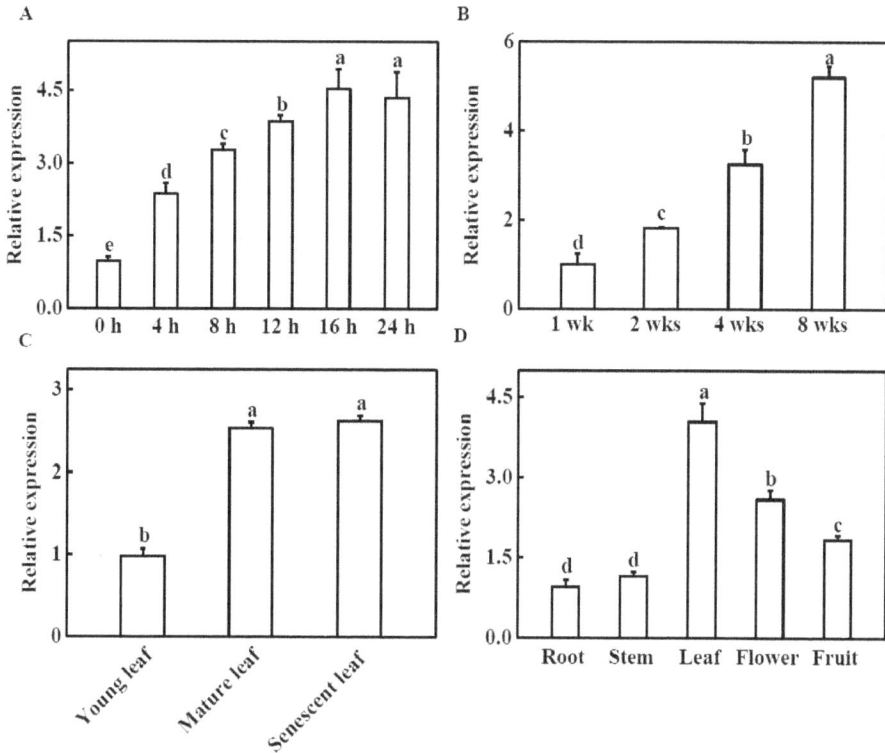

Figure 1. Expression profiles of *TGase*. (**A**) Light-induced *TGase* expression in cucumber. The cucumber seedlings were exposed to light 0, 4, 8, 12, 16 and 24 h and *TGase* transcript levels were analyzed by quantitative real-time PCR (qPCR). (**B**) Transcript levels of *TGase* in cucumber leaves at 1, 2, 4, 8 weeks of development. (**C**) Transcript levels of *TGase* in young, mature and old leaves of cucumber plants. (**D**) qPCR analysis of *TGase* transcript in roots, stems, leaves, flowers and fruit of cucumber. Each histogram represents a mean ± SE of four independent experiments (*n* = 4). Different letters indicate significant differences between treatments (*p* < 0.05) according to Duncan's multiple range test.

2.2. Immunolocalization of TGase Protein in Cucumber Leaves

Subcellular immunolocalization in cucumber leaf mesophyll cells provided details on the presence of TGase. The signal was localized in the chloroplasts and near the chloroplast grana (Figure 2A). The presence of TGase spots in the cell wall was detected (Figure 2B). These were not significantly localized to TGase in other cell organelles.

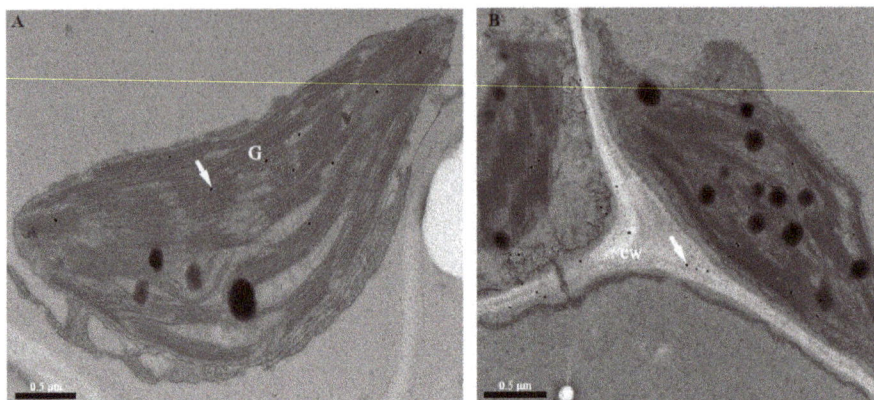

Figure 2. TEM immunolocalization of TGase in cucumber leaf chloroplasts using the monoclonal antibody (1:1000). (**A**) Signal in the granary of the chloroplasts. (**B**) Signal in the cell well. G, grana; cw, cell wall.

2.3. Effects of TGase on the Biomass and Photosynthetic Characteristics of Transgenic Tobacco Lines

To analyze the role of cucumber TGase in salt tolerance, we overexpressed the *TGase* gene in tobacco plants. As shown in Figure 3A, as compared with WT plants, the biomass was higher in the *CsTGase*-overexpressing (*CsTGase*OE) plants after salt stress. On the other hand, the *CsTGase*OE plants had a higher biomass relative to WT plants under normal conditions. As shown in Figure 3B, the proline content in all the plants increased after salt treatment, but this increase was much greater in *CsTGase*OE plants than in the WT plants. In WT plants, the chlorophyll a and chlorophyll b contents decreased by 62.0% and 51.3%, respectively, after salt treatment, whereas they maintained higher levels in *CsTGase*OE plants compared with those of the WT (Figure 3C,D).

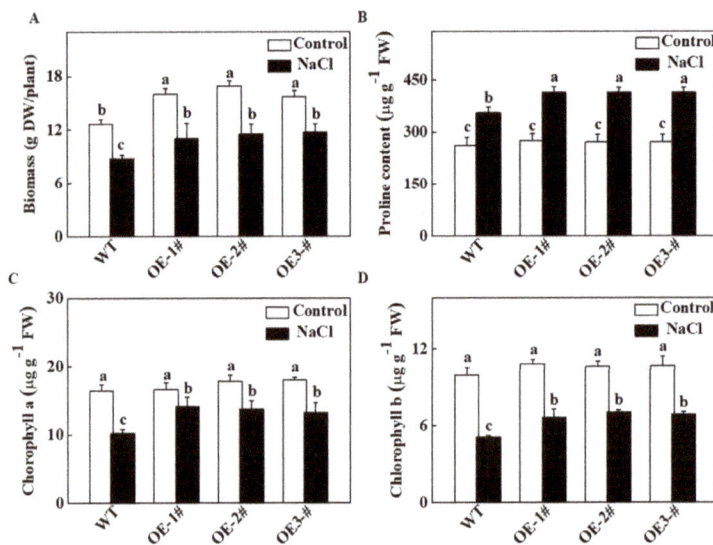

Figure 3. Effects of salt stress on biomass, proline and chlorophyll content in wild-type (WT) and *CsTGase*OE plants. (**A**) Biomass. (**B**) Proline content in leaves. (**C,D**) Chlorophyll a and b content. Each histogram represents a mean $\pm$ SE of four independent experiments ($n = 4$). Different letters indicate significant differences between treatments ($p < 0.05$) according to Duncan's multiple range test.

We also evaluated the effects of salt stress on photosynthetic gas exchange parameters. As shown in Figure 4A, under normal conditions, the net photosynthesis rate (Pn) of *CsTGase*OE plants was significantly higher than that of the WT plants. Salt stress resulted in a significant decrease in Pn, but this value was still maintained at a higher level in *CsTGase*OE plants than in the WT (Figure 4A). Similar results were observed for stomatal conductance (Gs) (Figure 4B) and transpiration rate (Tr) (Figure 4D). However, there were no significant differences in intercellular CO_2 concentration (Ci) in WT and *CsTGase*OE plants after salt stress (Figure 4C). These results suggest that TGase plays a critical role in the tobacco response to salt stress, especially in maintaining photosynthetic properties.

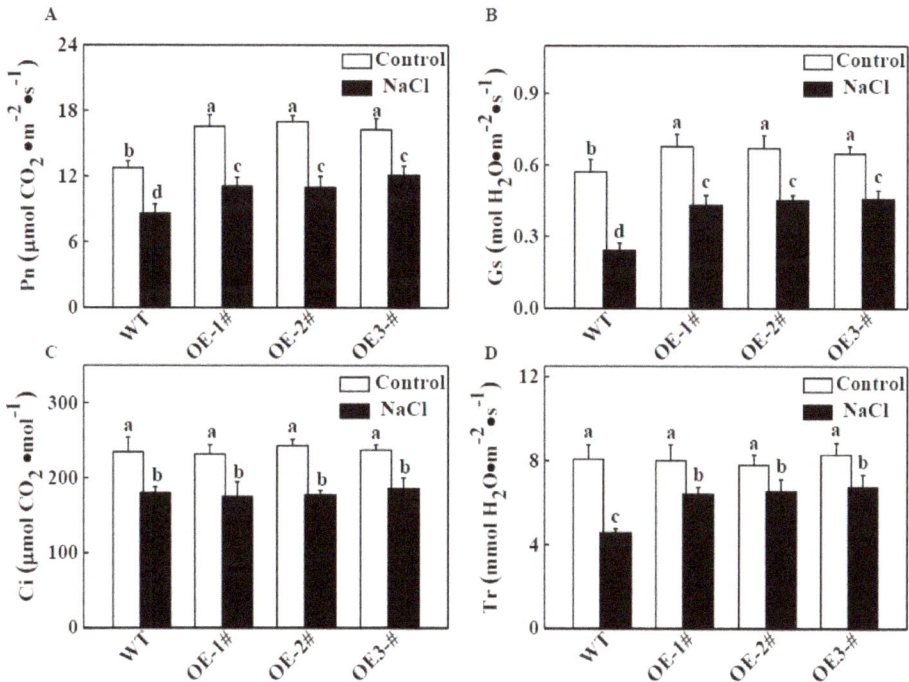

Figure 4. Photosynthetic parameters of WT and *CsTGase*OE plants in response to salt stress. (**A**) Net photosynthetic rate (Pn). (**B**) Stomatal conductance (Gs). (**C**) Intercellular CO_2 concentration (Ci). (**D**) Transpiration rate (Tr). Each histogram represents a mean ± SE of four independent experiments ($n = 6$). Different letters indicate significant differences between treatments ($p < 0.05$) according to Duncan's multiple range test.

2.4. Effects of TGase on Endogenous PA Content in Thylakoid Membranes

To determine whether PAs were involved in *TGase*-induced salt tolerance by protecting photosynthetic properties, we first measured the endogenous concentration of PAs in thylakoid membranes using a sensitive HPLC method. Under normal conditions, in *CsTGase*OE plants, the thylakoid associated PAs (Put, Spd and Spm) showed significantly higher levels compared with those of WT plants. Salt-induced bound Put, Spd, and Spm increased in comparison to the WT (Figure 5A–C). Meanwhile, the PA concentration increased by 161.8%, 155.9%, and 167.4% in the three *CsTGase*OE lines, respectively, compared with the WT under normal conditions (Figure 5D). PA accumulation rose by 28.7% in WT plants after salt treatment, but was still lower than in *CsTGase*OE plants.

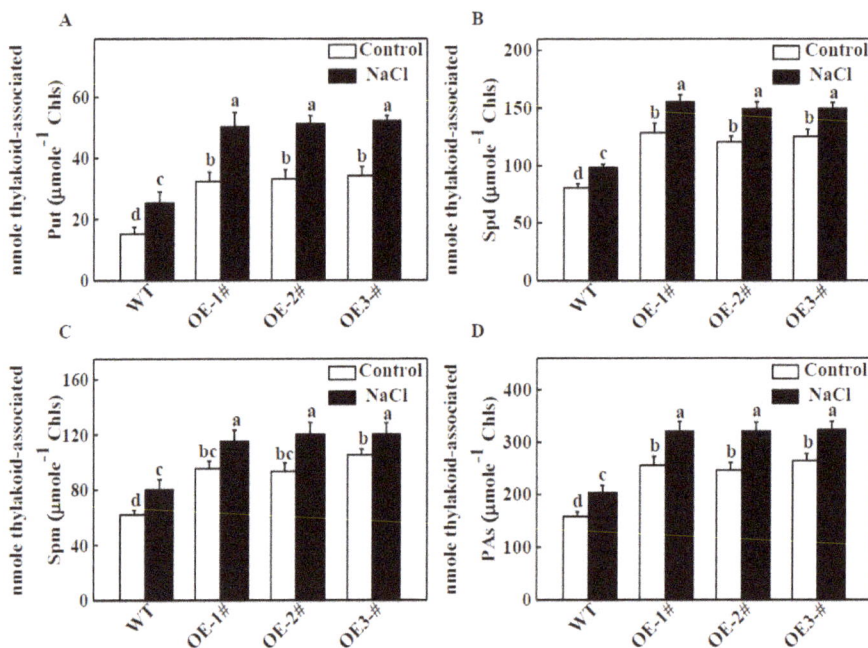

Figure 5. Effects of salt stress on the contents of endogenous putrescine (Put), spermidine (Spd), spermine (Spm) and total polyamines (PAs) in chloroplast of WT and *CsTGase*OE plants. (**A**) Thylakoid-associated Put content. (**B**) Thylakoid-associated Spd content. (**C**) Thylakoid-associated Spm content. (**D**) Total thylakoid-associated PAs content. Each histogram represents a mean ± SE of three independent experiments (*n* = 4). Different letters indicate significant differences between treatments (*p* < 0.05) according to Duncan's multiple range test.

2.5. Effects of TGase on the Ultrastructure of Thylakoids

PAs are a major positive factor in chloroplast ultrastructure [14]. To determine whether TGase regulates the ultrastructure of chloroplasts, we assayed the architecture of the thylakoid network using transmission electron microscopy (TEM). Under normal conditions, TEM revealed that the chloroplasts of WT plants had well-structured thylakoid membranes composed of grana connected by stroma lamellae (Figure 6A). Interestingly, overexpression of *TGase* resulted in chloroplasts having more grana and a larger size than those of the WT plants (Figure 6B–D). And in *CsTGase*OE plants, chloroplasts grana stacks reached up to 600 nm, whereas in the WT plants, chloroplast grana stacks were a maximum of 200 nm (Figure S1). These results suggest that TGase plays an important role in chloroplast development. Furthermore, under salt stress, chloroplasts were severely deformed into irregular shapes, and starch granules accumulated; moreover, a separation between cell membranes and chloroplasts was observed in WT plants (Figure 6E). However, the disintegration of grana thylakoids was significantly lower in *CsTGase*OE plants compared to the salt-treated WT plants (Figure 6E–H).

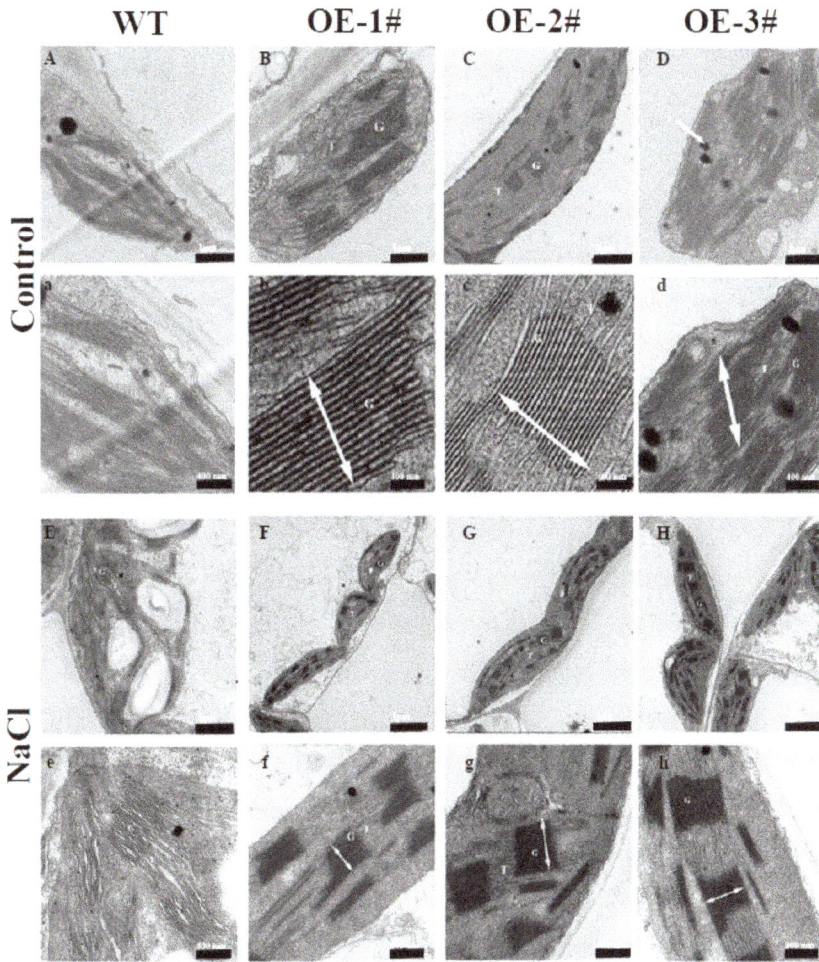

Figure 6. Electron microscopy in chloroplast of WT and *CsTGase*OE plants after salt stress. (**B–D, b–d,** **F–H** and **f–h**) shows an increased grana appression and a reduced stroma thylakoid network with respect to the WT (**A,a,E,e**) under normal conditions and salt stress. G, grana; T, thylakoid; SG, starch grana, P, plastoglobule. Grana height is indicated by white arrows. Scale bars for chloroplast and thylakoid are indicated. Three biological replicates were performed, and similar results were obtained.

2.6. Effect of TGase on Chl a Fluorescence Transients (OJIP)

A kinetic comparison was made of the raw OJIP transients measured in *CsTGase*OE and WT plants after salt stress. No significant difference was observed under the absence of NaCl in any of the plants. However, there were significant differences in *CsTGase*OE and WT plants under salt stress (Figure 7). In WT plants, salt stress resulted in a significant decrease in the intensities of fluorescence at J, I, and P levels with no major change in the minimal fluorescence (F_0) (Figure 7). Compared to the WT plants, the OJIP fluorescence transient and F_m values were near the normal levels in *CsTGase*OE plants after salt stress (Figure 7).

Int. J. Mol. Sci. **2019**, *20*, 894

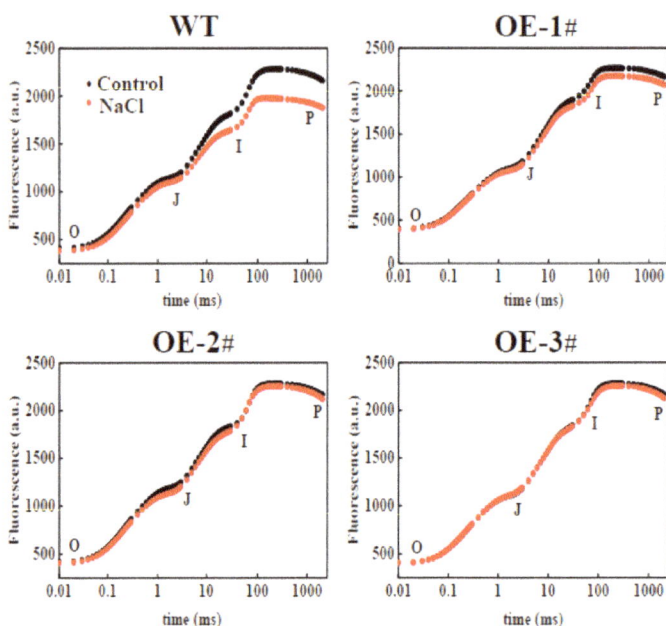

Figure 7. Change in Chl *a* fluorescence transient curves (OJIP) (log time scale) in leaves of WT and *CsTGase*OE plants under salt stress.

2.7. Effect of TGase on Nonphotochemical Quenching (NPQ) Induction

NPQ induction during light and dark transition periods was monitored. Transformed tobacco illuminated with 1 min light (500 µmol m^{-2} s^{-1}) and treat with 4 min dark showed NPQ values proximal to ~0.8, whereas there was little activation of photoprotection in the WT, with an NPQ of ~0.4 (Figure 8A). At a light intensity level of 500 µmol m^{-2} s^{-1} the NPQ value of the WT was significantly decreased by salt stress compared to that of the *CsTGase*OE plants (Figure 8B).

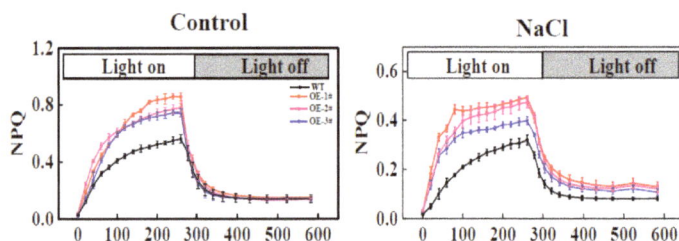

Figure 8. Nonphotochemical quenching (NPQ) induction and relaxation kinetics of WT and *CsTGase*OE plants under salt stress.

2.8. Effect of TGase on Quantum Yield of Energy Conversion in PSII and PSI

The F$_v$/F$_m$ values were not significantly different in all plants under normal conditions (Figure 9A). However, the F$_v$/F$_m$ values of the WT plants were significantly lower than those of *CsTGase*OE plants under salt stress (Figure 9A), indicating that photoinhibition was more severe in WT plants and the TGase has a positive role in photoprotection. The Y(II) in *CsTGase*OE plants was 19.8–24.0% higher than in WT plants, mainly due to an extremely higher photochemical quenching coefficient (qP) under normal conditions. After salt stress, the Y(NO) and Y(NPQ) values of *CsTGase*OE plants were lower compared with those of WT plants (Figure 9C–E). In *CsTGase*OE plants, the quantum yield of regulated energy dissipation (Y(NPQ)), an important process that consumes excess absorbed light energy and protects the photosynthetic apparatus, was higher than that in WT plants after salt stress (Figure 9G).

Similar to F_v/F_m and Y(II), P_m and Y(I) increased under normal conditions and were notably higher after salt stress in *CsTGase*OE plants compared with WT plants. Y(ND) showed a significant decrease in *CsTGase*OE plants compared to that of the WT plants after salt stress (Figure 9F). In addition, Y(NA) showed a significant increase in *CsTGase*OE leaves compared with those of the WT plants after salt stress (Figure 9H).

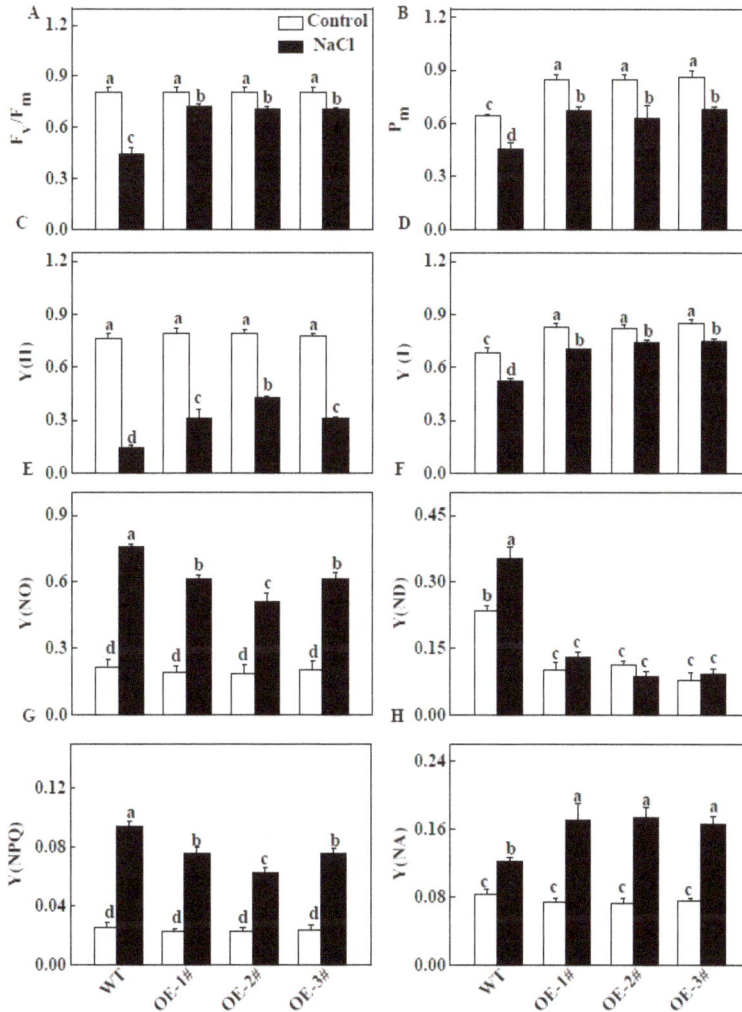

Figure 9. Changes in PSII and PSI in leaves of WT and *CsTGase*OE plants under salt stress. (**A**) The maximal quantum efficiency of PSII (F_v/F_m). (**B**) The maximum fluorescence of PSI (P_m). (**C**) The effective quantum efficiency of PSII (Y(II)). (**D**) The effective quantum efficiency of PSI (Y(I)). (**E**) The nonregulated energy dissipation (Y(NO)). (**F**) The oxidation status of PSI donor side (Y(ND)). (**G**) The regulated energy dissipation (Y(NPQ)). (**H**) The reduction status of PSI accept or side (Y(NA)). Each histogram represents a mean ± SE of four independent experiments ($n = 4$). Different letters indicate significant differences between treatments ($p < 0.05$) according to Duncan's multiple range test.

2.9. Effects of TGase on the Regulation of Photosynthesis Related Gene Expression

To analyze whether TGase is involved in affecting photosynthesis-related genes, we first examined the expression of photosynthesis-related genes in the WT and *TGase*OE plants, such as PSII-related genes (*NtpsbA/B/C/D/E*), PSI-related genes (*NtpsaA/B*), ATP synthesis-related genes (*NtatpA/B*), Calvin cycle-related genes (*NtrbcL/NtrbcS/NtFBPase*) and cytochrome-related genes (*NtpetA/B/D*). As shown in Figure 10, the transcript levels of photosynthesis-related genes were higher in *CsTGase*OE plants than the WT plants under normal conditions. We further analyzed the expression patterns of those genes in WT and *CsTGase*OE plants after 7 days of salt stress. Except for *NtpsbD*, the transcript levels of these genes in WT and *CsTGase*OE plants were suppressed under salt stress, but its levels in *CsTGase*OE plants were still higher than those in the WT (Figure 10).

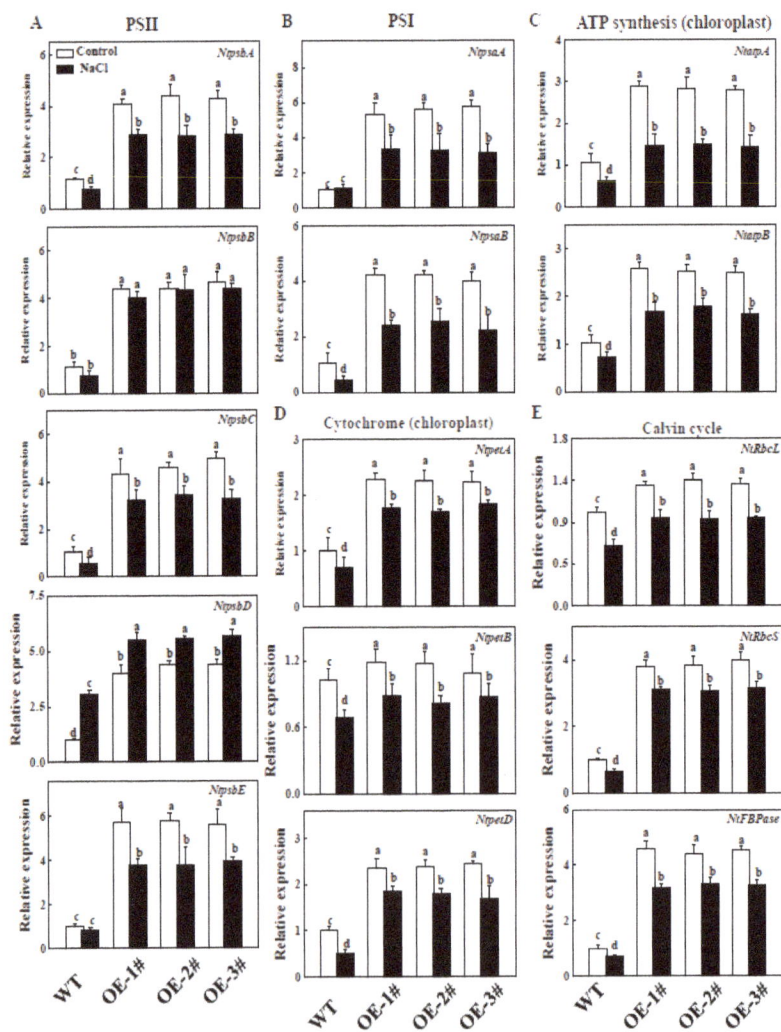

Figure 10. Expression of PSII (**A**); PSI (**B**); Cytochrome (**C**); ATP synthesis (**D**); Calvin cycle-related genes (**E**) in WT and *CsTGase*OE plants under salt stress. Each histogram represents a mean ± SE of four independent experiments (*n* = 4). Different letters indicate significant differences between treatments (*p* < 0.05) according to Duncan's multiple range test.

To confirm these results, we monitored the differential changes of some photosynthesis proteins using western blot (WB). As shown in Figure 11, in agreement with the results of gene expression, the levels of D2 proteins in *CsTGase*OE plants were significantly higher than that of WT plants under normal conditions. After 7 days of salt stress treatment, a significant reduction of D1, D2, and Cytf proteins was detected in the leaves of WT plants, but higher levels of these proteins were still present in the leaves of *CsTGase*OE plants (Figure 11 and Figure S2). In addition, LHCA1, and LHCB1 had showed no significant differences in WT and *CsTGase*OE plants before and after salt stress (Figure 11 and Figure S2).

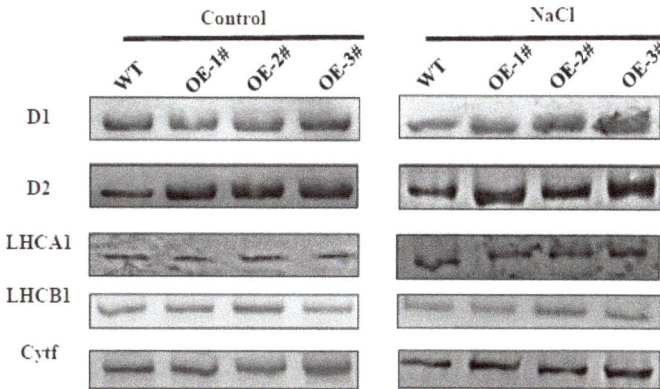

Figure 11. Analysis of thylakoid membrane protein changes in WT and *CsTGase*OE plants under salt stress. Thylakoid membrane proteins were separated by 12% SDS-urea-PAGE, transferred to PVDF membranes and probed with antisera against known thylakoid membrane proteins obtained from Agrisera company.

3. Discussion

CsTGase expression was dependent on light induction (Figure 1A), and *CsTGase* transcript levels increased with leaf development and aging, especially in mature and senescent leaves (Figure 1B,C), supporting the notion that TGase mainly functions in the senescence process [27]. Furthermore, cucumber TGase (CsTGase) was located not only in the chloroplast grana but also in the cell wall (Figure 2). These results indicate that TGase plays a vital role in early plant development.

Salt stress and other abiotic stresses can decrease photosynthetic capacity. Under salt conditions, plants overexpressing *CsTGase* showed enhanced salt tolerance displaying vigorous growth and higher Pn and Gs (Figures 3 and 4), suggesting that this gene might be particularly involved in the salt stress response. Additionally, some photosynthesis-related genes such as *NtpsbA, NtpsbB, NtpsbC, NtpsbD, NtpsbE, NtpsaA, NtpsaB, NtpetA, NtpetB, NtatpA,* and *NtatpB,* had higher transcript levels in *CsTGase* overexpressing plants (Figure 10), suggesting that the improved tolerance of transgenic plants over-expressing *CsTGase* might result from regulation of photosynthetic systems. Further, identification of the regulation of photosynthetic systems and unraveling this regulatory network may shed light on the mechanism underlying *CsTGase*-dependent tolerance to salt stress.

Many reports have shown that PAs play critical roles in regulating plant responses to abiotic stresses such as salinity, high temperature, and cold stresses, which are related to changes in endogenous PA levels and gene modifications [25,28,29]. *CsTGase* overexpressing lines had higher PA contents than the WT under normal conditions (Figure 5). Moreover, the overexpression of *CsTGase* increased endogenous PA levels in chloroplasts compared with those of the WT under salt stress (Figure 5), indicating that TGase may play positive roles in PA-dependent pathways to enhance plant resistance to salt stress.

The salt stress induced serious changes in photochemical efficiency and was often associated with the suppression of PSII and PSI activity [30]. In the present study, salt stress resulted in a significant decline in F_v/F_m and based on the analysis of the OJIP curves, levels at J, I, and P transients were gradually decreased in WT plants under salt stress (Figure 7). This inhibition resulted in strong fluorescence and pronounced suppression of total fluorescence emission, indicating that salt stress reduced PSII and PSI electron transport. However, overexpression of *CsTGase* interrupted the decrease in the OJIP curves and F_m values (Figure 7). These results are consistent with the fact that TGase plays a critical role in photoprotection [31]. Electron microscopy revealed that *CsTGase* overexpression resulted in an increase in grana stacking (Figure 6). Moreover, salt stress induced more severe damage to the ultrastructure of chloroplast and thylakoids in WT plants compared with *CsTGase*OE plants. Taken together, these results indicated that TGase plays a critical role in the protection of chloroplast under salt stress.

It was demonstrated that chlorophyll-a/b proteins (LHCII, CP29, CP26, and CP24) as substrates of TGase, as well as TGase itself, can catalyze the modification of light harvesting complex II by PAs in a light-dependent pathway [32,33]. However, the regulation of PSI through TGase has not been well studied. In this study, the decrease of Y(I) in the treated WT leaves resulted from the increase in the donor side limitation of PSI, as reflected by Y(ND), whereas the Y(ND) was not increased in the *CsTGase*OE plants (Figure 9). This finding indicates that salt stress releases an excessive amount of light energy to PSI in WT plants but not in *CsTGase*OE plants. The proportion of reduced electron carriers cannot be oxidized on the acceptor side of PSI by Y(NA) in WT plants, which is often used as an indicator of PSI photoinhibition. Our results showed that TGase could alleviate the salt stress caused the acceptor side limitation of PSI, as reflected by the higher value of Y(NA) in the *CsTGase*OE plants under salt stress than in the WT plants. An increase in Y(NA) indicates an increasing acceptor side limitation [34]. This phenomenon indicates that TGase may play a positive role in PSI under salt stress.

Apart from the effect on photosystems, we also found that the Calvin cycle is regulated by *CsTGase*. Indeed, the upregulation of some Calvin cycle genes, such as *NtRbcL* and *NtRbcS*, were observed in *CsTGase*OE plants under normal conditions. Furthermore, the expression of the *NtFBPase* gene was slightly enhanced in the leaves of *CsTGase*OE plants (Figure 10). However, the expression of Calvin cycle-related genes was inhibited in WT and *CsTGase*OE plants after salt stress, but these genes still maintained a high level in *CsTGase*OE plants (Figure 10). Rubisco and FBPase are sensitive to oxidative stress, as salt stress has been shown to induce ROS accumulation [35]. Hence, we hypothesize that TGase may be involved in the downregulation of ROS accumulation by enhancing endogenous PA content, especially in chloroplasts, thereby protecting the photosynthetic organs and the Calvin cycle.

A number of previous studies have established the role of TGase in maintaining the activity of chloroplast-related proteins, which is associated with enhancement of PA conjugation to light harvesting complex II (LHCII) proteins [21]. Proteomic analysis revealed that some proteins of photosystems were substrates of TGase, such as, LHCII, CP29, CP26, and CP24 [36]. In this study, the D1 and D2 proteins were strongly accumulated in *CsTGase*OE plants, while their levels were noticeably decreased in WT plants under salt stress. In addition, the LHCA1, LHCB1 and Cytf protein levels were not significantly different in WT and *CsTGase*OE plants (Figure 11 and Figure S2). Taken together, these results suggest a critical role for TGase in maintaining protein stability under salt stress.

In summary, this study provides compelling evidence to support our assumption that TGase participates in the enhancement of salt tolerance by modulating photosynthetic proteins; the manipulation of endogenous TGase activity could increase PA levels to alleviate salt-induced photoinhibition in tobacco plants. Meanwhile, TGase increased the levels of chloroplast proteins and protected these proteins under salt stress. Therefore, TGase mediates the processing of PA biosynthesis and the protection of chloroplast proteins to enhance salt tolerance, thereby increasing the survival of plants under salt stress.

4. Materials and Methods

4.1. Cucumber Plant Materials and Treatments for Expression Analysis

The cucumber *Cucumis sativus* L. cv. 9930 genotype was used in the experiment. Seeds were germinated and grown in 250 cm^3 plastic pots filled with peat. The plants were watered daily with Hoagland's nutrition solution in the chamber. The growing conditions were as follows: 14/10 light/dark cycle, 28/22 °C day/night temperatures and 600 µmol m^{-2} s^{-1} photosynthetic photon flux density (PPFD).

To analyze the possible influence of light on gene expression, 30-day-old cucumber plants growing under the normal 16/8 h photoperiod, were incubated to a 24 h illumination period, with or without being previously subjected to 24 h darkness. Leaf samples were taken at 0, 4, 8, 12, and 24 h and during the continuous illumination period.

To analyze the tissue-specific expression of *CsTGase*, roots, stems, leaves, flowers and fruits were collected. Young, mature, and senescent leaves were collected. We also collected the leaves at 1 week, 2 weeks, 4 weeks, and 8 weeks during plant growth.

4.2. TEM Observations: Immunogold Transmission Electron Microscopy

Cucumber leaf sections were fixed, dehydrated, and embedded in Lowicryl K4M resin (Pelco International, Redding, CA, USA), following the previously described procedures [37]. A monoclonal antibody, anti-plant TGase (produced in rabbit; Univ-bio, Shanghai, China) at 1:1000 dilutions was used. A gold AffiniPure anti rabbit IgG was used as a secondary antibody. For electron microcopy, a Jeol-JEM-1010 transmission electron microscope operating at 80 kV.

4.3. Generation and Selection of Transgenic Plants

To obtain the cucumber *TGase* overexpression (*TGase*OE) construct, the 1836 bp full-length coding DNA sequence (CDS) was amplified with the primer *TGase*OE-F (5′-CGAGCTCATGGATGATC GTGAGGCGTTTAAGA-3′) and *TGase*OE-R (5′-CGGGGTACCACGTTGCATGCAATTCCCGTAG-3′) using cucumber cDNA as the template. The PCR product was digested with *Sac*I and *Kpn*I and inserted behind the CaMV 35S promoter in the binary vector pCAMBIA1301-GUS. The *TGase*OE-GUS plasmid was transformed into *Agrobacterium tumefaciens* strain EHA105. NC89 tobacco plants were used for transformation, as described by Horsch [38].

4.4. Salt Tolerance Analysis of the Transgenic Plants

Three-week-old plants grown in vermiculite were treated with 200 mM NaCl for 7 days, and leaves were used to measure the biomass, proline content, and chlorophyll content. For biomass measurements, plants were dried for 48 h at 75 °C and then weighed. The proline content was measured according to Bates et al. [39]. The chlorophyll content was measured by UV spectrophotometry as described by Yang et al. [40].

Three-week-old plants grown in vermiculite were treated with 200 mM NaCl for 14 days, and the leaves were used to measure the photosynthetic parameters. The net photosynthetic rate (Pn), intercellular CO_2 concentration (Ci), and stomatal conductance (Gs) were measured according to Zhang et al. [41].

To avoid light impact on the TGase activity and PA content, potted plants in the experimental field were exposed to a 14/10 light/dark cycle. Then, we collected the samples and measured photosynthetic parameters at 12 h of illumination.

4.5. Thylakoid Isolation

Thylakoids were isolated as previously described with minor modifications [13]. Intact chloroplasts from the fully expanded leaves were homogenized in 50 mM KCl, 1 mM MgCl$_2$,

1 mM MnCl$_2$, 1 mM EDTA, 0.5 mM KH$_2$PO$_4$, 25 mM HEPES, pH 7.6, 330 mM sorbitol, 10 μM sodium ascorbate, and 0.2% (*w/w*) bovine serum albumin. The homogenates were filtered through a 300-μm and then 100-μm nylon mesh and debris was removed by centrifugation at 300× *g* for 1 min, followed by centrifugation at 4000× *g* for 10 min to collect the thylakoids. The pellet was separated from starch, resuspended and washed in 7 mM MgCl$_2$, 10 mM KCl, and 25 mM HEPES, pH 7.6 to break intact chloroplasts and removed free polyamines. Finally, for the polyamine analysis, thylakoids were resuspended in the medium containing 7 mM MgCl$_2$, 50 mM KCl, 25 mM HEPES, pH 7.6 and 330 mM sorbitol.

4.6. Analysis of Endogenous Polyamines in Thylakoid Membranes

The contents of endogenous PAs in thylakoid membranes were analyzed according to a method described by Zhang et al. [41] with some modifications. Briefly, for polyamine analysis, isolated thylakoids were incubated in 1.6 mL of 5% (*w/v*) cold perchloric acid (PCA) for 1 h on ice. After centrifugation for 20 min at 12,000× *g*, the pellet was used to determine bound PAs. PAs were analyzed using a high-performance liquid chromatography with a 1200 series system (Agilent Technologies, Santa Clara, CA, USA), a C18 reversed-phase column (4.6 mm by 250 mm, 5 μm Kromasil) and a two solvent system including a methanol gradient (36%–64%, *v/v*) at a flow rate of 0.8 mL min^{-1}.

4.7. Observation of the Ultrastructure of the Chloroplast

Tobacco leaves were collected and cut into pieces of approximately 1 mm^2 and fixed by vacuum infiltration with 3% glutaraldehyde and 1% formaldehyde in a 0.1 M phosphate buffer (pH 7.4) for 2 h (primary fixation). After washing, the sample were fixed for 2 h in osmium tetroxide at room temperature; then, the samples were dehydrated in acetone and embedding in Durcupan ACM, Ultrathin sections of the leaf pieces (70 nm) were cut, stained with uranium acetate and lead citrate in series and examined using a H7650 transmission electron microscope (Hitachi, Tokyo, Japan) at an accelerating voltage of 80 kV. A minimum of 50 chloroplasts of each type of plant were examined.

4.8. Chl a fluorescence Measurement and OJIP Transient Analyses

The chlorophyll *a* (Chl *a*) fluorescence induction kinetics were measured at 12 h of illumination using dual portable fluorescence (Dual-PAM-100, Walz, Germany). Measurements were analyzed using the automated induction program provided by the Dual-PAM software. PSII and PSI activities were quantified by chlorophyll fluorescence and P700$^+$ absorbance changes.

The OJIP curves showed a polyphasic rise. The initial fluorescence (F$_0$) (approx. 50 μs) was set as O, followed by the O to J phase (ends at approx. 2 ms), then the J to I phase (ends at approx. 30 ms) and I to P phase (at the peak of the OJIP curve). The JIP measurement is named after the basic steps in fluorescence transience when plotted on a logarithmic time scale [42]. Leaves were dark adapted for 30 min prior to the measurement or the NPQ measurement leaves were continuously illuminated for 270 s with 500 μmol photons m^{-2} s^{-1} using the Handy-PEA (multihit- mode). Every 30 s was given a 3000 μmol photons m^{-2} s^{-1} (duration 0.8 s) saturating pulse for maximal fluorescence, F$_m$'. To calculate the NPQ at the end of the actinic light phase we followed the equation, NPQ = F$_m$/ F$_m$' − 1 [43]. The tests were shown in the middle portion of infiltration on the ventral surface of the leaves. Measurements were taken in 6 replications.

4.9. Quantitative Real-Time PCR

Total RNA was extracted from three biological replicates of leaves using an RNA extraction kit (Tiangen, Beijing, China) according to the manufacturer's instructions. The first strand cDNA was synthesized from 1 μg of DNase-treated RNA using reverse transcriptase (Takara, Dalian, China) following the manufacturer's protocol. Quantitative real-time PCR (qPCR) was using gene specific primers (Table S1) in 20 μL reaction system using SYBR Premix Ex Taq II (Takara, Dalian, China).

Tobacco *β-actin* was used as a reference gene for tobacco; cucumber *actin* was used as a reference gene for cucumber. Relative gene expression was calculated according to Livak and Schmittgen [44].

4.10. Protein Extraction and Western Blotting

For the thylakoid membranes, the intact chloroplasts were reputed in low osmotic buff (50 mM HEPES–KOH (pH 7.6) and 2 mM $MgCl_2$) on the ice, then the thylakoid membranes were collected and the protein content was determined by a BCA Protein Assay Kit (Solarbio, Beijing, China). For immunoblot analysis, thylakoid proteins were solubilized and separated on 12% SDS-urea-PAGE gels. After electrophoresis, the proteins were transferred to polyvinylidene difluoride (PVDF) membranes (Millipore, Billerica, MA, USA) and probed using commercial antibodies specific for the PSII subunits (D1 and D2) (AS05084 and AS06146), light-harvesting antenna proteins (LHCA1 and LHCB1) (AS01005 and AS01004) and Cytb6f subunit (Cytf) (AS14169), at 1:5000 dilutions were used. These antibodies were from Agrisera (Vännäs, Sweden). At least three independent replicates were used for each determination. Accumulation of proteins were quantified using Quantity One software (Bio-Rad, Hercules, California, CA, USA).

4.11. Statistical Analysis

At least 4 independent replicates were used for each determination. Statistical analysis of the bioassays was performed using the SPSS 20 statistical package (SPSS Inc., Chicago, IL, USA). Experimental data were analyzed with a Duncan's multiple range test at $p < 0.05$.

Supplementary Materials: Supplementary materials can be found at http://www.mdpi.com/1422-0067/20/4/894/s1. Figure S1: The granum height of TGase in each transgenic line. Figure S2: The relative intensity of thylakoid membrane protein changes in WT and *CsTGase*OE plants under salt stress. Table S1: Sequences of primers used for qPCR assays.

Author Contributions: S.G. designed the research and proposed the research proceeding. M.Z. and Y.W. performed the experiments and wrote the main manuscript text. Y.Z. prepared all figures and modified this manuscript until submit. S.S. and J.S. improved the manuscript. All authors reviewed and approved the manuscript.

Funding: This work was supported by the National Natural Science Foundation of China (31672199 and 31801902), the China Earmarked Fund for Modern Agro-industry Technology Research System (CARS-23-B12).

Conflicts of Interest: The authors declare no conflict of interests.

References

1. Ioannidis, N.E.; Kotzabasis, K. Polyamines in chemiosmosis in vivo: A cunning mechanism for the regulation of ATP synthesis during growth and stress. *Front. Plant Sci.* **2014**, *5*, 71. [CrossRef] [PubMed]
2. Yang, Y.; Guo, Y. Unraveling salt stress signaling in plants. *J. Integr. Plant Biol.* **2018**, *60*, 796–804. [CrossRef] [PubMed]
3. Zhou, H.; Guo, S.; An, Y.; Shan, X.; Wang, Y.; Shu, S.; Sun, J. Exogenous spermidine delays chlorophyll metabolism in cucumber leaves (*Cucumis sativus* L.) under high temperature stress. *Acta Physiol. Plant.* **2016**, *38*, 224. [CrossRef]
4. Shu, S.; Yuan, Y.; Chen, J.; Sun, J.; Zhang, W.; Tang, Y.; Zhong, M.; Guo, S. The role of putrescine in the regulation of proteins and fatty acids of thylakoid membranes under salt stress. *Sci. Rep.* **2015**, *5*, 14390. [CrossRef] [PubMed]
5. Sagor, G.H.M.; Zhang, S.; Kojima, S.; Simm, S.; Berberich, T.; Kusano, T. Reducing cytoplasmic polyamine oxidase activity in *Arabidopsis* increases salt and drought tolerance by reducing reactive oxygen species production and increasing defense gene expression. *Front. Plant Sci.* **2016**, *7*, 214. [CrossRef] [PubMed]
6. Bortolotti, C.; Cordeiro, A.; Alcázar, R.; Borrell, A.; Culiañez-Macià, F.A.; Tiburcio, A.F.; Altabella, T. Localization of arginine decarboxylase in tobacco plants. *Physiol. Plant.* **2004**, *120*, 84–92. [CrossRef] [PubMed]
7. Del Duca, S.; Serafini-Fracassini, D.; Cai, G. Senescence and programmed cell death in plants: Polyamine action mediated by transglutaminase. *Front. Plant Sci.* **2014**, *5*, 120. [CrossRef] [PubMed]

8. Serafini-Fracassini, D.; Del Duca, S.; D'Orazi, D. First evidence for polyamine conjugation mediated by an enzymic activity in plants. *Plant Physiol.* **1988**, *87*, 757–761. [CrossRef]

9. Tiburcio, A.F.; Altabella, T.; Bitrián, M.; Alcázar, R. The roles of polyamines during the lifespan of plants: From development to stress. *Planta* **2014**, *240*, 1–18. [CrossRef] [PubMed]

10. Sfichi, L.; Loannidis, N.; Kotzabasis, K. Thylakoid-associated polyamines adjust the UV-B sensitivity of the photosynthetic apparatus by means of light-harvesting complex II changes. *Photochem. Photobiol.* **2004**, *80*, 499–506. [CrossRef]

11. Demetriou, G.; Neonaki, C.; Navakoudis, E.; Kotzabasis, K. Salt stress impact on the molecular structure and function of the photosynthetic apparatus—The protective role of polyamines. *Biochim. Biophys. Acta* **2007**, *1767*, 272–280. [CrossRef]

12. Dondini, L.; Del Duca, S.; Dall'Agata, L.; Bassi, R.; Gastaldelli, M.; Della Mea, M.; Di Sandro, A.; Claparols, I.; Serafini-Fracassini, D. Suborganellar localisation and effect of light on Helianthus tuberosus chloroplast transglutaminases and their substrates. *Planta* **2003**, *217*, 84–95. [PubMed]

13. Shu, S.; Yuan, L.-Y.; Guo, S.-R.; Sun, J.; Yuan, Y.-H. Effects of exogenous spermine on chlorophyll fluorescence, antioxidant system and ultrastructure of chloroplasts in *Cucumis sativus* L. under salt stress. *Plant Physiol. Biochem.* **2013**, *63*, 209–216. [CrossRef] [PubMed]

14. Hamdani, S.; Yaakoubi, H.; Carpentier, R. Polyamines interaction with thylakoid proteins during stress. *J. Photochem. Photobiol.* **2011**, *104*, 314–319. [CrossRef] [PubMed]

15. Ioannidis, N.E.; Cruz, J.A.; Kotzabasis, K.; Kramer, D.M. Evidence that putrescine modulates the higher plant photosynthetic proton circuit. *PLoS ONE* **2012**, *7*, e29864. [CrossRef]

16. Ioannidis, N.E.; Ortigosa, S.M.; Veramendi, J.; Pintó-Marijuan, M.; Fleck, I.; Carvajal, P.; Kotzabasis, K.; Santos, M.; Torné, J.M. Remodeling of tobacco thylakoids by over-expression of maize plastidial transglutaminase. *Biochim. Biophys. Acta* **2009**, *1787*, 1215–1222. [CrossRef] [PubMed]

17. Serafini-Fracassini, D.; Del Duca, S. Transglutaminases: Widespread cross-linking enzymes in plants. *Ann. Bot.* **2008**, *102*, 145–152. [CrossRef]

18. Lilley, G.R.; Skill, J.; Griffin, M.; Bonner, P.L. Detection of Ca^{2+}-dependent transglutaminase activity in root and leaf tissue of monocotyledonous and dicotyledonous plants. *Plant Physiol.* **1998**, *117*, 1115–1123. [CrossRef]

19. Della Mea, M.; Caparrós-Ruiz, D.; Claparols, I.; Serafini-Fracassini, D.; Rigau, J. *AtPng1p*. The first plant transglutaminase. *Plant Physiol.* **2004**, *135*, 2046–2054. [CrossRef]

20. Del Duca, S.; Faleri, C.; Iorio, R.A.; Cresti, M.; Serafini-Fracassini, D.; Cai, G. Distribution of transglutaminase in pear pollen tubes in relation to cytoskeleton and membrane dynamics. *Plant Physiol.* **2013**, *161*, 1706–1721. [CrossRef]

21. Del Duca, S.; Tidu, V.; Bassi, R.; Esposito, C.; Serafmi-Fracassini, D. Identification of chlorophyll-a/b proteins as substrates of transglutaminase activity in isolated chloroplasts of *Helianthus tuberosus* L. *Planta* **1994**, *193*, 283–289. [CrossRef]

22. Campos, N.; Castañón, S.; Urreta, I.; Santos, M.; Torné, J. Rice transglutaminase gene: Identification, protein expression, functionality, light dependence and specific cell location. *Plant Sci.* **2013**, *205*, 97–110. [CrossRef] [PubMed]

23. Villalobos, E.; Santos, M.; Talavera, D.; Rodrıguez-Falcón, M.; Torné, J. Molecular cloning and characterization of a maize transglutaminase complementary DNA. *Gene* **2004**, *336*, 93–104. [CrossRef] [PubMed]

24. Duan, J.; Li, J.; Guo, S.; Kang, Y. Exogenous spermidine affects polyamine metabolism in salinity-stressed Cucumis sativus roots and enhances short-term salinity tolerance. *J. Plant Physiol.* **2008**, *165*, 1620–1635. [CrossRef] [PubMed]

25. Sang, Q.; Shan, X.; An, Y.; Shu, S.; Sun, J.; Guo, S. Proteomic analysis reveals the positive effect of exogenous spermidine in tomato seedlings' response to high-temperature stress. *Front. Plant Sci.* **2017**, *8*, 120. [CrossRef] [PubMed]

26. Tang, Y.-Y.; Yuan, Y.-H.; Shu, S.; Guo, S.-R. Regulatory mechanism of NaCl stress on photosynthesis and antioxidant capacity mediated by transglutaminase in cucumber (*Cucumis sativus* L.) seedlings. *Sci. Hortic.* **2018**, *235*, 294–306. [CrossRef]

27. Sobieszczuk-Nowicka, E.; Zmienko, A.; Samelak-Czajka, A.; Łuczak, M.; Pietrowska-Borek, M.; Iorio, R.; Del Duca, S.; Figlerowicz, M.; Legocka, J. Dark-induced senescence of barley leaves involves activation of plastid transglutaminases. *Amino Acids* **2015**, *47*, 825–838. [CrossRef]

28. Yuan, Y.; Zhong, M.; Shu, S.; Du, N.; Sun, J.; Guo, S. Proteomic and physiological analyses reveal putrescine responses in roots of cucumber stressed by NaCl. *Front. Plant Sci.* **2016**, *7*, 1035. [CrossRef]

29. Zhuo, C.; Liang, L.; Zhao, Y.; Guo, Z.; Lu, S. A cold responsive ethylene responsive factor from Medicago falcata confers cold tolerance by up-regulation of polyamine turnover, antioxidant protection, and proline accumulation. *Plant Cell Environ.* **2018**, *41*, 2021–2032. [CrossRef]

30. Oukarroum, A.; Bussotti, F.; Goltsev, V.; Kalaji, H.M. Correlation between reactive oxygen species production and photochemistry of photosystems I and II in Lemna gibba L. plants under salt stress. *Environ. Exp. Bot.* **2015**, *109*, 80–88. [CrossRef]

31. Ioannidis, N.E.; Malliarakis, D.; Torné, J.M.; Santos, M.; Kotzabasis, K. The over-expression of the plastidial transglutaminase from maize in Arabidopsis increases the activation threshold of photoprotection. *Front. Plant Sci.* **2016**, *7*, 635. [CrossRef] [PubMed]

32. Del Duca, S.; Tidu, V.; Bassi, R.; Serafini-Fracassini, D.; Esposito, C. Identification of transglutaminase activity and its substrates in isolated chloroplast of Helianthus tuberosus. *Planta* **1994**, *193*, 283–289. [CrossRef]

33. Sobieszczuk-Nowicka, E.; Krzesłowska, M.; Legocka, J. Transglutaminases and their substrates in kinetin-stimulated etioplast-to-chloroplast transformation in cucumber cotyledons. *Protoplasma* **2008**, *233*, 187. [CrossRef] [PubMed]

34. Huang, W.; Yang, S.-J.; Zhang, S.-B.; Zhang, J.-L.; Cao, K.-F. Cyclic electron flow plays an important role in photoprotection for the resurrection plant Paraboearufescens under drought stress. *Planta* **2012**, *235*, 819–828. [CrossRef] [PubMed]

35. Miller, G.; Suzuki, N.; Ciftci-Yilmaz, S.; Mittler, R. Reactive oxygen species homeostasis and signalling during drought and salinity stresses. *Plant Cell Environ.* **2010**, *33*, 453–467. [CrossRef] [PubMed]

36. Campos, A.; Carvajal-Vallejos, P.; Villalobos, E.; Franco, C.; Almeida, A.; Coelho, A.; Torné, J.; Santos, M. Characterisation of *Zea mays* L. plastidial transglutaminase: Interactions with thylakoid membrane proteins. *Plant Biol.* **2010**, *12*, 708–716. [CrossRef] [PubMed]

37. Campos, N.; Villalobos, E.; Fontanet, P.; Torné, J.M.; Santos, M. A peptide of 17 aminoacids from the N-terminal region of maize plastidial transglutaminase is essential for chloroplast targeting. *Am. J. Mol. Biol.* **2012**, *2*, 245–257. [CrossRef]

38. Horsch, R.B.; Fry, J.E.; Hoffmann, N.L.; Eichholtz, D.; Rogers, S.G.; Fraley, R.T. A simple and general-method for transferring genes into plants. *Science* **1985**, *227*, 1229–1231.

39. Bates, L.; Waldren, R.; Teare, I. Rapid determination of free proline for water-stress studies. *Plant Soil* **1973**, *39*, 205–207. [CrossRef]

40. Yang, Q.; Chen, Z.-Z.; Zhou, X.-F.; Yin, H.-B.; Li, X.; Xin, X.-F.; Hong, X.-H.; Zhu, J.-K.; Gong, Z. Overexpression of *SOS* (Salt Overly Sensitive) genes increases salt tolerance in transgenic Arabidopsis. *Mol. Plant* **2009**, *2*, 22–31. [CrossRef]

41. Zhang, R.H.; Li, J.; Guo, S.R.; Tezuka, T. Effects of exogenous putrescine on gas-exchange characteristics and chlorophyll fluorescence of NaCl-stressed cucumber seedlings. *Photosynth. Res.* **2009**, *100*, 155–162. [CrossRef] [PubMed]

42. Force, L.; Critchley, C.; van Rensen, J.J. New fluorescence parameters for monitoring photosynthesis in plants. *Photosynth. Res.* **2003**, *78*, 17. [CrossRef] [PubMed]

43. Bilger, W.; Björkman, O. Relationships among violaxanthin deepoxidation, thylakoid membrane conformation, and nonphotochemical chlorophyll fluorescence quenching in leaves of cotton (*Gossypium hirsutum* L.). *Planta* **1994**, *193*, 238–246. [CrossRef]

44. Livak, K.J.; Schmittgen, T.D. Analysis of relative gene expression data using real-time quantitative PCR and the $2^{-\Delta\Delta CT}$ method. *Methods* **2001**, *25*, 402–408. [CrossRef] [PubMed]

International Journal of
Molecular Sciences

MDPI

Review

Progress in Understanding the Physiological and Molecular Responses of *Populus* to Salt Stress

Xiaoning Zhang [1,†], Lijun Liu [2,†], Bowen Chen [1], Zihai Qin [1], Yufei Xiao [1], Ye Zhang [1], Ruiling Yao [1], Hailong Liu [1,*] and Hong Yang [3,*]

1   Guangxi Key Laboratory of Superior Timber Trees Resource Cultivation, Guangxi Forestry Research Institute, 23 Yongwu Road, Nanning 530002, China; sflzxn@163.com (X.Z.); gfri_bwchen@163.com (B.C.); qinzihai@126.com (Z.Q.); xiaoyufei33@163.com (Y.X.); elaine.ye@163.com (Y.Z.); jullyudi@163.com (R.Y.)
2   Key Laboratory of State Forestry Administration for Silviculture of the lower Yellow River, College of Forestry, Shandong Agricultural University, Taian 271018, Shandong, China; lijunliu@sdau.edu.cn
3   Key Laboratory of Economic Plants and Biotechnology, Kunming Institute of Botany, Academy of Sciences, Yunnan Key Laboratory for Wild Plant Resources, Kunming 650201, China
*   Correspondence: hailon_liu@163.com (H.L.); yanghong@mail.kib.ac.cn (H.Y.); Tel.: +86-185-7786-9045 (H.L.); +86-153-9871-2027 (H.Y.)
†   These authors contribute equally to this work.

Received: 21 February 2019; Accepted: 9 March 2019; Published: 15 March 2019

Abstract: Salt stress (SS) has become an important factor limiting afforestation programs. Because of their salt tolerance and fully sequenced genomes, poplars (*Populus* spp.) are used as model species to study SS mechanisms in trees. Here, we review recent insights into the physiological and molecular responses of *Populus* to SS, including ion homeostasis and signaling pathways, such as the salt overly sensitive (SOS) and reactive oxygen species (ROS) pathways. We summarize the genes that can be targeted for the genetic improvement of salt tolerance and propose future research areas.

Keywords: poplars (*Populus*); salt tolerance; molecular mechanisms; SOS; ROS

---

1. Introduction

Poplars (*Populus* spp.), which include about 100 species [1], are widely distributed across a variety of climatic regions [2] and have become important species for global afforestation and shelterbelt projects because of their rapid growth and high biomass yields [3]. These traits, in combination with other characteristics, such as extensive and deep root systems, considerable genetic variation, small genome size, convenient asexual propagation, genetic transformability, and economic significance, have led to the use of poplars as model tree species [4–6].

The increasing salinization of soils has greatly limited the planting of salt-sensitive *Populus* species [7]. Salt stress (SS) induces water deficiency, osmotic stress, ion toxicity, and oxidative damage [8] and thereby reduces photosynthesis, respiration, transpiration, metabolism, and growth in poplars. Like most plants, poplars can adapt to SS by maintaining their cellular ion homeostasis, accumulating osmotic-adjustment substances, and activating scavengers of reactive oxygen species (ROS) via the initiation of an efficient signal transduction network [9]. Desert poplar (*Populus euphratica*) is one of the most salt-tolerant *Populus* species [10] and is often used to study the salt-response mechanisms of trees. *P. euphratica* was reported to be tolerant to up to 450 mM NaCl (about 2.63%) under hydroponic conditions and showed high recovery efficiency when NaCl was removed from the culture medium [11]. A previous study has shown that *P. euphratica* could grow in soils with up to 2.0% salinity and can survive in soils with up to 5.0% salinity [12]. As a non-halophyte, *P. euphratica* could activate salt secretion mechanisms when soil salinity concentrations are greater than 20%, which may be one of the reasons for its high salt-tolerance [13]. Most other *Populus* species are relatively salt-sensitive, including the grey poplar (*Populus* × *canescens*), *Populus* × *euramericana*, and *Populus popularis*, which are usually used as a salt-sensitive control in salt-response studies. *Populus alba* has a wide variation of salinity tolerance within the species: for example 'Guadalquivir F-21–38', 'Guadalquivir F-21–39', and 'Guadalquivir F-21–40' clones show salt tolerance, while most other clones have a common salt sensitivity. Considering the wide variation within the species, *P. alba* could be used as a model species to understand the mechanisms of SS [2].

Previous research primarily focused on the anatomical, physiological, and biochemical changes in poplars during SS; many recent studies have focused on the molecular mechanisms using new techniques, such as genome-scale transcript analysis [14], high-throughput sequencing [15], metabolite profiling [16], bioinformatic analyses [17–19], and a non-invasive micro-test technique (NMT) [20].

Here, we review the recent progress in understanding the physiological and molecular responses of *Populus* to SS, including SS injuries, the main mechanisms of salt tolerance, and the genes targeted for the genetic improvement of salt tolerance in *Populus*, with a major focus on ion homeostasis, osmotic adjustment, ROS scavenging, salt overly sensitive (SOS) signaling pathways, potential candidate genes, and transcription factor-mediated regulation of the SS response.

2. SS Injury

2.1. Inhibition of Poplar Growth by Salinity Stress

Salinity affects all stages of *Populus* growth, including germination [21], vegetation growth [15,22], and sexual reproduction [23]. The percentage of seeds that germinate and the extent of leaf expansion were both reported to decline as salt concentrations increase [21]. In addition, shoot growth is more sensitive to salt than root growth [8]. When salt-sensitive white poplar (*P. alba*) clones were exposed to SS (0.6% NaCl), striking reductions were observed in their leaf elongation rate and internode lengths, while significant increases were observed in the numbers of short branches and bud numbers, as well as in the levels of leaf epinasty, necrosis, and abscission [7].

Long-term SS also induces early leaf maturity, early flowering, and early tree maturation in *P. alba* clones, which display a greater architectural modification when exposed to high SS (0.6%) than lower SS (0.3%) [22]. When exposed to 0.6% NaCl, the heights, ground diameter, and leaf numbers of Poplar 107 were significantly reduced, while plants exposed to 0.3% NaCl stress had relatively minor phenotypic changes [15]. When *P. euphratica* was exposed to 300 mM (1.76%) NaCl stress, the three growth indexes (plant height, ground diameter, and leaf number) were reduced to 31%, 45.5%, and 20% of the control plants, respectively. The mean leaf area of these stressed trees was reduced by up to 60%, and the leaves began to wither and yellow after 10 days. By contrast, a treatment of 50 mM (about 0.29%) NaCl did not cause a significant reduction in these traits in *P. euphratica* [24].

2.2. Salt-Induced Physiological and Cellular Changes

The adverse effects of SS also result in physiological and microscopic anatomical changes. *P. euphratica* and *P. alba* trees exposed to SS have significantly reduced stomatal area, aperture, and conductance, but increased stomatal density and hydraulic conductance [25–27]. The salt-induced reduction of leaf area in *Populus* may be one of the reasons for the increased stomatal density and decreased stomatal area [24,27]. The percentage loss of hydraulic conductivity (PLC%) in *P. euphratica* increased from 31.81% at 0 mM NaCl to 83.83% at 150 mM NaCl (0.88%), causing a 40–80% decrease in hydraulic conductivity and ensuring high hydraulic efficiency [27]. *Populus* trees thus reduce their transpiration by decreasing their stomatal apertures and conductance and increasing their PLC% values to cope with the salt-induced water deficit.

Populus may adjust their xylems to adapt to salinity stress [28]; for example, when exposed to SS, salt-sensitive *P.* × *canescens* produces narrower xylem vessels in which it stores sodium ions (Na^+), reducing the effect of ion toxicity. In contrast, the salt-tolerant species *P. euphratica* produces narrow xylem vessels to reduce Na^+ uptake even under normal conditions, the abundance of which remains largely unaltered under moderate SS [29]. Overall, this may indicate an evolutionary adaptation of the xylem structure in *P. euphratica*.

2.3. Salt-Induced Damage to the Photosynthetic System

Chlorophyll (Chl) is the main pigment used for photosynthesis in plants and is often studied in the evaluation of salt damage. The contents of Chla and Chlb, as well as the relative electron transport rate, decreased significantly in poplar 107 (a superior variety selected from *Populus × euramericana* cv. '74/76' hybrids) under 0.6% NaCl but increased under 0.3% NaCl [15].

Almost all poplar trees show a decreased net photosynthetic rate (Pn) under high SS; for example, the Pn decreased by 48.3% in *P. euphratica* under a 300 mM NaCl treatment, in comparison with the unstressed plants [24]. SS has a negative effect on growth, possibly due to the Pn affecting the accumulation of biomass. The stomatal conductance (Gs), transpiration rate, and internal CO_2 (Ci) concentration of *P. euphratica* leaves also decreased by 48.5%, 42.1%, and 15.7% under 300 mM NaCl, indicating that the photosynthetic system was injured [24]. The fluorescence transient curve OJIP is highly sensitive to salinity stress; the OJ phase is the photochemical phase, leading to the reduction of QA to QA$^-$. The I level is related to the heterogeneity in the filling up of the plastoquinone (PQ) pool. The P level is reached when all the PQ molecules are reduced to PQH2 [30]. While in poplars treated with 0.3% NaCl the OJIP curve follows the same trend as in the control plants, it significantly decreases at the J, I, and P phases in plants treated with 0.6% NaCl [15]. Large decreases in the J and I phases are also observed in *P. euphratica* treated with 300 mM NaCl [24]. As the electron transfer between Pheo and QA occurs during the J phase, these results indicate the higher salt concentration had a significant influence on electron transfer, which led to a great degree of salt damage [15]. Besides, F_0 (minimal constant fluorescence of dark-adapted plants) and Fv/Fm (a surrogate of the maximum quantum efficiency of PSII) are also used as fluorescence parameters to identify salt tolerance, for example, to detect genotypic differences in the sensitivity of white poplar clones to SS [31].

3. Primary Mechanism of Salt Tolerance in *Populus*

3.1. Maintaining an Optimal K+/Na+ Ratio

Salt tolerance can be determined by the net Na$^+$ efflux capacity in *Populus*. The net uptake of Na$^+$ depends on its influx, exclusion, and sequestration, as well as on other Na$^+$ regulation processes, such as xylem Na$^+$ loading and unloading, and phloem Na$^+$ recirculation [32]. Na$^+$ moves into cells through non-selective cation channels (NSCCs) and high-affinity K$^+$ transporter (HKT) [33], while excessive Na$^+$ can be extruded into the apoplast across the plasma membrane (PM) by the Na$^+$/H$^+$ antiporter salt overly sensitive1 (SOS1; see Section 3.4) [34], which may also involve Na$^+$ loading in the xylem [35]. The net Na$^+$ efflux increased significantly in *P. euphratica* after 0.5–12 h under 100 mM NaCl but was reduced at 0.5 h in *P. popularis*, which showed an overall Na$^+$ influx after 6–12 h of SS [36]. This is consistent with results of X-ray microanalysis showing that *P. euphratica* had lower concentration of Na$^+$ in all subcellular compartments than salt-sensitive poplar species [37]. Moreover, salt tolerance can be improved by increasing Na$^+$ efflux and the uptake of mineral nutrients via interactions between mycorrhizal fungi and the roots of salt-sensitive *Populus* [10,38].

It is crucial for *Populus* to maintain an optimal K$^+$/Na$^+$ ratio in the cytoplasm when exposed to high salinity. An excessive uptake of Na$^+$ ions not only increases their abundance in plant cells, but also induces the loss of potassium ions (K$^+$) by depolarizing the cellular membranes. Moreover, Na$^+$ can compete with K$^+$ for the binding sites of the uptake system, resulting in an imbalance of the K$^+$/Na$^+$ ratio that eventually causes ion toxicity [32]. Under high SS, *P. euphratica* maintained an optimal Na$^+$/K$^+$ ratio by restricting the net Na$^+$ uptake and transport from roots to shoots and by maintaining higher K$^+$ uptake and transport [39,40]. X-ray microanalysis showed that high salinity reduced the K$^+$/Na$^+$ ratio by 93% in *P. popularis* but only by 69% in *P. euphratica* [40].

Populus improves salt tolerance by accumulating Na^+/K^+ ions in the vacuoles. Na^+ is sequestered into plant cell vacuoles by the tonoplast Na^+/H^+ antiporter NHX1 [41]. Besides detoxifying the cytoplasm, the accumulation of Na^+ ions in the vacuoles is used as an osmoticum to draw water into the cells [42]. Vacuolar H^+-ATPase and vacuolar H^+-PPase generate an electrochemical gradient across the vacuolar membrane for the tonoplast Na^+/H^+ antiporters in *P. euphratica* cells [42]. In addition, expressing the *Arabidopsis thaliana* Na^+/H^+ antiporter gene *AtNHX1* in transgenic poplars improves their salt resistance by improving their Na^+/H^+ exchange activity [43–45]. AtNHX3 was previously shown to act as the tonoplast K^+/H^+ antiporter for the transportation of K^+ and the maintenance of ion homeostasis [46]; however, AtNHX1 was also recently found to enhance the accumulation of K^+ in the vacuoles of transgenic poplars [47]. Yang reported that the constitutive expression of either *AtNHX1* or *AtNHX3* in transgenic *Populus* increased the vacuolar accumulation of Na^+ and K^+, leading to improved salt tolerance and drought tolerance [47].

K^+ and Na^+ uptake and transport mediated by HKT1 facilitate the rapid response of *Populus* to SS. The Na^+/K^+ transporter HKT1 is located on the plasma membrane and mediates the uptake and transport of K^+ and Na^+ in poplar. Furthermore, HKT is also involved in xylem Na^+ unloading in tomato (*Solanum lycopersicum*) [48], as well as in the phloem-mediated recirculation of Na^+ from the shoots to the roots of rice (*Oryza sativa*) [49], avoiding the excessive accumulation of Na^+ in the leaves. The expression of *HKT1* in *P. euphratica* is three times higher after 1 h of a 1% NaCl treatment, which facilitates its rapid response to SS by taking a certain amount of Na^+ ions into the cells to maintain their osmotic balance [50].

Restricting the K^+ efflux is important for the salt resistance of *Populus*. Plasma membrane H^+-ATPase restricts K^+ efflux, thus improving salt resistance of *Populus*. K^+ can be transported into cells via an inward-rectifying K^+ channel and the high-affinity K^+ transporter HKT1, while K^+ efflux from the roots is mediated by the activation (by depolarization) of outward-rectifying K^+ channels (DA-KORCs) and NSCCs (DA-NSCCs), which can be induced by salt and inhibited by the plasma membrane H^+-ATPase [51,52]. The concentration of K^+ is markedly reduced in *P. euphratica* callus cells exposed to SS when pretreated with vanadate, an inhibitor of H^+-ATPase, because of the enhanced efflux of K^+ [53,54].

H^+-ATPase activity plays a large role in salt tolerance of *Populus*. In addition to restricting K^+ efflux, H^+-ATPases can also maintain a proton gradient across the membrane, used by the Na^+/H^+ antiporters [8,55,56], which is closely associated with the salt sensitivity of *Populus*. The activity of plasma membrane H^+-ATPase was higher in salt-tolerant genotypes than in salt-sensitive genotypes of *P. alba* [57]. Similarly, H^+-ATPase genes were more highly expressed in the salt-tolerant species *P. euphratica* than in the salt-sensitive *Populus trichocarpa* [55]. This is further supported by the work of Ma et al., who showed that the *P. euphratica* genome contains more copies of the P-type H^+-ATPase genes than *P. trichocarpa* [58].

Overall, except the above mechanisms, *P. euphratica* owns more effective mechanisms to respond to SS, for example, develop smaller vessel lumina than other salt sensitive poplars to limit ion loading into the xylem and develop leaf succulence after a long time of SS to dilute salt as a plastic morphological adaptation. Up to one-fifth of Na^+ and one-third of Cl^- are stored in foliage in the harvest season and are eliminated as leaves are ultimately shed. Excessive Na^+(Cl^-) can also be extruded via phloem retranslocating into the roots [59]. Sequestering Cl^- in cortical vacuoles at high salinity is also important for restricting its transport into above-ground organs [60].

3.2. Accumulation of Osmotic-Adjustment Substances

The water potential of the soil and the availability of water to plant roots are lower in saline soils; therefore, salt-stressed plants first experience water deficiencies caused by osmotic stress [16]. Stress caused by 150 mM NaCl caused a drop of -0.68 MPa in the osmotic potential and a rapid decrease of 0.77 MPa in the shoot water potential in young *P. euphratica* trees [16]. Plant cells tend to accumulate soluble osmolytes to adjust their osmotic potential, such as proline, glycine betaine, soluble sugars, and proteins, which enable the plants to alleviate the osmotic stress and maintain cell turgor, water uptake, and metabolic activity [8,61]. Both salt-sensitive and salt-tolerant poplar species showed an accumulation of free amino acids under long-term SS [62]. Proline is an important osmotic-adjustment substance that exists in a free state in plant cells, has a low molecular weight, is highly soluble in water, is relatively non-toxic, and has no net charge in the physiological pH range [63]. Proline accumulation preserves the osmotic balance under salinity stress, and proline content can be used as a physiological index of plant resistance to SS [8,64]. Salt-tolerant *P. euphratica* increased proline accumulation by 50–90% when exposed to 150–300 mM NaCl [24], while salt-sensitive hybrid poplars, such as *P. alba* cv. *Pyramidalis* × *P. tomentosa*, showed no significant accumulation when exposed to 50 and 150 mM NaCl [65]. Sucrose and total soluble sugars increased with the elevation of foliar Na^+ and Cl^- concentrations in *P. euphratica* [59]. Except for Valine (Val) and Isoleucine (Ile), soluble carbohydrates, sugar alcohols, organic acids, and amino acids in *P. euphratica* leaves did not show significant changes after 24 h of SS. However, the changes of these amino acids were too low to significantly affect the total osmotic potential of leaves [16]. As a "cheap" osmolyte, the accumulation of sodium mainly contributes to osmotic recovery in *P. euphratica* [16,66].

3.3. ROS and Reactive Nitrogen Species (RNS)

ROS, including hydrogen peroxide (H_2O_2), superoxide anions ($O_2\cdot^-$), hydroxyl radicals ($\cdot OH$), and singlet oxygen ($^1O_2$), accumulate when plants are exposed to high SS [67,68]. At moderate levels, functioning as signaling molecules, ROS trigger signal transduction events and elicit specific cellular responses thus regulating plant growth and stress responses [69]. Some ROS can react with almost all the components of living cells leading to severe damage to lipids, proteins, and nucleic acids [70]. Excessive ROS can induce oxidative damage and might be detoxified through enzymatic and non-enzymatic antioxidant systems. Poplars expressing *TaMnSOD* show greatly improved tolerance to NaCl, with higher superoxide dismutase (SOD) activities, lower malondialdehyde (MDA) contents, and lower relative electrical conductivity (REC) than the wild-type lines [71]. The peroxidase (POD) activity of *P. euphratica* was 61.8% higher under 200 mM NaCl stress relative to the control [26], while 3,3'-diaminobenzidine (DAB) staining and H_2O_2 measurement demonstrated a sharply increased level of H_2O_2 in Chinese white poplar *(Populus tomentosa)* exposed to 200 mM NaCl for 24 h [72]. Most of the genes encoding glutathione peroxidases (GSH-Px), glutathione S-transferases (GST), and glutaredoxins were more highly expressed in salt-stressed *P. tomentosa* than in the control [72], while the transcription levels of genes encoding antioxidant enzymes were upregulated [69] in plants exposed to 150 mM NaCl for 24 h [55]. Genes encoding enzymes involved in the glutathione metabolism pathway, including GSH-Px, glucose-6-phosphate Dehydrogenase (G6PD), glucose phosphate dehydrogenase (GPD), and isocitrate dehydrogenase (IDH), were significantly upregulated in plants treated with NaCl, which facilitated the detoxification of the salinity-induced ROS [15].

Non-enzymatic antioxidants include ascorbate (AsA), glutathione (GSH), and carotenoids (Car). Carotenoids not only protect against active oxygen species by quenching the excited states of photosensitizing molecules and singlet oxygen and by scavenging free radicals, but also protect biomembranes against oxidative damage by modifying the structural and dynamic properties of lipid membranes [73]. Recently, a carotenoid-deficient mutant of bacteria Pantoea sp. YR343 was found, showing reduced colonization on *Populus deltoids* roots [74].

In addition to the antioxidant enzymes, heat-shock transcription factors (HSFs) play a role in scavenging ROS in plants under SS. The transgenic expression of *PeHSF* in tobacco enhanced the activities of ascorbate peroxidase, GSH-Px, and glutathione reductase [75], and *PtHSP17.8* expression in *Arabidopsis* increased the activation levels of antioxidative enzymes under SS [76].

RNS includes nitric oxide ($NO\cdot$), nitric dioxide ($NO_2\cdot$), nitrous acid (HNO_2), and dinitrogen tetroxide (N_2O_4), which can be produced when plants are subjected to SS. Like ROS, RNS also function as signaling molecules in the response to abiotic stress. NO is involved in plant growth, development, senescence, as well as stress response [69]. NO was also reported to enhance salt tolerance in plants [77]. NO reacts with GSH, forming S-nitrosoglutathione (GSNO); NO, GSNO, and peroxynitrite (ONOO$^-$) can produce covalent post-translational modifications (PTMs), such as S-nitrosylation and the protein nitration [78]. However, ONOO$^-$, generated from nitric oxide NO and superoxide anion($O_2^{\cdot-}$), can produce tyrosine nitration of plant proteins and originate nitrosative damage in plant cells [69].

3.4. Poplar Salt Stress (SS) Signaling Pathways

Calcium ions (Ca^{2+}) are an important secondary messenger in plants and mediate poplar salt tolerance by enhancing Na^+ exclusion, restricting K^+ efflux, and sustaining the selectivity of the cell membrane [40]. In higher plants, the Ca^{2+}- dependent SOS signaling pathway helps maintain ion homoeostasis and thus confers salt tolerance under saline conditions [79,80]. Upon NaCl exposure, the resulting elevated cytosolic Ca^{2+} levels are sensed by SOS3, which activates SOS2 and stimulates the membrane-localized Na^+/H^+ antiporter SOS1, resulting in Na^+ efflux into the apoplast of the root [79,81]. Similarly, in *Populus*, although *SOS* gene expression is generally ubiquitous, some studies have indicated that SOS2 functions upstream of SOS1 and downstream of SOS3 [82].

In addition to extruding Na^+ from the roots, SOS1 controls the long-distance transport of Na^+ and affects its partitioning in plant organs [34,35]. *PeSOS1* (Salt overly sensitive 1 from *P. euphratica*) expression was upregulated 5- to 10-fold in *P. euphratica* leaves treated with 200 mM NaCl for 24 h relative to the untreated controls, and *PeSOS1* partially suppressed salt sensitivity when transgenically expressed in the *Escherichia coli* mutant strain EP432 [83]. Similarly, *PabSOS1* expression was about five times higher after 12 h of NaCl treatment [82]. SOS2 not only acts as a central regulator of Na^+ extrusion but also is involved in the signaling node between the SOS pathway and other signaling pathways [80].

Tang identified two *CBL10* homologs, *PtCBL10A* and *PtCBL10B*, in the western balsam poplar (*P. trichocarpa*) genome, which may interact with the salt tolerance component PtSOS2 and may help accumulate Na^+ in vacuoles [84]. Like PtSOS3, PtCBL10s also interacts with PtSOS2 to stimulate the activity of PtSOS1. Whereas PtCBL10s primarily functions in green tissues such as the shoots and targets the downstream component PtSOS2 to the tonoplast, PtSOS3 functions in the roots and targets PtSOS2 to the plasma membrane [84].

H^+-ATPases not only provide the proton-motive force used to enhance Na^+/H^+ antiporter activity, but also can restrict the NaCl-induced efflux of K^+ through DA-KORCs and DA-NSCCs (Figure 1). Genes encoding plasma membrane H^+-ATPases are upregulated in *P. euphratica* [37], likely enhancing the exchange of Na^+ and H^+ across the plasma membrane.

As a hinge signal molecule for sensing and responding to SS, H_2O_2 is vital for K^+/Na^+ homeostasis. In response to SS, the salt-resistant species *P. euphratica* rapidly produces H_2O_2 in a process triggered by proton-coupled ion transporters such as the H^+-pumps and the Na^+/H^+ antiporters in the plasma membrane [53]. This H_2O_2 accumulation causes a net Ca^{2+} influx by activating non-selective cation channels, which enhances Ca^{2+} concentration in the cytosol [54] and stimulates the SOS signaling pathway [81,85,86] (Figure 1). In addition, H_2O_2 signaling results in the upregulation of plasma membrane H^+-ATPases, whose activity limits the NaCl-induced efflux of K^+ [40,87]. Overall, H_2O_2 is involved in salt resistance in *P. euphratica* by controlling Na^+ extrusion via the H_2O_2–cytosolic [Ca^{2+}]–SOS pathway and by reducing K^+ efflux to maintain ion homeostasis via the H_2O_2–Ca^{2+}–PM H^+-ATPases pathway (Figure 1).

NADPH oxidases are the main source of H_2O_2. During SS, plasma membrane H^+-ATPases enhance H^+ efflux, decreasing the pH and contributing to the activation of NADPH oxidases, which leads to H_2O_2 production and triggers the Ca^{2+}-dependent SOS signaling pathway [53,54,87] (Figure 1). NaCl induces a transient increase in extracellular ATP (eATP), which is sensed by purinoceptors in the plasma membrane (e.g., P2K$_1$) and causes a rapid H_2O_2 burst that in turn increases the concentration of Ca^{2+} in the cytosol of *Populus* cells [36,88,89]. Consequently, the salt-elicited eATP-H_2O_2-cytosolic [Ca^{2+}] cascade contributes to enhancing Na^+ extrusion through the Ca^{2+}-dependent SOS pathways and to reducing K$^+$ efflux by activating H$^+$-ATPase, thus controlling cellular K$^+$/Na$^+$ homeostasis (Figure 1).

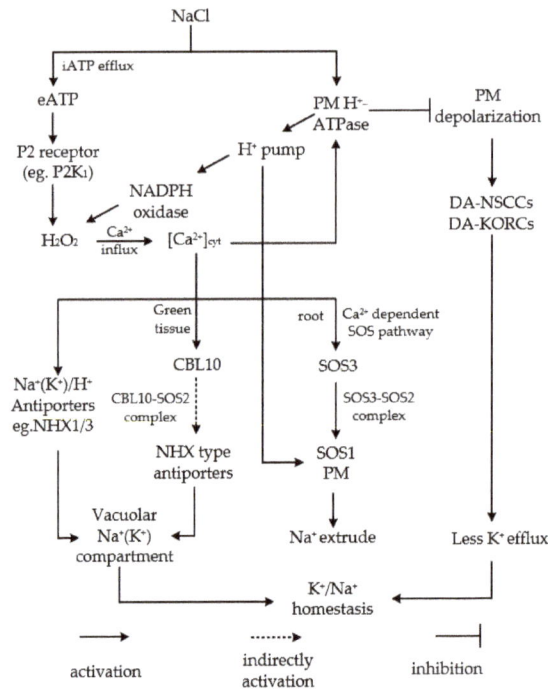

NaCl

iATP efflux

eATP

PM H$^+$-ATPase ──── PM depolarization

P2 receptor (eg. P2K$_1$)

H$^+$ pump

NADPH oxidase

DA-NSCCs
DA-KORCs

H$_2$O$_2$ $\xrightarrow[\text{influx}]{\text{Ca}^{2+}}$ [Ca^{2+}]$_{cyt}$

Green tissue

root | Ca^{2+} dependent SOS pathway

Na$^+$(K$^+$)/H$^+$ Antiporters eg.NHX1/3

CBL10

CBL10-SOS2 complex

SOS3

SOS3-SOS2 complex

NHX type antiporters

SOS1 PM

Vacuolar Na$^+$(K$^+$) compartment

Na$^+$ extrude

Less K$^+$ efflux

K$^+$/Na$^+$ homestasis

⟶ activation
·······▸ indirectly activation
⊣ inhibition

Figure 1. Schematic model showing multiple signaling networks active in *Populus* in response to NaCl stress. NaCl induces the efflux of intracellular ATP (iATP) and an increase in extracellular ATP (eATP), which is sensed by P2K$_1$ in the plasma membrane (PM) and leads to the induction of H$_2$O$_2$ production. This stimulates the movement of Ca^{2+} into the cells via Ca^{2+}-permeable channels. The elevated cytosolic Ca^{2+} concentration initiates the SOS pathway by stimulating Na$^+$/H$^+$ antiporters, such as SOS1, localized in the PM to extrude Na$^+$, or activates CBL10, forming the CBL10–SOS2 complex, which may indirectly target the NHX type antiporters to the tonoplast to compartmentalize Na$^+$ into vacuoles in green tissues. The elevated cytosolic Ca^{2+} also stimulates tonoplast-localized NHX1/3 to accumulate Na$^+$(K$^+$) into vacuoles. Besides, the elevated cytosolic Ca^{2+} increases H$^+$-ATPase activity in the PM, which activates a H$^+$ pump to supply a proton gradient for the Na$^+$/H$^+$ antiporters, stimulating the extrusion of Na$^+$. A proton gradient supplied by the H$^+$ pump contributes to the activation of NADPH oxidases, which leads to H$_2$O$_2$ production. H$^+$-ATPases can also inhibit the efflux of K$^+$ by further polarizing the PM. All these signaling components help to maintain K$^+$/Na$^+$ homeostasis in *Populus* cells.

NaCl-induced expression of *HSF* (Heat shock transcription factor) in *P. euphratica* is markedly restricted by inhibitors of NADPH oxidase and Ca^{2+}-permeable channels, suggesting that salt-induced H$_2$O$_2$ and cytosolic Ca^{2+} enhance the transcription of *HSFs*, which in turn upregulate genes encoding antioxidant enzymes for scavenging ROS under saline conditions [75].

In conclusion, eATP, Ca^{2+}, H$_2$O$_2$, NADPH, H$^+$-ATPase, and Na$^+$ (K$^+$)/H$^+$ transporters play important roles in mediating salt tolerance in *Populus* trees.

4. Candidate Genes Used for the Genetic Improvement of Salt Tolerance

Currently, many studies are focused on introducing known salt-response signaling genes into *Populus* and testing the performance of the transgenic plants in high-salinity conditions. Many of these genes confer a significantly improved salt tolerance, as described below. Transferring transcription factors (TFs) genes into poplars is generally a more efficient approach than transferring structural genes, because transcription factors usually regulate the expression of many target genes in related pathways. A total of 59 *ERF* (Ethylene response factor) genes are associated with SS in *Populus* [90]. *ERF76* from dihaploid *P. simonii* × *P. nigra* plants was transferred into the same *Populus* clone and significantly upregulated 16 genes encoding other transcription factors, as well as 45 stress-related genes [91]. When exposed to SS, *ERF76*-expressing transgenic plants were significantly taller and had increased root lengths, fresh weights, abscisic acid (ABA) and gibberellic acid (GA) contents compared to the control. Transgenic *ERF76* expression enhanced salt tolerance by upregulating the expression of stress-related genes and increasing ABA and GA biosynthesis [91]. The *PsnERF75* gene from *P. simonii* × *P. nigra* is induced by salt, drought, and ABA treatments [92] and confers salt tolerance when transgenically expressed in *Arabidopsis* [91].

The DREB (for dehydration-responsive element-binding protein) transcription factors, members of the ERF family, are vital regulatory nodes in the signaling pathways involved in the salt-stress response [93]. *PeDREB2a*, encoding a DREB transcription factor in *P. euphratica*, improved salt tolerance in *Arabidopsis* or birdsfoot trefoil (*Lotus corniculatus*) when transgenically expressed under the stress-inducible *rd29A* promoter [94]. The transgenic expression of *LbDREB* (a *DREB* gene from the halophyte *Limonium bicolor*) in *Populus ussuriensis* enhanced its resistance to salt, increasing its SOD and POD activities and the expression of the genes encoding these enzymes, reducing its MDA content, and enhancing its proline accumulation in the leaves [93]. The transgenic *P. ussuriensis* plants also had higher root/shoot ratios, higher relative water contents (RWC), and lower relative electrolytic leakage. Consistent with these changes, the genes encoding NAM (no apical meristem), GT-1(trihelix transcription factor), and WRKY70 (WRKY transcription factor 70) displayed inducible temporal expression patterns and are important components in the SS response signaling networks [93]. The LbDREB protein may inhibit the expression of *NAM*, *GT-1*, and *WRKY70* and induce the expression of *SOD* and *POD* in response to high salinity stress, but this requires further verification.

The GTPase RabE is located in the Golgi apparatus and the plasma membrane, where it plays an important role in vesicle transport [95]. The overexpression of constitutively active *PtRabE1b* conferred salt tolerance in poplar [96]. This gene is directly co-expressed with many genes involved in salt tolerance, such as *HSFA4a*, *SOS2*, *MPK19*, and the Ca^{2+} signaling-related genes *CAM7*, *CKL6*, and calcium exchanger. *HSFA4a* expression is regulated by oxidative stress and MPK3/MPK6 and positively influenced salt tolerance in *Arabidopsis* [97]. *CmHSFA4*, a *Chrysanthemum* homologue of this gene, positively regulates salt tolerance by regulating the activities of SOS1, HKT2, and the ROS scavengers [98].

Recently, Yoon et al. identified a novel gene, *PagSAP1*(stress-associated proteins), from the hybrid poplar *P. alba* × *Populus glandulosa*. *PagSAP1* negatively mediates salt-stress responses, and SS can in turn suppress the expression of this gene in poplar roots [99]. *PagSAP1* overexpression resulted in enhanced sensitivity to SS, while *PagSAP1* silencing via RNA interference (RNAi) significantly increased cytosolic Ca^{2+} in the roots. This increased cytosolic Ca^{2+} activated SOS signal transduction, resulting in high *SOS3* transcript levels in the RNAi-derived plants. *HKT1* expression is significantly reduced in all poplar genotypes under salt treatment; however, the lowest level is observed in the *PagSAP1*-overexpressing lines. HKT1 is responsible for Na^+ influx and xylem-mediated Na^+ recirculation from the shoot to the root [100]; therefore, the low HKT1 activity levels in the *PagSAP1*-overexpressing lines may explain the higher Na^+ accumulation in the leaves and the lower Na^+ levels in the roots compared with the control and RNAi-derived lines. By contrast, the excess Na^+ in the roots of the *PagSAP1*-RNAi lines was eliminated by increased SOS1 activity, which resulted in lower Na^+ levels in both the roots and the leaves of these lines. As a result, the salt tolerance of the *PagSAP1*-RNAi lines was improved through the upregulation of *SOS3*, *SOS1*, *HKT1*, *H⁺-ATPase*, *AAA-type ATPase*, and Arabidopsis K+ channel 2 (*AKT2*), all of which are essential for maintaining Na$^+$/K$^+$ homeostasis [99].

The poplars SOS proteins share high functional conservation with their *Arabidopsis* homologues [82]. SOS2 interacts with or regulates the activity of several tonoplast-localized transporters, such as the Ca^{2+}/H$^+$ antiporter [101], the vacuolar H$^+$- ATPase [102], and the Na$^+$/H$^+$ exchanger [103,104]. *PtSOS2.1*, *PtSOS2.2*, and *PtSOS2.3* (the *PtSOS2* genes in *P. trichocarpa*) overexpression improves the salt tolerance of poplars and increases the concentrations of proline and photosynthetic pigments, relative water content, and the activity of their antioxidant enzymes, while significantly decreasing the levels of MDA [105].

The mutant SOS2 protein PtSOS2TD, generated by mutating the 169th amino acid in the activation loop of PtSOS2 from threonine (T) to aspartic acid (D), is more active than PtSOS2 and can sufficiently activate PtSOS1 in a PtSOS3-independent manner [80]. *PtSOS2TD* overexpression in poplars significantly increased salt tolerance, causing higher plasma membrane Na$^+$/H$^+$ exchange activity, greater Na$^+$ efflux, decreased Na$^+$ accumulation in the leaves, and improved ROS scavenging capacity [80].

The transgenic expression of the *PtCBL10s* (Calcineurin B-like from *P. trichocarpa*) conferred greater salt tolerance to poplars by maintaining shoot ion homeostasis under SS. The doubling of the *CBL10* genes in poplar may represent an evolutionary adaptation to the adverse environment [84].

The genes that have been shown to increase the salt tolerance of transgenic *Populus* are presented in Table 1.

Table 1. Candidate genes for improving salt tolerance in *Populus*.

Candidate Genes and Source	Transgenic Species	Effect of SS in Transgenic Species Compared with WT	Reference
AtNHX1/3 (Vacuolar Na$^+$/H$^+$ antiporter from *Arabidopsis thaliana*)	*Populus davidiana* × *Populus bolleana*	1 Normal growth and morphology; 2 Promoted vacuolar Na$^+$ (K$^+$)/H$^+$ exchange activity; 3 Increased Na$^+$ and K$^+$ accumulation in the vacuoles; 4 Elevated the eaccumulation of proline.	[47]
AtNHX1 (See above)	*Populus* × *euramericana* 'Neva'	1 Enhanced plant growth and photosynthetic capacity; 2 Lowerd MDA and REC; 3 Increased Na$^+$ (K$^+$) accumulation in roots and leaves.	[44]
	Populus × *euramericana* 'Neva'	1 Reduced decrease in Chl, Car, PSII, Fv/Fm, and qP; 2 Smaller reduction of Pn, Gs, Ci, CE; 3 Greater increase of stem and leaf, smaller increase in root.	[43]
	Populus deltoides × *P. euramericana* CL 'NL895'	1 Higher content of sodium ions; 2 Decreased MDA content.	[45]
PtSOS2TD (Salt overly sensitive 2 from *Populus trichocarpa*)	*P. davidiana* × *P. bolleana* hybrid poplar clone Shanxin	1 More vigorous growth; 2 Greater biomass produced; 3 Less Na$^+$ in the leaves; 4 Higher Na$^+$/H$^+$ exchange activity and Na$^+$ efflux; 5 More scavenging of ROS.	[80]
PtSOS2 (See above)	*Populus tremula* × *Populus tremuloides* Michx clone T89	1 Improved PM Na$^+$/H$^+$ exchange activity, Na$^+$ efflux; 2 Higher proline activity; 3 Higher RWC and sustained decrease of water loss; 4 Increased SOD, POD, CAT activity; 5 Decreased MDA concentration.	[105]
PtCBL10A and *PtCBL10B* (Calcineurin B-like from *P. trichocarpa*)	*P. davidiana* × *P. bolleana* hybrid poplar clone Shanxin	1 Less impairment by SS with higher stature and greater shoot biomass; 2 Lower Na$^+$ in the leaves, more Na$^+$ in the stem.	[84]
PeCBL6, *PeCBL10* (Calcineurin B-like from *P. euphratica*)	triploid white poplar	1 Higher height growth rate; 2 Less wilted leaves; 3 Lower MDA content; 4 Higher chl content.	[106]
PtSOS3 (Salt overly sensitive 3 from *P. trichocarpa*)	*P. davidiana* × *P. bolleana* hybrid poplar clone Shanxin	1 Lower Na$^+$ in the root; 2 Higher K$^+$ content in the root; 3 More Na$^+$ in the stem.	[84]
TaMnSOD (Mn-superoxide dismutases from *Tamarix Androssowii*)	*P. davidiana* × *P. bolleana* hybrid poplar clone Shanxin	1 Higher SOD activity; 2 Lower MDA contents; 3 Lower REC; 4 More weight gains.	[71]

Table 1. *Cont.*

Candidate Genes and Source	Transgenic Species	Effect of SS in Transgenic Species Compared with WT	Reference
TaLEA (Late embryogenesis abundant from *T. androssowii*)	*Populus simonii* × *Populus nigra* Xiaohei poplar	1 Decrease in MDA content; 2 Decrease in relative electrolyte leakage; 3 Improved salt and drought resistance.	[107]
	P. davidiana × *P. bolleana*	1 Higher Survival percentages; 2 Higher Seedling height and photosynthetic capabilities; 3 Lower Na+ in young leaves but higher in yellow and withered leaves.	[108]
ERF76 (Ethylene response factor from di-haploid *P. simonii* × *P. nigra*)	*P. simonii* × *P. nigra* di-haploid	1 Higher plant height, root length, fresh weight; 2 Higher in ABA and GA concentration.	[91]
JERFs (Jasmonic ethylene responsive factor from the tomato)	*Populus alba* × *Populus berolinensis*	1 Lower reductions of height, basal diameter, and biomass; 2 Lower reduction in leaf water content and increase in root/crown ratio; 3 Greater increase of foliar proline concentration; 4 Higher foliar Na+ concentration.	[109]
LbDREB (dehydration responsive element binding TF from *Limonium bicolor*)	*Populus ussuriensis* Kom. Chinese Daqing poplar	1 Higher SOD, POD activity; 2 Less MDA accumulation in the leaves; 3 More proline accumulation; 4 Increased root/shoot ratio; 5 Reduced decrease of RWC; 6 Lower increase of relative electrolytic leakage.	[93]
AhDREB1 (dehydration responsive element binding-like TF from the halophyte *Atriplex hortensis*)	*Populus tomentosa*	1 Higher survival rate; 2 High proline content.	[110]
AtSTO1 (Salt tolerant1 from *Arabidopsis thaliana*)	*P. tremula* × *P. alba* Poplar 717-1B4	1 Higher aboveground biomass; 2 Higher root biomass; 3 Higher shoot height; 4 Higher chl content.	[111]
AtPLDα (Phospholipase Dα from *A. thaliana*)	*P. tomentosa*	1 Higher root rate and root length; 2 Reduced decrease of total chl content; 3 Lower REC and MDA content; 4 Higher SOD, POD, and CAT activities.	[112]
AtSRK2C, AtGolS2 (Stress responses, SNF1-related protein kinase 2C, galactinol synthase 2 from *A. thaliana*)	*P. tremula* × *tremuloides*	1 Reduced decrease of dry weight; 2 Reduced decrease of total adventitious root length.	[113]
PtRabE1b(Q74L) (Rab GTPase from *P. trichocarpa*)	*P. alba* × *P. glandulosa* clone 84 K	1 More adventitious roots; 2 Greater root growth status in seedlings.	[96]
PagSAP1 (stress-associated proteins from *P. alba* × *P. glandulosa*)	*P. alba* × *P. glandulosa*	RNAi plants accumulate more Ca2+, and K+ and less Na+.	[99]

SS, salt stress; PM, plasma membrane; SOD, superoxide dismutase; POD, peroxidase; CAT, catalase; MDA, malondialdehyde; Chl, chlorophyll; Car, carotenoid; PSII, actual quantum yield of PSII; Fv/Fm, maximum photochemical efficiency; qP, photochemical quenching coefficient; Pn, net photosynthetic rate; Gs, stomatal conductance; Ci, internal CO_2; CE, carboxylation efficiency concentration; ROS, reactive oxygen species; REC, relative electrical conductivity; RWC, relative water content; ABA, abscisic acid; GA, gibberellic acid.

5. Conclusions and Outlook

Soil salinization is increasingly problematic and is now a dominant factor limiting *Populus* growth [7]. Therefore, it is important to improve the salt tolerance of poplar trees. Plant salt tolerance is a typical quantitative trait affected by many physiological and biochemical factors [15]. Different *Populus* species, such as the salt-resistant poplar species *P. euphratica* and the salt-sensitive *Populus* species *P.* × *canescens*, have different SS responses [114,115]. Furthermore, trees such as the hybrid poplar 107 display different responses when exposed to different salinity levels [15]. Overall, *Populus* adapt to SS by maintaining suitable Na+/K+ ratios, accumulating osmotic-adjustment substances, activating antioxidative enzymes and antioxidants, and activating stress response signaling networks to reduce the negative effects of high salinity. With the innovation of transcriptomics technologies, a substantial number of stress-responsive and/or stress-regulated genes have been identified—in addition to the signal regulatory networks in which they function—and transferred between *Populus* and *Arabidopsis* or other species with great success. This has greatly elucidated the molecular mechanisms of the poplar stress responses [113].

ROS/RNS and hormones were identified as signaling molecules involved in the response to SS avoiding high salinity damage. The crosstalk between ROS, RNS, ABA, ethylene, and/or other hormones in poplar salt stress will be further studied. Molecular chaperones, especially dehydrins and osmotin, which contribute to protect proteins, are supposed key factors for coping with SS [16]. Nutrient fertilization with N and P was reported to reduce the accumulation of ROS (e.g., O_3) and enhance membrane stability, thus protecting from oxidative stress by activating a cross-talk between antioxidant and osmotic mechanisms [116]. The role of mycorrhization and polymer amendment in enhancing mineral nutrition and improving salt tolerance is also a topic of future study. All of the above topics need to be more deeply studied in the future to improve salt tolerance in poplar species as well as in other tree species.

Determining the key transcription factors and molecular mechanisms underlying salt tolerance is an important goal for future research and will facilitate the enhancement of salt tolerance in *Populus*.

Author Contributions: All authors worked on manuscript preparation.

Funding: This work was supported by the GuangXi Natural Science Foundation (2016GXNSFBA380224); the Department of Human Resources and Social Security of Guangxi Zhuang Autonomous Region, China (GuiCaiSheHan [2018]112), and the Fundamental Research Funds for Guangxi Forestry Research Institute No. LK201812).

Acknowledgments: We are grateful to Li Liu in KIB for the kind discussion and guidance to the project.

Conflicts of Interest: The authors declare that they have no conflict of interest.

Abbreviations

SOS	salt overly sensitive
ROS	reactive oxygen species
SOD	superoxide dismutase
POD	peroxidase
MDA	malondialdehyde
GSH-Px	glutathione peroxidases
GST	glutathione S-transferases
GR	glutathione reductase
GPX	glutathione peroxidases
GPD	glucose phosphate dehydrogenase
G6PD	glucose-6-phosphate dehydrogenase
APX	ascorbate peroxidase
IDH	Isocitrate dehydrogenase
GA	gibberellic acid
ABA	abscisic acid
REC	relative electrical conductivity
RWC	relative water content
TFs	transcription factors
Chl	chlorophyll
Pn	net photosynthetic rate
Gs	stomatal conductance
Ci	internal CO_2
CE	carboxylation efficiency concentration;
Car	carotenoid
PSII	actual quantum yield of PSII
qP	photochemical quenching coefficient
NPQ	non photochemical quenching
Fv/Fm	maximum photochemical efficiency
eATP	extracellular ATP
iATP	intracellular ATP
DAB	3,3'-diaminobenzidine
PM	plasma membrane

D A	depolarization activated
KORCs	K$^+$ outward rectifying channels
NSCCs	non-selective cation channels
SS	salt stress

References

1. Wu, Z.Y.; Raven, P.H. (Eds.) *Flora of China*; Beijing: Science Press: Beijing, China; Missouri Botanical Garden Press: St. Louis, MO, USA, 1999; Volume 4, pp. 139–162.
2. Sixto, H. Response to sodium chloride in different species and clones of genus *Populus* L. *Forestry* **2005**, *78*, 93–104. [CrossRef]
3. Jansson, S.; Douglas, C.J. *Populus*: A model system for plant biology. *Annu. Rev. Plant Biol.* **2007**, *58*, 435–458. [CrossRef]
4. Bradshaw, H.D.; Ceulemans, R.; Davis, J.; Stettler, R. Emerging Model Systems in Plant Biology: Poplar (*Populus*) as A Model Forest Tree. *J. Plant Growth Regul.* **2000**, *19*, 306–313. [CrossRef]
5. Polle, A.; Douglas, C. The molecular physiology of poplars: Paving the way for knowledge-based biomass production. *Plant Biol.* **2010**, *12*, 239–241. [CrossRef]
6. Taylor, G. *Populus*: Arabidopsis for Forestry. Do We Need a Model Tree? *Ann. Bot.* **2002**, *90*, 681–689. [CrossRef] [PubMed]
7. Wang, R.G.; Chen, S.L.; Deng, L.; Fritz, E.; Hüttermann, A.; Polle, A. Leaf photosynthesis, fluorescence response to salinity and the relevance to chloroplast salt compartmentation and anti-oxidative stress in two poplars. *Trees* **2007**, *21*, 581–591. [CrossRef]
8. Munns, R.; Tester, M. Mechanisms of salinity tolerance. *Annu. Rev. Plant Biol.* **2008**, *59*, 651–681. [CrossRef]
9. Hirayama, T.; Shinozaki, K. Research on plant abiotic stress responses in the post-genome era: Past, present and future. *Plant J.* **2010**, *61*, 1041–1052. [CrossRef]
10. Chen, S.L.; Hawighorst, P.; Sun, J.; Polle, A. Salt tolerance in *Populus*: Significance of stress signaling networks, mycorrhization, and soil amendments for cellular and whole-plant nutrition. *Environ. Exp. Bot.* **2014**, *107*, 113–124. [CrossRef]
11. Gu, R.S.; Fonseca, S.; PuskÁs, L.G.; Hackler, L.J.; Zvara, Á.; Dudits, D.; Pais, M.S. Transcript identification and profiling during salt stress and recovery of *Populus euphratica*. *Tree Physiol.* **2004**, *24*, 275–276. [CrossRef]
12. Wang, S.; Chen, B.; Li, H. *Euphrates Poplar Forest*; China Environmental Science Press: Beijing, China, 1996; pp. 43–52.
13. Fu, A.H.; Li, W.H.; Chen, Y.N. The threshold of soil moisture and salinity influencing the growth of *Populus euphratica* and *Tamarix ramosissima* in the extremely arid region. *Environ. Earth Sci.* **2012**, *66*, 2519–2529. [CrossRef]
14. Qiu, Q.; Ma, T.; Hu, Q.J.; Liu, B.B.; Wu, Y.X.; Zhou, H.H.; Wang, Q.; Wang, J.; Liu, J.Q. Genome-scale transcriptome analysis of the desert poplar, *Populus euphratica*. *Tree Physiol.* **2011**, *31*, 452–461. [CrossRef] [PubMed]
15. Chen, P.F.; Zuo, L.H.; Yu, X.Y.; Dong, Y.; Zhang, S.; Yang, M.S. Response mechanism in *Populus × euramericana* cv. '74/76' revealed by RNA-seq under salt stress. *Acta Physiol. Plant* **2018**, *40*. [CrossRef]
16. Brinker, M.; Brosche, M.; Vinocur, B.; Abo-Ogiala, A.; Fayyaz, P.; Janz, D.; Ottow, E.A.; Cullmann, A.D.; Saborowski, J.; Kangasjarvi, J.; et al. Linking the salt transcriptome with physiological responses of a salt-resistant *Populus* species as a strategy to identify genes important for stress acclimation. *Plant Physiol.* **2010**, *154*, 1697–1709. [CrossRef] [PubMed]
17. Yer, E.N.; Baloglu, M.C.; Ayan, S. Identification and expression profiling of all Hsp family member genes under salinity stress in different poplar clones. *Gene* **2018**, *678*, 324–336. [CrossRef] [PubMed]
18. Zhao, K.; Li, S.X.; Yao, W.J.; Zhou, B.R.; Li, R.H.; Jiang, T.B. Characterization of the basic helix-loop-helix gene family and its tissue-differential expression in response to salt stress in poplar. *PeerJ* **2018**, *6*, e4502. [CrossRef] [PubMed]
19. Zhao, K.; Zhang, X.M.; Cheng, Z.H.; Yao, W.J.; Li, R.H.; Jiang, T.B.; Zhou, B.R. Comprehensive analysis of the three-amino-acid-loop-extension gene family and its tissue-differential expression in response to salt stress in poplar. *Plant Physiol. Biochem.* **2019**, *136*, 1–12. [CrossRef]

20. Zhang, Y.N.; Wang, Y.; Sa, G.; Zhang, Y.H.; Deng, J.Y.; Deng, S.R.; Wang, M.J.; Zhang, H.L.; Yao, J.; Ma, X.Y.; et al. *Populus euphratica* J3 mediates root K$^+$/Na$^+$ homeostasis by activating plasma membrane H$^+$-ATPase in transgenic Arabidopsis under NaCl salinity. *Plant Cell Tiss Organ Cult.* **2017**, *131*, 75–88. [CrossRef]

21. Liu, J.P.; Li, Z.J.; He, L.R.; Zhou, Z.L.; Xu, Y.L. Salt-tolerance of *Populus euphratica* and *P. pruinosa* seed during germination. *Sci. Silvae Sin.* **2004**, *40*, 165–169. [CrossRef]

22. Abassi, M.; Mguis, K.; Béjaoui, Z.; Albouchi, A. Morphogenetic responses of *Populus alba* L. under salt stress. *J. For. Res.-JPN* **2014**, *25*, 155–161. [CrossRef]

23. Meilan, R.; Sabatti, M.; Ma, C.P.; Elena, K. An Early-Flowering Genotype of *Populus*. *J. Plant Biol.* **2004**, *47*, 52–56. [CrossRef]

24. Zhao, C.Y.; Si, J.H.; Feng, Q.; Deo, R.C.; Yu, T.F.; Li, P.D. Physiological response to salinity stress and tolerance mechanics of *Populus euphratica*. *Environ. Monit. Assess.* **2017**, *189*, 533. [CrossRef] [PubMed]

25. Abbruzzese, G.; Beritognolo, L.; Muleo, R.; Piazzai, M.; Sabatti, M.; Mugnozza, G.S.; Kuzminsky, E. Leaf morphological plasticity and stomatal conductance in three *Populus alba* L. genotypes subjected to salt stress. *Environ. Exp. Bot.* **2009**, *66*, 381–388. [CrossRef]

26. Rajput, V.D.; Chen, Y.; Ayup, M. Effects of high salinity on physiological and anatomical indices in the early stages of *Populus euphratica* growth. *Russ. J. Plant Physiol.* **2015**, *62*, 229–236. [CrossRef]

27. Rajput, V.D.; Chen, Y.N.; Ayup, M.; Minkina, T.; Sushkova, S.; Mandzhieva, S. Physiological and hydrological changes in *Populus euphratica* seedlings under salinity stress. *Acta Ecol. Sin.* **2017**, *37*, 229–235. [CrossRef]

28. Awad, H.; Barigah, T.; Badel, E.; Cochard, H.; Herbette, S. Poplar vulnerability to xylem cavitation acclimates to drier soil conditions. *Physiol. Plant.* **2010**, *139*, 280–288. [CrossRef] [PubMed]

29. Junghans, U.; Polle, A.; Duchting, P.; Weiler, E.; Kuhlman, B.; Gruber, F.; Teichmann, T. Adaptation to high salinity in poplar involves changes in xylem anatomy and auxin physiology. *Plant Cell Environ.* **2006**, *29*, 1519–1531. [CrossRef] [PubMed]

30. Strassert, R.J.; Srivastava, A. Polyphasic chlorophyll a fluorescence transient in plants and cyanobacteria. *Photochem. Photobiol.* **1995**, *61*, 32–42. [CrossRef]

31. Sixto, H.; Aranda, I.; Grau, J.M. Assessment of salt tolerance in *Populus alba* clones using chlorophyll fluorescence. *Photosynthetica* **2006**, *44*, 169–173. [CrossRef]

32. Wu, H.H. Plant salt tolerance and Na$^+$ sensing and transport. *Crop J.* **2018**, *6*, 215–225. [CrossRef]

33. Ward, J.M.; Hirschi, K.D.; Sze, H. Plants pass the salt. *Trends Plant Sci.* **2003**, *8*, 200–201. [CrossRef]

34. Olias, R.; Eljakaoui, Z.; Li, J.; de Morales, P.A.; Marin-Manzano, M.C.; Pardo, J.M.; Belver, A. The plasma membrane Na$^+$/H$^+$ antiporter SOS1 is essential for salt tolerance in tomato and affects the partitioning of Na$^+$ between plant organs. *Plant Cell Environ.* **2009**, *32*, 904–916. [CrossRef] [PubMed]

35. Shi, H.Z.; Quintero, F.J.; Pardo, J.M.; Zhu, J.K. The Putative Plasma Membrane Na$^+$/H$^+$ Antiporter SOS1 Controls Long-Distance Na$^+$ Transport in Plants. *Plant Cell Online* **2002**, *14*, 465–477. [CrossRef]

36. Zhao, N.; Wang, S.J.; Ma, X.J.; Zhu, H.P.; Sa, G.; Sun, J.; Li, N.F.; Zhao, C.J.; Zhao, R.; Chen, S.L. Extracellular ATP mediates cellular K$^+$/Na$^+$ homeostasis in two contrasting poplar species under NaCl stress. *Trees* **2015**, *30*, 825–837. [CrossRef]

37. Ma, X.Y.; Deng, L.; Li, J.K.; Zhou, X.Y.; Li, N.Y.; Zhang, D.C.; Lu, Y.J.; Wang, R.G.; Sun, J.; Lu, C.F.; et al. Effect of NaCl on leaf H$^+$-ATPase and the relevance to salt tolerance in two contrasting poplar species. *Trees* **2010**, *24*, 597–607. [CrossRef]

38. Li, J.; Bao, S.Q.; Zhang, Y.H.; Ma, X.J.; Mishra-Knyrim, M.; Sun, J.; Sa, G.; Shen, X.; Polle, A.; Chen, S.L. *Paxillus involutus* strains MAJ and NAU mediate K$^+$/Na$^+$ homeostasis in ectomycorrhizal *Populus* × *canescens* under sodium chloride stress. *Plant Physiol.* **2012**, *159*, 1771–1786. [CrossRef] [PubMed]

39. Chen, S.L.; Li, J.K.; Wang, S.S.; Fritz, E.; Hüttermann, A.; Altman, A. Effects of NaCl on shoot growth, transpiration, ion compartmentation, and transport in regenerated plants of *Populus euphratica* and *Populus tomentosa*. *Can. J. For. Res.* **2003**, *33*, 967–975. [CrossRef]

40. Sun, J.; Dai, S.X.; Wang, R.G.; Chen, S.L.; Li, N.Y.; Zhou, X.Y.; Lu, C.F.; Shen, X.; Zheng, X.J.; Hu, Z.M.; et al. Calcium mediates root K$^+$/Na$^+$ homeostasis in poplar species differing in salt tolerance. *Tree Physiol.* **2009**, *29*, 1175–1186. [CrossRef]

41. Mansour, M.M.F.; Salama, K.H.A.; Al Mutawa, M.M. Transport proteins and salt tolerance in plants. *Plant Sci.* **2003**, *164*, 891–900. [CrossRef]

42. Silva, P.; Façanha, A.R.; Tavares, R.M.; Gerós, H. Role of Tonoplast Proton Pumps and Na$^+$/H$^+$ Antiport System in Salt Tolerance of *Populus euphratica* Oliv. *J. Plant Growth Regul.* **2010**, *29*, 23–34. [CrossRef]

43. Jiang, C.Q.; Zheng, Q.S.; Liu, Z.P.; Liu, L.; Zhao, G.M.; Long, X.H.; Li, H.Y. Seawater-irrigation effects on growth, ion concentration, and photosynthesis of transgenic poplar overexpressing the Na$^+$/H$^+$ antiporter AtNHX1. *J. Plant Nutr. Soil Sci.* **2011**, *174*, 301–310. [CrossRef]

44. Jiang, C.Q.; Zheng, Q.S.; Liu, Z.P.; Xu, W.J.; Liu, L.; Zhao, G.M.; Long, X.H. Overexpression of *Arabidopsis thaliana* Na$^+$/H$^+$ antiporter gene enhanced salt resistance in transgenic poplar (*Populus* × *euramericana* 'Neva'). *Trees* **2012**, *26*, 685–694. [CrossRef]

45. Qiao, G.R.; Zhuo, R.Y.; Liu, M.Y.; Jiang, J.; Li, H.Y.; Qiu, W.M.; Pan, L.Y.; lin, S.; Zhang, X.G.; Sun, Z.X. Over-expression of the *Arabidopsis* Na$^+$/H$^+$ antiporter gene in *Populus deltoides* CL × *P. euramericana* CL "NL895" enhances its salt tolerance. *Acta Physiol. Plant* **2011**, *33*, 691–696. [CrossRef]

46. Liu, H.; Tang, R.J.; Zhang, Y.; Wang, C.T.; Lv, Q.D.; Gao, X.S.; Li, W.B.; Zhang, H.X. AtNHX3 is a vacuolar K$^+$/H$^+$ antiporter required for low-potassium tolerance in *Arabidopsis thaliana*. *Plant Cell Environ.* **2010**, *33*, 1989–1999. [CrossRef]

47. Yang, L.; Liu, H.; Fu, S.M.; Ge, H.M.; Tang, R.J.; Yang, Y.; Wang, H.H.; Zhang, H.X. Na$^+$/H$^+$ and K$^+$/H$^+$ antiporters AtNHX1 and AtNHX3 from *Arabidopsis* improve salt and drought tolerance in transgenic poplar. *Biol. Plant.* **2017**. [CrossRef]

48. Jaime-Perez, N.; Pineda, B.; Garcia-Sogo, B.; Atares, A.; Athman, A.; Byrt, C.S.; Olias, R.; Asins, M.J.; Gilliham, M.; Moreno, V.; et al. The sodium transporter encoded by the HKT1;2 gene modulates sodium/potassium homeostasis in tomato shoots under salinity. *Plant Cell Environ.* **2017**, *40*, 658–671. [CrossRef]

49. Kobayashi, N.I.; Yamaji, N.; Yamamoto, H.; Okubo, K.; Ueno, H.; Costa, A.; Tanoi, K.; Matsumura, H.; Fujii-Kashino, M.; Horiuchi, T.; et al. OsHKT1;5 mediates Na$^+$ exclusion in the vasculature to protect leaf blades and reproductive tissues from salt toxicity in rice. *Plant J.* **2017**, *91*, 657–670. [CrossRef]

50. Xu, M.; Sun, Z.M.; Liu, S.A.; Chen, C.H.; Xu, L.A.; Huang, M.R. Cloning and Expression Analysis of Salinity Stress Related Peu HKT1 Gene from *Populus euphratica*. *MPB* **2016**, in press.

51. Britto, D.T.; Kronzucker, H.J. Cellular mechanisms of potassium transport in plants. *Physiol. Plant.* **2008**, *133*, 637–650. [CrossRef]

52. Shabala, L.; Zhang, J.; Pottosin, I.; Bose, J.; Zhu, M.; Fuglsang, A.T.; Velarde-Buendia, A.; Massart, A.; Hill, C.B.; Roessner, U.; et al. Cell-Type-Specific H$^+$-ATPase Activity in Root Tissues Enables K$^+$ Retention and Mediates Acclimation of Barley (*Hordeum vulgare*) to Salinity Stress. *Plant Physiol.* **2016**, *172*, 2445–2458. [CrossRef]

53. Sun, J.; Li, L.S.; Liu, M.Q.; Wang, M.J.; Ding, M.Q.; Deng, S.R.; Lu, C.F.; Zhou, X.Y.; Shen, X.; Zheng, X.J.; et al. Hydrogen peroxide and nitric oxide mediate K$^+$/Na$^+$ homeostasis and antioxidant defense in NaCl-stressed callus cells of two contrasting poplars. *Plant Cell Tiss Organ Cult.* **2010**, *103*, 205–215. [CrossRef]

54. Sun, J.; Wang, M.J.; Ding, M.Q.; Deng, S.R.; Liu, M.Q.; Lu, C.F.; Zhou, X.Y.; Shen, X.; Zheng, X.J.; Zhang, Z.K.; et al. H$_2$O$_2$ and cytosolic Ca^{2+} signals triggered by the PM H$^+$-coupled transport system mediate K$^+$/Na$^+$ homeostasis in NaCl-stressed *Populus euphratica* cells. *Plant Cell Environ.* **2010**, *33*, 943–958. [CrossRef] [PubMed]

55. Ding, M.Q.; Hou, P.C.; Shen, X.; Wang, M.J.; Deng, S.R.; Sun, J.; Xiao, F.; Wang, R.G.; Zhou, X.Y.; Lu, C.F.; et al. Salt-induced expression of genes related to Na$^+$/K$^+$ and ROS homeostasis in leaves of salt-resistant and salt-sensitive poplar species. *Plant Mol. Biol.* **2010**, *73*, 251–269. [CrossRef] [PubMed]

56. Polle, A.; Chen, S. On the salty side of life: Molecular, physiological and anatomical adaptation and acclimation of trees to extreme habitats. *Plant Cell Environ.* **2015**, *38*, 1794–1816. [CrossRef] [PubMed]

57. Beritognolo, I.; Piazzai, M.; Benucci, S.; Kuzminsky, E.; Sabatti, M.; Mugnozza, G.S.; Muleo, R. Functional characterisation of three Italian *Populus alba* L. genotypes under salinity stress. *Trees* **2007**, *21*, 465–477. [CrossRef]

58. Ma, T.; Wang, J.Y.; Zhou, G.K.; Yue, Z.; Hu, Q.J.; Chen, Y.; Liu, B.B.; Qiu, Q.; Wang, Z.; Zhang, J.; et al. Genomic insights into salt adaptation in a desert poplar. *Nat. Commun.* **2013**, *4*, 2797. [CrossRef]

59. Zeng, F.J.; Yan, H.L.; Arndt, S.K. Leaf and whole tree adaptations to mild salinity in field grown *Populus euphratica*. *Tree Physiol.* **2009**, *29*, 1237–1246. [CrossRef]

60. Chen, S.L.; Li, J.K.; Fritzb, E.; Wang, S.S.; Huttermann, A. Sodium and chloride distribution in roots and transport in three poplar genotypes under increasing NaCl stress. *For. Ecol. Manag.* **2002**, *168*, 217–230. [CrossRef]

61. Verslues, P.E.; Agarwal, M.; Katiyar-Agarwal, S.; Zhu, J.; Zhu, J.K. Methods and concepts in quantifying resistance to drought, salt and freezing, abiotic stresses that affect plant water status. *Plant J.* **2006**, *45*, 523–539. [CrossRef]

62. Ottow, E.A.; Brinker, M.; Teichmann, T.; Fritz, E.; Kaiser, W.; Brosche, M.; Kangasjarvi, J.; Jiang, X.N.; Polle, A. *Populus euphratica* displays apoplastic sodium accumulation, osmotic adjustment by decreases in calcium and soluble carbohydrates, and develops leaf succulence under salt stress. *Plant Physiol.* **2005**, *139*, 1762–1772. [CrossRef]

63. Liang, W.J.; Ma, X.L.; Wan, P.; Liu, L.Y. Plant salt-tolerance mechanism: A review. *Biochem. Biophys. Res. Commun.* **2018**, *495*, 286–291. [CrossRef]

64. Hasanuzzaman, M.; Alam, M.M.; Rahman, A.; Hasanuzzaman, M.; Nahar, K.; Fujita, M. Exogenous proline and glycine betaine mediated upregulation of antioxidant defense and glyoxalase systems provides better protection against salt-induced oxidative stress in two rice (*Oryza sativa* L.) varieties. *Biomed. Res. Int.* **2014**, *2014*, 757219. [CrossRef]

65. Watanabe, S.; Kojima, K.; Ide, Y.; Sasaki, S. Effects of saline and osmotic stress on proline and sugar accumulation in *Populus euphratica* in vitro. *Plant Cell Tiss Organ Cult.* **2000**, *63*, 199–206. [CrossRef]

66. Janz, D.; Polle, A. Harnessing salt for woody biomass production. *Tree Physiol.* **2012**, *32*, 1–3. [CrossRef]

67. Apel, K.; Hirt, H. Reactive oxygen species: Metabolism, oxidative stress, and signal transduction. *Annu. Rev. Plant Biol.* **2004**, *55*, 373–399. [CrossRef]

68. Miller, G.; Suzuki, N.; Ciftci-Yilmaz, S.; Mittler, R. Reactive oxygen species homeostasis and signalling during drought and salinity stresses. *Plant Cell Environ.* **2010**, *33*, 453–467. [CrossRef]

69. Del Rio, L.A. ROS and RNS in plant physiology: An overview. *J. Exp. Bot.* **2015**, *66*, 2827–2837. [CrossRef]

70. Sies, H. Role of metabolic H_2O_2 generation: Redox signaling and oxidative stress. *J. Biol. Chem.* **2014**, *289*, 8735–8741. [CrossRef]

71. Wang, Y.C.; Qu, G.Z.; Li, H.Y.; Wu, Y.J.; Wang, C.; Liu, G.F.; Yang, C.P. Enhanced salt tolerance of transgenic poplar plants expressing a manganese superoxide dismutase from *Tamarix androssowii*. *Mol. Biol. Rep.* **2010**, *37*, 1119–1124. [CrossRef]

72. Zheng, L.Y.; Meng, Y.; Ma, J.; Zhao, X.L.; Cheng, T.L.; Ji, J.; Chang, E.M.; Meng, C.; Deng, N.; Chen, L.Z.; et al. Transcriptomic analysis reveals importance of ROS and phytohormones in response to short-term salinity stress in *Populus tomentosa*. *Front Plant Sci.* **2015**, *6*, 678. [CrossRef]

73. Gruszecki, W.I.; Strzalka, K. Carotenoids as modulators of lipid membrane physical properties. *Biochim. Biophys. Acta* **2005**, *1740*, 108–115. [CrossRef]

74. Bible, A.N.; Fletcher, S.J.; Pelletier, D.A.; Schadt, C.W.; Jawdy, S.S.; Weston, D.J.; Engle, N.L.; Tschaplinski, T.; Masyuko, R.; Polisetti, S.; et al. A Carotenoid-Deficient Mutant in *Pantoea* sp. YR343, a Bacteria Isolated from the Rhizosphere of *Populus deltoides*, Is Defective in Root Colonization. *Front. Microbiol.* **2016**, *7*, 491. [CrossRef]

75. Shen, Z.D.; Ding, M.Q.; Sun, J.; Deng, S.R.; Zhao, R.; Wang, M.J.; Ma, X.J.; Wang, F.F.; Zhang, H.L.; Qian, Z.Y.; et al. Overexpression of *PeHSF* mediates leaf ROS homeostasis in transgenic tobacco lines grown under salt stress conditions. *Plant Cell Tiss Organ Cult.* **2013**, *115*, 299–308. [CrossRef]

76. Li, J.B.; Zhang, J.; Jia, H.X.; Li, Y.; Xu, X.D.; Wang, L.J.; Lu, M.Z. The *Populus trichocarpa PtHSP17.8* involved in heat and salt stress tolerances. *Plant Cell Rep.* **2016**, *35*, 1587–1599. [CrossRef]

77. Poor, P.; Kovacs, J.; Borbely, P.; Takacs, Z.; Szepesi, A.; Tari, I. Salt stress-induced production of reactive oxygen-and nitrogen species and cell death in the ethylene receptor mutant Never ripe and wild type tomato roots. *Plant Physiol. Biochem.* **2015**, *97*, 313–322. [CrossRef]

78. Corpas, F.J.; Palma, J.M.; Del Rio, L.A.; Barroso, J.B. Protein tyrosine nitration in higher plants grown under natural and stress conditions. *Front Plant Sci.* **2013**, *4*, 29. [CrossRef]

79. Ji, H.T.; Pardo, J.M.; Batelli, G.; Van Oosten, M.J.; Bressan, R.A.; Li, X. The Salt Overly Sensitive (SOS) pathway: Established and emerging roles. *Mol. Plant* **2013**, *6*, 275–286. [CrossRef]

80. Yang, Y.; Tang, R.J.; Jiang, C.M.; Li, B.; Kang, T.; Liu, H.; Zhao, N.; Ma, X.J.; Yang, L.; Chen, S.L.; et al. Overexpression of the *PtSOS2* gene improves tolerance to salt stress in transgenic poplar plants. *Plant Biotechnol. J.* **2015**, *13*, 962–973. [CrossRef]

81. Zhu, J.K. Salt and drought stress signal transduction in plants. *Annu. Rev. Plant Biol.* **2002**, *53*, 247–273. [CrossRef]

82. Tang, R.J.; Liu, H.; Bao, Y.; Lv, Q.D.; Yang, L.; Zhang, H.X. The woody plant poplar has a functionally conserved salt overly sensitive pathway in response to salinity stress. *Plant Mol. Biol.* **2010**, *74*, 367–380. [CrossRef]

83. Wu, Y.X.; Ding, N.; Zhao, X.; Zhao, M.G.; Chang, Z.Q.; Liu, J.Q.; Zhang, L.X. Molecular characterization of PeSOS1: The putative Na$^+$/H$^+$ antiporter of *Populus euphratica*. *Plant Mol. Biol.* **2007**, *65*, 1–11. [CrossRef]

84. Tang, R.J.; Yang, Y.; Yang, L.; Liu, H.; Wang, C.T.; Yu, M.M.; Gao, X.S.; Zhang, H.X. Poplar calcineurin B-like proteins PtCBL10A and PtCBL10B regulate shoot salt tolerance through interaction with PtSOS2 in the vacuolar membrane. *Plant Cell Environ.* **2014**, *37*, 573–588. [CrossRef]

85. Zhu, J.K. Plant salt tolerance. *Trends Plant Sci.* **2001**, *6*, 66–71. [CrossRef]

86. Zhu, J.K. Regulation of ion homeostasis under salt stress. *Curr. Opin. Plant Biol.* **2003**, *6*, 441–445. [CrossRef]

87. Zhang, F.; Wang, Y.; Yang, Y.; Wu, H.; Wang, D.; Liu, J. Involvement of hydrogen peroxide and nitric oxide in salt resistance in the calluses from *Populus euphratica*. *Plant Cell Environ.* **2007**, *30*, 775–785. [CrossRef]

88. Choi, J.; Tanaka, K.; Cao, Y.R.; Qi, Y.; Qiu, J.; Liang, Y.; Lee, S.Y.; Stacey, G. Identification of a Plant Receptor for Extracellular ATP. *Science* **2014**, *343*. [CrossRef]

89. Sun, J.; Zhang, X.; Deng, S.R.; Zhang, C.L.; Wang, M.J.; Ding, M.Q.; Zhao, R.; Shen, X.; Zhou, X.Y.; Lu, C.F.; et al. Extracellular ATP signaling is mediated by H$_2$O$_2$and cytosolic Ca^{2+} in the salt response of *Populus euphratica* cells. *PLoS ONE* **2012**, *7*, e53136. [CrossRef]

90. Wang, S.J.; Zhou, B.R.; Yao, W.J.; Jiang, T.B. PsnERF75 Transcription Factor from *Populus simonii* × *P. nigra* Confers Salt Tolerance in Transgenic Arabidopsis. *J. Plant Biol.* **2018**, *61*, 61–71. [CrossRef]

91. Yao, W.J.; Wang, S.J.; Zhou, B.R.; Jiang, T.B. Transgenic poplar overexpressing the endogenous transcription factor *ERF76* gene improves salinity tolerance. *Tree Physiol.* **2016**, *36*, 896–908. [CrossRef]

92. Wang, Z.L.; Liu, J.; Guo, H.Y.; He, X.; Wu, W.B.; Du, J.C.; Zhang, Z.Y.; An, X.M. Characterization of two highly similar CBF/DREB1-like genes, PhCBF4a and PhCBF4b, in *Populus hopeiensis*. *Plant Physiol. Biochem.* **2014**, *83*, 107–116. [CrossRef]

93. Zhao, H.; Zhao, X.Y.; Li, M.Y.; Jiang, Y.; Xu, J.Q.; Jin, J.J.; Li, K.L. Ectopic expression of *Limonium bicolor* (Bag.) Kuntze *DREB (LbDREB)* results in enhanced salt stress tolerance of transgenic *Populus ussuriensis* Kom. *Plant Cell Tiss Organ Cult.* **2017**, *132*, 123–136. [CrossRef]

94. Zhou, M.L.; Ma, J.T.; Zhao, Y.M.; Wei, Y.H.; Tang, Y.X.; Wu, Y.M. Improvement of drought and salt tolerance in Arabidopsis and Lotus corniculatus by overexpression of a novel DREB transcription factor from *Populus euphratica*. *Gene* **2012**, *506*, 10–17. [CrossRef]

95. Speth, E.B.; Imboden, L.; Hauck, P.; He, S.Y. Subcellular Localization and Functional Analysis of the Arabidopsis GTPase RabE. *Plant Physiol.* **2009**, *149*, 1824–1837. [CrossRef]

96. Zhang, J.; Li, Y.; Liu, B.L.; Wang, L.J.; Zhang, L.; Hu, J.J.; Chen, J.; Zheng, H.Q.; Lu, M.Z. Characterization of the *Populus Rab* family genes and the function of *PtRabE1b* in salt tolerance. *BMC Plant Biol.* **2018**, *18*, 124. [CrossRef]

97. Perez-Salamo, I.; Papdi, C.; Rigo, G.; Zsigmond, L.; Vilela, B.; Lumbreras, V.; Nagy, I.; Horvath, B.; Domoki, M.; Darula, Z.; et al. The heat shock factor A4A confers salt tolerance and is regulated by oxidative stress and the mitogen-activated protein kinases MPK3 and MPK6. *Plant Physiol.* **2014**, *165*, 319–334. [CrossRef]

98. Li, F.; Zhang, H.; Zhao, H.; Gao, T.; Song, A.; Jiang, J.; Chen, F.; Chen, S. Chrysanthemum *CmHSFA4* gene positively regulates salt stress tolerance in transgenic chrysanthemum. *Plant Biotechnol. J.* **2018**, *16*, 1311–1321. [CrossRef]

99. Yoon, S.K.; Bae, E.K.; Lee, H.; Choi, Y.L.; Han, M.; Choi, H.; Kang, K.S.; Park, E.J. Downregulation of stress-associated protein 1 (*PagSAP1*) increases salt stress tolerance in poplar (*Populus alba* × *P. glandulosa*). *Trees* **2018**, *32*, 823–833. [CrossRef]

100. Xue, S.W.; Yao, X.; Luo, W.; Jha, D.; Tester, M.; Horie, T.; Schroeder, J.I. AtHKT1;1 mediates nernstian sodium channel transport properties in *Arabidopsis* root stelar cells. *PLoS ONE* **2011**, *6*, e24725. [CrossRef]

101. Cheng, N.H.; Pittman, J.K.; Zhu, J.K.; Hirschi, K.D. The protein kinase SOS2 activates the Arabidopsis H$^+$/Ca^{2+} antiporter CAX1 to integrate calcium transport and salt tolerance. *J. Biol. Chem.* **2004**, *279*, 2922–2926. [CrossRef]

102. Baxter, A.; Mittler, R.; Suzuki, N. ROS as key players in plant stress signalling. *J. Exp. Bot.* **2014**, *65*, 1229–1240. [CrossRef]

103. Huertas, R.; Olias, R.; Eljakaoui, Z.; Galvez, F.J.; Li, J.; De Morales, P.A.; Belver, A.; Rodriguez-Rosales, M.P. Overexpression of *SlSOS2 (SlCIPK24)* confers salt tolerance to transgenic tomato. *Plant Cell Environ.* **2012**, *35*, 1467–1482. [CrossRef]

104. Qiu, Q.S.; Guo, Y.; Quintero, F.J.; Pardo, J.M.; Schumaker, K.S.; Zhu, J.K. Regulation of vacuolar Na$^+$/H$^+$ exchange in *Arabidopsis thaliana* by the salt-overly-sensitive (SOS) pathway. *J. Biol. Chem.* **2004**, *279*, 207–215. [CrossRef]

105. Zhou, J.; Wang, J.J.; Bi, Y.F.; Wang, L.K.; Tang, L.Z.; Yu, X.; Ohtani, M.; Demura, T.; Zhu Ge, Q. Overexpression of *PtSOS2* Enhances Salt Tolerance in Transgenic Poplars. *Plant Mol. Biol. Rep.* **2014**, *32*, 185–197. [CrossRef]

106. Li, D.D.; Song, S.Y.; Xia, X.L.; Yin, W.L. Two CBL genes from *Populus euphratica* confer multiple stress tolerance in transgenic triploid white poplar. *Plant Cell Tiss Organ Cult.* **2012**, *109*, 477–489. [CrossRef]

107. Gao, W.; Bai, S.; Li, Q.; Gao, C.; Liu, G.; Li, G.; Tan, F. Overexpression of *TaLEA* Gene from *Tamarix androssowii* improves salt and drought tolerance in transgenic Poplar (*Populus simonii* × *P. nigra*). *PLoS ONE* **2013**, *8*, e67462. [CrossRef]

108. Sun, Y.S.; Chen, S.; Huang, H.J.; Jiang, J.; Bai, S.; Liu, G.F. Improved salt tolerance of *Populus davidiana* × *P. bolleana* overexpressed LEA from *Tamarix androssowii*. *J. For. Res.-JPN* **2014**, *25*, 813–818. [CrossRef]

109. Li, Y.L.; Su, X.H.; Zhang, B.Y.; Huang, Q.J.; Zhang, X.H.; Huang, R.F. Expression of jasmonic ethylene responsive factor gene in transgenic poplar tree leads to increased salt tolerance. *Tree Physiol.* **2009**, *29*, 273–279. [CrossRef]

110. Du, N.X.; Liu, X.; Li, Y.; Chen, S.Y.; Zhang, J.S.; Ha, D.; Deng, W.G.; Sun, C.K.; Zhang, Y.Z.; Pijut, P.M. Genetic transformation of *Populus tomentosa* to improve salt tolerance. *Plant Cell Tiss Organ Cult.* **2011**, *108*, 181–189. [CrossRef]

111. Lawson, S.S.; Michler, C.H. Overexpression of *AtSTO1* leads to improved salt tolerance in *Populus tremula* × *P. alba*. *Transgenic Res.* **2014**, *23*, 817–826. [CrossRef]

112. Zhang, T.T.; Song, Y.Z.; Liu, Y.D.; Guo, X.Q.; Zhu, C.X.; Wen, F.J. Overexpression of phospholipase Dα gene enhances drought and salt tolerance of *Populus tomentosa*. *Chin. Sci. Bull.* **2008**, *53*, 3656–3665. [CrossRef]

113. Yu, X.; Ohtani, M.; Kusano, M.; Nishikubo, N.; Uenoyama, M.; Umezawa, T.; Saito, K.; Shinozaki, K.; Demura, T. Enhancement of abiotic stress tolerance in poplar by overexpression of key Arabidopsis stress response genes, *AtSRK2C* and *AtGolS2*. *Mol. Breed.* **2017**, *37*. [CrossRef]

114. Chen, S.; Polle, A. Salinity tolerance of *Populus*. *Plant Biol.* **2010**, *12*, 317–333. [CrossRef] [PubMed]

115. Janz, D.; Behnke, K.; Schnitzler, J.P.; Kanawati, B.; Schmitt Kopplin, P.; Polle, A. Pathway analysis of the transcriptome and metabolome of salt sensitive and tolerant poplar species reveals evolutionary adaption of stress tolerance mechanisms. *BMC Plant Biol.* **2010**, *10*, 150. [CrossRef] [PubMed]

116. Podda, A.; Pisuttu, C.; Hoshika, Y.; Pellegrini, E.; Carrari, E.; Lorenzini, G.; Nali, C.; Cotrozzi, L.; Zhang, L.; Baraldi, R.; et al. Can nutrient fertilization mitigate the effects of ozone exposure on an ozone-sensitive poplar clone? *Sci. Total Environ.* **2019**, *657*, 340–350. [CrossRef] [PubMed]

International Journal of
Molecular Sciences

MDPI

Review

Melatonin: A Small Molecule but Important for Salt Stress Tolerance in Plants

Haoshuang Zhan [1,†], Xiaojun Nie [1,†], Ting Zhang [1], Shuang Li [1], Xiaoyu Wang [1], Xianghong Du [1], Wei Tong [1,*] and Weining Song [1,2,*]

[1] State Key Laboratory of Crop Stress Biology in Arid Areas, College of Agronomy and Yangling Branch of China Wheat Improvement Center, Northwest A&F University, Yangling 712100, China; zhanhaoshuang@nwsuaf.edu.cn (H.Z.); small@nwsuaf.edu.cn (X.N.); zhangting@nwsuaf.edu.cn (T.Z.); Lishuang@nwsuaf.edu.cn (S.L.); xiaoyuw@nwsuaf.edu.cn (X.W.); xianghongdu@nwsuaf.edu.cn (X.D.)
[2] ICARDA-NWSUAF Joint Research Center for Agriculture Research in Arid Areas, Yangling 712100, China
* Corresponding authors: tongw@nwsuaf.edu.cn (W.T.); sweining2002@nwsuaf.edu.cn or sweining2002@yahoo.com (W.S.); Tel.: +86-29-8708-2984 (W.S.); Fax: +86-29-8708-2203 (W.S.)
† These authors contributed equally to this work.

Received: 1 January 2019; Accepted: 4 February 2019; Published: 7 February 2019

Abstract: Salt stress is one of the most serious limiting factors in worldwide agricultural production, resulting in huge annual yield loss. Since 1995, melatonin (*N*-acetyl-5-methoxytryptamine)—an ancient multi-functional molecule in eukaryotes and prokaryotes—has been extensively validated as a regulator of plant growth and development, as well as various stress responses, especially its crucial role in plant salt tolerance. Salt stress and exogenous melatonin lead to an increase in endogenous melatonin levels, partly via the phyto-melatonin receptor CAND2/PMTR1. Melatonin plays important roles, as a free radical scavenger and antioxidant, in the improvement of antioxidant systems under salt stress. These functions improve photosynthesis, ion homeostasis, and activate a series of downstream signals, such as hormones, nitric oxide (NO) and polyamine metabolism. Melatonin also regulates gene expression responses to salt stress. In this study, we review recent literature and summarize the regulatory roles and signaling networks involving melatonin in response to salt stress in plants. We also discuss genes and gene families involved in the melatonin-mediated salt stress tolerance.

Keywords: antioxidant systems; ion homeostasis; melatonin; salt stress; signal pathway

1. Introduction

Salinity represents an environmental stress factor affecting plant growth and development, and a destructive threat to global agricultural production [1], which damages more than 400 million hectares of land—over 6% of the world's total land area. Of the irrigated farmland areas, currently 19.5% are salt-affected, with increasing numbers facing the threat of salinization (http://www.plantstress. com/Articles/index.asp). The effects of salt stress on plants mainly include osmotic stress, specific ion toxicity, nutritional imbalance, and reactive oxygen species [2]. Osmotic stress is a rapid process caused by salt concentrations around the roots, which is induced at the initial stage of salt stress [1–3]. Na^+ accumulation at a later stage causes nutrient imbalance, leading to specific ion toxicity [4]. Plants' exposure to salt stress induces overproduction of reactive oxygen species (ROS), which results in membrane injury [5,6].

Melatonin is a multi-regulatory molecule likely to be present in most plants and animals [7]. It was first identified in 1958, in the bovine pineal gland [8], and is a well-known animal hormone regulating various biological processes, such as the circadian rhythm [9,10], antioxidant activity [11], immunological enhancement [12], seasonal reproduction [13], emotional status, and physical conditions [14]. In 1995, melatonin was discovered in vascular plants [15,16], which initiated this field of study. Melatonin was found to have many physiological functions similar to indole-3-acetic acid (IAA), such as regulating plant photoperiod and protecting chlorophyll [17]. More importantly, it acts as a powerful antioxidant, thus protecting plants from various biotic/abiotic stresses [18,19].

In recent years, more functions of melatonin have been identified in higher plants, mainly its roles as a stress responses regulator. In this review, we systematically discuss the functional and potential regulatory mechanisms of melatonin in response to salt stress. We also focus on the putative genes involved in the melatonin-induced salt stress resistance. Furthermore, we summarized plant melatonin receptors, thus outlining the current situation and further directions for promoting the study of plant salt stress tolerance.

2. Function and Mechanism of Melatonin Effects on Plant Salt Tolerance

Extensive studies have revealed the crucial and indispensable roles that melatonin plays in increasing salt tolerance in diverse plant species (Table 1). These functions regulate antioxidant systems to protect plants from the salt stress-induced water deficits and physiological damages, improve photosynthetic efficiency and ion homeostasis, and behave as an activator mediating NO signaling and the polyamine metabolism pathway [7,17,33].

Table 1. The reported roles melatonin plays in response to salt and other stresses in plants.

Plant Species	Stress Condition	References
Actinidia deliciosa	Salt	[20]
Malus hupehensis	Salt	[21]
Arabidopsis thaliana	salt	[22]
Arabidopsis thaliana	Salt, drought and cold	[23]
Arabidopsis thaliana	Salt	[24]
Cynodon dactylon (L). Pers.	Salt, drought and cold	[25]
Chara australis	Salt	[26]
Chlamydomonas reinhardtii	Salt	[27]
Citrus aurantium L.	Salt	[28]
Cucumis sativus L.	Salt	[29]
Cucumis sativus L.	Salt	[17]
Cucumis sativus L.	Salt	[30]
Zea mays L.	Salt	[31]
Zea mays L.	Salt	[32]
Zea mays L.	Salt	[33]
Raphanus sativus L.	Salt	[34]
Raphanus sativus L.	Salt	[35]
Brassica napus L.	Salt	[36]
Brassica napus L.	Salt	[37]
Oryza sativa L.	Leaf senescence and salt	[38]
Oryza sativa L.	Salt	[39]
Glycine max	Salt and drought	[40]
Helianthus annuus	Salt	[41]
Helianthus annuus	Salt	[42]
Ipomoea batatas	Salt	[43]
Solanum lycopersicum	Salt	[44]
Vicia faba L.	Salt	[45]
Citrullus lanatus L.	Salt	[46]
Triticum aestivum L.	Salt	[47]

2.1. Melatonin Activates Antioxidant Systems in Response to Salt Stress

Salinity induces reactive oxygen species (ROS) production, including superoxide anion (O_2^-), hydrogen peroxide (H_2O_2), hydroxyl radical (OH^-), and singlet oxygen ($^1O_2$) [47]. Excess ROS usually leads to cell damage and oxidative stress [22]; it also acts as signaling molecules fundamentally involved in mediating salt tolerance [48]. Plants have developed two antioxidant systems to alleviate ROS-triggered damages: the enzymatic and non-enzymatic systems [49]. In response to salt stress, plants have evolved a complex antioxidant enzyme system, including superoxide dismutase (SOD), guaiacol peroxidase (POD), catalase (CAT), glutathione peroxidases (GPX), glutathione S-transferase (GST), dehydroascorbate reductase (DHAR), glutathione reductase (GR), and ascorbate peroxidase (APX) [17]. The non-enzymatic system, including ascorbic acid (AsA), α-tocopherols, glutathione (GSH), carotenoids, and phenolic compounds, is also essential for ROS elimination [50].

Exogenous melatonin treatment significantly reduced salinity-induced ROS. Following 12 days of salt stress, H_2O_2 concentration increased by 37.5%, while melatonin pre-treatment of cucumber maintained a low H_2O_2 concentration throughout the experiment [17]. Similar results were also observed in salt-stressed rapeseed seedlings, and the application of exogenous melatonin decreased H_2O_2 content by 11.2% [36]. Liang et al. [38] discovered inhibitory effects of melatonin resulting in an increased rate of H_2O_2 production in rice seedlings under salt stress, showing that melatonin works in a concentration-dependent manner. Melatonin scavenges ROS, mainly triggered by salt stress, via three pathways. Melatonin acts as a broad-spectrum antioxidant that interacts with ROS and directly scavenges it [51]. The primary function of melatonin is to act as a free radical scavenger and an antioxidant. Through the free radical scavenging cascade, a single melatonin molecule can scavenge up to 10 reactive oxygen species (ROS)/reactive nitrogen species (RNS), which differs from other conventional antioxidants [51]. Exogenous melatonin decreases H_2O_2 and O_2^- concentrations by activating antioxidant enzymes. This function has been confirmed in many plant species, such as rapeseed, radish, cucumber, rice, maize, bermudagrass, soybean, watermelon, kiwifruit, and *Malus hupehensis* [36]. In cucumber, the activity of major protective antioxidant enzymes—including SOD, CAT, POD, and APX—in melatonin pre-treated plants was significantly higher than control plants [17]. Under salt stress, exogenous melatonin application also significantly increased the activities of APX, CAT, SOD, POD, GR, and GPX in melatonin-treated seedlings compared to their non-treated counterparts [31,33]. Moreover, melatonin interacts with ROS by improving concentrations of antioxidants (AsA-GSH) [17]. In cucumber, AsA and GSH concentrations in melatonin pre-treated plants were 1.7- and 1.3-fold higher, respectively, compared to control plants [17]. Other studies have reported a marked melatonin-dependent induction of AsA and GSH in maize seedlings under salt stress [31]. These findings suggest that exogenous melatonin could activate enzymatic and non-enzymatic antioxidants to scavenge salt stress-induced ROS, thus improving salt stress tolerance in plants.

2.2. Melatonin Improves Plant Photosynthesis under Salt Stress

Photosynthesis, an important physio-chemical process responsible for energy production in higher plants, can be indirectly affected by salt stress [46,52]. For many plant species suffering salt stress, decline in productivity is often associated with lower photosynthesis levels [52]. There are two possible reasons for the salt-induced photosynthesis decline: stomatal closure and affected photosynthetic apparatus [52]. Salt stress can cause stomatal closure, and stomatal conductance (Gs) is one of the parameters for evaluating photosynthesis [52]. The parameters of chlorophyll fluorescence include maximum photochemical efficiency of PSII (Fv/Fm), photochemical quenching (qP), non-photochemical quenching [Y(NPQ)], and actual photochemical efficiency of PSII [Y(II), etc. [46].

In addition to its broad-spectrum antioxidant effects, melatonin participates in the regulation of plant photosynthesis under salt stress. Pretreatment with various concentrations (50–500 μM) of melatonin clearly improved salt tolerance in watermelons, where the leaf net photosynthetic rate (Pn), Gs, chlorophyll content, Y(II) and qP were significantly decreased under salt stress. However, this decrease was alleviated by melatonin pretreatment. Melatonin can also protect watermelon photosynthesis by alleviating stomatal limitation [46]. Similar results were observed in salt-stressed cucumber seedlings, where the photosynthetic capacity of cucumber was significantly improved by exogenous melatonin at 50–150 μM concentrations. Photosynthesis improvement is manifested by increased P_N, maximum quantum efficiency of PSII, and total chlorophyll content [17]. In radish seedling, chlorophyll a, chlorophyll b and total chlorophyll contents increased upon melatonin treatment under salt stress, and the 100 μM dose was the best [34]. Melatonin also enhanced rice seedlings' salt tolerance by decreasing chlorophyll's degradation rate [38]. Even though the chlorophyll content in melatonin-treated maize seedlings did not change, an obvious increase in Pn was observed under salt stress [33]. Exogenous melatonin's protective roles in photosynthesis were also observed in soybean, apple, and tomato [21,40,44]. Overall, exogenous melatonin improves photosynthesis by effectively alleviating chlorophyll degradation and stomatal closure caused by salt stress, therefore enhancing salt stress tolerance.

2.3. Melatonin Promotes Ion Homeostasis under Salt Stress

Ion homeostasis refers to the ability of living organisms to maintain stable ion concentrations in a defined space [53]. Na^+, K^+, Ca^{2+}, and H^+ are major intracellular ions [53,54]. In salt-stressed plants, Na^+ can enter into plant cells, which at high concentrations is harmful to cytosolic enzymes [55]. Therefore, regulation of K^+ and Na^+ concentrations to maintain high of K^+ and low Na^+ cytosolic levels has a significant impact on salt-stressed plants [54,55]. Restriction of Na^+ influx, active Na^+ efflux, and compartmentalization of Na^+ into the vacuole are three major mechanisms of preventing Na^+ accumulation in the cytoplasm [56]. The *NHX1* gene encodes a vacuolar Na^+/H^+ exchanger, whose homologue in *Arabidopsis*, *AtNHX1*, was upregulated by salt stress resulting in excess transfer of Na^+ into vacuolar [57]. Salt Overly Sensitive1 (*SOS1*) encodes a transmembrane protein, identified as a plasma membrane Na^+/H^+ antiporter. SOS signaling is responsible for transporting Na^+ out of the cells [37,56]. The *Arabidopsis SOS1* gene possesses 12 transmembrane domains. Similar to *AtNHX1*, *AtSOS1* was also upregulated by salt stress [56]. Besides Na^+/H^+ antiporters, the involvement of K^+ channels has also been reported in plants' salt stress response. The *AKT1* gene encoding a Shaker type K^+ channel protein is responsible for absorbing K^+ from the soil and transporting it into the roots [58]. Under salt stress, *NHX1*, *SOS1* and *AKT1* upregulated gene expression leads to an increase of K^+ and decreased Na^+ in plant cells, thereby improving plants' salt stress tolerance.

Recently, studies have shown that the exogenous application of melatonin improves plants' ion homeostasis under salt stress. Melatonin significantly increased K^+ and decreased Na^+ contents in shoots of maize seedlings, leading to a significantly higher K^+/Na^+ ratio in shoots under melatonin-mediated salinity [33]. Improved ion homeostasis may be related to the upregulation of several genes, such as *NHX*, *SOS* and *AKT*. Under salt stress, *MdNHX1* and *MdAKT1* transcript levels were greatly upregulated by melatonin, which is consistent with the relatively high K^+ levels and K^+/Na^+ ratio in melatonin pretreated *Malus hupehensis* seedlings [21]. Similarly, *NHX1* and *SOS2* expression was higher in melatonin-treated rapeseed seedlings compared to non-treated plants, which correlated with the lower Na^+/K^+ ratio [37]. Ca^{2+} signaling plays critical roles in plant biotic and abiotic stress responses; however, no evidence regarding the involvement of Ca^{2+} signaling in melatonin-triggered salinity tolerance exists.

2.4. Melatonin Regulates Plant Hormones Metabolism

Plant hormones are important signals for plant growth and development [30]. Melatonin widely participates in the metabolism of most plant hormones, such as indole-3-acetic acid (IAA), abscisic acid (ABA), gibberellic acids (GA), cytokinins (CK), and ethylene [59].

The melatonin molecule shares chemical similarities with IAA, both using tryptophan as a substrate in their biosynthesis pathways [60]. It is reported that melatonin acts as a growth regulator and exhibits auxin-like activities [61]. Melatonin promotes vegetative growth and root development in many plant species, such as wheat, barley, rice, *Arabidopsis*, soybean, maize, tomato, etc. [59]. Under stress conditions, the growth-promoting effects of melatonin are higher compared to those in control plants [59]. Melatonin has been proposed to regulate lateral root formation through an

IAA-independent pathway in *Arabidopsis* [61]. In contrast, others suggest a certain relationship between melatonin and IAA; for example, a slight increase in endogenous IAA content was observed in *Brassica juncea* [59,62] when treated with exogenous melatonin. Furthermore, application of low concentrations of IAA increases endogenous melatonin levels. At the same time, high concentrations of melatonin inhibit PIN1,3,7 expression and decrease IAA levels in *Arabidopsis* roots, suggesting that melatonin may regulate root growth in *Arabidopsis*, completely or partially, through auxin synthesis and polar auxin transport [60].

Abscisis acid (ABA) and gibberellic acids (GA) are important plant hormones in stress responses. The dynamic balance of endogenous ABA and GA levels is crucial for seed germination [30,63]. Genes related to ABA synthesis—such as *ZEP* and *NCED1*—were upregulated during abiotic stresses, resulting in increased endogenous ABA levels [64]. GA acts as an ABA antagonist [65], and plays essential roles in plant stress tolerance [66]. Studies show that melatonin mediates ABA biosynthesis and metabolism regulation, thus decreasing ABA content under stress conditions. For example, in two drought-stressed *Malus* species, melatonin selectively downregulates *MdNCED3*, a key ABA biosynthesis gene, and upregulates *MdCYP707A1* and *MdCYP707A2*, ABA catabolic genes [67]. Similarly, in perennial ryegrass, exogenous melatonin downregulates ABA biosynthesis genes under heat stress, thereby decreasing ABA content [64]. However, melatonin treatment has no effect on water stress-induced ABA accumulation in maize [68]. Under salt stress, melatonin increased endogenous ABA content in *Elymus nutans*, which was significantly suppressed by fluridone. ABA and fluridone pretreatments had no effect on endogenous melatonin concentration, indicating that ABA might act as a downstream signal that participates in the melatonin-induced cold tolerance. Interestingly, melatonin can also activate the expression of cold-responsive genes to improve plant cold-stress tolerance in an ABA-independent manner. This suggests that both ABA-dependent and ABA-independent pathways might be involved in melatonin-induced cold tolerance [69]. These data suggest that, similar to the heat-related results, under drought and cold stresses, exogenous melatonin can also alleviate salt stress by regulating ABA biosynthesis and catabolism. Under salt stress, *CsNCED1* and *CsNCED2*—ABA synthesis-related genes—transcript levels were reduced in melatonin-pretreated seeds, and genes related to ABA catabolism were significantly increased, thus leading to a decreased ABA content. On the contrary, *GA20ox* and *GA3ox*—genes involved in GA synthesis—were significantly upregulated by melatonin, which is consistent with the increased GA content [30]. Overall, hormone biosynthesis- and catabolism-related research is helpful for understanding melatonin's mechanisms in response to salt stress.

2.5. Melatonin Mediates NO Signaling Pathway

Nitric oxide (NO) is an important messenger and ubiquitous signaling molecule, which participates in various plant physiological processes [70], and responds to abiotic and biotic stresses [41,42,71,72]. In animals, NO is synthesized by NO synthase (NOS) [72], and whether NOS-like proteins exist in plants remains controversial. NOS-like proteins were first identified in plants by Ninnemann and Maier [73]. Initially, *Arabidopsis* nitric oxide associated 1 (*NOA1*) was characterized as a NOS-like gene with NOS activity. However, further research indicated that these proteins function as a GTPases, involved in binding RNA/ribosomes [74]. There are at least seven different NO biosynthetic pathways found in plants, which can be classified as oxidative or reductive based on the operation [75]. Oxidative routes of NO biosynthesis use L-arginine, polyamine, or droxylamine as substrates [75]. S-nitrosylation refers to the process of covalently binding a NO group to its target proteins via cysteine (Cys) residues, and producing an S-nitrosothiol [76]. S-nitrosylation, with NO, is widely used to explain NO signaling in both animals and plants [77,78].

Studies have shown that melatonin, through its interaction with NO, plays important roles in plant stress responses. For examples, NO acts as a downstream signal for melatonin mitigated sodic alkaline stress in tomato seedlings [79]. In addition, exogenous melatonin significantly induces the accumulation of polyamine-mediated NO in the roots of *Arabidopsis* under Fe deficiency conditions, and increases the plants' tolerance to Fe deficiency [80]. Melatonin-induced NO production is also involved in the innate immune response of *Arabidopsis* against P. syringe pv. tomato (Pst) DC3000 infection [81]. In rapeseed seedlings, the possible roles of NO in melatonin-enhanced salt stress tolerance have been reported. Salt stress firstly induces the increase in melatonin and NO serves as the downstream signal. In addition, both melatonin and sodium nitroprusside (SNP) increased salinity-induced S-nitrosylation. Increased S-nitrosylation could be partially impaired by 2-phenyl-1-4,4,5,5-tetramethylimidazoline-1-oxyl-3-oxide (PTIO), an NO scavenger. Application of melatonin increased *NHX1* and *SOS2* transcript levels, which was blocked by NO removal. These data suggest that NO is involved in the maintenance of ion homeostasis in plant salt stress tolerance. NO is also involved in the improvement of the antioxidant systems triggered by melatonin [37]. However, the above research still lacks S-nitrosylation target protein identification. In addition, the interactions between NO and other substances, such as hormones, chlorophyll, polyamines, etc., in melatonin-enhanced salt stress tolerance requires further exploration.

2.6. Melatonin Regulates Polyamine Metabolism

Polyamines (PAs) are small aliphatic polycations that have been found in almost all living organisms. They play important roles in plant growth and development, and responses to various biotic and abiotic stimuli [82–84]. Spermidine (Spd), putrescine (Put), and spermine (Spm) are three main polyamines in plants [84]. Both the application of exogenous polyamines and modulating endogenous polyamine contents effectively enhance plant stress tolerance [83,84].

Studies have shown that melatonin plays a key role in polyamine-mediated signaling pathways under various abiotic stresses, such as alkaline stress, cold, oxidative, and iron deficiency tolerance [7]. For example, polyamines mediate the melatonin-induced alkaline stress tolerance of *Malus hupehensis*. Under alkaline stress, melatonin application significantly upregulated the expression of six polyamine synthesis-related genes, including *SAMDC1, -3, -4*, and *SPDS1, -3, -5, -6*. Moreover, melatonin-treated *Malus hupehensis* exhibited more polyamine accumulation compared to the untreated seedlings [85]. Exogenous melatonin also modulates polyamine and ABA metabolisms of cucumber seedlings during chilling stress. The melatonin-related cold tolerance improvement is consistent with the increased PA content [24]. PA modulation by melatonin under a salt stress response was also described by Ke et al. [7], where they show that melatonin treatment increases PAs content by accelerating the conversion of arginine and methionine to polyamines in wheat seedlings. At the same time, melatonin suppresses PAO (polyamine oxidase) and DAO (diamine oxidase) activities—two enzymes involved in polyamines metabolism—which decrease melatonin-induced polyamine degradation, thus improving salt stress tolerance [7]. This provides initial evidence that exogenous melatonin treatment enhances plant salt tolerance by regulating PAs, whether the proposed mechanisms are applicable to other plant species requires further investigation. In addition, polyamines are involved in the melatonin-induced NO production in the roots of Fe deficient *Arabidopsis*, and increase the plant tolerance to Fe deficiency [80]. Thus, the interaction between PAs and NO in melatonin-induced salt stress tolerance of plants requires further confirmation.

3. Melatonin Correlated Genes and Gene Families in Plants

To further investigate melatonin's mechanism in regulating salt tolerance in plants, melatonin biosynthesis- and metabolism-related genes, transcription factors, and other related genes and gene families were summarized.

3.1. Putative Genes Involved in Melatonin-Mediated Salt Stress Tolerance

In a wide range of plant species, the melatonin biosynthesis pathway begins with tryptophan, which is converted to tryptamine by tryptophan decarboxylase. Subsequently, tryptamine is converted to serotonin by tryptamine 5-hydroxylase (T5H). In some of the other plant species, the first two steps of the melatonin biosynthesis pathway are reversed. Tryptophan is first converted into 5-hydroxytrytophan by tryptophan 5-hydroxylase (TPH), and then to serotonin by aromatic-L-amino-acid decarboxylase (TDC/AADC) [86]. Although no TPH enzyme been cloned, the presence of ^{14}C-5-hydroxytryptophan and ^{14}C-serotonin have been detected when using ^{14}C-tryptophan as substrate in *Hypericum perforatum* [87]. In the following two steps, three distinct enzymes and two inversed routes were involved. Serotonin N-acetyltransferase (SNAT) catalyzes serotonin into N-acetylserotonin, and N-acetylserotonin was then converted into melatonin by N-acetylserotonin methyl-transferase (ASMT) or caffeic acid O-methyltransferase (COMT). As ASMT/COMT exhibits substrate affinity towards serotonin, and SNAT has substrate affinity toward 5-methoxytryptamine, serotonin could have been first methylated to 5-methoxytryptamine by ASMT/COMT and then to melatonin by SNAT. Different steps involved in the melatonin biosynthesis pathways may occur in different subcellular locations d. In total, six enzymes are involved in plant melatonin biosynthesis, which are related to four different routes. In an *Arabidopsis AtSNAT* mutant, endogenous melatonin content was lower than that in wild-type *Arabidopsis* seedlings. Moreover, the *AtSNAT* mutant was salt hypersensitive compared to wild-type [22]. The possible functions of apple *MzASMT9* were investigated in *Arabidopsis*. Under salt stress, *MzASMT9* transcript levels were upregulated, and melatonin levels were also increased by the ectopic expression of *MzASMT9*, thus leading to an enhanced salt tolerance in transgenic *Arabidopsis* lines [88]. Although there is no direct evidence about the possible roles of TDC, T5H, and COMT in plant salt tolerance, overexpression and suppression of these genes obviously affected endogenous plant melatonin levels [89–92].

The catabolism of phyto-melatonin has also been reported in recent years. Unlike the biosynthesis of melatonin, the metabolism of phyto-melatonin is either through an enzymatic or non-enzymatic pathway [41]. The major melatonin metabolites in plants are N^1-acetyl-N^2-formyl-5-methoxykynuramine (AFMK) and melatonin hydroxylated derivatives, such as 2-hydroxymelatonin and cyclic-3-hydroxymelatonin (3-OHM) [41,93,94].

In rice, melatonin is catabolized into 2-hydroxymelatonin by melatonin 2-hydroxylase (M2H), which belongs to the 2-oxoglutarate-dependent dioxygenase (2-ODD) superfamily [95]. The first *M2H* gene was cloned from rice in 2015 [96].

Except for genes involved in the biosynthesis and catabolism of phyto-melatonin, transcription factors also play critical roles in the melatonin-mediated salt stress response. Under abiotic stress (salt, drought, and cold), exogenous melatonin significantly improves endogenous melatonin levels and upregulates the expression of C-repeat binding factors (CBFs)/Drought response element Binding 1 factors (DREB1s), thus leading to an increase in transcript levels of multiple stress-responsive genes, including *COR15A*, *RD22*, and *KIN1* [23]. RNA sequencing was performed in cucumber roots with or without melatonin treatment under salt stress. The results show that many transcription factors including WRKY, MYB, NAC, and the ethylene-responsive transcription factor were differentially expressed in melatonin-treated plants compared to control plants under NaCl-induced stress [97].

The effects of melatonin on the expression of genes involved in ROS scavenging under NaCl stress were investigated. The application of 1 mM melatonin induced the expression of *CsCu-ZnSOD*, *CsFe-ZnSOD*, *CsPOD*, and *CsCAT* in cucumber under salt stress [30]. Similar results were also observed in rapeseed, and studies showed that antioxidant defense-related genes such as *APX*, *Cu/ZnSOD* and *MnSOD* were involved in melatonin-induced salt stress tolerance [37]. In tomato seedlings under salt stress, melatonin significantly improved *TRXf* gene expression, which participates in the redox regulation of many physiological processes [44]. Genes responsible for maintaining ion homeostasis were also involved in melatonin-enhanced salt stress. *MdNHX1* and *MdAKT1*, two ion-channel genes, were upregulated by exogenous melatonin in *Malus hupehensis* under salinity [21]. *NHX1* and *SOS2 expression* was also modulated by melatonin in salt-stressed rapeseed. Several studies have shown that melatonin alleviates salinity stress by regulating hormone biosynthesis and metabolism gene expression. Melatonin induced the expression of GA biosynthesis genes (*GA20ox* and *GA3ox*). Meanwhile, the ABA catabolism genes, *CsCYP707A1* and *CsCYP707A2*, were obviously upregulated, whereas the ABA biosynthesis gene *CsNECD2* was downregulated by melatonin in salt-stressed cucumber seedlings [30].

3.2. Comparative and Phylogenetic Analysis of TDC, T5H, SNAT, and ASMT Gene Families in Plants

TDC, *T5H*, *SNAT*, and *ASMT* correlate with melatonin biosynthesis in most plant species [86]. Recently, a genome-wide expression, classification, phylogenetic, and expression profiles of the tryptophan decarboxylase (TDC) gene family was conducted in *Solanum lycopersicum* [98]. A total of five *TDC* genes were obtained from the tomato genome. Among the five candidate genes, *SlTDC3* was expressed in all the tested tissues, whereas *SlTDC1* and *SlTDC2* were specifically expressed in the fruit and leaves of the tomato plant, respectively. *SlTDC4* and *SlTDC5* are not expressed in tomato. The study of *TDC* genes in rice is relatively clearer compared to other plants. Rice has at least three *TDC* genes [89]. *OsTDC1* (AK31) and *OsTDC2* (AK53) were first identified by Kang et al. [99]. Heterologous expression of *OsTDC1* and *OsTDC2* in *Escherichia coli* showed that both genes exhibited TDC activity [99]. The expression profiles of *OsTDC1*, *OsTDC2*, and *OsTDC3* have also been investigated in rice. *OsTDC1* and *OsTDC2* have similar expression profiles, with low expression in seedling shoots, and relatively high levels in leafs, stems, roots and flowers. In comparison, *OsTDC3* expression was very low in almost all tested organs, except the roots [89]. These results indicated that different *TDC* genes might play different roles during plant growth and development. Overexpression of *OsTDC1*, *OsTDC2*, and especially *OsTDC3* leads to improved melatonin levels in transgenic rice [89]. The phylogenetic relationships and gene structures of TDCs from algae to higher plants showed that they are found throughout the high plant kingdom with a small family size. The evolution of *TDC* genes in plants was mainly through gene expansion and intron loss events. This is the first research of its kind on the TDC gene family; however, the expression profiles of TDCs were not investigated under the salt stress condition [98–100]. The ASMT gene family was also analyzed in *Solanum lycopersicum* [101]. There are 14 candidate *ASMT* genes involved in tomato, three of which may be pseudogenes. The expression patterns of *SlASMTs* suggested that four *SlASMTs* were involved in tomato plant response to biotic stresses [101].

TDC, *T5H*, *SNAT*, and genes have been identified and functionally analyzed in many plants, especially in rice [89,91,102–104]. A systematic analysis of the tomato TDC gene family has been conducted, and the phylogenetic relationships between *TDC* genes in plants have also been analyzed. In addition to the *ASMT* gene families in tomato, genome-wide analysis of *SNAT*, *ASMT*, and *T5H* families has not been reported. Based on the methods described by Pang et al. [98] and Liu et al. [101], we searched *TDC* genes in wheat genome, as well as *SNAT* and *ASMT* genes in 10 plant species from algae to higher plants. We further validated these *TDC* and *ASMT* genes using the previously reported main residues [105–107]. Only BLASTP (identity >70%, coverage >70%) was conducted for *T5H* genes identification, using rice *T5H* genes as the query. A total of eight *T5H*, 37 *SNAT*, and 140 *ASMT* candidate genes were obtained in 10 plant species (Supplementary Table S1). Furthermore, there are 33 candidate *TDC* genes in wheat. Phylogenetic relationships of *SNAT* and *ASMT* are shown in Figures 1 and 2. Based on the phylogenetic tree topology, the SNAT gene family could be divided into four groups (Group I to IV). *SNAT* members in Group I are highly conserved across all species. Similar numbers of *SNAT* genes were found in different species, and no obvious gene expansion was observed. *OsSNAT2* of rice belongs to Group I, whose function is already revealed [108]. The *ASMT* gene family phylogenetic tree is similar to that of the *TDC* gene family [100]. One member from *Volvox carteri* clustered into a separate branch, indicating that *ASMT* genes originated before the divergence of green algae and land plant species. The average gene number of *ASMT* in algae, pteridophyta, gymnosperms, and angiosperms is 1, 4, 25, and 18.3, respectively, suggesting that gene expansion occurred during the evolution from algae to higher plants.

Furthermore, we specially investigated the expression profiles of *TDC*, *T5H*, *SNAT*, and *ASMT* genes in wheat under salt stress. RNA-sequencing data were downloaded from the NCBI Sequence Read Archive (SRA) ddabase (https://www.ncbi.nlm.nih.gov/sra/). FPKM (fragments per kilobase of transcript per million fragments mapped) values for all candidate genes in wheat were calculated using Hisat2 and Stringtie, and the heat maps were generated using the geom_tile method in ggplot2 [109]. As shown in Figure 3, there are four *TDC* genes, two *T5H* genes, one *SNAT* gene, and 10 *ASMT* genes specifically expressed under salt stress, and lots of genes are upregulated under salt conditions, indicating that these genes could be involved in the salt stress tolerance of wheat.

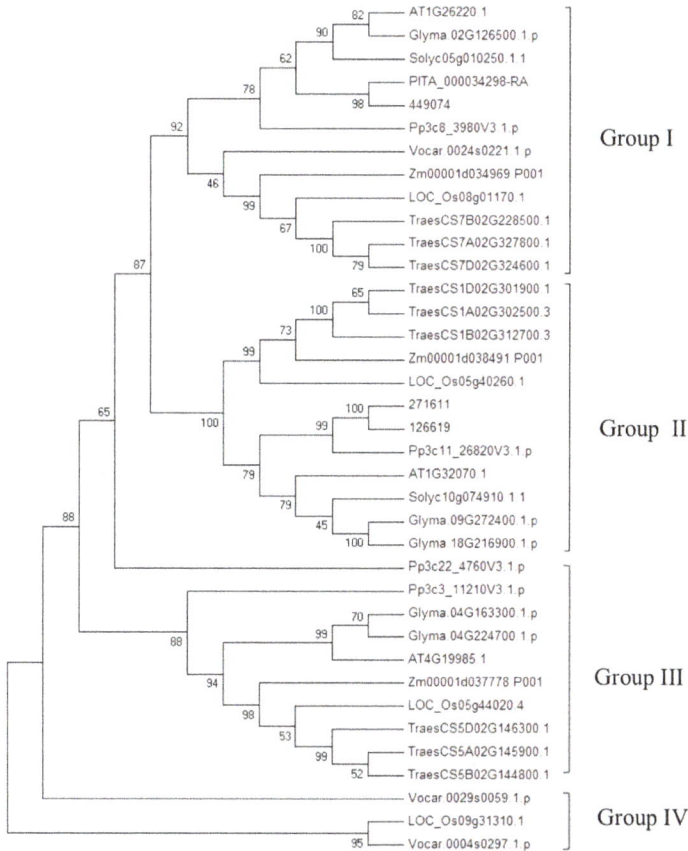

Figure 1. Phylogenetic relationship of the serotonin N-acetyltransferase (*SNAT*) genes from 10 plant species. The candidate *SNAT* genes involved in the phylogenetic tree include the dicots (*Arabidopsis.thaliana* (AT): AT1G26220.1, AT1G32070.1, and AT4G19985.1; *Solanum lycopersicum* (Solyc): Solyc05g010250.1.1, and Solyc10g074910.1.1; *Glyma max* (Glyma): Glyma.02G126500.1.p, Glyma.04G163300.1.p, Glyma.04G224700.1.p, Glyma.09G272400.1.p, and Glyma.18G216900.1.p), monocot (*Zea mays* (Zm): Zm00001d037778_P001, Zm00001d034969_P001, and Zm00001d038491_P001; *Oryza sativa* (LOC_Os): LOC_Os05g40260.1, LOC_Os05g44020.4, LOC_Os08g01170.1, and LOC_Os09g31310.1; *Triticum aestivum* (Traes): TraesCS5D02G146300.1, TraesCS7B02G228500.1, TraesCS7A02G327800.1, TraesCS7D02G324600.1, TraesCS5A02G145900.1, TraesCS1D02G301900.1, TraesCS5B02G144800.1, TraesCS1B02G312700.3, and TraesCS1A02G302500.3), Gymnospermae (*Pinus taeda* (PITA): PITA_000034298-RA), Pteridophyta (*Selaginella moellendorffii*: 271611, 449074, and 126619), Bryophyta (*Physcomitrella patens* (Pp)): Pp3c22_4760V3.1.p, Pp3c3_11210V3.1.p, Pp3c11_26820V3.1.p, and Pp3c8_3980V3.1.p), and algae (*Volvox carteri* (Vocar): Vocar.0029s0059.1.p, Vocar.0004s0297.1.p, and Vocar.0024s0221.1.p)

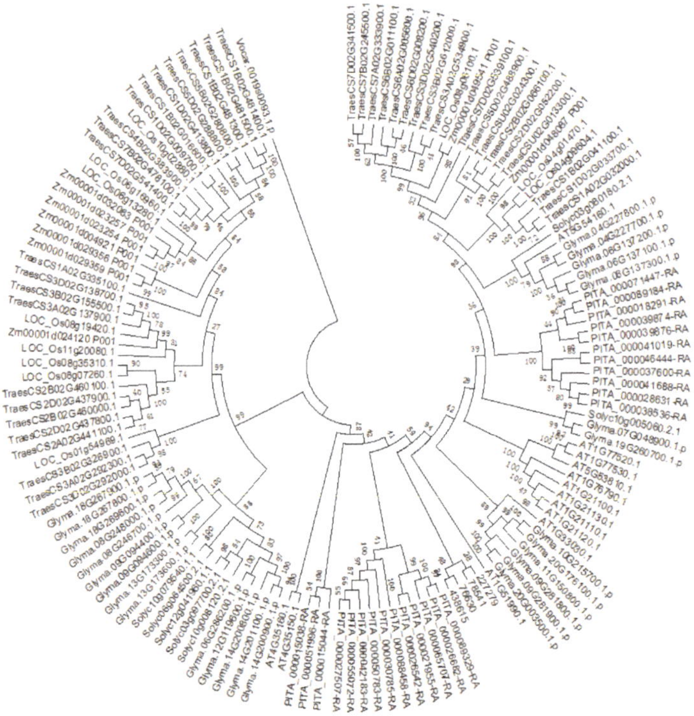

Figure 2. Phylogenetic relationship of the N-acetylserotonin methyl-transferase (*ASMT*) genes from 10 plant species. The 10 plant species include *A.thaliana*, *S.lycopersicum*, *G.max*, *Z.mays*, *O.sativa*, *T.aestivum*, *P.taeda*, *S.moellendorffii*, *P.patens*, and algae.

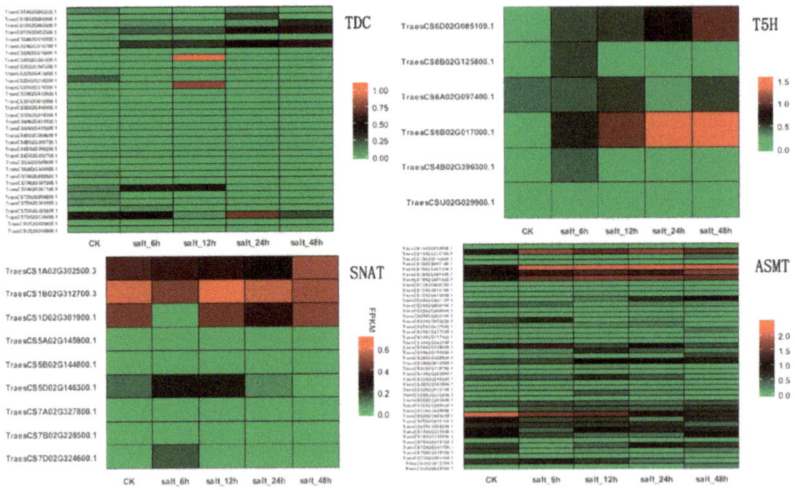

Figure 3. Expression profiles of *TDC*, *T5H*, *SNAT*, and *ASMT* genes in wheat under salt stress conditions. The red or green colors represent the higher or lower relative abundance of each transcript in each sample, respectively.

4. Phyto-Melatonin Receptor

It is clear that exogenous melatonin plays a considerable role during plant growth and development, and is associated with plant stress responses—including salt stress. However, the method by which plants perceive exogenous melatonin and convert it into downstream signals remains unknown. The phyto-melatonin receptor holds promise for better understanding melatonin's biological function and mechanism. Animal melatonin receptors were discovered earlier than the phyto-melatonin receptor. The first melatonin receptor (Mel1c) was cloned from frogs (*Xenopus laevis*) in 1994 [110]. Melatonin receptors belong to the G protein-coupled receptor (GPCR) superfamily, which possess seven transmembrane helices [111]. To date, a total of three melatonin receptor subtyoes have been reported in mammals; MT1 (Mel1a), MT2 (Mel1b), and MT3 (ML2) [112,113]. MT1 and MT2 are G protein-coupled receptors, which exhibit high-affinity for melatonin [112,114], while MT3 exhibits low affinity for melatonin and it belongs to the quinone reductases family [115].

AtCAND2/PMTR1, the first phyto-melatonin receptor, was recently discovered in *Arabidopsis*. When melatonin is perceived by CAND2/PMTR1, it triggers the dissociation of Gα form Gγβ, which activates the downstream H_2O_2 and Ca^{2+} signaling transduction cascade, leading to the phenotype of stomatal closure. Several studies have identified CAND2 as the first phyto-melatonin receptor. *AtCAND2* is a membrane protein with seven transmembrane helices. Interaction with the unique G proteinαsubunit (GPA1) of *Arabidopsis* proved that CAND2 is a G protein-coupled receptor. ^{125}I-melatonin can bind to CAND2 in a specific and saturated manner. *Arabidopsis AtCand2* mutant exhibits no changes in the stomatal aperture when treated with melatonin, while 10 μmol/L melatonin induced stomatal closure in the wild-type counterparts [114]. These data indicate that further research on CAND2/PMTR1-mediated signaling in salt stress is required. Moreover, the discovery of CAND2/PMTR1 provides a new method for finding other melatonin receptors in plants.

5. Conclusions and Future Perspectives

Melatonin, as an antioxidant and signaling molecule, modulates a wide range of physiological functions in bacteria, fungi, invertebrates, vertebrates, algae, and plants. It has been extensively studied in humans and other animals, while plant studies have lagged behind. In light of its importance and significance, more and more attention has focused on the biosynthesis and bio-function of melatonin in plants. It has become a research hotspot in the plant biology kingdom, with increasing research being conducted in recent years [116,117]. To promote related research in plant salt tolerance, we summarized the regulatory roles and mechanisms of melatonin in plants during salt stress resistance by reviewing recently published literature, and we finally propose a model (Figure 4).

First, salt stress or the application of exogenous melatonin improves endogenous melatonin levels in plants, which modulates the expression of genes involved in melatonin biosynthesis and metabolisms or assimilates exogenous melatonin directly [116]. Increased levels of endogenous melatonin occur mainly by upregulation of melatonin biosynthesis-related genes or absorption of exogenous melatonin by plants; both mechanisms require further investigation. Increased endogenous levels enhanced plant salt stress tolerance via several different pathways. The improvement of antioxidant capacity, ion homeostasis, photosynthetic capacity and the regulation of ROS, NO, hormone, and polyamine metabolism by melatonin in salt-stressed plants was discussed. Previous studies have shown that Ca^{2+} signaling plays important roles in salt stress tolerance [118]; however, little evidence of Ca^{2+} signaling was observed in the melatonin-induced salt stress tolerance. Therefore, whether melatonin enhances plants salinity resistance through Ca^{2+} signaling requires further investigation.

Genetic modification and RNA-sequencing analysis are effective tools in the identification of the putative target genes involved in melatonin-enhanced salt stress tolerance. TPH, a putative gene involved in serotonin biosynthesis, has not been cloned in plants yet. However, *TDC* and *T5H*, two genes involved in serotonin biosynthesis, have been identified in many plants, but have not

Int. J. Mol. Sci. **2019**, *20*, 709

been cloned in *Arabidopsis*. We suspect that other biosynthesis pathways of melatonin may also exist in plants.

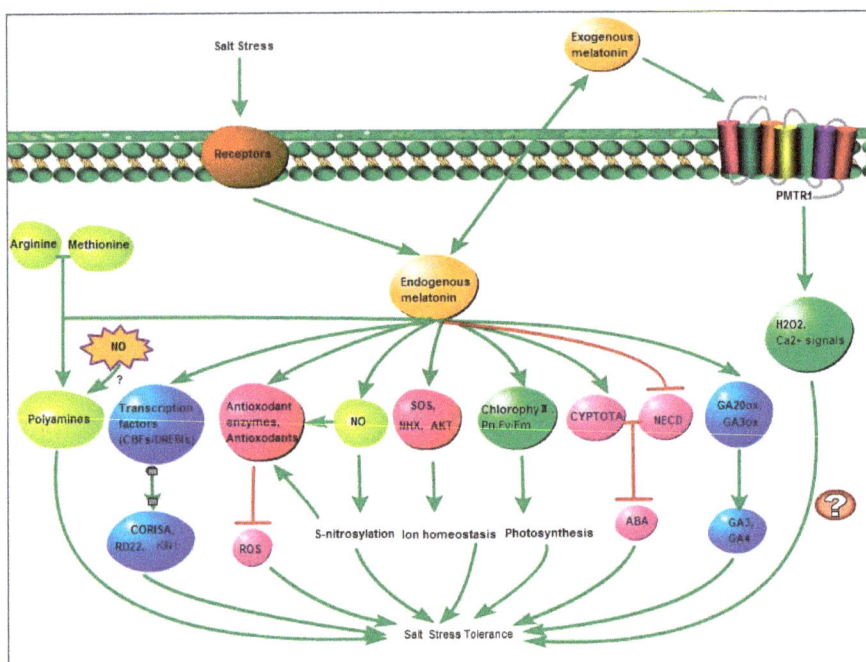

Figure 4. Melatonin-mediated salt stress response in plants. Abbreviation: NO, nitric oxide; ROS, reactive oxygen species; Pn, net photosynthetic rate; ABA, abscisic acid; GA, gibberellin acid. ⊥: represents inhibition; and →: represents promotion.

Plant melatonin receptors have been the bottleneck in the study of phyto-melatonin in the past few decades. With the first phytomelatonin receptor discovered recently in *Arabidopsis*, the involvement of PMTR1-mediated phytomelatonin signaling in salt stress response requires updated exploration. In addition, three melatonin receptors MT1, MT2, and MT3, have been identified in mammals, the identification of new phytomelatonin receptors is another exciting field to explore. Further studies in this field might deepen our understanding of the biological functions and molecular mechanisms governing melatonin's regulatory role during salt stress tolerance and beyond.

Supplementary Materials: Supplementary materials can be found at http://www.mdpi.com/1422-0067/20/3/709/s1.

Author Contributions: Conceptualization, X.N.; Formal Analysis, H.Z., T.Z., and S.L.; Resources, W.T. and W.S.; Data Curation, H.Z., X.W. and X.D.; Writing—Original Draft Preparation, H.Z.; Writing—Review and Editing, X.N. and W.S.; Supervision, W.S. and W.T.; Funding Acquisition, W.S. and X.N.

Funding: This work was mainly supported by the National Natural Science Foundation of China (Grant No. 31771778 and 31561143005). The funders had no role in study design, data collection and analysis, decision to publish, or preparation of the manuscript.

Acknowledgments: We are grateful to Hong Yue for her help with the phylogeny analysis, and Kewei Feng for his help on executing the figures.

Conflicts of Interest: The authors declare no conflict of interest.

References

1. Munns, R.; Tester, M. Mechanisms of salinity tolerance. *Annu. Rev. Plant Biol.* **2008**, *59*, 651–681. [CrossRef]
2. Abbasi, H.; Jamil, M.; Haq, A.; Ali, S.; Ahmad, R.; Malik, Z.; Parveen. Salt stress manifestation on plants, mechanism of salt tolerance and potassium role in alleviating it: A review. *Zemdirbyste-Agriculture* **2016**, *103*, 229–238. [CrossRef]
3. Rahnama, A.; James, R.; Poustini, K.; Munns, R. Stomatal conductance as a screen for osmotic stress tolerance in durum wheat growing in saline soil. *Funct. Plant Biol.* **2010**, *37*, 255–263. [CrossRef]
4. Ashraf, M.; Wu, L. Breeding for salinity tolerance in plants. *Crit. Rev. Plant Sci.* **1994**, *13*, 17–42. [CrossRef]
5. Shalata, A.; Mittova, V.; Volokita, M.; Guy, M.; Tal, M. Response of the cultivated tomato and its wild salt-tolerant relative Lycopersicon pennellii to salt-dependent oxidative stress: The root antioxidative system. *Physiol. Plant* **2001**, *112*, 487–494. [CrossRef] [PubMed]
6. Hasanuzzaman, M.; Oku, H.; Nahar, K.; Bhuyan, M.H.M.B.; Mahmud, J.A.; Baluska, F.; Fujita, M. Nitric oxide-induced salt stress tolerance in plants: ROS metabolism, signaling, and molecular interactions. *Plant Biotechnol. Rep.* **2018**, *12*, 77–92. [CrossRef]
7. Ke, Q.; Ye, J.; Wang, B.; Ren, J.; Yin, L.; Deng, X.; Wang, S. Melatonin mitigates salt stress in wheat seedlings by modulating polyamine metabolism. *Front. Plant Sci.* **2018**, *9*, 914. [CrossRef]
8. Lerner, A.B.; Case, J.D.; Takahashi, Y.; Lee, T.H.; Mori, W. Isolation of melatonin, the pineal gland factor that lightens melanocyteS1. *J. Am. Chem. Soc.* **1958**, *80*, 2587. [CrossRef]
9. Brainard, G.C.; Hanifin, J.P.; Greeson, J.M.; Byrne, B.; Glickman, G.; Gerner, E.; Rollag, M.D. Action spectrum for melatonin regulation in humans: evidence for a novel circadian photoreceptor. *J. Neurosci.* **2001**, *21*, 6405–6412. [CrossRef] [PubMed]
10. Mishima, K. Melatonin as a regulator of human sleep and circadian systems. *Nihon Rinsho* **2012**, *70*, 1139–1144, [Article in Japanese].
11. Rodriguez, C.; Mayo, J.C.; Sainz, R.M.; Antolín, I.; Herrera, F.; Martín, V.; Reiter, R.J. Regulation of antioxidant enzymes: a significant role for melatonin. *J. Pineal Res.* **2004**, *36*, 1–9. [CrossRef] [PubMed]
12. Calvo, J.R.; González-Yanes, C.; Maldonado, M.D. The role of melatonin in the cells of the innate immunity: a review. *J. Pineal Res.* **2013**, *55*, 103–120. [CrossRef] [PubMed]
13. Barrett, P.; Bolborea, M. Molecular pathways involved in seasonal body weight and reproductive responses governed by melatonin. *J. Pineal Res.* **2012**, *52*, 376–388. [CrossRef] [PubMed]
14. Dollins, A.B.; Zhdanova, I.V.; Wurtman, R.J.; Lynch, H.J.; Deng, M.H. Effect of inducing nocturnal serum melatonin concentrations in daytime on sleep, mood, body temperature, and performance. *Proc. Natl. Acad. Sci. USA* **1994**, *91*, 1824–1828. [CrossRef]
15. Hattori, A.; Migitaka, H.; Iigo, M.; Itoh, M.; Yamamoto, K.; Ohtani-Kaneko, R.; Hara, M.; Suzuki, T.; Reiter, R.J. Identification of melatonin in plants and its effects on plasma melatonin levels and binding to melatonin receptors in vertebrates. *Biochem. Mol. Biol. Int.* **1995**, *35*, 627–634. [PubMed]
16. Dubbels, R.; Reiter, R.J.; Klenke, E.; Goebel, A.; Schnakenberg, E.; Ehlers, C.; Schiwara, H.W.; Schloot, W. Melatonin in edible plants identified by radioimmunoassay and by high performance liquid chromatography-mass spectrometry. *J. Pineal Res.* **1995**, *18*, 28–31. [CrossRef] [PubMed]
17. Wang, L.Y.; Liu, J.L.; Wang, W.X.; Sun, Y. Exogenous melatonin improves growth and photosynthetic capacity of cucumber under salinity-induced stress. *Photosynthetica* **2016**, *54*, 19–27. [CrossRef]
18. Tan, D.-X.; Hardeland, R.; Manchester, L.C.; Korkmaz, A.; Ma, S.; Rosales-Corral, S.; Reiter, R.J. Functional roles of melatonin in plants, and perspectives in nutritional and agricultural science. *J. Exp. Bot.* **2012**, *63*, 577–597. [CrossRef] [PubMed]
19. Yu, Y.; Lv, Y.; Shi, Y.; Li, T.; Chen, Y.; Zhao, D.; Zhao, Z. The Role of Phyto-Melatonin and Related Metabolites in Response to Stress. *Molecules* **2018**, *23*, 1887. [CrossRef] [PubMed]
20. Xia, H.; Ni, Z.; Pan, D. Effects of exogenous melatonin on antioxidant capacity in Actinidia seedlings under salt stress. *IOP Conf. Ser. Earth Environ. Sci.* **2017**, *94*, 012024. [CrossRef]
21. Li, C.; Wang, P.; Wei, Z.; Liang, D.; Liu, C.; Yin, L.; Jia, D.; Fu, M.; Ma, F. The mitigation effects of exogenous melatonin on salinity-induced stress in Malus hupehensis. *J. Pineal Res.* **2012**, *53*, 298–306. [CrossRef] [PubMed]

22. Chen, Z.; Xie, Y.; Gu, Q.; Zhao, G.; Zhang, Y.; Cui, W.; Xu, S.; Wang, R.; Shen, W. The AtrbohF-dependent regulation of ROS signaling is required for melatonin-induced salinity tolerance in Arabidopsis. *Free Radic. Bio. Med.* **2017**, *108*, 465–477. [CrossRef] [PubMed]

23. Shi, H.; Qian, Y.; Tan, D.-X.; Reiter, R.J.; He, C. Melatonin induces the transcripts of CBF/DREB1s and their involvement in both abiotic and biotic stresses in Arabidopsis. *J. Pineal Res.* **2015**, *59*, 334–342. [CrossRef] [PubMed]

24. Zheng, X.; Tan, D.X.; Allan, A.C.; Zuo, B.; Zhao, Y.; Reiter, R.J.; Wang, L.; Wang, Z.; Guo, Y.; Zhou, J.; et al. Chloroplastic biosynthesis of melatonin and its involvement in protection of plants from salt stress. *Scientific reports* **2017**, *7*, 41236. [CrossRef] [PubMed]

25. Shi, H.; Jiang, C.; Ye, T.; Tan, D.-x.; Reiter, R.J.; Zhang, H.; Liu, R.; Chan, Z. Comparative physiological, metabolomic, and transcriptomic analyses reveal mechanisms of improved abiotic stress resistance in bermudagrass (*Cynodon dactylon* (L). Pers.) by exogenous melatonin. *J. Exp. Bot.* **2015**, *66*, 681–694. [CrossRef] [PubMed]

26. Beilby, M.J.; Al Khazaaly, S.; Bisson, M.A. Salinity-induced noise in membrane potential of Characeae chara australis: effect of exogenous melatonin. *J. Membrane Biol.* **2015**, *248*, 93–102. [CrossRef] [PubMed]

27. Zhang, Y.; Gao, W.; Lv, Y.; Bai, Q.; Wang, Y. Exogenous melatonin confers salt stress tolerance to Chlamydomonas reinhardtii (Volvocales, Chlorophyceae) by improving redox homeostasis. *Phycologia* **2018**, *57*, 680–691. [CrossRef]

28. Kostopoulou, Z.; Therios, I.; Roumeliotis, E.; Kanellis, A.K.; Molassiotis, A. Melatonin combined with ascorbic acid provides salt adaptation in *Citrus aurantium* L. seedlings. *Plant Physiol. Bioch.* **2015**, *86*, 155–165. [CrossRef]

29. Zhang, N.; Zhang, H.-J.; Sun, Q.-Q.; Cao, Y.-Y.; Li, X.; Zhao, B.; Wu, P.; Guo, Y.-D. Proteomic analysis reveals a role of melatonin in promoting cucumber seed germination under high salinity by regulating energy production. *Sci. Rep.* **2017**, *7*, 503. [CrossRef]

30. Zhang, H.-J.; Zhang, N.; Yang, R.-C.; Wang, L.; Sun, Q.-Q.; Li, D.-B.; Cao, Y.-Y.; Weeda, S.; Zhao, B.; Ren, S.; et al. Melatonin promotes seed germination under high salinity by regulating antioxidant systems, ABA and GA4 interaction in cucumber (*Cucumis sativus* L.). *J. Pineal Res.* **2014**, *57*, 269–279. [CrossRef]

31. Chen, Y.-E.; Mao, J.-J.; Sun, L.-Q.; Huang, B.; Ding, C.-B.; Gu, Y.; Liao, J.-Q.; Hu, C.; Zhang, Z.-W.; Yuan, S.; et al. Exogenous melatonin enhances salt stress tolerance in maize seedlings by improving antioxidant and photosynthetic capacity. *Physiol. Plant* **2018**, *164*, 349–363. [CrossRef] [PubMed]

32. Jiang, X.; Li, H.; Song, X. Seed priming with melatonin effects on seed germination and seedling growth in maize under salinity stress. *Pak. J. Bot.* **2016**, *48*, 1345–1352.

33. Jiang, C.; Cui, Q.; Feng, K.; Xu, D.; Li, C.; Zheng, Q. Melatonin improves antioxidant capacity and ion homeostasis and enhances salt tolerance in maize seedlings. *Acta Physiol. Plant.* **2016**, *38*, 82. [CrossRef]

34. Jiang, Y.; Liang, D.; Liao, M.A.; Lin, L. Effects of melatonin on the growth of radish Seedlings under salt stress. In Proceedings of the 3rd international conference on renewable energy and environmental technology (ICERE 2017), Hanoi, Vietnam, 25–27 February 2017.

35. Yao, H.; Wang, X.; Liao, M.A.; Lin, L. Effects of melatonin treated radish on the growth of following stubble lettuce under salt stress. In Proceedings of the 3rd international conference on renewable energy and environmental technology (ICERE 2017), Hanoi, Vietnam, 25–27 February 2017.

36. Zeng, L.; Cai, J.S.; Li, J.J.; Lu, G.Y.; Li, C.S.; Fu, G.P.; Zhang, X.K.; Ma, H.Q.; Liu, Q.Y.; Zou, X.L.; et al. Exogenous application of a low concentration of melatonin enhances salt tolerance in rapeseed (Brassica napus L.) seedlings. *J. Integ. Agr* **2018**, *17*, 328–335. [CrossRef]

37. Zhao, G.; Zhao, Y.; Yu, X.; Kiprotich, F.; Han, H.; Guan, R.; Wang, R.; Shen, W. Nitric oxide is required for melatonin-enhanced tolerance against salinity stress in rapeseed (*Brassica napus* L.) seedlings. *Int. J. MolSci.* **2018**, *19*, 1912. [CrossRef]

38. Liang, C.; Zheng, G.; Li, W.; Wang, Y.; Hu, B.; Wang, H.; Wu, H.; Qian, Y.; Zhu, X.-G.; Tan, D.-X.; et al. Melatonin delays leaf senescence and enhances salt stress tolerance in rice. *J. Pineal Res.* **2015**, *59*, 91–101. [CrossRef] [PubMed]

39. Li, X.; Yu, B.; Cui, Y.; Yin, Y. Melatonin application confers enhanced salt tolerance by regulating Na+ and Cl− accumulation in rice. *Plant Growth Regul.* **2017**, *83*, 441–454. [CrossRef]

40. Wei, W.; Li, Q.-T.; Chu, Y.-N.; Reiter, R.J.; Yu, X.-M.; Zhu, D.-H.; Zhang, W.-K.; Ma, B.; Lin, Q.; Zhang, J.-S.; et al. Melatonin enhances plant growth and abiotic stress tolerance in soybean plants. *J. Exp Bot.* **2015**, *66*, 695–707. [CrossRef] [PubMed]

41. Arora, D.; Bhatla, S.C. Melatonin and nitric oxide regulate sunflower seedling growth under salt stress accompanying differential expression of Cu/Zn SOD and Mn SOD. *Free Radical Biol. Med.* **2017**, *106*, 315–328. [CrossRef]

42. Kaur, H.; Bhatla, S.C. Melatonin and nitric oxide modulate glutathione content and glutathione reductase activity in sunflower seedling cotyledons accompanying salt stress. *Nitric Oxide* **2016**, *59*, 42–53. [CrossRef]

43. Yu, Y.; Wang, A.; Li, X.; Kou, M.; Wang, W.; Chen, X.; Xu, T.; Zhu, M.; Ma, D.; Li, Z.; et al. Melatonin-stimulated triacylglycerol breakdown and energy turnover under salinity stress contributes to the maintenance of plasma membrane H^+-ATPase activity and K^+/Na^+ homeostasis in sweet potato. *Front. Plant Sci.* **2018**, *9*, 256. [CrossRef] [PubMed]

44. Zhou, X.; Zhao, H.; Cao, K.; Hu, L.; Du, T.; Baluška, F.; Zou, Z. Beneficial roles of melatonin on redox regulation of photosynthetic electron transport and synthesis of D1 protein in tomato seedlings under salt stress. *Front. Plant Sci.* **2016**, *7*, 1823. [CrossRef] [PubMed]

45. Dawood, M.G.; El-Awadi, M.E. Alleviation of salinity stress on *Vicia faba* L. plants via seed priming with melatonin. *Acta Biológica Colombiana* **2015**, *20*, 223–235. [CrossRef]

46. Li, H.; Chang, J.; Chen, H.; Wang, Z.; Gu, X.; Wei, C.; Zhang, Y.; Ma, J.; Yang, J.; Zhang, X. Exogenous Melatonin Confers Salt Stress Tolerance to Watermelon by Improving Photosynthesis and Redox Homeostasis. *Front. Plant Sci.* **2017**, *8*, 295. [CrossRef] [PubMed]

47. El-Mashad, A.A.A.; Mohamed, H.I. Brassinolide alleviates salt stress and increases antioxidant activity of cowpea plants (*Vigna sinensis*). *Protoplasma* **2012**, *249*, 625–635. [CrossRef] [PubMed]

48. Zhang, M.; Smith, J.A.C.; Harberd, N.P.; Jiang, C. The regulatory roles of ethylene and reactive oxygen species (ROS) in plant salt stress responses. *Plant Mol. Biol.* **2016**, *91*, 651–659. [CrossRef] [PubMed]

49. Ahmad, P.; Abdul Jaleel, C.; A Salem, M.; Nabi, G.; Sharma, S. Roles of Enzymatic and non-enzymatic antioxidants in plants during abiotic stress. *Crit Rev Biotechnol.* **2010**, *30*, 161–175. [CrossRef]

50. Tan, D.X.; Manchester, L.C.; Terron, M.P.; Flores, L.J.; Reiter, R.J. One molecule, many derivatives: A never-ending interaction of melatonin with reactive oxygen and nitrogen species? *J. Pineal Res.* **2007**, *42*, 28–42. [CrossRef]

51. Campos, L.M.O.; Hsie, S.B.; Granja, A.J.A.; Correia, M.R.; Almeida-Cortez, J.; Pompelli, M.F. Photosynthesis and antioxidant activity in *Jatropha curcas* L. under salt stress. *Braz. J. Plant Physiol.* **2012**, *24*, 55–67. [CrossRef]

52. Meloni, D.A.; Oliva, M.A.; Martinez, C.A.; Cambraia, J. Photosynthesis and activity of superoxide dismutase, peroxidase and glutathione reductase in cotton under salt stress. *Environ. Exp. Bot.* **2003**, *49*, 69–76. [CrossRef]

53. Amtmann, A.; Leigh, R. Ion Homeostasis. In *Abiotic Stress Adaptation in Plants: Physiological, Molecular and Genomic Foundation*; Pareek, A., Sopory, S.K., Bohnert, H.J., Eds.; Springer: Dordrecht, The Netherlands, 2010.

54. Zhu, J.K. Regulation of ion homeostasis under salt stress. *Curr. Opin. Plant Biol.* **2003**, *6*, 441–445. [CrossRef]

55. Fukuda, A.; Nakamura, A.; Hara, N.; Toki, S.; Tanaka, Y. Molecular and functional analyses of rice NHX-type Na^+/H^+ antiporter genes. *Planta* **2011**, *233*, 175–188. [CrossRef] [PubMed]

56. Padan, E.; Venturi, M.; Gerchman, Y.; Dover, N. Na^+/H^+ antiporters. *BBA- Bioenergetics* **2001**, *1505*, 144–157. [CrossRef]

57. Shi, H.; Zhu, J.-K. Regulation of expression of the vacuolar Na^+/H^+ antiporter gene AtNHX1 by salt stress and abscisic acid. *Plant Mol. Biol.* **2002**, *50*, 543–550. [CrossRef] [PubMed]

58. Garriga, M.; Raddatz, N.; Véry, A.-A.; Sentenac, H.; Rubio-Meléndez, M.E.; González, W.; Dreyer, I. Cloning and functional characterization of HKT1 and AKT1 genes of *Fragaria* spp.—Relationship to plant response to salt stress. *J. Plant. Physiol.* **2017**, *210*, 9–17. [CrossRef] [PubMed]

59. Arnao, M.B.; Hernández-Ruiz, J. Melatonin and its relationship to plant hormones. *Ann. Bot* **2018**, *121*, 195–207. [CrossRef]

60. Wang, Q.; An, B.; Wei, Y.; Reiter, R.J.; Shi, H.; Luo, H.; He, C. Melatonin regulates root meristem by repressing auxin synthesis and polar auxin transport in Arabidopsis. *Front. Plant Sci.* **2016**, *7*, 1882. [CrossRef]

61. Pelagio-Flores, R.; Muñoz-Parra, E.; Ortiz-Castro, R.; López-Bucio, J. Melatonin regulates Arabidopsis root system architecture likely acting independently of auxin signaling. *J. Pineal Res.* **2012**, *53*, 279–288. [CrossRef]

62. Chen, Q.; Qi, W.-b.; Reiter, R.J.; Wei, W.; Wang, B.-m. Exogenously applied melatonin stimulates root growth and raises endogenous indoleacetic acid in roots of etiolated seedlings of Brassica juncea. *J. Plant Physiol.* **2009**, *166*, 324–328. [CrossRef]

63. Footitt, S.; Douterelo-Soler, I.; Clay, H.; Finch-Savage, W.E. Dormancy cycling in Arabidopsis seeds is controlled by seasonally distinct hormone-signaling pathways. *Proc. Natl. Acad. Sci. USA* **2011**, *108*, 20236–20241. [CrossRef]

64. Zhang, J.; Shi, Y.; Zhang, X.; Du, H.; Xu, B.; Huang, B. Melatonin suppression of heat-induced leaf senescence involves changes in abscisic acid and cytokinin biosynthesis and signaling pathways in perennial ryegrass (*Lolium perenne* L.). *Environ. Exp. Bot.* **2017**, *138*, 36–45. [CrossRef]

65. Yang, R.; Yang, T.; Zhang, H.; Qi, Y.; Xing, Y.; Zhang, N.; Li, R.; Weeda, S.; Ren, S.; Ouyang, B.; et al. Hormone profiling and transcription analysis reveal a major role of ABA in tomato salt tolerance. *Plant Physiol. Bioch.* **2014**, *77*, 23–34. [CrossRef] [PubMed]

66. Maggio, A.; Barbieri, G.; Raimondi, G.; De Pascale, S. Contrasting Effects of GA3 Treatments on Tomato Plants Exposed to Increasing Salinity. *J. Plant Growth Regul.* **2010**, *29*, 63–72. [CrossRef]

67. Li, C.; Tan, D.-X.; Liang, D.; Chang, C.; Jia, D.; Ma, F. Melatonin mediates the regulation of ABA metabolism, free-radical scavenging, and stomatal behaviour in two Malus species under drought stress. *J. Exp. Bot.* **2015**, *66*, 669–680. [CrossRef] [PubMed]

68. Jia, W.; Zhang, J. Water stress-induced abscisis acid accumulation in relation to reducing agents and sulfhydryl modifiers in maize plant. *Plant Cell Environ.* **2000**, *12*, 1389–1395. [CrossRef]

69. Fu, J.; Wu, Y.; Miao, Y.; Xu, Y.; Zhao, E.; Wang, J.; Sun, H.; Liu, Q.; Xue, Y.; Xu, Y.; et al. Improved cold tolerance in Elymus nutans by exogenous application of melatonin may involve ABA-dependent and ABA-independent pathways. *Scientific Reports* **2017**, *7*, 39865. [CrossRef] [PubMed]

70. Aydogan, S.; Yerer, M.B.; Goktas, A. Melatonin and nitric oxide. *J. Endocrinol. Invest.* **2006**, *29*, 281–287. [CrossRef] [PubMed]

71. Zhao, M.G.; Tian, Q.Y.; Zhang, W.H. Nitric oxide synthase-dependent nitric oxide production is associated with salt tolerance in Arabidopsis. *Plant Physiol.* **2007**, *144*, 206–217. [CrossRef]

72. Lozano-Juste, J.; León, J. Enhanced abscisic acid-mediated responses in nia1nia2noa1-2 triple mutant impaired in NIA/NR- and AtNOA1-dependent nitric oxide biosynthesis in Arabidopsis. *Plant Physiol.* **2010**, *152*, 891–903. [CrossRef]

73. Ninnemann, H.; Maier, J. Indications for the occurrence of nitric oxide synthases in fungi and plants and the involvement in photoconidiation of *Neurospora crassa*. *Photochem. Photobiol.* **1996**, *64*, 393–398. [CrossRef]

74. Corpas, F.J.; Palma, J.M.; Del Río, L.A.; Barroso, J.B. Evidence supporting the existence of L-arginine-dependent nitric oxide synthase activity in plants. *New Phytologist* **2009**, *184*, 9–14. [CrossRef] [PubMed]

75. Gupta, K.J.; Fernie, A.R.; Kaiser, W.M.; van Dongen, J.T. On the origins of nitric oxide. *Trends Plant Sci.* **2011**, *16*, 160–168. [CrossRef] [PubMed]

76. Astier, J.; Rasul, S.; Koen, E.; Manzoor, H.; Besson-Bard, A.; Lamotte, O.; Jeandroz, S.; Durner, J.; Lindermayr, C.; Wendehenne, D. S-nitrosylation: An emerging post-translational protein modification in plants. *Plant Sci.* **2011**, *181*, 527–533. [CrossRef] [PubMed]

77. Gupta, K.J. Protein S-nitrosylation in plants: photorespiratory metabolism and NO signaling. *Sci Signal.* **2011**, *4*, jc1. [CrossRef] [PubMed]

78. Jaffrey, S.R.; Erdjument-Bromage, H.; Ferris, C.D.; Tempst, P.; Snyder, S.H. Protein S-nitrosylation: A physiological signal for neuronal nitric oxide. *Nat. Cell Biol.* **2001**, *3*, 193. [CrossRef]

79. Liu, N.; Gong, B.; Jin, Z.; Wang, X.; Wei, M.; Yang, F.; Li, Y.; Shi, Q. Sodic alkaline stress mitigation by exogenous melatonin in tomato needs nitric oxide as a downstream signal. *J. Plant Physiol.* **2015**, *186-187*, 68–77. [CrossRef] [PubMed]

80. Zhou, C.; Liu, Z.; Zhu, L.; Ma, Z.; Wang, J.; Zhu, J. Exogenous melatonin improves plant iron deficiency tolerance via increased accumulation of polyamine-mediated nitric oxide. *Int. J. Mol. Sci.* **2016**, *17*, 1777. [CrossRef]

81. Shi, H.; Chen, Y.; Tan, D.-X.; Reiter, R.J.; Chan, Z.; He, C. Melatonin induces nitric oxide and the potential mechanisms relate to innate immunity against bacterial pathogen infection in Arabidopsis. *J. Pineal Res.* **2015**, *59*, 102–108. [CrossRef]

82. Masson, P.H.; Takahashi, T.; Angelini, R. Editorial: Molecular mechanisms underlying polyamine functions in plants. *Front. Plant Sci.* **2017**, *8*, 14. [CrossRef]
83. Gill, S.S.; Tuteja, N. Polyamines and abiotic stress tolerance in plants. *Plant Signal. Behav.* **2010**, *5*, 26–33.
84. Sánchez-Rodríguez, E.; Romero, L.; Ruiz, J.M. Accumulation of free polyamines enhances the antioxidant response in fruits of grafted tomato plants under water stress. *J. Plant Physiol.* **2016**, *190*, 72–78. [CrossRef] [PubMed]
85. Gong, X.; Shi, S.; Dou, F.; Song, Y.; Ma, F. Exogenous melatonin alleviates alkaline stress in *Malus hupehensis* Rehd. by regulating the biosynthesis of polyamines. *Molecules* **2017**, *22*, 1542. [CrossRef] [PubMed]
86. Zhao, H.; Zhang, K.; Zhou, X.; Xi, L.; Wang, Y.; Xu, H.; Pan, T.; Zou, Z. Melatonin alleviates chilling stress in cucumber seedlings by upregulation of CsZat12 and modulation of polyamine and abscisic acid metabolism. *Sci. Rep.* **2017**, *7*, 4998. [CrossRef]
87. Back, K.; Tan, D.-X.; Reiter, R.J. Melatonin biosynthesis in plants: Multiple pathways catalyze tryptophan to melatonin in the cytoplasm or chloroplasts. *J. Pineal Res.* **2016**, *61*, 426–437. [CrossRef] [PubMed]
88. Murch, S.J.; KrishnaRaj, S.; Saxena, P.K. Tryptophan is a precursor for melatonin and serotonin biosynthesis in in vitro regenerated St. John's wort (*Hypericum perforatum* L. cv. Anthos) plants. *Plant Cell Rep.* **2000**, *19*, 698–704. [CrossRef]
89. Byeon, Y.; Park, S.; Lee, H.Y.; Kim, Y.-S.; Back, K. Elevated production of melatonin in transgenic rice seeds expressing rice tryptophan decarboxylase. *J. Pineal Res.* **2014**, *56*, 275–282. [CrossRef]
90. Zhao, D.; Wang, R.; Liu, D.; Wu, Y.; Sun, J.; Tao, J. Melatonin and expression of tryptophan decarboxylase gene (TDC) in Herbaceous peony (*Paeonia lactiflora* Pall.) flowers. *Molecules* **2018**, *23*, 1164. [CrossRef]
91. Park, S.; Byeon, Y.; Back, K. Transcriptional suppression of tryptamine 5-hydroxylase, a terminal serotonin biosynthetic gene, induces melatonin biosynthesis in rice (*Oryza sativa* L.). *J. Pineal Res.* **2013**, *55*, 131–137. [CrossRef]
92. Byeon, Y.; Choi, G.-H.; Lee, H.Y.; Back, K. Melatonin biosynthesis requires *N*-acetylserotonin methyltransferase activity of caffeic acid *O*-methyltransferase in rice. *J. Exp. Bot.* **2015**, *66*, 6917–6925. [CrossRef]
93. Hardeland, R. Taxon- and site-specific melatonin catabolism. *Molecules* **2017**, *22*, 2015. [CrossRef]
94. Kanwar, M.K.; Yu, J.; Zhou, J. Phytomelatonin: Recent advances and future prospects. *J. Pineal Res.* **2018**, *65*, e12526. [CrossRef] [PubMed]
95. Wei, Y.; Zeng, H.; Hu, W.; Chen, L.; He, C.; Shi, H. Comparative transcriptional profiling of melatonin synthesis and catabolic genes indicates the possible role of melatonin in developmental and stress responses in rice. *Front. Plant Sci.* **2016**, *7*, 676. [CrossRef] [PubMed]
96. Byeon, Y.; Back, K. Molecular cloning of melatonin 2-hydroxylase responsible for 2-hydroxymelatonin production in rice (Oryza sativa). *J. Pineal Res.* **2015**, *58*, 343–351. [CrossRef] [PubMed]
97. Zhang, N.; Zhang, H.J.; Zhao, B.; Sun, Q.Q.; Cao, Y.Y.; Li, R.; Wu, X.X.; Weeda, S.; Li, L.; Ren, S.; et al. The RNA-seq approach to discriminate gene expression profiles in response to melatonin on cucumber lateral root formation. *J. Pineal Res.* **2014**, *56*, 39–50. [CrossRef] [PubMed]
98. Pang, X.; Wei, Y.; Cheng, Y.; Pan, L.; Ye, Q.; Wang, R.; Ruan, M.; Zhou, G.; Yao, Z.; Li, Z.; et al. The Tryptophan Decarboxylase in *Solanum lycopersicum*. *Molecules* **2018**, *23*, 998. [CrossRef] [PubMed]
99. Kang, S.; Kang, K.; Lee, K.; Back, K. Characterization of rice tryptophan decarboxylases and their direct involvement in serotonin biosynthesis in transgenic rice. *Planta* **2007**, *227*, 263–272. [CrossRef] [PubMed]
100. Fan, J.B.; Xie, Y.; Zhang, Z.C.; Chen, L. Melatonin: A Multifunctional Factor in Plants. *Int. J. Mol. Sci.* **2018**, *19*, 1528. [CrossRef]
101. Liu, W.; Zhao, D.; Zheng, C.; Chen, C.; Peng, X.; Cheng, Y.; Wan, H. Genomic analysis of the ASMT gene family in *Solanum lycopersicum*. *Molecules* **2017**, *22*, 1984. [CrossRef]
102. Kang, K.; Lee, K.; Park, S.; Byeon, Y.; Back, K. Molecular cloning of rice serotonin N-acetyltransferase, the penultimate gene in plant melatonin biosynthesis. *J. Pineal Res.* **2013**, *55*, 7–13. [CrossRef]
103. Byeon, Y.; Lee, H.Y.; Lee, K.; Park, S.; Back, K. Cellular localization and kinetics of the rice melatonin biosynthetic enzymes SNAT and ASMT. *J. Pineal Res.* **2014**, *56*, 107–114. [CrossRef]
104. Kang, S.; Kang, K.; Lee, K.; Back, K. Characterization of tryptamine 5-hydroxylase and serotonin synthesis in rice plants. *Plant Cell Rep.* **2007**, *26*, 2009–2015. [CrossRef]

105. Torrens-Spence, M.P.; Liu, P.; Ding, H.; Harich, K.; Gillaspy, G.; Li, J. Biochemical evaluation of the decarboxylation and decarboxylation-deamination activities of plant aromatic amino acid decarboxylases. *J. Biol. Chem.* **2013**, *288*, 2376–2387. [CrossRef] [PubMed]

106. Torrens-Spence, M.P.; Lazear, M.; von Guggenberg, R.; Ding, H.; Li, J. Investigation of a substrate-specifying residue within Papaver somniferum and Catharanthus roseus aromatic amino acid decarboxylases. *Phytochemistry* **2014**, *106*, 37–43. [CrossRef]

107. Kang, K.; Kong, K.; Park, S.; Natsagdorj, U.; Kim, Y.S.; Back, K. Molecular cloning of a plant N-acetylserotonin methyltransferase and its expression characteristics in rice. *J. Pineal Res.* **2011**, *50*, 304–309. [CrossRef] [PubMed]

108. Byeon, Y.; Lee, H.Y.; Back, K. Cloning and characterization of the serotonin N-acetyltransferase-2 gene (SNAT2) in rice (*Oryza sativa*). *J. Pineal Res.* **2016**, *61*, 198–207. [CrossRef] [PubMed]

109. Maag, J.L.V. gganatogram: An R package for modular visualisation of anatograms and tissues based on ggplot2. *F1000Research* **2018**, *7*, 1576. [CrossRef] [PubMed]

110. Ebisawa, T.; Karne, S.; Lerner, M.R.; Reppert, S.M. Expression cloning of a high-affinity melatonin receptor from Xenopus dermal melanophores. *Proc. Natl. Acad. Sci. USA* **1994**, *91*, 6133–6137. [CrossRef]

111. Ng, K.Y.; Leong, M.K.; Liang, H.; Paxinos, G. Melatonin receptors: distribution in mammalian brain and their respective putative functions. *Brain Struct. Funct.* **2017**, *222*, 2921–2939. [CrossRef]

112. Witt-Enderby, P.A.; Bennett, J.; Jarzynka, M.J.; Firestine, S.; Melan, M.A. Melatonin receptors and their regulation: biochemical and structural mechanisms. *Life Sci.* **2003**, *72*, 2183–2198. [CrossRef]

113. Dubocovich, M.L.; Delagrange, P.; Krause, D.N.; Sugden, D.; Cardinali, D.P.; Olcese, J. International union of basic and clinical pharmacology. LXXV. Nomenclature, classification, and pharmacology of G protein-coupled melatonin receptors. *Pharmacol. Rev.* **2010**, *62*, 343–380. [CrossRef]

114. Wei, J.; Li, D.-X.; Zhang, J.-R.; Shan, C.; Rengel, Z.; Song, Z.-B.; Chen, Q. Phytomelatonin receptor PMTR1-mediated signaling regulates stomatal closure in Arabidopsis thaliana. *J. Pineal Res.* **2018**, *65*, e12500. [CrossRef] [PubMed]

115. Nosjean, O.; Ferro, M.; Cogé, F.; Beauverger, P.; Henlin, J.-M.; Lefoulon, F.; Fauchère, J.-L.; Delagrange, P.; Canet, E.; Boutin, J.A. Identification of the Melatonin-binding SiteMT 3 as the Quinone Reductase 2. *J. Biol. Chem.* **2000**, *275*, 31311–31317. [CrossRef] [PubMed]

116. Zhang, N.; Sun, Q.; Zhang, H.; Cao, Y.; Weeda, S.; Ren, S.; Guo, Y.-D. Roles of melatonin in abiotic stress resistance in plants. *J. Exp. Bot.* **2015**, *66*, 647–656. [CrossRef] [PubMed]

117. Tan, D.X.; Manchester, L.C.; Liu, X.; Rosales-Corral, S.A.; Acuna-Castroviejo, D.; Reiter, R.J. Mitochondria and chloroplasts as the original sites of melatonin synthesis: a hypothesis related to melatonin's primary function and evolution in eukaryotes. *J. Pineal Res.* **2013**, *54*, 127–138. [CrossRef] [PubMed]

118. Park, S.-Y.; B Seo, S.; J Lee, S.; G Na, J.; Kim, Y.J. Mutation in PMR1, a Ca^{2+}-ATPase in Golgi, confers salt tolerance in Saccharomyces cerevisiae by inducing expression of PMR2, an Na$^+$-ATPase in plasma membrane. *J. Biol. Chem.* **2001**, *276*, 28694–28699. [CrossRef] [PubMed]

International Journal of
Molecular Sciences

MDPI

Article

Nitric Oxide Is Required for Melatonin-Enhanced Tolerance against Salinity Stress in Rapeseed (*Brassica napus* L.) Seedlings

Gan Zhao [1], Yingying Zhao [1], Xiuli Yu [1], Felix Kiprotich [1], Han Han [1], Rongzhan Guan [2], Ren Wang [3] and Wenbiao Shen [1,*]

[1] College of Life Sciences, Laboratory Center of Life Sciences, Nanjing Agricultural University, Nanjing 210095, China; 2015116111@njau.edu.cn (G.Z.); 2017116114@njau.edu.cn (Y.Z.); 2016116109@njau.edu.cn (X.Y.); kiprotichfelix@yahoo.com (F.K.); martinhan956@gmail.com (H.H.)
[2] National Key Laboratory of Crop Genetics and Germplasm Enhancement, Jiangsu Collaborative Innovation Center for Modern Crop Production, Nanjing Agricultural University, Nanjing 210095, China; guanrzh@njau.edu.cn
[3] Institute of Botany, Jiangsu Province and Chinese Academy of Sciences, Nanjing 210014, China; wangren@126.com
* Correspondence: wbshenh@njau.edu.cn; Tel./Fax: +86-258-439-6542

Received: 4 May 2018; Accepted: 27 June 2018; Published: 29 June 2018

Abstract: Although melatonin (*N*-acetyl-5-methoxytryptamine) could alleviate salinity stress in plants, the downstream signaling pathway is still not fully characterized. Here, we report that endogenous melatonin and thereafter nitric oxide (NO) accumulation was successively increased in NaCl-stressed rapeseed (*Brassica napus* L.) seedling roots. Application of melatonin and NO-releasing compound not only counteracted NaCl-induced seedling growth inhibition, but also reestablished redox and ion homeostasis, the latter of which are confirmed by the alleviation of reactive oxygen species overproduction, the decreases in thiobarbituric acid reactive substances production, and Na^+/K^+ ratio. Consistently, the related antioxidant defense genes, *sodium hydrogen exchanger* (*NHX1*), and *salt overly sensitive 2* (*SOS2*) transcripts are modulated. The involvement S-nitrosylation, a redox-based posttranslational modification triggered by NO, is suggested. Further results show that in response to NaCl stress, the increased NO levels are strengthened by the addition of melatonin in seedling roots. Above responses are abolished by the removal of NO by NO scavenger. We further discover that the removal of NO does not alter endogenous melatonin content in roots supplemented with NaCl alone or together with melatonin, thus excluding the possibility of NO-triggered melatonin production. Genetic evidence reveals that, compared with wild-type Arabidopsis, the hypersensitivity to NaCl in *nia1/2* and *noa1* mutants (exhibiting null nitrate reductase activity and indirectly reduced endogenous NO level, respectively) cannot be rescued by melatonin supplementation. The reestablishment of redox homeostasis and induction of *SOS* signaling are not observed. In summary, above pharmacological, molecular, and genetic data conclude that NO operates downstream of melatonin promoting salinity tolerance.

Keywords: Arabidopsis; *Brassica napus*; ion homeostasis; melatonin; NaCl stress; nitric oxide; redox homeostasis

1. Introduction

Soil salinity is a major factor that significantly influences global agricultural production [1]. High salinity (mainly NaCl) provokes two primary effects on plants, including ionic and oxidative effects [1–4]. In general, high NaCl stress disturbs the ionic environment of plant cells, notably forming a higher Na^+/K^+ ratio [5]. Plants usually remove excessive Na^+ by Na^+/H^+ antiporters, and genetic evidence revealed that overexpressing these antiporter genes can improve salt tolerance [6,7]. For example, *SOS* signaling is a well-known pathway responsible for initiating transport of Na^+ out of the cells, or activating an unknown transporter, thus leading to the sequestration of Na^+ in the vacuole [8–10]. Another type of Na^+/H^+ antiporter belongs to the Na^+/H^+ exchanger (NHX) family, and constitutive overexpression of *NHX* can increase Na accumulation in vacuoles, and thus, enhance salt tolerance [10]. Meanwhile, a large number of reactive oxygen species (ROS), such as superoxide anion, hydrogen peroxide, and hydroxy1 radicals, are induced under salinity conditions [11]. To combat salt-induced oxidative stress, the enzymatic antioxidant system provides a highly efficient and specific ROS scavenging approach for plants. For example, superoxide dismutase (SOD), catalase (CAT), and guaiacol peroxidase (POD), are very important parts of this enzymatic system, and normally, plants decrease ROS by upregulating activities of these enzymes [11–13].

Rapeseed (*Brassica napus* L.) is one of the most widely cultivated oil crops in the world because of the healthy fatty acid composition of its oil and high protein content of its meal. It is classified as a moderate salinity-tolerant crop [14]. During the growth period, rapeseed plants are challenged by salt stress, and ionic and redox imbalance are two major effects associated with salinity toxicity [11,14–16]. As more land becomes salinized, the studying of related mechanisms (the reestablishment of ionic and redox balance) and the application of effective methods (including reclamation of saline soils by use of chemicals or plant growth-promoting bacteria, or by growing salt tolerant cultivars in the saline soils, etc.) for improving salt tolerance in rapeseed plants [14–18], are becoming increasingly significant [19–25].

Melatonin (*N*-acetyl-5-methoxytryptamine) was discovered, and isolated from the bovine pineal gland in 1958 [26]. With a large set of functions in animals (circadian rhythms, seasonal rhythms, and alleviating oxidative stress; [27–30]), this compound was also detected in plants, and used as both a plant growth regulator and a biostimulator to alleviate abiotic and biotic stresses, including salinity, cold, drought, chemical pollutants, and defense against bacterial pathogen infection [31–33]. Previous results revealed that exogenous application of melatonin not only increased endogenous melatonin levels, but also improved the salt tolerance in Arabidopsis, soybean, Chinese crab apple, rice, cucumber, and bermudagrass [19,20,34–37]. Above beneficial roles of melatonin are normally associated with enhanced activities of antioxidant enzymes, as well as upregulating transcripts of ion channel genes, or sugar and glycolysis metabolism-related genes [36,37]. However, the corresponding detailed mechanism, especially the crosstalk with other signaling components and related transduction cascade, is still not fully characterized.

It is well known that nitric oxide (NO), one of the important gasotransmitters controlling a diverse range of physiological functions in plants [38–40], can also enhance salinity tolerance [21,41–44]. Previous reports revealed that there are at least two major enzymatic sources of NO: a nitrate/nitrite-dependent pathway and an L-Arg-dependent pathway [38,39]. Further experiments with Arabidopsis single and triple mutants, exhibiting null nitrate reductase (NR) activity (*nia1/2*) and indirectly reduced endogenous NO level (*nitric oxide associated1*; *noa1*), revealed that NO production is associated with salinity tolerance in Arabidopsis [42,44,45]. Importantly, protein post-translational modification by *S*-nitrosylation was preliminarily used to explain the physiological functions of NO in both animals and plants [46], including adaptation against biotic and abiotic stresses in plants [41,47,48]. However, it is not clear whether NO-dependent *S*-nitrosylation is also associated with melatonin responses in plants.

Although previous pharmacological data showed the interplay between melatonin and NO leading to plant tolerance against NaCl stress in sunflower seedlings [49,50], no genetic study has yet provided definitive proof of a role of endogenous NO in melatonin signaling governing salinity tolerance. In this study, we firstly evaluated the role of NO in melatonin-triggered salinity tolerance in rapeseed seedlings by using pharmacological and biochemical approaches. The reestablishment of redox and ion homeostasis was confirmed. The involvement of *S*-nitrosylation is also discovered, suggesting that NO is involved in melatonin signaling as a downstream messenger. Afterwards, both *nia1/2* and *noa1* Arabidopsis mutants were utilized to investigate the relationship between NO and melatonin in salinity tolerance. We thus concluded that NO acts downstream of melatonin signaling to enhance tolerance against salinity.

2. Results

2.1. Salt Stress Stimulates Melatonin and NO Production

To assess the sensitivity of rapeseed seedling growth to NaCl stress, the effects of varying concentrations (100, 150, 200, and 250 mM) of NaCl on root growth were investigated. As shown in Figure 1, the exposure of seedlings to NaCl resulted in dose-dependent decreases in the root elongation and root fresh weight. Since approximate 50% inhibition in above parameters was observed in 200 mM NaCl-treated seedlings, this concentration of NaCl was applied in the following experiments.

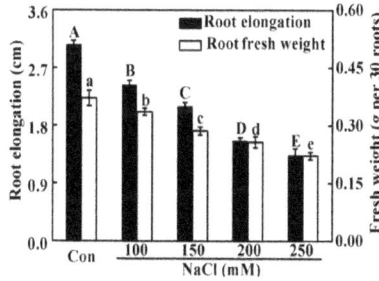

Figure 1. Growth inhibition of seedling roots upon NaCl stress. Three-day-old rapeseed seedlings were transferred to 100, 150, 200, and 250 mM NaCl for 2 days. Afterwards, the root elongation (left) and root fresh weight (right) were measured. The sample without chemicals was the control (Con). Values are means ± standard error (SE) of three independent experiments with at least three replicates for each. Bars with different letters are significant different at $p < 0.05$ according to Duncan's multiple range test.

Griess reagent (visible spectrophotography) and laser scanning confocal microscopy (LSCM) with the specific probe (4-amino-5-methylamino-2′,7′-difluorofluorescein diacetate; DAF-FM DA), are the most frequently used methods for the determination of NO production in plants. Subsequently, the time course experiments for 48 h revealed the rapid burst of endogenous melatonin (Figure 2A; detected by enzyme-linked immunosorbent assay) and NO (Figure 2B,C; respectively determined by visible spectrophotography and LSCM) accumulation in NaCl-treated root tissues, peaking at 6 h and 12 h of stress, compared to the control sample (Con). We also noticed that the increases of melatonin and NO were still evident until 48 h, although both of them were decreased after the peaking points.

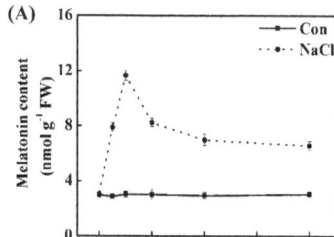

Figure 2. *Cont.*

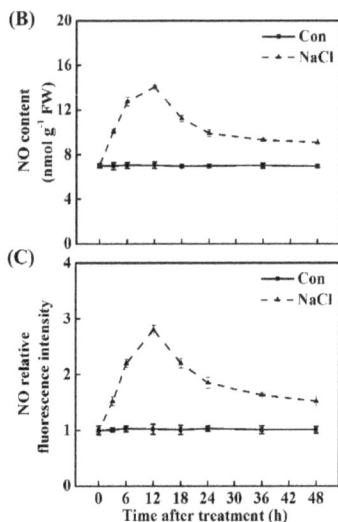

Figure 2. Changes in endogenous melatonin and nitric oxide (NO) levels in response to NaCl stress. Three-day-old rapeseed seedlings were transferred to 200 mM NaCl for 2 days. Meanwhile, melatonin (**A**) detected by enzyme-linked immunosorbent assay; and NO contents (**B**) determined by visible spectrophotography, and (**C**) determined by laser confocal scanning microscopy, and expressed as relative fluorescence intensity) in seedling roots were analyzed. The sample without chemicals was the control (Con). Values are means ± SE of three independent experiments with at least three replicates for each.

2.2. Melatonin and NO Alleviate NaCl-Induced Seedling Growth Inhibition

It was well known that to discern the role of melatonin in the alleviation of salt stress, a dose–response study of exogenous melatonin in vitro was firstly established. As shown in Figure 3, we observed that the addition of melatonin (0.1, 1, and 10 µM) not only promoted seedling root growth under the normal growth condition, but also differentially alleviated the growth inhibition in roots triggered by NaCl stress, while no significant rescuing effects were observed in 0.01 and 100 µM melatonin-pretreated seedlings. Among these pretreatments, the responses of 1 µM melatonin was maximal, and this concentration was further applied in the following test.

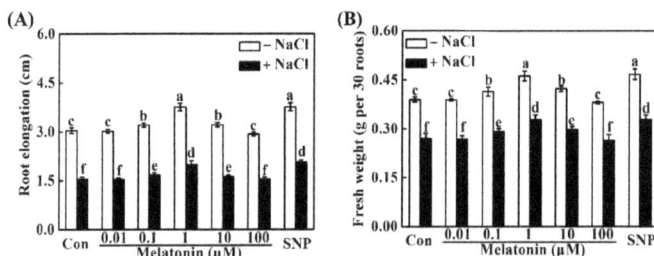

Figure 3. NaCl stress-triggered growth inhibition of seedling roots was alleviated by exogenous melatonin and sodium nitroprusside (SNP; a NO-releasing compound). Three-day-old rapeseed seedlings were pretreated with the indicated concentrations of melatonin or 10 µM SNP for 12 h, and then transferred to 200 mM NaCl for another 2 days. Afterwards, the root elongation (**A**) and root fresh weight (**B**) were measured. The sample without chemicals was the control (Con). Values are means ± SE of three independent experiments, with at least three replicates for each. Bars with different letters are significant different at $p < 0.05$ according to Duncan's multiple range test.

Meanwhile, the treatment with three types of NO-releasing compounds, namely sodium nitroprusside (SNP), diethylamine NONOate (NONOate), and S-nitrosoglutathione (GSNO), produced similar positive responses in the stressed condition (Figure 3 and Supplementary Materials Figure S1). While, old SNP (a negative control of SNP) failed to influence root growth inhibition. Above results thus suggested the beneficial role of exogenous NO in the plant tolerance against salinity stress. Considering the cost of chemicals, SNP was used as a NO-releasing compound in the following experiment.

2.3. PTIO-Dependent Removal of NO Production Impairs the Response of Melatonin

To assess the possible link between melatonin and NO in the alleviation of NaCl stress, the effects of the NO scavenger 2-phenyl-4,4,5,5-tetramethylimidazoline-1-oxyl-3-oxide (PTIO), on the abovementioned melatonin and SNP responses, were investigated and compared. The results shown in Figure 4 revealed that both melatonin- and SNP-alleviated root growth inhibition was greatly reduced in the presence of PTIO, which was similar to the phenotypes in NaCl-stressed alone conditions. In comparison with NaCl stress, the addition of PTIO aggravated root growth inhibition.

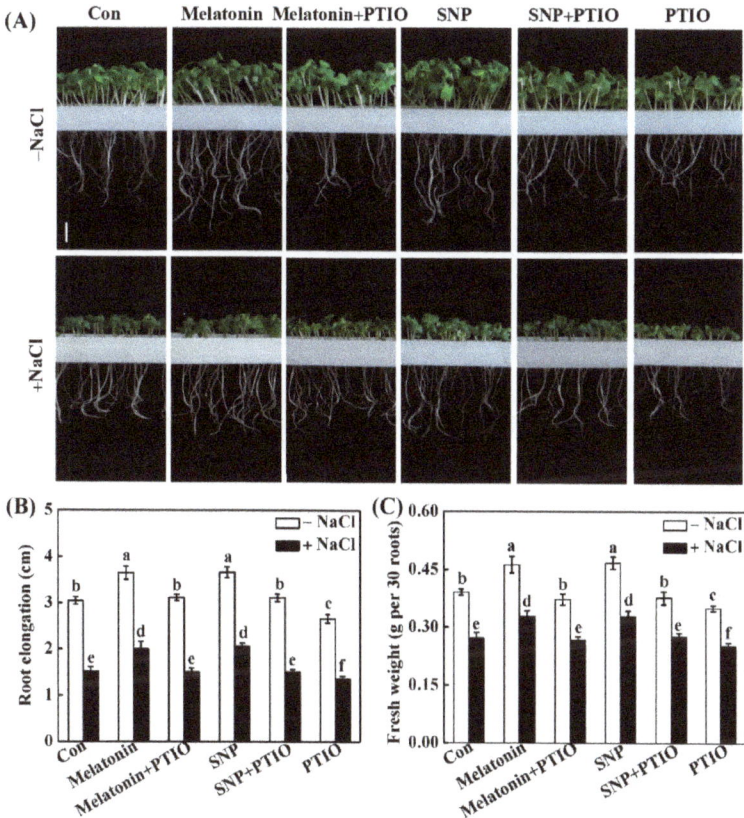

Figure 4. Exogenous melatonin-alleviated root growth inhibition caused by NaCl stress was sensitive to the removal of NO. Three-day-old rapeseed seedlings were pretreated with 1 μM melatonin, 10 μM SNP, 200 μM PTIO, alone or their combinations for 12 h, and then transferred to 200 mM NaCl for 2 days. Afterwards, corresponding photographs were taken ((**A**); bar: 1 cm). The root elongation (**B**) and root fresh weight (**C**) were measured. The sample without chemicals was the control (Con). Values are means ± SE of three independent experiments with at least three replicates for each. Bars with different letters are significant different at $p < 0.05$ according to Duncan's multiple range test.

The role of NO in melatonin-enhanced salinity tolerance was further examined by monitoring NO synthesis in response to applied melatonin and SNP in the presence or absence of PTIO. Similar to the response of SNP, a significant increase in NO-induced fluorescence was observed in stressed seedling roots compared with the control tissue, demonstrating melatonin-mediated NO production (Figure 5A,B). Importantly, melatonin-induced NO synthesis was abolished by co-incubation with PTIO, correlating these data with those from phenotypic analysis (Figure 4). The above results were further confirmed by Griess reagent method (Figure 5C). Together, the pharmacological evidence revealed that PTIO-dependent removal of NO production impairs the response of melatonin.

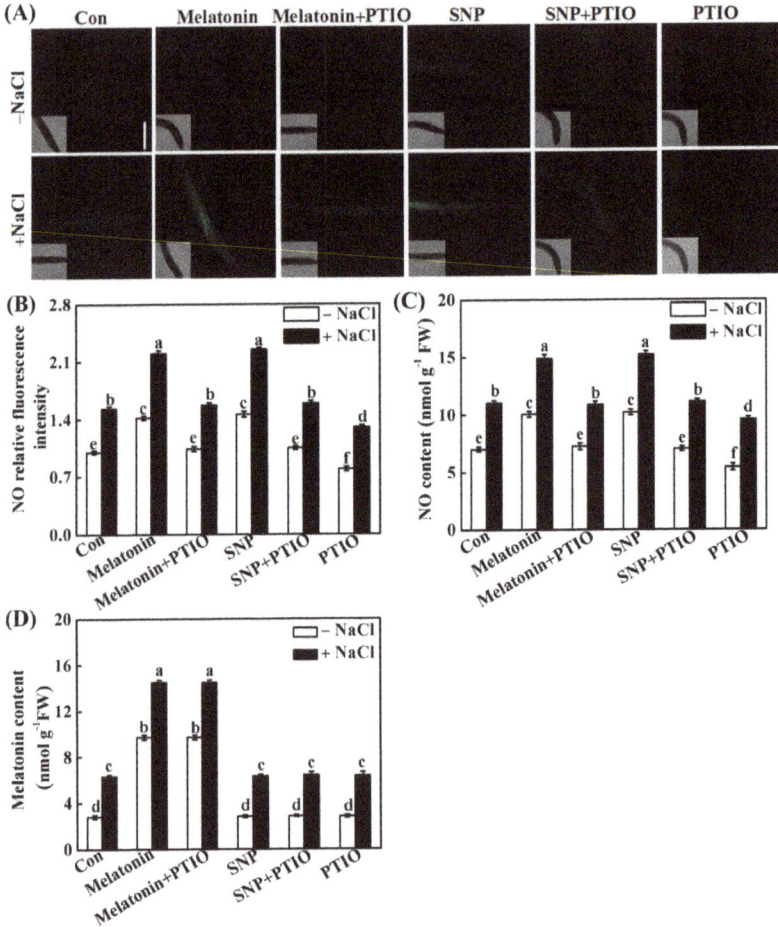

Figure 5. The removal of NO did not alter endogenous melatonin level, but melatonin triggered NO production. Three-day-old rapeseed seedlings were pretreated with 1 µM melatonin, 10 µM SNP, 200 µM PTIO, alone or their combinations for 12 h, and then transferred to 200 mM NaCl for another 2 days. Afterwards, NO ((**A**); determined by laser confocal scanning microscopy; (**C**); determined by visible spectrophotography) and melatonin contents ((**D**); detected by enzyme-linked immunosorbent assay) in root tissues were detected. Scale bar = 1 mm. DAF-FM DA fluorescence densities according to (**A**) were also given (**B**). The sample without chemicals was the control (Con). Values are means ± SE of three independent experiments with at least three replicates for each. Bars with different letters are significant different at $p < 0.05$ according to Duncan's multiple range test.

2.4. NO Does Not Alter Melatonin Synthesis

To further confirm above hypothesis, the effects of SNP and PTIO on endogenous melatonin levels were analyzed. Unlike the inducible responses of exogenous melatonin, the treatment with SNP had no effect on either basal or NaCl-induced melatonin production (Figure 5D). Interestingly, the co-incubation with PTIO did not influence melatonin levels in response to either melatonin or SNP when applied exogenously, no matter if seedlings were with or without the treatment of NaCl.

2.5. Redox Balance Is Reestablished by Melatonin via NO

To unravel the molecular mechanism underlying melatonin-triggered salinity tolerance, subsequent histochemical detection of hydrogen peroxide (H_2O_2; DAB staining) and superoxide anion (O_2^-; NBT staining) was applied. Similar to the positive responses of SNP, NaCl-induced H_2O_2 and O_2^- overproduction in roots, confirmed by the dark brown (Figure 6A) and purple-blue (Figure 6B) color precipitates, was differentially abolished by melatonin. Contrasting results were observed when PTIO was added together. These were in accordance with the results of TBARS contents (Figure 6C).

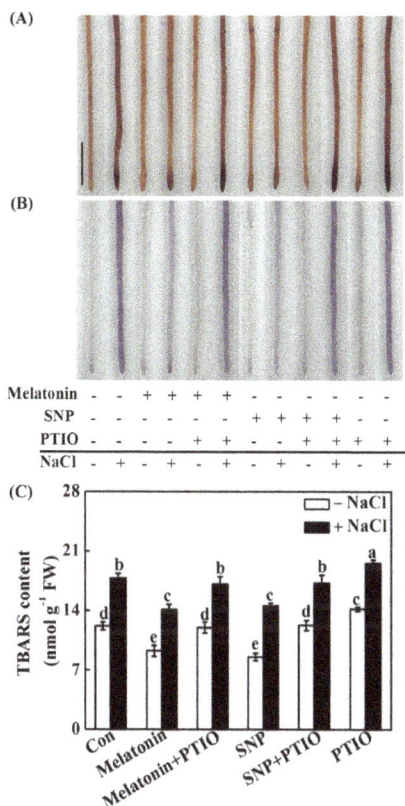

Figure 6. Redox balance was reestablished by melatonin via NO. Three-day-old rapeseed seedlings were pretreated with 1 µM melatonin, 10 µM SNP, 200 µM PTIO, alone or their combinations for 12 h, and then transferred to 200 mM NaCl for another 2 days. Afterwards, seedling roots were stained with DAB (**A**) and NBT (**B**) to detect H_2O_2 and O_2^-. Scale bar = 1 cm. TBARS content (**C**) were also determined. The sample without chemicals was the control (Con). Values are means ± SE of three independent experiments with at least three replicates for each. Bars with different letters are significant different at $p < 0.05$ according to Duncan's multiple range test.

Molecular and biochemical experiments revealed that treatment with PTIO almost completely blocked the increases in the expression of the antioxidant genes *APX*, *MnSOD*, *Cu/ZnSOD* (Figure 7A–C), and the activities of APX and SOD (Figure 7D,E) in NaCl-stressed root tissues. Combined with the results in histochemical detection and TBARS content analysis (Figure 6), these clearly suggested the requirement of NO in melatonin-reestablished redox balance.

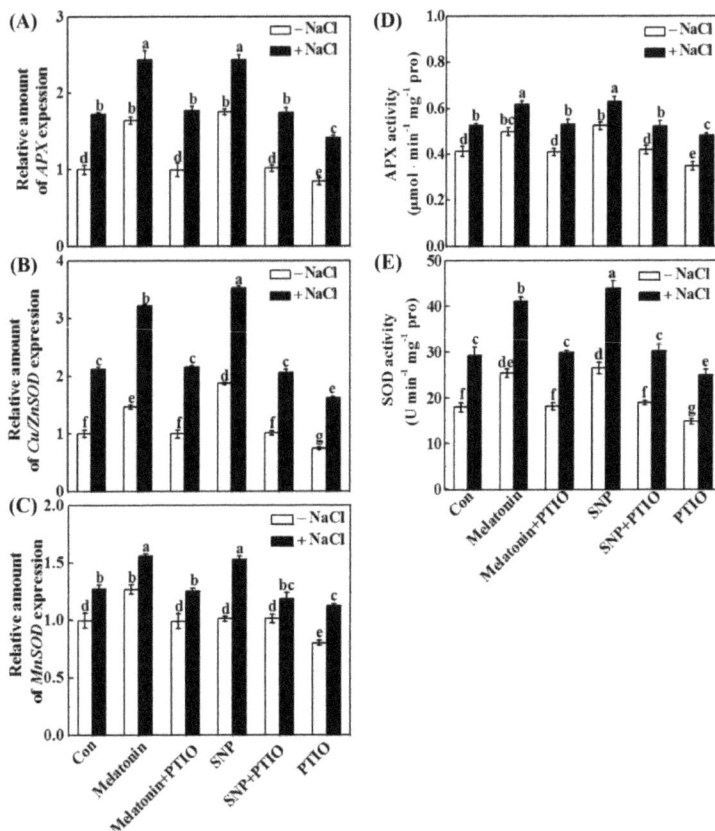

Figure 7. Antioxidant genes and corresponding enzymatic activities were modulated by melatonin-mediated NO. Three-day-old rapeseed seedlings were pretreated with 1 μM melatonin, 10 μM SNP, 200 μM PTIO, alone or their combinations for 12 h, and then transferred to 200 mM NaCl for another 12 h (**A–C**) or 2 days (**D,E**). Then, the mRNA expression of *APX* (**A**), *Cu/ZnSOD* (**B**), and *MnSOD* (**C**) in root tissues was analyzed by qPCR. The activities of ascorbate peroxidase (APX; (**D**)) and superoxide dismutase (SOD; (**E**)) were determined. The sample without chemicals was the control (Con). Values are means ± SE of three independent experiments with at least three replicates for each. Bars with different letters are significant different at $p < 0.05$ according to Duncan's multiple range test.

2.6. Melatonin Modulates Ion Homeostasis via NO

The keeping of ion homeostasis is essential for plants to resist salt stress. Therefore, the effects of melatonin and NO, as well as their interplay on ion homeostasis, were investigated. Compared with the control, NaCl stress significantly increased the Na⁺ accumulation and decreased the K⁺ accumulation, thus leading to a higher Na⁺/K⁺ ratio (Figure 8A) in seedling roots. By contrast, the addition of melatonin and SNP helped seedling roots to reduce the accumulation of Na⁺ and improve K$^+$ assimilation, resulting in lower Na$^+$/K$^+$ ratio, compared with NaCl stressed alone, both of which could be reversed by PTIO.

To confirm the cause of this phenomenon, transcripts of Na$^+$ transporter *NHX1* and *SOS2* were analyzed. As expected, changes in *NHX1* and *SOS2* were consistent with the results in Na$^+$/K$^+$ ratio. For example, the removal of endogenous NO by PTIO completely impaired the effects of melatonin and SNP on activating *NHX1* and *SOS2* mRNA (Figure 8B,C).

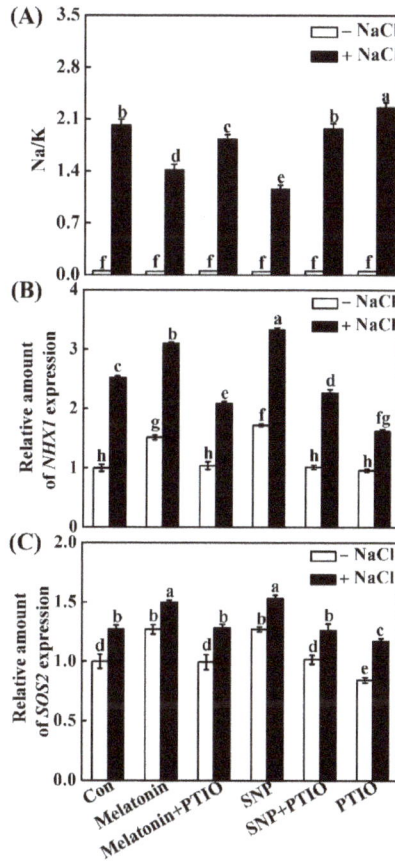

Figure 8. Melatonin modulated ion homeostasis via NO. Three-day-old rapeseed seedlings were pretreated with 1 μM melatonin, 10 μM SNP, 200 μM PTIO, alone or their combinations for 12 h, and then transferred to 200 mM NaCl for another 2 days (**A**) or 12 h (**B,C**). Afterwards, Na$^+$ to K$^+$ ratio (**A**) in seedling roots were detected by ICP-OES. The mRNA expression of *NHX1* (**B**) and *SOS2* (**C**) were analyzed by qPCR. The sample without chemicals was the control (Con). Values are means ± SE of three independent experiments with at least three replicates for each. Bars with different letters are significant different at $p < 0.05$ according to Duncan's multiple range test.

2.7. The Possible Involvement of NO-Dependent S-Nitrosylation

An important bioactivity of NO is implemented by regulating the activity of targeted proteins through S-nitrosylation, a redox-based posttranslational modification. For further confirming the involvement of NO in above melatonin responses, the profiles in S-nitrosylation were analyzed by using the modified biotin switch technique. Figure 9 showed that both melatonin and SNP increased

S-nitrosylation under NaCl stress, which could be partially blocked by the PTIO-dependent removal of NO production. Alone, PTIO supplementation slightly decreased S-nitrosylation, compared to the control samples.

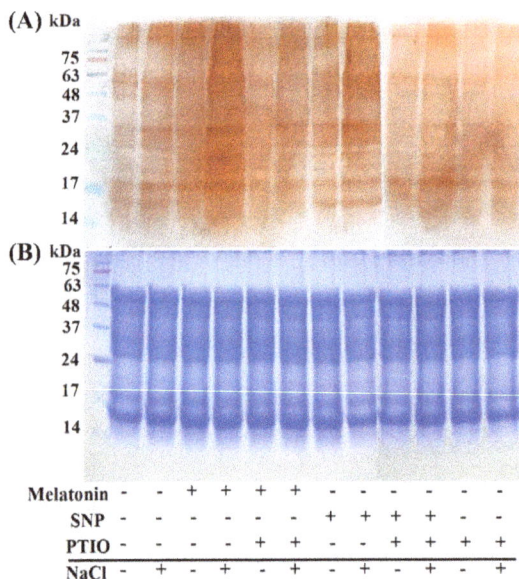

Figure 9. Immunoblot analysis of the total S-nitrosylated protein. Three-day-old rapeseed seedlings were pretreated with 1 μM melatonin, 10 μM SNP, 200 μM PTIO alone, or in various combinations, for 12 h, and then transferred to 200 mM NaCl for another 2 days. The sample without chemicals was the control (Con). Afterwards, proteins were extracted from seedling roots, and subjected to the modified biotin switch method. The labelled proteins were detected using protein blot analysis with antibodies against biotin (A). Numbers on the left of the panels indicate the position of the protein markers in kDa. A Coomassie Brilliant Blue-stained gel (B) is present to show that equal amounts of proteins were loaded.

2.8. Genetic Evidence Reveals That NO Is Required for Melatonin-Induced Salinity Tolerance

To complement above results, we evaluated the role of NO in melatonin-triggered salinity tolerance by using Arabidopsis wild-type (WT) and mutants exhibiting null nitrate reductase (NR) activity (*nia1/2*) and indirectly reduced endogenous NO level (*nitric oxide associated1*; *noa1*). Figure 10A shows WT and *nia1/2* and *noa1* mutants challenged by NaCl stress in the presence or absence of melatonin. As expected, the *nia1/2* and *noa1* mutants exhibited more sensitivity to salinity stress than WT, as measured by the responses in primary root elongation (Figure 10B) and chlorophyll content (Figure 10C) in seeding leaves. It was subsequently observed that the NaCl-induced toxicity in WT plants was obviously rescued by the pretreatment with melatonin. By contrast, no significant rescuing responses were observed in the stressed *nia1/2* and *noa1* mutants with melatonin. Since there are at least two distinct pathways responsible for NO synthesis in plants, the NR- and L-Arg-dependent pathways, our genetic evidence further confirmed the central role of NO in salt tolerance triggered by melatonin.

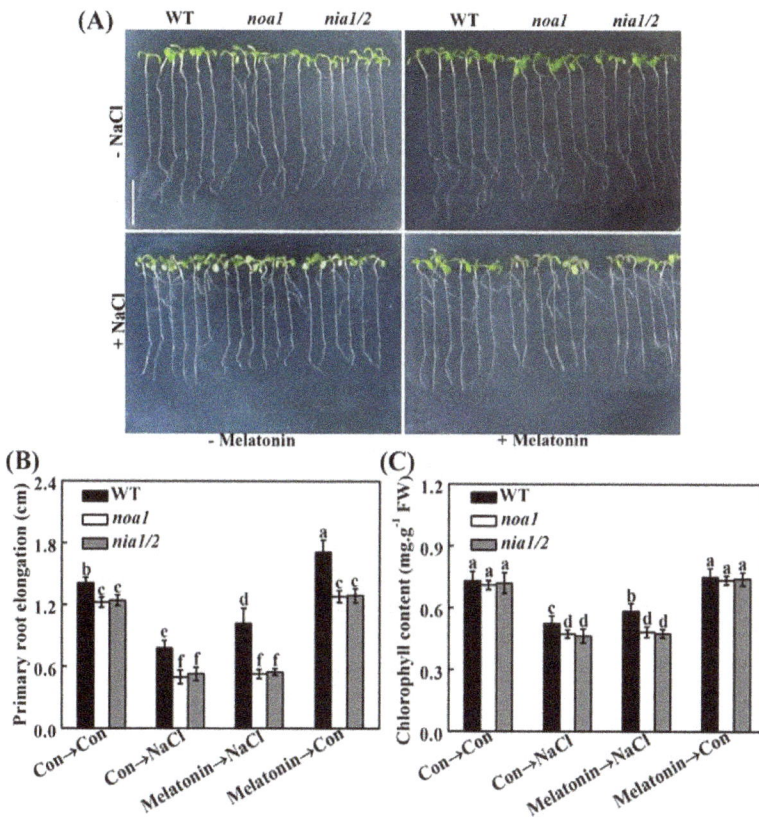

Figure 10. Genetic evidence supported the requirement of NO in melatonin-alleviated NaCl stress. Five-day-old wild-type (WT), and *noa1* and *nia1/2* mutant plants were grown on MS medium supplemented with 1.0 µM melatonin for 5 days, and then transplanted to medium in the presence or absence of 125 mM NaCl for another 5 days. Primary root elongation (**B**) and total chlorophyll content in leaves (**C**) were then determined to assess changes in salt tolerance described in (**A**). Control seedlings (Con) were grown in MS medium alone. Scale bar = 1 cm. Data are means ± SE of three independent experiments with at least three replicates for each. Bars with different letters are significant different at $p < 0.05$ according to Duncan's multiple range test.

In order to further assess whether melatonin-reestablished redox and ion homeostasis is associated with NO signaling, histochemical detection and molecular approach were adopted. As shown in Figure 11, unlike the WT plants, the NaCl-triggered H_2O_2 and O_2^- overproduction in *nia1/2* and *noa1* mutants was largely insensitive to melatonin supplementation. Consistently, NaCl-induced *APX1*, *APX2*, *CAT1*, and *FSD1* transcripts were strengthened by melatonin in wild-type (Figure 12). In comparison, no such induction conferred by melatonin appeared in the mutant plants upon NaCl stress. Molecular evidence further showed that *SOS1*, *SOS2*, and *SOS3* transcripts were upregulated in wild-type and mutant seedlings upon NaCl stress (Figure 13). In the presence of melatonin, above induction was more pronounced in wild-type, but not in *nia1/2* and *noa1* mutants. Combined with the changes in phenotypes (Figure 10), the above genetic evidence thus suggested that endogenous NO mainly produced by NR and NOA1 is required for melatonin-triggered salinity tolerance in Arabidopsis, and the reestablished redox and ion homeostasis were suggested.

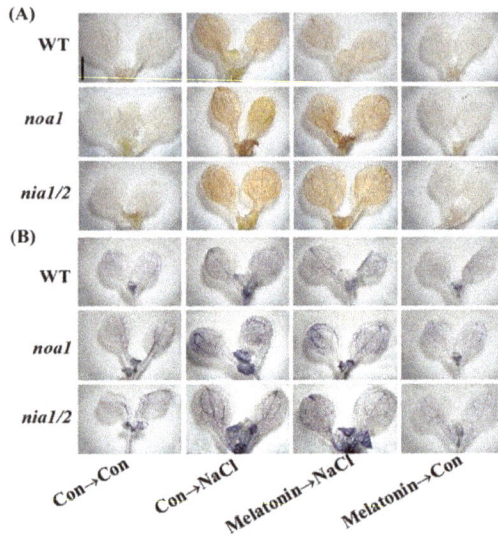

Figure 11. Genetic evidence revealed that redox balance was reestablished by melatonin via NO. Five-day-old wild-type (WT), *noa1*, and *nia1/2* mutant plants were grown on MS medium supplemented with 1.0 μM melatonin for 5 days, and then transplanted to medium in the presence or absence of 125 mM NaCl for another 5 days. Control seedlings (Con) were grown in MS medium alone. Afterwards, seedlings were stained with DAB (**A**) and NBT (**B**) to detect H_2O_2 and O_2^-. Scale bar = 1 mm.

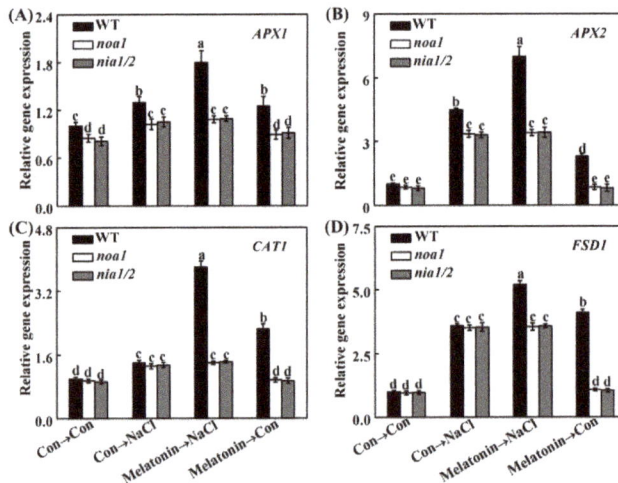

Figure 12. Changes in antioxidant gene expression. Five-day-old wild-type (WT), *noa1*, and *nia1/2* mutant plants were grown on MS medium supplemented with 1.0 μM melatonin for 5 days, and then transplanted to medium in the presence or absence of 125 mM NaCl for another 24 h. The mRNA expression of *APX1* (**A**), *APX2* (**B**), *CAT1* (**C**), and *FSD1* (**D**) in root tissues were analyzed by qPCR. Control seedlings (Con) were grown in MS medium alone. Data are means ± SE of three independent experiments with at least three replicates for each. Bars with different letters are significant different at $p < 0.05$ according to Duncan's multiple range test.

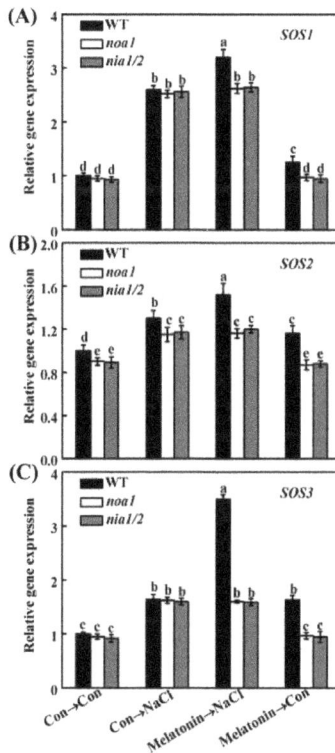

Figure 13. Gene evidence indicated that *SOS* signaling pathway was regulated by melatonin via NO. Five-day-old wild-type (WT), *noa1*, and *nia1/2* mutant plants were grown on MS medium supplemented with 1.0 μM melatonin for 5 days, and then transplanted in medium in the presence or absence of 125 mM NaCl for another 24 h. The mRNA expression of *SOS1* (**A**), *SOS2* (**B**), and *SOS3* (**C**) in root tissues were analyzed by qPCR. Control seedlings (Con) were grown in MS medium alone. Data are means ± SE of three independent experiments with at least three replicates for each. Bars with different letters are significant different at $p < 0.05$ according to Duncan's multiple range test.

3. Discussion

In this work, the molecular basis of melatonin-mediated plant salinity tolerance was investigated. The pharmacological and genetic data presented here show that rapeseed seedlings respond to salinity stress by increasing the concentration of melatonin and NO successively, and that NO is required for the melatonin-mediated plant tolerance against salinity stress.

First, the result shows that an increase in the concentration of melatonin followed by the induction of NO production, is one of the earliest responses involved in the signaling transduction elicited by NaCl stress in rapeseed seedlings (Figure 2). Although the major source(s) of NO have not been investigated in rapeseed seedlings, a cause-effect relationship between melatonin and NO production in salinity tolerance was further established.

Compared to the beneficial role of melatonin in the terms of the alleviation of rapeseed root growth inhibition caused by NaCl stress (Figure 3 and Figure S1), the application of SNP, NONOate, and GSNO, three well-known NO-releasing compounds, could result in the similar responses. The above effect was not found in seedlings supplemented with old SNP, a negative control of SNP, which contains no NO, but nitrate, nitrite, and ferrocyanide [51,52]. These results confirmed the previous conclusion that NO acts as an important gaseous molecule with multiple biological functions in plants, especially in salt tolerance [34–38].

Previously, the interplay between melatonin and NO in plant responses against stress has been controversial [53]. For example, pharmacological data showed that exogenous melatonin induces the production of NO in alkaline stressed tomato seedlings, and NO might be a downstream signal involved in the enhancement of tomato tolerance against alkaline stress triggered by melatonin [54]. A similar conclusion was obtained in Arabidopsis innate immunity against bacterial pathogen (Pst DC3000) [33]. In the present study, mimicking the action of SNP, the increased NO synthesis in response to melatonin in NaCl-stressed rapeseed seedling roots is demonstrated (Figure 5A–C). The removal of endogenous NO by PTIO, however, impaired the effects of melatonin and SNP. Most importantly, the above processes were correlated to the phenotypic responses, showing that the enhancement of plant tolerance against NaCl stress by melatonin is associated with endogenous NO levels (Figure 4). The above results were consistent with reported by Wen et al. [55], in which they found that the increased NO levels caused by the inhibition of S-nitrosoglutathione reductase (GSNOR), which could negatively regulate the NO accumulation in plants [56], were involved in melatonin-triggered adventitious rooting in cucumber plants. Since abiotic stress could trigger adventitious root formation, an important phenotype of the stress-induced morphogenic response (SIMR) in plants [57], our results further confirmed the biological roles of NO in both development and plant responses against stress [38]. Meanwhile, the removal of endogenous NO by PTIO does not alter endogenous melatonin synthesis in roots supplemented with melatonin or NaCl, neither alone or in different combinations (Figure 5D), thus excluding the possibility of NO-triggered melatonin production. By contrast, several previous reports have provided pharmacological evidence indicating that NO could stimulate endogenous melatonin accumulation in sunflower seedling cotyledons as a long-distance signaling response under salt stress [49,50]. NO-dependent melatonin synthesis in cadmium-stressed rice seedlings were also observed [58]. The above differences may reflect the complexity of melatonin signaling in plants [52], and the interplay between melatonin and NO might be dependent on the doses of stresses and exposure times, and even different plant species.

Lozano-Juste and Leŏn [45] reported that *nia1/2* and *noa1* mutants were more sensitive to NaCl-inhibited germination than wild-type seeds. Thus, we obtained these mutants to further study the role of endogenous NO in Arabidopsis salinity tolerance achieved by melatonin. Figure 10 shows that *nia1/2* and *noa1* seedlings seemed to be more sensitive than wild-type to NaCl stress, in terms of primary root elongation and chlorophyll degradation (in particular). Most importantly, exogenous melatonin-enhanced salt tolerance was markedly impaired in above Arabidopsis mutants, which were impaired in NIA/NR- and AtNOA1-dependent NO biosynthesis [42,44,45,51,59]. Thus, this genetic evidence supported the requirement of NO in melatonin-alleviated NaCl stress, at least in our experimental conditions, although the possibility of the direct scavenging or the inhibition of NO synthesis by melatonin could not be easily ruled out [60–62].

Upon NaCl stress, redox imbalance caused by ROS overproduction normally occurred in plants, thus leading to growth stunt and even cell death [1–4]. Further genetic evidence strongly revealed that plant tolerance against NaCl stress is closely associated with a more efficient antioxidant defense [13,20,35,63]. Previously, the antioxidant properties of melatonin against the overproduction of ROS have been confirmed in plant responses against salinity stress [34–37]. Similarly to the above discoveries, our results revealed that melatonin counteracted NaCl stress-induced oxidative damage in rapeseed seedlings, which could be confirmed by the decreased ROS production and lipid peroxidation (Figure 6), as well as the induction of representative antioxidant gene expression, including *APX*,

Cu/ZnSOD, and *MnSOD* (Figure 7A–C). By using biochemical determination, it was further suggested that melatonin was able to increase APX and SOD activities (Figure 7D,E), not only at the transcriptional levels. By contrast, the above changes were prevented by PTIO, an NO scavenger [41,42], indicating that NO is involved in the reestablishment redox balance triggered by melatonin in NaCl stressed rapeseed seedlings. Consistently, we further showed that melatonin was able to reestablish redox balance in Arabidopsis wild-type seedlings upon NaCl stress, but was ineffective in *nia1/2* and *noa1* plants, two NO-deficient mutants (Figures 11 and 12), both of which were used to dissect physiological function of NO in plants [42,44,45]. Together, the above genetic evidence further confirmed that the counteracting effect of melatonin on oxidative damage induced by NaCl stress is NO-dependent.

Maintenance of ion homeostasis, in particular, the Na^+ to K^+ ratio, is another important approach for plants to resist NaCl stress [5,63]. Previous results showed that SOS1, a Na^+/H^+ antiporter, mainly mediated Na efflux, and the activation of SOS2/SOS3 complex can regulate *SOS1* gene expression [63,64]. The reestablishment of ion homeostasis by NO was previously confirmed in NaCl-stressed reed plants, showing that the NaCl-enhanced Na^+ to K^+ ratio is decreased by NO [41]. The application with melatonin exhibited a similar action [34,35], which was confirmed in rapeseed seedlings upon NaCl stress (Figure 8A). Interestingly, changes in rapeseed *NHX1* and *SOS2* transcripts were consistent with the results in Na^+/K^+ ratio (Figure 8B,C). However, the above changes were impaired by the removal of NO, suggesting the NHX-mediated Na^+-sequestration [10] and SOS-mediated Na^+ efflux [64] triggered by NO, might be two important strategies for plant tolerance against NaCl stress when supplemented with melatonin. The above conclusion was partially supported by the changes of *SOS1*, *SOS2*, and *SOS3* transcripts in wild-type, and *nia1/2* and *noa1* Arabidopsis mutants (Figure 13).

NO-mediated *S*-nitrosylation, a redox-related modification of cysteine thiol, is regarded as one of the important post-translational modifications to regulate enzyme activity and interactions among proteins in animals [46]. In higher plants, the formation of *S*-nitrosylation is associated with a wide range of physiological responses, including stress tolerance and root organogenesis [40,47,48,62]. Although a previous paper points to the possible involvement of NO-dependent *S*-nitrosylation in melatonin signaling [33], no study has yet provided definitive proof. In the subsequent work, we discovered that NaCl-induced *S*-nitrosylation was intensified by melatonin and SNP, but impaired by the removal of NO, respectively (Figure 9). Although we have not investigated the detailed target proteins of *S*-nitrosylation, our results strengthened the role of NO in melatonin responses.

Overall, we here presented the first genetic and pharmacological evidence showing the involvement of NO in melatonin signaling in salinity tolerance, and the model was shown in Figure 14. This model proposes that, upon NaCl stress, the increased melatonin triggers a signaling cascade that leads to an induction in NR- and NOA1-dependent NO concentration, thus resulting in the enhancement of salt stress tolerance. Meanwhile, the reestablishment of redox and *SOS*-mediated ion homeostasis are regarded as the important mechanism. NO-dependent *S*-nitrosylation was also illustrated in melatonin responses, and this is a new finding. Thus, the possibility that *miRNA398* modulated antioxidant gene expression [65] and *S*-nitrosylation targeted APX [48], in the above melatonin-mediated NO action, could not be easily ruled out. Further investigation may reveal melatonin-mediated NO-targeted proteins for genetic modification following biotechnological approaches, ultimately aiming to enhance plant tolerance against NaCl stress. Additionally, related studies using potted or field grown rapeseed plants incubated in soil/potting media with a prolonged experimental time would further help in understanding the melatonin and NO application in improving abiotic stress tolerance, thus providing potential benefits in agriculture.

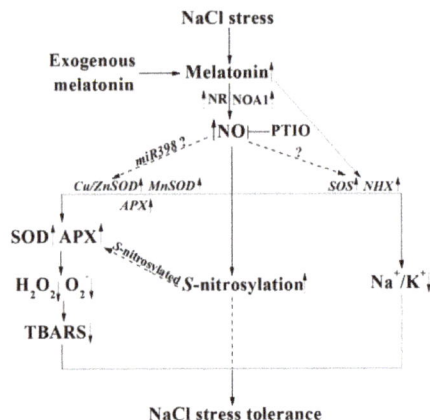

Figure 14. A model depicting the requirement of NR- and NOA1-dependent NO in melatonin-enhanced tolerance against salinity. The reestablishment of redox and ion homeostasis was involved. Dashed lines denote indirect or still undescribed pathways, including *miRNA398*-modulated gene expression [65] and *S*-nitrosylated antioxidant enzymes (APX, [48]).

4. Materials and Methods

4.1. Chemicals

All chemicals were purchased from Sigma-Aldrich (St. Louis, MO, USA) unless stated otherwise. The chemicals used for treatment were melatonin, sodium nitroprusside (SNP, a NO-releasing compound), 2-phenyl-4,4,5,5,-tetramethylimidazoline-1-oxyl-3-oxide (PTIO, a scavenger of NO), diethylamine NONOate (NONOate, a NO donor), and *S*-nitrosoglutathione (GSNO, a NO donor). An SNP solution was used as a negative control of SNP, by maintaining a separated solution of SNP for at least 10 days in light, in a specific open tube, to eliminate all NO [66]. The concentrations of chemicals used in this study were determined in pilot experiments from which maximal induced responses were obtained.

4.2. Plant Materials, Growth Condition, and Experimental Design

Rapeseed (*Brassica napus* L. zhongshuang 11) were kindly supplied by Chinese Academy of Agricultural Sciences. Rapeseed were surface-sterilized with 5% NaClO for 10 min, and rinsed comprehensively in distilled water, then germinated for 2 days at 25 °C in the darkness. Subsequently, the uniform seedlings were cultured with half-strength Hoagland solution in an illuminating incubator (16 h light with a light intensity of 200 $\mu mol \cdot m^{-2} \cdot s^{-1}$, 25 ± 1 °C, and 8 h dark, 23 ± 1 °C). Three-day-old rapeseed seedlings were treated with the indicated chemicals as shown in legends. Three independent experiments with at least three replicates for each were performed, and 30 samples were included in each replicate. Afterwards, rapeseed plants were photographed, and seedling roots were measured immediately after various treatments, or sampled for other analysis.

Arabidopsis thaliana noa1 (CS6511, Col-0) and *nia1/2* (CS2356, Col-0) mutants were obtained from the Arabidopsis Biological Resource Center (http://www.arabidopsis.org/, 5 June 2018). Seeds were surface-sterilized by sodium hypochlorite and rinsed three times with sterile water, then cultured on the solid Murashige and Skoog (MS, pH 5.8) medium containing 1% (*w/v*) agar and 1% (*w/v*) sucrose. For culturing NR-related mutant, the nitrogen in the MS medium included 1 mM NH_4^+ and 1.94 mM NO_3^- [44]. Plates containing seeds were kept at 4 °C for 2 days, and then transferred into the growth chamber with a 16/8 h (23/21 °C) day/light regimes at 150 $\mu mol \cdot m^{-2} \cdot s^{-1}$ irradiation. Five-day-old wild-type (WT), and *noa1*, and *nia1/2* mutant plants grown on MS medium were treated with the indicated chemicals, as shown in legends. Three independent experiments with at least three replicates for each were performed, and about 80 samples were included in each replicate. Afterwards, Arabidopsis plants were photographed, and seedling leaf and root parts were measured immediately, or sampled for other analysis.

4.3. Determination of Melatonin by Enzyme-Linked Immunosorbent Assay (ELISA)

Melatonin was extracted from root tissues by using an acetone-methanol method, and determined by enzyme-linked immunosorbent assay (ELISA) [33,35]. After centrifugation at $12,000 \times g$ for 15 min at 4 °C, the extract was used for quantification of melatonin using the Melatonin ELISA Kit (Jiangsu Baolai Biotechnology, Nanjing, China).

4.4. Determination of NO by Griess Reagent

NO production in root tissues was determined by using Griess reagent [66,67]. Identical filtrate pretreated with 2-(4-carboxyphenyl)-4,4,5,5-tetramethylimidazoline-1-oxyl-3-oxide potassium salt (cPTIO), the specific scavenger of NO, for 15 min, was used as blanks. Absorbance was assayed at 540 nm, and NO content was calculated by comparison to a standard curve of $NaNO_2$.

4.5. Laser Confocal Determination of Endogenous NO Production

Using a fairly specific NO fluorescent probe 4-amino-5-methylamino-2′,7′-difluorofluorescein diacetate (DAF-FM DA), the endogenous NO level of root tissues was detected by a TCS-SP2 confocal laser scanning microscopy (Leica Lasertechnik GmbH, Heidelberg, Germany; excitation at 488 nm, emission at 500–530 nm) [67,68]. Results are from five replicates per experiment. Fluorescence was expressed as relative fluorescence units using the Leica Confocal Software 2.5 (Leica Lasertechnik GmbH, Heidelberg, Germany).

4.6. ROS Detection

H_2O_2 and $O_2{}^-$ were histochemically detected by 3,3′-diaminobenzidine (DAB) and nitroblue tetrazolium (NBT) staining [35]. Finally, all samples were observed using a light microscope (model Stemi 2000-C; Carl Zeiss, Oberkochen, Germany).

4.7. Assay of Thiobarbituric Acid Reactive Substances (TBARS) Content

Oxidative damage was estimated by measuring the concentration of TBARS as previously described [69].

4.8. Determination of Antioxidant Enzyme Activities

Root tissues were crushed into fine powder in a mortar and pestle under liquid N_2. Soluble proteins were extracted and homogenized in 50 mM PBS (pH 7.0) containing 1 mM EDTA and 1% polyvinylpyrrolidone (PVP), or with the addition of 1 mM ascorbic acid (ASA) in the case of ascorbate peroxidase (APX) activity determination. APX and SOD activities were measured as described previously [35,70]. Protein content was determined according to the method described by Bradford [71].

4.9. Real-Time Quantitative RT-PCR (qPCR) Analysis

According to our pervious method [35], total RNA was extracted from seedling roots using a Tranzol up kit (TransGen Biotech, Beijing, China). RNA concentration and quality were determined using the NanoDrop 2000 (Thermo Fisher Scientific, Wilmington, DE, USA), and then incubated with RNase-free DNase (TaKaRa Bio Inc., Dalian, China) to eliminate traces of DNA. cDNAs were then synthesized using an oligo(dT) primer and a SuperScript First-Strand Synthesis System (Invitrogen, Carlsbad, CA, USA).

By using the gene-specific primers (Supplementary Materials Tables S1 and S2), qPCR was conducted using a Mastercycler ep® *realplex* real-time PCR system (Eppendorf, Hamburg, Germany) with *TransStart*® Green qPCR SuperMix (TransGen Biotech, Beijing, China) [35]. In rapeseed plants, the expression levels of genes were normalized to two internal control gene *Actin* and *GADPH* transcript levels, and presented as values relative to corresponding control samples in NaCl-free conditions. In Arabidopsis, the expression levels of genes were normalized to two internal control gene *Actin* 2 and *GADPH* transcript levels, and presented as values relative to corresponding control of wild-type samples in NaCl-free conditions.

4.10. Determination of Ion Contents

Fresh seedling roots were harvested and washed four times by deionized water after treatments. According to the previous method [21,35], Na and K element contents were measured with an Inductively Coupled Plasma Optical Emission Spectrometer (ICP-OES, Perkin Elmer Optima 2100 DV; PerkinElmer, Shelton, CT, USA).

4.11. Quantification of Chlorophyll Content

Total chlorophyll was extracted using 95% (v/v) ethanol for 24 h in darkness, and then calculated by examining the absorbance at 649 nm and 665 nm [35].

4.12. Modified Biotin Switch Method

Modified biotin switch method was carried out according to the previous method [40,62]. After treatment, total protein was extracted from fresh seedling roots, and S-nitrosylated and biotin-labelled proteins were separated by using non-reducing sodium dodecyl sulfate polyacrylamide gel electrophoresis (SDS-PAGE; 12%). Western blotting was then performed to detect the proteins. Anti-biotin antibody horseradish peroxidase (Abcam antibodies, Cambridge, UK) was diluted 1:6000. Coomassie Brilliant Blue-stained gels were used to show that equal amounts of proteins were loaded.

4.13. Statistical Analysis

Values are means ± SE of three independent experiments with at least three replicates for each. Statistical analysis was performed using SPSS 16.0 software (IBM Corporation, Armonk, NY, USA). Differences among treatments were analyzed by one-way analysis of variance (ANOVA), taking $p < 0.05$ as significant according to Duncan's multiple range test.

Supplementary Materials: Supplementary materials can be found at http://www.mdpi.com/1422-0067/19/7/1912/s1. Supplementary data associated with this article can be found in the online version. Table S1: The sequences of primers for qPCR used in rapeseed. Table S2: The sequences of primers for qPCR used in Arabidopsis. Figure S1: NaCl-induced inhibition of root elongation was rescued by three types of NO-releasing compounds, but aggravated by the scavenger of NO.

Author Contributions: G.Z., Y.Z., X.Y., and W.S. designed and refined the research; G.Z., Y.Z., X.Y., H.H., R.G., and R.W. performed research; G.Z. and Y.Z. prepared the mutant materials; G.Z., Y.Z., F.K., and W.S. analyzed data; G.Z., Y.Z., R.G., and W.S. wrote the article. All authors discussed the results and comments on the manuscript.

Acknowledgments: This work was partly supported by the National Key Research and Development Plan (2016YFD0101306), the Fundamental Research Funds for the Central Universities (KYTZ201402), and the Priority Academic Program Development of Jiangsu Higher Education Institutions (PAPD).

Conflicts of Interest: The authors declare no conflict of interest.

Abbreviations

ASA	Ascorbic acid
CAT	Catalase
cPTIO	2-(4-carboxyphenyl)-4,4,5,5-tetramethylimidazoline-1-oxyl-3-oxide potassium salt
DAB	3,3′-diaminobenzidine
GSNO	*S*-nitrosoglutathione
NBT	Nitroblue tetrazolium
NHX1	Sodium hydrogen exchanger
noa1	Nitric oxide associated1
NO	Nitric oxide
NONOate	Diethylamine
NR	Nitrate reductase
POD	Guaiacol peroxidase
PTIO	2-phenyl-4,4,5,5,-tetramethylimidazoline-1-oxyl-3-oxide
ROS	Reactive oxygen species
SNP	Sodium nitroprusside
SOD	Superoxide dismutase
SOS	Salt overly sensitive

References

1. Zhu, J.K. Plant salt tolerance. *Trends Plant Sci.* **2001**, *6*, 66–71. [CrossRef]
2. Zhu, J.K. Salt and drought stress signal transduction in plants. *Annu. Rev. Plant Biol.* **2002**, *53*, 247–273. [CrossRef] [PubMed]
3. Parida, A.K.; Das, A.B. Salt tolerance and salinity effects on plants: A review. *Ecotoxicol. Environ. Saf.* **2005**, *60*, 324–349. [CrossRef] [PubMed]
4. Hasegawa, P.M.; Bressan, R.A.; Zhu, J.K.; Bohnert, H.J. Plant cellular and molecular responses to high salinity. *Annu. Rev. Plant Biol.* **2000**, *51*, 463–499. [CrossRef] [PubMed]
5. Zhu, J.K. Regulation of ion homeostasis under salt stress. *Curr. Opin. Plant Biol.* **2003**, *6*, 441–445. [CrossRef]
6. Ohta, M.; Hayashi, Y.; Nakashima, A.; Hamada, A.; Tanaka, A.; Nakamura, T.; Hayakawa, T. Introduction of a Na^+/H^+ antiporter gene from *Atriplex gmelini* confers salt tolerance to rice. *FEBS Lett.* **2002**, *532*, 279–282. [CrossRef]
7. Shi, H.; Lee, B.; Wu, S.J.; Zhu, J.K. Overexpression of a plasma membrane Na^+/H^+ antiporter gene improves salt tolerance in *Arabidopsis thaliana*. *Nat. Biotechnol.* **2003**, *21*, 81–85. [CrossRef] [PubMed]
8. Shi, H.; Ishitani, M.; Kim, C.; Zhu, J.K. The *Arabidopsis thaliana* salt tolerance gene *SOS1* encodes a putative Na^+/H^+ antiporter. *Proc. Natl. Acad. Sci. USA* **2000**, *97*, 6896–6901. [CrossRef] [PubMed]
9. Shi, H.; Quintero, F.J.; Pardo, J.M.; Zhu, J.K. The putative plasma membrane Na^+/H^+ antiporter SOS1 controls long-distance Na^+ transport in plants. *Plant Cell* **2002**, *14*, 465–477. [CrossRef] [PubMed]
10. Munns, R.; Tester, M. Mechanisms of salinity tolerance. *Annu. Rev. Plant Biol.* **2008**, *59*, 651–681. [CrossRef] [PubMed]
11. Ashraf, M.; Ali, Q. Relative membrane permeability and activities of some antioxidant enzymes as the key determinants of salt tolerance in canola (*Brassica napus* L.). *Environ. Exp. Bot.* **2008**, *63*, 266–273. [CrossRef]
12. Miller, G.; Suzuki, N.; Ciftci-Yilmaz, S.; Mittler, R. Reactive oxygen species homeostasis and signalling during drought and salinity stresses. *Plant Cell Environ.* **2010**, *33*, 453–467. [CrossRef] [PubMed]
13. Mittler, R.; Vanderauwera, S.; Suzuki, N.; Miller, G.; Tognetti, V.B.; Vandepoele, K.; Gollery, M.; Shulaev, V.; Breusegem, F.V. ROS signaling: The new wave? *Trends Plant Sci.* **2011**, *16*, 300–309. [CrossRef] [PubMed]
14. Naeem, M.S.; Jin, Z.L.; Wan, G.L.; Liu, D.; Liu, H.B.; Yoneyama, K.; Zhou, W.J. 5-Aminolevulinic acid improves photosynthetic gas exchange capacity and ion uptake under salinity stress in oilseed rape (*Brassica napus* L.). *Plant Soil* **2010**, *332*, 405–415. [CrossRef]
15. Ruiz, J.M.; Blumwald, E. Salinity-induced glutathione synthesis in *Brassica napus*. *Planta* **2002**, *214*, 965–969. [CrossRef] [PubMed]

16. Dai, Q.; Chen, C.; Feng, B.; Liu, T.; Tian, X.; Gong, Y.; Sun, Y.; Wang, J.; Du, S. Effects of different NaCl concentration on the antioxidant enzymes in oilseed rape (*Brassica napus* L.) seedlings. *Plant Growth Regul.* **2009**, *59*, 273–278. [CrossRef]

17. Kagale, S.; Divi, U.K.; Krochko, J.E.; Keller, W.A.; Krishna, P. Brassinosteroid confers tolerance in *Arabidopsis thaliana* and *Brassica napus* to a range of abiotic stresses. *Planta* **2007**, *225*, 353–364. [CrossRef] [PubMed]

18. Jalili, F.; Khavazi, K.; Pazira, E.; Nejati, A.; Rahmani, H.A.; Sadaghiani, H.R.; Miransarf, M. Isolation and characterization of ACC deaminase-producing fluorescent pseudomonads, to alleviate salinity stress on canola (*Brassica napus* L.) growth. *J. Plant Physiol.* **2009**, *166*, 667–674. [CrossRef] [PubMed]

19. Wei, W.; Li, Q.T.; Chu, Y.N.; Reiter, R.J.; Yu, X.M.; Zhu, D.H.; Zhang, W.K.; Ma, B.; Lin, Q.; Zhang, J.S.; et al. Melatonin enhances plant growth and abiotic stress tolerance in soybean plants. *J. Exp. Bot.* **2015**, *66*, 695–707. [CrossRef] [PubMed]

20. Liang, C.; Zheng, G.; Li, W.; Wang, Y.; Hu, B.; Wang, H.; Wu, H.; Qian, Y.; Zhu, X.G.; Tan, D.X.; et al. Melatonin delays leaf senescence and enhances salt stress tolerance in rice. *J. Pineal Res.* **2015**, *59*, 91–101. [CrossRef] [PubMed]

21. Xie, Y.; Ling, T.; Liu, K.; Zheng, Q.; Huang, L.; Yuan, X.; He, Z.; Hu, B.; Fang, L.; Shen, Z.; et al. Carbon monoxide enhances salt tolerance by nitric oxide-mediated maintenance of ion homeostasis and up-regulation of antioxidant defence in wheat seeding roots. *Plant Cell Environ.* **2008**, *31*, 1864–1881. [CrossRef] [PubMed]

22. Kumar, M.; Choi, J.; An, G.; Kim, S.R. Ectopic expression of *OsSta2* enhances salt stress tolerance in rice. *Front. Plant Sci.* **2017**, *8*, 316. [CrossRef] [PubMed]

23. Li, H.; Chang, J.; Chen, H.; Wang, Z.; Gu, X.; Wei, C.; Zhang, Y.; Ma, J.; Yang, J.; Zhang, X. Exogenous melatonin confers salt stress tolerance to watermelon by improving photosynthesis and redox homeostasis. *Front. Plant Sci.* **2017**, *8*, 295. [CrossRef] [PubMed]

24. Zeng, L.; Cai, J.; Li, J.; Lu, G.; Li, C.; Fu, G.; Zhang, X.; Ma, H.; Liu, Q.; Zou, X.; et al. Exogenous application of a low concentration of melatonin enhances salt tolerance in rapeseed (*Brassica napus* L.) seedlings. *J. Integr. Agric.* **2018**, *17*, 328–335. [CrossRef]

25. Kumar, M.; Choi, J.Y.; Kumari, N.; Pareek, A.; Kim, S.R. Molecular breeding in *Brassica* for salt tolerance: Importance of microsatellite (SSR) markers for molecular breeding in *Brassica*. *Front. Plant Sci.* **2015**, *6*, 688. [CrossRef] [PubMed]

26. Lerner, A.B.; Case, J.D.; Takahashi, Y.; Lee, T.H.; Mori, W. Isolation of melatonin, the pineal gland factor that lightens melanocytes. *J. Am. Chem. Soc.* **1958**, *80*, 2587. [CrossRef]

27. Reiter, R.J.; Tan, D.X.; Galano, A. Melatonin: Exceeding expectations. *Physiology* **2014**, *29*, 325–333. [CrossRef] [PubMed]

28. Jung-Hynes, B.; Reiter, R.J.; Ahmad, N. Melatonin and circadian rhythms: Building a bridge between aging and cancer. *J. Pineal Res.* **2010**, *48*, 9–19. [CrossRef] [PubMed]

29. Cozzi, B.; Morei, G.; Ravault, J.P.; Chesneau, D.; Reiter, R.J. Circadian and seasonal rhythms of melatonin production in mules (*Equus asinus* × *Equus caballus*). *J. Pineal Res.* **1991**, *10*, 130–135. [CrossRef] [PubMed]

30. Santofimia-Castaño, P.; Ruy, D.C.; Garcia-Sanchez, L.; Jimenez-Blasco, D.; Fernandez-Bermejo, M.; Bolaños, J.P.; Salido, G.M.; Gonzalez, A. Melatonin induces the expression of Nrf2-regulated antioxidant enzymes via PKC and Ca^{2+} influx activation in mouse pancreatic acinar cells. *Free Radic. Biol. Med.* **2015**, *87*, 226–236. [CrossRef] [PubMed]

31. Reiter, R.J.; Tan, D.X.; Zhou, Z.; Cruz, M.H.C.; Fuentes-Broto, L.; Galano, A. Phytomelatonin: Assisting plants to survive and thrive. *Molecules* **2015**, *20*, 7396–7437. [CrossRef] [PubMed]

32. Bajwa, V.S.; Shukla, M.R.; Sherif, S.M.; Murch, S.J.; Saxena, P.K. Role of melatonin in alleviating cold stress in *Arabidopsis thaliana*. *J. Pineal Res.* **2014**, *56*, 238–245. [CrossRef] [PubMed]

33. Shi, H.; Chen, Y.; Tan, D.X.; Reiter, R.J.; Chan, Z.; He, Z. Melatonin induces nitric oxide and the potential mechanisms relate to innate immunity against bacterial pathogen infection in *Arabidopsis*. *J. Pineal Res.* **2015**, *59*, 102–108. [CrossRef] [PubMed]

34. Li, C.; Wang, P.; Wei, Z.; Liang, D.; Liu, C.; Yin, L.; Jia, D.; Fu, M.; Ma, F. The mitigation effects of exogenous melatonin on salinity-induced stress in *Malus hupehensis*. *J. Pineal Res.* **2012**, *53*, 298–306. [CrossRef] [PubMed]

35. Chen, Z.; Xie, Y.; Gu, Q.; Zhao, G.; Zhang, Y.; Cui, W.; Xu, S.; Wang, R.; Shen, W. The *AtrbohF*-dependent regulation of ROS signaling is required for melatonin-induced salinity tolerance in *Arabidopsis*. *Free Radic. Biol. Med.* **2017**, *108*, 465–477. [CrossRef] [PubMed]

36. Zhang, H.J.; Zhang, N.; Yang, R.C.; Wang, L.; Sun, Q.Q.; Li, D.B.; Cao, Y.Y.; Wedda, S.; Zhao, B.; Ren, S.; et al. Melatonin promotes seed germination under high salinity by regulating antioxidant systems, ABA and GA$_4$ interaction in cucumber (*Cucumis sativus* L.). *J. Pineal Res.* **2014**, *57*, 269–279. [CrossRef] [PubMed]

37. Shi, H.; Jiang, C.; Ye, T.; Tan, D.X.; Reiter, R.J.; Zhang, H.; Liu, R.; Chan, R. Comparative physiological, metabolomic, and transcriptomic analyses reveal mechanisms of improved abiotic stress resistance in bermudagrass [*Cynodon dactylon* (L.). Pers.] by exogenous melatonin. *J. Exp. Bot.* **2015**, *66*, 681–694. [CrossRef] [PubMed]

38. Besson-Bard, A.; Pugin, A.; Wendehenne, D. New insights into nitric oxide signaling in plants. *Annu. Rev. Plant Biol.* **2008**, *59*, 21–39. [CrossRef] [PubMed]

39. Gupta, K.J.; Fernie, A.R.; Kaiser, W.M.; Dengon, J. On the origins of nitric oxide. *Trends Plant Sci.* **2011**, *16*, 160–168. [CrossRef] [PubMed]

40. Su, J.; Zhang, Y.; Nie, Y.; Chen, D.; Wang, R.; Hu, H.; Chen, J.; Zhang, J.; Du, Y.; Shen, W. Hydrogen-induced osmotic tolerance is associated with nitric oxide-mediated proline accumulation and reestablishment of redox balance in alfalfa seedlings. *Environ. Exp. Bot.* **2018**, *147*, 249–260. [CrossRef]

41. Zhao, L.; Zhang, F.; Guo, J.; Yang, Y.; Li, B.; Zhang, L. Nitric oxide functions as a signal in salt resistance in the calluses from two ecotypes of reed. *Plant Physiol.* **2004**, *134*, 849–857. [CrossRef] [PubMed]

42. Zhao, M.G.; Tian, Q.Y.; Zhang, W.H. Nitric oxide synthase-dependent nitric oxide production is associated with salt tolerance in *Arabidopsis*. *Plant Physiol.* **2007**, *144*, 206–217. [CrossRef] [PubMed]

43. Wang, Y.; Li, L.; Cui, W.; Xu, S.; Shen, W.; Wang, R. Hydrogen sulfide enhances alfalfa (*Medicago sativa*) tolerance against salinity during seed germination by nitric oxide pathway. *Plant Soil* **2012**, *351*, 107–119. [CrossRef]

44. Xie, Y.; Mao, Y.; Lai, D.; Zhang, W.; Zheng, T.; Shen, W. Roles of NIA/NR/NOA1-dependent nitric oxide production and HY1 expression in the modulation of *Arabidopsis* salt tolerance. *J. Exp. Bot.* **2013**, *64*, 3045–3060. [CrossRef] [PubMed]

45. Lozano-Juste, J.; León, J. Enhanced abscisis acid-mediated responses in *nia1nia2noa1-2* triple mutant impaired in NIA/NR- and AtNOA1-dependent nitric oxide biosynthesis in Arabidopsis. *Plant Physiol.* **2010**, *152*, 891–903. [CrossRef] [PubMed]

46. Jaffrey, S.R.; Erdjument-Bromage, H.; Ferris, C.D.; Tempst, P.; Snyder, S.H. Protein *S*-nitrosylation: A physiological signal for neuronal nitric oxide. *Nat. Cell Biol.* **2001**, *3*, 193–197. [CrossRef] [PubMed]

47. Yun, B.W.; Feechan, A.; Yin, M.; Saidi, N.B.B.; Bihan, T.L.; Yu, M.; Moore, J.W.; Kang, J.G.; Kwon, E.; Spoel, S.H.; et al. *S*-nitrosylation of NADPH oxidase regulates cell death in plant immunity. *Nature* **2011**, *478*, 264–268. [CrossRef] [PubMed]

48. Lindermayr, C.; Saalbach, G.; Durner, J. Proteomic identification of *S*-nitrosylated proteins in Arabidopsis. *Plant Physiol.* **2005**, *137*, 921–930. [CrossRef] [PubMed]

49. Kaur, H.; Bhatla, S.C. Melatonin and nitric oxide modulate glutathione content and glutathione reductase activity in sunflower seedling cotyledons accompanying salt stress. *Nitric Oxide* **2016**, *59*, 42–53. [CrossRef] [PubMed]

50. Arora, D.; Bhatla, S.C. Melatonin and nitric oxide regulate sunflower seedling growth under salt stress accompanying differential expression of Cu/Zn SOD and Mn SOD. *Free Radic. Biol. Med.* **2017**, *106*, 315–328. [CrossRef] [PubMed]

51. Han, B.; Yang, Z.; Xie, Y.; Nie, L.; Cui, J.; Shen, W. *Arabidopsis* HY1 confers cadmium tolerance by decreasing nitric oxide production and improving iron homeostasis. *Mol. Plant* **2014**, *7*, 388–403. [CrossRef] [PubMed]

52. Chen, M.; Cui, W.; Zhu, K. Hydrogen-rich water alleviates aluminum-induced inhibition of root elongation in alfalfa via decreasing nitric oxide production. *J. Hazard. Mater.* **2014**, *267*, 40–47. [CrossRef] [PubMed]

53. Arnao, M.B.; Hernández-Ruiz, J. Melatonin and its relationship to plant hormones. *Ann. Bot. Lond.* **2018**, *121*, 195–207. [CrossRef] [PubMed]

54. Liu, N.; Gong, B.; Jin, Z.; Wang, X.; Wei, M.; Yang, F.; Li, Y.; Shi, Q. Sodic alkaline stress mitigation by exogenous melatonin in tomato needs nitric oxide as a downstream signal. *J. Plant Physiol.* **2015**, *186*, 68–77. [CrossRef] [PubMed]

55. Wen, D.; Gong, B.; Sun, S.; Liu, S.; Wang, X.; Wei, M.; Yang, F.; Li, Y.; Shi, H. Promoting roles of melatonin in adventitious root development of *Solanum lycopersicum* L. by regulating auxin and nitric oxide signaling. *Front. Plant Sci.* **2016**, *7*, 718. [CrossRef] [PubMed]

56. Corpas, F.J.; Alché, J.D.; Barroso, J.B. Current overview of *S*-nitrosoglutathione (GSNO) in higher plants. *Front. Plant Sci.* **2013**, *4*, 126. [CrossRef] [PubMed]

57. Potters, G.; Pasternak, T.P.; Guisez, Y.; Palme, J.K.; Jansen, M.A.K. Stress-induced morphogenic responses: Growing out of trouble? *Trends Plant Sci.* **2007**, *12*, 98–105. [CrossRef] [PubMed]

58. Lee, K.; Choi, G.H.; Back, K. Cadmium-induced melatonin synthesis in rice requires light, hydrogen peroxide, and nitric oxide: Key regulatory roles for tryptophan decarboxylase and caffeic acid *O*-methyltransferase. *J. Pineal Res.* **2017**, *63*, e12441. [CrossRef] [PubMed]

59. Xie, Y.; Mao, Y.; Zhang, W.; Lai, D.; Wang, Q.; Shen, W. Reactive oxygen species-dependent nitric oxide production contributes to hydrogen-promoted stomatal closure in Arabidopsis. *Plant Physiol.* **2014**, *165*, 759–773. [CrossRef] [PubMed]

60. Noda, Y.; Mori, A.; Liburdy, R.; Packer, L. Melatonin and its precursors scavenge nitric oxide. *J. Pineal Res.* **1999**, *27*, 159–163. [CrossRef] [PubMed]

61. Kang, Y.S.; Kang, Y.G.; Park, H.J.; Wee, H.J.; Jang, H.O.; Bae, M.K.; Bae, S.K. Melatonin inhibits visfatin-induced inducible nitric oxide synthase expression and nitric oxide production in macrophages. *J. Pineal Res.* **2013**, *55*, 294–303. [CrossRef] [PubMed]

62. Guerrero, J.M.; Reiter, R.J.; Ortiz, G.G.; Pablos, M.I.; Sewerynek, E.; Chuang, J.I. Melatonin prevents increases in neural nitric oxide and cyclic GMP production after transient brain ischemia and reperfusion in the Mongolian gerbil (*Meriones unguiculatus*). *J. Pineal Res.* **1997**, *23*, 24–31. [CrossRef] [PubMed]

63. Deinlein, U.; Stephan, A.B.; Horie, T.; Luo, W.; Xu, G.; Schroeder, J.I. Plant salt-tolerance mechanisms. *Trends Plant Sci.* **2014**, *19*, 371–379. [CrossRef] [PubMed]

64. Qiu, Q.S.; Guo, Y.; Dietrich, M.A.; Schumaker, K.S.; Zhu, J.K. Regulation of SOS1, a plasma membrane Na$^+$/H$^+$ exchanger in *Arabidopsis thaliana*, by SOS2 and SOS3. *Proc. Natl. Acad. Sci. USA* **2002**, *99*, 8436–8441. [CrossRef] [PubMed]

65. Sunkar, R.; Kapoor, A.; Zhu, J.K. Posttranscriptional induction of two Cu/Zn superoxide dismutase genes in *Arabidopsis* is mediated by downregulation of miR398 and important for oxidative stress tolerance. *Plant Cell* **2006**, *18*, 2051–2065. [CrossRef] [PubMed]

66. Zhou, B.; Guo, Z.; Xing, J.; Huang, B. Nitric oxide is involved in abscisic acid-induced antioxidant activities in *Stylosanthes guianensis*. *J. Exp. Bot.* **2005**, *56*, 3223–3228. [CrossRef] [PubMed]

67. Balcerczyk, A.; Soszynski, M.; Bartosz, G. On the specificity of 4-amino-5-methylamino-2′,7′-difluorofluorescein as a probe for nitric oxide. *Free Radic. Biol. Med.* **2005**, *39*, 327–335. [CrossRef] [PubMed]

68. Qi, F.; Xiang, Z.; Kou, N.; Cui, W.; Wang, R.; Zhu, D.; Shen, W. Nitric oxide is involved in methane-induced adventitious root formation in cucumber. *Physiol. Plant.* **2017**, *159*, 366–377. [CrossRef] [PubMed]

69. Han, Y.; Zhang, J.; Chen, X.; Gao, Z.; Xuan, W.; Xu, S.; Ding, X.; Shen, W. Carbon monoxide alleviates cadmium-induced oxidative damage by modulating glutathione metabolism in the roots of *Medicago sativa*. *New Phytol.* **2008**, *177*, 155–166. [CrossRef] [PubMed]

70. Nakano, Y.; Asada, K. Hydrogen peroxide is scavenged by ascorbate-specific peroxidase in spinach chloroplasts. *Plant Cell Physiol.* **1981**, *22*, 867–880.

71. Bradford, M.M. A rapid and sensitive method for the quantitation of microgram quantities of protein utilizing the principle of protein-dye binding. *Anal. Biochem.* **1976**, *72*, 248–254. [CrossRef]

International Journal of
Molecular Sciences

MDPI

Article

Variations in Physiology and Multiple Bioactive Constituents under Salt Stress Provide Insight into the Quality Evaluation of Apocyni Veneti Folium

Cuihua Chen [1], Chengcheng Wang [1], Zixiu Liu [1], Xunhong Liu [1,2,3,*], Lisi Zou [1], Jingjing Shi [1], Shuyu Chen [1], Jiali Chen [1] and Mengxia Tan [1]

[1] College of Pharmacy, Nanjing University of Chinese Medicine, Nanjing 210023, China; cuihuachen2013@163.com (C.C.); ccw199192@163.com (C.W.); liuzixiu3221@126.com (Z.L.); zlstcm@126.com (L.Z.); shijingjingquiet@163.com (J.S.); 18305172513@163.com (S.C.); 18994986833@163.com (J.C.); 18816250751@163.com (M.T.)
[2] Collaborative Innovation Center of Chinese Medicinal Resources Industrialization, Nanjing 210023, China
[3] National and Local Collaborative Engineering Center of Chinese Medicinal Resources Industrialization and Formulae Innovative Medicine, Nanjing 210023, China
* Correspondence: liuxunh1959@163.com; Tel./Fax: +86-25-8581-1524

Received: 25 August 2018; Accepted: 3 October 2018; Published: 5 October 2018

Abstract: As one of the major abiotic stresses, salinity stress may affect the physiology and biochemical components of *Apocynum venetum* L. To systematically evaluate the quality of Apocyni Veneti Folium (AVF) from the perspective of physiological and the wide variety of bioactive components response to various concentrations of salt stress, this experiment was arranged on the basis of ultra-fast liquid chromatography tandem triple quadrupole mass spectrometry (UFLC-QTRAP-MS/MS) technology and multivariate statistical analysis. Physiological characteristics of photosynthetic pigments, osmotic homeostasis, lipid peroxidation product, and antioxidative enzymes were introduced to investigate the salt tolerance mechanism of AVF under salinity treatments of four concentrations (0, 100, 200, and 300 mM NaCl, respectively). Furthermore, a total of 43 bioactive constituents, including 14 amino acids, nine nucleosides, six organic acids, and 14 flavonoids were quantified in AVF under salt stress. In addition, multivariate statistical analysis, including hierarchical clustering analysis, principal component analysis (PCA), and gray relational analysis (GRA) was employed to systematically cluster, distinguish, and evaluate the samples, respectively. Compared with the control, the results demonstrated that 200 mM and 100 mM salt stress contributed to maintain high quality of photosynthesis, osmotic balance, antioxidant enzyme activity, and the accumulation of metabolites, except for total organic acids, and the quality of AVF obtained by these two groups was better than others; however, under severe stress, the accumulation of the oxidative damage and the reduction of metabolite caused by inefficiently scavenging reactive oxygen species (ROS) lead to lower quality. In summary, the proposed method may provide integrated information for the quality evaluation of AVF and other salt-tolerant Chinese medicines.

Keywords: Apocyni Veneti Folium; salt stress; multiple bioactive constituents; physiological changes; multivariate statistical analysis

1. Introduction

As a major abiotic factors constraint on agriculture, salinity affects about 20% of the cultivated lands in the world and nearly 50% of all irrigated lands [1,2]. In China, about 34.6 million hectares lands are suffering from salinity interference. The medicinal plants among the major and vital groups of crops that exert a significant role in disease prevention and treatment are also being threatened by this constraint [3].

The effects of salt stress on plant growth are mainly revealed in ion toxicity, osmotic stress, and secondary oxidative stress. Plants subjected to salt stress form a series of physiological and molecular mechanisms that respond to salt stress, including ion transport and distribution to maintain ion balance [4,5], osmotic adjustment substances, and metabolites formation to maintain osmotic balance, antioxidant enzyme accumulation, and activity enhancement to resist oxidative stress, signal transduction factors, and salt stress-related genes regulation [6–10].

Apocyni Veneti Folium (AVF), Luobumaye in Chinese, is the dried leaf of *Apocynum venetum* L. (Apocynaceae) [11]. For centuries, AVF has been used to treat cardiac disease, hypertension, nephritis, and neurasthenia, and processed into tea due to its health benefits. Documentaries demonstrated that AVF has functions of antihypertension, antidepression, hepatoprotection, antianxiety, antioxidation, and diuresis [12,13]. It is well established that metabolites of medicinal plants, as sources of natural antioxidant and immune enhancer, are involved in the treatment of human diseases and health disorders [14]. However, their synthesis and accumulation depend on the growing conditions and are vitalized under abiotic stresses [5]. Therefore, a systematic quality assessment method is required for the quality control of herbal medicines.

For the assessment of contents of bioactive components in AVF, many analytical methods have been established by using high performance liquid chromatography (HPLC) coupled to an UV detector [15] or high performance capillary electrophoresis method with diode array detection (HPCE-DAD) [16]. However, these methods have been limited to the quantification of only a few flavones. Additionally, ion trap-time-of-flight (IT-TOF) MS system has been applied to sensitively detect phenolic acids and flavonoids in AVF [17], but not simultaneously detect a variety of bioactive compounds. Recently, ultra-fast liquid chromatography tandem triple quadrupole mass spectrometry (UFLC-QTRAP-MS/MS) is useful for the qualitative and quantitative analysis of bioactive compounds with the advantages of great separation efficiency, high peak capacity, and high sensitivity [18,19] and has emerged as a good tool for the sensitive and selective analysis of various constituents [20].

Apocynum venetum L., as a medicinal halophytic plant, is able to grow in most regions of China, but widely distributed among saline-alkali wasteland, desert edge, and the Gobi desert. According to the Chinese Pharmacopoeia (2015), hyperoside is used as a basis for assessing the quality of AVF and its content should be no less than 0.3% [13]. However, documents have shown that the content of particular class of components is not efficient to evaluate AVF, considering its growing environment [17,21], and there is no published comparative study reported for the quantification of bioactive constituents combined with physiological changes simultaneously.

Choosing of optimum environmental condition to elevate the metabolites has been reported on medicinal plants [22]. In this study, the effects of salt stress on the physiological characteristics and multiple bioactive constituents of AVF were studied, and quantitative results were then further interpreted by multivariate statistical analysis to evaluate its quality. Our investigation may provide a theoretical basis for the quality evaluation of AVF in respect of the increasing in metabolites under superior salt stressed condition and the mechanism of salt tolerance.

2. Results

2.1. Physiological Changes Affected by NaCl

2.1.1. Effects of Salt Stress on Photosynthetic Pigments

Salt-treated plant did not show significant change from the perspective of plant phenotypes even at the end of the 40 days experiment (Figure S1), as compared to the control. Chlorophyll a, total chlorophyll, chlorophyll a/b, and carotenoids were significantly increased in the presence of 100 and 200 mM salt, but the changes were not significant under severe stress (Table 1). However, chlorophyll b had little change compared to the control throughout the salt stressed experiments with value ranging from 0.43 to 0.55 mg g^{-1} FW.

Table 1. Effects of different NaCl concentrations on the chlorophyll (Chl) and carotenoids contents of AVF.

Treatments	Pigment Content				
	Chl a (mg g^{-1} FW)	Chl b (mg g^{-1} FW)	total Chl (mg g^{-1} FW)	Chl a/b	Carotenoids (mg g^{-1} FW)
0 mM	2.35 ± 0.11 c	0.49 ± 0.06 ab	2.84 ± 0.05 c	4.77 ± 0.82 c	0.86 ± 0.07 c
100 mM	3.66 ± 0.19 a	0.55 ± 0.04 a	4.21 ± 0.22 a	6.63 ± 0.14 a	1.25 ± 0.01 a
200 mM	3.00 ± 0.20 b	0.48 ± 0.01 ab	3.47 ± 0.20 b	6.30 ± 0.49 ab	1.04 ± 0.02 b
300 mM	2.31 ± 0.12 c	0.43 ± 0.01 b	2.74 ± 0.11 c	5.33 ± 0.40 bc	0.83 ± 0.01 c

Data are the mean ± SD (*n* = 3). Different letters following values in the same column indicate significant difference among salt treatments using Duncan's multiple-range test at $p < 0.05$.

2.1.2. Effects of Salt Stress on Osmolytes and Lipid Peroxidation

Compared to the control, the content of soluble sugars, proline and soluble proteins under salt treatment showed significant changes, which increased first and then decreased throughout the experiment (Figure 1). It is worth noting that the first two increased about 3.76 and 2.11-fold, respectively, and the osmolytes were affected significantly under 200 mM NaCl treatment. The accumulation of malondiadehyde (MDA) was significantly increased with the elevated salt treatments ranging from 74.42 to 91.21 nmol g^{-1} FW (Table S1).

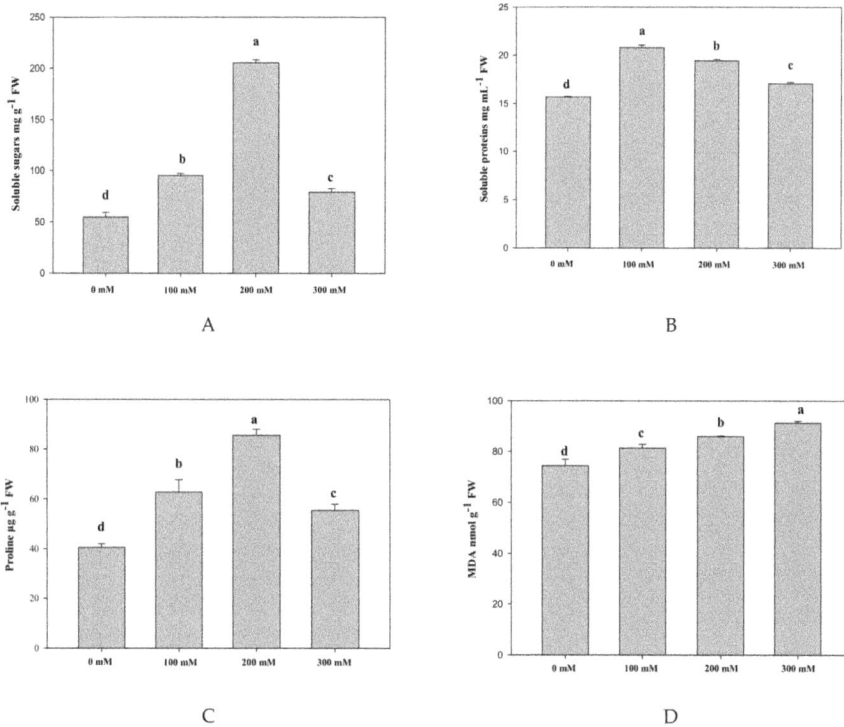

Figure 1. Effects of different concentrations of NaCl on soluble sugars (**A**), soluble proteins (**B**), proline (**C**), and MDA (**D**). Bars are expressed as the mean ± SD (*n* = 3). Bars carrying different letters are significantly different at $p < 0.05$.

2.1.3. Effects of Salt Stress on Antioxidant Enzyme and Ascorbic Acid

It can be observed from Figure 2 that the activity of superoxide dismutase (SOD) was significantly increased under moderate stress and severe stress with respect to the control. Peroxidase (POD) activity was noticeably enhanced, specifically under severe stress, increased to 38.66 U mg^{-1} prot, about 2.2-fold compared with the control (Table S1). However, catalase (CAT) activity in salt-treated AVF was significantly declined compared to the control, but it was elevated with the increasing salt concentrations from 100 to 300 mM (Table S1). Significant change was shown between any two groups of ascorbic acid ranging from 980.9 to 1095 μg mL^{-1}, but it was moderately higher in the salt-treated groups than in the control.

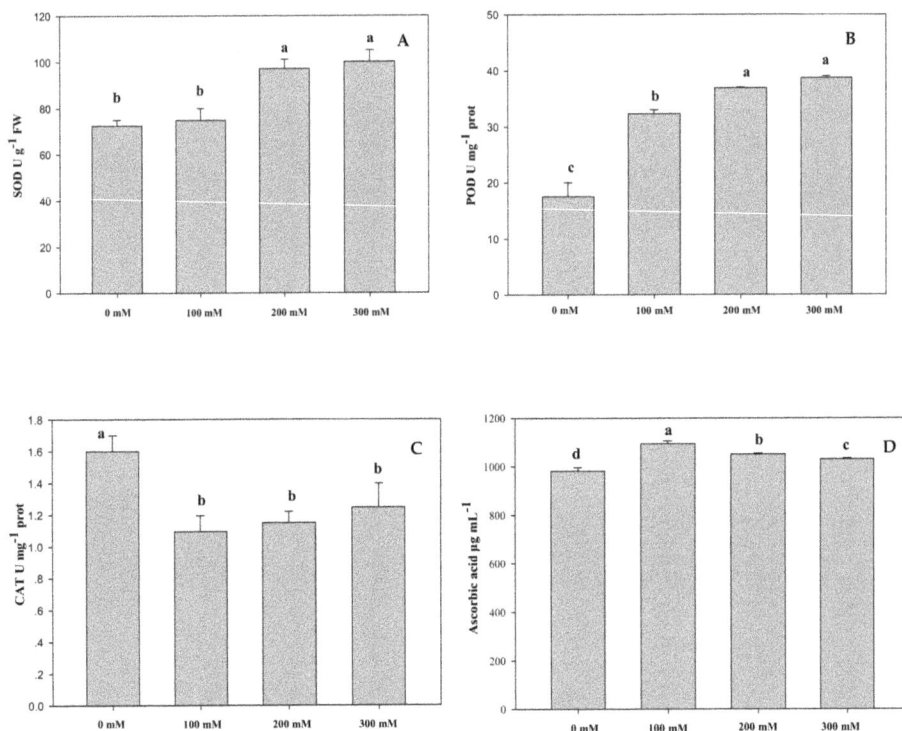

Figure 2. Effects of different NaCl concentrations on the activities of superoxide dismutase (**A**), peroxidase (**B**) catalase (**C**), and ascorbic acid contents (**D**) in Apocyni Veneti Folium (AVF). Bars are expressed as the mean ± SD (*n* = 3). Bars carrying different letters are significantly different at *p* < 0.05 among NaCl treatments.

2.2. Determination of Multiple Bioactive Components

2.2.1. Optimization of Sample Preparation and UFLC-QTRAP-MS/MS Conditions

In order to obtain the optimal extraction efficiency, extraction methods were optimized. Relatively speaking, ultrasonic extraction of samples with a ratio of water volume (mL) to sample weight (g) in accordance with 100:1 for 45 min under 30 °C was the appropriate condition. Then, four kinds of standard compounds with low and high contents, uracil, phenylalanine, neochlorogenic acid, and hyperoside were used to optimize the UFLC-QTRAP-MS/MS conditions. By using the UFLC system with a XBridge® C$_{18}$ column (100 mm × 4.6 mm, 3.5 μm) in the case of the mobile phase of

0.1% formic acid in water—0.1% formic acid in acetonitrile, the flow rate of 0.8 mL min^{-1}, and the column temperature of 30 °C; the analytes were well separated.

MRM (multiple reaction monitoring) technology mainly targeted selection of data for mass spectrometry signal acquisition, recorded signal of the regular ion pairs, and removed the interference ion signal. Only the MS/MS^2 ions selected for mass spectrometry acquisition, in order to achieve more specific, sensitive and accurate analysis of the target molecules [20]. Representative extracted ion chromatograms of 43 analytes in the MRM mode were presented in Figure S2, the detailed information about MS/MS condition for each analyte was listed in Table S2, and the characteristic total ion chromatograms (TIC) was displayed in Figure S3.

2.2.2. Method Validation

Quantitative analysis was performed using the UFLC-QTRAP-MS/MS technique. Calibration curves were constructed by injecting each analyte three times and over six suitable concentrations into UFLC-QTRAP-MS/MS system. As listed in Table S3, the limits of detection (LODs) and limits of quantitation (LOQs) were measured in the range of 0.91–6.15 ng mL^{-1} and 3.03–20.5 ng mL^{-1}, respectively. Each RSD was less than 5%, and the recovery ranged from 95.19 to 103.91%.

2.2.3. Sample Determination

Compared with the control, there was no significant change in asparagine, a higher content of amino acids in AVF, except for 200 mM of salinity stressed group (Table S4). The content of glutamine, proline, and glutamic acid changed notably between any two groups, while the change in the total amino acids was non-significant (Figure 3). Little change was detected on the total nucleosides except for the 300 mM NaCl stressed group. By comparison with the control, the accumulation of inosine and thymidine were considerably reduced. The content of neochlorogenic acid and the total organic acids under salinity stress was significantly reduced, but the change in caffeic acid was different. Compared to the control, kaempferol 3-O-rutinoside, gallocatechin, and epigallocatechin were significantly increased under salt stress, and significant differences were shown from each other. The accumulations of hyperoside, isoquercitrin, astragalin, trifolin, and total flavonoids were changed similarly, increasing first and then decreasing with the increasing salt concentrations.

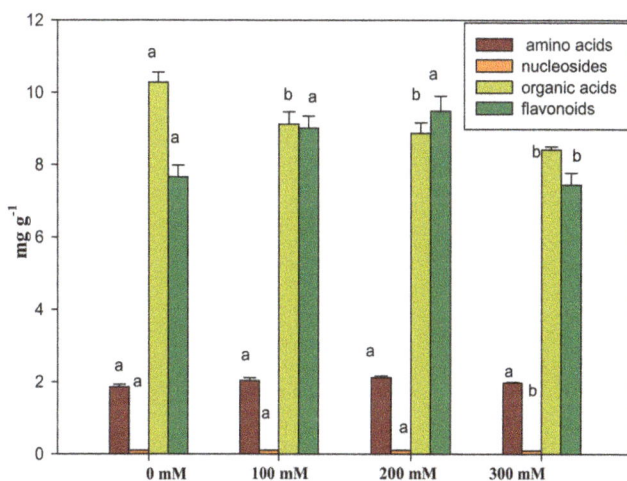

Figure 3. The accumulation of four kinds of constituents in AVF under salt tolerance. Bars are expressed as the mean ± SD ($n = 3$). Bars carrying different letters are significantly different at $p < 0.05$ among NaCl treatments.

2.2.4. Multivariate Statistical Analysis of Samples

A heat map derived from hierarchical clustering analysis intuitively displayed the changes of the accumulation of 43 bioactive components under salinity stress (Figure 4A), and on the other hand, the clustering of samples. In detail, 0 and 100 mM salt treated samples, and 200 and 300 mM salt treated ones were clustered separately and then gathered together. Principal component analysis (PCA) scores plot (Figure 4B) exhibited a statistical distinction based on 43 compositions under salinity stress with R2X [1] and R2X [2] accounted for 47.0% and 16.6% of the total variance, respectively [11]. In the PCA loading plot, chemical markers possessing large loading values of ions, such askaempferol 3-*O*-rutinoside, hypoxanthine, and thymidine strongly contribute to sample classification. Additionally, gray relational analysis (GRA) is part of the grey system theory and is suitable for solving problems with complicated interrelationships between multiple factors and variables. It provides a reliable guarantee for the quality evaluation of traditional Chinese medicines. The relative correlation degree (r_i) derived from GRA is proportional to the sample quality. Thus, the quality order of AVF under different NaCl treatments was: 200 mM salinity stressed group > 100 mM salinity stressed group > 300 mM salinity stressed group > control, and the corresponding values of r_i were 0.6363, 0.5253, 0.4827, and 0.3984, respectively. These directly revealed that the accumulations of metabolites were affected under saline condition in AVF (Table 2).

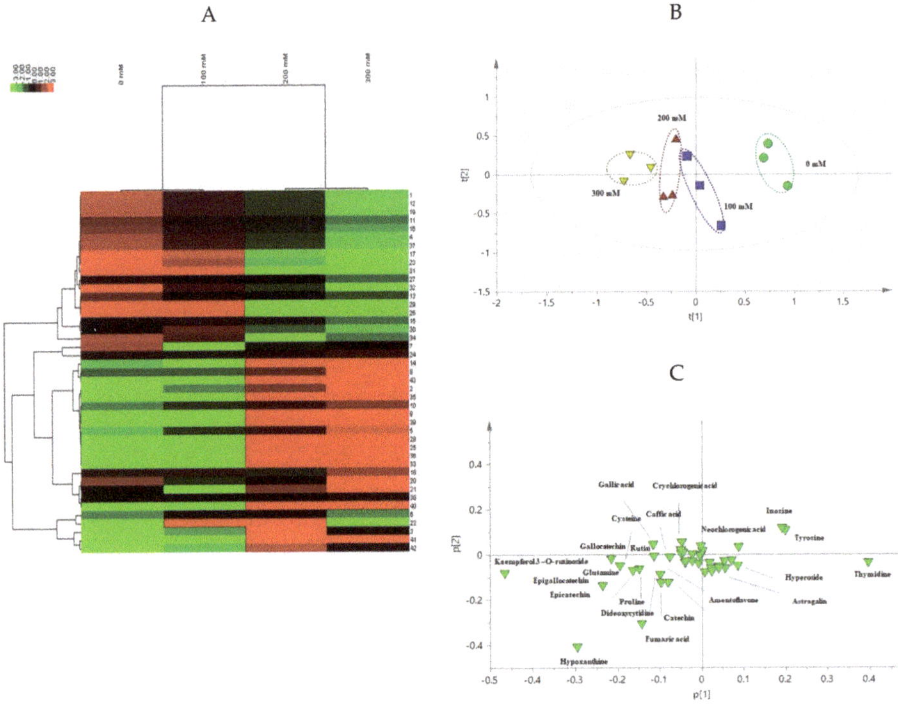

Figure 4. Multivariate statistical analysis of AVF under different salt treatments. Heat map derived from hierarchical clustering analysis (**A**), principal component analysis (PCA) scores plot (**B**), and PCA loading plot (**C**) of AVF.

Table 2. Quality sequencing of the tested samples affected by NaCl.

Treatments	r_i	Quality-Ranking
0 mM	0.3984	4
100 mM	0.5253	2
200 mM	0.6363	1
300 mM	0.4827	3

3. Discussion

Plants exposed to salt stress undergo physiological and biochemical adaptations to help maintain protoplasmic viability in response to salinity. It has been reported that salt stress can cause oxidative stress through increased reactive oxygen species (ROS). Though ROS molecules are the by-products of vital metabolisms, the built-in antioxidant system maintains the ROS under the controlled level. Temporal and spatial-localization of ROS is vital for the regulation of signaling mechanisms [23]. Highly-accumulated ROS, generated as a result of the decreased gas exchange processes and impairment in protective mechanisms, could damage the cellular components, such as lipids, proteins, and nucleic acids [24]. The increased salinity affects primary carbon metabolism, plant growth, and development by ion toxicity, and induces nutritional deficiency, water deficits, and oxidative stress [5,25]. Moreover, it modulates the levels of secondary metabolites, which are physiologically important particularly under stress tolerance [26]. To mitigate ROS-mediated oxidative damage, plants have developed a complex antioxidant defense system that includes osmotic homeostasis, antioxidant enzymes, and metabolites [27–29].

Photosynthesis is involved in the energy metabolism of all the plant systems. When higher plants suffer from salt stress, a growth disorder normally occurs, such as membrane damage and toxic compound accumulation. It can lead to a reduction of the chlorophyll content, disintegration of chloroplast membranes, disruption of photosystem biochemical reactions, and the reduction of photosynthetic activity [30]. The decrease in the chlorophyll content under environmental stress could be attributed to the enhancement of chlorophyll degradation [31]. Carotenoids are accessory light harvesting pigments, preventing the photosynthetic pigments from photo-damage, stabilizing the phospholipids, and scavenging various ROS generated during stressful salinity.

Compared with the control, the content of pigments shown in this study was increased under low and moderated stress, but decreased under severe stress. It indicates that salt stress at specific concentrations might promote the photosynthesis of AVF and severe stress might have a certain inhibitory effect, probably due to the impact of salt on disturbing photosynthesis process, photosynthetic enzymes, chlorophylls, and carotenoids, respectively [5]. The change of chlorophyll a/b reflects the photosynthetic activity of the leaves, and a reasonable value can prevent the excessive light energy in the leaves from inducing the generation of free radicals and photo oxidation of pigment molecules. In summary, it indicated a full utilization of light energy and enhancement of metabolic activity of AVF exposed to low concentrations of NaCl.

The accumulation of osmotic adjustments is one of salt-tolerant mechanisms. It may increase cellular concentrations, help maintain ion homeostasis and water relations, alleviate the negative effects of high ion concentrations on the enzymes and proteins under stressed conditions [32]. Soluble sugars are involved in biosynthetic process and can balance the osmotic strength of cytosol with that of vacuole [33]. The fluctuation in this study under salt stress might be caused by the changes of CO_2 assimilation, source-sink carbon partitioning, and/ or the activity of related enzymes [34]. Soluble proteins in AVF clearly increased depending on the rising of NaCl, and then significantly reduced, indicating that cell turgor maintaining and water acquisition regulation perhaps was affected by salt stress [35,36]. The results were consistent with the literature reported on *Salvia miltiorrhiza* [37]. The increased osmolytes in AVF might be responsible for maintaining homeostasis under low and moderate salt stress, but under severe salt stress, it was notably affected, possibly due to the inability to effectively keep osmotic balance.

Proline accumulation is one of the adaptations of plants to salinity. It has a wide range of biological functions in plants, such as scavenging free-radical by quenching of singlet oxygen, protecting macromolecules against denaturation [38–40], reducing the acidity in the cell, and helping rapid growth after stress [41]. In this study, proline alleviated NaCl stressed induction, but it was significantly reduced caused by osmotic tolerance due to the severe salt stress [42].

Under severe salinity stress, loss of the membrane integrity and stability is a common symptom developed in plants [43] due to excessively formation of free radicals and lipid peroxidation. As one of the lipid peroxidation products, MDA plays an important role in modifying core proteins and in many stressed plants, and it is considered as a useful oxidative marker to indicate the chloroplast lipid peroxidation [44]. A significant increase in the MDA content with salt stress elevated in our study might indicate the oxidative degradation of chloroplast membranes [33,45].

SOD, CAT, and POD, which are involved in antioxidation processes, protect plants from oxidative damage caused by abiotic stresses [46]. SOD catalyzes the dismutation of O_2^- into O_2 and H_2O_2, whereas CAT dis-mutates mostly photorespiratory/respiratory H_2O_2 into H_2O and O_2, and POD is responsible for the removal of H_2O_2 by oxidation of co-substrate, such as phenolic compounds [47]. Integral coordination of antioxidant enzymes could be vital for the redox homeostasis mechanism under the oxidative stress, as previously reported in wheat [48] and barley [49]. In the present study, elevated NaCl increased the activities of SOD and POD, but not CAT. Furthermore, POD might be efficient in clearing the excess H_2O_2 with significantly increased activity and slightly increased SOD activity but significantly decreased in CAT activity appeared under 50 mM NaCl stress. Besides H_2O_2 stress, salt stress might simultaneously enhance superoxide production in cells [50].

As a non-enzymatic free radical scavenger and a key substance in the network of antioxidants, ascorbic acid has been shown to play multiple roles in plant growth. It has also been seen in regulating the normal reactive oxygen species in plant cells together with other small molecules [51,52]. A significant increase was shown in ascorbic acid to protect the body from endogenous damage of oxygen free radicals and then a decrease with the increasing NaCl concentrations, which might be due to the consumption as an enzyme-catalyzed substrate for scavenging ROS.

The increased level of free amino acids in the cell cytoplasm plays an important role in osmotic adjustments, which are also involved in the stability and integrity of cellular membranes in saline environment [53]. The amino acid content was found increased in *Aloe vera* during salinity stress [54]. In this study, the elevated level of amino acids, such as glutamic acid, cysteine, and proline helped provide osmotic protection for AVF. Nucleosides and their derivatives have significant physiological functions. Higher salinity had been reported to induce changes in protein structure, increase in cytoplasmic RNAase activity, leading to decrease in DNA synthesis and creating many cellular menaces to activity required for development processes in plants [55]. Under adverse conditions, transcription factors associated with stress resistance can regulate the simultaneous expression of multiple stress-tolerant genes and the transmission of stress signals [56]. Lower content of nucleosides was shown under severe stress due to the salt stress.

Abiotic stress promotes the synthesis of various secondary metabolites possessing antioxidant activity. Organic acids and flavonoids are ubiquitous in plants and are generally accumulated in response to salinity stress [57]. As mentioned above, phenolic metabolites can cooperate with POD in H_2O_2 scavenging. The phenylpropanoid pathway is the main metabolic route for the synthesis of phenolics and flavonoids [58]. The accumulation of organic acids may vary in different plant species in response to salinity tolerance. In previous reports, they were observed increased in buckwheat sprout [59], but decreased in baby Romaine lettuce [60]. In this experiment, they were significantly depressed under salinity conditions, possibly because of the consumption of scavenging or detoxifying excess free radicals [61].

Flavonoids have a wide array of physiological functions in plants, e.g., involvement in UV filtration and symbiotic nitrogen fixation, and acting as chemical signal messengers for initiating plant-microbe symbiotic associations [62,63], and they also contribute significantly for human by virtue of antioxidative, antiviruses, antiangiogenic, and neuropharmacological effects [64,65]. Flavonoid biosynthesis can be stimulated by the variation of the cellular redox homeostasis and lipid peroxidation of membranes of the plant cell [33,66]. In our study, the increase in flavonoid content under low and moderate might be associated with the increases in chlorophyll, and this enhanced synthesis of secondary metabolites under stressful conditions was believed to protect the cellular structures from oxidative damage and osmotic stress [67]. However, it decreases under severe salt stress, which might be due to the oxidative damage caused by the imbalance between antioxidants formation and ROS scavenging.

According to the results of hierarchical clustering analysis showed in the heat map, the control and low concentrations of salt stressed AVF, and the moderate and high concentrations of salt stressed were aggregated together orderly. This was consistent with PCA results that AVF samples exposed to salt stress (0, 100, 200, and 300 mM, respectively) were sequentially distributed and could be distinguished from each other from the positive to the negative axis of PC1. Moreover, chemical markers in the PCA loading plot provided the possibility of sample differentiation. In all, the results of multivariate statistical analysis indicated that it was not only provided relevant basis for the classification of various salt treatment of AVF, but also comprehensively evaluated the quality of it, that is, moderate salt treated samples were superior to others, and salt treated ones were better than the control.

Generally speaking, in plants under salt stress, photosynthesis, osmotic balance, and metabolic processes are deeply affected [5]. Impairment in the photosynthetic process leads to the higher lipid peroxidation and excessive accumulation of reactive oxygen species (ROS) [24]. The productions of ROS are much higher than its detoxification in abiotic stress conditions. Osmolytes play key roles in maintaining normal osmotic potential, and antioxidant systems of enzymes and metabolites protect plants from oxidative damage and efficiently retain more tolerance against abiotic stresses. In the subsequent experiments, we will use multi-omics approaches to comprehensively interpret the effects of salt stress on the quality of AVF and the salt tolerance mechanism at the transcription, protein, and metabolic levels [68].

4. Materials and Methods

4.1. Plant Materials and Salinity Treatments

The experimental samples of AVF were obtained by the following steps. Firstly, the site for salt stress test was selected in Medicinal Botanical Garden of Nanjing University of Traditional Chinese Medicine (latitude 118°57′1″, East longitude 32°6′5″). The experiment was carried out in the shelter covered by a transparent film that blocked rainwater, while other conditions were similar to the open-air environment. Secondly, the materials and methods of salt stress were given as follow: The botanical origins of the materials were identified by Professor Xunhong Liu (Department for Authentication of Chinese Medicines, Nanjing University of Chinese Medicine); the main root of *Apocynum Venetum* L., two years old from the same plant, and the number of bud and head close to each other, was excavated from the garden in December 2016, and then planted in pots (50 cm height, 34 cm of top diameter, and 26 cm of bottom diameter). Each pot was filled with 25 kg of dry soil and 3 roots, and was placed in the open air before the salt-treated experiments.

Salt stress tests had been conducted since 20 May 2017 when *Apocynum venetum* L. was in normal growth (about 30 cm height). Four levels of salt treatment concentrations, 0 (control, watering), 100 (low stress), 200 (moderate stress), and 300 mM (severe stress) NaCl treatments were designed with 3 replicates at each concentration level and 3 pots per replicate. According to the previous research, by calculating the amount of water, the final determination of the solution per pot was 2 L. In order to prevent osmotic shock, salt concentrations increased gradually by 50 mM NaCl every four days until the designated concentration was reached and lasted 6 times. The photograph of changes of plant phenotypes upon treatment of different concentrations of NaCl was seen in Figure S1. At last, experimental samples were harvested on 30 June 2017. The collected samples of four groups were immediately frozen at $-80\ °C$ for subsequent experiments, some for physiological experiments, some for quantitative analysis, and the rest for the voucher specimens deposited at the Herbarium in School of Pharmacy, Nanjing University of Chinese Medicine.

4.2. Physiological Experiment

4.2.1. Extraction and Assay of Pigment

Four groups of fresh AVF samples (0.2 g) under salt stress were homogenized with ethanol (95%, v/v), filtered, and made up to 2 mL, respectively. Photosynthetic pigments (chlorophyll a, b, total chlorophyll and carotenoids) concentrations were calculated from the absorbance of extract at 665, 649 and 470 nm using the formula [37,69], given as follow: Chlorophyll a (mg g^{-1} FW) = (13.95 × A665 − 6.88 × A649) × 2/(1000 × 0.2); Chlorophyll b (mg g^{-1} FW) = (24.96 × A649 − 7.32 × A665) × 2/(1000 × 0.2); Carotenoids (mg g^{-1} FW) = ((1000 × A470) − (2.05 × Chl a) − (114.8 × Chl b)) × 2/(245 × 1000 × 0.2)

4.2.2. Osmolytes and MDA Assay

Osmolytes, including soluble proteins, soluble sugars and proline, were assayed in this study to measure the salt tolerance of AVF. The content of soluble proteins was determined according to the ultraviolet absorption method. 0.5 g of fresh sample of AVF under salt stress was homogenized and extracted by 8 mL of PBS (phosphate buffer saline, 0.1 mM Na_2HPO_4 and NaH_2PO_4, pH 7.4), respectively. After centrifugation, 1 μL of the supernatant was subjected to detect using UV-Vis spectrophotometer (DENOVIX DS-11, Wilmington, DE, USA), which can directly display the concentration of protein in the solution by detecting the absorbance of the solution at 260 nm and 280 nm with 1 μL of PBS as a control. The content of soluble sugars and proline was quantified by colorimetric method. To measure the content of soluble sugars, 0.5 g of fresh leaves was homogenized with 4.5 mL of PBS in an ice water bath and centrifuge the homogenate at 3500 rpm for 10 min. 0.5 mL of extract was mixed with 3 mL of anthrone solution (75 mg anthrone in 50 mL of 72% sulphuric acid (w/w)), and was immediately placed in a boiling water bath for 10 min. The light absorption was estimated at 620 nm. The content of soluble sugars was determined by using glucose as a standard and expressed as mg g^{-1} FW. With regard to the free proline assay, the procedure was as follows: 0.5 g of harvested leaf fragments were extracted with 4.5 mL of aqueous sulfosalicylic acid (3%, w/v) in boiled water for 10 min and then centrifuged at 3500 rpm for 10 min. After that, 2 mL of glacial acetic acid and 4 mL of acidninhydrin agent (1.25 g acidninhydrin in 30 mL glacial acetic acid and 20 mL of 6 mol L^{-1} H_3PO_4) were added to the homogenate in a test tube. The mixture was incubated in boiling water for 30 min, and then the test tube was placed in the cold water to terminate the reaction. Each test tube was added to 4 mL of toluene and vortexed for 30 s. The supernatant was taken and centrifuged at 3000 rpm for 5 min. Proline content was quantified by using the Bate's method at 520 nm [69,70].

Lipid peroxidation was measured in terms of MDA by thiobarbituric acid (TBA) method [69,71]. Fresh leaf (0.5 g) fragment were homogenized with 5 mL of trichloroacetic acid (3%, m/v), and then centrifuged at 3000 rpm for 10 min. Took 2.0 mL aliquot of the supernatant to the test tube and added 2.0 mL of thiobarbituric acid (0.67%, m/v). The mixture was heated in boiling water for 30 min and then quickly cooled in an ice bath. After centrifugation at 3000 rpm for 10 min, the absorbance of the samples was recorded at 530 nm.

4.2.3. Enzyme Activities and Ascorbic Acid Assay

To determine the antioxidant enzyme activities, 0.5 g of fresh AVF under the stress of different salt concentrations were homogenized with 4.5 mL of PBS in an ice water bath and centrifuged at 3500 rpm for 10 min. The supernatant was collected to determine the antioxidant enzyme activities. SOD activity was assayed by hydroxylamine method, CAT activity was determined by ammonium molybdate method, and POD activity was measured according to the colorimetric method [69,70].

As for the determination of ascorbic acid, it was assayed based on the oxidation of ascorbic acid by iron (III) in the presence of 1,10-phenanthroline with subsequent formation of ferroin and a suitable anion associate according to the Zenki et al. method [72]. All of them were tested by assay kits bought from Nanjing Jiancheng Bioengineering Institute (Nanjing, China). UV-visible absorptions were measured by multi-mode microplate reader (SpectraMax M5, San Jose, CA, USA) and the detection amount of the reaction solution was 200 μL.

4.3. Multiple Bioactive Constituents Assay

4.3.1. Chemicals and Reagents

Ultrapure water was prepared using a Milli-Q purifying system (Millipore, Bedford, MA, USA). Methanol and acetonitrile of HPLC grade were purchased from Merck (Damstadt, Germany). Standard compounds of histidine (1), arginine (3), cysteine (4), asparagine (5), serine (6), lysine (7), glutamine (8), proline (9), cytidine (10), hypoxanthine (11), deoxycytidine (12), uridine (13), tyrosine (14), guanine (15), guanosine(16), inosine(17), deoxyguanosine (19), isoleucine (20), leucine (21), thymidine (23), phenylalanine (24), tryptophan (27), epicatechin (33), rutin (34), hyperoside (35), and quercitrin (37) were purchased from Shanghai Yuanye Biotechnology (Shanghai, China); glutamic acid (2), gallic acid (22) and apigenin (43) were obtained from Chinese National Institute of Control of Pharmaceutical and Biological Products (Beijing, China); fumaric acid (18), gallocatechin (25), epigallocatechin (28), cryptochlorogenic acid (31), kaempferol 3-*O*-rutinoside (39) and amentoflavone (42) were acquired from Chengdu Chroma Biotechnology (Chengdu, China); neochlorogenic acid (26), chlorogenic acid (29), catechin (30), caffeic acid (32), isoquercitrin (36), avicularin (38), trifolin (40), and astragalin (41) were bought from Baoji Chenguang Biotechnology Co., LTD. (Baoji, China) with the purity greater than 98% and their structures were presented in Figure S4.

4.3.2. Sample Preparation

Four groups of fresh AVF harvested under salt treatments were naturally dried, and then powdered and passed through a 60-mesh sieve. 0.3 g of sample was weighed accurately and ultrasonically extracted with 30 mL of water for 45 min, supplemented with water to compensate for the lost weight, and centrifuged at 12000 rpm for 15 min [11,73]. The supernatant was stored at 4 °C and filtered through a 0.22 μm membrane (Jinteng laboratory equipment Co., Ltd., Tianjin, China) before being subjected to UFLC-MS/MS analysis.

4.3.3. Chromatographic and Mass Spectrometric Conditions

The mobile phase of AB Sciex QTRAP® 4500 UFLC-MS/MS spectrometry consisted of water containing 0.1% formic acid (v/v, A) and acetonitrile containing 0.1% formic acid (v/v, B). The analytes were eluted using a linear gradient program: 1–3 min, 5% B; 3–6 min, 5–15% B; 6–15 min, 15–20% B; 15–17 min, 20–70% B, 17–17.5 min, 70–5% B, and 17.5–23 min, 5% B. The flow rate was 0.80 mL/min. The column temperature was 30 °C. The injection volume was 1 μL. According to our previous reports [11], the standard solution of each analyte was injected separately into the electrospray ionization (ESI) source in the direct infusion mode of MS to acquire the fragmentor voltage and collision energy in both positive and negative modes. Next, ESI source operates in both ion modes using the MRM transition acquiring the spectra and the Analyst 1.6.3 software analyzing data, respectively. In the same ion mode, isomers with the same ion pairs, such as catechin/epicatechin, chlorogenic acid/neochlorogenic acid/cryptochlorogenic acid, gallocatechin/epigallocatechin, hyperoside/isoquercetin, and leucine/isoleucine were separately and injected into UFLC-QTRAP-MS/MS to find the accurate t_R for identification and quantification. The operating parameters were set as follows: GS1 flow, 65 L min^{-1}; GS2 flow, 65 L min^{-1}; and CUR flow, 30 L min^{-1}; gas temperature, 650 °C; pressure of the nebulizer, 5500 V for the positive ion mode, and −4500 V for the negative ion mode, respectively.

4.3.4. Method Validation and Sample Determination

The standard solution containing 43 reference substances was prepared and diluted with water to appropriate concentrations for the construction of calibration curves. The concentrations of 43 analytes in mixed solution were seen in Table S3. The LODs and LOQs of constituents were measured at signal-to-noise (S/N) ratios of 3 and 10, respectively. Precision of the intra and inter-day was expressed as relative standard deviation (RSD). Repeatability was achieved by six different analytical sample solutions prepared by the same sample, and stability was performed by analyzing the variations at 0, 2, 4, 8, 12, and 24 h, respectively. While the recovery test was performed by adding a known amount of corresponding constituents in triplicate at low, medium, or high levels to 0.5 g of 100 mM NaCl treated samples, respectively. The quantitative determination of the bioactive compounds of AVF under different sat stress was performed under the optimal condition by UFLC-QTRAP-MS/MS.

4.3.5. Multivariate Statistical Analysis

Hierarchical cluster analysis is a method of cluster analysis, which seeks to build a hierarchy of clusters. PCA is a statistical procedure that uses an orthogonal transformation to convert a set of observations of possibly correlated variables (entities each of which takes on various numerical values) into a set of values of linearly uncorrelated variables called principal components [11]. Hierarchical clustering analysis and PCA were introduced to cluster and classify samples based on the content of constituents by Java Treeview 3.0 software and SIMCA-P 13.0 software, respectively. Then, GRA was performed according to the contents of 43 bioactive components by Microsoft Excel 2010 for Window 10 to evaluate the quality of AVF under different concentration of salt stress. Specifically, through the establishment of sample dataset and normalization treatment of raw data, the optimal and the worst reference sequences were conducted. After establishing dimension of the differences between comparing sequences and reference sequences, correlation coefficient and correlation degree were calculated, followed by the weight value of the evaluation samples (r_i).

4.4. Data Processing

The mean values of all parameters were taken from the measurements of three replicates with the standard deviation calculated. One-way ANOVA followed by Duncan's multiple-range test was used to compare the means with the significance level set as 0.05 by SPSS 19.0

5. Conclusions

In this study, our aim was to use the changes in physiology and biochemical components as references to study the quality control of AVF response to salt stress. Thus, an efficient analytical method of simultaneous determination of multiple bioactive constituents combined with physiological analysis was established for the quality evaluation based on the multivariate statistical analysis. Investigations into the physiological changes of photosynthetic pigments, osmotic homeostasis, lipid peroxidation, antioxidative enzymes, and ascorbic acid could provide comprehensive insights into the response mechanisms induced by salt stress. Furthermore, a total of 43 bioactive constituents, including amino acids, nucleosides, organic acids, and flavonoids, were successfully identified and quantified in different salinity-treated AVF with the application of UFLC-QTRAP-MS/MS technology. Multivariate statistical analysis was performed for the group classification and quality evaluation. Overall, the quality of AVF subjected to NaCl was superior to the control and AVF treated with 200 mM NaCl had the best quality. In general, this study was conducted to the quality evaluation of AVF concerning the impacts caused by salinity on the physiology and bioactive constituents. The results might provide a valuable reference for the quality assessment of other herbal medicines and the development of salt-tolerant plants in saline soils.

Supplementary Materials: Supplementary materials can be found at www.mdpi.com/1422-0067/19/10/3042/s1.

Int. J. Mol. Sci. **2018**, *19*, 3042

Author Contributions: C.C. and X.L. conceived and designed the experiments. C.C., C.W. and Z.L. performed the experiments. C.C., L.Z., J.S., S.C., J.C., and M.T. analyzed the data and drafted the manuscript. All authors contributed to the revision of this manuscript and approved the final manuscript.

Acknowledgments: This research was supported by the Priority Academic Program Development of Jiangsu Higher Education Institutions of China (NO. ysxk-2014) and Postgraduate Research & Practice Innovation Program of Jiangsu Province (KYCX18_1606).

Conflicts of Interest: The authors declare no conflicts of interest.

References

1. Zhao, G.M.; Han, Y.; Sun, X.; Li, S.H.; Shi, Q.M.; Wang, C.H. Salinity stress increases secondary metabolites and enzyme activity in safflower. *Ind. Crop. Prod.* **2015**, *64*, 175–181.
2. Zhou, Y.; Tang, N.Y.; Huang, L.J.; Zhao, Y.J.; Tang, X.Q.; Wang, K.C. Effects of Salt Stress on Plant Growth, Antioxidant Capacity, Glandular Trichome Density, and Volatile Exudates of Schizonepeta tenuifolia Briq. *Int. J. Mol. Sci.* **2018**, *19*, 252. [CrossRef] [PubMed]
3. Aghaei, K.; Komatsu, S. Crop and medicinal plants proteomics in response to salt stress. *Front. Plant Sci.* **2013**, *4*, 8. [CrossRef] [PubMed]
4. Liu, A.L.; Xiao, Z.X.; Li, M.; Wong, F.; Yung, W.S.; Ku, Y.S.; Wang, Q.W.; Wang, X.; Xie, M.; Yim, A.K.; et al. Transcriptomic reprogramming in soybean seedlings under salt stress. *Plant Cell Environ.* **2018**, 1–17.
5. Munns, R.; Tester, M. Mechanisms of salinity tolerance. *Annu. Rev. Plant Biol.* **2008**, *59*, 651–681. [CrossRef] [PubMed]
6. Arbona, V.; Manzi, M.; Ollas, C.; Gómez-Cadenas, A. Metabolomics as a Tool to Investigate Abiotic Stress Tolerance in Plants. *Int. J. Mol. Sci.* **2013**, *14*, 4885–4911. [CrossRef] [PubMed]
7. Zhang, B.; Liu, K.; Zheng, Y.; Wang, Y.; Wang, J.; Liao, H. Disruption of AtWNK8 Enhances Tolerance of Arabidopsis to Salt and Osmotic Stresses via Modulating Proline Content and Activities of Catalase and Peroxidase. *Int. J. Mol. Sci.* **2013**, *14*, 7032–7047. [CrossRef] [PubMed]
8. Jaleel, C.A.; Riadh, K.; Gopi, R.; InèsHameed, M.; Inès, J.; Al-Juburi, H.; Zhao, C.X.; Shao, H.B.; Rajaram, P. Antioxidant defense responses: Physiological plasticity in higher plants under abiotic constraints. *Acta Physiol. Plant.* **2009**, *31*, 427–436. [CrossRef]
9. Tran, L.S.; Urao, T.; Qin, F.; Maruyama, K.; Kakimoto, T.; Shinozaki, K.; Yamaguchi-Shinozaki, K. Functional analysis of AHK1/ATHK1 and cytokinin receptor histidine kinases in response to abscisic acid, drought, and salt stress in Arabidopsis. *Proc. Natl. Acad. Sci. USA* **2007**, *104*, 20623–20638. [CrossRef] [PubMed]
10. Tang, X.L.; Mu, X.M.; Shao, H.B.; Wang, H.; Brestic, M. Global plant-responding mechanisms to salt stress: Physiological and molecular levels and implications in biotechnology. *Crit. Rev. Biotechnol.* **2015**, *35*, 425–437. [CrossRef] [PubMed]
11. Chen, C.H.; Liu, Z.X.; Zou, L.X.; Liu, X.H.; Chai, C.; Zhao, H.; Yan, Y.; Wang, C.C. Quality evaluation of Apocyni Veneti Folium from different habitats and commercial herbs based on simultaneous determination of multiple bioactive constituents combined with multivariate statistical analysis. *Molecules* **2018**, *23*, 573. [CrossRef] [PubMed]
12. The Pharmacopoeia Committee of the Health Ministry of People's Republic of China. *Pharmacopoeia of People's Republic of China*; Guangdong Scientific Technologic Publisher: Guangzhou, China, 1995; p. 182.
13. Pharmacopoeia Commission of the Ministry of Health of the People's Republic of China. *Pharmacopoeia of the People's Republic of China*; Part I; Medical Science and Technology Press: Beijing, China, 2015; pp. 211–212.
14. Ksouri, R.; Megdiche, W.; Debez, A.; Falleh, H.; Grignon, C.; Abdelly, C. Salinity effects on polyphenol content and antioxidant activities in leaves of the halophyte *Cakile maritima*. *Plant Physiol. Biochem.* **2007**, *45*, 244–249. [CrossRef] [PubMed]
15. Zhou, C.Z.; Gao, G.H.; Zhou, X.M.; Yu, D.; Chen, X.H.; Bi, K.S. Simultaneous determination of five active components in traditional Chinese medicine *Apocynum venetum* L. by RP-HPLC–DAD. *J. Med. Plants Res.* **2011**, *5*, 735–742.
16. Liu, X.H.; Zhang, Y.C.; Li, S.J.; Wang, M.; Wang, L.J. Simultaneous Determination of Four Flavonoids in Folium Apocyni Veneti by HPCE-DAD. *Chin. Pharmacol. J.* **2010**, *45*, 464–467.
17. An, H.J.; Wang, H.; Lan, Y.X.; Hashi, Y.; Chen, S.Z. Simultaneous qualitative and quantitative analysis of phenolic acids and flavonoids for the quality control of *Apocynum venetum* L. leaves by

HPLC–DAD–ESI–IT–TOF–MS and HPLC–DAD. *J. Pharmaceut. Biomed.* **2013**, *85*, 295–304. [CrossRef] [PubMed]

18. Chen, F.; Zhang, F.S.; Yang, N.Y.; Liu, X.H. Simultaneous Determination of 10 Nucleosides and Nucleobases in *Antrodia camphorata* Using QTRAP LC–MS/MS. *J. Chromatogr. Sci.* **2014**, *52*, 852–861. [CrossRef] [PubMed]

19. Yuan, M.; Breitkopf, S.B.; Yang, X.; Asara, J.M. A positive/negative ion-switching, targeted mass spectrometry-based metabolomics platform for bodily fluids, cells, and fresh and fixed tissue. *Nat. Protoc.* **2012**, *7*, 872–881. [CrossRef] [PubMed]

20. March, R.E. An introduction to quadrupole ion trap mass spectrometry. *J. Mass Spectrom.* **1997**, *32*, 351–369. [CrossRef]

21. Shi, J.Y.; Li, G.L.; Zhang, R.; Zheng, J.; Suo, Y.R.; You, J.M.; Liu, Y.J. A validated HPLC-DAD-MS method for identifying and determining the bioactive components of two kinds of luobuma. *J. Liq. Chromatogr. Relat. Technol.* **2011**, *34*, 537–547. [CrossRef]

22. Wahid, A.; Ghazanfar, A. Possible involvement of some secondary metabolites in salt tolerance of sugarcane. *J. Plant Physiol.* **2006**, *163*, 723–730. [CrossRef] [PubMed]

23. Mateos-Naranjo, E.; Andrades-Moreno, L.; Davy, A.J. Silicon alleviates deleterious effects of high salinity on the halophytic grass Spartina densiflora. *Plant Physiol. Biochem.* **2013**, *63*, 115–121. [CrossRef] [PubMed]

24. Kim, Y.H.; Khan, A.L.; Kim, D.H.; Lee, S.Y.; Kim, K.M.; Waqas, M.; Jung, H.Y.; Shin, J.H.; Kim, J.G.; Lee, I.J. Silicon mitigates heavy metal stress by regulating P-type heavy metal ATPases, Oryza sativa low silicon genes, and endogenous phytohormones. *BMC Plant Biol.* **2014**, *14*, 13. [CrossRef] [PubMed]

25. Flowers, T.J.; Colmer, T.D. Salinity tolerance in halophytes. *New Phytol.* **2008**, *179*, 945–963. [CrossRef] [PubMed]

26. Parihar, P.; Singh, S.; Singh, R.; Singh, V.P.; Prasad, S.M. Effect of salinity stress on plants and its tolerance strategies: A review. *Environ. Sci. Pollut. Res. Int.* **2015**, *22*, 4056–4075. [CrossRef] [PubMed]

27. Deinlein, U.; Stephan, A.B.; Horie, T.; Luo, W.; Xu, G.; Schroeder, J.I. Plant salt-tolerance mechanisms. *Trends Plant Sci.* **2014**, *19*, 371–379. [CrossRef] [PubMed]

28. Golldack, D.; Li, C.; Mohan, H.; Probst, N. Tolerance to drought and salts tress in plants: Unraveling the signaling networks. *Front. Plant Sci.* **2014**, *5*, 151. [CrossRef] [PubMed]

29. Roy, S.J.; Negrão, S.; Tester, M. Salt resistant crop plants. *Curr. Opin. Biotechnol.* **2014**, *26*, 115–124. [CrossRef] [PubMed]

30. Gururani, M.A.; Venkatesh, J.; Tran, L.S.P. Regulation of photosynthesis during abiotic stress-induced photoinhibition. *Mol. Plant* **2015**, *8*, 1304–1320. [CrossRef] [PubMed]

31. Gururani, M.A.; Mohanta, T.K.; Bae, H. Current understanding of the interplay between phytohormones and photosynthesis under environmental stress. *Int. J. Mol. Sci.* **2015**, *16*, 19055–19085. [CrossRef] [PubMed]

32. Flowers, T.J.; Munns, R.; Colmer, T.D. Sodium chloride toxicity and the cellular basis of salt tolerance in halophytes. *Ann. Bot.* **2015**, *115*, 419–431. [CrossRef] [PubMed]

33. D'Souza, M.R.; Devaraj, V.R. Biochemical responses of Hyacinth bean (*Lablab purpureus*) to salinity stress. *Acta Physiol. Plant.* **2010**, *32*, 341–353.

34. Rosa, M.; Prado, C.; Podazza, G.; Interdonato, R.; Gonzalez, J.A.; Hilal, M.; Prado, F.E. Soluble sugars: Metabolism, sensing and abiotic stress. A complex network in the life of plants. *Plant Signal. Behav.* **2009**, *4*, 388–393. [CrossRef] [PubMed]

35. Abbaspour, H.; Afshari, H.; Abdel-Wahhab, A. Influence of salt stress on growth, pigments, soluble sugars and ion accumulation in three pistachio cultivars. *J. Med. Plants Res.* **2012**, *6*, 2468–2473. [CrossRef]

36. Mittal, S.; Kumari, N.; Sharma, V. Differential response of salt stress on *Brassica juncea*: Photosynthetic performance, pigment, proline, D1 and antioxidant enzymes. *Plant Physiol. Biochem.* **2012**, *54*, 17–26. [CrossRef] [PubMed]

37. Gengmao, Z.; Quanmei, S.; Yu, H.; Shihui, L.; Changhai, W. The physiological and biochemical responses of a medicinal plant (*Salvia miltiorrhiza* L.) to stress caused by various concentrations of NaCl. *PLoS ONE* **2014**, *9*, e89624. [CrossRef] [PubMed]

38. Sharma, S.; Verslues, P.E. Mechanisms independent of abscisic acid (ABA) or proline feedback have a predominant role in transcriptional regulation of proline metabolism during low water potential and stress recovery. *Plant Cell Environ.* **2010**, *33*, 1838–1851. [CrossRef] [PubMed]

39. Szabados, L.; Savoure, A. Proline: A multifunctional amino acid. *Trends Plant Sci.* **2010**, *15*, 89–97. [CrossRef] [PubMed]

40. Kumar, S.G.; Reddy, A.M.; Sudhakar, C. NaCl effects on proline metabolism in two high yielding genotypes of mulberry (*Morus alba* L.) with contrasting salt tolerance. *Plant Sci.* **2003**, *165*, 1245–1251. [CrossRef]
41. Cuin, T.A.; Shabala, S. Compatible solutes reduce ROS-induced potassium efflux in Arabidopsis roots. *Plant Cell Environ.* **2007**, *30*, 875–885. [CrossRef] [PubMed]
42. Çoban, Ö.; Baydar, N.G. Brassinosteroid effects on some physical and biochemical properties and secondary metabolite accumulation in peppermint (*Mentha piperita* L.) under salt stress. *Ind. Crops Prod.* **2016**, *86*, 251–258. [CrossRef]
43. Bajji, M.; Kinet, J.M.; Lutts, S. The use of the electrolyte leakage method for assessing cell membrane stability as a water stress tolerance test in durum wheat. *Plant Growth Regul.* **2002**, *36*, 61–70. [CrossRef]
44. Yamauchi, Y.; Sugimoto, Y. Effect of protein modification by malondialdehyde on the interaction between the oxygen-evolving complex 33 kDa protein and photosystem II core proteins. *Planta* **2010**, *231*, 1077–1088. [CrossRef] [PubMed]
45. Wang, Q.H.; Liang, X.; Dong, Y.J.; Xu, L.L.; Zhang, X.W.; Kong, J.; Liu, S. Effects of exogenous salicylic acid and nitric oxide on physiological characteristics of perennial ryegrass under cadmium stress. *J. Plant Growth Regul.* **2013**, *32*, 721–731. [CrossRef]
46. Tasgin, E.; Atici, O.; Nalbantoglu, B.; Popova, L.P. Effects of salicylic acid and cold treatments on protein levels and on the activities of antioxidant enzymes in the apoplast of winter wheat leaves. *Phytochemistry* **2006**, *67*, 710–715. [CrossRef] [PubMed]
47. Kang, G.Z.; Wang, C.H.; Sun, G.C.; Wang, Z.X. Salicylic acid changes activities of H2O2-metabolizing enzymes and increases the chilling tolerance of banana seedlings. *Environ. Exp. Bot.* **2003**, *50*, 9–15. [CrossRef]
48. Nwugo, C.C.; Huerta, A.J. The effect of silicon on the leaf proteome of rice (*Oryza sativa* L.) Plants under Cadmium-Stress. *J. Proteome Res.* **2011**, *10*, 518–528. [CrossRef] [PubMed]
49. Melo, A.M.P.; Roberts, T.H.; Moller, I.M. Evidence for the presence of two rotenone-insensitive NAD(P)H dehydrogenases on the inner surface of the inner membrane of potato tuber mitochondria. *Biochim. Biophys. Acta* **1996**, *1276*, 133–139. [CrossRef]
50. Apel, K.; Hirt, H. Reactive oxygen species: Metabolism, oxidative stress, and signal transduction. *Annu. Rev. Plant Biol.* **2004**, *55*, 373–399. [CrossRef] [PubMed]
51. Wang, Z.Y.; Xiong, L.; Li, W.; Zhu, J.K.; Zhu, J. The plant cuticle is required for osmotic stress regulation of abscisic acid biosynthesis and osmotic stress tolerance in Arabidopsis. *Plant Cell* **2011**, *23*, 1971–1984. [CrossRef] [PubMed]
52. Moradi, F.; Ismail, A.M. Responses of photosynthesis, chlorophyll fluorescence and ROS-Scavenging systems to salt stress during seedling and reproductive stages in rice. *Ann. Bot.* **2007**, *99*, 1161–1173. [CrossRef] [PubMed]
53. Mishra, A.; Patel, M.K.; Jha, B. Non-targeted metabolomics and scavenging activity of reactive oxygen species reveal the potential of *Salicornia brachiata* as a functional food. *J. Funct. Foods* **2015**, *13*, 21–31. [CrossRef]
54. Murillo-Amador, B.; Córdoba-Matson, M.V.; Villegas-Espinoza, J.A.; Hernández-Montiel, L.G.; Troyo-Diéguez, E.; García-Hernández, J.L. Mineral content and biochemical variables of *Aloe vera* L. under salt stress. *PLoS ONE* **2014**, *9*, e9487. [CrossRef] [PubMed]
55. Niu, X.; Bressan, R.A.; Hasegawa, P.M.; Pardo, J.M. Ion homeostasis in NaCl stress environments. *Plant Physiol.* **1995**, *109*, 735–742. [CrossRef] [PubMed]
56. Qi, Z.; Xiong, L. Characterization of a Purine Permease Family Gene OsPUP7 Involved in Growth and Development Control in Rice. *Chin. Bull. Botany* **2013**, *55*, 1119–1135.
57. Zhao, X.; Wang, W.; Zhang, F.; Deng, J.; Li, Z.; Fu, B. Comparative metabolite profiling of two rice genotypes with contrasting salt stress tolerance at the seedling stage. *PLoS ONE* **2014**, *29*, e108020. [CrossRef] [PubMed]
58. Yokozawa, T.; Kashiwada, Y.; Hattori, M.; Chung, H.Y. Study on the components of Luobuma with peroxynitrite-scavenging activity. *Biol. Pharm. Bull.* **2002**, *25*, 748–752. [CrossRef] [PubMed]
59. Lim, J.H.; Park, K.J.; Kim, B.K.; Jeong, J.W.; Kim, H.J. Effect of salinity stress on phenolic compounds and carotenoids in buckwheat (*Fagopyrum esculentum* M.) sprout. *Food Chem.* **2012**, *135*, 1065–1070. [CrossRef] [PubMed]
60. Chisari, M.; Todaro, A.; Barbagallo, R.N.; Spagna, G. Salinity effects on enzymatic browning and antioxidant capacity of fresh-cut baby Romaine lettuce (*Lactuca sativa* L. cv. Duende). *Food Chem.* **2010**, *119*, 1502–1506. [CrossRef]

61. Nichenametla, S.N.; Taruscio, T.G.; Barney, D.L.; Exon, J.H. A review of the effects and mechanisms of polyphenolics in cancer. *Crit. Rev. Food Sci. Nutr.* **2006**, *46*, 161–183. [CrossRef] [PubMed]
62. Chen, C.H.; Xu, H.; Liu, X.H.; Zou, L.S.; Wang, M.; Liu, Z.X.; Fu, X.S.; Zhao, H.; Yan, Y. Site-specific accumulation and dynamic change of flavonoids in Apocyni Veneti Folium. *Microsc. Res. Tech.* **2017**, *80*, 1315–1322. [CrossRef] [PubMed]
63. Lee, B.H.; Jeong, S.M.; Lee, J.H.; Kim, J.H.; Yoon, I.S.; Lee, J.H.; Choi, S.H.; Lee, S.M.; Chang, C.G.; Kim, H.C.; et al. Quercetin inhibits the 5-hydroxytryptamine type 3 receptor-mediated ion current by interacting with pre-transmembrane domain I. *Mol. Cells* **2005**, *20*, 69–73. [PubMed]
64. Kim, Y.H.; Lee, Y.J. RAIL apoptosis is enhanced by quercetin through Akt dephosphorylation. *J. Cell Biochem.* **2007**, *100*, 998–1009. [CrossRef] [PubMed]
65. Winkel-Shirley, B. Biosynthesis of flavonoids and effects of stress. *Curr. Opin. Plant Biol.* **2002**, *5*, 218–223. [CrossRef]
66. Bettaieb, I.; Knioua, S.; Hamrouni, I.; Limam, F.; Marzouk, B. Water-deficit impact on fatty acid and essential oil composition and antioxidant activities of cumin (*Cuminum cyminum* L.) aerial parts. *J. Agr. Food Chem.* **2011**, *59*, 328–334. [CrossRef] [PubMed]
67. Close, D.C.; McArthor, C. Rethinking the role of many plant phenolics—Protection against photodamage not herbivores? *OIKOS* **2002**, *99*, 166–172. [CrossRef]
68. Hirayama, T.; Shinozaki, K. Research on plant abiotic stress responses in the post-genome era: Past, present and future. *Plant J.* **2010**, *61*, 1041–1052. [CrossRef] [PubMed]
69. Wang, X.H.; Huang, J.L. *Principles and Techniques of Plant Physiological Biochemical Experiment*, 3rd ed.; Higher Education Press: Beijing, China, 2015.
70. Zeng, J.W.; Chen, A.M.; Li, D.D.; Yi, B.; Wu, W. Effects of Salt Stress on the Growth, Physiological Responses, and Glycoside Contents of *Stevia rebaudiana* Bertoni. *J. Agric. Food Chem.* **2013**, *61*, 5720–5726. [CrossRef] [PubMed]
71. Wu, F.B.; Zhang, G.P.; Dominy, P. Four barley genotypes respond differently to cadmium: Lipid peroxidation and activities of antioxidant capacity. *Environ. Exp. Bot.* **2003**, *50*, 67–78. [CrossRef]
72. Zenki, M.; Tanishita, A.; Yokoyama, T. Repetitive determination of ascorbic acid using iron(III)-1.10-phenanthroline-peroxodisulfate system in a circulatory flow injection method. *Talanta* **2004**, *64*, 1273–1277. [CrossRef] [PubMed]
73. Hua, Y.J.; Wang, S.N.; Chai, C.; Liu, Z.S.; Liu, X.H.; Zou, L.S.; Wu, Q.N.; Zhao, H.; Yan, Y. Quality Evaluation of Pseudostellariae Radix Based on Simultaneous Determination of Multiple Bioactive Components Combined with Grey Relational Analysis. *Molecules* **2016**, *22*, 13. [CrossRef] [PubMed]

International Journal of
Molecular Sciences

MDPI

Article

Root Abscisic Acid Contributes to Defending Photoinibition in Jerusalem Artichoke (*Helianthus tuberosus* L.) under Salt Stress

Kun Yan [1,*], Tiantian Bian [1,2], Wenjun He [1], Guangxuan Han [1,*], Mengxue Lv [1], Mingzhu Guo [3] and Ming Lu [3]

[1] Key Laboratory of Coastal Environmental Processes and Ecological Remediation, Yantai Institute of Coastal Zone Research, Chinese Academy of Sciences, Yantai 264003, China; czlbtt@163.com (T.B.); wjhe@yic.ac.cn (W.H.); xuhualing1981@163.com (M.L.)
[2] School of Life Sciences, Ludong University, Yantai 264025, China
[3] College of Life Sciences, Yantai University, Yantai 264005, China; m17616156626@163.com (M.G.); 17865561399@163.com (M.L.)
* Correspondence: kyan@yic.ac.cn (K.Y.); gxhan@yic.ac.cn (G.H.); Tel.: +86-0535-2109279 (K.Y.)

Received: 15 November 2018; Accepted: 4 December 2018; Published: 7 December 2018

Abstract: The aim of the study was to examine the role of root abscisic acid (ABA) in protecting photosystems and photosynthesis in Jerusalem artichoke against salt stress. Potted plants were pretreated by a specific ABA synthesis inhibitor sodium tungstate and then subjected to salt stress (150 mM NaCl). Tungstate did not directly affect root ABA content and photosynthetic parameters, whereas it inhibited root ABA accumulation and induced a greater decrease in photosynthetic rate under salt stress. The maximal photochemical efficiency of PSII (Fv/Fm) significantly declined in tungstate-pretreated plants under salt stress, suggesting photosystem II (PSII) photoinhibition appeared. PSII photoinhibition did not prevent PSI photoinhibition by restricting electron donation, as the maximal photochemical efficiency of PSI ($\Delta MR/MR_0$) was lowered. In line with photoinhibition, elevated H_2O_2 concentration and lipid peroxidation corroborated salt-induced oxidative stress in tungstate-pretreated plants. Less decrease in $\Delta MR/MR_0$ and Fv/Fm indicated that PSII and PSI in non-pretreated plants could maintain better performance than tungstate-pretreated plants under salt stress. Consistently, greater reduction in PSII and PSI reaction center protein abundance confirmed the elevated vulnerability of photosystems to salt stress in tungstate-pretreated plants. Overall, the root ABA signal participated in defending the photosystem's photoinhibition and protecting photosynthesis in Jerusalem artichoke under salt stress.

Keywords: chlorophyll fluorescence; lipid peroxidation; Na^+; photosynthesis; photosystem

1. Introduction

Soil salinity is a serious problem for agricultural cultivation because of the detrimental effects on crop growth and yield. Under salt stress, plants have to tolerate osmotic stress, ionic toxicity, and secondary oxidative stress, and the metabolisms may be disrupted with damaged biological macromolecules [1–3]. Correspondently, plants have evolved some physiological adaption measures such as Na^+ exclusion, osmolyte synthesis, and antioxidant induction, however, signal molecules which sensitively perceive external stresses are required to activate these protective mechanisms [4,5].

Abscisic acid (ABA) is defined as a stress hormone, because ABA can mediate extrinsic stress signals to improve expression of resistance genes [6–8]. As well documented, ABA plays an important role in regulating stomatal closure to limit water loss from transpiration, which assists in plant acclimatization to osmotic tolerance [8–10]. The positive role of ABA in plant salt tolerance also has been reviewed, and besides stomatal closure, osmolyte synthesis and antioxidant induction usually associate with ABA signal under salt stress despite some inconsistent reports due to species difference [7,11–14]. Na^+ is the primary toxic component for plants upon salt stress [2]. To date, it is still ambiguous whether ABA signal contributes to controlling Na^+ long-distance transportation and exclusion [4,15]. Particularly, Cabot et al. [16] reported that leaf ABA accumulation resulted in higher leaf Na^+ concentration in *Phaseolus vulgaris* under salt stress due to lowered leaf Na^+ exclusion and increased Na^+ translocation from root to shoot. Therefore, ABA function in defending salt-induced ionic toxicity seems not definite in contrast to its role in osmotic tolerance. Moreover, it remains unknown whether root ABA or leaf ABA has a greater effect on plant salt tolerance.

As one of the most important metabolisms for plant growth, photosynthesis is sensitive to salt stress. Photosynthetic analysis seems to be an effective and convenient way for diagnosing plant salt tolerance, because photosynthetic capacity in susceptible cultivars is more liable to be inhibited than tolerant ones [17–22]. Salt-induced stomatal closure initially depressed photosynthesis by lowering CO_2 availability [23,24], and subsequently, the negative effect on Rubisco can further restrict CO_2 fixation [25,26]. Eventually, the declined CO_2 assimilation will elevate excitation pressure in chloroplast through feedback inhibition on photosynthetic electron transport and then bring about photosystems photoinhibition or even irreversible damage with excess ROS production [27,28]. At present, photosystems photoinhibition and interaction under salt stress have been reported. In addition to PSII, PSI is also a crucial photoinhibition site and PSI photoinhibition poses a great threat to the entire photosynthetic apparatus by inducing PSII photoinhibition [20,29]. However, the relationship between the ABA signal and photosystem photoinhibition remains to be disclosed. ABA-induced stomatal limitation may trigger photosystem photoinhibition, but ABA-induced antioxidant activity can prevent from photoinhibition by scavenging reactive oxygen species (ROS). Particularly, the ambiguous function of ABA for regulating Na^+ transportation make it more complex.

Jerusalem artichoke (*Helianthus tuberosus* L.) is a valuable energy crop with high fructose and inulin concentrations in the tuber [30]. Jerusalem artichoke has certain salt tolerance and serves as a promising crop for utilizing coastal marginal land in China [30,31]. According to previous studies, salt stress could induce photosynthetic stomatal limitation, oxidative injury, chlorophyll loss, and ABA accumulation in Jerusalem artichoke [32–34]. However, the importance of ABA for salt tolerance in Jerusalem artichoke has not been tested. At present, gas exchange combined with modulated chlorophyll fluorescence has become a traditional method to examine plant stress tolerance. Recently, a simultaneous measurement of chlorophyll fluorescence transients and modulated 820 nm reflection has been applied to investigate PSII and PSI performance and their coordination, which enriches the traditional photosynthetic analysis [20,35–38]. In this study, we aimed to verify ABA function for salt adaptability in Jerusalem artichoke by photosynthetic analysis after applying a specific ABA synthesis inhibitor to the roots. Simultaneous measurement of chlorophyll fluorescence transients and modulated 820 nm reflection was carried out to complement traditional gas exchange analysis for revealing photosystems performance and coordination. Particularly, the abundance of PSII and PSI reaction center proteins was detected by immunoblot analysis to confirm salt-induced damage on photosystems. We hypothesized that root ABA accumulation helped prevent photosystems photoinhibition and protect photosynthesis by alleviating water loss and ionic toxicity. Our study can deepen the knowledge of salt tolerance in Jerusalem artichoke and may provide a reference for the cultivation in coastal saline land.

2. Results

2.1. Leaf Na^+, Relative Water, Malondialdehyde (MDA) and H_2O_2 Content, and Root Na^+ Flux

After four days of salt stress, leaf Na^+ and H_2O_2 content were significantly increased, whereas leaf relative water content was significantly decreased (Table 1). Leaf Na^+, MDA and H_2O_2 content, and root Na^+ flux were not directly affected by tungstate. Upon four days of salt stress, tungstate had no effect on the decreased amplitude of leaf relative water content but amplified the increase in leaf Na^+ and H_2O_2 content (Table 1). Salt-induced significant increase in leaf MDA content was found in tungstate-pretreated plants rather than non-pretreated plants (Table 1). Root Na^+ efflux was significantly elevated by salt stress, but salt-induced increase in Na^+ efflux was greatly reduced in tungstate-pretreated plants (Table 1).

Table 1. H_2O_2, malondialdehyde (MDA), Na^+ and relative water contents in the leaf and average root Na^+ flux in Jerusalem artichoke after four days of salt stress. Data in the table indicate the mean of five replicates (±SD). Within each column, means followed by the same letters are not significantly different at $p < 0.05$. FW indicates fresh weight. CP indicates control plants without pretreatment and NaCl stress; T1 indicates tungstate-pretreated plants without NaCl stress; T2 indicates non-pretreated plants under 150 mM NaCl stress; T3 indicates tungstate-pretreated plants under 150 mM NaCl stress.

Treatments	H_2O_2 Content ($\mu mol \cdot g^{-1}$ FW)	MDA Content ($nmol \cdot g^{-1}$ FW)	Na^+ Content ($mg \cdot g^{-1}$ FW)	Root Na^+ Efflux ($pmol \cdot cm^{-2} s^{-1}$)	Relative Water Content (%)
CP	0.11 ± 0.01c	53.00 ± 5.86b	1.08 ± 0.20c	1.80 ± 0.70c	91.41 ± 3.55a
T1	0.10 ± 0.02c	54.13 ± 5.64b	1.23 ± 0.44c	3.84 ± 1.09c	90.30 ± 1.92a
T2	0.18 ± 0.04b	52.83 ± 4.11b	3.58 ± 0.25b	130.13 ± 23.59a	62.93 ± 5.78b
T3	0.27 ± 0.03a	69.01 ± 6.50a	7.22 ± 0.59a	33.79 ± 6.59c	62.49 ± 4.01b

2.2. ABA Content in Leaf and Root

Single tungstate pretreatment did not affect root ABA content (Figure 1). After two days of salt stress, root ABA content was significantly increased by 47.8%, and the increase was dampened by tungstate pretreatment (Figure 1). After four days of salt stress, root ABA content was still remarkably lower in tungstate-pretreated plants than non-pretreated plants under salt stress (Figure 1). In all treatment groups, root ABA content after four days was lower than that after two days (Figure 1), which might originate from root development or ABA translocation from root to leaf.

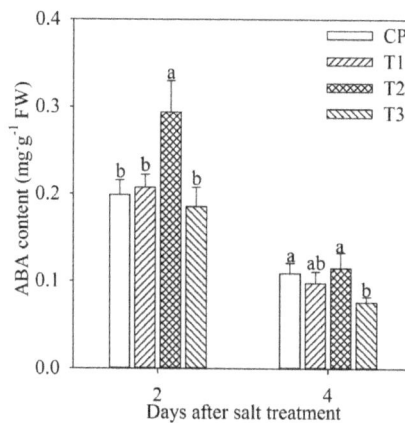

Figure 1. Changes in root abscisic acid (ABA) content in Jerusalem artichoke after salt stress for two days (a) and four days (b). Data in the figure indicate mean of five replicates (±SD), and different letters on error bars indicate significant difference at $p < 0.05$. CP indicates control plants without pretreatment and NaCl stress; T1 indicates tungstate-pretreated plants without NaCl stress; T2 indicates non-pretreated plants under 150 mM NaCl stress; T3 indicates tungstate-pretreated plants under 150 mM NaCl stress. The symbols, CP, T1, T2, and T3 are also used in the following figures.

2.3. Gas Exchange and Modulated Chlorophyll Fluorescence Parameters

Tungstate did not obviously influenced photosynthetic rate (Pn), stomatal conductance (g_s) and transpiration rate (Tr) (Figure 2a–c). Pn, g_s, and Tr significantly decreased in non-pretreated plants after one day of salt stress, and the decrease was up to 56.54%, 74.31% and 76.86% after four days of salt stress. In contrast, the decrease in Pn, g_s and Tr was remarkably higher in tungstate-pretreated plants upon salt stress (Figure 2a–c).

Under salt stress, decreased actual photochemical efficiency of PSII (ΦPSII) was noted with increased non-photochemical quenching (NPQ) in non-pretreated plants, whereas PSII excitation pressure (1-qP) did not show obvious change (Figure 2d–f). Tungstate did not significantly influenced ΦPSII, 1-qP and NPQ, and salt-induced decrease in ΦPSII was greater in tungstate-pretreated plants than non-pretreated plants. After two and three days of salt stress, 1-qP was significantly increased in tungstate-pretreated plants, but the increase became slight after four days of salt stress (Figure 2e). NPQ was significantly increased in tungstate-pretreated plants after one day of salt stress, but the increase disappeared after 3 days of salt stress (Figure 2f).

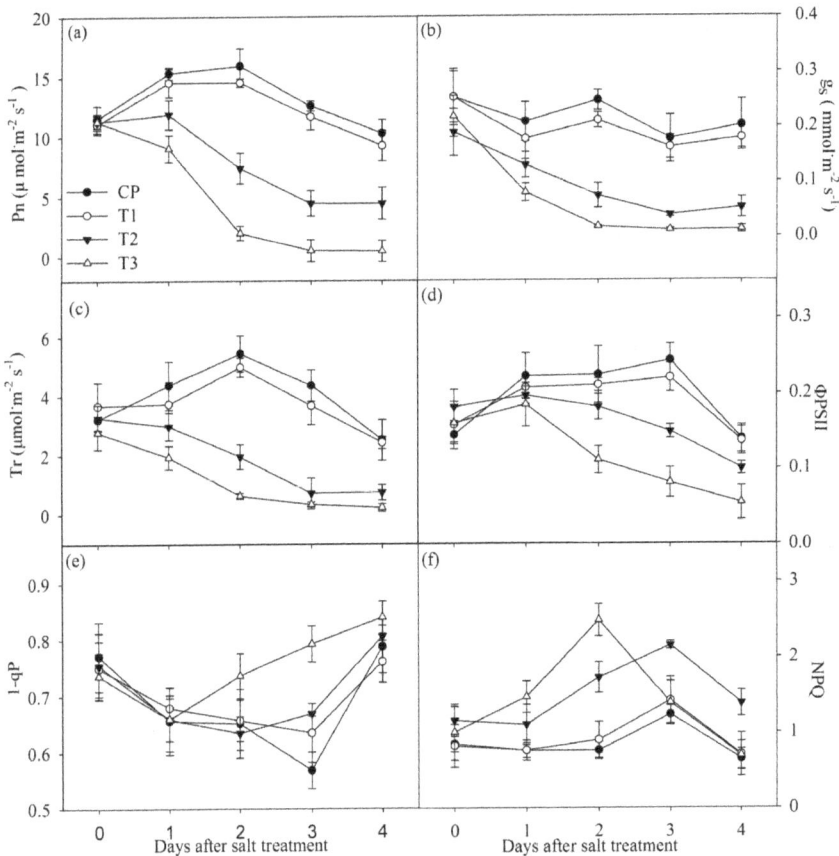

Figure 2. Changes in photosynthetic rate (Pn, (**a**)), stomatal conductance (g_s, (**b**)), transpiration (Tr, (**c**)), actual photochemical efficiency of PSII (ΦPSII, (**d**)), PSII excitation pressure (1-qP, (**e**)) and non-photochemical quenching (NPQ, (**f**)) in Jerusalem artichoke under salt stress. Data in the figure indicate the mean of five replicates (±SD).

2.4. Chlorophyll Fluorescence and Modulated 820 nm Reflection Transients

After two days of salt stress, chlorophyll fluorescence and modulated 820 nm reflection transients did not exhibit obvious change. The initial decrease in 820 nm reflection signal indicated PSI oxidation process, and the subsequent increase suggested that PSI was gradually re-reduced. After two days of salt stress, chlorophyll fluorescence transient descended in tungstate-pretreated plants (Figure 3a), suggesting PSII capacity was negatively affected. After salt stress for two days, the 820 nm reflection transient also changed in tungstate-pretreated plants, indicated by prolonged PSI oxidation process and lowered PSI re-reduction level (Figure 3b).

After four days of salt stress, chlorophyll fluorescence transient declined, while the PSI oxidation process was shortened (Figure 3c,d). Tungstate pretreatment never induced any change in chlorophyll fluorescence and 820 nm reflection transients, but their variations under salt stress were amplified by tungstate pretreatment (Figure 3c,d).

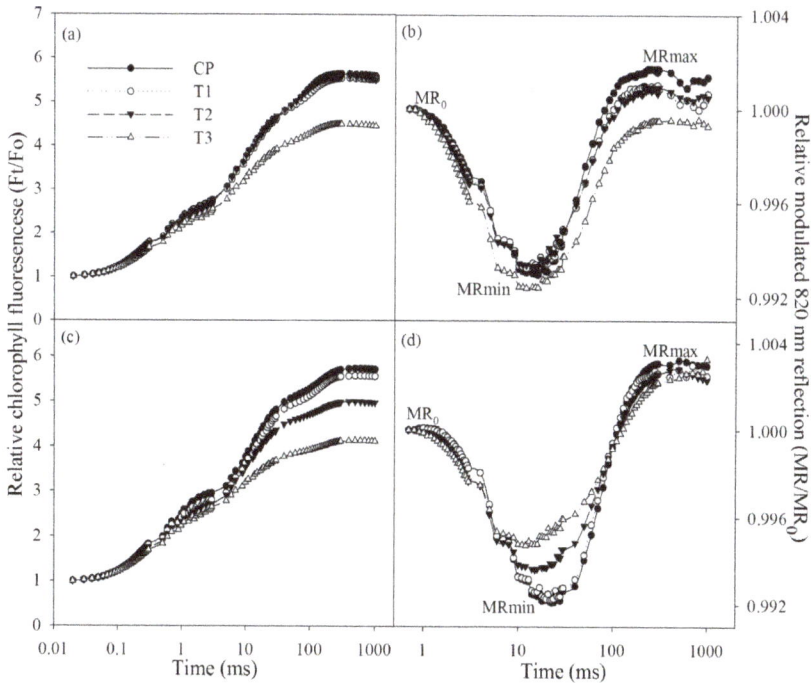

Figure 3. Chlorophyll fluorescence transients and 820 reflection transients during the first 1 s red illumination in Jerusalem artichoke under salt stress for two days (**a,b**) and four days (**c,d**). Ft is chlorophyll fluorescence intensity during the 1 s of red illumination, and Fo is fluorescence intensity at 20 µs, when all reaction centers of PSII are open. MR is the reflection signal during the 1 s of red illumination, and MR_0 is the value of modulated 820 nm reflection at the onset of red light illumination (0.7 ms, the first reliable MR measurement). MRmin and MRmax indicate the maximal point during PSI oxidation and the maximal point during PSI re-reduction, respectively. Data in the figure indicate the mean of five replicates.

2.5. PSII Performance, the Maximal Photochemical Capacity of PSI, and Immunoblot Analysis

Tungstate had no direct effect on the maximal photochemical capacity of PSI ($\Delta MR/MR_0$), the maximal quantum yield of PSII (Fv/Fm), probability that an electron moves further than primary acceptor of PSII (ETo/TRo) and quantum yield for electron transport (ETo/ABS) (Figure 4c–f). After two days of salt stress, Fv/Fm, $\Delta MR/MR_0$, ETo/TRo and ETo/ABS did not obviously change in non-pretreated plants, but significant decrease in Fv/Fm was observed in tungstate-pretreated plants (Figure 4c–f). After four days of salt stress, significant decrease in $\Delta MR/MR_0$ appeared with slightly lowered Fv/Fm, ETo/TRo and ETo/ABS in non-pretreated plants, but the decrease was greater in tungstate-pretreated plants (Figure 4c–f).

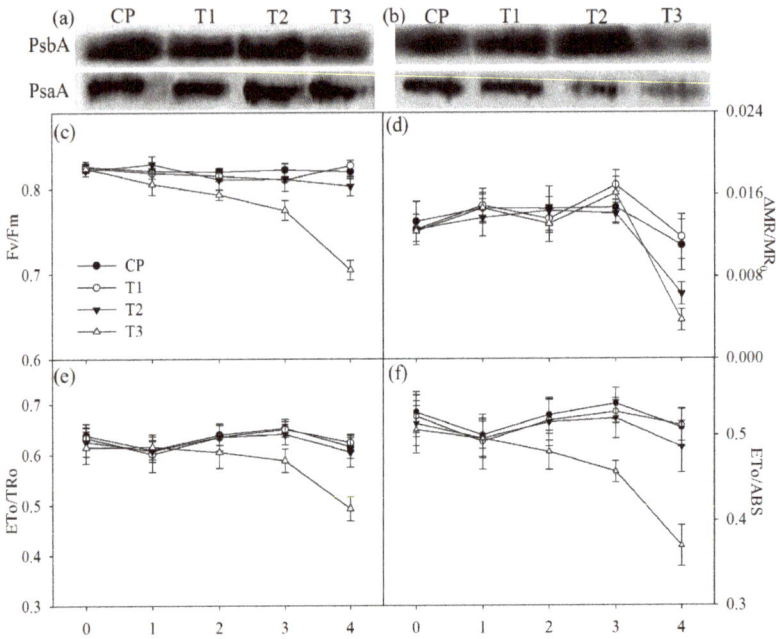

Figure 4. Immunoblot analysis of PSII reaction center protein (PsbA) and PSI reaction center protein (PsaA) abundance after two days (**a**) and four days (**b**) of salt stress and salt-induced changes in the maximal photochemical efficiency of PSII (Fv/Fm, (**c**)) and PSI ($\Delta MR/MR_0$, (**d**)), probability that an electron moves further than Q_A (ETo/TRo, (**e**)) and quantum yield for electron transport (REo/ETo, (**f**)) in Jerusalem artichoke. Data in the figure indicate the mean of five replicates (±SD).

The amount of PSII reaction center protein (PsbA) was reduced in tungstate-pretreated plants after two days of salt stress, and the reduction became more obvious after four days of salt stress (Figure 4a,b). In contrast, PsbA abundance was not affected by salt stress in plants without tungstate pretreatment (Figure 4a,b). After four days of salt stress, PSI reaction center protein (PsaA) abundance was decreased, and the decrease was greater in tungstate-pretreated than non-pretreated plants (Figure 4b).

3. Discussion

As with common knowledge, salt stress elevated root ABA concentration in Jerusalem artichoke, and tungstate pretreatment prevented salt-induced root ABA accumulation (Figure 1). The salt-induced greater decrease in Pn and ΦPSII in tungstate-pretreated plants than non-pretreated plants suggested that root ABA aided in protecting photosynthetic process in Jerusalem artichoke against salt stress (Figure 2a,d). Under salt stress, leaf stomatal closure reduced water loss from transpiration in Jerusalem artichoke (Figure 2b,c), but could inevitably induce stomatal limitation on photosynthesis. Tungstate-pretreated plants should encounter stronger photosynthetic stomatal limitation under salt stress due to the greater decrease in g_s compared with non-pretreated plants (Figure 2b). Lowered CO_2 assimilation can elevate PSII excitation pressure by feedback inhibition on photosynthetic electron transport and cause oxidative injury with excessive ROS production [28,39]. Under salt stress, PSII excitation pressure did not obviously change in spite of lowered CO_2 assimilation in non-pretreated plants, as the excessive excitation energy was effectively dissipated as heat (Figure 2e,f). In contrast, elevated PSII excitation pressure due to greater lowered CO_2 assimilation and insufficient heat dissipation could bring about photosystems photoinhibition in tungstate-pretreated plants upon salt stress. Notably, elevated PSII excitation pressure disappeared in tungstate-pretreated plants after four days of salt stress (Figure 2e), implying tremendous decrease in trapped energy in reaction center due to severe PSII photoinhibition.

In line with the above deduction, PSII photoinhibition actually occurred in tungstate-pretreated plants upon salt stress, indicated by declined Fv/Fm and chlorophyll fluorescence transient (Figure 3a,c and Figure 4c). Thus, considering slight change in Fv/Fm and chlorophyll fluorescence transient in non-pretreated plants (Figure 3a,c and Figure 4c), root ABA should participate in protecting PSII against photoinhibition in Jerusalem artichoke under salt stress. This positive role of root ABA was corroborated by immunoblot analysis, as lowered and unchanged PsbA abundance appeared, respectively, in tungstate-pretreated and non-pretreated plants upon salt stress (Figure 4a,b). Similar to PSII, PSI photoinhibition also derives from oxidative injury on reaction center proteins [27,28,36]. Along with elevated ROS production and lipid peroxidation (Table 1), PSI photoinhibition appeared after four days of salt stress, indicated by the significant decrease in $\Delta MR/MR_0$ (Figure 3d). In agreement with our recent study on waterlogging [40], PSI was also more vulnerable than PSII in Jerusalem artichoke under salt stress according to less decrease in Fv/Fm than $\Delta MR/MR_0$ (Figure 3c,d). Nonetheless, inhibition on root ABA synthesis led to higher PSII susceptibility to salt stress compared with PSI, as earlier significant decrease was observed in Fv/Fm rather than $\Delta MR/MR_0$ in tungstate-pretreated plants (Figure 3c,d). Prolonged PSI oxidation and lowered PSI re-reduction level in 820 nm reflection transients after two days of salt stress verified greater PSII vulnerability (Figure 2b). Thus, contrary to recent studies [20,29], PSII photoinhibition was not induced by PSI photoinhibition in tungstate-pretreated plants under salt stress. We inferred that photoprotective mechanisms were not adequately induced by salt stress in tungstate-pretreated plants and, as a result, lower heat dissipation appeared with greater excitation pressure on PSII (Figure 2e,f). As a traditional viewpoint, PSII photoinhibition can protect PSI against photoinhibition by restricting photosynthetic electron transport to PSI. In this study, PSII photoinhibition declined electron flow to PSI in tungstate-pretreated plants under salt stress, but PSI photoinhibition was still exacerbated according to greater decrease in $\Delta MR/MR_0$ compared with non-pretreated plants (Figure 4c–f). After four days of salt stress, greater shortened PSI oxidation also implied more severe PSI photoinhibition in tungstate-pretreated plants (Figure 3d), and this result was confirmed by salt-induced greater reduction in PsaA abundance in tungstate-pretreated plants (Figure 4b). Overall, root ABA signal helped defend salt-induced PSII and PSI photoinhibition in Jerusalem artichoke, and the protective way for PSI did not depend on PSII inactivation.

Although osmotic pressure can rapidly depress photosynthesis through stomatal limitation, Na^+ toxicity is more hazardous under salt stress. Na^+ can irreversibly inactivate PSII and PSI by inducing secondary oxidative injury or through direct damage on photosynthetic proteins [41–44]. Particularly, severe PSII photoinhibition without elevated excitation pressure in tungstate-pretreated plants after four days of salt stress may result from the direct effect of Na^+ in large part. In this study, inhibition on root ABA synthesis did not influence leaf water status in Jerusalem artichoke under salt stress, as similar relative leaf water content existed in tungstate-pretreated and non-pretreated plants (Table 1). In contrast, inhibited root ABA accumulation declined Na^+ exclusion from roots and led to prominent increase in leaf Na^+ concentration (Table 1). Therefore, Na^+ toxicity should be responsible for more severe PSII and PSI photoinhibition in tungstate-pretreated plants. However, the signal pathway for regulating Na^+ transport and uptake needs to be revealed in future study.

In agreement with the hypothesis, root ABA signal contributed to defending photosystems photoinhibition and protecting photosynthesis in Jerusalem artichoke under salt stress, but this positive role of root ABA was actualized mainly by reducing Na^+ toxicity.

4. Materials and Methods

4.1. Plant Material and Treatment

Tubers of Jerusalem artichoke were collected in Laizhou Bay, China. The tubers were planted in plastic pots filled with vermiculite (one tuber in each pot) and placed in an artificial climatic room

(Qiushi, China). The vermiculite was kept wet by watering. In the room, day/night temperature and humidity were controlled at 25/18 °C and 70%, and photon flux density was 400 μmol·m^{-2} s^{-1} for 12 h per day from 07:00 to 19:00. After one month, the tubers germinated and were daily watered with Hoagland nutrient solution (pH 5.7). One month later, health and uniform plants were selected and separated to four groups. In the first group, plants without tungstate sodium pretreatment were not subjected to NaCl stress. In the second group, plants were pretreated with tungstate sodium but not subjected to NaCl stress. In the third group, plants were exposed to 150 mM NaCl for four days without tungstate sodium pretreatment. In the fourth group, plants were pretreated with tungstate sodium and then subjected to 150 mM NaCl for four days. NaCl was added to nutrient solution incrementally by 50 mM step every day to reach the final concentration. The solution was refreshed every two days, and before refreshing solution, the culture substrate was thoroughly leached using nutrient solution for avoiding ion accumulation. One day before salt treatment, tungstate sodium (1 mM), a specific inhibitor of ABA synthesis, was added to nutrient solution for pretreatment.

4.2. Measurements of Na$^+$, Relative Water Content, and Root Na$^+$ Flux

The extraction of Na$^+$ was performed according to Song et al. [45]. Deionized H$_2$O (25 mL) was added to 0.1 g dried leaf powder and boiled for 2 h. The supernatant was diluted 50 times with deionized H$_2$O for measuring Na$^+$ content by using an atomic absorption spectrophotometer (TAS-990, Beijing, China). Net Na$^+$ flux was measured using NMT (Younger, Amherst, MA, USA) and the principle and protocol for measuring root Na$^+$ flux have been elucidated in detail in our recent study [20]. In this experiment, newly developed root segments were sampled and a vigorous Na$^+$ flux was identified at 500 μm from the root apex. The measured root position can be visualized under microscope, and tungstate pretreatment dampened salt-induced increase in root Na$^+$ efflux (Supplemental Figure S1). The average value of Na$^+$ flux is presented in Table 1.

Fresh leaves were harvested and weighed (fresh weight, FW), and then were immersed in distilled water for 4 h at room temperature to determine saturated fresh weight (SW). Subsequently, the leaves were dried completely in an oven at 70 °C and weighed (dry weight, DW). Relative water content (RWC) was calculated as: RWC = (FW − DW)/(SW − DW) × 100%.

4.3. Measurements of MDA, H$_2$O$_2$, and ABA Content

Leaf tissues (0.5 g) were ground under liquid nitrogen and homogenized in 5 mL 0.1% TCA. The homogenate was centrifuged at 12,000× *g* and 4 °C for 10 min to collect the supernatant for measurements of MDA and H$_2$O$_2$ content. The supernatant (0.5 mL) was mixed with 10 mM potassium phosphate buffer (0.5 mL, pH 7.0) and 1 M KI (1 mL), and the absorbance at 390 nm was recorded for calculating H$_2$O$_2$ content [46]. MDA content was determined by thiobarbituric acid reaction to reflect the extent of lipid peroxidation [47].

ABA content was analyzed according to Lopez-Carbonell and Jauregui [48] with some modification. Root and leaf tissues (0.5 g) were ground under liquid nitrogen and homogenized in 3 mL 80% methanol containing 0.1% acetic acid. After agitation for 30 min at 4 °C, the homogenate was centrifuged at 12,000× *g* and 4 °C for 10 min. The supernatant was filtered through a 0.45 μm polytetrafluoroethylene membrane, and the filtrate (10 μL) was injected into a high performance liquid chromatography instrument equipped with mass spectrometer (Thermo, Waltham, MA, USA). A hypersil C18 column (4.6 mm × 150 mm; particle size, 5.0 μm) was used in the liquid chromatography system, and the mobile phase consisted of water with 0.1% HCO$_2$H (A) and MeOH with 0.1% HCO$_2$H (B). A gradient elution program was applied, and the initial gradient of methanol was kept at 30% for 2 min and increased linearly to 100% at 20 min. All the analyses of mass spectrum (MS) were performed using ionspray source in negative ion mode, and MS/MS product ions were produced by collision-activated dissociation of selected precursor ions. Since many compounds could present the same nominal molecular mass, MS/MS method was required to selectively monitor ABA in crude plant extracts by identifying parent mass and unique fragment ion. In this study, MS/MS method was used for the quantitation of ABA by monitoring 263/153 transition, and ABA concentration was determined by using a standard curve plotted with known concentrations of the standards.

4.4. Measurements of Gas Exchange and Modulated Chlorophyll Fluorescence

Gas exchange and modulated chlorophyll fluorescence were simultaneously measured by using an open photosynthetic system (LI-6400XTR, Li-Cor, Lincoln, NE, USA) equipped with a fluorescence leaf chamber (6400-40 LCF, Li-Cor). Temperature, CO_2 concentration and actinic light intensity were, respectively, set at 25 °C, 400 $\mu mol \cdot mol^{-1}$ and 1000 $\mu mol \cdot m^{-2} s^{-1}$ in the leaf cuvette. Pn, g_s and Tr were simultaneously noted. After steady-state fluorescence yield was recorded, a saturating actinic light pulse of 8000 $\mu mol \cdot m^{-2} s^{-1}$ for 0.7 s was used to produce maximum fluorescence yield by temporarily inhibiting PSII photochemistry for measuring ΦPSII. Photochemical quenching coefficient was also recorded for calculating 1-qP. Thereafter, the leaves were dark-adapted for 30 min, and a saturating actinic light pulse of 8000 $\mu mol \cdot m^{-2} s^{-1}$ for 0.7 s was applied to measure the maximal fluorescence for calculating NPQ [49].

4.5. Simultaneous Measurements of Chlorophyll Fluorescence and Modulated 820 nm Reflection Transients

A multifunctional plant efficiency analyzer (MPEA, Hansatech, Norfolk, UK) was used for the measurements, and its operating mechanism has been described in detail [37]. The leaves were dark-adapted for 30 min, and the leaves were orderly illuminated with 1 s red light (627 nm, 5000 μmol photons$\cdot m^{-2} s^{-1}$), 10 s far red light (735 nm, 200 μmol photons$\cdot m^{-2} s^{-1}$) and 2 s red light (627 nm, 5000 μmol photons$\cdot m^{-2} s^{-1}$). Chlorophyll fluorescence and modulated 820 nm reflection were simultaneously detected during the illumination. Chlorophyll fluorescence and modulated 820 nm reflection transients were simultaneously recorded during the illumination. Fv/Fm, ETo/TRo, and ETo/ABS were calculated according to chlorophyll fluorescence transients [20], and $\Delta MR/MR_0$ was determined from modulated 820 nm reflection signal [50–52].

4.6. Isolation of Thylakoid Membranes and Western Blot

Five grams of leaf discs were ground under liquid nitrogen and homogenized in a solution containing 400 mM sucrose, 50 mM HEPES-KOH (pH 7.8), 10 mM NaCl, and 2 mM $MgCl_2$ [53]. The homogenate was filtered through two layers of cheesecloth and then centrifuged at 5000× *g* and 4 °C for 10 min to collectthylakoid pellets. The pellets were resuspended in the homogenization buffer, and chlorophyll content was measured.

Thylakoid membranes with 10 μg chlorophyll were separated by a 12% (*w*/*w*) SDS-PAGE gel. Proteins from the gel were transferred onto polyvinylidene fluoride membrane by semi dry method. After blocking with 5% skimmed milk for 1 h, the membranes were incubated for 2 h with the primary anti-PsbA or anti-PsaA antibodies (PhytoAB, San Francisco, CA, USA) and then incubated with horseradish peroxidase-conjugated anti-rabbit IgG antibody (PhytoAB, USA) for 2 h. BeyoECL Plus substrate (Beyotime Biotechnology, Shanghai, China) was used to test immunoreaction, and the chemiluminescence was detected by a Tanon-5500 cooled CCD camera (Tanon, Shanghai, China).

4.7. Statistical Analysis

One-way ANOVA was carried out by using SPSS 16.0 (SPSS Inc., Chicago, IL, USA) for all sets of data. The values presented are the means of measurements with five replicate plants, and comparisons of means were determined through LSD test. The difference was considered significant at $p < 0.05$.

Supplementary Materials: Supplementary material is available online at http://www.mdpi.com/1422-0067/19/12/3934/s1.

Author Contributions: K.Y. designed the experiment, performed data analysis and wrote the manuscript. T.B., W.H., M.L. (Mengxue Lv), M.G., and M.L. (Ming Lu) participated in the experiment. G.H. reviewed the manuscript and proposed some critical suggestions.

Funding: This research was jointly supported by National Natural Science Foundation of China (41201292), Shandong Provincial Natural Science Foundation, China (ZR2017QC005), Key Deployment Project of Chinese Academy of Sciences (KFZD-SW-112), the Science and Technology Service Network Initiative (KFJ-STS-ZDTP-023), and the Opening Foundation of the State Key Laboratory of Crop Biology, Shandong Agricultural University (2016KF07).

Conflicts of Interest: The authors declare that the research was conducted in the absence of any commercial or financial relationships that could be construed as a potential conflict of interest.

Abbreviations

ETo/ABS	quantum yield for electron transport
ETo/TRo	probability that an electron moves further than primary acceptor of PSII
Fv/Fm	the maximal quantum yield of PSII
g_s	stomatal conductance
MDA	malondialdehyde
NPQ	non-photochemical quenching
Pn	photosynthetic rate
PSI	Photosystem I
PSII	Photosystem II
Q_A	primary quinone
ROS	reactive oxygen species
Tr	transpiration rate
1-qP	excitation pressure of PSII
$\Delta MR/MR_0$	the maximal photochemical capacity of PSI
$\Phi PSII$	actual photochemical efficiency of PSII

References

1. Hossain, M.S.; Dietz, K.J. Tuning of redox regulatory mechanisms, reactive oxygen species and redox homeostasis under salinity stress. *Front. Plant Sci.* **2016**, *7*, 548. [CrossRef] [PubMed]
2. Munns, R.; Tester, M. Mechanisms of salinity tolerance. *Annu. Rev. Plant Biol.* **2008**, *59*, 651–681. [CrossRef] [PubMed]
3. Zhu, J.K. Regulation of ion homeostasis under salt stress. *Curr. Opin. Plant Biol.* **2003**, *6*, 441–445. [CrossRef]
4. Zhu, J.K. Abiotic Stress Signaling and Responses in Plants. *Cell* **2016**, *167*, 313–324. [CrossRef] [PubMed]
5. Julkowska, M.M.; Testerink, C. Tuning plant signaling and growth to survive salt. *Trends Plant Sci.* **2015**, *20*, 586–594. [CrossRef] [PubMed]
6. Hong, J.H.; Seah, S.W.; Xu, J. The root of ABA action in environmental stress response. *Plant Cell Rep.* **2013**, *32*, 971–983. [CrossRef] [PubMed]
7. Sah, S.K.; Reddy, K.R.; Li, J. Abscisic acid and abiotic stress tolerance in crop plants. *Front. Plant Sci.* **2016**, *7*, 571. [CrossRef] [PubMed]
8. Zhang, J.; Jia, W.; Yang, J.; Ismail, A.M. Role of ABA in integrating plant responses to drought and salt stresses. *Field Crop Res.* **2006**, *97*, 111–119. [CrossRef]
9. Zhang, F.P.; Sussmilch, F.; Nichols, D.S.; Cardoso, A.A.; Brodribb, T.J.; McAdam, S.A.M. Leaves, not roots or floral tissue, are the main site of rapid, external pressure-induced ABA biosynthesis in angiosperms. *J. Exp. Bot.* **2018**, *69*, 1261–1267. [CrossRef]
10. Zhu, J.K. Salt and drought stress signal transduction in plants. *Annu. Rev. Plant Biol.* **2002**, *53*, 247–273. [CrossRef]
11. Ryu, H.; Cho, Y.G. Plant hormones in salt stress tolerance. *J. Plant Biol.* **2015**, *58*, 147–155. [CrossRef]
12. Hong, C.Y.; Chao, Y.Y.; Yang, M.Y.; Cheng, S.Y.; Cho, S.C.; Kao, C.H. NaCl-induced expression of glutathione reductase in roots of rice (*Oryza sativa* L.) seedlings is mediated through hydrogen peroxide but not abscisic acid. *Plant Soil* **2009**, *320*, 103–115. [CrossRef]
13. Kalinina, E.B.; Keith, B.K.; Kern, A.J.; Dyer, W.E. Salt- and osmotic stress-induced choline monooxygenase expression in Kochia scoparia is ABA-independent. *Biol. Plant.* **2012**, *56*, 699–704. [CrossRef]

14. Per, T.S.; Khan, N.A.; Reddy, P.S.; Masood, A.; Hasanuzzaman, M.; Khan, M.I.R.; Anjum, N.A. Approaches in modulating proline metabolism in plants for salt and drought stress tolerance: Phytohormones, mineral nutrients and transgenics. *Plant Physiol. Biochem.* **2017**, *115*, 126–140. [CrossRef]

15. Osakabe, Y.; Yamaguchi-Shinozaki, K.; Shinozaki, K.; Tran, L.S. ABA control of plant macroelement membrane transport systems in response to water deficit and high salinity. *New Phytol.* **2014**, *202*, 35–49. [CrossRef] [PubMed]

16. Cabot, C.; Sibole, J.V.; Barceló, J.; Poschenrieder, C. Abscisic acid decreases leaf Na$^+$ exclusion in salt-treated *Phaseolus vulgaris* L. *J. Plant Growth Regul.* **2009**, *28*, 187–192. [CrossRef]

17. Chen, P.; Yan, K.; Shao, H.; Zhao, S. Physiological mechanisms for high salt tolerance in wild soybean (*Glycine soja*) from Yellow River Delta, China: Photosynthesis, osmotic regulation, ion flux and antioxidant capacity. *PLoS ONE* **2013**, *8*, e83227. [CrossRef]

18. Kalaji, H.M.; Govindjee Bosa, K.; Koscielniak, J.; Zuk-Golaszewska, K. Effects of salt stress on photosystem II efficiency and CO_2 assimilation of two Syrian barley landraces. *Environ. Exp. Bot.* **2011**, *73*, 64–72. [CrossRef]

19. Stepien, P.; Johnson, G.N. Contrasting responses of photosynthesis to salt stress in the glycophyte *Arabidopsis* and the halophyte *Thellungiella*: Role of the plastid terminal oxidase as an alternative electron sink. *Plant Physiol.* **2009**, *149*, 1154–1165. [CrossRef]

20. Yan, K.; Wu, C.; Zhang, L.; Chen, X. Contrasting photosynthesis and photoinhibition in tetraploid and its autodiploid honeysuckle (*Lonicera japonica* Thunb.) under salt stress. *Front. Plant Sci.* **2015**, *6*, 227. [CrossRef]

21. Yan, K.; Xu, H.; Cao, W.; Chen, X. Salt priming improved salt tolerance in sweet sorghum by enhancing osmotic resistance and reducing root Na$^+$ uptake. *Acta Physiol. Plant.* **2015**, *37*, 203. [CrossRef]

22. Aparicio, C.; Urrestarazu, M.; Cordovilla, M.D. Comparative physiological analysis of salinity effects in six olive genotypes. *Hortscience* **2014**, *49*, 901–904.

23. Chaves, M.M.; Flexas, J.; Pinheiro, C. Photosynthesis under drought and salt stress: Regulation mechanisms from whole plant to cell. *Ann. Bot.* **2009**, *103*, 551–560. [CrossRef] [PubMed]

24. Loreto, F.; Centritto, M.; Chartzoulakis, K. Photosynthetic limitations in olive cultivars with different sensitivity to salt stress. *Plant Cell. Environ.* **2003**, *26*, 595–601. [CrossRef]

25. Feng, L.L.; Han, Y.J.; Liu, G.; An, B.G.; Yang, J.; Yang, G.H.; Li, Y.S.; Zhu, Y.G. Overexpression of sedoheptulose-1,7-bisphosphatase enhances photosynthesis and growth under salt stress in transgenic rice plants. *Funct. Plant Biol.* **2007**, *34*, 822–834. [CrossRef]

26. Lu, K.X.; Cao, B.H.; Feng, X.P.; He, Y.; Jiang, D.A. Photosynthetic response of salt-tolerant and sensitive soybean varieties. *Photosynthetica* **2009**, *47*, 381–387. [CrossRef]

27. Sonoike, K. Photoinhibition of photosystem I. *Physiol. Plant.* **2011**, *142*, 56–64. [CrossRef] [PubMed]

28. Takahashi, S.; Murata, N. How do environmental stresses accelerate photoinhibition? *Trends Plant Sci.* **2008**, *13*, 178–182. [CrossRef] [PubMed]

29. Yan, K.; Zhao, S.; Liu, Z.; Chen, X. Salt pretreatment alleviated salt-induced photoinhibition in sweet sorghum. *Theor. Exp. Plant Physiol.* **2015**, *27*, 119–129. [CrossRef]

30. Long, X.H.; Liu, L.P.; Shao, T.Y.; Shao, H.B.; Liu, Z.P. Developing and sustainably utilize the coastal mudflat areas in China. *Sci. Total Environ.* **2016**, *569–570*, 1077–1086. [CrossRef]

31. Dias, N.S.; Ferreira, J.F.S.; Liu, X.; Suarez, D.L. Jerusalem artichoke (*Helianthus tuberosus*, L.) maintains high inulin, tuber yield, and antioxidant capacity under moderately-saline irrigation waters. *Ind. Crops Prod.* **2016**, *94*, 1009–1024. [CrossRef]

32. Huang, Z.; Long, X.; Wang, L.; Kang, J.; Zhang, Z.; Zed, R.; Liu, Z. Growth, photosynthesis and H+-ATPase activity in two Jerusalem artichoke varieties under NaCl-induced stress. *Process Biochem.* **2012**, *47*, 591–596. [CrossRef]

33. Li, L.; Shao, T.; Yang, H.; Chen, M.; Gao, X.; Long, X.; Shao, H.; Liu, Z.; Rengel, Z. The endogenous plant hormones and ratios regulate sugar and dry matter accumulation in Jerusalem artichoke in salt-soil. *Sci. Total Environ.* **2017**, *578*, 40–46. [CrossRef]

34. Long, X.; Huang, Z.; Zhang, Z.; Li, Q.; Zed, R.; Liu, Z. Seawater stress differentially affects germination, growth, photosynthesis, and ion concentration in genotypes of Jerusalem Artichoke (*Helianthus tuberosus* L.). *J. Plant Growth Regul.* **2009**, *29*, 223–231. [CrossRef]

35. Li, P.M.; Ma, F.W. Different effects of light irradiation on the photosynthetic electron transport chain during apple tree leaf dehydration. *Plant Physiol. Biochem.* **2012**, *55*, 16–22. [CrossRef] [PubMed]

36. Oukarroum, A.; Bussotti, F.; Goltsev, V.; Kalaji, H.M. Correlation between reactive oxygen species production and photochemistry of photosystems I and II in *Lemna gibba* L. plants under salt stress. *Environ. Exp. Bot.* **2015**, *109*, 80–88. [CrossRef]

37. Strasser, R.J.; Tsimilli-Michael, M.; Qiang, S.; Goltsev, V. Simultaneous in vivo recording of prompt and delayed fluorescence and 820 nm reflection changes during drying and after rehydration of the resurrection plant *Haberlea rhodopensis*. *Biochim. Biophys. Acta* **2010**, *1797*, 122. [CrossRef]

38. Zivcak, M.; Brestic, M.; Kunderlikova, K.; Olsovska, K.; Allakhverdiev, S.I. Effect of photosystem I inactivation on chlorophyll a fluorescence induction in wheat leaves: Does activity of photosystem I play any role in OJIP rise? *J. Photochem. Photobiol. B Biol.* **2015**, *152*, 318–324. [CrossRef] [PubMed]

39. Gill, S.S.; Tuteja, N. Reactive oxygen species and antioxidant machinery in abiotic stress tolerance in crop plants. *Plant Physiol. Biochem.* **2010**, *48*, 909–930. [CrossRef] [PubMed]

40. Yan, K.; Zhao, S.; Cui, M.; Han, G.; Wen, P. Vulnerability of photosynthesis and photosystem I in Jerusalem artichoke (*Helianthus tuberosus* L.) exposed to waterlogging. *Plant Physiol. Biochem.* **2018**, *125*, 239–246. [CrossRef] [PubMed]

41. Allakhverdiev, S.I.; Murata, N. Salt stress inhibits photosystems II and I in cyanobacteria. *Photosynth Res.* **2008**, *98*, 529–539. [CrossRef] [PubMed]

42. Munns, R. Comparative physiology of salt and water stress. *Plant Cell Environ.* **2002**, *25*, 239–250. [CrossRef] [PubMed]

43. Murata, N.; Takahashi, S.; Nishiyama, Y.; Allakhverdiev, S.I. Photoinhibition of photosystem II under environmental stress. *Biochim. Biophys. Acta* **2007**, *1767*, 414–421. [CrossRef]

44. Yang, C.; Zhang, Z.S.; Gao, H.Y.; Fan, X.L.; Liu, M.J.; Li, X.D. The mechanism by which NaCl treatment alleviates PSI photoinhibition under chilling-light treatment. *J. Photochem. Photobiol. B Biol.* **2014**, *140*, 286–291. [CrossRef] [PubMed]

45. Song, J.; Shi, G.W.; Gao, B.; Fan, H.; Wang, B.S. Waterlogging and salinity effects on two *Suaeda salsa* populations. *Physiol. Plant.* **2011**, *141*, 343–351. [CrossRef] [PubMed]

46. Velikova, V.; Yordanov, I.; Edreva, A. Oxidative stress and some antioxidant systems in acid rain-treated bean plants—Protective role of exogenous polyamines. *Plant Sci.* **2000**, *151*, 59–66. [CrossRef]

47. Yan, K.; Cui, M.; Zhao, S.; Chen, X.; Tang, X. Salinity stress is beneficial to the accumulation of chlorogenic acids in honeysuckle (*Lonicera japonica* Thunb.). *Front. Plant Sci.* **2016**, *7*, 1563. [CrossRef]

48. Lopez-Carbonell, M.; Jauregui, O. A rapid method for analysis of abscisic acid (ABA) in crude extracts of water stressed Arabidopsis thaliana plants by liquid chromatography—Mass spectrometry in tandem mode. *Plant Physiol. Biochem.* **2005**, *43*, 407–411. [CrossRef]

49. Maxwell, K.; Johnson, G.N. Chlorophyll fluorescence—A practical guide. *J. Exp. Bot.* **2000**, *51*, 659–668. [CrossRef]

50. Schansker, G.; Srivastava, A.; Strasser, R.J. Characterization of the 820-nm transmission signal paralleling the chlorophyll a fluorescence rise (OJIP) in pea leaves. *Funct. Plant Biol.* **2003**, *30*, 785–796. [CrossRef]

51. Yan, K.; Chen, P.; Shao, H.B.; Zhao, S.J. Characterization of photosynthetic electron transport chain in bioenergy crop Jerusalem artichoke (*Helianthus tuberosus* L.) under heat stress for sustainable cultivation. *Ind. Crop. Prod.* **2013**, *50*, 809–815. [CrossRef]

52. Yan, K.; Han, G.; Ren, C.; Zhao, S.; Wu, X.; Bian, T. *Fusarium solani* infection depressed photosystem performance by inducing foliage wilting in apple seedlings. *Front. Plant Sci.* **2018**, *9*, 479. [CrossRef] [PubMed]

53. Zhang, Z.S.; Jin, L.Q.; Li, Y.T.; Tikkanen, M.; Li, Q.M.; Ai, X.Z.; Gao, H.Y. Ultraviolet-B Radiation (UV-B) Relieves Chilling-light-induced PSI photoinhibition and accelerates the recovery Of CO_2 assimilation in cucumber (*Cucumis sativus* L.) leaves. *Sci. Rep.* **2016**, *6*, 34455. [CrossRef] [PubMed]

International Journal of
Molecular Sciences

MDPI

Article

SNF1-Related Protein Kinases SnRK2.4 and SnRK2.10 Modulate ROS Homeostasis in Plant Response to Salt Stress

Katarzyna Patrycja Szymańska *, Lidia Polkowska-Kowalczyk, Małgorzata Lichocka, Justyna Maszkowska and Grażyna Dobrowolska *

Institute of Biochemistry and Biophysics, Polish Academy of Sciences, Pawińskiego 5a, 02-106 Warsaw, Poland; lidekp@ibb.waw.pl (L.P.-K.); mlichocka@ibb.waw.pl (M.L.); j.maszkowska@ibb.waw.pl (J.M.)
* Correspondence: kszymanska@ibb.waw.pl (K.P.S.); dobrowol@ibb.waw.pl (G.D.);
   Tel.: +48-22-592-5715 (K.P.S. & G.D.)

Received: 12 December 2018; Accepted: 24 December 2018; Published: 2 January 2019

Abstract: In response to salinity and various other environmental stresses, plants accumulate reactive oxygen species (ROS). The ROS produced at very early stages of the stress response act as signaling molecules activating defense mechanisms, whereas those produced at later stages in an uncontrolled way are detrimental to plant cells by damaging lipids, DNA, and proteins. Multiple systems are involved in ROS generation and also in ROS scavenging. Their level and activity are tightly controlled to ensure ROS homeostasis and protect the plant against the negative effects of the environment. The signaling pathways responsible for maintaining ROS homeostasis in abiotic stress conditions remain largely unknown. Here, we show that in *Arabidopsis thaliana*, two abscisic acid- (ABA)-non-activated SNF1-releted protein kinases 2 (SnRK2) kinases, SnRK2.4 and SnRK2.10, are involved in the regulation of ROS homeostasis in response to salinity. They regulate the expression of several genes responsible for ROS generation at early stages of the stress response as well as those responsible for their removal. Moreover, the SnRK2.4 regulate catalase levels and its activity and the level of ascorbate in seedlings exposed to salt stress.

Keywords: antioxidant enzymes; *Arabidopsis thaliana*; ascorbate cycle; hydrogen peroxide; reactive oxygen species; salinity; SnRK2

1. Introduction

Plants growing in nature are exposed to ever changing environmental conditions. They experience various abiotic stresses, such as drought, temperature extremes, and salinity. Salinity and drought are among the most detrimental factors limiting plant growth and development. Salinity causes ion-related stress, limitations in nutrient uptake as well as osmotic stress.

A secondary effect of salt stress and several other stresses is the accumulation of reactive oxygen species (ROS) in plant cells. Various ROS, such as singlet oxygen ($^1O_2$), superoxide radical ($O_2{}^-$), hydroxyl radical ($-OH$), and hydrogen peroxide (H_2O_2), are produced at low levels in chloroplasts, mitochondria, peroxisomes, and the apoplast during plant growth in optimal conditions [1,2]. They are involved in the regulation of plant growth and development, acting as signaling molecules. Upon stress, however, ROS play a double role [1,3–5]. ROS production at a low level is needed at the first stages of the stress response for induction of the plant defense, e.g., activation of signaling cascades, expression of stress response genes encoding enzymes involved in the synthesis of osmoprotectants, and some enzymes responsible for ROS scavenging [6–8]. At the later stages, ROS that accumulate in a non-controlled way have a widespread toxic effect, causing peroxidation of lipids, and damaging proteins and DNA, eventually leading to cell death. In response to stress, ROS are produced by diverse enzymes, e.g., NADPH oxidases, glycolate oxidases, oxalate oxidase, xanthine oxidase, and some peroxidases [9,10]. In *Arabidopsis thaliana* subjected to salinity stress, mainly two NADPH oxidases, respiratory burst oxidases, AtRbohD and AtRbohF, are involved in ROS production. They generate O_2^- free radicals in the apoplastic space by transferring electrons from NADPH to O_2. Then, the O_2^- is dismutated to H_2O_2 by superoxide dismutase (SOD) and the H_2O_2 molecules diffuse to adjacent cells, where they can play a role of signaling molecules, inducing plant defense, or they cause oxidative stress and cell damage. To achieve ROS homeostasis, which is required for efficient defense against the negative effects of harmful environmental conditions, plants have evolved several systems for ROS removal, both enzymatic, such as catalases (CATs), SODs, and various peroxidases (PRXs), and non-enzymatic, i.e., the ascorbate-glutathione cycle [9,11–16]. The enzymes involved in the ROS production and removal are encoded by multiple genes and are strictly regulated in response to stress, at the transcriptional, protein, and activity levels.

There are several reports showing that kinases from the SNF1-releted protein kinases 2 (SnRK2s) family are major regulators of the plant response to osmotic stress (drought, salinity). SnRK2s are plant-specific kinases activated in response to osmotic stress and some of them additionally in response to abscisic acid (ABA). They have been found in every plant species analyzed [17]. The SnRK2s are classified into three groups based on their phylogeny. The classification correlates well with their response to ABA: Group 1 comprises ABA-non-activated kinases; in group 2, are kinases weakly activated or non-activated by ABA (depending on plant species); and group 3 contains kinases strongly activated by ABA [18,19]. SnRK2s play a crucial role in the induction of defense mechanisms against drought [20,21] and salinity [22–24] via ABA-dependent and ABA-independent pathways. So far, the role of ABA-dependent SnRK2s (SnRK2.2, SnRK2.3, and SnRK2.6) has been mostly studied and found to be crucial for ABA signaling [25,26]. In response to drought, they regulate stomatal closure in an ABA-dependent manner by phosphorylating several ion channels [27–29], expression of stress-response genes by phosphorylating transcription factors activated in response to ABA [30–32], and the activity of aquaporins [33]. SnRK2.6 is involved in the ABA-dependent ROS production indispensable for stomatal closure [34], possibly by phosphorylating the NADPH oxidase, RbohF [35,36].

Much less is known about the involvement of SnRK2s in response to salinity, even though these kinases are known to be strongly activated in response to this stress. It has been shown that ABA-non-responsive kinases (belonging to group 1) are activated rapidly and transiently in response to salt stress [23,37,38]. McLoughlin et al. [23] showed that two kinases from group 1, SnRK2.4 and SnRK2.10, are fully active within seconds in roots of Arabidopsis plants after treatment with NaCl. Both kinases were found to be required for plant tolerance to salinity stress by regulating root growth and architecture [23]. Moreover, it has been shown that SnRK2s from group 1 influence the plant tolerance to salt stress via regulation of mRNA decay. They phosphorylate VARICOSE (VCS), a protein regulating mRNA decapping [24]. Very recently, using the phosphoproteomic approach, several potential ABA-non-activated SnRK2s' targets that phosphorylated in response to salinity have been found [39]. Among them there were several proteins, e.g., RNA- and DNA- binding proteins, protein kinases, phosphatases, and dehydrins, Early Responsive to Dehydration 10 (ERD10) and ERD14, whose phosphorylation likely affects the plant tolerance to salt stress. It has also been suggested that kinases from this group could be involved in the regulation of tolerance to salt stress via regulation of oxidative stress generated in response to salinity. Diédhiou et al. [22] showed that transgenic rice overexpressing Stress-Activated Protein Kinase 4 (SAPK4), the rice ABA-non-activated SnRK2, exhibited improved tolerance to salt stress. Their results indicated that SAPK4 regulates Na^+ and Cl^- accumulation and ROS homeostasis; the enhanced level of the kinase caused up-regulation of the *CatA* gene encoding catalase. Additionally, it has been shown that SnRK2.4 positively regulates the accumulation of ROS in response to stress induced by cadmium ions [40].

These data suggest that at least some members of the group 1 of the SnRK2 family are likely to be involved in the regulation of plant tolerance to osmotic stress by controlling the ROS level. The aim of the present study was to establish the role of SnRK2.4 and SnRK2.10 in the regulation of the ROS homeostasis in response to salt stress in *Arabidopsis thaliana*.

2. Results

2.1. SnRK2.4 and SnRK2.10 Kinases Are Involved in H_2O_2 Accumulation in Response to Salt Stress

To check whether SnRK2.4 and/or SnRK2.10 are involved in ROS accumulation in the early response to salt stress, we compared the accumulation of H_2O_2 in leaves of four-week-old plants of the *snrk2.4* and *snrk2.10* knockout mutants and wild type Col-0 plants exposed to 150 mM NaCl for various time using a luminol-based assay. The mutant lines accumulated significantly less H_2O_2 than the wild type (wt) plants did in response to salinity (Figure 1A). The maximal level of H_2O_2 in Col-0 leaves was observed 30 min after the stressor application (over a three-fold increase in respect to the control level), whereas in the *snrk2.4* and *snrk2.10* mutants, the maximal H_2O_2 accumulation was only two-fold and occurred at 60 min. Notably, the level of H_2O_2 in control conditions was by ca. 30% lower in three out of four mutants tested relative to wt. The lower initial H_2O_2 content combined with the smaller increase resulted in the mutant plants having less than half of the H_2O_2 level found in the wt at the peak of the response to salt stress. Since the two independent lines of both mutants showed similar behavior for further studies, we decided to use only one line of each mutant, *snrk2.4-1* and *snrk2.10-1* (later referred to as *snrk2.4* and *snrk2.10* mutants), previously well characterized [23,39,40]. To verify the observed differences between the wt plants and the *snrk2* mutants in ROS accumulation in response to salt stress, we monitored their level in Arabidopsis roots using the fluorescent dye, dichlorofluorescin diacetate (H_2DCFDA). We analyzed the accumulation of ROS in roots of five-day-old seedlings of wt plants, and the *snrk2.4* and *snrk2.10* mutants (Figure 1B,C) exposed to 250 mM NaCl for 15 min. Similarly to what was observed for Arabidopsis leaves exposed to the salt treatment, in roots of the *snrk2* mutant lines, the basal level of ROS was lower than in the roots of the wt plants and the ROS accumulation after the stress application was lower in those mutants in comparison with the one observed for wt seedlings. Thus, the obtained results strongly suggested the role of SnRK2.4 and SnRK2.10 in the regulation of ROS accumulation at the early stages of plant response to salinity.

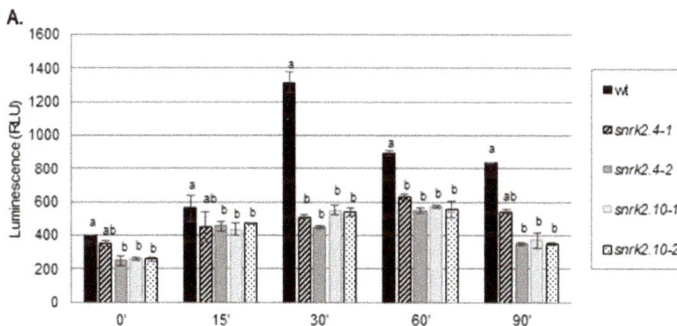

Figure 1. *Cont.*

B.

C.

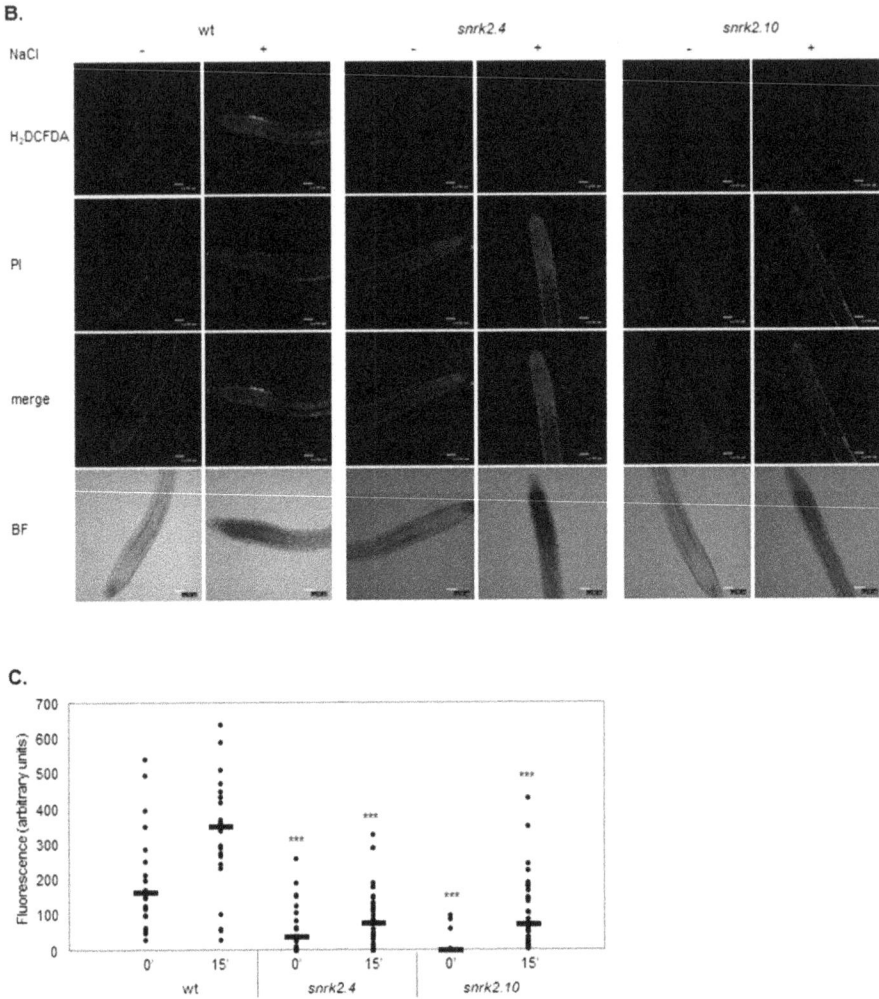

Figure 1. SnRK2.4 and SnRK2.10 affect ROS level in plants subjected to salt stress. (**A**). Leaves of wt plants and *snrk2.4* and *snrk2.10* mutant lines were subjected to 150 mM NaCl for the indicated time and H_2O_2 was determined using a luminol-based assay. Letters represent statistical differences in respect to the wt plants where a means no significant difference, and b means a significant difference [one way analysis of variance (ANOVA). Error bars represent standard deviation (SD). Three independent biological replicates, each with four samples per data point were performed. Results of all combined experiments are shown. B and C. Roots of five-day-old Arabidopsis seedlings (wt plants and *snrk2.4* and *snrk2.10* mutant lines) were stained with propidium iodide (PI; 20 μg/mL) and 2',7'-Dichlorofluorescin diacetate (H_2DCFDA; 30 μg/mL) and then treated for 15 min with 250 mM NaCl in $\frac{1}{2}$ MS (+) or $\frac{1}{2}$ MS only (−). (**B**) The production of ROS was monitored by imaging of H_2DCFDA fluorescence in the roots using confocal microscopy; BF – bright field image, scale bars =50 μm (**C**) Fluorescence intensity of H_2DCFDA was calculated from well-defined region of interest (4000 μm^2) in the root meristematic zone on each single confocal section; stars represent statistically significant differences in respect to the wt plants (Mann-Whitney U test) where *** $p < 0.0001$; results represent data collected from at least 30 seedlings/line/conditions where each dot represents the sample value and a dash represents the median of measurements.

2.2. SnRK2.4 and SnRK2.10 Regulate Expression of Genes Involved in ROS Generation in Response to Salinity

To determine the mechanism by which SnRK2.4 and SnRK2.10 affect the H_2O_2 homeostasis in salt-stressed plants, we investigated their impact on the enzymes involved in ROS production and scavenging. Since mitogen-activated protein kinase (MAPK, MPK) cascades regulate the ROS homeostasis by controlling the expression of genes encoding enzymes involved in ROS production [41,42], scavenging, as well as genes involved in ROS signaling [43], we studied the expression of several genes playing a role in the regulation of ROS levels. As the first approach, we analyzed the expression of genes whose products are responsible for ROS generation. Several cellular ROS generating enzymes, especially NADPH oxidases and apoplastic peroxidases, are involved in the plant response to environmental stresses [12]. Since, in the Arabidopsis, in response to salinity ROS, are generated mainly by the RbohD and RbohF oxidases [8,44,45], we analyzed transcript levels of *RbohD* and *RbohF*. For the studies, we used two-week-old Arabidopsis seedlings of the *snrk2.4* and *snrk2.10* mutant lines and wt plants treated with 150 mM NaCl up to 24 h. At the first stage of the response to salt stress (1 h), the transcript levels of *RbohD* and *RbohF* were significantly lower in the *snrk2.4* mutant than in the wt plants (Figure 2A,B) in agreement with the lower level of H_2O_2 found in this mutant compared to the wt. At the later time points, this difference was reversed and the expression of *RbohD* and *RbohF* became higher in the mutant than in the wt plants. Notably, the two genes assayed showed markedly different kinetics and extents of activation during the salt treatment. The expression pattern of *RbohD* and *RbohF* in the *snrk2.10* mutant in response to the stress differed significantly from the one observed for the *snrk2.4* mutant. *RbohD* expression was hardly affected by the mutation (it was lower by only 10–15% than in the wt plants throughout the experiment), while the expression of *RbohF* was enhanced markedly at the early stages (up to 3 h) of the response and then fell to slightly below that in the wt. This behavior was opposite to that shown by the *snrk2.4* mutant.

Additionally, we analyzed the expression of genes encoding two apoplastic peroxidases (PRX), PRX33 and PRX34, known to play an important role in the oxidative burst in response to biotic stresses [46,47]. The expression of both *PRXs* was induced in the wt plants in response to salinity stress, indicating their involvement in an abiotic stress response as well (Figure 2C,D). Notably, their expression patterns were affected in a complex manner in both *snrk2* mutants indicating an involvement of SnRK2.4 and SnRK2.10 in their regulation upon salt stress. As for the two *Rboh* genes described above, also here, the two mutant lines showed contrasting responses. In the *snrk2.4* mutant, the *PRX33* gene showed a delayed induction compared to the wt, and at 24 h, its transcript level was double that in the wt. The induction of the *PRX34* gene was enhanced several fold at 3 h and 6 h and was only slightly higher at 24 h relative to the wt. In the *snrk2.10* mutant, *PRX33* showed about a two-fold higher expression level than in the wt at 1 h, nearly identical ones at 3 and 6 h, and again much higher (over three-fold) at 24 h. One should note that also in control conditions, the expression of *PRX33* was markedly up-regulated in the *snrk2.10* mutant. The *SnRK2.10* mutation had a negligible effect on the expression of the other *PRX* gene studied, *PRX34*.

These results indicate that both kinases have an impact on the expression of *RbohD/F* and *PRX33/34* upon salt stress and that their roles are markedly different—they cannot substitute each other in this respect.

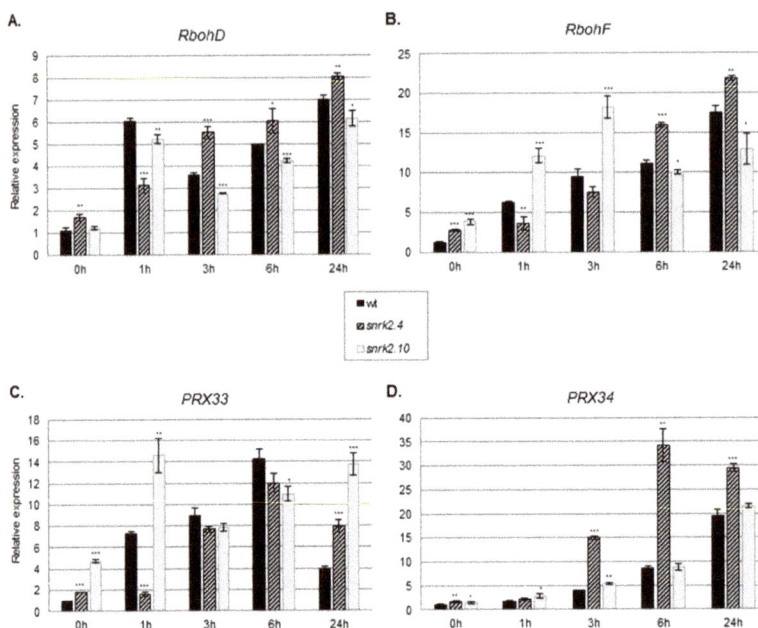

Figure 2. SnRK2.4 and SnRK2.10 affect the expression of genes involved in ROS homeostasis during response to salt stress. Expression (mRNA level) of (**A**). *RbohD—respiratory burst oxidase homolog protein D*; (**B**) *RbohF—respiratory burst oxidase homolog protein F*; (**C**) *PRX33—peroxidase 33*; and (**D**) *PRX34—peroxidase 34* was determined by RT-qPCR in wt plants and *snrk2* mutant lines subjected to treatment with 150 mM NaCl at times indicated (h); error bars represent SD; stars represent statistically significant differences in comparison with the wt plants (Student *t*-test) where * $p < 0.05$; ** $p < 0.001$; *** $p < 0.0001$. At least two independent biological replicates of the experiment were performed. Results of one representative experiment are shown.

2.3. SnRK2.4 and SnRK2.10 Are Involved in Regulation of ROS Scavenging in Response to Salt Stress

Plants have evolved several ROS scavenging pathways, both enzymatic and non-enzymatic. We analyzed the involvement of SnRK2.4 and SnRK2.10 in the regulation of some of them upon salinity stress by comparing the responses of two-week-old *snrk2.4* and *snrk2.10* mutants and wt seedlings exposed to 150 mM NaCl.

2.3.1. SnRK2s Affect CAT1 Gene Expression, Catalase Level, and Activity

Among the most prominent ROS scavenging enzymes are catalases. The Arabidopsis genome encodes three catalases—CAT1, CAT2, and CAT3. Expression of all of them is induced by salinity, however, the strongest changes were observed for the *CAT1* gene [48], therefore, we focused our studies on this gene. Our analysis revealed significant differences in the pattern of *CAT1* expression in response to salt stress between the *snrk2.4* and *snrk2.10* seedlings and the wt ones (Figure 3A).

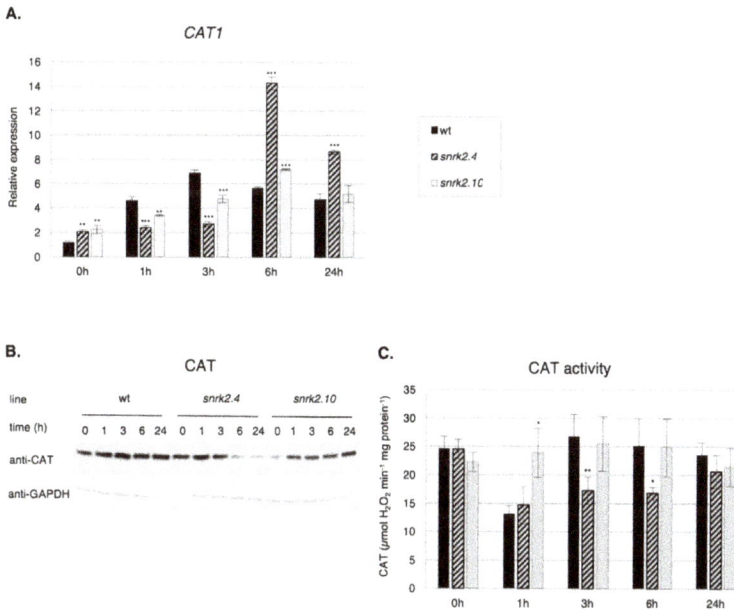

Figure 3. SnRK2.4 and SnRK2.10 modulate catalase (CAT) on multiple levels during response to salt stress. Wild type and *snrk2* mutants' seedlings were subjected to treatment with 150 mM NaCl for times indicated. *CAT1* expression was determined by RT-qPCR (**A**), total catalase protein was determined by immunoblot analysis (**B**), and total catalase activity assay was performed (**C**); error bars represent SD; stars represent statistically significant differences in comparison with the wt plants (Student *t*-test) where * $p < 0.05$; ** $p < 0.001$; *** $p < 0.0001$. After exposure, membranes were stripped and reused for glyceraldehyde 3-phosphate dehydrogenase (GAPDH) detection as a loading control. At least two independent biological replicates of the experiment were performed, each with four samples per data point. Results of one representative experiment are shown.

At the first stages of the response (up to 3 h), the *CAT1* transcript level was significantly lower in both mutants than in the wt, whereas at the later stages, the reverse was true, especially in *snrk2.4*. This suggests that at first, the *CAT1* expression is positively regulated by SnRK2.4 and SnRK2.10, whereas at the later stages of the stress response, the SnRK2s, especially SnRK2.4, exert an inhibitory action. Notably, the *CAT1* transcript level was up-regulated ca. two-fold in the both mutants in control conditions. In that, the effect of a lack of either SnRK2s resembled the situation observed earlier for *PRX33* and *RbohD/RbohF* genes. Unexpectedly, the differences in *CAT1* transcript levels were not reflected by the amount of catalase protein (Figure 3B) or activity (Figure 3C). The catalase protein level was in fact lower in the *snrk2* mutants exposed to salinity stress than in wt plants (Figure 3B); the lowest level was observed in the *snrk2.4* mutant after salt treatment. These data apparently indicate discrepancies between the transcript and protein levels. However, the immunoblot analysis was performed using antibodies recognizing all three isoforms of catalase—CAT1, CAT2, and CAT3. It should be mentioned that in Arabidopsis rosettes, *CAT2* and *CAT3* transcripts are much more abundant than that for *CAT1* [49,50]. Possibly the same is true for the corresponding proteins, therefore, the changes in the amount of the least abundant CAT1 were likely obscured by changes in the other two catalase isoenzymes. The latter, however, suggests that CAT2 and CAT3 could be also under the control of SnRK2s.

The catalase activity, reflecting the combined activity of the three isoenzymes, showed a different pattern for each line studied (Figure 3C); two mutants differed substantially from each other and also from the wt. Strikingly, at 1 h of salt treatment, there was a substantial drop of the catalase activity in the *snrk2.4* and the wt, whereas in the *snrk2.10* mutant, the activity basically did not change.

Taken together, our results show that during the plant response to salinity, SnRK2.4 and SnRK2.10 (especially SnRK2.4) regulate catalase at various levels, including gene expression, catalase protein level, and probably also its enzymatic activity.

2.3.2. SnRK2.4 and SnRK2.10 Regulate the Ascorbate Cycle

Ascorbate is a major antioxidant in plants. To check whether SnRK2.4 and SnRK2.10 play a role in the regulation of the ascorbate cycle in plants subjected to salinity, we compared the expression of genes and their protein products involved in the ascorbate cycle (ascorbate peroxidases, APXs, and dehydroascorbate reductase 1, DHAR1) as well as the APX activity in the *snrk2.4* and *snrk2.10* mutants and wt seedlings exposed to salt stress (Figure 4A–D). Expression of all the genes encoding cytoplasmic *APXs* (*APX1*, *APX2*, and *APX6*) and *DHAR1* was highly induced in response to salinity and that response was markedly and in a complex manner affected in both *snrk2* mutants. Notably, in control conditions, their expression was significantly higher in both mutants than in the wt.

APX1 expression was induced rapidly in response to salinity in wt seedlings, reached a maximum at 1 h and declined below control level at 6 h and 24 h, whereas in the *snrk2.4* mutant, it increased gradually between 3 h and 24 h; in *snrk2.10*, the pattern was similar except for the presence of an early (1 h) peak of expression.

A different situation was observed for the *APX2* gene. It underwent progressive very strong induction in wt plants (1000-fold increase at 24 h of salt treatment), and a similarly progressive, but much less pronounced, induction in the mutants.

APX6 expression was rapidly induced up to six-fold in response to salinity in the wt to reach the maximum at 3 h followed by a slight drop to four-times value at time 0. In contrast, in the *snrk2.4* and *snrk2.10* mutants, where the *APX6* expression was only slightly (*snrk2.10*) or not at all (*snrk2.4*) induced upon salt application, and dropped below the initial level at 24 h in both mutants.

Immunoblot analysis (with antibodies recognizing all three APXs) showed a slight decrease of the APX amount during salt stress in the wt and substantial accumulation at the later stages (6 h and 24 h) in the *snrk2* mutants, especially in the *snrk2.4* (Figure 4E). This accumulation roughly parallels the transcript pattern of *APX1* and *APX2* and suggests that the SnRK2.4 and SnRK2.10 kinases are negative regulators of APX accumulation at the later, but not the early stages of the plant response to salinity.

DHAR1 expression was highly and progressively induced in wt seedlings exposed to salt stress, and to a lesser extent also in the mutants. Notably, in the latter, its level fell between 6 h and 24 h of salt stress. DHAR1 protein showed strong and rapid accumulation in the wt line in response to the stress, slightly lower in the *snrk2.10* mutant, whereas in the *snrk2.4* mutant, it was barely detectable in control conditions and then grew gradually until the end of the treatment (Figure 4F), but it was significantly lower than in the wt and the *snrk2.10* mutant plants. These results suggest that the two SnRK2s, especially SnRK2.4, positively regulate DHAR1 accumulation.

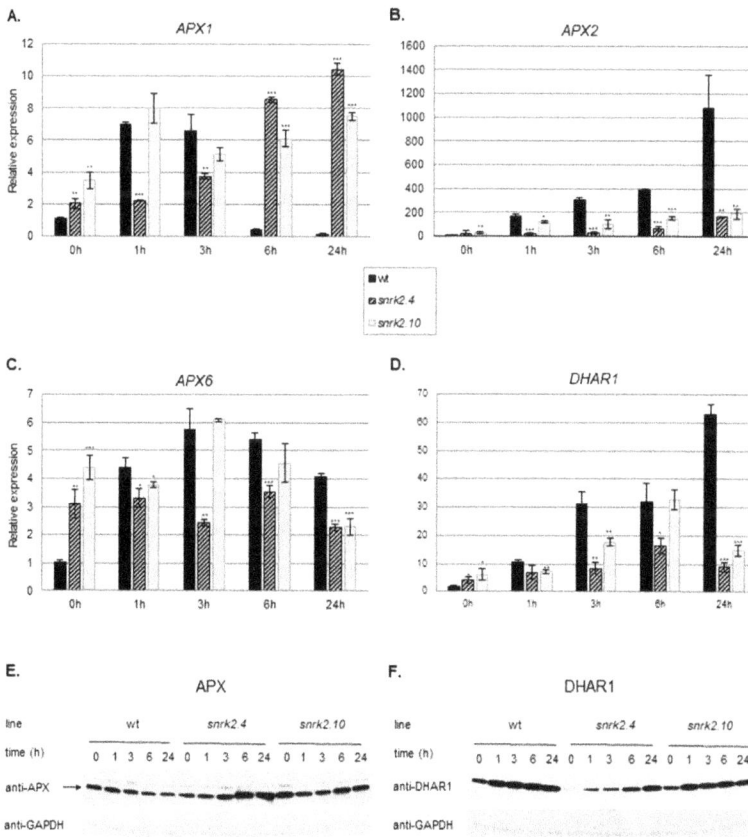

Figure 4. SnRK2.4 and SnRK2.10 regulate enzymes of the ascorbate cycle during response to salt stress. Wild type and *snrk2* mutant seedlings were subjected to treatment with 150 mM NaCl for times indicated. Expression of (**A**) *APX1—Ascorbate Peroxidase 1*, (**B**) *APX2—Ascorbate Peroxidase 2*, (**C**) *APX6—Ascorbate Peroxidase 6*, and (**D**) *DHAR1—Dehydroascorbate Reductase 1* was monitored by RT-qPCR; error bars represent SD; stars represent statistically significant differences in comparison with the wt plants (Student *t*-test) where * $p < 0.05$; ** $p < 0.001$; *** $p < 0.0001$. Total protein level of (**E**) APX and (**F**) DHAR1 was monitored with immunoblot analysis; after exposure, membranes were stripped and reused for GAPDH detection as a loading control. At least two independent biological replicates of the experiment were performed. Results of one representative experiment are shown.

The above results indicate that SnRK2.4, and to lesser extent also SnRK2.10, regulate the level of the enzymes of the ascorbate cycle during the plant response to salinity.

Next, we compared the total ascorbate (Asc) content, APX activity, and the ascorbate/dehydroascorbate (Asc/DHAsc) ratio in the *snrk2.4* and *snrk2.10* mutants and wt seedlings subjected to salinity (Figure 5).

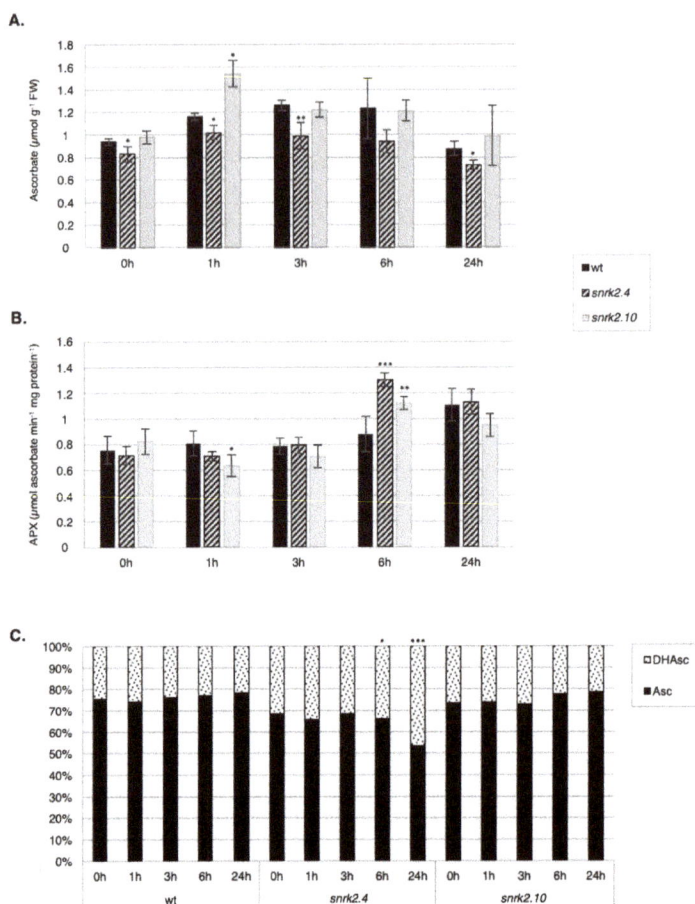

Figure 5. SnRK2.4 and SnRK2.10 regulate ascorbate cycle during response to salt stress. Wild type and *snrk2* mutant seedlings were subjected to treatment with 150 mM NaCl for times indicated and (**A**) ascorbate content, (**B**) ascorbate peroxidase (APX) activity, and (**C**) ascorbate redox status were monitored. Asc—ascorbate, DHAsc—dehydroascorbate; error bars represent SD; stars represent statistically significant differences from wt plants (Student *t*-test for Asc and APX activity, Chi-square test for Asc/DHAsc ratio) where * $p < 0.05$; ** $p < 0.001$; *** $p < 0.0001$. At least two independent biological replicates of the experiment were performed, each with four samples per data point. Results of one representative experiment are shown.

The total Asc content in wt plants was slightly increased in response to salt stress (from 0.9 μmol g^{-1} FW in control conditions to 1.3 μmol g^{-1} FW after 6 h of the treatment) (Figure 5A). A similar pattern was observed for the *snrk2.10* mutant, but with stronger Asc accumulation after 1 h of treatment (from 1 μmol g^{-1} FW in control conditions to 1.55 μmol g^{-1} FW after 1 h and 1.2 μmol g^{-1} FW after 3 h and 6 h of salt stress), while in the *snrk2.4* mutant, the Asc content only slightly increased in response to salt stress (0.83–1.0 μmol g^{-1} FW).

The ratio between oxidized (DHAsc) and the reduced form of Asc was virtually identical in the wt and the *snrk2.10* mutant and did not change upon salt stress (Figure 5C). In the *snrk2.4* mutant line, the fraction of the reduced form of Asc was lower than that in the other lines at all time points of the treatment and additionally showed a significant decrease after 6 h and 24 h of the stress (from 69% in control to 53% after 24 h), suggesting an increased APX activity. Indeed, measurements of the APX activity confirmed this conjecture, albeit the activity pattern did not exactly match the DHAsc/Asc pattern (Figure 5B). Generally, the APX activity reflected the APX protein level (compare Figure 4E).

3. Discussion

Salinity imposes ion and osmotic stresses on plant cells and leads to accumulation of ROS. The understanding of the signaling pathways controlling redox homeostasis during salt stress remains limited. A majority of the data concerning this subject pertains to the plant responses to biotic stresses. Several kinases regulating ROS production in response to pathogen infection have been identified. Some of them act as positive regulators of ROS generation via direct or indirect regulation of RbohD and/or RbohF activity or of their transcript accumulation (e.g., MPK3/6, Flagellin-sensitive 2, EF-Tu receptor, Brassinosteroid insensitive 1 associated receptor kinase 1, Botrytis-induced kinase 1, CPK5), some positively regulate PRX activity (like ZmMPK7), whereas others inhibit ROS accumulation (e.g., MPK4, AtCPK28) [51–60]. Much less is known regarding the protein kinases involved in ROS production or scavenging in response to abiotic stresses. In response to salinity, ROS are generated in different cellular compartments: Chloroplasts (by the photosynthetic electron transport), mitochondria (by the respiratory electron transport), peroxisomes, and, in the apoplast, by the action of oxidases present in the plasma membrane. The ROS generated in the various cellular compartments cross talk with each other. Recently it has been shown that, similarly to animals, the ROS-induced ROS release (RIRR) process (e.g., ROS produced in one cellular organelle or compartment induce ROS production in another one) takes place also in plants [61]. An RIRR-generated ROS wave leads to ROS amplification and signal transduction to neighboring compartments and cells. Therefore, enzymes localized to the cytoplasm or nucleus (as is in the case of the SnRK2s studied here) can indirectly affect the ROS level also in other subcellular compartments.

3.1. Role of SnRK2.4/SnRK2.10 in ROS Accumulation in Response to Salinity Stress

Previously published results have indicated that ABA-non-responsive SnRK2s (from group 1) regulate ROS levels in response to abiotic stresses. Diédhiou et al. [22] showed that an ABA-non-activated SnRK2, SAPK4, regulates ROS homeostasis in rice in response to salt stress, and Kulik et al. [40] that SnRK2.4 and SnRK2.10 are involved in positive regulation of H_2O_2 accumulation in Arabidopsis roots in the early response to heavy metal stress.

Surprisingly, studies performed on Arabidopsis, the *snrk2.2/3/6* triple, the septuple, and the decuple mutants defective in several SnRK2s did not show any correlation between SnRK2s level and H_2O_2 accumulation in response to osmotic stress (polyethylene glycol (PEG) treatment of Arabidopsis seedlings) [62]. However, first, the measurement was done 12 h after PEG addition. Moreover, it seems likely that individual SnRK2s might differently affect ROS production and scavenging and their defects in the multiple *snrk2* mutants could effectively cancel out, resulting in no net change in H_2O_2 accumulation. Finally, the role of various SnRK2s might be different in response to ABA, PEG, and salinity.

Our present studies focused on the role of two ABA-non-responsive SnRK2 kinases, SnRK2.4 and SnRK2.10, in the regulation of ROS homeostasis in Arabidopsis exposed to salinity stress. Since these kinases localize to the cytoplasm (SnRK2.10) and the cytoplasm and nucleus (SnRK2.4), we studied those enzymes involved in ROS homeostasis that in principle could be regulated by cytoplasmic and nuclear kinases: Directly by phosphorylation or indirectly at the transcriptional level. In respect to ROS production, we studied two plasma membrane NADPH oxidases, RbohD and RbohF, and two apoplastic peroxidases, PRX33 and PRX34. It has been established that RbohD and RbohF are regulated at the activity and expression levels. Their activity is tightly controlled by phosphorylation and by Ca^{2+} binding. Several kinases capable of phosphorylating RbohD/F have been identified, but the role of these phosphorylations is not fully clear. Drerup et al. [63] showed that Calcineurin B-like protein 1/9 (CBL1/9)-CIPK26 (from CBL-interacting protein kinase 26) complexes phosphorylate and activate RbohF. Similarly, Han et al. [36] presented that CIPK11 and CIPK26 phosphorylate RbohF, constituting alternative paths for RbohF activation, whereas Kimura et al. [64] suggest that the binding of CIPK26 to RbohF decreases ROS production. It has also been shown that RbohD/F are regulated also by Ca^{2+}-independent kinases, like MPK8, which inhibits RbohD activity in response to wounding [65]. Some data indicate an involvement of SnRK2s in the regulation of NADPH oxidase activity. OST1/SnRK2.6, an ABA-dependent kinase, regulates the ROS level required for the stomatal closure [34]. Sirichandra et al. [35] showed that OST1 phosphorylates RbohF in vitro and suggested that this phosphorylation plays a role in its activation and possibly in the regulation of stomatal movement in response to ABA. Recently, Han et al. [36] showed that OST1 together with CIPKs is involved in RbohF activation.

Our present results revealed that SnRK2.4 and SnRK2.10 positively regulate the ROS production at the early stages of the response to salinity; we observed significantly lower H_2O_2 levels in the *snrk2.4* and *snrk2.10* mutants than in wt plants salt-treated for up to 90 min. These results suggest that SnRK2.4 and SnRK2.10 might phosphorylate RbohD and/or RbohF in response to salinity and thereby regulate the ROS level. The phosphorylation of RbohD/F by SnRK2.4/10 is highly plausible since the substrate specificities of SnRK2s, CIPKs, and calcium-dependent protein kinases (CDPKs) are quite similar; all of them belong to the CDPK-SnRK superfamily [66]. The SnRK2.4 and SnRK210 kinases are activated rapidly in response to salinity, within seconds after the stressor application. Therefore, it is likely that they are involved in the earliest events of the response to salt stress, i.e., activation of the Rbohs and production of ROS responsible for triggering the defense mechanisms. However, this hypothesis needs further studies.

Our results pointed out to a role of SnRK2.4 and SnRK2.10 in the regulation of *RbohD* and *RbohF* expression. In *Arabidopsis thaliana* seedlings, the expression of *RbohD* and *RbohF* is induced in response to salinity [14,44,45,67]. At the first stages of the response (in our case, the treatment with NaCl for 1 h), the transcript levels of *RbohD* and *RbohF* were significantly lower in the *snrk2.4* mutant than in wt plants, which is in agreement with the lower level of H_2O_2 found in the mutants. Later, during the salt treatment, this correlation was no longer sustained and the *RbohD/F* expression in the mutant became elevated above the level observed for the wt plants. The effect of disruption of the *SnRK2.10* gene was more complex, as it has little effect on induction of *RbohD* expression, but actually enhanced that of *RbohF*.

Additionally, we analyzed the expression of genes encoding apoplastic peroxidases, *PRX33* and *PRX34*, whose involvement in the response to salt stress has not been considered so far. Expression of both *PRXs* was induced in response to salinity stress, which indicates their role in the abiotic stress response, and this induction was apparently regulated by SnRK2s. Expression of *PRX33* was significantly lower in the *snrk2.4* mutant early in the response to salinity, but became highly elevated relative to the wt plants after prolonged salt treatment. For *PRX34*, the effect of the *snrk2.4* mutation was visible only at later stages of the response and manifested as a several fold enhancement of the induction. This suggests an inhibitory role of SnRK2.4 on the salt-induced *PRX34* expression.

As for the *Rboh* genes, also here, SnRK2.10 turned out to act differently to SnRK2.4. The *snrk2.10* mutation had virtually no effect on *PRX34* expression, but greatly stimulated the expression of *PRX33* in control conditions and also at early and late response to salt stress.

These data indicate that even though in both the *snrk2.4* and *snrk2.10* mutants, the level of H_2O_2 produced early in response to salt stress is significantly lower than in wt plants, the mechanisms of the regulation of ROS accumulation by the two kinases seem to be different. It should be stressed here, that unlike SnRK2.4, SnRK2.10 does not localize to the nucleus, therefore, the different modes of regulation of gene expression by these kinases are not surprising (Figure 6).

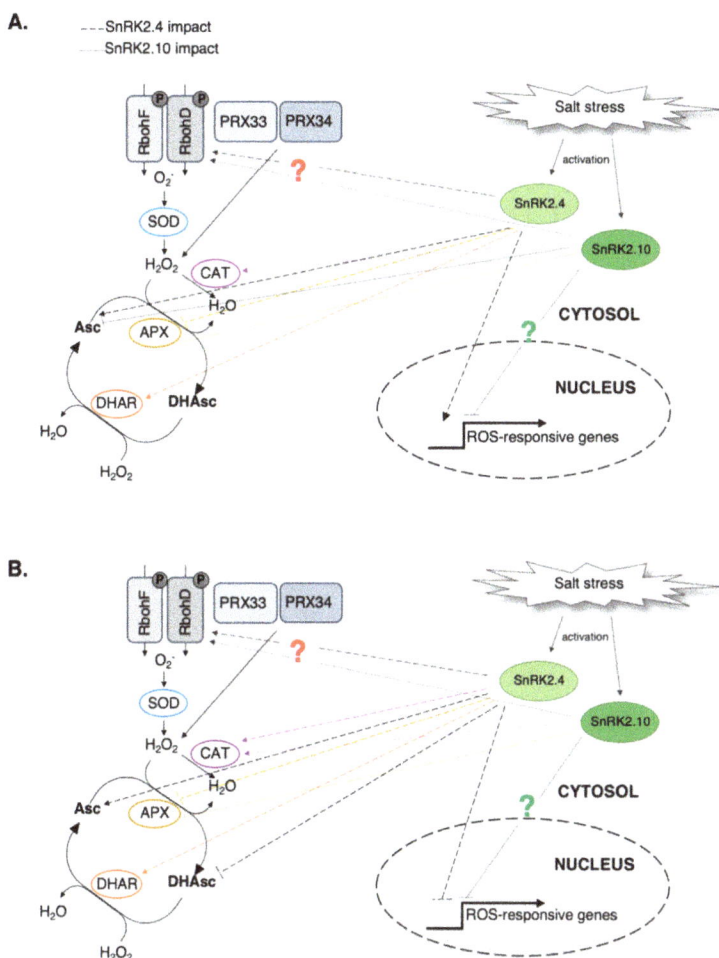

Figure 6. Possible roles of SnRK2.4 and SnRK2.10 in the regulation of the ROS homeostasis in Arabidopsis seedlings exposed to the salt stress. Proposed role of the SnRK2s in (**A**) early response and (**B**) late response to the salt stress. In response to salinity, SnRK2.4 along with SnRK2.10 regulate the ROS production/accumulation as well as ROS scavenging at the transcription as well as protein and/or activity levels. Detailed description in the text; dash lines—SnRK2.4 impact; dotted lines—SnRK2.10 impact; green question mark—probably indirect regulation; red question mark—plausible direct regulation by phosphorylation.

The expression of genes encoding ROS producing enzymes in the *snrk2.4* and *snrk2.10* (especially *snrk2.4*) is very different at early and late stages of the response. We propose that at the later stages the expression of genes studied might be regulated by other signaling pathways, for example, MAPK cascade(s), which are involved in controlling ROS homeostasis. Those pathways might be triggered to compensate for the low ROS level in the *snrk2s* mutants. In response to several stimuli MAPK cascade(s) control *RbohD, RbohF, PRX33,* and *PRX34* expression [52,68]. MPK3/MPK6 phosphorylate and thus activate the ERF6 transcription factor, whose targets are *RbohD* and *PRX33* in response to fungal pathogen [69–71]. Moreover, in *Nicotiana benthamiana* during the ETI (from effector-triggered immunity) and Elicitin-(INF1)-triggered PTI (from pattern-triggered immunity), salicylic acid induced protein kinase (SIPK, orthologue of Arabidopsis MPK6) phosphorylates four

W-box binding transcription factors (WRKYs), which are responsible for the expression of *RBOHB* (ortholog of Arabidopsis *RbohD*), and positively regulates the *RBOHB* transcript level [72]. Since MPK6 and SIPK are activated in response to salinity and water deficits [37,73] and ERF6 is involved in response to the water limitation [74], it seems likely that the MPK6 pathway, and presumably some others, could overcompensate for the low expression of *RbohD* and *PRX33* at early stages of the response in the *snrk2.4* mutant.

3.2. Involvement of SnRK2.4/SnRK2.10 in ROS Removal under Salinity Stress Conditions

Data on the signaling pathways and protein kinases involved in the regulation of the antioxidant systems engaged in ROS scavenging in response to salinity or osmotic stress are scarce. It has been shown that GSK3 kinase (ASKα) regulates salt stress tolerance of Arabidopsis by phosphorylation and activation of glucose-6-phosphate dehydrogenase, an enzyme important for maintaining the cellular redox balance [75]. Zong et al. [76] have shown that ectopic expression of *ZmMPK7* in *Nicotiana tabaccum* enhances peroxidase activity, which results in lower accumulation of H_2O_2 in response to osmotic stress. It has been reported that several protein kinases of MAPK cascades as well as CIPK are involved in the regulation of expression and/or activity of CAT1 [43,48,77].

We analyzed here the impact of SnRK2.4 and SnRK2.10 on several enzymes involved in ROS scavenging (CATs, APXs, and DHAR1). A comparison of the changes in the *CAT1* transcript level in the *snrk2.4*, *snrk2.10*, and wt plants exposed to salinity indicated that SnRK2.4, and to a lesser extent also SnRK2.10, positively regulate the expression of *CAT1* during the first stages of the stress response. However, similar to what was observed for *RbohD*, at the later stages of the response, the impact of the *snrk2* mutations became just the opposite, which results in a higher *CAT1* expression than in the wt plants. We conjecture that this effect was not due to a direct regulation of *CAT1* expression by SnRK2.4/SnRK2.10, but rather because of reflected activation of some other signaling pathway(s) in order to compensate for the low *CAT1* expression. One such pathway might be again the MPK6 cascade, known to mediate *CAT1* expression and H_2O_2 production [77].

Besides the regulation of the *CAT1* expression, SnRK2.4 and, less markedly, also SnRK2.10 positively regulated catalase protein accumulation and activity during salt stress, as they were both significantly lower in the *snrk2* mutants exposed to the stress than in the wt. The discrepancy between the enhanced expression of *CAT1* later in the response and the lower catalase protein level and activity indicates that in response to salinity, the SnRK2s affect not only CAT1, but most likely also the CAT2 and CAT3 levels, in opposing directions. Since CAT2 and CAT3 are more abundant than CAT1 in Arabidopsis seedlings, it is likely that the catalase activity, which we measured, represented mainly the activity of CAT2 and CAT3. It is plausible that SnRK2s regulate also the expression of *CAT2* and/or *CAT3* genes. We observed lower catalase activity in the *snrk2.4* mutant exposed for up to 6 h to the salt stress. It is not clear whether SnRK2s modulate the enzyme(s) specific activity or only the catalases' protein level or through the phosphorylation they impact the targeting of catalases into the peroxisomes (their final destination). Phosphorylation of catalases by SnRK2s is quite feasible, since it has been shown that salt overly sensitive 2 (SOS2), a kinase belonging to the SnRK3 subfamily, interacts with CAT2 and CAT3 and possibly influences H_2O_2 accumulation in response to salinity [78]. SOS2 localizes to the plasma membrane and cytoplasm and its substrate specificity is nearly the same as that of the SnRK2's.

The available information on the regulation of the ascorbate cycle in response to abiotic stresses and the protein kinases is very limited. It has been suggested that in response to strong light, the SnRK2.6/OST1 kinase activates *APX2* expression [79]. Pitzschke and Hirt [80] have speculated that MAPK cascades could be involved in ascorbate cycle regulation based on transcriptomic analysis performed on *mapk* mutants and wt plants. They revealed that several genes encoding enzymes involved in ascorbate biosynthesis and metabolism were differentially expressed compared to wt plants [43], but those results have not been confirmed so far.

Our data regarding the expression of genes encoding selected enzymes of the ascorbate cycle, APX (APX1, 2, 6) and DHAR1, indicate that SnRK2.4 and SnRK2.10 also modulate the ascorbate cycle. The two kinases had a significant impact on *APXs'* expression upon salt stress, but the different *APX* genes were regulated differently. The effect of the *snrk2* mutations on the *APX1* expression profile was similar to those observed for *RbohD* or *PRX33*. At present, we do not know how the *APX2* and *APX6* up-regulation by SnRK2s affects the overall APX activity. The combined APX protein level in the mutants and wt plants exposed to the stress correlated well with the level of expression of *APX1*, but not of *APX2* or *APX6*. These data indicate that most likely APX1 has the largest share in the overall cytoplasmic APX pool, and importantly, SnRK2.4 and SnRK2.10 play a role in its regulation. Furthermore, another enzyme of the ascorbate cycle that regenerates DHAsc to Asc, DHAR1, was strongly up-regulated at both the transcript and protein levels by SnRK2.4 and slightly less by SnRK2.10 in plants exposed to salinity.

The total Asc level and the Asc/DHAsc ratio were significantly lower in the *snrk2.4* mutant in comparison with the wt plants in response to salinity, which suggests that SnRK2.4 kinase positively regulates Asc accumulation. In agreement with these data, the APX activity was higher in *snrk2.4* than in the two other lines studied. Taken together, these data indicate that SnRK2.4 plays a substantial role in the regulation of the ascorbate cycle in response to salt stress, by direct or indirect regulation of APX and DHAR1 (Figure 6).

Discussing our results, one issue should be pointed out—the role of the circadian clock in the regulation of the redox homeostasis. Circadian clocks regulate the plant growth and development as well as responses to multiple environmental cues, both biotic as well as abiotic [81,82]. Numerous genes involved in the response to osmotic stress and ABA signaling have been identified as circadian clock-dependent, including *SnRK2.6* and several genes encoding stress-responsive transcription factors (for review see [83]). It has been shown that the redox homeostasis (ROS production, scavenging, and expression of ROS-responsive genes) is tuned with diurnal and circadian rhythms, for example, the expression of *CAT1* and *CAT3* is highest at noon, whereas *CAT2* is highest at dawn [84]. On the other hand, in the feedback response, ROS signals affect clock responses [84,85]. Since our results show that SnRK2.4 and SnRK2.10 regulate the ROS level, it is highly likely that the kinases have some impact on the circadian clock. We also conjecture that expression of SnRK2.4 and/or SnRK2.10 might be regulated by a circadian rhythm. It should be stressed at this point that the expression of some genes studied by us might be affected not only by the salt, but also, to some extent, by the diurnal/circadian rhythms.

It has been reported that the transcriptomic analysis using a circadian-guided network approach might be used for identification of the genes involved in the early sensing of mild drought [86], indicating again a close relation between the stress responses and the circadian clock. Salt stress and dehydration signaling pathways have several common elements, including SnRK2s. Importantly, dehydration accompanies salt stress and it has been shown that not only ABA-activated SnRK2s, but also SnRK2, which are not activated in response to ABA, e.g., SnRK2.10 are involved in the plant response to a water deficit [39]. Therefore, when analyzing the plant response to salinity stress, as well as all other stresses, one should be aware of circadian/diurnal rhythms, which play a role in tuning those responses [84–86].

Regulation of the plant response to salt stress by SnRK2s is complex, and our knowledge on this subject is very limited. To provide the full picture, presenting the role of SnRK2s in the salt stress response, additional extensive work is required, e.g., the elucidation of the interplay between SnRK2s, ROS, circadian clock, and various signaling pathways.

A model summarizing our knowledge on the involvement of SnRK2.4 and SnRK2.10 in the plant response to salt stress is presented in Figure 7. In response to salinity stress, the kinases regulate root growth and architecture [23], mRNA decay (by phosphorylation of VCS) [24], have an impact on dehydrin ERD14 localization and likely interactions with plant membranes [39], and on ROS homeostasis (the results described here).

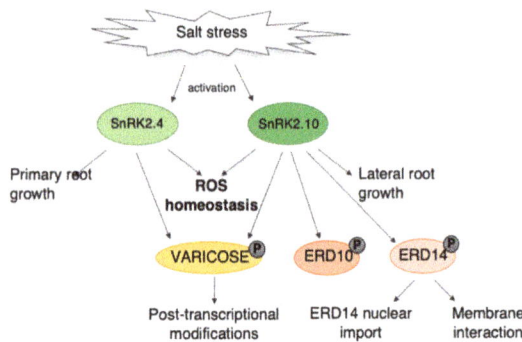

Figure 7. Schematic model illustrating the role of SnRK2.4 and SnRK2.10 in Arabidopsis' response to salt stress. SnRK2.4 and SnRK2.10 modulate root growth under the salinity conditions. Moreover, in response to salt stress, the ABA-non-activated SnRK2s phosphorylate VARICOSE (VCS), a protein participating in mRNA decay, and two dehydrins, Early Responsive to Dehydration 10 (ERD10) and ERD14. Our results presented here revealed that SnRK2.4 and SnRK2.10 regulate the ROS homeostasis in the response to salinity.

In conclusion, our data described here show that SnRK2.4 along with SnRK2.10 positively regulate the first ROS wave that transduces the salt stress signal. The kinases regulate ROS accumulation as well as ROS scavenging, by modulating the catalase level and the ascorbate cycle (Figure 6). These results suggest that the two studied SnRK2s are involved in the fine tuning of the ROS level and thus contribute to the regulation of ROS homeostasis required for the plant acclimation to unfavorable environmental conditions.

4. Materials and Methods

4.1. Plant Material, Growth, and Treatment Conditions

The following *Arabidopsis thaliana* lines were used: Arabidopsis Col-0 ecotype ("wild type"; wt); homozygous T-DNA insertion lines *snrk2.4-1* (SALK_080588), *snrk2.4-2* (SALK_146522), and *snrk2.10-1* (WiscDsLox233E9) kindly provided by Prof. C. Testerink (University of Amsterdam, The Netherlands), and *snrk2.10-2* (SAIL_698_C05) from the Nottingham Arabidopsis Stock Center (NASC). Seedlings were grown in a sterile hydroponic culture as described Kulik et al. [40] for two weeks.

For luminol-based H_2O_2 determination, plants were grown for four weeks on Jiffy pods (Jiffy-7, Jiffy Group) in a growth chamber under 8 h of light /16 h dark conditions at 21 °C/18 °C.

For ROS production measurements with H_2DCFDA, Arabidopsis seedling were grown on $\frac{1}{2}$ MS plates supplemented with 0.8% agar for five days in a growth chamber under 8 h of light/16 h dark conditions at 21 °C/18 °C.

Two- or four-week-old plants were treated with 150 mM NaCl for the indicated time (as described in the results section; stress was applied 2 h after the light was turn on), harvested by sieving, and immediately frozen in liquid nitrogen. Plant material was kept at −80 °C until further analysis.

4.2. Determination of H_2O_2

Luminol-based assay for H_2O_2 was performed according to Rasul et al. [87]. Discs of 2-mm diameter were excised from leaves of four-week-old plants using a cork borer, from 5 leaves per sample per condition, and placed into assay vials with 200 μL of MQ water, sealed with parafilm, and incubated at RT overnight. Next, stress conditions were applied (either 150 mM NaCl or MQ water as a control) and 4 μL of luminol solution [3 mM luminol dissolved in dimethyl sulfoxide (DMSO); final concentration 60 μM] was added at appropriate time points. Vials were gently mixed and luminescence was measured using a luminometer for a total time of 120 s. The measurements

Int. J. Mol. Sci. 2019, 20, 143

were performed at selected time points up to 90 min post treatment. Statistical analysis was performed using one way analysis of variance (ANOVA).

ROS detection with H_2DCFDA was performed as described previously by Kulik et al. [40] and Srivastava et al. [88] with minor modifications. Staining of five-day-old Arabidopsis seedlings roots with PI and H_2DCFDA was performed before treatment with 250 mM NaCl in $\frac{1}{2}$ MS or $\frac{1}{2}$ MS only. Single confocal sections were collected with a 20× (NA 0.75) Plan Fluor multiimmersion objective mounted on an inverted epifluorescence TE 2000E microscope (Nikon, Tokyo, Japan) coupled with an EZ-C1 confocal laser-scanning head (Nikon). H_2DCFDA fluorescence was excited with blue light at 488 nm emitted by a Sapphire 488 nm laser (Coherent, Santa Clara, CA, USA) and detected with a 515/30-nm band-pass-filter and rendered in false green, PI fluorescence was excited with green light at 543 nm emitted by a 1 mW He-Ne laser (Melles Griot, Carlsbad, CA, USA) and detected with a 610 nm long-pass filter and rendered in false magenta. All confocal parameters (laser power, gain, etc.) and conditions were the same during the experiment. EZ-C1 FreeViewer software was used to quantify the fluorescence intensity from the 4000 μm^2 area of the root meristematic zone in each Arabidopsis seedling. Each experimental variant was repeated at least twice with a total of 30 single images collected. Statistical analysis was performed using the Mann-Whitney U test.

4.3. RNA Extraction and RT-qPCR Analysis

Total RNA was extracted with TRI Reagent® according to the manufacturer's protocol (MRC). Approximately 150–200 mg of frozen ground plant material was used. DNA contamination was removed from the obtained RNA using a RapidOut DNA Removal kit (Thermo Scientific, Waltham, MA, USA). cDNA was synthesized from 4 µg of purified RNA using an Enhanced Avian HS RT-PCR Kit (Sigma-Aldrich, St. Louis, MO, USA) following the manufacturer's protocol. RT-qPCR was performed on 50 ng of the cDNA using LightCycler® 480 SYBR Green I Master Mix (Roche, Basel, Switzerland) and a Roche LightCycler® 480 machine. Relative transcript levels were calculated according to Livak and Schmittgen [89] with UBQ10 (AT4G05320) and UBC (AT5G25760) as reference genes [90,91]. Statistical analysis was performed using Student t-test. All primers used in this study are listed in Table S1.

4.4. Protein Extraction and Immunoblot Analysis

Total proteins were extracted from frozen plant samples in two volumes of extraction buffer: 100 mM HEPES, pH7.5; 5 mM EDTA; 5 mM EGTA; 10 mM DTT; 1 mM Na_3VO_4; 10 mM NaF; 50 mM β-glycerophosphate; 10 mM pyridoxal 5-phosphate; 10% glycerol; and 1 × Complete protease inhibitors (EDTA-free, Roche) on a rotator for 30 min at 4 °C and then centrifuged at 12,000 rpm for 30 min at 4 °C. Protein concentration in the supernatant was measured using a Bradford Protein Assay. The extracts were used immediately or flash-frozen and kept at −80 °C for further analysis. The immunoblot blot analysis was based on a standard procedure described by Sambrook [92]. Protein samples (7–15 µg) were separated on 12% SDS-polyacrylamide gels and transferred to Immobilon®® P membrane by electroblotting in transfer buffer, TB (25 mM Tris base, 192 mM glycine), overnight at 18 V. Transferred proteins were visualized by staining the membranes with Ponceau S (2% Ponceau S in 3% trichloroacetic acid). Immunodetection with anti-APX rabbit IgG (AS08 368, Agrisera, Vännäs, Sweden), anti-CAT rabbit IgG (AS09 501, Agrisera), and anti-DHAR1 rabbit IgG (AS11 1746, Agrisera) was performed as described in the manufacturer's protocols. Anti-glyceraldehyde-3-phosphate dehydrogenase (GAPDH) rabbit IgG (raised against the CYDDIKAAIKEESEG peptide of GAPDH; BioGenes, Berlin, Germany) was used as described previously in Wawer et al. [93]. Secondary anti-rabbit antibodies (alkaline phosphatase (AP) conjugated—AS09 607, Agrisera; horseradish peroxidase (HRP) conjugated—AS09 602, Agrisera) were visualized using appropriate substrates—5-bromo-4-chloro-3-indolyl-phosphate/nitroblue tetrazolium (BCIP/NBT, Roche) for AP, and—ECL detection reagent (Pierce™ ECL Western Blotting Substrate, Thermo Scientific) for HRP according to the manufacturer's protocol. Membranes were reused for

329

GAPDH protein detection used as a loading control for Western blots. Stripping of the membranes was performed according to Abcam online protocols.

For APX and CAT activity assays (see further), proteins were extracted from frozen plant samples (0.5 g FW) with 1 mL of ice-cold 50 mM sodium phosphate buffer, pH 7.5 or 100 mM potassium phosphate buffer, pH 7.0, respectively, containing 1 mM polyethylene glycol, 1 mM phenylmethylsulfonyl fluoride, 8% (*w/v*) polyvinylpolypyrolydone, and 0.01% (*v/v*) Triton X-100, according to Venisse et al. [94].

4.5. Determination of Ascorbate and Ascorbate/Dehydroascorbate Ratio

Ascorbate (Asc) and dehydroascorbate (DHAsc) was determined using a modified bipyridyl method described in detail by Polkowska-Kowalczyk et al. [95]. Statistical analysis was performed using the Student *t*-test and Chi-square test.

4.6. Determination of APX and CAT Activity

Ascorbate peroxidase (APX, EC 1.11.1.11) activity was assayed as described previously in Polkowska-Kowalczyk et al. [95]. Enzyme activity was expressed as μmol of oxidized ascorbate per min per mg of protein.

Catalase (CAT, EC 1.11.1.6) activity was assayed at 25 °C following the decomposition of H_2O_2 at 240 nm (extinction coefficient 0.036 mM^{-1} cm^{-1}) according to a modified method of Aebi [96]. The reaction mixture contained 50 μL of plant extract in 1 mL 50 mM potassium phosphate buffer (pH 7.0) and 9.8 mM H_2O_2. Enzyme activity was expressed as μmol H_2O_2 decomposed per min per mg of protein.

Statistical analysis was performed using the Student *t*-test.

Supplementary Materials: The following are available online at http://www.mdpi.com/1422-0067/20/1/143/s1, Table S1: List of primers used in this study.

Author Contributions: Conceptualization, K.P.S. and G.D.; Methodology, K.P.S. and L.P.-K.; Formal Analysis, K.P.S.; Investigation, K.P.S., L.P.-K. and M.L.; Resources, K.P.S. and J.M.; Visualization K.P.S.; Writing—Original Draft Preparation, G.D. and K.P.S.; Writing—Review & Editing K.P.S., L.P.-K., J.M. and G.D.; Funding Acquisition K.P.S.

Funding: This research was funded by the National Science Centre, grant number 2013/11/N/NZ1/02417 and 2015/16/T/NZ1/00164 to K.S. and Foundation for Polish Science (Program No. MPD/2009-3/2).

Acknowledgments: We kindly thank Christa Testerink for providing the *snrk2* knockout mutants seeds.

Conflicts of Interest: The authors declare no conflict of interest.

Abbreviations

APX	Ascorbate peroxidase
Asc	Ascorbate
CAT	Catalase
CDPK	Calcium-dependent protein kinase
CIPK	CBL-interacting protein kinase
DAsc	Dehydroascorbate
DHAR	Dehydroascorbate reductase
H_2O_2	Hydrogen peroxide
MPK	Mitogen activated protein kinase
$O_2{}^-$	Superoxide radical
PRX	Peroxidase
Rboh	Respiratory burst oxidase homologue
ROS	Reactive oxygen species
SnRK2	SNF-1 Related Protein Kinases type 2

References

1. Mittler, R. ROS Are Good. *Trends Plant Sci.* **2017**, *22*, 11–19. [CrossRef] [PubMed]
2. Mhamdi, A.; Van Breusegem, F. Reactive Oxygen Species in Plant Development. *Development* **2018**, *145*, dev164376. [CrossRef] [PubMed]
3. Foyer, C.H.; Ruban, A.V.; Noctor, G. Viewing Oxidative Stress through the Lens of Oxidative Signalling Rather than Damage. *Biochem. J.* **2017**, *474*, 877–883. [CrossRef] [PubMed]
4. Waszczak, C.; Carmody, M.; Kangasjärvi, J. Reactive Oxygen Species in Plant Signaling. *Annu. Rev. Plant Biol.* **2018**, *69*, 209–236. [CrossRef] [PubMed]
5. Noctor, G.; Reichheld, J.P.; Foyer, C.H. ROS-Related Redox Regulation and Signaling in Plants. *Semin. Cell Dev. Biol.* **2018**, *80*, 3–12. [CrossRef]
6. Foyer, C.H.; Noctor, G. Redox Homeostasis and Antioxidant Signaling: A Metabolic Interface between Stress Perception and Physiological Responses. *Plant Cell Online* **2005**, *17*, 1866–1875. [CrossRef] [PubMed]
7. Sofo, A.; Scopa, A.; Nuzzaci, M.; Vitti, A. Ascorbate Peroxidase and Catalase Activities and Their Genetic Regulation in Plants Subjected to Drought and Salinity Stresses. *Int. J. Mol. Sci.* **2015**, *16*, 13561–13578. [CrossRef] [PubMed]
8. Ben Rejeb, K.; Lefebvre-De Vos, D.; Le Disquet, I.; Leprince, A.S.; Bordenave, M.; Maldiney, R.; Jdey, A.; Abdelly, C.; Savouré, A. Hydrogen Peroxide Produced by NADPH Oxidases Increases Proline Accumulation during Salt or Mannitol Stress in *Arabidopsis thaliana*. *New Phytol.* **2015**, *208*, 1138–1148. [CrossRef] [PubMed]
9. Mittler, R. Oxidative Stress, Antioxidants and Stress Tolerance. *Trends Plant Sci.* **2002**, *7*, 405–410. [CrossRef]
10. Baxter, A.; Mittler, R.; Suzuki, N. ROS as Key Players in Plant Stress Signalling. *J. Exp. Bot.* **2014**, *65*, 1229–1240. [CrossRef]
11. Mittler, R.; Vanderauwera, S.; Gollery, M.; Van Breusegem, F. Reactive Oxygen Gene Network of Plants. *Trends Plant Sci.* **2004**, *9*, 490–498. [CrossRef] [PubMed]
12. Miller, G.; Suzuki, N.; Ciftci-Yilmaz, S.; Mittler, R. Reactive Oxygen Species Homeostasis and Signalling during Drought and Salinity Stresses. *Plant Cell Environ.* **2010**, *33*, 453–467. [CrossRef] [PubMed]
13. Choudhury, S.; Panda, P.; Sahoo, L.; Panda, S.K. Reactive Oxygen Species Signaling in Plants under Abiotic Stress. *Plant Signal. Behav.* **2013**, *8*, e23681. [CrossRef] [PubMed]
14. Hossain, M.S.; Dietz, K.-J. Tuning of Redox Regulatory Mechanisms, Reactive Oxygen Species and Redox Homeostasis under Salinity Stress. *Front. Plant Sci.* **2016**, *7*, 548. [CrossRef] [PubMed]
15. Inupakutika, M.A.; Sengupta, S.; Devireddy, A.R.; Azad, R.K.; Mittler, R. The Evolution of Reactive Oxygen Species Metabolism. *J. Exp. Bot.* **2016**, *67*, 5933–5943. [CrossRef] [PubMed]
16. Czarnocka, W.; Karpiński, S. Friend or Foe? Reactive Oxygen Species Production, Scavenging and Signaling in Plant Response to Environmental Stresses. *Free Radic. Biol. Med.* **2018**, *122*, 4–20. [CrossRef] [PubMed]
17. Kulik, A.; Wawer, I.; Krzywińska, E.; Bucholc, M.; Dobrowolska, G. SnRK2 Protein Kinases-Key Regulators of Plant Response to Abiotic Stresses. *OMICS* **2011**, *15*, 859–872. [CrossRef] [PubMed]
18. Boudsocq, M.; Barbier-Brygoo, H.; Laurière, C. Identification of Nine Sucrose Nonfermenting 1-Related Protein Kinases 2 Activated by Hyperosmotic and Saline Stresses in *Arabidopsis thaliana*. *J. Biol. Chem.* **2004**, *279*, 41758–41766. [CrossRef]
19. Kobayashi, Y.; Yamamoto, S.; Minami, H.; Kagaya, Y.; Hattori, T. Differential Activation of the Rice Sucrose Nonfermenting1-Related Protein Kinase2 Family by Hyperosmotic Stress and Abscisic Acid. *Plant Cell* **2004**, *16*, 1163–1177. [CrossRef]
20. Fujii, H.; Zhu, J.K. *Arabidopsis* Mutant Deficient in 3 Abscisic Acid-Activated Protein Kinases Reveals Critical Roles in Growth, Reproduction, and Stress. *Proc. Natl. Acad. Sci. USA* **2009**, *106*, 8380–8385. [CrossRef]
21. Fujita, Y.; Nakashima, K.; Yoshida, T.; Katagiri, T.; Kidokoro, S.; Kanamori, N.; Umezawa, T.; Fujita, M.; Maruyama, K.; Ishiyama, K.; et al. Three SnRK2 Protein Kinases Are the Main Positive Regulators of Abscisic Acid Signaling in Response to Water Stress in Arabidopsis. *Plant Cell Physiol.* **2009**, *50*, 2123–2132. [CrossRef] [PubMed]
22. Diédhiou, C.J.; Popova, O.V.; Dietz, K.J.; Golldack, D. The SNF1-Type Serine-Threonine Protein Kinase SAPK4 Regulates Stress-Responsive Gene Expression in Rice. *BMC Plant Biol.* **2008**, *8*, 49. [CrossRef] [PubMed]

23. McLoughlin, F.; Galvan-Ampudia, C.S.; Julkowska, M.M.; Caarls, L.; Van Der Does, D.; Laurière, C.; Munnik, T.; Haring, M.A.; Testerink, C. The Snf1-Related Protein Kinases SnRK2.4 and SnRK2.10 Are Involved in Maintenance of Root System Architecture during Salt Stress. *Plant J.* **2012**, *72*, 436–449. [CrossRef] [PubMed]

24. Soma, F.; Mogami, J.; Yoshida, T.; Abekura, M.; Takahashi, F.; Kidokoro, S.; Mizoi, J.; Shinozaki, K.; Yamaguchi-Shinozaki, K. ABA-Unresponsive SnRK2 Protein Kinases Regulate MRNA Decay under Osmotic Stress in Plants. *Nat. Plants* **2017**, *3*, 16204. [CrossRef] [PubMed]

25. Hubbard, K.E.; Nishimura, N.; Hitomi, K.; Getzoff, E.D.; Schroeder, J.I. Early Abscisic Acid Signal Transduction Mechanisms: Newly Discovered Components and Newly Emerging Questions. *Genes Dev.* **2010**, *24*, 1695–1708. [CrossRef] [PubMed]

26. Umezawa, T.; Nakashima, K.; Miyakawa, T.; Kuromori, T.; Tanokura, M.; Shinozaki, K.; Yamaguchi-Shinozaki, K. Molecular Basis of the Core Regulatory Network in ABA Responses: Sensing, Signaling and Transport. *Plant Cell Physiol.* **2010**, *51*, 1821–1839. [CrossRef] [PubMed]

27. Geiger, D.; Scherzer, S.; Mumm, P.; Stange, A.; Marten, I.; Bauer, H.; Ache, P.; Matschi, S.; Liese, A.; Al-Rasheid, K.A.S.; et al. Activity of Guard Cell Anion Channel SLAC1 Is Controlled by Drought-Stress Signaling Kinase-Phosphatase Pair. *Proc. Natl. Acad. Sci. USA* **2009**, *106*, 21425–21430. [CrossRef] [PubMed]

28. Lee, S.C.; Lan, W.; Buchanan, B.B.; Luan, S. A Protein Kinase-Phosphatase Pair Interacts with an Ion Channel to Regulate ABA Signaling in Plant Guard Cells. *Proc. Natl. Acad. Sci. USA* **2009**, *106*, 21419–21424. [CrossRef] [PubMed]

29. Sato, A.; Sato, Y.; Fukao, Y.; Fujiwara, M.; Umezawa, T.; Shinozaki, K.; Hibi, T.; Taniguchi, M.; Miyake, H.; Goto, D.B.; et al. Threonine at Position 306 of the KAT1 Potassium Channel Is Essential for Channel Activity and Is a Target Site for ABA-Activated SnRK2/OST1/SnRK2.6 Protein Kinase. *Biochem. J.* **2009**, *424*, 439–448. [CrossRef]

30. Furihata, T.; Maruyama, K.; Fujita, Y.; Umezawa, T.; Yoshida, R.; Shinozaki, K.; Yamaguchi-Shinozaki, K. Abscisic Acid-Dependent Multisite Phosphorylation Regulates the Activity of a Transcription Activator AREB1. *Proc. Natl. Acad. Sci. USA* **2006**, *103*, 1988–1993. [CrossRef]

31. Wang, P.; Xue, L.; Batelli, G.; Lee, S.; Hou, Y.-J.; Van Oosten, M.J.; Zhang, H.; Tao, W.A.; Zhu, J.-K. Quantitative Phosphoproteomics Identifies SnRK2 Protein Kinase Substrates and Reveals the Effectors of Abscisic Acid Action. *Proc. Natl. Acad. Sci. USA* **2013**, *110*, 11205–11210. [CrossRef]

32. Umezawa, T.; Sugiyama, N.; Takahashi, F.; Anderson, J.C.; Ishihama, Y.; Peck, S.C.; Shinozaki, K. Genetics and Phosphoproteomics Reveal a Protein Phosphorylation Network in the Abscisic Acid Signaling Pathway in *Arabidopsis thaliana*. *Sci. Signal.* **2013**, *6*, rs8. [CrossRef]

33. Grondin, A.; Rodrigues, O.; Verdoucq, L.; Merlot, S.; Leonhardt, N.; Maurel, C. Aquaporins Contribute to ABA-Triggered Stomatal Closure through OST1-Mediated Phosphorylation. *Plant Cell* **2015**, *27*, 1945–1954. [CrossRef]

34. Mustilli, A.-C.; Merlot, S.; Vavasseur, A.; Fenzi, F.; Giraudat, J. *Arabidopsis* OST1 Protein Kinase Mediates the Regulation of Stomatal Aperture by Abscisic Acid and Acts Upstream of Reactive Oxygen Species Production. *Plant Cell* **2002**, *14*, 3089–3099. [CrossRef]

35. Sirichandra, C.; Gu, D.; Hu, H.C.; Davanture, M.; Lee, S.; Djaoui, M.; Valot, B.; Zivy, M.; Leung, J.; Merlot, S.; et al. Phosphorylation of the *Arabidopsis* AtrbohF NADPH Oxidase by OST1 Protein Kinase. *FEBS Lett.* **2009**, *583*, 2982–2986. [CrossRef]

36. Han, J.-P.; Köster, P.; Drerup, M.M.; Scholz, M.; Li, S.; Edel, K.H.; Hashimoto, K.; Kuchitsu, K.; Hippler, M.; Kudla, J. Fine Tuning of RBOHF Activity Is Achieved by Differential Phosphorylation and Ca^{2+} Binding. *New Phytol.* **2018**. [CrossRef]

37. Mikołajczyk, M.; Awotunde, O.S.; Muszyńska, G.; Klessig, D.F.; Dobrowolska, G. Osmotic Stress Induces Rapid Activation of a Salicylic Acid-Induced Protein Kinase and a Homolog of Protein Kinase ASK1 in Tobacco Cells. *Plant Cell* **2000**, *12*, 165–178. [CrossRef]

38. Burza, A.M.; Pekala, I.; Sikora, J.; Siedlecki, P.; Małagocki, P.; Bucholc, M.; Koper, L.; Zielenkiewicz, P.; Dadlez, M.; Dobrowolska, G. *Nicotiana tabacum* Osmotic Stress-Activated Kinase Is Regulated by Phosphorylation on Ser-154 and Ser-158 in the Kinase Activation Loop. *J. Biol. Chem.* **2006**, *281*, 34299–34311. [CrossRef]

39. Maszkowska, J.; Dębski, J.; Kulik, A.; Kistowski, M.; Bucholc, M.; Lichocka, M.; Klimecka, M.; Sztatelman, O.; Szymańska, K.P.; Dadlez, M.; et al. Phosphoproteomic Analysis Reveals That Dehydrins ERD10 and ERD14 Are Phosphorylated by SNF1-Related Protein Kinase 2.10 in Response to Osmotic Stress. *Plant Cell Environ.* **2018**. [CrossRef]

40. Kulik, A.; Anielska-Mazur, A.; Bucholc, M.; Koen, E.; Szymanska, K.; Zmienko, A.; Krzywinska, E.; Wawer, I.; McLoughlin, F.; Ruszkowski, D.; et al. SNF1-Related Protein Kinases Type 2 Are Involved in Plant Responses to Cadmium Stress. *Plant Physiol.* **2012**, *160*, 868–883. [CrossRef]

41. Nakagami, H.; Soukupová, H.; Schikora, A.; Zárský, V.; Hirt, H. A Mitogen-Activated Protein Kinase Kinase Kinase Mediates Reactive Oxygen Species Homeostasis in Arabidopsis. *J. Biol. Chem.* **2006**, *281*, 38697–38704. [CrossRef]

42. Yang, L.; Ye, C.; Zhao, Y.; Cheng, X.; Wang, Y.; Jiang, Y.Q.; Yang, B. An Oilseed Rape WRKY-Type Transcription Factor Regulates ROS Accumulation and Leaf Senescence in Nicotiana Benthamiana and *Arabidopsis* through Modulating Transcription of RbohD and RbohF. *Planta* **2018**, *247*, 1323–1338. [CrossRef]

43. Pitzschke, A.; Djamei, A.; Bitton, F.; Hirt, H. A Major Role of the MEKK1-MKK1/2-MPK4 Pathway in ROS Signalling. *Mol. Plant* **2009**, *2*, 120–137. [CrossRef]

44. Ben Rejeb, K.; Benzarti, M.; Debez, A.; Bailly, C.; Savouré, A.; Abdelly, C. NADPH Oxidase-Dependent H_2O_2 Production Is Required for Salt-Induced Antioxidant Defense in *Arabidopsis thaliana*. *J. Plant Physiol.* **2015**, *174*, 5–15. [CrossRef]

45. Ma, L.; Zhang, H.; Sun, L.; Jiao, Y.; Zhang, G.; Miao, C.; Hao, F. NADPH Oxidase AtrbohD and AtrbohF Function in ROS-Dependent Regulation of Na^+/K^+ Homeostasis in *Arabidopsis* under Salt Stress. *J. Exp. Bot.* **2012**, *63*, 305–317. [CrossRef]

46. O'Brien, J.A.; Daudi, A.; Finch, P.; Butt, V.S.; Whitelegge, J.P.; Souda, P.; Ausubel, F.M.; Bolwell, G.P. A Peroxidase-Dependent Apoplastic Oxidative Burst in Cultured *Arabidopsis* Cells Functions in MAMP-Elicited Defense. *Plant Physiol.* **2012**, *158*, 2013–2027. [CrossRef]

47. Daudi, A.; Cheng, Z.; O'Brien, J.A.; Mammarella, N.; Khan, S.; Ausubel, F.M.; Bolwell, G.P. The Apoplastic Oxidative Burst Peroxidase in *Arabidopsis* Is a Major Component of Pattern-Triggered Immunity. *Plant Cell* **2012**, *24*, 275–287. [CrossRef]

48. Xing, Y.; Jia, W.; Zhang, J. AtMEK1 Mediates Stress-Induced Gene Expression of CAT1 Catalase by Triggering H_2O_2 Production in Arabidopsis. *J. Exp. Bot.* **2007**, *58*, 2969–2981. [CrossRef]

49. Frugoli, J.A.; Zhong, H.H.; Nuccio, M.L.; McCourt, P.; McPeek, M.A.; Thomas, T.L.; McClung, C.R. Catalase Is Encoded by a Multigene Family in *Arabidopsis thaliana* (L.) Heynh. *Plant Physiol.* **1996**, *112*, 327–336. [CrossRef]

50. Mhamdi, A.; Queval, G.; Chaouch, S.; Vanderauwera, S.; Van Breusegem, F.; Noctor, G. Catalase Function in Plants: A Focus on *Arabidopsis* Mutants as Stress-Mimic Models. *J. Exp. Bot.* **2010**, *61*, 4197–4220. [CrossRef]

51. Gao, M.; Liu, J.; Bi, D.; Zhang, Z.; Cheng, F.; Chen, S.; Zhang, Y. MEKK1, MKK1/MKK2 and MPK4 Function Together in a Mitogen-Activated Protein Kinase Cascade to Regulate Innate Immunity in Plants. *Cell Res.* **2008**, *18*, 1190–1198. [CrossRef]

52. Asai, S.; Ohta, K.; Yoshioka, H. MAPK Signaling Regulates Nitric Oxide and NADPH Oxidase-Dependent Oxidative Bursts in Nicotiana Benthamiana. *Plant Cell Online* **2008**, *20*, 1390–1406. [CrossRef]

53. Dubiella, U.; Seybold, H.; Durian, G.; Komander, E.; Lassig, R.; Witte, C.-P.; Schulze, W.X.; Romeis, T. Calcium-Dependent Protein Kinase/NADPH Oxidase Activation Circuit Is Required for Rapid Defense Signal Propagation. *Proc. Natl. Acad. Sci. USA* **2013**, *110*, 8744–8749. [CrossRef]

54. Kadota, Y.; Sklenar, J.; Derbyshire, P.; Stransfeld, L.; Asai, S.; Ntoukakis, V.; Jones, J.D.; Shirasu, K.; Menke, F.; Jones, A.; et al. Direct Regulation of the NADPH Oxidase RBOHD by the PRR-Associated Kinase BIK1 during Plant Immunity. *Mol. Cell* **2014**, *54*, 43–55. [CrossRef]

55. Li, L.; Li, M.; Yu, L.; Zhou, Z.; Liang, X.; Liu, Z.; Cai, G.; Gao, L.; Zhang, X.; Wang, Y.; et al. The FLS2-Associated Kinase BIK1 Directly Phosphorylates the NADPH Oxidase RbohD to Control Plant Immunity. *Cell Host Microbe* **2014**, *15*, 329–338. [CrossRef]

56. Monaghan, J.; Matschi, S.; Shorinola, O.; Rovenich, H.; Matei, A.; Segonzac, C.; Malinovsky, F.G.G.; Rathjen, J.P.P.; Maclean, D.; Romeis, T.; et al. The Calcium-Dependent Protein Kinase CPK28 Buffers Plant Immunity and Regulates BIK1 Turnover. *Cell Host Microbe* **2014**, *16*, 605–615. [CrossRef]

57. Monaghan, J.; Matschi, S.; Romeis, T.; Zipfel, C. The Calcium-Dependent Protein Kinase CPK28 Negatively Regulates the BIK1-Mediated PAMP-Induced Calcium Burst. *Plant Signal. Behav.* **2015**, *10*, e1018497. [CrossRef]

58. Liu, Y.; He, C. A Review of Redox Signaling and the Control of MAP Kinase Pathway in Plants. *Redox Biol.* **2017**, *11*, 192–204. [CrossRef]

59. Kawasaki, T.; Yamada, K.; Yoshimura, S.; Yamaguchi, K. Chitin Receptor-Mediated Activation of MAP Kinases and ROS Production in Rice and Arabidopsis. *Plant Signal. Behav.* **2017**, *12*, e1361076. [CrossRef]

60. Zhang, M.; Chiang, Y.H.; Toruño, T.Y.; Lee, D.H.; Ma, M.; Liang, X.; Lal, N.K.; Lemos, M.; Lu, Y.J.; Ma, S.; et al. The MAP4 Kinase SIK1 Ensures Robust Extracellular ROS Burst and Antibacterial Immunity in Plants. *Cell Host Microbe* **2018**, *24*, 379–391.e5. [CrossRef]

61. Zandalinas, S.I.; Mittler, R. ROS-Induced ROS Release in Plant and Animal Cells. *Free Radic. Biol. Med.* **2018**, *122*, 21–27. [CrossRef]

62. Fujii, H.; Verslues, P.E.; Zhu, J.-K. *Arabidopsis* Decuple Mutant Reveals the Importance of SnRK2 Kinases in Osmotic Stress Responses in Vivo. *Proc. Natl. Acad. Sci. USA* **2011**, *108*, 1717–1722. [CrossRef]

63. Drerup, M.M.; Schlücking, K.; Hashimoto, K.; Manishankar, P.; Steinhorst, L.; Kuchitsu, K.; Kudla, J. The Calcineurin B-like Calcium Sensors CBL1 and CBL9 Together with Their Interacting Protein Kinase CIPK26 Regulate the *Arabidopsis* NADPH Oxidase RBOHF. *Mol. Plant* **2013**, *6*, 559–569. [CrossRef]

64. Kimura, S.; Kawarazaki, T.; Nibori, H.; Michikawa, M.; Imai, A.; Kaya, H.; Kuchitsu, K. The CBL-Interacting Protein Kinase CIPK26 Is a Novel Interactor of *Arabidopsis* NADPH Oxidase AtRbohF That Negatively Modulates Its ROS-Producing Activity in a Heterologous Expression System. *J. Biochem.* **2013**, *153*, 191–195. [CrossRef]

65. Takahashi, F.; Mizoguchi, T.; Yoshida, R.; Ichimura, K.; Shinozaki, K. Calmodulin-Dependent Activation of MAP Kinase for ROS Homeostasis in Arabidopsis. *Mol. Cell* **2011**, *41*, 649–660. [CrossRef]

66. Hrabak, E.M.; Chan, C.W.; Gribskov, M.; Harper, J.; Choi, J.; Halford, N.; Kudla, J.; Luan, S.; Nimmo, H.; Sussman, M.; et al. The *Arabidopsis* CDPK-SnRK Superfamily of Protein Kinases. *Plant Physiol.* **2003**, *132*, 666–680. [CrossRef]

67. Xie, Y.J.; Xu, S.; Han, B.; Wu, M.Z.; Yuan, X.X.; Han, Y.; Gu, Q.; Xu, D.K.; Yang, Q.; Shen, W.B. Evidence of *Arabidopsis* Salt Acclimation Induced by Up-Regulation of HY1 and the Regulatory Role of RbohD-Derived Reactive Oxygen Species Synthesis. *Plant J.* **2011**, *66*, 280–292. [CrossRef]

68. Arnaud, D.; Lee, S.; Takebayashi, Y.; Choi, D.; Choi, J.; Sakakibara, H.; Hwang, I. Cytokinin-Mediated Regulation of Reactive Oxygen Species Homeostasis Modulates Stomatal Immunity in Arabidopsis. *Plant Cell* **2017**, *29*, 543–559. [CrossRef]

69. Meng, X.; Xu, J.; He, Y.; Yang, K.-Y.; Mordorski, B.; Liu, Y.; Zhang, S. Phosphorylation of an ERF Transcription Factor by *Arabidopsis* MPK3/MPK6 Regulates Plant Defense Gene Induction and Fungal Resistance. *Plant Cell* **2013**, *25*, 1126–1142. [CrossRef]

70. Sewelam, N.; Kazan, K.; Thomas-Hall, S.R.; Kidd, B.N.; Manners, J.M.; Schenk, P.M. Ethylene Response Factor 6 Is a Regulator of Reactive Oxygen Species Signaling in Arabidopsis. *PLoS ONE* **2013**, *8*, e70289. [CrossRef]

71. Wang, P.; Du, Y.; Zhao, X.; Miao, Y.; Song, C.-P. The MPK6-ERF6-ROS-Responsive Cis-Acting Element7/GCC Box Complex Modulates Oxidative Gene Transcription and the Oxidative Response in Arabidopsis. *Plant Physiol.* **2013**, *161*, 1392–1408. [CrossRef]

72. Adachi, H.; Nakano, T.; Miyagawa, N.; Ishihama, N.; Yoshioka, M.; Katou, Y.; Yaeno, T.; Shirasu, K.; Yoshioka, H. WRKY Transcription Factors Phosphorylated by MAPK Regulate a Plant Immune NADPH Oxidase in Nicotiana Benthamiana. *Plant Cell* **2015**, *27*, 2645–2663. [CrossRef]

73. Ichimura, K.; Mizoguchi, T.; Yoshida, R.; Yuasa, T.; Shinozaki, K. Various Abiotic Stresses Rapidly Activate *Arabidopsis* MAP Kinases ATMPK4 and ATMPK6. *Plant J.* **2000**, *24*, 655–665. [CrossRef]

74. Dubois, M.; Skirycz, A.; Claeys, H.; Maleux, K.; Dhondt, S.; De Bodt, S.; Vanden Bossche, R.; De Milde, L.; Yoshizumi, T.; Matsui, M.; et al. Ethylene Response FACTOR6 Acts as a Central Regulator of Leaf Growth under Water-Limiting Conditions in Arabidopsis. *Plant Physiol.* **2013**, *162*, 319–332. [CrossRef]

75. Dal Santo, S.; Stampfl, H.; Krasensky, J.; Kempa, S.; Gibon, Y.; Petutschnig, E.; Rozhon, W.; Heuck, A.; Clausen, T.; Jonak, C. Stress-Induced GSK3 Regulates the Redox Stress Response by Phosphorylating Glucose-6-Phosphate Dehydrogenase in Arabidopsis. *Plant Cell* **2012**, *24*, 3380–3392. [CrossRef]

76. Zong, X.J.; Li, D.P.D.Q.; Gu, L.K.; Li, D.P.D.Q.; Liu, L.X.; Hu, X.L. Abscisic Acid and Hydrogen Peroxide Induce a Novel Maize Group C MAP Kinase Gene, ZmMPK7, Which Is Responsible for the Removal of Reactive Oxygen Species. *Planta* **2009**, *229*, 485–495. [CrossRef]

77. Xing, Y.; Jia, W.; Zhang, J. AtMKK1 Mediates ABA-Induced CAT1 Expression and H2O2 production via AtMPK6-Coupled Signaling in Arabidopsis. *Plant J.* **2008**, *54*, 440–451. [CrossRef]

78. Verslues, P.E.; Batelli, G.; Grillo, S.; Agius, F.; Kim, Y.-S.; Zhu, J.-K.; Agarwal, M.; Katiyar-Agarwal, S.; Zhu, J.-K. Interaction of SOS$_2$ with Nucleoside Diphosphate Kinase 2 and Catalases Reveals a Point of Connection between Salt Stress and H$_2$O$_2$ Signaling in *Arabidopsis thaliana*. *Mol. Cell. Biol.* **2007**, *27*, 7771–7780. [CrossRef]

79. Galvez-Valdivieso, G.; Fryer, M.J.; Lawson, T.; Slattery, K.; Truman, W.; Smirnoff, N.; Asami, T.; Davies, W.J.; Jones, A.M.; Baker, N.R.; et al. The High Light Response in *Arabidopsis* Involves ABA Signaling between Vascular and Bundle Sheath Cells. *Plant Cell Online* **2009**, *21*, 2143–2162. [CrossRef]

80. Pitzschke, A.; Hirt, H. Disentangling the Complexity of Mitogen-Activated Protein Kinases and Reactive Oxygen Species Signaling. *Plant Physiol.* **2009**, *149*, 606–615. [CrossRef]

81. Hotta, C.T.; Gardner, M.J.; Hubbard, K.E.; Baek, S.J.; Dalchau, N.; Suhita, D.; Dodd, A.N.; Webb, A.A.R. Modulation of Environmental Responses of Plants by Circadian Clocks. *Plant Cell Environ.* **2007**, *33*, 333–349. [CrossRef]

82. Bhardwaj, V.; Meier, S.; Petersen, L.N.; Ingle, R.A.; Roden, L.C. Defence Responses of *Arabidopsis thaliana* to Infection by Pseudomonas Syringae Are Regulated by the Circadian Clock. *PLoS ONE* **2011**, *6*, e26968. [CrossRef] [PubMed]

83. Seung, D.; Risopatron, J.P.M.; Jones, B.J.; Marc, J. Circadian Clock-Dependent Gating in ABA Signalling Networks. *Protoplasma* **2012**, *249*, 445–457. [CrossRef] [PubMed]

84. Lai, A.G.; Doherty, C.J.; Mueller-Roeber, B.; Kay, S.A.; Schippers, J.H.M.; Dijkwel, P.P. Circadian Clock-Associated 1 Regulates ROS Homeostasis and Oxidative Stress Responses. *Proc. Natl. Acad. Sci. USA* **2012**, *109*, 17129–17134. [CrossRef] [PubMed]

85. Li, Z.; Bonaldi, K.; Uribe, F.; Pruneda-Paz, J.L. A Localized Pseudomonas Syringae Infection Triggers Systemic Clock Responses in Arabidopsis. *Curr. Biol.* **2018**, *28*, 630–639.e4. [CrossRef] [PubMed]

86. Greenham, K.; Guadagno, C.R.; Gehan, M.A.; Mockler, T.C.; Weinig, C.; Ewers, B.E.; McClung, C.R. Temporal Network Analysis Identifies Early Physiological and Transcriptomic Indicators of Mild Drought in Brassica Rapa. *Elife* **2017**, *6*, e29655. [CrossRef] [PubMed]

87. Rasul, S.; Dubreuil-Maurizi, C.; Lamotte, O.; Koen, E.; Poinssot, B.; Alcaraz, G.; Wendehenne, D.; Jeandroz, S. Nitric Oxide Production Mediates Oligogalacturonide-Triggered Immunity and Resistance to Botrytis Cinerea in *Arabidopsis thaliana*. *Plant Cell Environ.* **2012**, *35*, 1483–1499. [CrossRef]

88. Srivastava, A.K.; Sablok, G.; Hackenberg, M.; Deshpande, U.; Suprasanna, P. Thiourea Priming Enhances Salt Tolerance through Co-Ordinated Regulation of MicroRNAs and Hormones in Brassica Juncea. *Sci. Rep.* **2017**, *7*, 1–15. [CrossRef] [PubMed]

89. Livak, K.J.; Schmittgen, T.D. Analysis of Relative Gene Expression Data Using Real-Time Quantitative PCR and the 2-ΔΔCT Method. *Methods* **2001**, *25*, 402–408. [CrossRef]

90. Czechowski, T. Genome-Wide Identification and Testing of Superior Reference Genes for Transcript Normalization in Arabidopsis. *Plant Physiol.* **2005**, *139*, 5–17. [CrossRef]

91. Remans, T.; Smeets, K.; Opdenakker, K.; Mathijsen, D.; Vangronsveld, J.; Cuypers, A. Normalisation of Real-Time RT-PCR Gene Expression Measurements in *Arabidopsis thaliana* Exposed to Increased Metal Concentrations. *Planta* **2008**, *227*, 1343–1349. [CrossRef] [PubMed]

92. Sambrook, J.; Fritsch, E.F.; Maniatis, T. *Molecular Cloning: A Laboratory Manual*; Cold Spring Harbor Laboratory Press: Cold Spring Harbor, NY, USA, 1989.

93. Wawer, I.; Bucholc, M.; Astier, J.; Anielska-Mazur, A.; Dahan, J.; Kulik, A.; Wysłouch-Cieszynska, A.; Zaręba-Kozioł, M.; Krzywinska, E.; Dadlez, M.; et al. Regulation of *Nicotiana tabacum* Osmotic Stress-Activated Protein Kinase and Its Cellular Partner GAPDH by Nitric Oxide in Response to Salinity. *Biochem. J.* **2010**, *429*, 73–83. [CrossRef] [PubMed]

94. Venisse, J.S.; Gullner, G.; Brisset, M.N. Evidence for the Involvement of an Oxidative Stress in the Initiation of Infection of Pear by Erwinia Amylovora. *Plant Physiol.* **2001**, *125*, 2164–2172. [CrossRef] [PubMed]

95. Polkowska-Kowalczyk, L.; Wielgat, B.; Maciejewska, U. Changes in the Antioxidant Status in Leaves of Solanum Species in Response to Elicitor from Phytophthora Infestans. *J. Plant Physiol.* **2007**, *164*, 1268–1277. [CrossRef] [PubMed]
96. Aebi, H. Catalase in Vitro. *Methods Enzymol.* **1984**, *105*, 121–126. [PubMed]

International Journal of
Molecular Sciences

MDPI

Article

The *Arabidopsis* Ca^{2+}-Dependent Protein Kinase CPK12 Is Involved in Plant Response to Salt Stress

Huilong Zhang [1,†], Yinan Zhang [1,†], Chen Deng [1,†], Shurong Deng [1], Nianfei Li [1], Chenjing Zhao [1], Rui Zhao [1,*], Shan Liang [2,*] and Shaoliang Chen [1]

[1] Beijing Advanced Innovation Center for Tree Breeding by Molecular Design, College of Biological Sciences and Technology, Beijing Forestry University, Beijing 100083, China; hlzhang2018@126.com (H.Z.); xhzyn007@163.com (Y.Z.); ced501@163.com (C.D.); danceon@126.com (S.D.); nl1669@nyu.edu (N.L.); 1120170396@mail.nankai.edu.cn (C.Z.); lschen@bjfu.edu.cn (S.C.)

[2] Beijing Advanced Innovation Center for Food Nutrition and Human Health, School of Food and Chemical Engineering, Beijing Technology and Business University, Beijing 100048, China

[*] Correspondence: ruizhao926@126.com (R.Z.); liangshan@btbu.edu.cn (S.L.); Tel.: +86-10-6233-8129 (R.Z.); +86-10-6233-8129 (S.L.)

[†] These authors contributed equally to this work.

Received: 1 December 2018; Accepted: 12 December 2018; Published: 14 December 2018

Abstract: CDPKs (Ca^{2+}-Dependent Protein Kinases) are very important regulators in plant response to abiotic stress. The molecular regulatory mechanism of CDPKs involved in salt stress tolerance remains unclear, although some CDPKs have been identified in salt-stress signaling. Here, we investigated the function of an *Arabidopsis* CDPK, CPK12, in salt-stress signaling. The *CPK12*-RNA interference (RNAi) mutant was much more sensitive to salt stress than the wild-type plant GL1 in terms of seedling growth. Under NaCl treatment, Na$^+$ levels in the roots of *CPK12*-RNAi plants increased and were higher than levels in GL1 plants. In addition, the level of salt-elicited H$_2$O$_2$ production was higher in *CPK12*-RNAi mutants than in wild-type GL1 plants after NaCl treatment. Collectively, our results suggest that CPK12 is required for plant adaptation to salt stress.

Keywords: *Arabidopsis*; CDPK; ion homeostasis; NMT; ROS; salt stress

1. Introduction

Saline soil cannot be used for agriculture and forestry production [1], and soil salinity is a major abiotic stress for plants worldwide [2,3]. When plants suffer from salt environments, the accumulation of sodium and chloride ions breaks the ion balance and causes secondary stress, such as oxidative bursts [4,5].

Plants have evolved sophisticated regulatory mechanisms to avoid and acclimate to salt stress and repair related damage, processes based on morphological, physiological, biochemical and molecular changes [6]. Salt overly sensitive (SOS) signaling is the most important pathway for regulating plant adaptation to salt stress [4,7]. In *Arabidopsis*, salt-induced increases in cytoplasmic calcium (Ca^{2+}) are sensed by the EF-hand–type Ca^{2+}-binding protein SOS3. Ca^{2+} together with SOS3 activates SOS2, a serine/threonine protein kinase. Activated SOS2 phosphorylates and stimulates the activity of SOS1, a plasma membrane–localized Na$^+$/H$^+$ antiporter, leading to regulation of ion homeostasis during salt stress [8–11]. A Na$^+$/H$^+$ exchanger, which is localized to plasma membrane, also plays an important role in *Populus euphratica*, the roots of which exhibit a strong capacity to extrude Na$^+$ under salt stress; furthermore, the protoplasts from root display enhanced Na$^+$/H$^+$ transport activity [12]. In addition, wheat *Nax1* and *Nax2* affect activity and expression levels of the SOS1-like Na$^+$/H$^+$ exchanger in both root cortical and stellar tissues [13].

Salt stress increases the production of reactive oxygen species (ROS), which plays a dual role in plants: they function as toxic byproducts of metabolism and as important signal transduction molecules [14–16]. Peroxisomes and chloroplasts are the major organelles of ROS generation [17–19], and plants eliminate ROS through non-enzymatic and enzymatic scavenging mechanisms [20]. Non-enzymatic antioxidants include the major cellular redox buffers glutathione and ascorbate, as well as flavonoids, carotenoids, alkaloids, and tocopherol [21]. Enzymatic ROS scavenging pathways in plants include superoxide dismutase (SOD), ascorbate peroxidase (APX), glutathione peroxidase (GPX), and catalase (CAT) [20]. H_2O_2 is the end-product of SOD, which is harmful to DNA, proteins, and lipids [20]. Halophytes can send stress signals quickly through H_2O_2, and have an efficient antioxidant ability to scavenge H_2O_2 upon completion of signaling [22]. In addition, H_2O_2 is a signaling molecule in the plant response to salt stress [23,24]. *P. euphratica* responds to salt stress with rapid H_2O_2 production, and exogenous H_2O_2 application enhances the Na^+/H^+ exchange [25]. Pharmacological experiments have strongly indicated that NaCl-induced Na^+/H^+ antiport is inhibited when H_2O_2 is absent [25], and H_2O_2-regulated K^+/Na^+ homeostasis in the salt-stressed plant is Ca^{2+}-dependent [25]. Exogenous H_2O_2 causes elevated cytosolic Ca^{2+} [25], which stimulates plasma membrane–localized Na^+/H^+ antiporters through the SOS signaling pathway [2,4,5]. Furthermore, H_2O_2 mediates increased *SOS1* mRNA stability in *Arabidopsis* and may therefore contribute to cellular Na^+ protection [26].

Ca^{2+} is a conserved second messenger in plant growth and development pathways and contributes to plant adaptations to environmental challenges [27,28]. In plants, calmodulin (CAM), calcineurin B-like proteins (CBL), and Ca^{2+}-dependent protein kinases (CDPKs) are important Ca^{2+} sensors [29–33]. For CDPKs, *Arabidopsis* has 34 members, rice (*Oryza sativa*) has 29 members, wheat (*Triticum aestivum*) has 20 members, and poplar (*Populus trichocarpa*) has 30 members [34–37]. In recent years, CDPKs have been characterized as playing an important function in mediating stress-signaling networks [38–40].

Genetic and biochemical evidence implicates several CDPKs in plant adaptations to environmental stress. *Arabidopsis* CPK32 (Ca^{2+}-Dependent Protein Kinase 32) phosphorylates ABF4 (ABRE Binding Factor 4) to participate in abscisic acid (ABA) signaling [41]. CPK4 and CPK11 are important positive regulators mediating ABA signaling pathways [42], but their homolog, CPK12, plays a negative role in this signaling [43,44]. CPK10 interacts with HSP1, which contributes to plant drought responses by modulating signaling through ABA and Ca^{2+} [45]. CPK23 responds to drought and salt stresses, and together with CPK21 constitutes a pair of critical Ca^{2+}-dependent regulators of the guard cell anion channel SLAC1 (Slow Anion Channel-Associated 1) in ABA signaling [46–48]. CPK3 and CPK6 positively regulate ABA signaling in stomatal movement [49–51], and CPK6 functions as a positive regulator of methyl jasmonate signaling in guard cells [52]. CPK13 inhibits opening of the stomata through its inhibition of guard cell–expressed KAT2 (K^+ transporter 2) and KAT1 (K^+ transporter 1) channels [53]. The expression of *CPK27* is induced by NaCl, and the *cpk27-1* mutant is much more sensitive to salt stress than wild-type plants in terms of seed germination and post-germination seedling growth [54].

Overexpression of rice *CPK21* enhances rice's capacity to tolerate high salinity [55]. OsCPK12 reduces ROS levels to regulate salt tolerance [56] and plays a positive role in plant responses to drought, osmotic stress, and dehydration [57]. In maize (*Zea mays*), ZmCCaMK is required for ABA-induced antioxidant defense, and the ABA-induced activation of ZmCCaMK is required for H_2O_2-dependent nitric oxide production [58]. ZmCPK4 positively regulates ABA signaling and enhances drought stress tolerance in *Arabidopsis* [59], while ZmCPK11 functions upstream of ZmMPK5 and regulates ABA-induced antioxidant defense [60]. The expression of *PeCPK10*, a gene cloned from *P. euphratica*, is induced by salt, drought and cold treatments, overexpression of *PeCPK10* in *Arabidopsis* improves the plant's tolerance of freezing [61]. In grape berry, ABA stimulates ACPK1 (ABA-stimulated calcium-dependent protein kinase1), which is potentially involved in ABA signaling [62]. Heterologous overexpression of ACPK1 in *Arabidopsis* promotes significant plant growth and enhances ABA sensitivity in seed germination, early seedling growth, and stomatal movement, providing evidence that ACPK1 is involved in ABA signal transduction as a positive regulator [63].

Although the functions of CDPKs in plant response to environmental stress have been demonstrated, the molecular biological mechanisms of CDPKs remain unclear. Previously, we reported that *Arabidopsis* CPK12 negatively regulates ABA signaling [43,44]. Here, we show that CPK12 mediates salt stress tolerance by regulating ion homeostasis and H_2O_2 production. Down-regulation of *CPK12* results in salt hypersensitivity in seedling growth and accumulation of higher levels of Na^+ and H_2O_2. Our results show that CPK12 may modulate salt stress tolerance in *Arabidopsis*.

2. Results

2.1. Identification of RNA Interference (RNAi) Mutants of CPK12

We previously identified the function of *Arabidopsis CPK12*, generated *CPK12*-RNAi lines, and observed that down-regulation of *CPK12* results in ABA hypersensitivity in seed germination and post-germination growth. CPK12 interacted with and phosphorylated and stimulated the type 2C protein phosphatase ABI2. In addition, CPK12 together with ABI2 negatively regulates ABA signal transduction [43]. Thus, we wondered whether *CPK12* was involved in salt-stress signaling. To address this question, we re-generated *CPK12*-RNAi lines and selected four (lines *R1*, *R4*, *R7*, *R8*) as examples for this study. Expression of *CPK12* was down-regulated in these RNAi lines, and the level of *CPK12* mRNA gradually decreased from line *R8* to line *R1*, creating a gradient of *CPK12* expression levels (Figure 1). In addition, the expression of a control gene *EF-1α* (*Elongation Factor-1α*), which is not related to salt-stress, was not affected in those *CPK12*-RNAi lines (Figure 1).

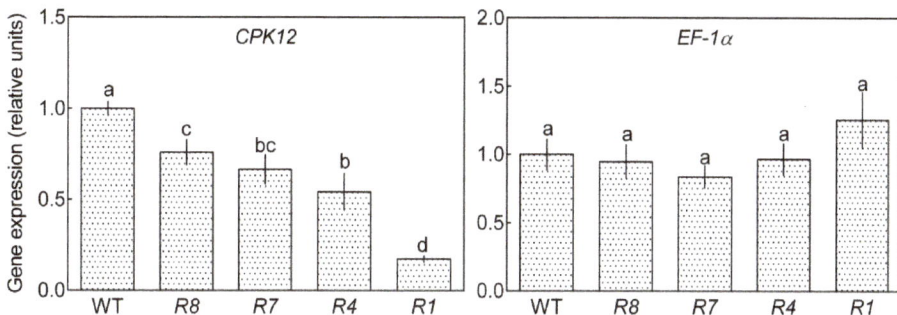

Figure 1. The expression of *CPK12* (*Ca²⁺-Dependent Protein Kinase 12*) and *EF-1α* (*Elongation Factor-1α*) in WT and *CPK12*-RNAi mutant. The mRNA levels (relative units, normalized relative to the mRNA level of the wild-type GL1 taken as 100%) of *CPK12* and *EF-1α*, estimated by qRT-PCR, in the non-transgenic GL1 (WT) and four different transgenic *CPK12*-RNAi lines (indicated by *R8*, *R7*, *R4*, and *R1*). Values are mean ± standard error from three independent experiments. Columns labeled with different letters indicate significant differences at *p* < 0.05.

2.2. Down-Regulation of CPK12 Results in NaCl Hypersensitivity in Seedling Growth

We next examined whether CPK12 affects seedling growth under salt stress. Seeds from the *CPK12*-RNAi mutant and GL1 plants were sown on medium containing various concentrations of NaCl. The presence of 110–150 mM NaCl inhibited *CPK12*-RNAi mutant growth. In addition, the cotyledons of the *CPK12*-RNAi mutants were chlorotic compared with GL1 seedlings (Figure 2).

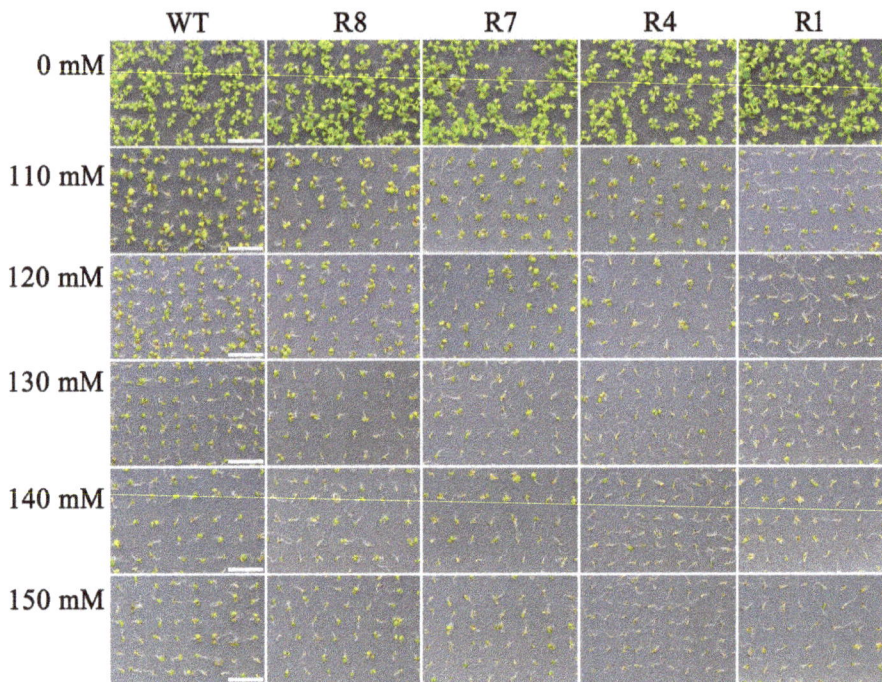

Figure 2. Down-regulation of *CPK12* results in NaCl-hypersensitive seedling growth. Seeds were planted in NaCl-free (0 mM) medium or media containing 110, 120, 130, 140, or 150 mM NaCl, and seedling growth was investigated 10 days after stratification. Scale bars, 1 cm.

CPK12-RNAi mutants exhibited the same post-germination seedling growth status as GL1 plants in the free of NaCl medium; however, compared with GL1 plants, NaCl suppressed root growth of *CPK12*-RNAi mutants more strongly, the reduction in the growth of salt-stressed *CPK12*-RNAi seedlings was more pronounced than for GL1 seedling (Figure 3). Taken together, these results suggest that *CPK12* is involved in salt-stress tolerance in *Arabidopsis*.

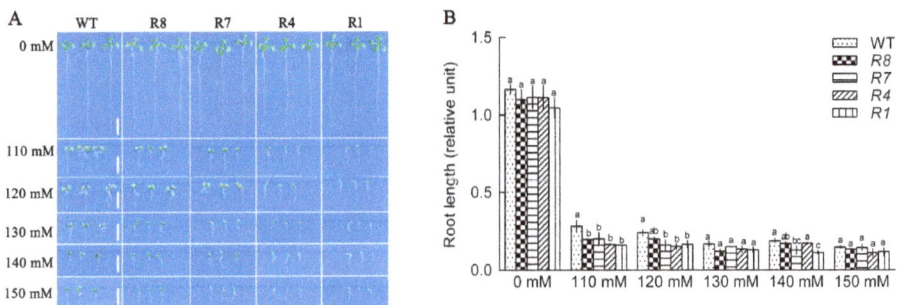

Figure 3. Down-regulation of *CPK12* results in NaCl-hypersensitive root growth. (**A**) Seeds were planted in NaCl-free (0 mM) medium or media containing 110, 120, 130, 140, 150, or mM NaCl, and root growth was detected 10 days after stratification. Scale bars, 0.5 cm. (**B**) Root lengths are mean ± standard error from three independent experiments. Thirty plants were measured for each genotype in each treatment. The mean values of root lengths are labeled with letters in the same group to denote significant differences (*p* < 0.05).

2.3. Salt Stress Induced the Ca^{2+} Elevation in Root Tissue

We examined the Ca^{2+} level in the *CPK12*-RNAi plants and wild type plants GL1 after salt stress using the Ca^{2+} specific probe, Rhod-2 AM. In the absence of salt treatment, the relative fluorescence intensity was not significantly different between GL1 and *CPK12*-RNAi plants, except the R1 line, probably due to the lowest expression of *CPK12* in R1 line. Under salt treatment, as expected, the Ca^{2+} levels in the roots of *CPK12*-RNAi plants and GL1 increased (Figure 4).

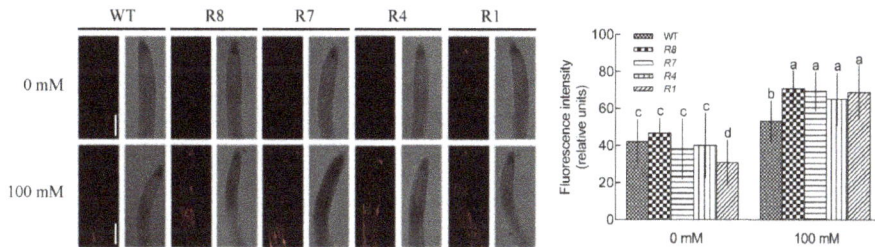

Figure 4. Ca^{2+} levels within roots of GL1 and *CPK12*-RNAi plants. (**A**) Seven-day-old seedlings were transferred to MS medium supplemented with (100 mM) or without NaCl (0 mM) for 12 h, then stained with the Ca^{2+}-specific fluorescent probe Rhod-2 AM for 1 h at room temperature. Orange-red fluorescence within cells was detected at the apical region of roots under a Leica confocal microscope. Representative confocal images show cytosolic Ca^{2+} content in plant roots. Scale bars, 100 μm. (**B**) The relative fluorescence intensity (±SD) represents the mean of 10 independent seedlings. The mean values of Ca^{2+} fluorescence are labeled with letters in the same group to denote significant differences ($p < 0.05$).

2.4. Down-Regulation of CPK12 Leads to Na^+ Accumulation in Root Tissue

To investigate the cause of the observed hypersensitivity of *CPK12*-RNAi mutants to salt stress, sodium accumulation in root cells was examined using the sodium-specific dye CoroNa-Green. Under no salt stress, CoroNa-Green fluorescence was almost undetectable in the root cells of *CPK12*-RNAi and wild-type plants because of low Na^+ content in root cells. Under NaCl treatment, Na^+ levels in the roots of *CPK12*-RNAi plants increased and were higher than levels in GL1 plants (Figure 5).

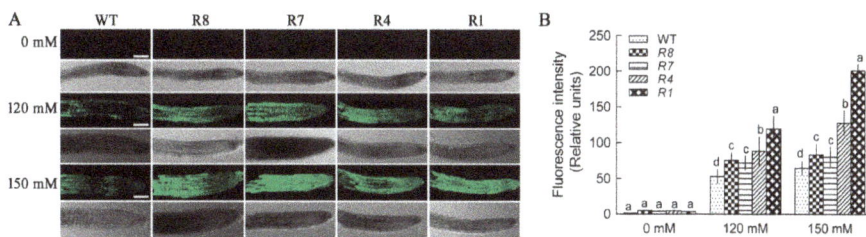

Figure 5. Na^+ levels in root cells of wild-type (WT) GL1 and *CPK12*-RNAi plants under salt stress. (**A**) Seven-day-old seedlings were transferred to MS medium supplemented with (120 mM, 150 mM) or without NaCl (0 mM) for 12 h, then seedlings were treated with CoroNa-Green AM (green fluorescence, sodium-specific) for 1 h. Green fluorescence in root cells was observed at the apical region of roots using a Leica confocal microscope. Typical images show Na^+ content in plant roots. Scale bars, 100 μm. (**B**) The mean relative fluorescence values marked with letters in the same group represent significant differences ($p < 0.05$).

To determine whether CPK12 contributed to the regulation of ion homeostasis, NMT (Non-invasive Micro-test Technique) was used to record root Na^+ fluxes in the *CPK12*-RNAi plants and GL1 plants under a long-term NaCl treatment (0, 110, 120, 130 mM; 7 d). In the absence of salt stress, Na$^+$ efflux in the apical region of roots was not significant between *CPK12*-RNAi plants and wild type GL1 plants. However, long-term salt treatment caused a rise in Na$^+$ efflux in GL1 plants and *CPK12*-RNAi mutants, but this was more pronounced in GL1 plants than in *CPK12*-RNAi plants (Figure 6). These observations indicate that CPK12 participates in salt-stress tolerance by regulating ion balance in root tissue.

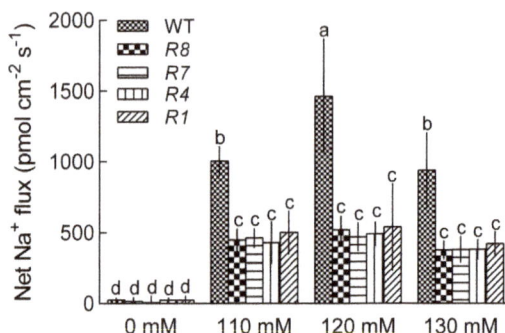

Figure 6. Na$^+$ flux in GL1 and *CPK12*-RNAi plants. Seeds were germinated for one week in a vertical direction on MS agar medium containing 0, 110, 120, 130 mM NaCl. Continuous NMT recording were applied at the meristem region of the root tips. Each column is the mean of six independent seedlings; bars show the standard error of the mean. Columns marked with letters in the same group indicate significant differences at $p < 0.05$.

2.5. Down-Regulation of CPK12 Results in H$_2$O$_2$ Burst and Accumulation

ROS accumulates when plants are exposed to salt stress, so we investigated H$_2$O$_2$ levels in *CPK12*-RNAi plants using a H$_2$O$_2$-specific fluorescent probe, H$_2$DCF-DA. In the NaCl shock condition, the levels of H$_2$O$_2$ in *CPK12*-RNAi plants were higher than GL1 plants (Figure 7). The salt stress-induced H$_2$O$_2$ accumulation in *Arabidopsis* was also detected after 12 h or 24 h treatment; compared with GL1 plants, the level of H$_2$O$_2$ in *CPK12*-RNAi plants was significantly higher after exposure to high NaCl concentrations (Figure 7).

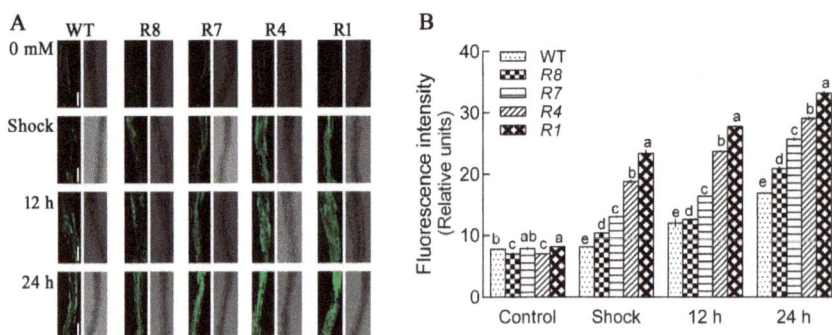

Figure 7. Accumulation of H$_2$O$_2$ in the root tips of GL1 and *CPK12*-RNAi plants exposed to salt stress. (**A**) After germinating seven days, the *Arabidopsis* seedlings were transferred to MS medium containing 0 or 100 mM NaCl for 10 min, 12 h, or 24 h. These seedlings were incubated with H$_2$DCF-DA for 5 min. The green fluorescence within cells at the apical region of roots was detected using a Leica confocal microscope. Scale bars, 100 μm. (**B**) The relative fluorescence intensity (±SD) represents the mean of 10 *Arabidopsis* seedlings. The mean values of H$_2$O$_2$ fluorescence are labeled with letters in the same group to denote significant differences ($p < 0.05$).

We measured the activities of antioxidant enzymes, such as SOD, CAT, and APX, in GL1 and *CPK12*-RNAi plants. In the absence of salt treatment, the activities of SOD and CAT were not significantly different between GL1 and *CPK12*-RNAi plants, but the activity of APX in *CPK12*-RNAi plants was lower than GL1. Under salt stress conditions, the activity of SOD in GL1 was higher than *CPK12*-RNAi plants, but CAT and APX were lower in *CPK12*-RNAi plants, when compared with GL1 (Figure 8). Taken together, these data imply that CPK12 is involved in the elimination of H_2O_2 under salt stress.

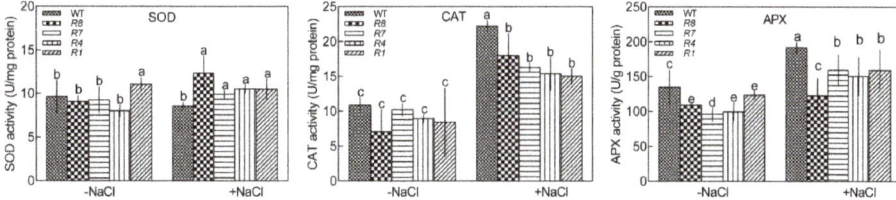

Figure 8. Effect of NaCl on activities of superoxide dismutase (SOD), catalase (CAT), and ascorbate peroxidase (APX) in wild-type (GL1) and *CPK12*-RNAi lines. Seven-day-old seedlings were transferred to MS medium supplemented with or without 100 mM NaCl for 10 d. The activities of antioxidant enzymes were analyzed. Each column shows the mean of three replicated experiments and bars represent the standard error of the mean. Columns labeled with letters in the same group denote significant difference at $p < 0.05$.

2.6. Down-Regulation of CPK12 Suppressed Cell Viability in Arabidopsis Roots

Previous studies showed that a high level of NaCl could reduce viability and increase programmed cell death in plants [64–67]. Cell viability was assayed with fluorescein diacetate (FDA) to determine whether salt stress could induce cell death in *CPK12*-RNAi plants. FDA staining showed an effect of salt treatment on cell viability in the elongation zone of roots. Wild-type GL1 and *CPK12*-RNAi plants grown in control conditions (NaCl-free Murashige–Skoog (MS) medium) showed clear FDA fluorescence with the cytoplasm of root cells, features indicating that the cells were viable. However, in salt-stressed *CPK12*-RNAi plants, the FDA fluorescence was undetectable in a number of root cells, and the fluorescence intensity was reduced (Figure 9). Compared to wild-type GL1 plants, the *CPK12*-RNAi plants exhibited lower cell viability during the period of salt stress.

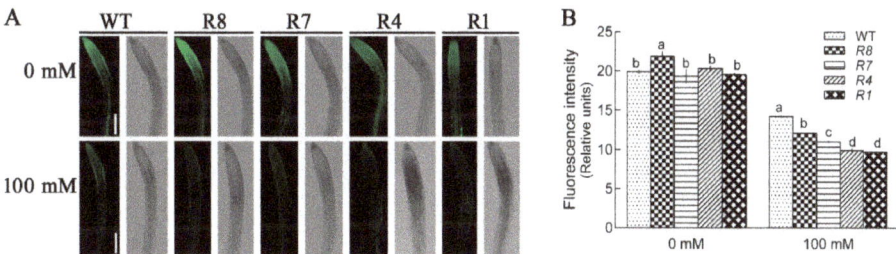

Figure 9. Effect of salt stress on cell viability in wild-type (GL1) and *CPK12*-RNAi lines. (**A**) Seven-day-old seedlings were transferred to MS medium supplemented with or without 100 mM NaCl for 12 h. Cell viability was assayed with fluorescein diacetate (FDA, green) stain. Representative images of apical region of roots are shown. Scale bars, 100 μm. (**B**) The fluorescence intensity (±SD) represents the mean of 10 independent seedlings. The mean values of FDA fluorescence are labeled with letters in the same group to denote significant differences ($p < 0.05$).

2.7. Down-Regulation of CPK12 Alters the Expression of Some Salt-Responsive Genes

We tested the expression of the following salt-related genes in the GL1 and *CPK12*-RNAi plants: *SOS1*, *SOS2*, and *SOS3* [4,7–11], *AHA1* and *AHA2* [68], *PER1* [69,70], *SOD*, *CAT*, and *APX*. In the absence of salt treatment, the expression of *AHA1* was not significantly difference between wild-type GL1 plants and *CPK12*-RNAi plants, but the expression of *SOS1*, *SOS2*, *SOS3 AHA2*, and *PER1* was down-regulated in *CPK12*-RNAi plants (Figure 10). Under salt stress, down-regulation of *CPK12* did not affect expression of *AHA2* and *PER1*, but significantly reduced expression of *SOS1* in *R8*, *R7*, and *R4* plants, *SOS2* in *R7*, *R4*, and *R1* plants, *AHA1* in *R8*, *R7*, *R4*, and *R1* plants (Figure 10). In the absence of salt treatment, the expression of *SOD* and *CAT* was down-regulated in *CPK12*-RNAi plants, and *APX* was down-regulated in *R8*, *R7*, and *R4* plants. Under salt stress, the expression of *APX* was down-regulated in *R1*. It is interesting that the expression of *SOD*, *CAT*, and *APX* was nearly undetectable in *R8* and *R7* plants, whether under salt or no-salt stress (Figure 10).

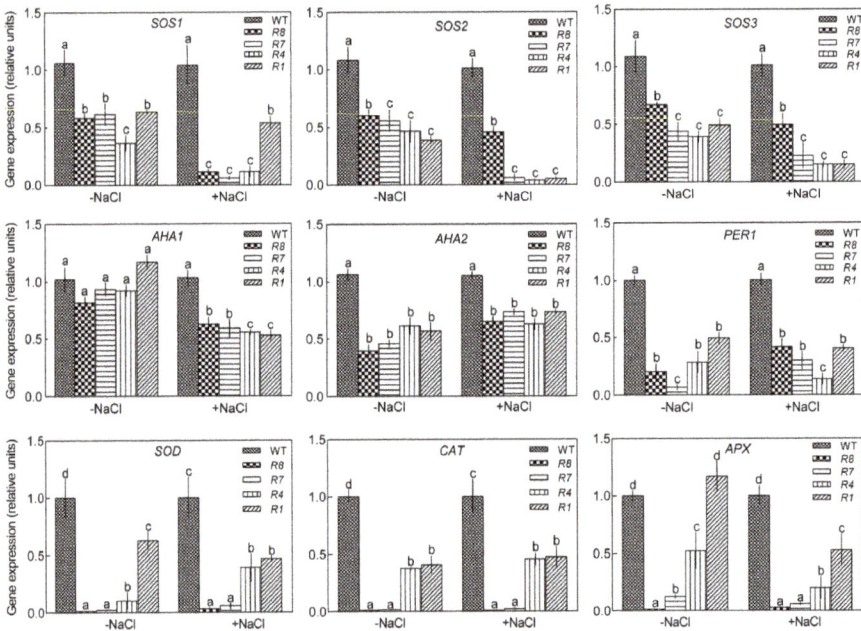

Figure 10. Changes in *CPK12* expression alter expression of a subset of genes involved in salt stress responses. The mRNA levels in the seedling of wild-type GL1, *CPK12*-RNAi mutants were determined by qRT-PCR. One-week-old seedlings were transferred to MS medium with and without the addition of 100 mM NaCl for ten days. The expression of salt stress responsive genes was analyzed. The gene expression levels were normalized relative to the value of the GL1 plants. Each value is the mean of the three independent determinations; columns labeled with letters in the same group indicate significant differences ($p < 0.05$).

3. Discussion

3.1. CPK12 Is Involved in Salt Stress Tolerance in Plants

In this investigation, *Arabidopsis* CPK12 was identified and characterized as a regulatory component involved in salt tolerance in terms of seedling growth. Previously we reported that CPK12 negatively regulates ABA signal transduction [43,44]. These results together imply an important function of CPK12 in regulating plant salt stress tolerance. Of note, although the expression of some salt-related genes were down-regulated in *CPK12*-RNAi plants without salt stress treatment (except *AHA1*), down-regulated *CPK12* did not influence post-germination seedling growth, and the expression of a control gene *EF-1α* that is not related to salt-stress, indicating that CPK12 is involved in salt stress signal transduction, but is not related to seedling development. Previously we reported that CPK27's function in salt stress tolerance [54], although CPK12 and CPK27 had similar functions. Our results suggest that the function of CPK12 may not be redundant for CPK27, because independent down-regulation of CPK12 or CPK27 can change the responses of *Arabidopsis* seedlings to salt stress. Current evidence suggests that CDPKs regulate plant tolerance to abiotic stress through ABA and Ca^{2+} pathways. For example, CPK10 regulates plants response to drought stress through ABA- and Ca^{2+}-mediated stomatal movements [45]; CPK4 and CPK11 are positive regulators in Ca^{2+}-mediated ABA signaling [42]. Thus, CPK12 and other CDPK members constitute a complicated regulation network, which functions in plant adaptation to salt and drought stresses.

3.2. CPK12 Regulates Na+ Balance in Salt-Stressed Plants

The ability to retain ion balance is very important for plant survival in saline environments [1,5,71]. Here, under salt stress, wild-type plants GL1 and *CPK12*-RNAi absorbed and accumulated Na^+ in roots; when compared with GL1, Na^+ accumulation was significantly higher in the roots of *CPK12*-RNAi plants, and net Na^+ efflux was reduced in *CPK12*-RNAi roots compared to GL1 plants. There are many CDPKs that can interact and regulate the activity of ion transporters. Under drought stress, AtCPK23 phosphorylates the guard cell anion channel SLAC1, which is collaborated with activation of the potassium-release channel GORK to regulate stomatal movement [47]. AtCPK3 and AtCPK6 are specifically expressed in guard-cell, regulate guard cell S-type anion channels and contribute to stomatal movement [49]. AtCPK13 phosphorylates two inward K^+ channels, KAT1 and KAT2, to restrict the stomatal aperture, and AtCPK3 phosphorylates and activates a two pore K^+ channel TPK1 [53,72]. Thus, like other CDPK members, it is speculated that CPK12 may regulate Na^+ balance in salt-stressed *Arabidopsis* seedlings. Compared with previous studies wherein some CDPKs were localized in guard cells to regulate stomatal aperture, our results give a new insight and show that CPK12 may function in root systems to regulate ion balance. CDPKs therefore constitute a network, which at the whole-plants level in roots and shoots, uses interaction and phosphorylation to regulate ion transporters or channels to improve plant tolerance to drought or salt stress over the short term.

3.3. CPK12 Regulates ROS Homeostasis in Salt-Stressed Plants

ROS, such as the superoxide anion, accumulates in stressed plants. Plasma membrane NADPH oxidases generate superoxide anions, which are transformed into H_2O_2 by superoxide dismutase [20]. In this study, compared with GL1 plants, *CPK12*-RNAi plants accumulated more H_2O_2 in roots, irrespective of the duration of NaCl treatment. Without salt-stress, compared with GL1, the expression of *CAT* gene was down-regulated in *CPK12*-RNAi plants, but the activity of CAT was not reduced, and the level of H_2O_2 was not significantly different between GL1 and *CPK12*-RNAi plants. Furthermore, under salt stress treatment, the activity of SOD was higher in *CPK12*-RNAi plants than GL1, but the activity of CAT was lower in *CPK12*-RNAi plants than GL1. The enhanced activity of SOD in *CPK12*-RNAi plants results in H_2O_2 production, but the reduced activity of CAT lead to H_2O_2 accumulation; these results suggest that down-regulated *CPK12* cannot scavenge NaCl-induced H_2O_2 bursts. Previous studies showed that down-regulated expression of CDPKs, which are related to antioxidases, may cause the accumulation of H_2O_2. *Arabidopsis cpk27-1* mutants accumulate more H_2O_2 in roots [54]. Potato StCDPK4 and StCDPK5 regulate the production of ROS [73]. In rice, *OsCPK12*-OX plants accumulate less H_2O_2 under conditions of high salinity, and this accumulation is more pronounced in *oscpk12* mutants and *OsCPK12* RNAi plants [56]. Similarly, excess H_2O_2 leads to oxidative damage and growth inhibition in *CPK12*-RNAi plants under salinity conditions. In contrast, *Arabidopsis cpk5 cpk6 cpk11 cpk4* quadruple mutants harbor decreased ROS content, suggesting that these CDPKs regulate ROS production potentially by phosphorylating the NADPH oxidase RBOHB [74]. Therefore, CDPKs play key roles in regulating ROS production and accumulation in plants [75].

3.4. CPK12 Functions with Potential Substrates in Salt Stress Signaling

In recent years, many substrates of CDPKs were identified. The transcription factor ABF4 is a substrate of CPK4/11 in *Arabidopsis* [42]. CPK12 can phosphorylate type-2C protein phosphatase ABI2 [43]. HSP1 interacts with CPK10 [45], SLAC1 is an interacting partner of CPK21 and CPK23 [47], and CPK13 specifically phosphorylates KAT2 and KAT1 [53]. In nutrient signaling, CPK10, CPK30, and CPK32 could potentially phosphorylate and activate all NLPs and possibly other transcription factors with overlapping or distinct target genes to support transcriptional, metabolic, and system-wide nutrient-growth regulations [76]. Potato CDPK5 phosphorylates the N-terminal region of plasma membrane RBOH (respiratory burst oxidase homolog) protein and participates in RBOHB-mediated ROS bursts, conferring resistance to near-obligate pathogens but increasing susceptibility to necrotrophic pathogens [77]. In this work, CPK12 was involved in plant adaptation to salt stress by regulating Na^+ and H_2O_2 homeostasis. These results indicate that CPK12 may interact with and phosphorylate several salt stress-related proteins as potential substrates in its regulatory function. To deeply demonstrate the regulatory mechanism of CPK12, the downstream components of CPK12 need to be identified, and their relationship with the whole complex CDPK regulation network need to be elucidated. Although the functions of CDPKs are widely identified in recent years, the complete CDPK signal transduction pathway is still not clearly illustrated, as the CDPK transduction network is very complex. Future progress is likely to identify sensors, channels, and other regulators involved in generating complex Ca^{2+} signatures in plant responses to hormones and environmental cues.

4. Materials and Methods

4.1. Plant Materials, Constructs, and Arabidopsis Transformation

Arabidopsis thaliana GL1 (Col-5) was used in this work for generating the *CPK12*-RNAi plants. A specific 242-bp fragment of *CPK12* (At5g23580) corresponding to the region of nt 6 to 247 was amplified with forward primer 5'-ACGCGTCGACGAACAAACCAAGAACCAGATGGGTT-3' and reverse primer 5'-CCGCTCGAGCGTTGGGGTATTCAGACAAGTGATG-3'. This fragment was inserted into the pSK-int vector, which was digested with the *Xho*I and *Sal*I. The same fragment, amplified with forward primer 5'-AACTGCAGGAACAAACCAAGAACCAGATGGGTT-3' and reverse primer 5'-GGACTAGTCGTTGGGGTATTCAGACAAGTGATG-3', was inserted into the pSK-int carrying the previous fragment. The entire RNAi cassette linked with actin 11 intron was excised from pSK-int vector, and inserted into the *Sac*I *Apa*I digested vector pSUPER1300(+) [78]. The construct was introduced into *Agrobacterium tumefaciens* GV3101 and transformed into GL1 by the floral dip method [79]. Transgenic plants were grown on MS agar plates containing hygromycin (50 µg/mL) to screen for positive seedlings. The homozygous T3 seeds of the transgenic plants were used for further analysis. Plants were grown in a growth chamber at 20–21 °C on MS medium at about 80 µmol photons $m^{-2} \cdot s^{-1}$, or in compost soil at about 120 µmol photons $m^{-2} \cdot s^{-1}$ over a 16-h photoperiod.

4.2. qRT-PCR Analysis

To assay the gene expression in the transgenic plants, quantitative real-time PCR analysis was performed with the RNA samples isolated from 10-day-old seedlings. For *CPK12*, qRT-PCR amplification was performed with forward primer 5'-CGAAACCCTCAAAGAAATAA-3' and reverse primer 5'-TGGTGTCCTCGTACGCACTCTC-3'. The primers specific for salt-related genes were: forward 5'-CACAAACATTTACCGAAAACCA-3' and reverse 5'-CAAATTTGCAAAGCTCATATCG-3' for *AHA1* (At2g18960); forward 5'-TGACTGATCTTCGATCCTCTCA-3' and reverse 5'-GAGAATGT GCATGTGCCAAA-3' for *AHA2* (At4g30190); forward 5'-CGTGCCCTTCATATTGTTGG-3' and reverse 5'-GACGCCATCAACAACGAGTC-3' for *PER1* (At1g48130); forward 5'-GTGAAGCAATCAAGC GGAAA-3' and reverse 5'-TGCGAAGAAGGCGTAGAACA-3' for *SOS1* (At2g01980); forward 5'-GCGAACTCAATGGGTTTTAAGT-3' and reverse 5'-CTTACGTCTACCATGAAAAGCG-3' for *SOS2*

(At5g35410); forward 5′-CCGGTCCATGAAAAAGTCAAAT-3′ and reverse 5′-CTCTTTCAATTCTT
CTCGCTCG-3′ for *SOS3* (At5g24270); forward 5′-AGGAAACATCACTGTTGGAGAT-3′ and reverse
5′-GAGTTTGGTCCAGTAAGAGGAA-3′ for *SOD* (At1g08830); forward 5′-AGGATCAAACTTT
GAGGGGTAG-3′ and reverse 5′-CTTGTGGTTCCTGGAATCTACT-3′ for *CAT* (At1g20620); forward
5′-GATGTCTTTGCTAAGCAGATGG-3′ and reverse 5′-GAGTTGTCGAAGATTAGAGGGT-3′ for *APX*
(At1g07890); forward 5′-CACCACTGGAGGTTTTGAGG-3′ and reverse 5′-TGGAGTATTTGGGG
GTGGT-3′ for *EF-1α* (At5g60390). Amplification of *ACTIN2/8* (forward primer 5′-GGTAACATTGTG
CTCAGTGGTGG-3′, reverse primer 5′-AACGACCTTAATCTTCATGCTGC-3′) gene was used as an
internal control.

4.3. Phenotype Identification

For the seedling growth experiment, seeds were germinated after stratification on MS medium
supplemented with various concentrations of NaCl. Seedling growth was examined 10 days
after stratification.

4.4. Measurement of Cytosolic Ca^{2+} Concentrations

For cytosolic Ca^{2+} concentration analysis, the Ca^{2+}-specific fluorescent probe Rhod-2 AM
(Invitrogen, Carlsbad, CA, USA) was used to measure the concentration of Ca^{2+} as previously
described [80]. Briefly, CPK12-RNAi mutants and GL1 seedlings were treated with MS liquid solution
supplemented with or without 100 mM NaCl for 12 h. Then, control and salinized plants roots
were 2 μM Rhod-2 AM (prepared in MS liquid solution, pH 5.8) incubated in the dark for 1 h at
room temperature. Then, the Arabidopsis plants were washed four to five times with distilled water.
The image of Ca^{2+} fluorescence in the probe-loaded roots was measured with a Leica SP5 confocal
microscope (Leica. Microsystems GmbH, Wetzlar, Germany), with emission at 570–590 nm and
excitation at 543 nm [80].

4.5. Detection of Cytosolic Na$^+$ Concentrations

Root cellular Na$^+$ levels were detected with Na$^+$-specific fluorescent probe, CoroNa™ Green,
AM (Invitrogen, Carlsbad, CA, USA). Seedlings of wild-type GL1 and *CPK12*-RNAi mutants were
treated with 0, 120, or 150 mM NaCl in MS liquid solution for 12 h. Then, control and salinized
seedlings were incubated with CoroNa in the dark for 1 h, and washed 3–4 times with distilled
water subsequently. Na$^+$ fluorescence was observed with a Leica SP5 confocal microscope (excitation:
488 nm; emission: 510–530 nm, Microsystems GmbH, Wetzlar, Germany) [25,80,81]. ImageJ software
(Version 1.48, National Institutes of Health, Bethesda, MD, USA) was used to quantify relative
fluorescence intensity. It is worth noting that there are several commercial Na$^+$ specific probes;
for example, SBFI, Sodium Green, CoroNa etc., Sodium Green displays a modest fluorescence increase
in response to Na$^+$ binding [82,83], while CoroNa is more suitable for detecting Na$^+$ in a wider range
of concentrations; and the selectivity of CoroNa is 4 times higher to Na$^+$ than to K$^+$ binding [82], but
CoroNa is not suitable for the detection of relatively low Na$^+$ changes in cells [83]. In this work, after
NaCl treatment, the roots absorbed and accumulated high levels of Na$^+$, and Na$^+$ efflux was inhibited.
Our previous studies showed that CoroNa is suitable for detecting Na$^+$ level after NaCl treatment in
tobacco, *Arabidopsis*, and *Glycyrrhiza uralensis* [84–86]; thus, we selected CoroNa Green to detect the
cytosolic Na$^+$ level in this work.

4.6. Net Fluxes Measurements of Na$^+$

Net Na$^+$ flux was measured using the NMT technique (NMT-YG-100, Younger, Amherst,
Massachusetts, USA) as described previously [12,80]. One-week-old seedling grown on MS medium
containing 0, 110, 120, 130 mM NaCl was washed 4–5 times with ddH$_2$O and transferred to the
measuring chamber containing 10–15 mL measuring solution, which included 0.1 mM NaCl, 0.5 mM
KCl, 0.1 mM CaCl$_2$, 0.1 mM MgCl$_2$, and 2.5% sucrose, pH 5.8. After the roots were immobilized to

the bottom of the chamber, Na flux measurements were started at 200–300 μm from the root apex. The Na$^+$ flux was continuously recorded for 17–20 min. The Na$^+$ flux was detected by shifting the ion-selective microelectrode between two sites close the roots over a preset length (30 μm for intact roots in this experiment) at a frequency in the range of 0.3–0.5 Hz. The electrode was stepped from one site to another in a predesigned sampling routine, while the sample was also scanned with a 3-D microstepper motor manipulator (CMC-4). Pre-pulled and salinized glass micropipettes (4–5 μm aperture, XYPG120-2; Xuyue Sci. and Tech. Co., Ltd., Beijing, China) were processed with a backfilling solution (Na: 250 mM NaCl) to a length of approximately 1 cm from the tip, then front-filled with about 10 μm columns of selective liquid ion-exchange cocktails (Na: Fluka 71178). An Ag/AgCl wire electrode holder (XYEH01-1; Xuyue Sci. and Tech. Co., Ltd., Beijing, China) was used to make electrical contact with the electrolyte solution. The reference electrode was DRIREF-2 (World Precision Instruments, www.wpiinc.com). Ion-selective electrodes were calibrated prior to flux measurements (The concentration of Na$^+$ was usually 0.1 mM in the measuring buffer for root samples). Na$^+$ flux was calculated by Fick's law of diffuseon: $J = -D(dc/dx)$, where J represents the Na$^+$ flux in the x direction, dc/dx is the Na$^+$-concentration gradient, and D is the Na$^+$ diffusion constant.

4.7. H$_2$O$_2$ Production with Root Cells

A specific fluorescent probe, 2′,7′-dichlorodihydrofluorescein diacetate (H$_2$DCF-DA; Molecular Probes) was used for H$_2$O$_2$ detection in the roots of GL1 plants and *CPK12*-RNAi plants. Shock and short-term responses of H$_2$O$_2$ to NaCl exposure were examined in this study.

Seven-day-old seedlings (GL1 and *CPK12*-RNAi) grown on MS medium were exposed to 0 or 100 mM NaCl for 10 min, 12 h, and 24 h and then incubated with 10 μM H$_2$DCF-DA (prepared in liquid MS medium, pH 5.7) for 5 min at room temperature in the dark. The H$_2$DCF-DA-loaded seedlings were washed 3–4 times with liquid MS solution. DCF-specific fluorescence was examined under a Leica SP5 confocal microscope (Leica Microsystem GmbH, Wetzlar, Germany), with excitation at 488 nm and emission at 510–530 nm. Relative H$_2$DCF-DA fluorescence intensities in root cells were measured with ImageJ 1.48 (National Institutes of Health, Bethesda, MD, USA).

4.8. Activity Analyses of Antioxidant Enzymes

For the detection of antioxidant enzyme activities, after salt treatment for ten days, the *Arabidopsis* seedling (0.2 g) was ground to a fine powder in liquid nitrogen, and then 2 mL ice-cold 50 mM potassium phosphate buffer (pH 7.0) was added. After centrifugation at 12,000 g for 20 min, the supernatant was used to detect the enzymatic activities of the antioxidant enzymes SOD, CAT, and APX. The activities of antioxidant enzymes were determined using commercial kits (Nanjing Jiancheng Bioengineering Institute, Nanjing, China). SOD activity was measured using the Superoxide Dismutase WST-1 Assay Kit, which, based on the xanthine/xanthine oxidase method, depended on the production of O^{2-} anions. CAT activity was measured by analyzing the yellowish complex produced by the reaction between H$_2$O$_2$ and ammonium molybdate and calculating CAT activity by measuring OD value at 405 nm. APX activity was estimated based on the reaction of ASA with H$_2$O$_2$ to oxidize ASA to MDASA; APX activity was calculated by measuring the reduced OD value at 290 nm. Activities of SOD and CAT are expressed as units per milligrams of protein (U/mg protein). The activity of APX is expressed as (U/g protein).

4.9. Cell Viability Analyses

Seven-day-old seedlings (GL1 and *CPK12*-RNAi) grown on MS medium were exposed to 0 or 100 mM NaCl for 12 h, and cell viability was measured by staining seedlings with FDA (Invitrogen, Carlsbad, CA, USA). The confocal parameters were set as described in previous studies: the excitation wavelength was 488 nm, and the emission wavelengths were 505 to 525 nm. Relative FDA fluorescence intensities in root cells were measured with ImageJ 1.48 (National Institutes of Health, Bethesda, MD, USA).

4.10. Data Analysis

All experimental data were analyzed with SPSS version 17.0 software (IBM China Company Ltd., Beijing, China) for statistical evaluations. Statistical analysis were performed using one-way ANOVA. Differences were considered significant at $p < 0.05$, unless otherwise stated.

Author Contributions: Conceptualization, R.Z., S.L. and S.C.; Investigation, H.Z., Y.Z., C.D., N.L. and C.Z.; Resources, R.Z. and S.D.; Writing—original draft preparation, H.Z., R.Z. and S.L.; Writing—review and editing, H.Z., R.Z. and S.L.; All authors discussed the results and comments on the manuscript.

Funding: This research was supported by the Beijing Natural Science Foundation (Grant No. 6172024, 6182030), Fundamental Research Funds for the Central Universities (Grant No. 2017ZY07), National Natural Science Foundation of China (Grant Nos. 31600205, 31770643, and 31570587), the Research Project of the Chinese Ministry of Education (Grant No. 113013A), the Program of Introducing Talents of Discipline to Universities (111 Project, Grant No. B13007).

Conflicts of Interest: The authors declare no conflict of interest.

Abbreviations

ABA	Abscisic Acid
APX	Ascorbate Peroxidase
CAT	Catalase
CDPK	Ca^{2+}-Dependent Protein Kinases
FDA	Fluorescein Diacetate
H_2DCF-DA	2′,7′-Dichlorodihydrofluorescein Diacetate
HSP	Heat Shock Protein
MS	Murashige–Skoog Medium
NADPH	Nicotinamide Adenine Dinucleotide Phosphate
NLP	NIN-Like Protein
NMT	Non-Invasive Micro-Test Technique
qRT-PCR	Quantitative Reverse Transcription PCR
RBOHB	Respiratory Burst Oxidase Homolog Protein B
ROS	Reactive Oxygen Species
SOD	Superoxide Dismutase
SOS	Salt overly sensitive

References

1. Polle, A.; Chen, S. On the salty side of life: Molecular physiological and anatomical adaptation and acclimation of trees to extreme habitats. *Plant Cell Environ.* **2015**, *38*, 1794–1816. [CrossRef] [PubMed]
2. Zhu, J.K. Plant salt tolerance. *Trends Plant Sci.* **2001**, *6*, 66–71. [CrossRef]
3. Janz, D.; Polle, A. Harnessing salt for woody biomass production. *Tree Physiol.* **2012**, *32*, 1–3. [CrossRef] [PubMed]
4. Zhu, J.K. Salt and drought stress signal transduction in plants. *Annu. Rev. Plant Biol.* **2002**, *53*, 247–273. [CrossRef] [PubMed]
5. Zhu, J.K. Regulation of ion homeostasis under salt stress. *Curr. Opin. Plant Biol.* **2003**, *6*, 441–445. [CrossRef]
6. Acosta-Motos, J.R.; Ortuño, M.F.; Bernal-Vicente, A.; Diaz-Vivancos, P.; Sanchez-Blanco, M.J.; Hernandez, J.A. Plant responses to salt stress: Adaptive mechanisms. *Agronomy* **2017**, *7*, 18. [CrossRef]
7. Xiong, L.; Zhu, J.K. Salt tolerance. *Arabidopsis Book* **2002**, *1*, e0048. [CrossRef]
8. Qiu, Q.S.; Guo, Y.; Dietrich, M.A.; Schumaker, K.S.; Zhu, J.K. Regulation of SOS1 a plasma membrane Na^+/H^+ exchanger in *Arabidopsis thaliana* by SOS2 and SOS3. *Proc. Natl. Acad. Sci. USA* **2002**, *99*, 8436–8441. [CrossRef]
9. Quan, R.; Lin, H.; Mendoza, I.; Zhang, Y.; Cao, W.; Yang, Y.; Shang, M.; Chen, S.; Pardo, J.M.; Guo, Y. SCABP8/CBL10 a putative calcium sensor interacts with the protein kinase SOS2 to protect *Arabidopsis* shoots from salt stress. *Plant Cell* **2007**, *19*, 1415–1431. [CrossRef]

10. Zhu, J.; Fu, X.; Koo, Y.D.; Zhu, J.K.; Jenney, F.E., Jr.; Adams, M.W.; Zhu, Y.; Shi, H.; Yun, D.J.; Hasegawa, P.M.; et al. An enhancer mutant of *Arabidopsis* salt overly sensitive 3 mediates both ion homeostasis and the oxidative stress response. *Mol. Cell Biol.* **2007**, *27*, 5214–5224. [CrossRef]

11. Yang, Q.; Chen, Z.Z.; Zhou, X.F.; Yin, H.B.; Li, X.; Xin, X.F.; Hong, X.H.; Zhu, J.K.; Gong, Z. Overexpression of SOS (Salt Overly Sensitive) genes increases salt tolerance in transgenic *Arabidopsis*. *Mol. Plant* **2009**, *2*, 22–31. [CrossRef] [PubMed]

12. Sun, J.; Chen, S.; Dai, S.; Wang, R.; Li, N.; Shen, X.; Zhou, X.; Lu, C.; Zheng, X.; Hu, Z.; et al. NaCl-induced alternations of cellular and tissue ion fluxes in roots of salt-resistant and salt-sensitive poplar species. *Plant Physiol.* **2009**, *149*, 1141–1153. [CrossRef] [PubMed]

13. Zhu, M.; Shabala, L.; Cuin, T.A.; Huang, X.; Zhou, M.; Munns, R.; Shabala, S. *Nax* loci affect SOS1-like Na^+/H^+ exchanger expression and activity in wheat. *J. Exp. Bot.* **2016**, *67*, 835–844. [CrossRef] [PubMed]

14. Mittler, R. Oxidative stress antioxidants and stress tolerance. *Trends Plant Sci.* **2002**, *7*, 405–410. [CrossRef]

15. Miller, G.; Shulaev, V.; Mittler, R. Reactive oxygen signaling and abiotic stress. *Physiol. Plant* **2008**, *133*, 481–489. [CrossRef] [PubMed]

16. Miller, G.; Suzuki, N.; Ciftci-Yilmaz, S.; Mittler, R. Reactive oxygen species homeostasis and signalling during drought and salinity stresses. *Plant Cell Environ.* **2010**, *33*, 453–467. [CrossRef] [PubMed]

17. Corpas, F.J.; Barroso, J.B.; del Rio, L.A. Peroxisomes as a source of reactive oxygen species and nitric oxide signal molecules in plant cells. *Trends Plant Sci.* **2001**, *6*, 145–150. [CrossRef]

18. Asada, K. Production and scavenging of reactive oxygen species in chloroplasts and their functions. *Plant Physiol.* **2006**, *141*, 391–396. [CrossRef]

19. Palma, J.M.; Corpas, F.J.; del Rio, L.A. Proteome of plant peroxisomes: New perspectives on the role of these organelles in cell biology. *Proteomics* **2009**, *9*, 2301–2312. [CrossRef]

20. Apel, K.; Hirt, H. Reactive oxygen species: Metabolism, oxidative stress, and signal transduction. *Annu. Rev. Plant Biol.* **2004**, *55*, 373–399. [CrossRef]

21. Creissen, G.; Firmin, J.; Fryer, M.; Kular, B.; Leyland, N.; Reynolds, H.; Pastori, G.; Wellburn, F.; Baker, N.; Wellburn, A.; et al. Elevated glutathione biosynthetic capacity in the chloroplasts of transgenic tobacco plants paradoxically causes increased oxidative stress. *Plant Cell* **1999**, *11*, 1277–1292. [CrossRef] [PubMed]

22. Bose, J.; Rodrigo-Moreno, A.; Shabala, S. ROS homeostasis in halophytes in the context of salinity stress tolerance. *J. Exp. Bot.* **2014**, *65*, 1241–1257. [CrossRef] [PubMed]

23. Rentel, M.C.; Knight, M.R. Oxidative stress-induced calcium signaling in *Arabidopsis*. *Plant Physiol.* **2004**, *135*, 1471–1479. [CrossRef] [PubMed]

24. Chen, S.; Polle, A. Salinity tolerance of *Populus*. *Plant Biol.* **2010**, *12*, 317–333. [CrossRef] [PubMed]

25. Sun, J.; Wang, M.J.; Ding, M.Q.; Deng, S.R.; Liu, M.Q.; Lu, C.F.; Zhou, X.Y.; Shen, X.; Zheng, X.J.; Zhang, Z.K.; et al. H_2O_2 and cytosolic Ca^{2+} signals triggered by the PM H-coupled transport system mediate K^+/Na^+ homeostasis in NaCl-stressed *Populus euphratica* cells. *Plant Cell Environ.* **2010**, *33*, 943–958. [CrossRef] [PubMed]

26. Chung, J.S.; Zhu, J.K.; Bressan, R.A.; Hasegawa, P.M.; Shi, H. Reactive oxygen species mediate Na^+-induced SOS1 mRNA stability in *Arabidopsis*. *Plant J.* **2008**, *53*, 554–565. [CrossRef] [PubMed]

27. Sanders, D.; Pelloux, J.; Brownlee, C.; Harper, J.F. Calcium at the crossroads of signaling. *Plant Cell* **2002**, *14* (Suppl. S1), 401–417. [CrossRef]

28. Hepler, P.K. Calcium: A central regulator of plant growth and development. *Plant Cell* **2005**, *17*, 2142–2155. [CrossRef]

29. Zielinski, R.E. Calmodulin and calmodulin-binding proteins in plants. *Annu. Rev. Plant Physiol. Plant Mol. Biol.* **1998**, *49*, 697–725. [CrossRef]

30. Cheng, S.H.; Willmann, M.R.; Chen, H.C.; Sheen, J. Calcium signaling through protein kinases. The *Arabidopsis* calcium-dependent protein kinase gene family. *Plant Physiol.* **2002**, *129*, 469–485. [CrossRef]

31. Luan, S.; Kudla, J.; Rodriguez-Concepcion, M.; Yalovsky, S.; Gruissem, W. Calmodulins and calcineurin B-like proteins: Calcium sensors for specific signal response coupling in plants. *Plant Cell* **2002**, *14* (Suppl. 1), 389–400. [CrossRef]

32. Harper, J.F.; Breton, G.; Harmon, A. Decoding Ca^{2+} signals through plant protein kinases. *Annu. Rev. Plant Biol.* **2004**, *55*, 263–288. [CrossRef] [PubMed]

33. Bouche, N.; Yellin, A.; Snedden, W.A.; Fromm, H. Plant-specific calmodulin-binding proteins. *Annu. Rev. Plant Biol.* **2005**, *56*, 435–466. [CrossRef] [PubMed]

34. Hrabak, E.M.; Chan, C.W.; Gribskov, M.; Harper, J.F.; Choi, J.H.; Halford, N.; Kudla, J.; Luan, S.; Nimmo, H.G.; Sussman, M.R.; et al. The *Arabidopsis* CDPK-SnRK superfamily of protein kinases. *Plant Physiol.* **2003**, *132*, 666–680. [CrossRef] [PubMed]

35. Asano, T.; Tanaka, N.; Yang, G.; Hayashi, N.; Komatsu, S. Genome-wide identification of the rice calcium-dependent protein kinase and its closely related kinase gene families: Comprehensive analysis of the CDPKs gene family in rice. *Plant Cell Physiol.* **2005**, *46*, 356–366. [CrossRef]

36. Li, A.L.; Zhu, Y.F.; Tan, X.M.; Wang, X.; Wei, B.; Guo, H.Z.; Zhang, Z.L.; Chen, X.B.; Zhao, G.Y.; Kong, X.Y.; et al. Evolutionary and functional study of the CDPK gene family in wheat (*Triticum aestivum* L.). *Plant Mol. Biol.* **2008**, *66*, 429–443. [CrossRef]

37. Zuo, R.; Hu, R.; Chai, G.; Xu, M.; Qi, G.; Kong, Y.; Zhou, G. Genome-wide identification classification and expression analysis of CDPK and its closely related gene families in poplar (*Populus trichocarpa*). *Mol. Biol. Rep.* **2013**, *40*, 2645–2662. [CrossRef]

38. Boudsocq, M.; Sheen, J. CDPKs in immune and stress signaling. *Trends Plant Sci.* **2013**, *18*, 30–40. [CrossRef]

39. Schulz, P.; Herde, M.; Romeis, T. Calcium-dependent protein kinases: Hubs in plant stress signaling and development. *Plant Physiol.* **2013**, *163*, 523–530. [CrossRef]

40. Hamel, L.P.; Sheen, J.; Seguin, A. Ancient signals: Comparative genomics of green plant CDPKs. *Trends Plant Sci.* **2014**, *19*, 79–89. [CrossRef]

41. Choi, H.I.; Park, H.J.; Park, J.H.; Kim, S.; Im, M.Y.; Seo, H.H.; Kim, Y.W.; Hwang, I.; Kim, S.Y. *Arabidopsis* calcium-dependent protein kinase AtCPK32 interacts with ABF4, a transcriptional regulator of abscisic acid-responsive gene expression, and modulates its activity. *Plant Physiol.* **2005**, *139*, 1750–1761. [CrossRef] [PubMed]

42. Zhu, S.Y.; Yu, X.C.; Wang, X.J.; Zhao, R.; Li, Y.; Fan, R.C.; Shang, Y.; Du, S.Y.; Wang, X.F.; Wu, F.Q.; et al. Two calcium-dependent protein kinases CPK4 and CPK11 regulate abscisic acid signal transduction in *Arabidopsis*. *Plant Cell* **2007**, *19*, 3019–3036. [CrossRef] [PubMed]

43. Zhao, R.; Sun, H.L.; Mei, C.; Wang, X.J.; Yan, L.; Liu, R.; Zhang, X.F.; Wang, X.F.; Zhang, D.P. The *Arabidopsis* Ca^{2+}-dependent protein kinase CPK12 negatively regulates abscisic acid signaling in seed germination and post-germination growth. *New Phytol.* **2011**, *192*, 61–73. [CrossRef] [PubMed]

44. Zhao, R.; Wang, X.F.; Zhang, D.P. CPK12: A Ca^{2+}-dependent protein kinase balancer in abscisic acid signaling. *Plant Signal. Behav.* **2011**, *6*, 1687–1690. [CrossRef] [PubMed]

45. Zou, J.J.; Wei, F.J.; Wang, C.; Wu, J.J.; Ratnasekera, D.; Liu, W.X.; Wu, W.H. *Arabidopsis* calcium-dependent protein kinase CPK10 functions in abscisic acid- and Ca^{2+}-mediated stomatal regulation in response to drought stress. *Plant Physiol.* **2010**, *154*, 1232–1243. [CrossRef] [PubMed]

46. Ma, S.Y.; Wu, W.H. AtCPK23 functions in *Arabidopsis* responses to drought and salt stresses. *Plant Mol. Biol.* **2007**, *65*, 511–518. [CrossRef] [PubMed]

47. Geiger, D.; Scherzer, S.; Mumm, P.; Marten, I.; Ache, P.; Matschi, S.; Liese, A.; Wellmann, C.; Al-Rasheid, K.A.; Grill, E.; et al. Guard cell anion channel SLAC1 is regulated by CDPK protein kinases with distinct Ca^{2+} affinities. *Proc. Natl. Acad. Sci. USA* **2010**, *107*, 8023–8028. [CrossRef]

48. Franz, S.; Ehlert, B.; Liese, A.; Kurth, J.; Cazale, A.C.; Romeis, T. Calcium-dependent protein kinase CPK21 functions in abiotic stress response in *Arabidopsis thaliana*. *Mol. Plant* **2011**, *4*, 83–96. [CrossRef]

49. Mori, I.C.; Murata, Y.; Yang, Y.; Munemasa, S.; Wang, Y.F.; Andreoli, S.; Tiriac, H.; Alonso, J.M.; Harper, J.F.; Ecker, J.R.; et al. CDPKs CPK6 and CPK3 function in ABA regulation of guard cell S-type anion- and Ca^{2+}-permeable channels and stomatal closure. *PLoS Biol.* **2006**, *4*, e327. [CrossRef]

50. Mehlmer, N.; Wurzinger, B.; Stael, S.; Hofmann-Rodrigues, D.; Csaszar, E.; Pfister, B.; Bayer, R.; Teige, M. The Ca^{2+}-dependent protein kinase CPK3 is required for MAPK-independent salt-stress acclimation in *Arabidopsis*. *Plant J.* **2010**, *63*, 484–498. [CrossRef]

51. Xu, J.; Tian, Y.S.; Peng, R.H.; Xiong, A.S.; Zhu, B.; Jin, X.F.; Gao, F.; Fu, X.Y.; Hou, X.L.; Yao, Q.H. AtCPK6 a functionally redundant and positive regulator involved in salt/drought stress tolerance in *Arabidopsis*. *Planta* **2010**, *231*, 1251–1260. [CrossRef] [PubMed]

52. Ye, W.; Muroyama, D.; Munemasa, S.; Nakamura, Y.; Mori, I.C.; Murata, Y. Calcium-dependent protein kinase CPK6 positively functions in induction by yeast elicitor of stomatal closure and inhibition by yeast elicitor of light-induced stomatal opening in *Arabidopsis*. *Plant Physiol.* **2013**, *163*, 591–599. [CrossRef] [PubMed]

53. Ronzier, E.; Corratge-Faillie, C.; Sanchez, F.; Prado, K.; Briere, C.; Leonhardt, N.; Thibaud, J.B.; Xiong, T.C. CPK13 a noncanonical Ca²⁺-dependent protein kinase specifically inhibits KAT2 and KAT1 shaker K⁺ channels and reduces stomatal opening. *Plant Physiol.* **2014**, *166*, 314–326. [CrossRef] [PubMed]

54. Zhao, R.; Sun, H.M.; Zhao, N.; Jing, X.S.; Shen, X.; Chen, S.L. The *Arabidopsis* Ca²⁺-dependent protein kinase CPK27 is required for plant response to salt-stress. *Gene* **2015**, *563*, 203–214. [CrossRef] [PubMed]

55. Asano, T.; Hakata, M.; Nakamura, H.; Aoki, N.; Komatsu, S.; Ichikawa, H.; Hirochika, H.; Ohsugi, R. Functional characterisation of OsCPK21, a calcium-dependent protein kinase that confers salt tolerance in rice. *Plant Mol. Biol.* **2011**, *75*, 179–191. [CrossRef] [PubMed]

56. Asano, T.; Hayashi, N.; Kobayashi, M.; Aoki, N.; Miyao, A.; Mitsuhara, I.; Ichikawa, H.; Komatsu, S.; Hirochika, H.; Kikuchi, S.; et al. A rice calcium-dependent protein kinase OsCPK12 oppositely modulates salt-stress tolerance and blast disease resistance. *Plant J.* **2012**, *69*, 26–36. [CrossRef] [PubMed]

57. Wei, S.; Hu, W.; Deng, X.; Zhang, Y.; Liu, X.; Zhao, X.; Luo, Q.; Jin, Z.; Li, Y.; Zhou, S.; et al. A rice calcium-dependent protein kinase OsCPK9 positively regulates drought stress tolerance and spikelet fertility. *BMC Plant Biol.* **2014**, *14*, 133. [CrossRef]

58. Ma, F.; Lu, R.; Liu, H.; Shi, B.; Zhang, J.; Tan, M.; Zhang, A.; Jiang, M. Nitric oxide-activated calcium/calmodulin-dependent protein kinase regulates the abscisic acid-induced antioxidant defence in maize. *J. Exp. Bot.* **2012**, *63*, 4835–4847. [CrossRef]

59. Jiang, S.; Zhang, D.; Wang, L.; Pan, J.; Liu, Y.; Kong, X.; Zhou, Y.; Li, D. A maize calcium-dependent protein kinase gene, ZmCPK4, positively regulated abscisic acid signaling and enhanced drought stress tolerance in transgenic *Arabidopsis*. *Plant Physiol. Biochem.* **2013**, *71*, 112–120. [CrossRef]

60. Ding, Y.; Cao, J.; Ni, L.; Zhu, Y.; Zhang, A.; Tan, M.; Jiang, M. ZmCPK11 is involved in abscisic acid-induced antioxidant defence and functions upstream of ZmMPK5 in abscisic acid signalling in maize. *J. Exp. Bot.* **2013**, *64*, 871–884. [CrossRef]

61. Chen, J.; Xue, B.; Xia, X.; Yin, W. A novel calcium-dependent protein kinase gene from *Populus euphratica*, confers both drought and cold stress tolerance. *Biochem. Biophys. Res. Commun.* **2013**, *441*, 630–636. [CrossRef] [PubMed]

62. Yu, X.C.; Li, M.J.; Gao, G.F.; Feng, H.Z.; Geng, X.Q.; Peng, C.C.; Zhu, S.Y.; Wang, X.J.; Shen, Y.Y.; Zhang, D.P. Abscisic acid stimulates a calcium-dependent protein kinase in grape berry. *Plant Physiol.* **2006**, *140*, 558–579. [CrossRef]

63. Yu, X.C.; Zhu, S.Y.; Gao, G.F.; Wang, X.J.; Zhao, R.; Zou, K.Q.; Wang, X.F.; Zhang, X.Y.; Wu, F.Q.; Peng, C.C.; et al. Expression of a grape calcium-dependent protein kinase ACPK1 in *Arabidopsis thaliana* promotes plant growth and confers abscisic acid-hypersensitivity in germination postgermination growth and stomatal movement. *Plant Mol. Biol.* **2007**, *64*, 531–538. [CrossRef] [PubMed]

64. Lin, J.S.; Wang, T.; Wang, G.X. Salt stress-induced programmed cell death via Ca²⁺-mediated mitochondrial permeability transition in tobacco protoplasts. *Plant Growth Regul.* **2005**, *45*, 243–250. [CrossRef]

65. Lin, J.S.; Wang, Y.; Wang, G.X. Salt stress-induced programmed cell death in tobacco protaplasts is mediated by reactive oxygen species and mitochondrial permeability transition pore status. *J. Plant Physiol.* **2006**, *163*, 731–739. [CrossRef] [PubMed]

66. Li, J.Y.; Jiang, A.; Zhang, W. Salt stress-induced programmed cell death in rice root tip cells. *J. Integr. Plant Biol.* **2007**, *49*, 481–486. [CrossRef]

67. Shabala, S. Salinity and programmed cell death: Untavelling mechanisms for ion specific signalling. *J. Exp. Bot.* **2009**, *60*, 709–712. [CrossRef]

68. Bose, J.; Xie, Y.; Shen, W.; Shabala, S. Haem oxygenase modifies salinity tolerance in Arabidopsis by controlling K⁺ retention via regulation of the plasma membrane H⁺-ATPase and by altering SOS1 transcript levels in roots. *J. Exp. Bot.* **2013**, *64*, 471–481. [CrossRef]

69. Lee, S.; Lee, H.J.; Jung, J.H.; Park, C.M. The *Arabidopsis thaliana* RNA-binding protein FCA regulates thermotolerance by modulating the detoxification of reactive oxygen species. *New Phytol.* **2015**, *205*, 555–569. [CrossRef]

70. Ha, J.H.; Kim, J.H.; Kim, S.G.; Sim, H.J.; Lee, G.; Halitschke, R.; Baldwin, I.T.; Kim, J.I.; Park, C.M. Shoot phytochrome B modulates reactive oxygen species homeostasis in roots via abscisic acid signaling in *Arabidopsis*. *Plant J.* **2018**, *94*, 790–798. [CrossRef]

71. Adams, E.; Shin, R. Transport, signaling, and homeostasis of potassium and sodium in plants. *J. Integr. Plant Biol.* **2014**, *56*, 231–249. [CrossRef] [PubMed]

72. Latz, A.; Mehlmer, N.; Zapf, S.; Mueller, T.D.; Wurzinger, B.; Pfister, B.; Csaszar, E.; Hedrich, R.; Teige, M.; Becker, D. Salt stress triggers phosphorylation of the *Arabidopsis* vacuolar K^+ channel TPK1 by calcium-dependent protein kinases (CDPKs). *Mol. Plant* **2013**, *6*, 1274–1289. [CrossRef] [PubMed]

73. Kobayashi, M.; Ohura, I.; Kawakita, K.; Yokota, N.; Fujiwara, M.; Shimamoto, K.; Doke, N.; Yoshioka, H. Calcium-dependent protein kinases regulate the production of reactive oxygen species by potato NADPH oxidase. *Plant Cell* **2007**, *19*, 1065–1080. [CrossRef] [PubMed]

74. Boudsocq, M.; Willmann, M.R.; McCormack, M.; Lee, H.; Shan, L.; He, P.; Bush, J.; Cheng, S.H.; Sheen, J. Differential innate immune signalling via Ca^{2+} sensor protein kinases. *Nature* **2010**, *464*, 418–422. [CrossRef] [PubMed]

75. Romeis, T.; Herde, M. From local to global: CDPKs in systemic defense signaling upon microbial and herbivore attack. *Curr. Opin. Plant Biol.* **2014**, *20c*, 1–10. [CrossRef] [PubMed]

76. Liu, K.H.; Niu, Y.; Konishi, M.; Wu, Y.; Du, H.; Sun Chung, H.; Li, L.; Boudsocq, M.; McCormack, M.; Maekawa, S.; et al. Discovery of nitrate-CPK-NLP signalling in central nutrient-growth networks. *Nature* **2017**, *545*, 311–316. [CrossRef]

77. Kobayashi, M.; Yoshioka, M.; Asai, S.; Nomura, H.; Kuchimura, K.; Mori, H.; Doke, N.; Yoshioka, H. StCDPK5 confers resistance to late blight pathogen but increases susceptibility to early blight pathogen in potato via reactive oxygen species burst. *New Phytol.* **2012**, *196*, 223–237. [CrossRef]

78. Ni, M.; Cui, D.; Einstein, J.; Narasimhulu, S.; Vergara, C.E.; Gelvin, S.B. Strength and tissue specificity of chimeric promoters derived from the octopine and mannopine synthase genes. *Plant J.* **1995**, *7*, 661–676. [CrossRef]

79. Clough, S.J.; Bent, A.F. Floral dip: A simplified method for *Agrobacterium*-mediated transformation of *Arabidopsis thaliana*. *Plant J.* **1998**, *16*, 735–743. [CrossRef]

80. Sun, J.; Zhang, X.; Deng, S.; Zhang, C.; Wang, M.; Ding, M.; Zhao, R.; Shen, X.; Zhou, X.; Lu, C.; et al. Extracellular ATP signaling is mediated by H_2O_2 and cytosolic Ca^{2+} in the salt response of *Populus euphratica* cell. *PLoS ONE* **2012**, *7*, e53136. [CrossRef]

81. Sun, J.; Li, L.; Liu, M.; Wang, M.; Ding, M.; Deng, S.; Lu, C.; Zhou, X.; Chen, X.; Zheng, X.; et al. Hydrogen peroxide and nitric oxide mediate K^+/Na^+ homeostasis and antioxidant defense in NaCl-stressed callus cells of two contrasting poplars. *Plant Cell Tiss. Organ. Cult.* **2010**, *103*, 205–215. [CrossRef]

82. Martin, V.V.; Rothe, A.; Gee, K.R. Fluorescent metal ion indicators based on benzoannelated crown systems: A green fluorescent indicator for intracellular sodium ions. *Bioorg. Med. Chem. Lett.* **2005**, *15*, 1851–1855. [CrossRef]

83. Iamshanova, O.; Mariot, P.; Lehen'kyi, V.; Prevarskaya, N. Comparison of fluorescence probes for intracellular sodium imaging in prostate cancer cell lines. *Eur. Biophys. J.* **2016**, *45*, 765–777. [CrossRef] [PubMed]

84. Han, Y.; Wang, W.; Sun, J.; Ding, M.; Zhao, R.; Deng, S.; Wang, F.; Hu, Y.; Wang, Y.; Lu, Y.; et al. *Populus euphratica* XTH overexpression enhances salinity tolerance by the development of leaf succulence in transgenic tobacco plants. *J. Exp. Bot.* **2013**, *64*, 4225–4238. [CrossRef] [PubMed]

85. Zhang, Y.N.; Wang, Y.; Sa, G.; Zhang, Y.H.; Deng, J.Y.; Deng, S.R.; Wang, M.J.; Zhang, H.L.; Yao, J.; Ma, X.Y.; et al. *Populus euphratica* J3 mediates root K^+/Na^+ homeostasis by activating plasma membrane H^+-ATPase in transgenic Arabidopsis under NaCl salinity. *Plant Cell Tiss. Organ Cult.* **2017**, *131*, 75–88. [CrossRef]

86. Lang, T.; Deng, S.; Zhao, N.; Deng, C.; Zhang, Y.; Zhang, Y.; Zhang, H.; Sa, G.; Yao, J.; Wu, C.; et al. Salt-sensitive signaling networks in the mediation of K^+/Na^+ homeostasis gene expression in *Glycyrrhiza uralensis* roots. *Front. Plant Sci.* **2017**, *8*, 1403. [CrossRef] [PubMed]

International Journal of
Molecular Sciences

MDPI

Review

Role and Functional Differences of HKT1-Type Transporters in Plants under Salt Stress

Akhtar Ali [1], Albino Maggio [2], Ray A. Bressan [3] and Dae-Jin Yun [1,*]

[1] Department of Biomedical Science & Engineering, Konkuk University, Seoul 05029, Korea; gultkr@yahoo.com
[2] Department of Agriculture, University of Naples Federico II, Via Universita 100, I-80055 Portici, Italy; almaggio@unina.it
[3] Department of Horticulture and Landscape Architecture, Purdue University, West Lafayette, IN 47907-2010, USA; bressan@purdue.edu
* Correspondence: djyun@konkuk.ac.kr; Tel.: +02-450-0583

Received: 30 January 2019; Accepted: 25 February 2019; Published: 1 March 2019

Abstract: Abiotic stresses generally cause a series of morphological, biochemical and molecular changes that unfavorably affect plant growth and productivity. Among these stresses, soil salinity is a major threat that can seriously impair crop yield. To cope with the effects of high salinity on plants, it is important to understand the mechanisms that plants use to deal with it, including those activated in response to disturbed Na^+ and K^+ homeostasis at cellular and molecular levels. HKT1-type transporters are key determinants of Na^+ and K^+ homeostasis under salt stress and they contribute to reduce Na^+-specific toxicity in plants. In this review, we provide a brief overview of the function of HKT1-type transporters and their importance in different plant species under salt stress. Comparison between HKT1 homologs in different plant species will shed light on different approaches plants may use to cope with salinity.

Keywords: abiotic stresses; high salinity; HKT1; halophytes; glycophytes

1. Introduction

Plants are sessile organisms, which are continuously challenged by various biotic and abiotic environmental stresses, such as soil salinity, extreme temperatures, drought, nutrients deficiency or pathogen attack. These stresses have a tremendous impact on agricultural crops, reducing their potential yields by more than half [1]. Soil salinization is one of the most serious causes of stress for world's agriculture, and it is progressing in most agricultural regions [2–4]. In saline soils, the ability of plants to grow and complete their life cycle can be severely compromised. Increased salinity leads to cytosolic osmotic stress and sodium ion specific toxicity that exert a combined inhibitory effect on physiological, biochemical and developmental pathways [5–7]. To deal with toxic levels of Na^+, plants may restrict Na^+ influx, compartmentalize Na^+ to vacuoles and/or mobilize the un-avoidable influx of Na^+ outside the cell and/or in different regions/organs of the plants [8–11]. In addition, the ability to take up K^+, a plant essential nutrient, is also crucial under salinity stress [12–14].

In dealing with the potentially detrimental effects of Na^+, sodium transporters play a pivotal role in plant protection in saline environments. These include antiporters that extrude Na^+ from root cells and/or re-distribute Na^+ throughout different tissues (Salt-Overly-Sensitive or SOS pathway) so as to reduce toxicity in critical cellular regions and reestablish to some degree water homeostasis [15–17]. Symporters, known as HKT1-type transporters (high-affinity potassium transporter1) also contribute to Na^+ detoxification by retrieving/diverting Na^+ from the xylem stream to protect the shoots from Na^+ toxicity [18–21]. the function of this mechanism is to confine toxic ions to the roots, thus protecting above ground tissues from damage [8,9,22]. the critical role of HKT1 transporters under salt stress has been well characterized in a number of plant species including the model plant *Arabidopsis*, wheat, rice, sorghum, tomato, as well as in extremophile models such as *Eutrema parvula* and *Eutrema salsuginea* [8,23–28]. HKT1-type transporters mediate the balance between Na$^+$ and K$^+$ ions under salt stress, a function that has recently been reported also for HKT1 transporters in monocots [12–14,29]. In cereal crops such as wheat and rice, which contain multiple *HKT1*-type genes, some members of this transporters family have been identified as key components of plant salt stress tolerance [12,30,31].

2. Na$^+$ Homeostasis

There are two main drawbacks of salt stress: osmotic stress and ion imbalance. Osmotic stress is caused by a decrease of the available water in the soil due to a reduced osmotic potential which makes more difficult for a plant to extract water [32]. Ion imbalance is mostly caused by excessive accumulation of Na$^+$ ions, which can inhibit normal cellular functions [33]. To achieve protection against high salinity, plants need to activate mechanisms that regulate both Na$^+$ uptake and homoeostasis [34,35]. the orchestrated distribution of Na$^+$ ions throughout the entire plant body represents one crucial activity to keep Na$^+$ away from sites of metabolism [36]. Moreover, the management of Na$^+$ must be balanced against the control of specific ion toxicity and the uptake of K$^+$, which is essential for normal plant growth and development [36]. High Na$^+$/K$^+$ ratios in the cytosol are toxic to plants, inhibiting various processes such as K$^+$ absorption, vital enzyme reactions, protein synthesis and photosynthesis [5,37]. In order to control the adverse effects of salt stress, plants have evolved various adaptive mechanisms to control ion homeostasis regulated by several proteins, working alone or in a group. Among them, HKT1-type transporters regulate sodium homeostasis by keeping a balance between Na$^+$ and K$^+$ in the cytoplasm [14,19,21,22,38]. To further explain and sheds light on the role of HKT1-type transporters, we will briefly discuss their discovery and classification based on their cation selectivity.

3. Discovery of HKT1-Type Transporters

HKT1-type transporters have an important role in mediating the distribution of Na$^+$ within the plant by a repeated pattern of Na$^+$ removal from the xylem, particularly in the roots, so that the amount of Na$^+$ reaching the shoot becomes more easily manageable [20,21,39]. Since the discovery of HKT1 in the early 90s [8,40], many more HKT transporter homologs from other species and with different cation transport properties have been isolated, which opened a new area of salt stress signaling in plants [7,23,25,27,28,41–48]. Interestingly, HKT1-type transporters from various species received the same name independently of their specific transport characteristics. For example, HKT1 from wheat was named TaHKT2;1 whereas its homolog from *Arabidopsis* was named AtHKT1. Nevertheless, the cation transport properties of these two HKT1 transporters are different from each other, AtHKT1 is a Na$^+$ transporter whereas TaHKT2;1 is a K$^+$/Na$^+$ symporter [8,23]. Subsequently, due to their different cation selectivity, HKT1-transporters have been divided into different classes based on their cation transport properties [30,49].

4. Classification HKT1-Type Transporters

Homologs of *HKT1* genes and proteins have been identified in a number of plant species, including *Arabidopsis*. Typically, their ion selectivity has been characterized in yeast and/or *Xenopus oocytes* [23,41,43]. Based on protein structure and ion selectivity, HKT1-type transporters have been divided into two sub-classes with differences in the amino acids of the first pore domain (PD) of the protein as the main distinguishing feature. This was the basis of an international agreement for the nomenclature of HKT1-type transporters established back in 2006 [30]. Accordingly, members of class-1 contain a Ser (serine) residue at the first pore-loop domain (pA) and show higher selectivity for Na$^+$ than K$^+$ (Figure 1). In contrast, members of class-2 contain Gly (glycine) residues at the same position and are considered to function as Na$^+$/K$^+$ co-transporters (Figure 1) [24,30,50]. *Arabidopsis* contains a single copy *HKT1* gene, *AtHKT1*, that encodes for a member of class-1 and shows highly specific sodium influx when expressed in *Xenopus laevis* oocytes and *Saccharomyces cerevisiae* [23]. Monocots such as rice and wheat contain more than one copy of the *HKT1* gene and their coding proteins belong to both class-1 as well as class-2 [3,12,24,25].

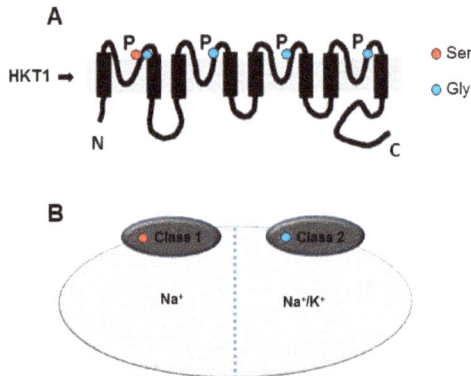

Figure 1. Classification and structure of HKT1 proteins. (**A**) Structural analysis of HKT1 protein containing Ser or Gly residues in their conserved regions. Members of class-1 transporters carry a Ser in the selectivity filter position while on the same position class-2 transporters contain a Gly residue. P denotes pore-loop domain while N and C indicate N-terminus and C-terminus of HKT-proteins. (**B**) Members of class-1 HKT1 that carry a Ser residue transport Na^+ while members of class-2 that carry a Gly residue can transport both Na^+ as well as K^+

5. Role of HKT1-Type Transporters in Glycophytes and Halophytes

HKT1-type transporters play a crucial role in Na^+ homeostasis. Knock-out of *HKT1* leads to NaCl sensitivity in *Arabidopsis* [19,20]. Homologs of *HKT1* have been isolated from different glycophytic species including wheat [8,40], *Arabidopsis* [19,21–23], rice [12,38,48], eucalyptus [43], barley [51], tomato [26], sorghum [27], strawberry [52], pumpkin [53], poplar [54]. the *Arabidopsis* genome contains a single *HKT1* gene that encodes for AtHKT1, a member of class-1 transporters. AtHKT1 acts as a high-affinity selective Na^+ transporter in heterologous systems such as *Xenopus oocytes* and yeast [19,23]. AtHKT1 has been shown to retrieve Na^+ from the xylem stream to reduce its transport and accumulation to the shoots [19,20]. This process prevents Na^+ toxicity in the shoots through recirculation of Na^+ to the roots from which it could be exported again [8,9,22]. On the other hand, members of class-2 transporters contribute to maintain a balanced Na^+/K^+ ratio in the cytoplasm under salt stress. the mode of action of class-2 transporters depends on the external Na^+ concentration. *TaHKT1* from wheat, a member of subclass-2, at low concentrations of Na^+ works as Na^+/K^+ symporter, but at high concentration of Na^+ TaHKT1 act as a Na^+ uniporter [8]. It has been recently demonstrated, via transgenic analysis, that over-expression of *AtHKT1* contributes to maintain optimal K^+/Na^+ in tobacco plants and to improve plant biomass under salt stress [55]. However, different modifications of HKT1 transporters may also cause variations of leaf Na^+ exclusion and salt tolerance in maize. Therefore, the exact mechanism through which HKT1 transporters confer salt tolerance deserves further attention. New insights in glycophytes have been obtained by comparative analysis of cultivated plants with wild relatives. In rice, different salt-tolerance during seed germination and seedling vegetative growth in weedy and cultivated plants has been associated to variants of *HKT1*-mediated transport and regulatory mechanisms that affect Na^+/K^+ balance [56].

Considering the activity of *HKT1* genes in glycophytic species, their functions in naturally salt-tolerant plants (halophytes) was also investigated [6,13,57]. Halophytes also control Na^+ toxicity based on efflux and re-distribution of Na^+ ions into various tissues to reduce its toxicity in specific plant organs and on Na^+ sequestration in the vacuole [13,58]. the *Arabidopsis* close relative *Eutrema salsuginea* (previously *Thellungiella halophila*) is a model halophyte [6,7,13]. *E. salsuginea*'s genome has been recently sequenced and it provides a resource to characterize the function of different genes in this species [11,59]. Although the precise nature of mechanisms that regulate halophytism is not fully understood [6,13,60,61], much progress has been made based on a comparative analysis of halophyte with glycophytes. Similar to glycophytes, halophytes also rely on genes coding for salt overly sensitive (SOS), vacuolar Na^+/H^+ antiporter (NHX) and sodium transporter (HKT1) proteins to cope with high salinity [7,28,59,60,62]. However, growing evidence indicates that these genes may have temporal and spatial expression patterns under normal and stress conditions that differentiate halophytes vs. glycophytes [7,11,28,58].

6. Functional Differences in Halophytic and Glycophytic HKTs

One close relative of *Arabidopsis* is the halophyte *Eutrema salsuginea* (previously *Thellungiella halophile* or *Thulengiella salsuginea*) [6,11,63]. *E. salsuginea*'s genome sequence is known and its juxta-positioning with the *Arabidopsis* genome provides a genetic blueprint that highlights similarities as well as differences between these genomes [59]. the genome of *E. salsuginea* includes three copies of *HKT1* genes in a tandem array [59]. Of the three *HKT1* homologs only *EsHKT1;2* is dramatically induced at the transcript level following salt stress [7]. When expressed in yeast, EsHKT1;2 shows high affinity for potassium whereas EsHKT1;1 more likely behaves as AtHKT1, with high specificity for sodium uptake [7]. Another *Arabidopsis* halophytic relative is *Eutrema parvula* (now *Schrenkiella parvula*). the genome of *Eutrema parvula* contains two *HKT1* genes, *EpHKT1;1* and *EpHKT1;2* [62]. *EpHKT1;2* is induced very rapidly upon salt stress [28]. All of these halophytic *HKT1* genes (three from *E. salsuginea* and two from *E. parvula*) contain a Ser residue at the selectivity filter in the first pore-loop domain and have therefore been grouped as class-1 transporters [30]. Members of class-1 HKT1-transporters lack the ability to uptake K^+. However, both EsHKT1;2 and EpHKT1;2 possess conserved Asp (aspartate) residue in the second pore-loop domain [7] and show K^+ uptake ability which makes them functionally different from other members of this class such as AtHKT1 (Figure 2).

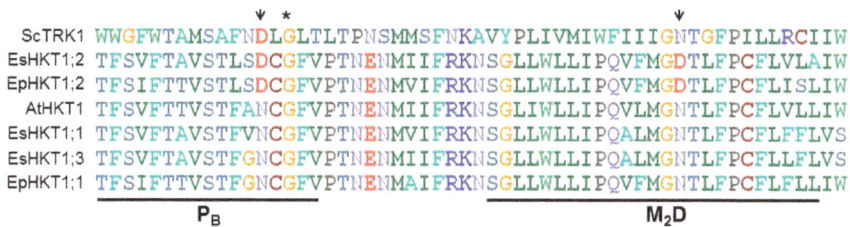

```
                ↓  *                                          ↓
ScTRK1   WWGFWTAMSAFNDLGLTLTPNSMMSFNKAVYPLIVMIWFIIIGNTGFPILLRCIIW
EsHKT1;2 TFSVFTAVSTLSDCGFVPTNENMIIFRKNSGLLWLLIPQVFMGDTLFPCFLVLAIW
EpHKT1;2 TFSIFTTVSTLSDCGFVPTNENMVIFRKNSGLLWLLIPQVFMGDTLFPCFLISLIW
AtHKT1   TFSVFTTVSTFANCGFVPTNENMIIFRKNSGLIWLLIPQVLMGNTLFPCFLVLLIW
EsHKT1;1 TFSVFTAVSTFVNCGFVPTNENMVIFRKNSGLLWLLIPQALMGNTLFPCFLFFLVS
EsHKT1;3 TFSVFTAVSTFGNCGFVPTNENMIIFRKNSGLLWLLIPQALMGNTLFPCFLLFLVS
EpHKT1;1 TFSIFTTVSTFGNCGFVPTNENMAIFRKNSGLLWLLIPQVFMGNTLFPCFLFLLIW
         ‾‾‾‾‾‾‾‾‾‾‾‾‾‾‾‾                 ‾‾‾‾‾‾‾‾‾‾‾‾‾‾‾‾‾‾‾‾
               PB                              M2D
```

Figure 2. Sequence comparison of HKT homologs from *Arabidopsis*, *E. salsuginea* and *E. parvula*. Amino acid sequences in the second pore loop region (PB) and the adjacent transmembrane domain (M2B) are aligned by clustalw (http://www.ebi.ac.uk/Tools/msa/clustalw2/). the conserved Gly residues in the PB region [49] are indicated by asterisks. the Asp residues specific for EsHKT1;2 (D207) and EpHKT1;2 (D205) are indicated by arrows.

Excess of $[Na^+]$ in the cytosol impairs the optimal cytosolic Na^+/K^+ ratio, which is recognized by the plant as K^+ deficiency indicating that potassium homeostasis is important for plants during salt stress [7,13,38,64]. Induction of *HKT1* under K^+ shortage would be detrimental if HKT1 was Na^+ specific [7,24,28,65]. Under salt stress, when the cytosolic sodium concentration reaches a toxic level, plants activate high-affinity potassium transporters to re-establish an optimal $[Na^+]/[K^+]$ cellular balance [12,14,27]. One example of high-affinity potassium transporters is EsHKT1;2 (and possibly EpHKT1;2). Down-regulation of *EsHKT1;2* in *E. salsuginea* leads to hypersensitive phenotypes under K^+-deficient conditions (Figures 3 and 4). Based on these findings and on the activation of EsHKT1;2 and EpHKT1;2 in response to high salinity, these genes can be considered major contributors to the halophytic nature of *E. salsuginea* and *E. parvula* [28].

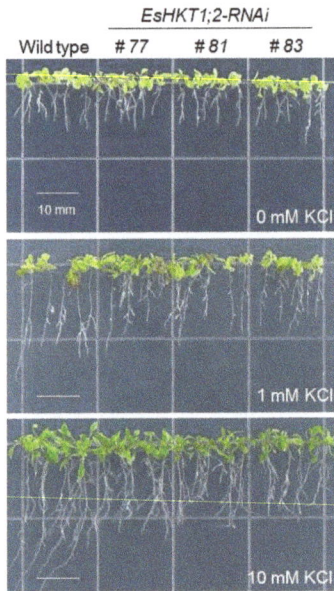

Figure 3. *EsHKT-RNAi* plants are sensitive to low K⁺-limiting conditions. Wild type and knock-down lines of *EsHKT1;2* (*EsHKT1;2-RNAi*) were grown on MS-medium for 10-days and then transferred to K⁺-deficient media with 0, 1 and 10 mM KCl (see Ali et al. 2012 for the detailed methodology) and allowed to grow for further 10-days. *EsHKT1;2-RNAi* lines were more sensitive to K⁺-limiting conditions as compared with the wild type Control. A gradual increase of K⁺ concentration greatly promotes the growth of wild type plants, whereas *EsHKT1;2-RNAi* lines were still sensitive. This result demonstrates the crucial role of EsHKT1;2 for K⁺-uptake.

Figure 4. *EsHKT-RNAi* plants are sensitive to salt stress. Wild type and knock-down lines of *EsHKT1;2* (*EsHKT-RNAi*) were grown on MS-medium for 2-weeks and then transferred to inert soil (porous soil, see Ali et al. 2012 for details) and grown for further 3-weeks. Plants were then treated with 300 mM NaCl for another 3-weeks period, twice a week (control represents untreated plants). *EsHKT-RNAi* lines were more sensitive to salt stress as compared with wild type Control.

7. Importance of Conserved Amino Acids in the 2nd Pore-Loop of HKT1 in Glycophytes and Halophytes

As discussed earlier, certain residues in the HKT transporters have a crucial role in the functioning of the transporter. Alignment of most published HKTs with ScTRK1 showed that both EsHKT1;2 and EpHKT1;2 contain conserved Asp residues in their second pore-loop domains (Asp207 and Asp205, respectively), (Figure 2) [14]. Yeast ScTRK1, a known high-affinity potassium transporter [66], also carries an Asp in the second pore-loop position (Figure 2). However, in most HKT1 proteins an Asn (asparagine) is present at the above said position (Asn211 in AtHKT1), while SlHKT1;1 and SbHKT1-4 both carry Ser residues (Ser264 and Ser277, respectively) [12,14,26]. These considerations were confirmed by showing that Asp207 and Asp205, could impart selectivity to subclass-1 HKT1 transporters. Asp207 to Asn207 in EsHKT1;2 and Asp205 to Asn205 in EpHKT1;2 were able to abolish potassium uptake and generate canonical subclass-1 Na$^+$-selective transporters [7,28]. In addition, changing the Asn residue in the 2nd pore-loop domain of AtHKT1 to Asp (N211D) resulted in a transporter that resembled EsHKT1;2 with high affinity for potassium transport. More importantly, transgenic *Arabidopsis* plants expressing AtHKT1^{N211D} tolerate salt stress more effectively than the wild type AtHKT1 and show exactly the same phenotype as *EsHKT1;2*- or *EpHKT1;2*-overexpressing Arabidopsis plants [14,28]. This means that HKTs from dicots can be differentiated from each other with respect to their monovalent cation selectivity by the presence of either Asp or Asn residues in the 2nd pore loop domain.

8. Substitution of Conserved Residues in the Pore-Region Affects the Cation Selectivity of HKT1-Transporters

The cation selectivity of HKT1 transporters is convertible by exchanging a single amino acid. Ser in the 1st pore loop domain appears not to be the only essential amino acid favoring K$^+$ uptake (at least in *Arabidopsis* and *Eutrema* species), but it possibly functions as a supporting residue. Nevertheless, Ser or Gly at the first pore-loop differentiates class-1 and class-2 HKT-transporters based on their Na$^+$ or Na$^+$/K$^+$ co-transport activity. However, this hypothesis failed to differentiate EsHKT1;2 which, although it contains a Ser residue at 1st pore-loop region and is a member of class-1 transporters, it unexpectedly functions as a K$^+$ transporter. In addition, EsHKT1;2 and EpHKT1;2 both contain a conserved Asp residue in the 2nd pore-loop region which is the key residue for their cation selectivity [14,28]. In contrast, it has been shown as indispensable the presence of an Asp (D) replacing an Asn residue (N211) to convert the Na$^+$ uniporter AtHKT1 into a Na$^+$/K$^+$ symporter [14]. Furthermore, the replacement of Asp with Asn in the EsHKT1;2 protein abolishes its potassium transport properties and it converts EsHKT1;2 into a Na$^+$ uniporter. Substitution of the corresponding Asn in AtHKT1 to Asp (N211D) confirmed the importance of this residue by expression in yeast cells, *Xenopus oocytes* and *Arabidopsis* (Figure 5) [14]. More recently, EpHKT1;2 was also shown to carry an Asp (D205) in the 2nd pore-loop domain. When Asp205 was substituted by Asn, EpHKT1;2 lost its ability to tolerate sodium stress in the presence of potassium [28]. According to these reports, HKT-type transporters possess several key amino acids which define their transport properties. In this regard, the presence of only Ser or Gly residues might not be sufficient to assign HKT-transporters to a specific class.

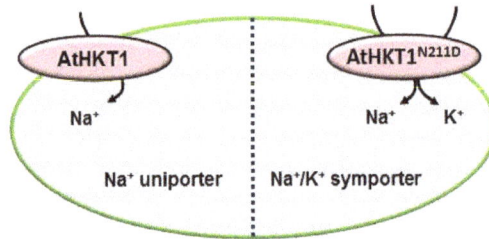

Figure 5. Functional properties of AtHKT1 and AtHKT1^{N211D} and differential selectivity for Na$^+$ and K$^+$ based on the Asn/Asp variance in the pore region. Wild type AtHKT1 is a sodium uniporter and does not confer salt stress tolerance. An altered version of AtHKT1 with a mutation of the Asn to Asp (AtHKT1^{N211D}) is also able to uptake potassium and confers salt stress tolerance. It has already been shown by Ali et al. 2016 that *athkt1-1* plants complemented by *AtHKT1^{N211D}* showed higher tolerance to salt stress than lines complemented by the wild type *AtHKT1*. Thus, the introduction of Asp, replacing Asn, in HKT1-type transporters established altered cation selectivity and uptake dynamics.

9. Contribution of HKT1 in Plant Na$^+$ Homeostasis and Salinity Tolerance

A balance between Na$^+$ and K$^+$ ions under salt stress is crucial for plant survival [64], but it is not clear how such balance can be established under conditions of (hyper-) accumulation of Na$^+$ (and to some degree Cl$^-$) leading to osmotic stress and ionic imbalance [67]. the localization of AtHKT1 in the xylem parenchyma cells appears to provide an answer because its activity can reduce the flux of Na$^+$ to the shoot tip in the (for most plants extremely rare) conditions of excess Na$^+$ in the root-zone. It is believed that high-affinity potassium transporters will be active during salt stress. the presence and stress-induced activity of Na$^+$/H$^+$ antiporters have also been shown [11]. Yet other transporters are active in partitioning Na$^+$ into vacuoles which can act as ultimate sinks for sodium ions [68].

For plants that are exposed to an excess of Na$^+$, the function of HKT1 isoforms seems to have changed from a distribution role that curtails Na$^+$ flux throughout the plant into a supporting role as K$^+$ transporters. the ability of *Thellungiella* species to maintain a low cytosolic Na$^+$/K$^+$ ratio in the presence of high salinity stress has been shown [69]. Suppression of *HKT1* expression in *E. salsuginea* by RNAi leads to hyper-accumulation of sodium in the shoots, reduced sodium content in the roots and, consequently, a disturbed Na$^+$/K$^+$ ratio in the plant. Shoot sodium hyper-accumulation brings salt sensitivity suggesting that *EsHKT1;2* in *E. salsuginea* is one of the major components of its halophytic behavior (Figures 3 and 4). While the RNAi targeted all *EsHKT1* copies, it is likely that the phenotype of the RNAi lines mirrored the function of *EsHKT1;2*, which is the most highly expressed copy among the three tandem duplicated *HKT1* paralogs in the *E. salsuginea* genome [59].

Our recent findings showed that AtHKT1^{N-D}, like the native EsHKT1;2 transporter, contributed to salt tolerance. This was demonstrated based on the phenotype of AtHKT1^{N-D} both in yeast and transgenic Arabidopsis lines [14]. More importantly the current generated by AtHKT1^{N211D} in *xenopus oocytes* were more similar to that of EsHKT1;2 rather than AtHKT1, indicating that enhanced uptake of K$^+$ can reduce Na$^+$ toxicity [26]. Molecular and structural studies of both AtHKT1 and EsHKT1;2 and their mutated versions further explained that HKT1-proteins contain different charge distributions at the pore site [14].

10. Concluding Remarks

Salinity tolerance in plants is a very complex process whose components are only partially known. A comparative analysis of halophytic and glycophytic systems has helped to understand the nature and function of critical genes in salt stress adaptation [11,19,29,70]. It has been shown that ThSOS1 and EsHKT1;2, and more recently EpHKT1;2 are essential determinants of the halophytic behavior of *E. salsuginea* and *E. pavula* [7,11,28]. Unlike their *Arabidopsis* counterparts, strong induction and activity of EsHKT1;2 [7] and EpHKT1;2 [28] under salt stress might suggest that the co-evolution of these ion transporters play a critical role in shaping halophytic lifestyles of *E. salsuginea* and *E. parvula*. Functional studies in *Arabidopsis* homologs of these ion transporters and genetic duplication in halophytes may help us to understand how multiple ion tolerance has been acquired to support survival in an environment characterized by high levels of Na^+ [11,59,62,71,72]. Therefore, to better understand salt tolerance in crop plants, additional studies must be directed towards defining the regulatory mechanisms operating differently in glycophytic and halophytic HKT1 so to reconcile expression and protein location with the extremely high salt stress tolerance of these species. Improving our understanding of key functional mechanisms that halophytes use to cope with high salinity could help us in designing applications and strategies to improve salt stress tolerance in crop plants in the future.

Author Contributions: D.-J.Y. designed research, A.A. performed experiments, A.A., A.M., R.A.B. and D.-J.Y. analyzed the data, collected information from literature and wrote the paper.

Funding: This work was supported by grants from the National Research Foundation of Korea (NRF), funded by the Korean Government (MSIP No. 2016R1A2A1A05004931), Global Research Lab (2017K1A1A2013146), Next-Generation BioGreen 21 Program (SSAC, Grant # PJ01318201).

Conflicts of Interest: The author declares no conflict of interest.

References

1. Bray, E.A.; Bailey-Serres, J.; Weretilnyk, E. Responses to abiotic stresses. In *Biochemistry and Molecular Biology of Plants*; Gruissem, W., Buchannan, B., Jones, R., Eds.; American Society of Plant Physiologists: Rockville, MD, USA, 2000; pp. 1158–1249.

2. Wang, W.; Vinocur, B.; Altman, A. Plant responses to drought, salinity and extreme temperatures: Towards genetic engineering for stress tolerance. *Planta* **2003**, *218*, 1–14. [CrossRef] [PubMed]

3. Huang, S.; Spielmeyer, W.; Lagudah, E.S.; Munns, R. Comparative mapping of HKT1 genes in wheat, barley and rice, key determinants of Na^+ transport, and salt tolerance. *J. Exp. Bot.* **2008**, *59*, 927–937. [CrossRef] [PubMed]

4. Cirillo, V.; Masin, R.; Maggio, A.; Zanin, G. Crop-weed interactions in saline environments. *Eur. J. Agron.* **2018**, *99*, 51–61. [CrossRef]

5. Murguía, J.R.; Bellés, J.M.; Serrano, R. A salt-sensitive 3′(2′),5′-bisphosphate nucleotidase involved in sulfate activation. *Science* **1995**, *267*, 232–234. [CrossRef] [PubMed]

6. Inan, G.; Zhang, Q.; Li, P.; Wang, Z.; Cao, Z.; Zhang, H.; Zhang, C.; Quist, T.M.; Goodwin, S.M.; Zhu, J.; et al. Salt cress. A halophyte and cryophyte *Arabidopsis* relative model system and its applicability to molecular genetic analyses of growth and development of extremophiles. *Plant Physiol.* **2004**, *135*, 1718–1737. [CrossRef] [PubMed]

7. Ali, Z.; Park, H.C.; Ali, A.; Oh, D.H.; Aman, R.; Kropornicka, A.; Hong, H.; Choi, W.; Chung, W.S.; Kim, W.Y.; et al. TsHKT1;2, a HKT1 Homolog from the Extremophile *Arabidopsis* Relative *Thellungiella salsuginea*, Shows K^+ Specificity in the Presence of NaCl. *Plant Physiol.* **2012**, *158*, 1463–1474. [CrossRef] [PubMed]

8. Rubio, F.; Gassmann, W.; Schroeder, J.I. Sodium-driven potassium uptake by the plant potassium transporter HKT1 and mutations conferring salt tolerance. *Science* **1995**, *270*, 1660–1663. [CrossRef] [PubMed]

9. Rus, A.; Yokoi, S.; Sharkhuu, A.; Reddy, M.; Lee, B.H.; Matsumoto, T.K.; Koiwa, H.; Zhu, J.K.; Bressan, R.A.; Hasegawa, P.M. AtHKT1 is a salt tolerance determinant that controls Na^+ entry into plant roots. *Proc. Natl. Acad. Sci. USA* **2001**, *98*, 14150–14155. [CrossRef] [PubMed]

10. Quintero, F.J.; Ohta, M.; Shi, H.Z.; Zhu, J.K.; Pardo, J.M. Reconstitution in yeast of the *Arabidopsis* SOS signaling pathway for Na^+ homeostasis. *Proc. Natl. Acad. Sci. USA* **2002**, *99*, 9061–9066. [CrossRef] [PubMed]

11. Oh, D.H.; Leidi, E.; Zhang, Q.; Hwang, S.M.; Li, Y.; Quintero, F.J.; Jiang, X.; D'Urzo, M.P.; Lee, S.Y.; Zhao, Y.; et al. Loss of halophytism by interference with SOS1 expression. *Plant Physiol.* **2009**, *151*, 210–222. [CrossRef] [PubMed]

12. Yao, X.; Horie, T.; Xue, S.; Leung, H.Y.; Katsuhara, M.; Brodsky, D.E.; Wu, Y.; Schroeder, J.I. Differential sodium and potassium transport selectivities of the rice OsHKT2;1 and OsHKT2;2 transporters in plant cells. *Plant Physiol.* **2010**, *152*, 341–355. [CrossRef] [PubMed]

13. Shao, Q.; Han, N.; Ding, T.; Zhou, F.; Wang, B. SsHKT1;1 is a potassium transporter of the C3 halophyte *Suaeda salsa* that is involved in salt tolerance. *Funct. Plant Biol.* **2014**, *41*, 790–802. [CrossRef]

14. Ali, A.; Raddatz, N.; Aman, R.; Kim, S.; Park, H.C.; Jan, M.; Baek, D.; Khan, I.U.; Oh, D.H.; Lee, S.Y.; et al. A single amino acid substitution in the sodium transporter HKT1 associated with plant salt tolerance. *Plant Physiol.* **2016**, *171*, 2112–2126. [CrossRef] [PubMed]

15. Qiu, Q.S.; Guo, Y.; Dietrich, M.A.; Schumaker, K.S.; Zhu, J.K. Regulation of SOS1, a plasma membrane Na$^+$/H$^+$ exchanger in *Arabidopsis thaliana*, by SOS2 and SOS3. *Proc. Natl. Acad. Sci. USA* **2002**, *99*, 8436–8441. [CrossRef] [PubMed]

16. Oh, D.H.; Lee, S.Y.; Bressan, R.A.; Yun, D.J.; Bohnert, H.J. Intracellular consequences of SOS1 deficiency during salt stress. *J. Exp. Bot.* **2010**, *61*, 1205–1213. [CrossRef] [PubMed]

17. Quintero, F.J.; Martinez-Atienza, J.; Villalta, I.; Jiang, X.; Kim, W.Y.; Ali, Z.; Hiroaki, F.; Imelda, M.; Yun, D.J.; Zhu, J.K.; et al. Activation of the plasma membrane Na/H antiporter Salt-Overly-Sensitive 1 (SOS1) by phosphorylation of an auto-inhibitory C-terminal domain. *Proc. Natl. Acad. Sci. USA* **2011**, *108*, 2611–2616. [CrossRef] [PubMed]

18. Maser, P.; Eckelman, B.; Vaidyanathan, R.; Horie, T.; Fairbairn, D.J.; Kubo, M.; Yamagami, M.; Yamaguchi, K.; Nishimura, M.; Uozumi, N.; et al. Altered shoot/root Na$^+$ distribution and bifurcating salt sensitivity in *Arabidopsis* by genetic disruption of the Na$^+$ transporter AtHKT1. *FEBS Lett.* **2002**, *531*, 157–161. [CrossRef]

19. Berthomieu, P.; Conejero, G.; Nublat, A.; Brackenbury, W.J.; Lambert, C.; Savio, C.; Uozumi, N.; Oiki, S.; Yamada, K.; Cellier, F.; et al. Functional analysis of AtHKT1 in *Arabidopsis* shows that Na$^+$ recirculation by the phloem is phloem is crucial for salt tolerance. *EMBO J.* **2003**, *22*, 2004–2014. [CrossRef] [PubMed]

20. Sunarpi, H.T.; Horie, T.; Motoda, J.; Kubo, M.; Yang, H.; Yoda, K.; Horie, R.; Chan, W.Y.; Leung, H.Y.; Hattori, K.; et al. Enhanced salt tolerance mediated by AtHKT1 transporter-induced Na unloading from xylem vessels to xylem parenchyma cells. *Plant J.* **2005**, *44*, 928–938. [CrossRef] [PubMed]

21. Møller, I.S.; Gilliham, M.; Jha, D.; Mayo, G.M.; Roy, S.J.; Coates, J.C.; Haseloff, J.; Tester, M. Shoot Na$^+$ exclusion and increased salinity tolerance engineered by cell type-specific alteration of Na$^+$ transport in *Arabidopsis*. *Plant Cell* **2009**, *21*, 2163–2178. [CrossRef] [PubMed]

22. Davenport, R.J.; Muñoz-Mayor, A.; Jha, D.; Essah, P.A.; Rus, A.; Tester, M. the Na$^+$ transporter AtHKT1;1 controls retrieval of Na$^+$ from the xylem in *Arabidopsis*. *Plant Cell Environ.* **2007**, *30*, 497–507. [CrossRef] [PubMed]

23. Uozumi, N.; Kim, E.J.; Rubio, F.; Yamaguchi, T.; Muto, S.; Tsuboi, A.; Bakker, E.P.; Nakamura, T.; Schroeder, J.I. the *Arabidopsis* HKT1 gene homolog mediates inward Na$^+$ currents in *Xenopus laevis* oocytes and Na$^+$ uptake in *Saccharomyces cerevisiae*. *Plant Physiol.* **2000**, *122*, 1249–1259. [CrossRef] [PubMed]

24. Horie, T.; Yoshida, K.; Nakayama, H.; Yamada, K.; Oiki, S.; Shinmyo, A. Two types of HKT transporters with different properties of Na$^+$ and K$^+$ transport in *Oryza sativa*. *Plant J.* **2001**, *27*, 129–138. [CrossRef] [PubMed]

25. Golldack, D.; Su, H.; Quigley, F.; Kamasani, U.R.; Munoz-Garay, C.; Balderas, E.; Popova, O.V.; Bennett, J.; Bohnert, H.J.; Pantoja, O. Characterization of a HKT-type transporter in rice as a general alkali cation transporter. *Plant J.* **2002**, *31*, 529–542. [CrossRef] [PubMed]

26. Asins, M.J.; Villalta, I.; Aly, M.M.; Olias, R.; Morales, P.A.D.; Huertas, R.; Li, J.; Jaime-perez, N.; Haro, R.; Raga, V.; et al. Two closely linked tomato HKT coding genes are positional candidates for the major tomato QTL involved in Na$^+$/K$^+$ homeostasis. *Plant Cell Environ.* **2012**, *36*, 1171–1191. [CrossRef] [PubMed]

27. Wang, T.T.; Ren, Z.J.; Liu, Z.Q.; Feng, X.; Guo, R.Q.; Li, B.G.; Li, L.G.; Jing, H.-C. SbHKT1;4, a member of the high-affinity potassium transporter gene family from *Sorghum bicolor*, functions to maintain optimal Na$^+$/K$^+$ balance under Na$^+$ stress. *J. Integr. Plant Biol.* **2014**, *56*, 315–332. [CrossRef] [PubMed]

28. Ali, A.; Khan, I.U.; Jan, M.; Khan, H.A.; Hussain, S.; Nisar, M.; Chung, W.S.; Yun, D.-J. the High-Affinity Potassium Transporter EpHKT1;2 From the Extremophile Eutrema parvula Mediates Salt Tolerance. *Front. Plant Sci.* **2018**, *9*, 1108. [CrossRef] [PubMed]

29. Ali, A.; Park, H.C.; Aman, R.; Ali, Z.; Yun, D.J. Role of HKT1 in *Thellungiella salsuginea*, a model extremophile plant. *Plant Signal. Behav.* **2013**, *8*, e25196. [CrossRef] [PubMed]

30. Platten, J.D.; Cotsaftis, O.; Berthomieu, P.; Bohnert, H.J.; Davenport, R.J.; Fairbairn, D.J.; Horie, T.; Leigh, R.A.; Lin, H.X.; Luan, S.; et al. Nomenclature for HKT transporters, key determinants of plant salinity tolerance. *Trends Plant Sci.* **2006**, *11*, 372–374. [CrossRef] [PubMed]

31. Munns, R.; James, R.A.; Xu, B.; Athman, A.; Conn, S.J.; Jordans, C.; Byrt, C.S.; Hare, R.A.; Tyerman, S.D.; Tester, M.; et al. Wheat grain yield on saline soils is improved by an ancestral Na$^+$ transporter gene. *Nat. Biotechnol.* **2012**, *30*, 360–366. [CrossRef] [PubMed]

32. Cheong, M.S.; Yun, D.J. Salt-stress signaling. *J. Plant Biol.* **2008**, *50*, 148–155. [CrossRef]

33. Munns, R.; Tester, M. Mechanisms of salinity tolerance. *Annu. Rev. Plant Biol.* **2008**, *59*, 651–681. [CrossRef] [PubMed]

34. Hasegawa, P.M.; Bressan, R.A.; Zhu, J.K.; Bohnert, H.J. Plant Cellular and Molecular Responses to High Salinity. *Annu. Rev. Plant Physiol. Plant Mol. Biol.* **2000**, *51*, 463–499. [CrossRef] [PubMed]

35. Tester, M.; Davenport, R. Na$^+$ tolerance and Na$^+$ transport in higher plants. *Ann. Bot.* **2003**, *91*, 503–527. [CrossRef] [PubMed]

36. Hauser, F.; Horie, T. A conserved primary salt tolerance mechanism mediated by HKT transporters: A mechanism for sodium exclusion and maintenance of high K$^+$/Na$^+$ ratio in leaves during salinity stress. *Plant Cell Environ.* **2009**, *33*, 552–565. [CrossRef] [PubMed]

37. Tsugane, K.; Kobayashi, K.; Niwa, Y.; Ohba, Y.; Wada, K.; Kobayashi, H. A recessive *Arabidopsis* mutant that grows photo-autotrophically under salt stress shows enhanced active oxygen detoxification. *Plant Cell* **1999**, *11*, 1195–1206. [CrossRef] [PubMed]

38. Ren, Z.H.; Gao, J.P.; Li, L.G.; Cai, X.L.; Huang, W.; Chao, D.Y.; Zhu, M.Z.; Wang, Z.Y.; Luan, S.; Lin, H.X. A rice quantitative trait locus for salt tolerance encodes a sodium transporter. *Nat. Genet.* **2005**, *37*, 1141–1146. [CrossRef] [PubMed]

39. Chen, Z.H.; Zhou, M.X.; Newman, I.A.; Mendham, N.J.; Zhang, G.P.; Shabala, S. Potassium and sodium relations in salinised barley tissues as a basis of differential salt tolerance. *Funct. Plant Biol.* **2007**, *34*, 150–162. [CrossRef]

40. Schachtman, D.P.; Schroeder, J.I. Structure and transport mechanism of a high-affinity potassium uptake transporter from higher-plants. *Nature* **1994**, *370*, 655–658. [CrossRef] [PubMed]

41. Rubio, F.; Schwarz, M.; Gassmann, W.; Schroeder, J.I. Genetic selection of mutations in the high affinity K$^+$ transporter HKT1 that define functions of a loop site for reduced Na$^+$ permeability and increased Na$^+$ tolerance. *J. Biol. Chem.* **1999**, *274*, 6839–6847. [CrossRef] [PubMed]

42. Gassmann, W.; Rubio, F.; Schroeder, J.I. Alkali cation selectivity of the wheat root high-affinity potassium transporter HKT1. *Plant J.* **1996**, *10*, 869–882. [CrossRef]

43. Fairbairn, D.J.; Liu, W.H.; Schachtman, D.P.; Gomez-Gallego, S.; Day, S.R.; Teasdale, R.D. Characterization of two distinct HKT1-like potassium transporters from *Eucalyptus camaldulensis*. *Plant Mol. Biol.* **2000**, *43*, 515–525. [CrossRef] [PubMed]

44. Garciadeblás, B.; Senn, M.E.; Banuelos, M.A.; Rodriguez-Navarro, A. Sodium transport and HKT transporters: the rice model. *Plant J.* **2003**, *34*, 788–801. [CrossRef] [PubMed]

45. Su, H.; Balderas, E.; Vera-Estrella, R.; Golldack, D.; Quigley, F.; Zhao, C.S.; Pantoja, O.; Bohnert, J.H. Expression of the cation transporter McHKT1 in a halophyte. *Plant Mol. Biol.* **2003**, *52*, 967–980. [CrossRef] [PubMed]

46. Haro, R.; Banuelos, M.A.; Senn, M.A.E.; Barrero-Gil, J.; Rodríguez-Navarro, A. HKT1 mediates sodium uniport in roots: Pitfalls in the expression of HKT1 in yeast. *Plant Physiol.* **2005**, *139*, 1495–1506. [CrossRef] [PubMed]

47. Takahashi, R.; Liu, S.; Takano, T. Cloning and functional comparison of a high-affinity K$^+$ transporter gene PhaHKT1 of salt-tolerant and saltsensitive reed plants. *J. Exp. Bot.* **2007**, *58*, 4387–4395. [CrossRef] [PubMed]

48. Jabnoune, M.; Espeout, S.; Mieulet, D.; Fizames, C.; Verdeil, J.L.; Conejero, G.; Rodriguez-Navarro, A.; Sentenac, H.; Guiderdoni, E.; Abdelly, C.; et al. Diversity in expression patterns and functional properties in the rice HKT transporter family. *Plant Physiol.* **2009**, *150*, 1955–1971. [CrossRef] [PubMed]

49. Maser, P.; Hosoo, Y.; Goshima, S.; Horie, T.; Eckelman, B.; Yamada, K.; Yoshida, K.; Bakker, E.P.; Shinmyo, A.; Oiki, S.; et al. Glycine residues in potassium channel-like selectivity filters determine potassium selectivity in four-looper-subunit HKT transporters from plants. *Proc. Natl. Acad. Sci. USA* **2002**, *99*, 6428–6433. [CrossRef] [PubMed]

50. Kato, Y.; Sakaguchi, M.; Mori, Y.; Saito, K.; Nakamura, T.; Bakker, E.P.; Sato, Y.; Goshima, S.; Uozumi, N. Evidence in support of a four transmembrane-pore-transmembrane topology model for the *Arabidopsis thaliana* Na$^+$/K$^+$ translocating AtHKT1 protein, a member of the superfamily of K$^+$ transporters. *Proc. Natl. Acad. Sci. USA* **2001**, *98*, 6488–6493. [CrossRef] [PubMed]

51. Mian, A.; Oomen, R.J.; Isayenkov, S.; Sentenac, H.; Maathuis, F.J.; Very, A.A. Over-expression of an Na^+ -and K^+ -permeable HKT transporter in barley improves salt tolerance. *Plant J.* **2011**, *68*, 468–479. [CrossRef] [PubMed]

52. Garriga, M.; Raddatz, N.; Véry, A.A.; Sentenac, H.; Rubio-Meléndez, M.E.; González, W.; Dreyer, I. Cloning and functional characterization of HKT1 and AKT1 genes of Fragaria spp.—Relationship to plant response to salt stress. *J. Plant Physiol.* **2017**, *210*, 9–17. [CrossRef] [PubMed]

53. Sun, J.; Cao, H.; Cheng, J.; He, X.; Sohail, H.; Niu, M.; Huang, Y.; Bie, Z. Pumpkin CmHKT1;1 controls shoot Na^+ accumulation via limiting Na^+ transport from rootstock to scion in grafted cucumber. *Int. J. Mol. Sci.* **2018**, *19*, 2648. [CrossRef] [PubMed]

54. Xu, M.; Chen, C.; Cai, H.; Wu, L. Overexpression of pehkt1;1 improves salt tolerance in populous. *Genes* **2018**, *9*, 475. [CrossRef] [PubMed]

55. Wang, L.; Liu, Y.; Feng, S.; Wang, Z.; Zhang, J.; Zhang, J.; Wang, D.; Gan, Y. AtHKT1 gene regulating K^+ state in whole plant improves salt tolerance in transgenic tobacco plants. *Sci. Rep.* **2018**, *8*, 16585. [CrossRef] [PubMed]

56. Zhang, Y.; Fang, J.; Wu, X.; Dong, L. Na^+ /K^+ balance and transport regulatory mechanisms in weedy and cultivated rice (*Oryza sativa* L.) under salt stress. *BMC Plant Biol.* **2018**, *18*, 375. [CrossRef] [PubMed]

57. Vinocur, B.; Altman, A. Recent advances in engineering plant tolerance to abiotic stress: Achievements and limitations. *Curr. Opin. Biotechnol.* **2005**, *16*, 123–132. [CrossRef] [PubMed]

58. Gong, Q.; Li, P.; Ma, S.; Indu Rupassara, S.; Bohnert, H.J. Salinity stress adaptation competence in the extremo-phile *Thellungiella halophila* in comparison with its relative *Arabidopsis thaliana*. *Plant J.* **2005**, *44*, 826–839. [CrossRef] [PubMed]

59. Wu, H.J.; Zhang, Z.; Wang, J.Y.; Oh, D.H.; Dassanayake, M.; Liu, B.; Huang, Q.; Sun, H.X.; Xia, R.; Wu, Y.; et al. Insights into salt tolerance from the genome of *Thellungiella salsuginea*. *Proc. Natl. Acad. Sci. USA* **2012**, *109*, 12219–12224. [CrossRef] [PubMed]

60. Oh, D.H.; Hong, H.; Lee, S.Y.; Yun, D.J.; Bohnert, H.J.; Dassanayake, M. Genome Structures and Transcriptomes Signify Niche Adaptation for the Multiple-Ion-Tolerant Extremophyte *Schrenkiella parvula* (*Thellungiella parvula*). *Plant Physiol.* **2014**, *164*, 2123–2138. [CrossRef] [PubMed]

61. Vera-Estrella, R.; Barkla, B.J.; Pantoja, O. Comparative 2D-DIGE analysis of salinity responsive microsomal proteins from leaves of salt-sensitive *Arabidopsis thaliana* and salt-tolerant *Thellungiella salsuginea*. *J. Proteom.* **2014**, *111*, 113–127. [CrossRef] [PubMed]

62. Dassanayake, M.; Oh, D.H.; Haas, J.S.; Hernandez, A.; Hong, H.; Ali, S.; Yun, D.J.; Bressan, R.A.; Zhu, J.K.; Bohnert, H.J.; et al. the genome of the extremophile crucifer *Thellungiella parvula*. *Nat. Genet.* **2011**, *43*, 913–918. [CrossRef] [PubMed]

63. Oh, D.H.; Dassanayake, M.; Haas, J.S.; Kropornika, A.; Wright, C.; d'Urzo, M.P.; Hong, H.; Ali, S.; Hernandez, A.; Lambert, G.M.; et al. Genome structures and halophyte-specific gene expression of the extremophile *Thellungiella parvula* in comparison with *Thellungiella salsuginea* (*Thellungiella halophila*) and *Arabidopsis*. *Plant Physiol.* **2010**, *154*, 1040–1052. [CrossRef] [PubMed]

64. Qi, Z.; Spalding, E.P. Protection of plasma membrane K^+ transport by the salt overly sensitive1 Na^+/H^+ antiporter during salinity stress. *Plant Physiol.* **2004**, *136*, 2548–2555. [CrossRef] [PubMed]

65. Kader, M.A.; Seidel, T.; Golldack, D.; Lindberg, S. Expressions of OsHKT1, OsHKT2, and OsVHA are differentially regulated under NaCl stress in salt-sensitive and salt-tolerant rice (*Oryza sativa* L.) cultivars. *J. Exp. Bot.* **2006**, *57*, 4257–4268. [CrossRef] [PubMed]

66. Ko, C.H.; Gaber, R.F. TRK1 and TRK2 encode structurally related K^+ transporters in *Saccharomyces cerevisiae*. *Mol. Cell. Biol.* **1991**, *11*, 4266–4273. [CrossRef] [PubMed]

67. Shabala, S. Ionic and osmotic components of salt stress specifically modulate net ion fluxes from bean leaf mesophyll. *Plant Cell Environ.* **2000**, *23*, 825–837. [CrossRef]

68. Kim, B.G.; Waadt, R.; Cheong, Y.H.; Pandey, G.K.; Dominguez-Solis, J.R.; Schultke, S.; Lee, S.C.; Kudla, J.; Luan, S. the calcium sensorCBL10 mediates salt tolerance by regulating ion homeostasis in *Arabidopsis*. *Plant J.* **2007**, *52*, 473–484. [CrossRef] [PubMed]

69. Orsini, F.; D'Urzo, M.P.; Inan, G.; Serra, S.; Oh, D.H.; Mickelbart, M.V.; Consiglio, F.; Li, X.; Jeong, J.C.; Yun, D.J.; et al. A comparative study of salt tolerance parameters in 11 wild relatives of *Arabidopsis thaliana*. *J. Exp. Bot.* **2010**, *61*, 3787–3798. [CrossRef] [PubMed]

70. Shi, H.; Ishitani, M.; Kim, C.; Zhu, J.K. the *Arabidopsis thaliana* salt tolerance gene SOS1 encodes a putative Na$^+$/H$^+$ antiporter. *Proc. Natl. Acad. Sci. USA* **2000**, *97*, 6896–6901. [CrossRef] [PubMed]
71. Volkov, V.; Amtmann, A. *Thellungiella halophila*, a salt-tolerant relative of *Arabidopsis thaliana*, has specific root ion-channel features supporting K$^+$/Na$^+$ homeostasis under salinity stress. *Plant J.* **2006**, *48*, 342–353. [CrossRef] [PubMed]
72. Volkov, V.; Wang, B.; Dominy, P.J.; Fricke, W.; Amtmann, A. *Thellungiella* halophila, a salt-tolerant relative of *Arabidopsis thaliana*, possesses effective mechanisms to discriminate between potassium and sodium. *Plant Cell Environ.* **2003**, *27*, 1–14. [CrossRef]

International Journal of
Molecular Sciences

MDPI

Article

CaDHN5, a Dehydrin Gene from Pepper, Plays an Important Role in Salt and Osmotic Stress Responses

Dan Luo, Xiaoming Hou, Yumeng Zhang, Yuancheng Meng, Huafeng Zhang, Suya Liu, Xinke Wang and Rugang Chen *

College of Horticulture, Northwest A&F University, Yangling 712100, China; danluonwafu@163.com (D.L.); 15230286139@163.com (X.H.); Kexuanzhangyumeng@163.com (Y.Z.); YuanchengMeng07@126.com (Y.M.); 18848966687@163.com (H.Z.); YaSuLiu@126.com (S.L.); W1942399775@126.com (X.W.)
* Correspondence: rugangchen@nwsuaf.edu.cn; Tel./Fax: +86-29-87082613

Received: 15 April 2019; Accepted: 20 April 2019; Published: 23 April 2019

Abstract: Dehydrins (*DHNs*), as a sub-family of group two late embryogenesis-abundant (LEA) proteins, have attracted considerable interest owing to their functions in enhancing abiotic stress tolerance in plants. Our previous study showed that the expression of *CaDHN5* (a dehydrin gene from pepper) is strongly induced by salt and osmotic stresses, but its function was not clear. To understand the function of *CaDHN5* in the abiotic stress responses, we produced pepper (*Capsicum annuum* L.) plants, in which *CaDHN5* expression was down-regulated using VIGS (Virus-induced Gene Silencing), and transgenic *Arabidopsis* plants overexpressing *CaDHN5*. We found that knock-down of *CaDHN5* suppressed the expression of manganese superoxide dismutase (*MnSOD*) and peroxidase (*POD*) genes. These changes caused more reactive oxygen species accumulation in the VIGS lines than control pepper plants under stress conditions. *CaDHN5*-overexpressing plants exhibited enhanced tolerance to salt and osmotic stresses as compared to the wild type and also showed increased expression of salt and osmotic stress-related genes. Interestingly, our results showed that many salt-related genes were upregulated in our transgenic *Arabidopsis* lines under salt or osmotic stress. Taken together, our results suggest that *CaDHN5* functions as a positive regulator in the salt and osmotic stress signaling pathways.

Keywords: *Capsicum annuum* L.; *CaDHN5*; salt stress; osmotic stress; dehydrin

1. Introduction

Plants live in an open environment and cannot move from one place to another. As a result, plants are exposed to various biotic and abiotic stresses. These stresses individually, or in combination, result in huge losses in terms of growth, development, and yield, and sometimes threaten the survival of the plant. Amongst the abiotic factors, water stress is the most important [1,2]. Some earlier studies treated drought and salinity as similar stresses because plants respond in a similar manner to salt and drought stresses, and signaling mechanisms overlap [3].

Dehydrin, a highly hydrophilic plant protein, belongs to the second sub-family of the late embryogenesis developmental protein family (LEA II) [4]. The protein sequence has conserved K-segments (consisting of EKKGIMDKIKEKLPG located near the C-terminus, rich in lysine), Y-segments (consisting of T/VDEYGNP located close to the N-terminus), and S-segments (rich in serine) [5]. Dehydrins are classified into five categories: Y_nSK_n, K_n, SK_n, Y_nK_n, and K_nS [6]. Every type of dehydrin has a different function. For example, SK_n dehydrins can not only bind phospholipids, protect enzyme stability, and prevent heat-induced degeneration, but they are also crucial for plant growth, development, and resistance to low temperature stress responses [7]. Various abiotic stresses and hormones can strongly influence expression of dehydrin [8]. Studies have shown that Y_nSK_n is an alkaline or neutral protein that is highly upregulated by cold stress [9], and has a unique RRKK motif (a nuclear localization signal), which is key for the localization of Y_nSK_n-type dehydrins in the nucleus. Many studies have shown that there is a positive interaction between the expression of dehydrins and resistance to abiotic stresses [10]. *Cicer pinnatifidum* Y_2K-type dehydrin *CpDHN1* and white spruce S_8K_4-type dehydrin *PgDHN1* were induced by methyl jasmonate (MeJA) and salicylic acid (SA) [11,12]. *CaDHN5* belongs to the YSK_2 category of dehydrins. It is a neutral or basic protein that can be induced by osmotic or salt stresses, and exogenous abscisic acid (ABA), as shown in our previous study [13].

In recent years, many studies have explored the functional importance of dehydrin in plant stress resilience [14–20]. In addition, since the expression of various dehydrin genes can be induced by exogenous ABA treatment, dehydrins are also considered as ABA-responsive proteins (ABR) [14]. In tomato plants overexpressing dehydrin gene, drought resistance was enhanced without influencing tomato growth traits [15]. In barley, cold acclimation was due to faster *DHN5* accumulation rates in the winter lines compared to that of spring lines [16]. In *Physcomitrella*, under salt and mannitol stresses, the expression of *PpDHNA* and *PpDHNB* were strongly up-regulated [17]. Studies in transgenic *Arabidopsis thaliana* plants showed that overexpression of *AmDHN* (*Ammopiptanthus mongolicus* dehydrin) improved osmotic stress tolerance and drought resistance [18]. *MusaDHN-1*, an SK_3-type dehydrin gene in banana, contributes positively towards drought and salt stress tolerance, and responses to abscisic acid, ethylene, and methyl jasmonate [19]. Similarly, in *Boea crassifolia*, overexpression of $YNSK_2$-type dehydrin, *BcDh2* enhanced tolerance to mechanical stress, mediated by salicylic acid and jasmonic acid [20].

In a previous study, it was found that specific *DHNs* in pepper are differentially induced in response to different stresses [21]. Among the seven *Capsicum annuum* dehydrin genes, *CaDHN5* was significantly up-regulated under salt and osmotic stress treatments [13]. Therefore, in this study, we further explored the relationship between *CaDHN5* and salt and osmotic stresses via overexpressing and gene silencing techniques. The results showed that *CaDHN5*-silenced pepper plants were less tolerant to salt and osmotic stress, while *CaDHN5*-overexpressing *Arabidopsis* plants showed significantly increased tolerance to these stresses. These results suggest that *CaDHN5* functions as a positive regulator in salt and osmotic stress signaling pathways.

2. Results

2.1. Analysis of Silencing Efficiency of CaDHN5 in Pepper

Virus-Induced Gene Silencing (VIGS) technique was used to investigate the function of *CaDHN5* under salt and osmotic stresses [22,23]. About 310 bp specific sequences from *CaDHN5* were used to construct the vector pTRV2:*CaDHN5*. Phytoene desaturase (*PDS*) was used as a marker of gene silencing, due to its ability to induce a bleached phenotype after successfully silencing plants [22]. Approximately four weeks after the induction of TRV-mediated gene silencing, pepper plants induced with TRV2:*CaPDS* began to show an albino phenotype, while plants with the TRV2 empty vector and TRV2:*CaDHN5* showed no difference in phenotype (Figure 1a). We detected the expression level of the other six genes within the dehydrin family, and found that only *DHN5* expression level was down-regulated (Figure 1b). From the result of *CaDHN5* expression, it can be seen that the expression decreased by about 80% in the fourth week after induction in *CaDHN5*-silenced plants.

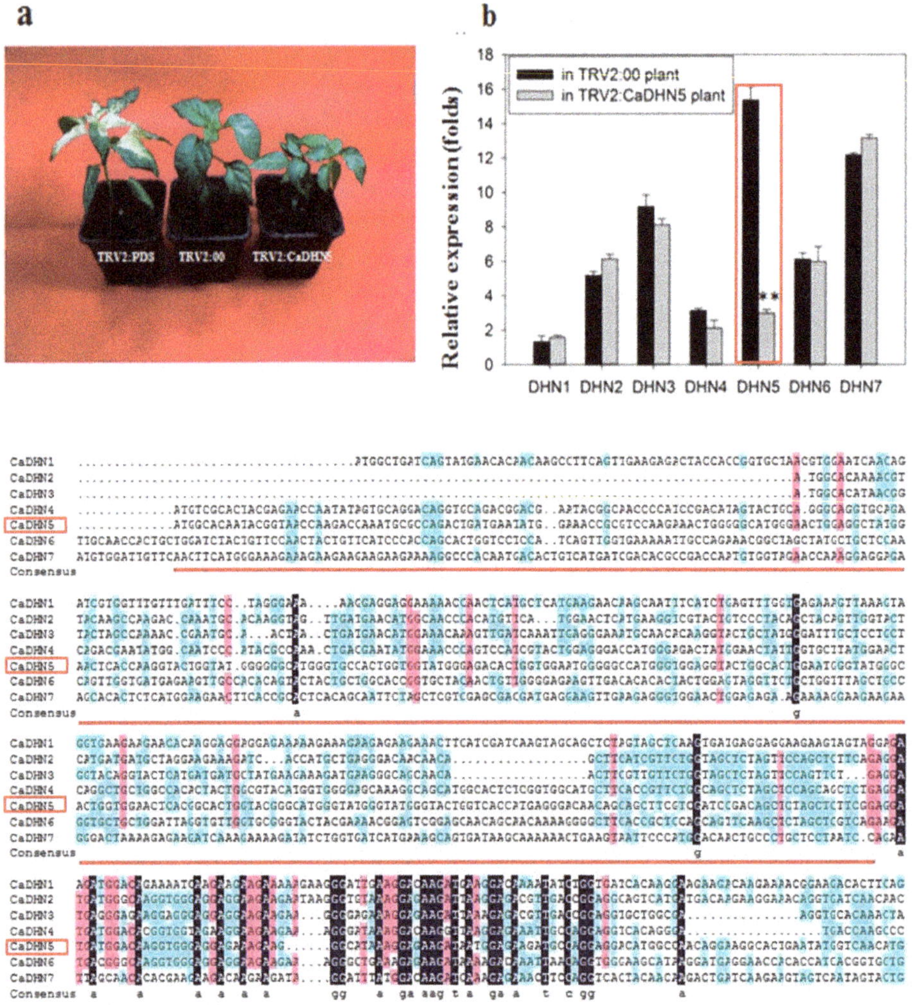

a

b

Figure 1. Phenotypes of *CaDHN5*-silenced plants and detection of gene silencing efficiency. (**a**) The phenotypes of silenced pepper plants (about four weeks after injection); (**b**) The relative expression of *CaDHN5* in silenced pepper plants and silencing sequence analysis of *CaDHN5*. The red frame is expression of *CaDHN5* in silenced and control plants. The black areas represent homology level 100%. The pink areas represent a level of homology greater than or equal to 75%. The blue areas represent a level of homology greater than or equal to 50%. The underlined part is a sequence of silencing. The results are the means ± standard deviation (S.D.), replicated three times. The means were compared using Student's test. Note: * indicates significant differences compared with the control at $p < 0.01$.

2.2. Influence of Silencing CaDHN5 on Tolerance of Salt and Osmotic Stresses in Pepper

To investigate the effect of *CaDHN5* silencing on the osmotic tolerance of pepper plants, control and *CaDHN5*-silenced plants were treated with 250 mM mannitol under continuous lighting conditions for three days. *CaDHN5*-silenced plants wilted considerably more than control plants, and some leaves became yellow after three days in mannitol-treated plants (Figure 2a).

In order to compare the differences between *CaDHN5*-silenced and control plants under mannitol treatment, we measured relative electrolyte leakage, rate of water loss, malondialdehyde (MDA), and chlorophyll content of these plants (Figure 2b–f). Following mannitol treatment, MDA levels in the control plants increased about four-fold, compared to control water treated plants, while in *CaDHN5*-silenced pepper plants, MDA levels increased by about six-fold (Figure 2b). After mannitol treatment, the proline content of silenced pepper plants increased two times as much as that seen in control plants (Figure 2c).

The rate of water loss and relative electrolyte leakage are indicators of the degree of membrane injury [24]. As can be seen from the relative electrolyte leakage measurements, the degree of membrane injury of pepper plants was significantly higher under mannitol treatment compared to controls (Figure 2d). Under 250 mM mannitol treatment, total chlorophyll content in control and silenced plants were both significantly decreased, and the difference between control and silenced plants was not significant (Figure 2e). In silenced plants, the rate of water loss increased, and the rate of water loss was three-fold lower than control plants after mannitol treatment (Figure 2f).

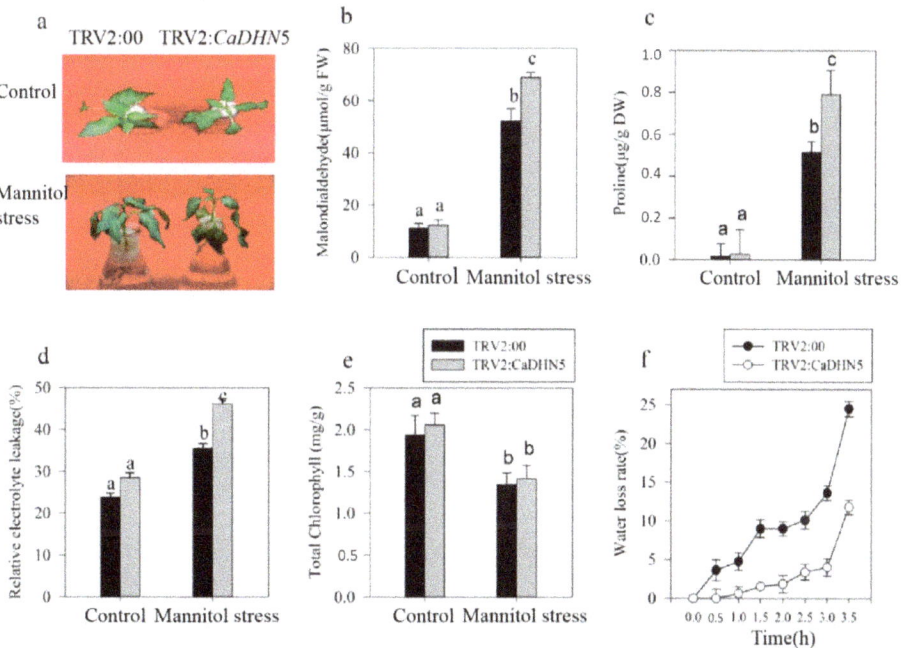

Figure 2. Effects of osmotic stress on plant phenotypes. (**a**) The phenotype of *CaDHN5*-silenced pepper plants under osmotic stress; (**b**) MDA content; (**c**) proline levels; (**d**) relative electrolytic leakage; (**e**) total Chlorophyll content; (**f**) water loss. Data that are significantly different are indicated with letters above the error bars (±S.D.). The different letters with the bars indicate significant differences as determined using Tukey HSD's multiple range tests ($p < 0.05$).

In normal condition the activities of superoxide dismutase (SOD) and peroxidase (POD) were not significantly different between control and silenced plants. However, after mannitol treatment, the activities of SOD and POD increased to scavenge superoxide anions and H_2O_2 produced in the plant. Therefore, the enzyme activities could reflect the ability of the plant to scavenge superoxide anions and H_2O_2. From Figure 3b,c, it can be clearly seen that in the *CaDHN5* silenced plant, the increase in enzyme activity was significantly lower than that of the control plant. The staining results of NBT also showed that after gene silencing, more superoxide anions accumulated in the leaves (Figure 3a).

We also analyzed the expression of the stress and antioxidant system-related genes (*MnSOD*, *POD*, and *ERD15* [23]) in control and silenced lines. There was no significant difference in the expression of

POD between control and silenced lines before treatment. After mannitol treatment, the expression of *POD* in control and silenced lines both increased, but in silenced plants it only increased two-fold, while in control plants this increase was four-fold (Figure 3d). A similar result was found for the expression of *MnSOD*. After *CaDHN5* silencing, the increased expression of *MnSOD* in silenced plants was only half compared to control plants (Figure 3e). In addition, the expression of *ERD15* was significantly higher in control plants treated with mannitol, while in silenced plants, increased expression of *ERD15* was significantly higher than control but less than mannitol-treated silenced pepper plants (Figure 3f).

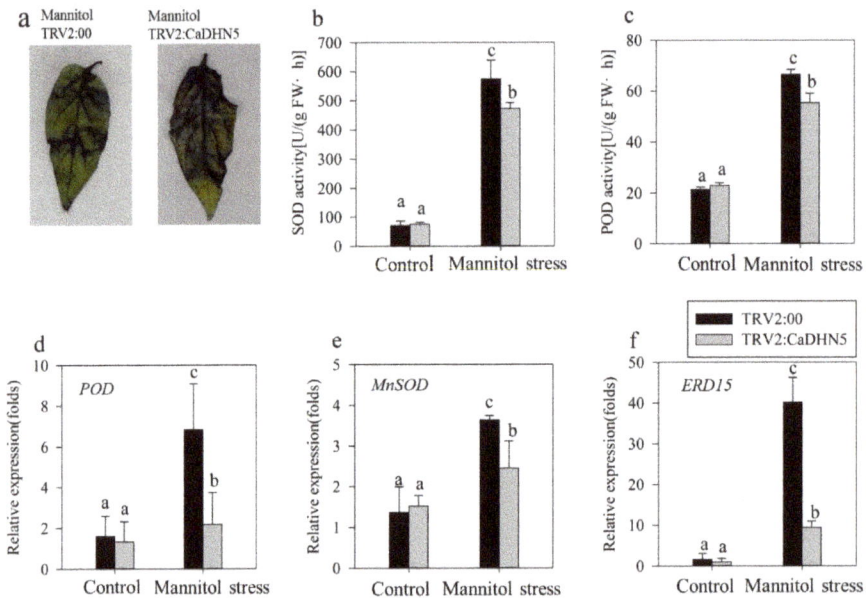

Figure 3. Determination of oxidative stress resistance and stress-related gene expression in mannitol treatment. (**a**) Results of pepper plants stained with NBT under mannitol treatment; (**b**) SOD activity under osmotic stress; (**c**) POD activity under mannitol treatment; (**d**) relative expression of *POD* under mannitol treatment; (**e**) *MnSOD* relative expression under mannitol treatment; (**f**) relative expression of *ERD15* under mannitol treatment. Data that are significantly different are indicated with letters above the error bars (±S.D.). The different letters with the bars indicate significant differences as determined using Tukey HSD's multiple range tests ($p < 0.05$).

In order to investigate the effect of silencing *CaDHN5* on salt stress in pepper, we measured the same physiological indices as those measured for mannitol stress, and performed the same analysis. Silenced and control plants were treated with 250 mM NaCl solution. Regarding the phenotype, the wilting conditions of silenced pepper plants were more evident under NaCl treatment (Figure 4a). Under normal conditions, there was no significant difference in the MDA content between control and silenced plants. However, after NaCl treatment, MDA levels in both control and silenced plants increased significantly; this increase in silenced plants was one and a half times more than that in control plants. Therefore, it appears that *CaDHN5*-silenced pepper plants experienced more serious membrane lipid peroxidation than control plants (Figure 4b). After NaCl treatment, the proline content of silenced pepper plants increased five-fold compared to control plants (Figure 4c). The chlorophyll content of *CaDHN5*-silenced pepper plants decreased more rapidly (Figure 4e). In silenced plants under NaCl treatment, the rate of water loss was four-fold faster than control plants (Figure 4f).

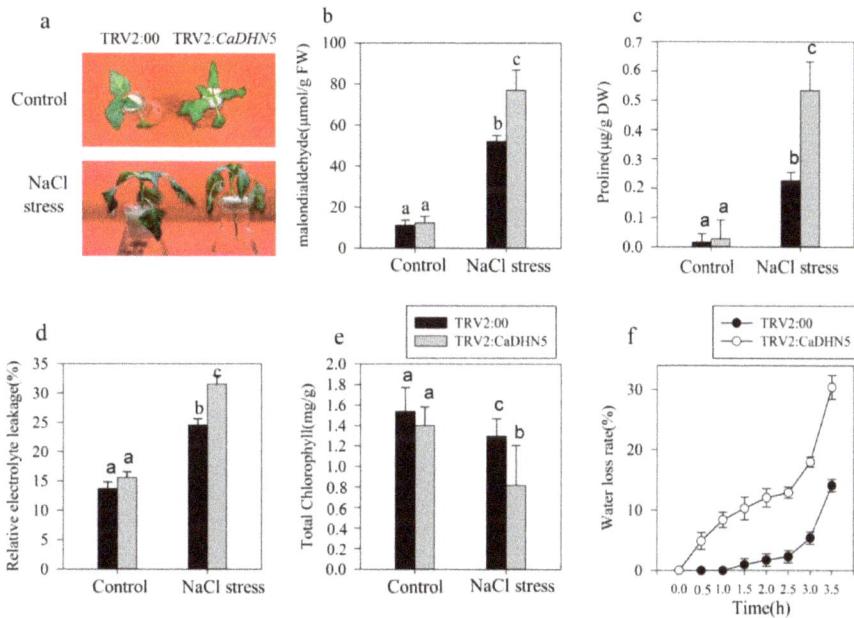

Figure 4. Effects of salt stress on plant phenotypes. (**a**) The phenotype of *CaDHN5*-silenced pepper plants under NaCl treatment; (**b**) MDA content; (**c**) proline levels; (**d**) relative electrolytic leakage; (**e**) total chlorophyll content; (**f**) water loss. Data that are significantly different are indicated with letters above the error bars (±S.D.). The different letters with the bars indicate significant differences as determined using Tukey HSD's multiple range tests ($p < 0.05$).

The activities of SOD and POD were measured under NaCl treatment. Results show that the activities increased in both silenced and control plants following NaCl treatment, but were slightly lowered in silenced plants compared to control pepper plants. These results suggested that the ability of plants to remove superoxide anions and H_2O_2 decreased slightly in silenced plants (Figure 5b,c). The NBT staining data reflected these observations (Figure 5a). In silenced plants treated with NaCl, NBT-stained leaves were more than that in the control lines. These data suggested that more superoxide anion accumulation occurred in silenced plants. We monitored the expression of POD and ERD15 (Figure 5d,f), which showed a significant increase in control plants treated with NaCl. MnSOD also exhibited a significant increase in expression, although this increase was less than that of POD and ERD15.

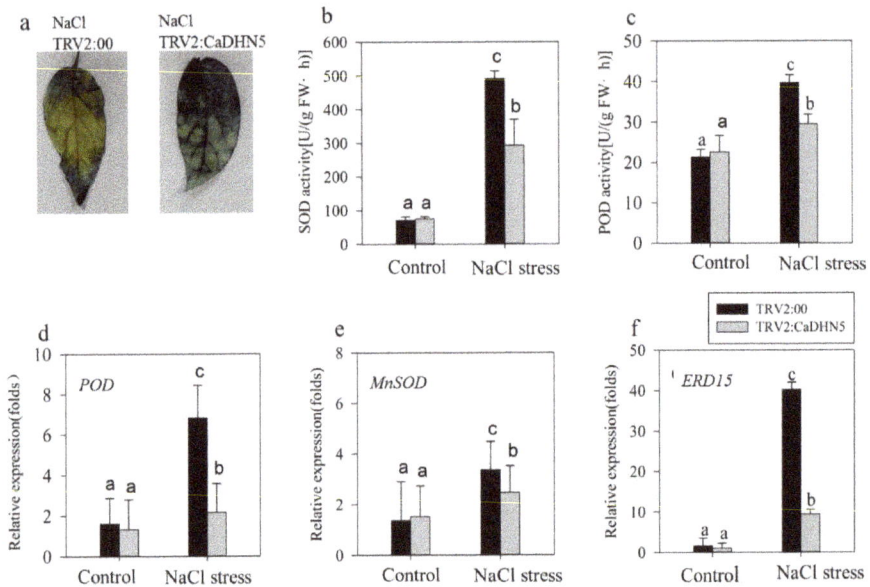

Figure 5. Determination of oxidative stress resistance and stress-related gene expression in NaCl treatment. (**a**) Results of NBT-stained pepper plants under NaCl treatment; (**b**) SOD activity under NaCl stress; (**c**) POD activity under NaCl treatment; (**d**) *POD* relative expression under NaCl treatment; € *MnSOD* relative expression under NaCl treatment; (**f**) *ERD15* relative expression under NaCl treatment. Data significantly different are indicated with letters above the error bars (±S.D.). The different letters with the bars indicate significant differences as determined using Tukey HSD's multiple range tests (*p* < 0.05).

2.3. Analysis of CaDHN5-Overexpression Arabidopsis

We constructed the overexpression vector pVBG2307:*CaDHN5*. The schematic diagram of the vector is shown in Figure 6a. After *Agrobacterium*-mediated transformation, the expression level of *CaDHN5* was estimated by qRT-PCR (Figure 6b). Expression of *CaDHN5* was higher in the lines D6 and D16 than other lines, so their homozygous T3 generation plants were chosen for further physiological analyses.

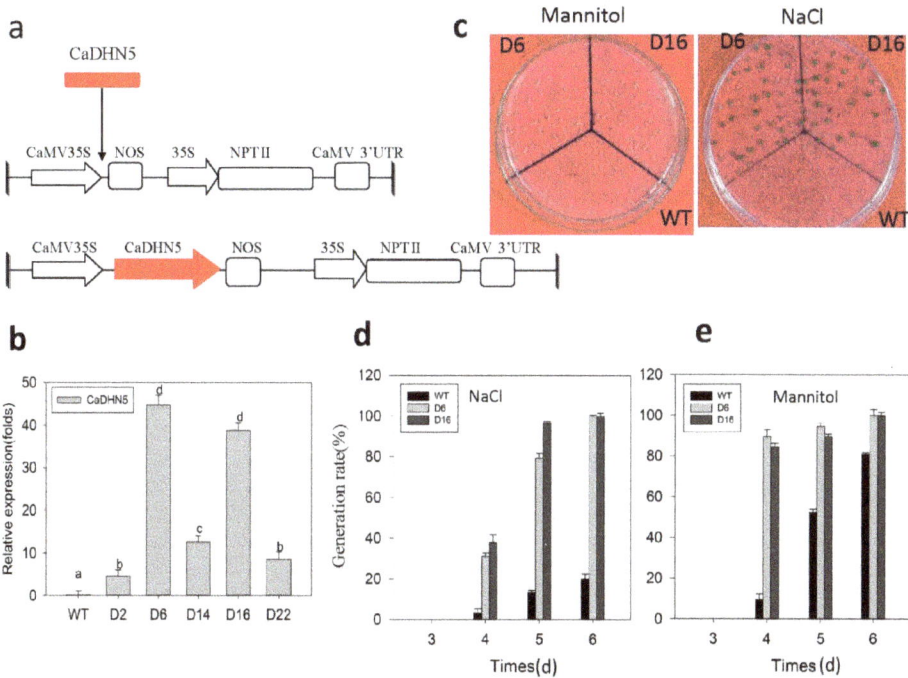

Figure 6. Assay of the transgenic *CaDHN5*-overexpressing lines. (**a**) Schematic representation of the pVBG2307:*CaDHN5* construct; (**b**) qRT-PCR analysis of *CaDHN5* expression in *Arabidopsis* transgenic lines (D2, D6, D14, D16, D22), with WT as control; (**c**) the phenotype of seed germination in wild type and two transgenic *Arabidopsis* plants (D6 and D16) subjected to salt (NaCl) and osmotic (mannitol) stress for five days; (**d**,**e**) seed germination rates of different lines subjected to salt (NaCl) and osmotic (mannitol) stress. Data significantly that are different are indicated with letters above the error bars (±S.D.). The different letters with the bars indicate significant differences as determined using Tukey HSD's multiple range tests ($p < 0.05$).

2.4. Seed Germination under Osmotic and Salt Stress Conditions

Transgenic *Arabidopsis* seeds were germinated on MS/2 agar medium containing 200 mM NaCl or mannitol solutions, and the germination rate was calculated (Figure 6c). Under NaCl treatment, at five days, almost all transgenic-seeds were germinated. However, only about 13% of WT seeds germinated (Figure 6d). Meanwhile, after six days, almost all transgenic seeds were germinated, while only 20% of the WT seeds were germinated. A similar trend was followed in the presence of mannitol, where the transgenic lines showed better germination compared to the WT seeds. At five days, almost all transgenic-seeds were germinated, while only 13% of WT seeds were germinated. In the following days, the germination rate of WT gradually increased, eventually reaching 80%, and transgenic seeds reached 100% (Figure 6e). These data show that the transgenic D6 and D16 lines displayed a better rate of seed germination than WT under salinity or osmotic stress.

2.5. Increased Tolerance of CaDHN5-Overexpressing Transgenic Arabidopsis Plants towards Salt and Osmotic Stresses

After three days treatment with 250 mM NaCl or mannitol, we observed the phenotype of *CaDHN5*-overexpressing transgenic *Arabidopsis* plants and measured physiological parameters (Figure 7).With mannitol treatment, the phenotypes of all *Arabidopsis thaliana* plants showed varying degrees of water loss, which occurred in all *Arabidopsis thaliana* plants, and the whole plants became brittle. After three days of mannitol treatment, WT leaves had suffered severe water loss and became brittle, while the leaves of transgenic plants remained moist (Figure 7a).

As can be seen from the content of MDA and chlorophyll (Figure 8b,c), the injury of wild-type plants was more serious under 250 mM mannitol treatment. Due to the influence of mannitol and salt, the chlorophyll content decreased to a similar extent in both WT and transgenic lines compared to controls (Figure 7b). The MDA content in WT following mannitol treatment increased by 15-fold compared to the control conditions, while in two transgenic lines D6 and D16, the increase was recorded to be about five-fold compared to controls (Figure 7c). After 250 mM NaCl treatment, the MDA content in WT increased by 12-fold, whereas in the two transgenic lines this increase was only about four-fold.

Based on previous research, we also selected several stress-related genes, *AtDREB2A*, *AtDREB2B* [25], *AtERD7*, and *AtMYC2* [26], to assess responses to osmotic stress, and *AtATR1/MYB34* [26], *AtSOS1* [25], *AtRITF1* [27], and *AtRSA1* for responses to salt stress in *Arabidopsis*. The relative expression levels of the above-mentioned genes were measured in WT and *CaDHN5*-overexpressing plants under stress and mannitol stresses (Figure 8). Only genes with significant changes are shown in Figure 8. *AtATR1/MYB34*, *AtSOS1*, and *AtRSA1* were up-regulated in both osmotic and salt stresses.

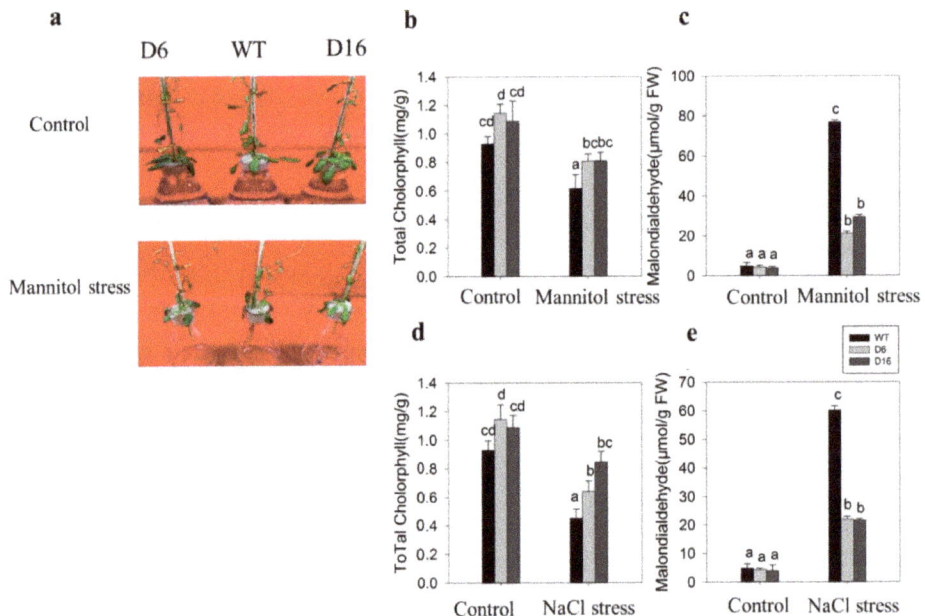

Figure 7. Related-physiological indices of *Arabidopsis thaliana* under salt and osmotic stress treatments. (a) The phenotypes of wild type (WT) and *CaDHN5*-overexpressing transgenic plants (D6 and D16) under mannitol treatment; (b) effects of mannitol treatment on total chlorophyll in transgenic *Arabidopsis* plants; (c) effects of mannitol treatment chlorophyll content in transgenic *Arabidopsis* plants; (d) effects of NaCl treatment on MDA content in transgenic *Arabidopsis* plants; (e) effects of NaCl treatment on total chlorophyll content in transgenic *Arabidopsis* plants. Data that are significantly different are indicated with letters above the error bars (±S.D.). The different letters with the bars indicate significant differences as determined using Tukey HSD's multiple range tests ($p < 0.05$).

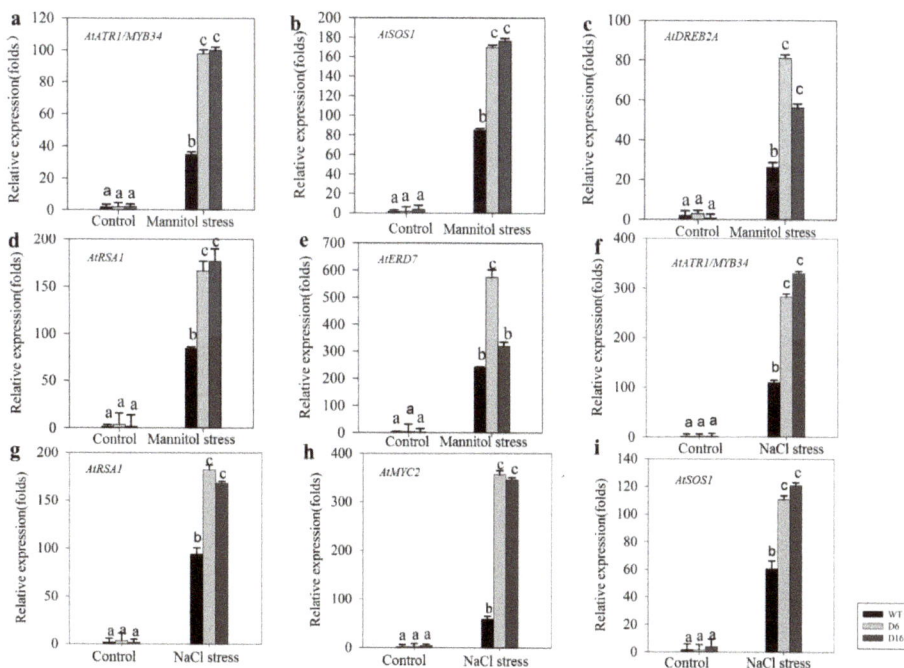

Figure 8. Expression of salt and osmotic related-genes in wild-type and transgenic plants. (**a**,**f**) *AtATR/MYB34* from *Arabidopsis* under mannitol and NaCl treanment; (**b**,**i**) *AtSOS1* from *Arabidopsis* under mannitol and NaCl treatments; (**c**) *AtDREB2A* from *Arabidopsis* under NaCl treatments; (**d**,**g**) *AtRSA1* from *Arabidopsis* under mannitol and NaCl treatments; (**e**) *AtERD7* from *Arabidopsis* under mannitol treatments; (**h**) *AtMYC2* from *Arabidopsis* under NaCl treanment. Data that are significantly different are indicated with letters above the error bars (±S.D.). The different letters with the bars indicate significant differences as determined using Tukey HSD's multiple range tests ($p < 0.05$).

3. Discussion

The LEA family of proteins were originally thought to be induced during seed maturation and drying [28]. In this study, *CaDHN5* cDNA was isolated from pepper leaves. Our results indicate *CaDHN5* shows a strong response to salt and osmotic stresses. In the experiments, NaCl and mannitol treatment were used to simulate salt and osmotic stresses. *CaDHN5* silenced and overexpressing transgenic plants were used to verify the function of *CaDHN5*. Silenced pepper plants were more sensitive to the effects of high salt and osmotic stresses, and *CaDHN5*-over-expressed plants were more tolerant than the WT plants.

We silenced *CaDHN5* in the pepper plant cultivar "P70". We first examined expression of *CaDHN5* in silenced pepper plants to ensure that subsequent experiments were carried out on the premise of successful gene silencing. Plant tolerance to stress is closely related to some physiological indices. It is well known that plants with strong stress tolerance usually have higher chlorophyll content and lower content of electrolyte leakage, proline, and MDA under stress situations. Under salt and osmotic stress conditions, the different trends in the decrease or increase of chlorophyll content, MDA, and conductivity suggest that *CaDHN5* may be involved in salt and osmotic stress responses. Meanwhile, these results indicated that the membrane damage and leaf senescence of the silenced *CaDHN5* pepper plants were higher under salt and osmotic stresses. Other studies in different plants have shown similar results [29,30]. Under salt stress and osmotic stresses, many *DHNs* were up-regulated in transgenic *Arabidopsis* plants, which showed high tolerance to these stresses [29]. It has also been found that the barley dehydrin *DHN3* responds to various stresses [30]. *POD* and *MnSOD* are important genes that function in the process of scavenging ROS. We found that when *CaDHN5* was silenced in pepper, the expression levels of these two genes were significantly lower than those of the control plants under salt and osmotic stress. This indicates that *CaDHN5* positively regulates the expression of these genes. In addition, results of enzyme activity and staining with NBT indicated that gene-silenced plants had higher levels of superoxide anion. *CaDHN5*-silenced pepper plants had lower tolerance to salt and osmotic stresses.

Further, we generated *CaDHN5* over-expressing transgenic *Arabidopsis*. Under high salinity and osmotic stress, it was found that when *CaDHN5* was overexpressed in *Arabidopsis*, it resulted in increased tolerance to salt and osmotic stress. Previous reports have described the increased anti-stress ability of different LEA genes in various plants, such as rice, wheat, and *Arabidopsis* [31,32]. The *MusaDHN1* gene of banana is not only induced by drought, salt, cold, oxidation, and heavy metal stress, but can also be induced by abscisic acid, ethylene, and methyl jasmonate [19]. In this study, *CaDHN5* transgenic lines are more tolerant under high concentrations of NaCl and mannitol. Transgenic seeds germinate rapidly under 200 mM mannitol compared to WT. Studies have shown that the dehydrin in Chinese cabbage has a similar function [33]. When we studied the influence of *CaDHN5* on salt stress tolerance, we found significant differences in MDA content between WT and transgenic plants. Meanwhile, overexpression of *CaDHN5* in *Arabidopsis* resulted in decreasing Chlorophyll degradation under NaCl treatment, but had no significant effect under osmotic stress. This could have resulted from the low accumulation of MDA in transgenic lines. In salt and osmotic stresses, the germination rates of *CaDHN5*-overexpressing *Arabidopsis* plants in the presence of NaCl and mannitol were significantly higher than those of WT plants. Monitoring the expression of other salts and osmotic stress-related genes showed that when *CaDHN5* was overexpressed, the expression levels of these stress-related genes also increased to varying degrees. Other studies have also shown similar results. When transgenic *Arabidopsis thaliana* transformed with wheat *TaDHN1* and *TaDHN3* genes were treated with salt and mannitol, the transgenic plants grew better and the root lengths were longer than wild type [34]. *HbDHN1*, *HbDHN2* were also transformed into *Arabidopsis thaliana*. *Arabidopsis thaliana* was transformed with *HbDHN1*, and *HbDHN2* reduced electrolyte leakage of cells and accumulation of ROS by increasing the SOD and POD activity, thereby resisting salt and osmotic stress [35]. In fact, among the expression of the eight related genes that we recorded, five genes (*AtATR1/MYB34*, *AtSOS1*, *AtDERB2A*, *AtRSA1*, and *AtERD7*) were up-regulated under osmotic stress, and four (*AtATR1/MYB34*, *AtRSA1*, *AtMYC2*, and *AtSOS1*) were up-regulated under salt stress. As mentioned earlier, signaling pathways in response to salt and osmotic stresses overlap [3]. The expression of *AtSOS1*, which encoded a plasma membrane Na^+/H^+ antiporter essential for salt tolerance. [27], was significantly increased in transgenic lines, both under salt and osmotic stresses. The expression of the transcription factor *AtDREB2A* in the ABA signaling pathway was also significantly increased. *AtRSA1* and *AtRITF1* are interacting genes that not only participate in the regulation of the transcription of several genes in the ROS scavenging system, but also regulate the expression of *AtSOS1*. It is worth noting that although the expression of *AtRSA1* gene was increased under salt and osmotic stresses, the interacting partners of *AtRSA1* and *AtRITF1* were only up-regulated under salt treatment.

4. Materials and Methods

4.1. Plant Materials, Growth Conditions

Seeds (wild-type: Columbia ecotype) and pepper (*Capsicum annuum* L.) cultivar "P70" were used in the current work, which were provided by Vegetable Plant Biotechnology and Germplasm Innovation laboratory, Northwest A&F University-China. The *Arabidopsis thaliana* seeds were treated as per Brini's method [32]. The pepper seedlings were cultured in a growth chamber by maintaining them in 16 h/8 h light/dark at 25 °C/20 °C [23]. The control plants were grown in the same environment and treated with corresponding solvents.

4.2. Isolation CaDHN5

According to the full-length CaDHN5 ORF sequence (GenBank accession No.: XM016705201), forward and reverse primers were designed as 5′-AGGAGATGGCACAATACGGT-3′AND5′-ATCCTTTGTTTTCATTTTCAGC-3′, respectively. PCR products were cloned into the pMD19-T vector (TaKaRa, Dalian, China) and sequenced (Xi'an AuGCT Biotechnologies Co. Xi'an, China).

4.3. Silencing Efficiency Analysis of CaDHN5 in Pepper

The pTRV2: *CaDHN5* construct was engineered to include a 310 bp sequence in *CaDHN5* cloned from a pepper cDNA template, using the forward primers 5′-ATGGCACAATACGGTAACC-3′and the reverse primers 5′-CCGAAGAGCTAGAGCTGTC-3′. The recombinant plasmid pTRV2: *CaDHN5* was constructed by combining CaDHN5 and pTRV2. *Agrobacterium tumefaciens* GV3101 containing pTRV2:*CaDHN5* was injected into pepper plants after combining GV3101 with pTRV1, and plants were grown as described previously. Fifty plants were used for the silencing assay [21].

4.4. Generation of Transgenic Arabidopsis Plants

The vector pVBG2307:*CaDHN5* contains the kanamycin resistance gene as a selectable marker between the 35S promoter and terminator (Figure 8a). Agrobacterium-mediated transformation was performed via the floral dipping technique of *Arabidopsis thaliana* (ecotype Columbia) [36]. Over-expressing transgenic plants were selected by growing seeds on MS/2 agar medium containing 50 mg/L kanamycin, which were grown up to the T3 generation to identify plants homozygous for the transgene.

4.5. Isolation of RNA, qRT-PCR

Total RNA was extracted from 200 mg of young leaves from *Arabidopsis transgenic* lines or silenced pepper plants using the RNeasy total RNA isolation kit (TianGen, Beijing, China). The cDNA was made by using PrimScript RT Kit (TaKaRa, Dalian, China). Primers are presented in Supplementary Table S1. The qRT-PCR was carried out as described previously [23]. The CaUbi3 gene (GenBank Accession No. AY486137.1) encoding the ubiquitin-conjugating protein was amplified from pepper plants as a reference gene for normalization of the *CaDHN5* cDNA samples [37], and the Atactin gene (GenBank Accession No. AY572427.1) was used as an internal control in *Arabidopsis* [38]. The relative fold difference in mRNA levels was determined using the $2^{-\Delta\Delta CT}$ method.

4.6. Measurement of Correlative Physiological Indices

4.6.1. Determination of MDA Content

Approximately 0.5 g of pepper leaves were weighed and rapidly ground with pre-chilled 10% trichloroacetic acid solution. Finally, to the mixed solution, 10% trichloroacetic acid was added to reconstitute the solution to 10 mL, and centrifuged at 4000 rpm for 10 min at 4 °C. A volume of 2 mL supernatant was taken and mixed with 2 mL 0.6% thiobarbituric acid solution. The mixed solution was heated in boiling water for 15 min and rapidly cooled. Following centrifugation at 4000 rpm for 10 min at 4 °C, the absorbance of the supernatant was measured at 532 nm, 450 nm, and 600 nm, according to the method described previously [39].

4.6.2. Total Chlorophyll Content

Pepper leaves (0.1 g) were immersed in 95% ethanol. After the leaves were completely decolored, the absorbance of the supernatant was measured at 470 nm, 649 nm, and 665 nm, as described previously [40].

4.6.3. Relative Electrolyte Leakage

Electrolyte leakage was measured according to the method described previously [41]. Leaves from treated and control plants were selected; 10 leaf discs were made by using a perforator, and the leaf discs were placed in a 50 mL centrifuge tube containing 10 mL of distilled water. After being kept at room temperature for 2 h, electrolyte leakage (EC1) was measured. The centrifuge tubes were heated in boiling water for 30 min after cooling, and the conductivity measurement value (EC2) was measured. Relative electrolyte leakage was calculated as (EC1/EC2) × 100.

4.6.4. Enzyme Activity

The SOD and POD activities were measured according to a previously described method [42]. Fresh leaves (0.5 g) were mixed with 8 mL PBS pH 7.8 and the mixture was centrifuged at 10,000 rpm for 15 min. The supernatant was considered as the crude enzyme extract. In the presence of hydrogen peroxide, POD can oxidize guaiacol to produce colored substances; the product concentration was calculated and POD activity was measured. SOD activity was determined by a similar principle, with NBT as the reaction substance.

4.6.5. NBT Staining

The NBT staining method was as used as described previously [43]. The plant leaves were immersed in a 0.1 mg/L NBT solution in Tris-HCl, pH 7.8, and vacuum infiltrated for about 1 min. After being incubated for 1 h in the dark, the leaves were placed in 80% ethanol, which was changed twice. After complete removal of chlorophyll, the degree of leaf staining was observed.

4.6.6. Water Loss Rate

Isolated plant leaves were placed on the laboratory bench (20–22 °C, humidity 45–60%) and their weight was measured every 30 min, as described previously [44]. The initial fresh weight of the leaves was recorded as W0, and thereafter weighed every 30 min. The leaf weight after 4 h was recorded as Wt. The water loss rate per 30 min was calculated as: (W0−Wt)/W0 × 100.

4.6.7. Proline Content

Approximately 0.5 g leaves were mixed with 5 mL of 3% sulfosalicylic acid; the mixture was placed in a 100 °C water bath for 10 min. After cooling, the mixture was centrifuged at 3000 rpm for 10 min. The supernatant (extraction solution, 2 mL) was mixed with a color rendering agent, indene (2 mL), and glacial acetic acid (2 mL). The mixed solution was heated in boiling water for about 40 min. A volume of 5 mL toluene was added into the mixing solution after cooling and the absorbance value was measured at 520 nm, according to a previously described method [45].

4.7. Statistical Analysis

The qRT-PCR data analysis was carried out using SPSS (Chicago, IL, USA). The relative expression levels of CaDHN5 under salt and osmotic stress are shown as mean ± SD of three biological replicate samples. Each replicate sample was a composite of leaves from three individual seedlings. Statistical analyses were performed using the SPSS (Chicago, IL, USA), and the means were compared using Tukey's HSD multiple range test, taking $p < 0.05$ as a significant difference.

5. Conclusions

In conclusion, although the physiological function of *CaDHN5* at a molecular level has not yet been identified, here we show that *Arabidopsis* plants overexpressing *CaDHN5* have higher survival rates in salt and osmotic stress conditions. These results suggest a functional role for *CaDHN5* in response to salt and osmotic stress. *Arabidopsis* plants overexpressing *CaDHN5* were significantly superior to WT in various physiological indices measured under salt and osmotic stresses. After gene silenced pepper plants, the tolerance of pepper plants to salt and mannitol were significantly decreased, and the above two factors jointly proved the effect of *CaDHN5* on plant tolerance to salt and osmotic stress.

Supplementary Materials: Supplementary materials can be found at http://www.mdpi.com/1422-0067/20/8/1989/s1.

Author Contributions: D.L., X.H., Y.Z., and R.C. conceived and designed the experiments. D.L., X.J., Y.M., and H.Z. performed the experiments. D.L., S.L., and X.W. analyzed the data. R.C. contributed reagents, materials, analysis tools. D.L. wrote the paper. All authors read and approved the final manuscript.

Funding: This research was funded by the National Natural Science Foundation of China (#31672146, #31201615) and the Natural Science Foundation of Shaanxi Province (2018JM3023).

Conflicts of Interest: The authors declare no conflict of interest.

References

1. Jaspers, P.; Kangasjarvi, J. Reactive oxygen species in abiotic stress signaling. *Physiol. Plant.* **2010**, *138*, 405–413. [CrossRef]
2. Ismail, A.M.; Hall, A.E.; Close, T.J. Purification and partial characterization of a dehydrin involved in chilling tolerance during seedling emergence of cowpea. *Plant Physiol.* **1999**, *120*, 237–244. [CrossRef] [PubMed]
3. Zhu, J.K. Salt and drought stress signal transduction in plants. *Annu. Rev. Plant. Biol.* **2002**, *53*, 247–273. [CrossRef] [PubMed]
4. Dure, L.; Crouch, M.; Harada, J. Common amino acid se-quence domains among the Lea proteins of higher plants. *Plant Mol. Biol.* **1989**, *12*, 475–486. [CrossRef]
5. Close, T.J. Dehydrins: A commonalty in the response of plants to dehydration and low temperature. *Physiol. Plant.* **1997**, *100*, 291–296. [CrossRef]
6. Kosova, K.; Vitamvas, P.; Prasil, I.T. Wheat and barley dehydrins under cold, drought, and salinity—what can LEA-II proteins tell us about plant stress response? *Front Plant Sci.* **2014**, *5*, 343. [CrossRef]
7. Kovacs, D.; Kalmar, E.; Torok, Z.; Tompa, P. Chaperone activity of ERD10 and ERD14, two disordered stress-related plant proteins. *Plant Physiol.* **2008**, *147*, 381–390. [CrossRef] [PubMed]
8. Close, T.J. Dehydrins: Emergence of a biochemical role of a family of plant dehydration proteins. *Physiol. Plant.* **1996**, *97*, 795–803. [CrossRef]
9. Zolotarov, Y.; Strmvik, M. De novo regulatory motif discovery identifies significant motifs in promoters of five classes of plandehydrin genes. *PLoS ONE* **2015**, *10*, 1522–1529. [CrossRef] [PubMed]
10. Riera, M.; Figueras, M.; López, C.; Goday, A.; Pagès, M. Protein kinase CK2 modulates developmental functions of the abscisic acidresponsive protein Rab17 from maiz. *Proc. Natl. Acad. Sci. USA* **2004**, *101*, 9879–9884. [CrossRef]
11. Richard, S.; Morency, M.J.; Drevet, C.; Jouanin, L.; Séguin, A. Isolation and characterization of a dehydrin gene from white spruce induced upon wounding, drought and cold stresses. *Plant Mol. Biol.* **2000**, *43*, 1–10. [CrossRef] [PubMed]
12. Bhattarai, T.; Fettig, S. Isolation and characterization of a dehydrin gene from Cicer pinnatifidum, a drought resistant wild relative of chickpea. *Physiol. Plant.* **2005**, *123*, 452–458. [CrossRef]
13. Jing, H.; Li, C.; Ma, F.; Ma, J.H.; Khan, A.; Wang, X. Genome-Wide Identification, Expression Diversication of Dehydrin Gene Family and Characterization of CaDHN3 in Pepper (*Capsicum annuum* L.). *PLoS ONE* **2016**, *11*, e0161073. [CrossRef]
14. Eriksson, S.K.; Kutzer, M.; Procek, J.; Gröbnercand, G.; Harryson, P. Tunable membrane binding of the intrinsically disordered dehydrin lti30, a cold-induced plant stress protein. *Plant Cell* **2011**, *23*, 2391–2404. [CrossRef]
15. Gerszberg, K.; HnatuszkoKonka, K. Tomato tolerance to abiotic stress: A review of most often engineered target sequences. *Plant Growth Regul.* **2017**, *83*, 175–198. [CrossRef]
16. Kosova, K.; Tom Prasil, I.; Prasilova, P.; Vitamvas, P.; Chrpova, J. The development of frost tolerance and DHN5 protein accumulation in barley (*Hordeum vulgare*) doubled haploid lines derived from Atlas 68 x Igri cross during cold acclimation. *J. Plant Physiol* **2010**, *67*, 343–350. [CrossRef]

17. Ruibal, C.; Salamó, I.P.; Carballo, V.; Castro, A.; Bentancor, M.; Borsani, O.; Szabados, L.; Vidal, S. Differential contribution of individual dehydrin genes from Physcomitrella patens to salt and osmotic stress tolerance. *Plant Sci.* **2012**, *190*, 89–102. [CrossRef]

18. Sun, J.; Nie, L.Z.; Sun, G.Q. Cloning and characterization of dehydrin gene from ammopiptanthus mongolicus. *Mol. Biol. Rep.* **2013**, *40*, 2281–2291. [CrossRef]

19. Shekhawat, U.K.; Srinivas, L.; Ganapathi, T.R. MusaDHN-1, a novel multiple stress-inducible SK3-type dehydrin gene, contributes affirmatively to drought and salt stress tolerance in banana. *Planta* **2011**, *234*, 915–932. [CrossRef]

20. Shen, Y.; Tang, M.J.; Hu, Y.L.; Lin, Z.P. Isolation and characterization of a dehydrin-like gene from drought tolerant Boea crassifolia. *Plant Sci.* **2004**, *166*, 1167–1175. [CrossRef]

21. Guo, W.L.; Chen, R.G.; Gong, Z.H.; Yin, Y.X.; Li, D.W. Suppression subtractive hybridization analysis of genes regulated by application of exogenous abscisic acid in pepper plant (*Capsicum annuum* L.) leaves under chilling stress. *PLoS ONE* **2013**, *8*, e66667. [CrossRef]

22. Wang, J.E.; Liu, K.K.; Li, D.W.; Zhang, Y.L.; Zhao, Q.; He, Y.M.; Gong, Z.H. A novel peroxidase CanPOD gene of pepper is involved in defense responses to Phytophthora capsici infection as well as abiotic stress tolerance. *Int. J. Mol. Sci.* **2013**, *14*, 3158–3177. [CrossRef]

23. Chen, R.G.; Jing, H.; Guo, W.L.; Wang, S.B.; Ma, F.; Pan, B.G. Silencing of dehydrin CaDHN1 diminishes tolerance to multiple abiotic stresses in *Capsicum annuum* L. *Plant Cell Rep.* **2015**, *34*, 2189–2200. [CrossRef]

24. Griffith, M.; Mclntyre, H.C.H. The interrelationship of growth and frost tolerance in winter rye. *Physiol. Plant.* **1993**, *87*, 335–344. [CrossRef]

25. Zhou, G.A.; Chang, R.Z.; Qiu, L.J. Overexpression of soybean ubiquitin-conjugating enzyme gene *GmUBC2* confers enhanced drought and salt tolerance through modulating abiotic stress-responsive gene expression in *Arabidopsis*. *Plant Mol. Biol.* **2010**, *72*, 357–367. [CrossRef]

26. Yuan, Y.; Fang, L.; Karungo, S.K. Overexpression of *VaPAT1*, a GRAS transcription factor from Vitis amurensis, confers abiotic stress tolerance in *Arabidopsis*. *Plant Cell Rep.* **2015**, *35*, 655. [CrossRef]

27. Guan, Q.; Wu, J.; Yue, X. A Nuclear Calcium-Sensing Pathway Is Critical for Gene Regulation and Salt Stress Tolerance in *Arabidopsis*. *PLoS Genetics* **2013**, *9*, e1003755. [CrossRef]

28. Bray, E.A.; BaileySerres, J.; Weretilnyk, E. Responses to abiotic stresses. In *Biochemistry and Molecular Biology of Plants*; Buchanan, B., Gruissem, W., Jones, R., Eds.; American Society of Plant Physiologists: Rockville, MD, USA, 2000; pp. 1158–1176.

29. Santos, A.B.; Mazzafera, P. Dehydrins are highly expressed in water-stressed plants of two coffee species. *Tropical Plant Biol.* **2012**, *5*, 218–232. [CrossRef]

30. Choi, D.W.; Zhu, B.; Close, T.J. The barley (Horderum vulgare L.) dehydrin multigene family: Sequences, allele types, chromosome assignments, and expression characteristics of 11 Dhn genes of cv Dicktoo. *Theor. Appl. Genet.* **1999**, *98*, 1234–1247. [CrossRef]

31. Sivamani, E.; Bahieldin, A.; Wraith, J.M.; AlNiemi, T.; Dyer, W.E.; Ho, T.H.D.; Wu, R. Improved biomass productivity and water use efficiency under water-deficit conditions in transgenic wheat constitutively expressing the barley HVA1 gene. *Plant Sci.* **2000**, *155*, 1–9. [CrossRef]

32. Brini, F.; Hanin, M.; Lumbreras, V.; Amara, I.; Khoudi, H.; Hassairi, A.; Pages, M.; Masmoudi, K. Overexpression of wheat dehydrin DHN-5 enhances tolerance to salt and osmotic stress in Arabidopsis thaliana. *Plant Cell Rep.* **2007**, *26*, 2017–2026. [CrossRef] [PubMed]

33. Park, B.J.; Liu, Z.; Kanno, A.; Kameya, T. Genetic improvement of Chinese cabbage for salt and drought tolerance by constitutive expression of a B. napus LEA gene. *Plant Sci.* **2005**, *169*, 553–558. [CrossRef]

34. Qin, Y.X.; Qin, F. Dehydrins from wheat x Thinopyrum ponticum amphiploid increase salinity and drought tolerance under their own inducible promoters without growth retardation. *Plant Physiol. Bioch.* **2016**, *99*, 142–149. [CrossRef] [PubMed]

35. Cao, Y.; Zhai, J.; Wang, Q.; Yuan, H.; Huang, X. Function of Hevea brasiliensis NAC1 in dehydration-induced laticifer differentiation and latex biosynthesis. *Planta* **2017**, *245*, 31–44. [CrossRef] [PubMed]

36. Clough, S.J.; Bent, A.F. Floral dip: A simplified method for Agrobacterium-mediated transformation of Arabidopsis thaliana. *Plant J.* **1998**, *16*, 735–743. [CrossRef]

37. Wan, H.J.; Yuan, W.; Ruan, M.; Ye, Q.; Wang, R.; Li, Z.; Zhou, G.; Yao, Z.; Zhao, J.; Liu, S.; et al. Identification of reference genes for reverse transcription quantitative real-time PCR normalization in pepper (*Capsicum annuum* L.). *Biochem. Biophy. Res. Commun.* **2011**, *416*, 24–30. [CrossRef] [PubMed]

38. Gutierrez, L.; Mauriat, M.; Gue'nin, S.; Pelloux, J.; Lefebvre, J.F.; Louvet, R.; Rusterucci, C.; Moritz, T.; Guerineau, F.; Bellini, C.; et al. The lack of asystematic validation of reference genes: A serious pitfall undervalued in reverse transcription-polymerase chain reaction (RT-PCR) analysis in plants. *Plant Biotechnol. J.* **2008**, *6*, 609–618. [CrossRef] [PubMed]

39. Dhindsa, R.S.; Plumb-Dhindsa, P.; Thorpe, T.A. Leaf senescence: Correlated with increased levels of membrane permeability and lipid peroxidation, and decreased levels of superoxide dismutase and catalase. *J. Exp. Bot.* **1981**, *32*, 93–101. [CrossRef]

40. Arkus, K.A.J.; Cahoon, E.B.; Jez, J.M. Mechanistic analysis of wheat chlorophyllase. *Arch. Biochem. Biophys.* **2005**, *438*, 146–155. [CrossRef]

41. Danyluk, J.; Perron, A.; Houde, M.; Limin, A.; Fowler, B.; Benhamou, N.; Sarhan, F. Accumulation of an acidic dehydrin in the vicinity of the plasma membrane during cold acclimation of wheat. *Plant Cell* **1998**, *10*, 623–638. [CrossRef]

42. Liang, J.G.; Tao, R.X.; Hao, Z.N.; Wang, L.P.; Zhang, X. Induction of resistance in cucumber against seedling damping-off by plant growth-promoting rhizobacteria (PGPR) Bacillus megaterium strain L8. *Afr. J. Biotechnol.* **2011**, *10*, 6920–6927.

43. Jabs, T.; Dietrich, R.A.; Dangl, J.L. Initiation of runaway cell death in an Arabidopsis mutant by extracellular superoxide. *Science* **1996**, *273*, 1853–1856. [CrossRef]

44. Zhang, L.N.; Zhang, L.C.; Xia, C.; Zhao, G.Y.; Liu, J.; Jia, J.Z.; Kong, X.Y. A novel wheat bZIP transcription factor, Tab ZIP60, confers multiple abiotic stress tolerances in transgenic Arabidopsis. *Physiol. Plant.* **2014**, *153*, 538–554. [CrossRef]

45. Bates, L.S.; Waldren, R.P.; Teeare, I.D. Rapid determination of free Pro for water-stress studies. *Plant Soil* **1973**, *39*, 205–207. [CrossRef]

International Journal of
Molecular Sciences

MDPI

Article

iTRAQ-Based Protein Profiling and Biochemical Analysis of Two Contrasting Rice Genotypes Revealed Their Differential Responses to Salt Stress

Sajid Hussain [1,†], Chunquan Zhu [1,†], Zhigang Bai [1], Jie Huang [1], Lianfeng Zhu [1], Xiaochuang Cao [1], Satyabrata Nanda [1], Saddam Hussain [2], Aamir Riaz [1], Qingduo Liang [1], Liping Wang [1], Yefeng Li [1], Qianyu Jin [1,*] and Junhua Zhang [1,*]

[1] State Key Laboratory of Rice Biology, China National Rice Research Institute, Hangzhou 310006, Zhejiang, China; sajid_2077uaf@yahoo.com (S.H.); zhuchunquan@caas.cn (C.Z.); baizg1989@163.com (Z.B.); huangjie67179484@163.com (J.H.); zlfnj@163.com (L.Z.); caoxiaochuang@126.com (X.C.); sbn.satyananda@gmail.com (S.N.); aamirriaz33@gmail.com (A.R.); 15550883578@163.com (Q.L.); 664948431@163.com (L.W.); m13067998118@163.com (Y.L.)
[2] Department of Agronomy, University of Agriculture Faisalabad, Punjab 38000, Pakistan; sadamhussainuaf@gmail.com
*   Correspondence: jinqianyu@caas.cn (Q.J.); zhangjunhua@caas.cn (J.Z.); Tel.: +86-571-63370122 (Q.J. & J.Z.)
†   Equal Contribution.

Received: 29 December 2018; Accepted: 24 January 2019; Published: 28 January 2019

Abstract: Salt stress is one of the key abiotic stresses causing huge productivity losses in rice. In addition, the differential sensitivity to salinity of different rice genotypes during different growth stages is a major issue in mitigating salt stress in rice. Further, information on quantitative proteomics in rice addressing such an issue is scarce. In the present study, an isobaric tags for relative and absolute quantitation (iTRAQ)-based comparative protein quantification was carried out to investigate the salinity-responsive proteins and related biochemical features of two contrasting rice genotypes—Nipponbare (NPBA, *japonica*) and Liangyoupeijiu (LYP9, *indica*), at the maximum tillering stage. The rice genotypes were exposed to four levels of salinity: 0 (control; CK), 1.5 (low salt stress; LS), 4.5 (moderate salt stress; MS), and 7.5 g of NaCl/kg dry soil (high salt stress, HS). The iTRAQ protein profiling under different salinity conditions identified a total of 5340 proteins with 1% FDR in both rice genotypes. In LYP9, comparisons of LS, MS, and HS compared with CK revealed the up-regulation of 28, 368, and 491 proteins, respectively. On the other hand, in NPBA, 239 and 337 proteins were differentially upregulated in LS and MS compared with CK, respectively. Functional characterization by KEGG and COG, along with the GO enrichment results, suggests that the differentially expressed proteins are mainly involved in regulation of salt stress responses, oxidation-reduction responses, photosynthesis, and carbohydrate metabolism. Biochemical analysis of the rice genotypes revealed that the Na^+ and Cl^- uptake from soil to the leaves via the roots was increased with increasing salt stress levels in both rice genotypes. Further, increasing the salinity levels resulted in increased cell membrane injury in both rice cultivars, however more severely in NPBA. Moreover, the rice root activity was found to be higher in LYP9 roots compared with NPBA under salt stress conditions, suggesting the positive role of rice root activity in mitigating salinity. Overall, the results from the study add further insights into the differential proteome dynamics in two contrasting rice genotypes with respect to salt tolerance, and imply the candidature of LYP9 to be a greater salt tolerant genotype over NPBA.

Keywords: Salt stress; *Oryza sativa*; proteomics; iTRAQ quantification; cell membrane injury; root activity

Int. J. Mol. Sci. **2019**, *20*, 547

1. Introduction

To satisfy the food demands of a population of more than nine billion people by 2050, the world's food productivity needs to be increased by 50% above current production [1,2]. The current growth trends of the major food crops, including wheat, rice, maize, and soybean, suggest that crop production will not be sufficient to meet these ever-rising food demands [3]. Further, the occurrence of abiotic stresses owing to climate change is one of the major reasons for the productivity gap [4]. Soil salinity is considered to be a major problem in the productivity of rice (*Oryza sativa* L.) worldwide [4]. Rice is highly sensitive to salt stress; however, the range of sensitivity varies with rice ecotypes, genotypes, and growth stages [5,6]. Salt tolerance in rice is correlated with variations in the translocation of sodium (Na^+) and chloride (Cl^-) ions in the aboveground plant organs, including the shoot and panicles [7–12]. Salinity affects rice physiology and growth by causing osmotic stress, nutrient imbalance, ionic toxicity, oxidative damage, alteration of metabolic processes, reduced cell division, genotoxicity, decline of growth and yield, and even the death of the plant [8,9,13–17]. In rice, salinity tolerance is usually achieved as a result of a cocktail of physiological and genetic reprogramming, including selective ion uptake and exclusion, preferential compartmentation of Na^+, alternation in stomatal closure, reactive oxygen species (ROS) signaling, and expression of salt-stress responsive genes and transcription factors [18–22].

Alterations in physiological and biochemical processes lead to changes in the protein pool in plants. In recent times, proteomic analysis has emerged as a significant molecular technique for the profiling and identification of proteins expressed in response to various abiotic stresses [23]. Isobaric tags for relative and absolute quantitation (iTRAQ)-based protein profiling and analysis has been performed in several crops, including rice [24], maize [25], wheat [26], tomatoes [27], and cotton [23], in response to abiotic stresses. Differential protein expressions in the areal tissues of rice subjected to salt stress have been reported by a few studies [28–30]. However, most of these studies have employed the 2D gel electrophoresis method to quantify the protein dynamics in rice. The 2D gel electrophoresis technique lacks efficiency in identifying the low abundant proteins, including extreme-acidic or basic proteins, proteins with molecular weights <15 kDa or >150 kDa, and hydrophobic proteins [23]. Furthermore, most of these works have been performed using the *japonica* rice genotype "Nipponbare" as the plant material. Therefore, in this study, we explored the proteomic dynamics of rice under salt stress in both *japonica* (Nipponbare, NPBA) and *indica* (Liangyoupeijiu, LYP9) rice genotypes by employing an iTRAQ-based proteomic study.

In the current study, the iTRAQ-based proteomic technique was used to identify the differentially expressed proteins in two rice genotypes of contrasting salt tolerance levels. The *indica* rice LYP9 has a higher salt tolerance level than the *japonica* rice NPBA [13]. Therefore, the proteomic analysis was performed with the aim of elucidating and comparing the effects of salt stress in these rice genotypes. Further, the physiological responses, such as cell membrane injury (CMI) and rice root activity of the NPBA and LYP9 genotypes, were assessed in response to various salt stress levels at the maximum tillering stage. Additionally, the Na^+ and Cl^- uptake from soil to leaf via root under the subjected salt stress levels were determined in both rice genotypes. The results from this study will help us to achieve better insights into the salt stress resistance mechanisms in rice.

2. Results

2.1. Na⁺ and Cl⁻ in the Soil

The soil Na^+ concentrations for LYP9 rice were recorded to be 0.17, 0.95, 1.7, and 2.0 mg·g⁻¹ for the control (no salt stress, CK), low salt stress (LS), moderate salt stress (MS), and high salt stress (HS) treatments, respectively. In NPBA, the soil Na^+ was recorded to be 0.18, 1.0, 1.6, and 2.15 mg·g⁻¹ for the CK, LS, MS, and HS treatments, respectively. The Na^+ concentration was found to be the highest in the HS treatment for NPBA rice, as most of the rice seedlings died under the HS condition before attaining the maximum tillering stage. Furthermore, the soil Na^+ concentration was lower for the LYP9 rice than the NPBA rice (Table 1). On the other hand, the soil Cl$^-$ concentrations were recorded to be 0.04, 0.59, 2.17, and 2.43 mg·g^{-1} for the CK, LS, MS, and HS treatments in LYP9 rice, respectively. In NPBA, the soil Cl$^-$ was found to be 0.01, 0.66, 1.64, and 3.03 mg·g^{-1} for the CK, LS, MS, and HS treatments, respectively.

Table 1. Differential Na$^+$ and Cl$^-$ uptake from soil to leaf via root in LYP9 and NPBA under different salt stress levels at rice maximum tillering stage.

Cultivars	Treatments	Na$^+$ (mg/g)		Cl$^-$ (mg/g)		Na$^+$ (mg/g)	Cl$^-$ (mg/g)
		Root	Leaf	Root	Leaf	Soil	Soil
LYP9	CK	0.7 ± 0.05d	0.2 ± 0.03c	0.5 ± 0.3d	6.8 ± 0.4d	0.2 ± 0.01d	0.04 ± 0.01e
	LS	1.1 ± 0.03bc	0.5 ± 0.08b	1.6 ± 0.7cd	12.4 ± 1.7bcd	1.0 ± 0.03c	0.6 ± 0.06de
	MS	1.5 ± 0.04b	0.8 ± 0.10a	7.9 ± 1.3ab	17.6 ± 2.4ab	1.7 ± 0.04b	2.2 ± 0.06bc
	HS	1.6 ± 0.09a	0.9 ± 0.11a	9.5 ± 1.6a	19.1 ± 2.9a	2.0 ± 0.07a	2.4 ± 0.33b
NPBA	CK	0.7 ± 0.03d	0.15 ± 0.01cd	1.0 ± 0.3d	9.9 ± 0.9cd	0.2 ± 0.01d	0.01 ± 0.01e
	LS	1.0 ± 0.07c	0.3 ± 0.01c	3.8 ± 0.4c	14.7 ± 2.8abc	1.0 ± 0.02c	0.7 ± 0.08e
	MS	1.3 ± 0.07b	0.9 ± 0.02a	6.3 ± 0.6b	18.7 ± 2.4ab	1.6 ± 0.12b	1.6 ± 0.035d
	HS	-	-	-	-	2.2 ± 0.07a	3.0 ± 0.23a

Values are denoted as mean ± SE (*n* = 3). Values followed by different letters denote significant difference (*p* ≤ 0.05) according to LSD test. Abbreviations: control (no salt stress, CK), low salt stress (LS), moderate salt stress (MS), and high salt stress (HS), Liangyoupeijiu (LYP9), Nipponbare (NPBA). The similar lettering within rice genotype shows the significant and different lettering mean non-significance within treatment levels.

2.2. Na$^+$ and Cl$^-$ in the Rice Plants

The concentration of Na$^+$ was found to increase in rice in proportion to rice growth. At the time of rice transplanting, the Na$^+$ concentration in the LYP9 and NPBA roots was 0.44 and 0.37 mg·g^{-1}, respectively. However, at the maximum tillering stage, Na$^+$ concentrations in rice roots was increased in both rice genotypes, with the increase in subjected salt stress levels. In LYP9 rice, LS, MS, and HS levels of salt stress resulted in the increase of Na$^+$ concentrations in rice roots amounting to 67.2%, 126.9%, and 138.8%, respectively, as compared with the CK treatment. Similarly, in NPBA rice, Na$^+$ concentration in the roots was increased by 42.9% for LS and 128.6% for MS as compared with the CK treatment. However, the NPBA rice could not survive under HS salinity conditions. These results indicated that the uptake of Na$^+$ is higher in rice in the maximum tillering stage as compared to the seedling stage (Table 1). Similar proportions were observed for Na$^+$ concentration in rice leaves, where the Na$^+$ concentrations were found to be increased by 163.2%, 305.3%, and 357.9% under LS, MS, and HS conditions, respectively, as compared with the CK condition in LYP9 rice, and by 86.7% and 480% under LS and MS conditions, respectively, as compared with the CK condition in NPBA rice (Table 1). The Na$^+$ uptake from root to shoot was found to be higher in LYP9 than NPBA. These results suggest that LYP9 has an enhanced ability to uptake Na$^+$ in the plant parts than compared to NPBA, which might aid in improved salt tolerance in LYP9 compared with NPBA. Likewise, at the maximum tillering stage, the Cl$^-$ uptake by the rice roots and leaves was increased with the increase in the salt stress levels (Table 1). Moreover, these increases in the Cl$^-$ ion uptakes were found to be higher in LYP9 leaves and roots than those of NPBA.

2.3. Cell Membrane Injury (CMI) in Rice Flag Leaves

Evaluations of cell membrane injury (CMI) in both LYP9 and NPBA rice revealed that salt concentrations and CMI are directly proportional, where higher salt concentrations cause severe cell membrane damage. The CMI was found to be higher in the HS condition as compared with MS, LS, and CK conditions in both rice cultivars (Figure 1). CMI was recorded as 5% for CK, 6.7% for LS, 7% for MS, and 15.2% for HS in LYP9. However, CMI in NPBA was recorded as 9.8% for CK, 10.6% for LS, and 11.9% for MS. Compared with the control (CK), the CMI in the LYP9 rice cultivar was increased by 34%, 40%, and 204% under LS, MS, and HS, respectively. On the other hand, CMI was increased by 8.1% (LS), and 21.4% (MS) in the NPBA rice, whilst rice seedlings died under HS conditions before reaching the maximum tillering stage in this genotype of rice (Figure 2). These results strongly suggest that salt stress negatively affects the cell membrane stability, and cell membrane integrity was found to be higher in LYP9 as compared with NPBA. Collectively, these results indicated that LYP9 is more tolerant to salt stress than NPBA.

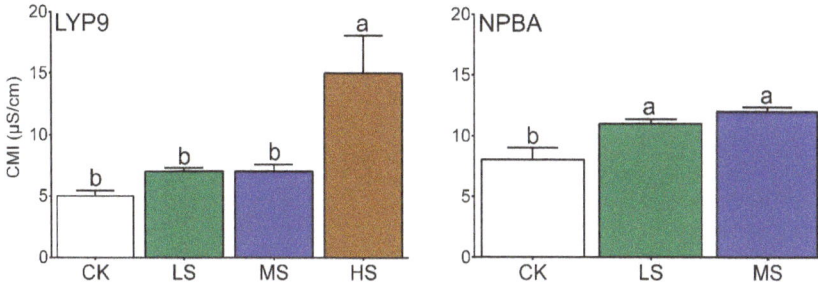

Figure 1. Evaluation of cell membrane injury under the subjected salt stress in LYP9 and NPBA. Bars denoted mean values ± SE (*n* = 3). Values followed by different letters denote significant difference ($p \leq 0.05$) according to LSD test. The similar lettering within rice genotype shows the significant and different lettering mean non-significance within treatment levels.

CK-LYP9   LS-LYP9   MS-LYP9   HS-LYP9   CK-NPBA   LS-NPBA   MS-NPBA   HS-NPBA

Figure 2. Effects of different levels of salt stress on the rice growth at the early stage in both LYP9 and NPBA.

2.4. Rice Root Activity

High root activity is an indicator of resistance against stress [31]. Rice root activity was increased by 2.1% for LS, 50.2% for MS, and 173.7% for HS as compared with CK in LYP9. In the case of NPBA, the rice root activity was decreased by 3.3% for LS, while it increased by 111.4% for MS, as compared to CK. In this study, the rice root activity was higher in LYP9 compared with NPBA under various salt stress levels, inferring the role of root activity in salt tolerance (Figure 3).

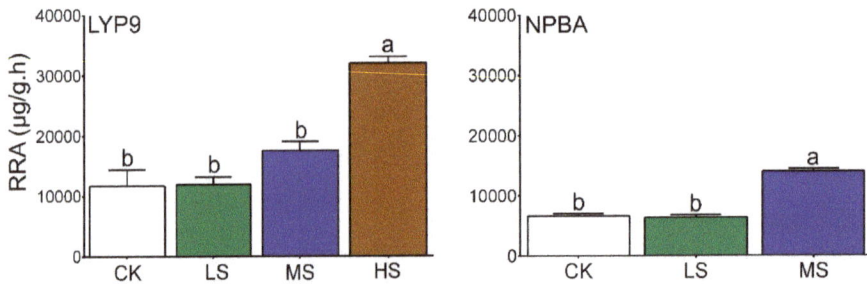

Figure 3. Rice root activity under different salt stress in LYP9 and NPBA. Bars denoted mean values ± SE (*n* = 3). Bars denoted mean values ±SE (*n* = 3). Values followed by different letters denote significant difference ($p \leq 05$) according to LSD test. The similar lettering within rice genotype shows the significant and different lettering mean non-significance within treatment levels.

2.5. iTRAQ-Based Protein Identification at the Rice Maximum Tillering Stage

Quantitative proteomic analysis of three leaf samples (CK, LS, and MS) from NPBA rice and four leaf samples (CK, LS, MS, and HS) from LYP9 rice were performed using the iTRAQ method. In total, 5340 proteins were identified with 1% FDR (Table 2). In LYP9, 28, 368, and 491 proteins were found to be up-regulated under LS, MS, and HS treatments, respectively, as compared with the CK treatment. On the other hand, in NPBA, 239 and 337 up-regulated proteins were detected under the LS and MS treatments as compared with the CK treatment (Table 3). The longest length of enriched peptides was 7 to 18, with the mass error below 0.025 to 1.00 and with a high performing Pearson correlation coefficient with repeated samples, showing a high quality of the mass spectroscopy data and sample preparation. Proteins with a 1.2 fold change and Q-value of >0.05 were considered as differentially expressed proteins.

Table 2. Overview of the total protein identification in both rice genotypes.

Total Spectra	Spectra	Unique Spectra	Peptides	Unique Peptide
402,823	71,146	53,833	21,741	18,899

Table 3. Differentially expressed proteins in NPBA and LYP9 rice under different salt levels with 1.2 fold change and Q-value > 0.05.

Protein ID	NCBI Accession	Protein Name	NPBA		LYP9		
			LS vs. CK	MS vs. CK	LS vs. CK	MS vs. CK	HS vs. CK
Salt responsive							
tr|B9FWE4|B9FWE4_ORYSJ	gi|222636749	Uncharacterized protein	1.516	1.415	0.906	1.234	1.255
tr|A2Y7R4|A2Y7R4_ORYSI	gi|115465579	Malate dehydrogenase	1.393	2	1.014	1.488	1.573
tr|B8BBS3|B8BBS3_ORYSI	gi|115476908	Os08g0478200 protein	1.389	1.593	0.951	1.402	2.706
tr|A2WT84|A2WT84_ORYSI	gi|115438875	Malate dehydrogenase	1.897	2.835	1.027	1.871	2.006
tr|A0A0P0VS15|A0A0P0VS15_ORYSJ	gi|115450217	Nascent polypeptide-associated complex subunit β (Fragment)	2.523	2.558	1.017	1.594	1.384
tr|A2XA10|A2XA10_ORYSI	gi|46805452	Os02g0768600 protein	1.506	2.225	1.071	2.213	2.212
tr|A0A190X658|A0A190X658_ORYSI	gi|115477769	L-isoaspartate methyltransferase	1.575	2.403	0.901	1.591	1.69
sp|Q43008|SODM_ORYSJ	gi|115463191	Superoxide dismutase	1.775	2.06	1.071	1.534	1.828
sp|Q9FE01|APX2_ORYSJ	gi|115474285	Ascorbate peroxidase	1.308	1.26	0.966	1.227	1.119
sp|Q07661|NDK1_ORYSJ	gi|61679782	Nucleoside diphosphate kinase 1	1.295	1.816	0.909	1.068	1.435
sp|Q5N725|ALFC3_ORYSJ	gi|297598143	Fructose-bisphosphate aldolase 3	1.399	1.639	1.023	1.089	1.532
sp|Q7XDC8|MDHC_ORYSJ	gi|115482534	Malate dehydrogenase	1.37	1.749	1.004	1.284	1.523
tr|A2X753|A2X753_ORYSI	gi|115447273	Os02g0612900 protein	1.441	1.552	1.036	1.506	1.597
tr|A2X7X9|A2X7X9_ORYSI	gi|125540544	Putative uncharacterized protein	1.152	1.502	0.882	1.378	1.25
tr|A0A0P0VTX8|A0A0P0VTX8_ORYSJ	gi|108706531	Os03g0182600 protein	0.852	1.876	0.908	0.949	1.367
tr|E0X6V4|E0X6V4_ORYSJ	gi|306415973	Triosephosphate isomerase	1.003	1.232	1.027	1.107	1.256
tr|A2ZAA7|A2ZAA7_ORYSI	gi|115483468	Nucleoside diphosphate kinase	1.197	2.406	0.882	1.08	1.768
tr|Q9ATR3|Q9ATR3_ORYSA	gi|13249140	Glucanase	1.063	2.009	0.876	0.912	1.39
tr|A2ZIH2|A2ZIH2_ORYSI	gi|115487556	Expressed protein	1.048	1.561	0.957	1.204	1.424
tr|B9FV80|B9FV80_ORYSJ	gi|222636335	Peroxidase	0.888	1.812	0.966	1.298	1.814
tr|B8B893|B8B893_ORYSI	gi|218199240	Plasma membrane ATPase	1.404	1.435	0.906	0.716	0.86
tr|A2XA20|A2XA20_ORYSI	gi|115448935	Proteasome subunit β type	0.862	1.109	0.946	1.067	1.059
tr|A2Y628|A2Y628_ORYSI	gi|125552829	Cysteine proteinase inhibitor	0.96	1.775	1.056	1.438	2.205
tr|Q9ZNZ1|Q9ZNZ1_ORYSA	gi|4097938	Beta-1,3-glucanase	0.795	1.711	1.003	0.932	1.619
tr|A2ZCK1|A2ZCK1_ORYSI	gi|148762354	Alcohol dehydrogenase 2	0.63	0.866	1.055	1.019	1.019
sp|A2XFC7|APX1_ORYSI	gi|158512874	L-ascorbate peroxidase 1	1.216	1.343	0.942	1.21	1.327
tr|A2X822|A2X822_ORYSI	gi|125540587	Glutathione peroxidase	0.717	0.617	0.924	1.77	1.567
tr|A2XFD1|A2XFD1_ORYSI	gi|125543402	Putative uncharacterized protein	1.1	1.543	0.943	1.23	1.554
tr|A2YLI3|A2YLI3_ORYSI	gi|115472191	Os07g0495200 protein	1.159	1.821	0.971	1.505	1.818
tr|B8ADI1|B8ADI1_ORYSI	gi|218187601	NADH-cytochrome b5 reductase	0.771	0.72	1.081	2.182	2.316
tr|A2YSB2|A2YSB2_ORYSI	gi|115475275	Os08g0205400 protein	1.587	2.217	0.908	1.597	1.194
tr|B8AY35|B8AY35_ORYSI	gi|218196772	Fructose-bisphosphate aldolase	0.458	0.214	0.964	0.74	1.505
tr|B8AY17|B8AY17_ORYSI	gi|218196757	Putative uncharacterized protein	0.725	0.849	0.996	1.174	1.649
tr|Q9ZNZ1|Q9ZNZ1_ORYSA	gi|4097938	Beta-1,3-glucanase	0.795	1.711	1.003	0.932	1.619
sp|Q94IZ0|NQR1_ORYSJ	gi|115442299	Putative uncharacterized protein	0.686	0.766	0.984	0.931	1.369
tr|A2WWV4|A2WWV4_ORYSI	gi|125528336	Putative uncharacterized protein	0.518	0.55	1.031	1.159	1.304

Table 3. *Cont.*

Protein ID	NCBI Accession	Protein Name	NPBA			LYP9	
			LS vs. CK	MS vs. CK	LS vs. CK	MS vs. CK	HS vs. CK
sp|P93438|METK2_ORYSJ	gi|3024122	S-adenosylmethionine synthase	1.282	1.017	1.013	1.226	1.092
tr|A2XUB9|A2XUB9_ORY1	gi|90265194	B0812A04.3 protein	1.074	1.225	1.215	1.186	1.437
tr|A2Z2Z0|A2Z2Z0_ORYSI	gi|125564321	Putative uncharacterized protein	1.01	0.776	0.921	1.202	1.102
tr|B8AEU4|B8AEU4_ORYSI	gi|218191814	Putative uncharacterized protein	0.954	1.212	0.948	0.908	1.134
tr|A0A0P0VTX8|A0A0P0VTX8_ORYSJ	gi|108706531	Os03g0182600 protein	0.852	1.876	0.908	0.949	1.367
tr|Q688M9|Q688M9_ORYSJ	gi|51854423	putative endo-1,31,4-β-D-glucanase	1.16	1.14	0.992	1.111	1.144
tr|B8ATW7|B8ATW7_ORYSI	gi|115460338	Os04g0602100 protein	1.386	1.494	1.115	1.448	1.482
sp|Q7FAH2|G3PC2_ORYSJ	gi|115459078	Glyceraldehyde-3-phosphate dehydrogenase 2	0.887	1.03	1.004	0.996	1.196
tr|Q0IG30|Q0IG30_ORYSJ	gi|297598314	Os01g0946500 protein	0.95	0.844	0.995	0.799	0.959
tr|Q6L5I4|Q6L5I4_ORYSJ	gi|47900421	Putative aldehyde dehydrogenase	0.735	0.737	0.911	1.167	1.008
sp|A2XW22|DHE2_ORYSI	gi|81686712	Glutamate dehydrogenase 2	1.177	1.142	0.912	0.8	1.184
sp|Q7FAY6|RGP2_ORYSJ	gi|115461086	Amylogenin	1.357	1.021	0.776	0.558	0.683
sp|Q259G4|PMM_ORYSI	gi|115461390	Phosphomannomutase	0.836	1.19	1.007	0.924	1.216
Photosynthesis related							
tr|A2YW57|A2YW57_ORYSI	gi|115477166	Os08g0504500 protein	1.317	2.22	0.883	1.852	1.77
tr|Q2QWM7|Q2QWM7_ORYSJ	gi|108862278	Os12g0190200 protein	1.053	1.535	0.91	1.363	1.279
tr|B8BCC6|B8BCC6_ORYSI	gi|115477246	Os08g0512500 protein	2.311	3.409	1.022	1.551	1.349
tr|A2ZK1|A2ZK1_ORYSI	gi|125550052	Putative uncharacterized protein	1.246	1.015	1.301	2.61	2.439
tr|B8AAX3|B8AAX3_ORYSI	gi|115440559	Os01g0805300 protein	1.418	2.178	0.986	1.943	1.703
tr|Q0D6V8|Q0D6V8_ORYSJ	gi|297607127	Os07g0433300 protein	2.246	3.387	0.982	2.305	2.053
tr|Q7XHS1|Q7XHS1_ORYSJ	gi|115472141	2Fe-2S iron-sulfur cluster protein-like	1.016	1.414	0.936	1.62	1.548
tr|A2X7M2|A2X7M2_ORYSI	gi|115447507	Os02g0638300 protein	1.096	1.611	1.14	1.73	1.825
tr|B0FFP0|B0FFP0_ORYSJ	gi|115470529	Chloroplast 23 kDa polypeptide of PS II (Fragment)	1.319	1.705	0.997	1.747	1.609
tr|Q7M1U9|Q7M1U9_ORYSA	gi|218186547	Photosystem I 9K protein	1.832	3.172	1.027	2.206	2.383
tr|A0A0P0XF80|A0A0P0XF80_ORYSJ	gi|38636895	Os08g0347500 protein	1.642	2.347	0.926	1.756	1.81
tr|Q7M1Y7|Q7M1Y7_ORYSA	gi|164375543	Photosystem II oxygen-evolving complex protein 2 (Fragment)	1.77	2.373	0.989	2.015	1.756
tr|B8AJX7|B8AJX7_ORYSI	gi|115455221	Serine hydroxymethyltransferase	2.24	2.885	1.078	1.525	1.334
tr|B8AY24|B8AY24_ORYSI	gi|218196765	Putative uncharacterized protein	1.288	1.56	1.026	1.639	1.326
sp|Q6ZT6|CHLP_ORYSJ	gi|297599916	Geranylgeranyl reductase	0.956	1.173	0.973	0.957	1.174
sp|P0C420|PSBH_ORYSA	gi|11466818	Photosystem II reaction center protein H	0.795	0.694	1.029	0.864	1.14
Oxidation reduction responsive							
tr|A3BVS6|A3BVS6_ORYSJ	gi|125604340	Superoxide dismutase	1.512	1.903	0.96	1.618	1.684
sp|Q6H7E4|TRXM1_ORYSJ	gi|115447527	Putative uncharacterized protein	0.941	1.681	1.049	1.75	2.489
sp|Q9SDD6|PRX2F_ORYSJ	gi|115435844	Peroxiredoxin-2F, mitochondrial	1.363	1.772	1.021	1.642	1.687
tr|B7FAE9|B7FAE9_ORYSJ	gi|215769368	Glutathione peroxidase	0.98	1.347	0.965	1.38	1.176
tr|A2Y043|A2Y043_ORYSI	gi|125550744	Peroxidase	1.232	2.046	0.87	0.633	1.271
tr|Q9FTN6|Q9FTN6_ORYSJ	gi|115434034	Os01g0106300 protein	0.732	1.977	0.788	0.606	1.476

Table 3. *Cont.*

Protein ID	NCBI Accession	Protein Name	NPBA			LYP9	
			LS vs. CK	MS vs. CK	LS vs. CK	MS vs. CK	HS vs. CK
tr|A2X2T0|A2X2T0_ORYSI	gi|55700921	Peroxidase	0.775	1.122	0.913	0.85	1.697
tr|O22440|O22440_ORYSA	gi|115474063	Peroxidase	1.763	2.554	0.963	2.051	1.612
tr|A3A7Y3|A3A7Y3_ORYSJ	gi|125582491	Uncharacterized protein	1.099	1.361	1.101	1.555	2.4
tr|B9FL20|B9FL20_ORYSJ	gi|115464801	Uncharacterized protein	1.175	1.416	0.965	1.159	1.356
tr|Q9AS12|Q9AS12_ORYSJ	gi|115436300	Peroxidase	4.654	5.188	0.78	1.948	2.334
tr|B8ATW7|B8ATW7_ORYSI	gi|115460338	Os04g0602100 protein	1.386	1.494	1.115	1.448	1.482
tr|B9FCM4|B9FCM4_ORYSJ	gi|116309795	OSIGBa0148A10.12 protein	2.208	2.05	1.017	1.365	1.123
tr|Q0JB49|Q0JB49_ORYSJ	gi|115459848	Glutathione peroxidase	1.449	1.435	0.933	1.537	1.271
tr|Q43006|Q43006_ORYSA	gi|20286|emb	Peroxidase	4.58	4.923	1.16	1.421	1.233
tr|Q5Z7I7|Q5Z7I7_ORYSJ	gi|55701041	Peroxidase	5.025	5.222	0.802	2.469	2.659
tr|Q25AK7|Q25AK7_ORYSA	gi|90265065	H0510A06.15 protein	1.326	1.047	0.91	1.209	1.022
tr|Q6K4J4|Q6K4J4_ORYSJ	gi|115479691	Peroxidase	1.23	1.049	0.988	1.16	0.919
tr|A2WJQ7|A2WJQ7_ORYSI	gi|115434036	Os01g0106400 protein	0.884	2.12	0.973	1.313	2.057
sp|P41095|RLA0_ORYSJ	gi|115474653	60S acidic ribosomal protein	1.312	1.231	1.073	0.882	0.83
sp|B8AUI3|GLO3_ORYSI	gi|115460650	Peroxisomal (S)-2-hydroxy-acid oxidase GLO3	0.627	0.615	1.305	0.833	0.972
tr|A0A0N7KI36|A0A0N7KI36_ORYSJ	gi|55700967	Peroxidase	0.895	0.817	0.934	1.403	1.041
tr|B8B5W7|B8B5W7_ORYSI	gi|218200254	Peroxidase	1.11	1.51	0.996	2.708	1.966
tr|A2WPA1|A2WPA1_ORYSI	gi|125525683	Peroxidase	1.258	1.625	1.07	2.133	3.577
tr|A2ZAA6|A2ZAA6_ORYSI	gi|115483466	Putative peptide methionine sulfoxide reductase	1.121	1.184	0.902	1.739	1.32
tr|A2XVK6|A2XVK6_ORYSI	gi|125549044	Putative uncharacterized protein	0.844	0.93	0.946	1.321	1.19
tr|B9F688|B9F688_ORYSJ	gi|222624472	Uncharacterized protein	2.063	3.091	1.018	2.407	2.351
tr|B8AU10|B8AU10_ORYSI	gi|218194884	Putative uncharacterized protein	1.206	0.747	1.145	1.386	1.226
tr|Q7F1J9|Q7F1J9_ORYSJ	gi|115477368	Os08g0522400 protein	1.225	1.347	1.072	1.309	1.121
sp|Q6K471|FTRC_ORYSJ	gi|75125055	Ferredoxin-thioredoxin reductase	1.28	2.03	0.905	1.598	1.668
tr|A0A0B4U1V7|A0A0B4U1V7_ORYSA	gi|115467518	Aldehyde dehydrogenase ALDH2b	1.178	1.029	1.016	1.006	1.233
sp|Q6AV34|ARGC_ORYSJ	gi|218193315	Probable N-acetyl-gamma-glutamyl-phosphate reductase	1.046	1.075	0.971	0.966	1.237
tr|Q2QV45|Q2QV45_ORYSJ	gi|115487998	70 kDa heat shock protein	1.415	1.387	1.068	1.391	1.254
sp|Q84VG0|CML7_ORYSJ	gi|115474531	Putative uncharacterized protein	1.351	1.579	0.859	1.35	1.392
tr|A2Y8A8|A2Y8A8_ORYSI	gi|115465902	Os06g0104300 protein	0.877	2.193	1.078	1.336	1.654
tr|A0A0P0X7V0|A0A0P0X7V0_ORYSJ	gi|115472943	Os07g0573800 protein (Fragment)	1.61	1.898	0.916	1.161	1.207
tr|B8BAM3|B8BAM3_ORYSI	gi|115474739	Os08g0139200 protein	1.096	1.539	0.841	0.9	1.207
sp|Q69TY4|PR2E1_ORYSJ	gi|115469028	Putative uncharacterized protein	1.289	1.212	0.944	1.336	1.361
sp|Q8W3D9|PORB_ORYSJ	gi|75248671	Protochlorophyllide reductase B	0.881	1.621	0.891	1.192	2.065
tr|B8AGN1|B8AGN1_ORYSI	gi|115445869	Os02g0328300 protein	1.63	2.925	0.927	1.814	1.808
tr|B9F604|B9F604_ORYSJ	gi|222625905	Uncharacterized protein	1.474	1.82	1.086	1.62	1.613
tr|Q7F229|Q7F229_ORYSJ	gi|115471449	Os07g0260300 protein	0.924	1.084	0.949	1.375	2.104
tr|A6N0B2|A6N0B2_ORYSI	gi|149391329	Mitochondrial formate dehydrogenase 1 (Fragment)	0.993	1.084	0.931	0.99	1.212

Table 3. *Cont.*

Protein ID	NCBI Accession	Protein Name	NPBA		LYP9		
			LS vs. CK	MS vs. CK	LS vs. CK	MS vs. CK	HS vs. CK
sp|Q10L32|MSRB5_ORYSJ	gi|115453111	Putative uncharacterized protein	1.116	1.471	0.84	1.272	1.479
tr|Q94IT6|Q94IT6_ORYSJ	gi|15408884	Os01g0847700 protein	1.028	0.871	1.18	1.416	1.504
tr|B8B2F2|B8B2F2_ORYSI	gi|218198209	Formate dehydrogenase	1.014	1.19	0.97	0.898	1.278
sp|Q7XPL2|HEM6_ORYSJ	gi|75232919	OSICBa0152L12.9 protein	0.993	1.224	0.873	0.963	1.253
sp|P0C5D4|PRXQ_ORYSI	gi|115466906	Peroxiredoxin Q, chloroplastic	1.215	1.691	0.909	1.71	2.077
tr|A0A0P0WR9|A0A0P0WWR9_ORYSJ	gi|300681235	Os06g0472000 protein	1.308	1.269	1.032	1.532	1.593
tr|A2WL79|A2WL79_ORYSI	gi|125524611	Peroxidase	0.826	0.852	1.047	1.182	1.233
sp|P37834|PER1_ORYSJ	gi|115464711	Peroxidase	0.702	0.999	0.819	0.742	1.764
tr|Q01LB1|Q01LB1_ORYSA	gi|115458104	OSJNBa0072K14.5 protein	1.175	1.243	0.937	1.124	1.221
tr|P0C01.1|APX6_ORYSJ	gi|115487636	Putative uncharacterized protein	1.127	1.213	1.031	1.282	1.477
sp|Q7X8R5|TRXM2_ORYSJ	gi|115459582	B1011H02.3 protein	1.557	2.198	0.916	1.577	3.486
tr|B7E4J4|B7E4J4_ORYSJ	gi|215704355	Putative uncharacterized protein	0.853	1.062	0.762	0.579	1.105
tr|Q7XV08|Q7XV08_ORYSJ	gi|38567882	OSJNBa0036B21.10 protein	1.159	1.382	0.971	1.084	1.397
Carbohydrate metabolism							
sp|Q8L7J2|BGL06_ORYSJ	gi|218192323	Beta-glucosidase 6	0.177	0.383	1.004	0.424	1.741
sp|Q76BW5|XTH8_ORYSJ	gi|115475445	Xyloglucan endotransglycosylase/hydrolase protein 8	0.953	2.101	0.939	1.074	1.369
tr|Q01JC3|Q01JC3_ORYSA	gi|116310134	Malate dehydrogenase	0.795	0.74	0.995	0.591	0.889
tr|Q0DCB1|Q0DCB1_ORYSJ	gi|115467998	Os06g0356700 protein	0.849	1.073	0.912	1.227	2.764
tr|Q10CU4|Q10CU4_ORYSJ	gi|115455353	GH family 3 N terminal domain containing protein, expressed	0.72	2.799	0.66	0.663	2.234
tr|Q9ZNZ1|Q9ZNZ1_ORYSA	gi|4097938	Beta-1,3-glucanase	0.795	1.711	1.003	0.932	1.619
tr|H2KWT0|H2KWT0_ORYSJ	gi|108863034	HIPL1 protein, putative, expressed	1.106	2.099	0.908	1.231	2.014
tr|B8AIS2|B8AIS2_ORYSI	gi|218191593	Putative uncharacterized protein	0.773	0.837	0.875	1.437	1.452
sp|Q0INM3|BGA15_ORYSJ	gi|115488372	Beta-galactosidase 15	1.348	1.602	0.924	1.187	1.56
tr|B9FWS5|B9FWS5_ORYSJ	gi|222636680	Uncharacterized protein	0.838	1.177	1.122	0.866	1.141
tr|Q0JG30|Q0JG30_ORYSJ	gi|297598314	Os01g0946500 protein	0.95	0.844	0.995	0.799	0.959
tr|Q0J0Q9|Q0J0Q9_ORYSJ	gi|115479865	Os09g0487600 protein	0.829	1.294	0.88	1.272	1.594
tr|A2XM08|A2XM08_ORYSI	gi|115455349	GH family 3 N terminal domain containing protein, expressed	0.859	1.124	0.866	0.78	1.387
sp|Q10NX8|BGAL6_ORYSJ	gi|152013362	Beta-galactosidase 6	1.063	1.684	0.938	1.413	1.814
tr|B8AII1|B8AII1_ORYSI	gi|218190145	Putative uncharacterized protein	0.904	1.55	0.954	1.167	1.478
tr|Q01IH0|Q01IH0_ORYSA	gi|116310092	H0502G05.3 protein	0.728	0.783	0.894	1.001	1.193
tr|Q01JK3|Q01JK3_ORYSA	gi|116310050	Aldose 1-epimerase	0.823	1.226	0.939	1.515	1.515
tr|B8BHM7|B8BHM7_ORYSI	gi|10140702	Alpha-galactosidase	0.723	1.32	1.006	1.331	1.544
tr|A229V6|A229V6_ORYSI	gi|125532825	Uncharacterized protein	0.73	2.017	0.876	1.276	1.194
tr|Q0DTS9|Q0DTS9_ORYSJ	gi|297600575	Os03g0227400 protein (Fragment)	1.101	1.226	0.804	1.06	1.306
tr|A2XME9|A2XME9_ORYSI	gi|115455637	Malate dehydrogenase	1.049	1.261	1.151	1.502	1.548
tr|Q6Z8F4|Q6Z8F4_ORYSJ	gi|115448091	Phosphoribulokinase	1.143	1.318	1.066	1.144	1.249

Table 3. *Cont.*

Protein ID	NCBI Accession	Protein Name	NPBA			LYP9	
			LS vs. CK	MS vs. CK	LS vs. CK	MS vs. CK	HS vs. CK
tr|A2YIJ5|A2YIJ5_ORYSJ	gi|50509727	Os07g0168600 protein	0.779	0.93	0.952	1.118	1.317
sp|Q75I93|BGL07_ORYSJ	gi|115454825	Beta-glucosidase	1.201	1.066	0.95	1.276	1.58
tr|Q7XIV4|Q7XIV4_ORYSJ	gi|115474081	Alpha-galactosidase	0.786	1.367	0.919	1.115	1.491
tr|A3A285|A3A285_ORYSJ	gi|115443693	Uncharacterized protein	0.83	1.101	0.843	1.151	1.262
tr|A0A0P0XVT5|A0A0P0XVT5_ORYSJ	gi|297610712	Alpha-galactosidase (Fragment)	0.72	1.115	0.848	1.16	1.321
tr|B7I946|B7I946_ORYSJ	gi|297605789	Os06g0356800 protein	0.681	1.016	0.7	1.159	2.984
Stress responsive							
tr|Q9AQU0|Q9AQU0_ORYSJ	gi|13486733	Peptidyl-prolyl cis-trans isomerase	1.249	1.825	0.965	1.578	1.772
tr|Q8GTB0|Q8GTB0_ORYSJ	gi|27476086	Putative heat shock 70 KD protein, mitochondrial	1.294	1.354	0.92	1.027	1.208
tr|Q84S20|Q84S20_ORYSJ	gi|28971968	CHP-rich zinc finger protein-like	2.605	2.416	0.869	1.374	1.439
tr|Q5JKK9|Q5JKK9_ORYSJ	gi|115442153	Os01g0940700 protein	1.897	3.948	0.955	1.04	0.959
sp|Q75I1Q00|BIP4_ORYSJ	gi|115464027	Heat shock 70 kDa protein BIP4	10	10	1.058	0.8	0.72
tr|Q53NM9|Q53NM9_ORYSJ	gi|115486793	DnaK-type molecular chaperone hsp70-rice	1.87	1.487	1.009	0.821	0.793
tr|Q10NA9|Q10NA9_ORYSJ	gi|115452223	70 kDa heat shock protein	2.198	1.668	1.086	0.907	0.816
sp|Q5VRY1|HSP18_ORYSJ	gi|115434946	17.5 kDa heat shock protein	1.413	3.508	1.045	1.084	1.043
tr|Q6YUA7|Q6YUA7_ORYSJ	gi|115476792	Os08g0464000 protein	1.323	1.3	1.041	0.866	1.034
tr|A2YK26|A2YK26_ORYSJ	gi|115471453	Os07g0262200 protein	1.096	1.252	0.994	0.995	1.314
tr|B9FK56|B9FK56_ORYSJ	gi|222631026	Uncharacterized protein	1.028	1.106	1.031	1.314	1.25
tr|A2Z3L9|A2Z3L9_ORYSJ	gi|115480445	Os09g0541700 protein	1.1	1.218	0.999	1.144	1.342
tr|O82143|O82143_ORYSJ	gi|115451853	26S proteasome regulatory particle	1.138	1.146	0.981	1.293	1.427
tr|Q5ZAV7|Q5ZAV7_ORYSJ	gi|115440349	Os01g0783500 protein	1.066	1.434	1.03	1.71	2.189
tr|A2Y628|A2Y628_ORYSJ	gi|125552829	Cysteine proteinase inhibitor	0.96	1.775	1.056	1.438	2.205
Osmotic stress responsive							
tr|A2XHR1|A2XHR1_ORYSJ	gi|125544232	Sucrose synthase	0.82	1.102	0.545	0.366	1.005
tr|B8B835|B8B835_ORYSJ	gi|115473055	NADH-dehydrogenase	0.992	1.182	0.879	1.282	1.538
tr|Q2RBD1|Q2RBD1_ORYSJ	gi|115483847	Non-specific lipid-transfer protein	0.988	1.244	0.894	1.274	2.009
tr|Q0IQK7|Q0IQK7_ORYSJ	gi|297612544	Non-specific lipid-transfer protein	1.226	2.979	0.78	1.023	2.235
tr|B8B936|B8B936_ORYSI	gi|218201512	Putative uncharacterized protein	0.871	1.281	0.93	0.976	1.51
tr|B8AII1|B8AII1_ORYSI	gi|218190145	Putative uncharacterized protein	0.904	1.55	0.954	1.167	1.478
sp|Q10LR9|DCUP2_ORYSJ	gi|6006382	Putative SAM-protoporphyrin IX methyltransferase	0.935	0.974	0.965	1.05	1.216
tr|A2X8B7|A2X8B7_ORYSI	gi|115452897	Uroporphyrinogen decarboxylase 2	1.265	1.835	0.902	0.95	1.425
tr|Q2RBD1|Q2RBD1_ORYSJ	gi|242062934	2-C-methyl-D-erythritol 2,4-cyclodiphosphate synthase	1.362	1.399	0.746	1.202	1.538
tr|Q2RBD1|Q2RBD1_ORYSJ	gi|115483847	Non-specific lipid-transfer protein	0.988	1.244	0.894	1.274	2.009
Ethylene responsive							
tr|B9G3V3|B9G3V3_ORYSJ	gi|222641669	Uncharacterized protein	1.837	2.313	1.837	1.825	1.982
sp|Q8W3D9|PORB_ORYSJ	gi|75248671	Protochlorophyllide reductase B	0.881	1.621	0.891	1.192	2.065
tr|Q0IQK7|Q0IQK7_ORYSJ	gi|297612544	Non-specific lipid-transfer protein (Fragment)	1.226	2.979	0.78	1.023	2.235
tr|Q2RBD1|Q2RBD1_ORYSJ	gi|115483847	Non-specific lipid-transfer protein	0.988	1.244	0.894	1.274	2.009

Table 3. *Cont.*

Protein ID	NCBI Accession	Protein Name	NPBA				LYP9	
			LS vs. CK	MS vs. CK	LS vs. CK	MS vs. CK	HS vs. CK	
Metabolic responsive								
tr|Q0D572|Q0D572_ORYSJ	gi|297607511	Os07g0577300 protein	1.28	1.719	1.105	0.899	2.422	
tr|A2YIJ5|A2YIJ5_ORYSI	gi|50509727	Os07g0168600 protein	0.779	0.93	0.952	1.118	1.317	
tr|B9F240|B9F240_ORYSJ	gi|222622048	Uncharacterized protein	0.739	1.149	1.372	1.24	1.422	
tr|B9F7T1|B9F7T1_ORYSJ	gi|222624734	Uncharacterized protein	1.389	1.083	1.317	0.854	0.954	

2.6. Identification of Differential Expressive Proteins in LYP9 and NPBA Subjected to Different Salt Stress Levels

From the iTRAQ-based identified proteins in both rice genotypes, the proteins that showed a relative abundance of >1.2 fold or <0.8 fold in the salt stressed plants, as compared to the control, were considered to be differential expressive proteins (DEPs). In LYP9 rice, 1927 DEPs were identified under various salt levels. For instance, 93 (28 up-regulated, 65 down-regulated) DEPs were identified in the LS condition, 782 (368 up-regulated, 414 down-regulated) DEPs were identified in the MS condition, and 1052 (561 up-regulated, 491 down-regulated) DEPs were identified in the HS plants, as compared to the control (Figure 4A). On the other hand, 1154 DEPs were identified in the NPBA rice under the applied salt stress levels. Briefly, 432 (239 up-regulated, 193 down-regulated) DEPs were identified in the LS condition and 722 (385 up-regulated, 337 down-regulated) DEPs were identified in the MS plants, as compared with the control (Figure 4B). Identification of the DEPs in both rice genotypes indicated that, with an increase in the salt levels, the number of DEPs was also increased in both rice types. Further, under LS stress levels, the number of DEPs was significantly less in the salt tolerant LYP9 genotype than in the salt sensitive NBPA rice.

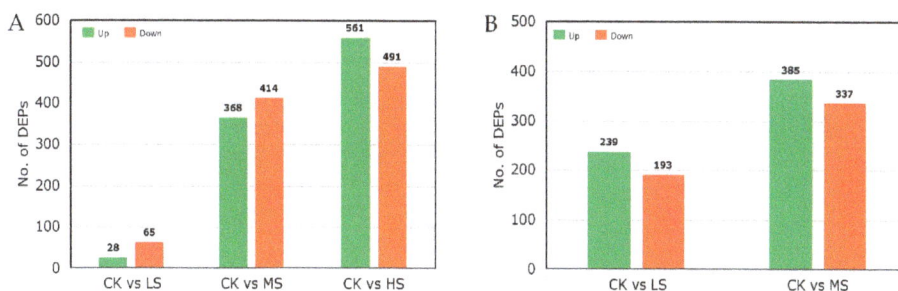

Figure 4. Identification of the differential expressive proteins (DEPs). (**A**) DEPs in LYP9 rice under various salt stress levels as compared with the control plants. (**B**) DEPs in NPBA under various salt stress levels as compared with the control plants. CK: control, LS: low salt, MS: medium salt, HS: high salt.

2.7. Gene Ontology (GO) and Kyoto Encyclopedia of Genes and Genomes (KEGG) Enrichment of the DEPs

To deduce the functionality and biological processes associated with the identified DEPs in the rice genotypes, GO analysis, Clusters of Orthologous Group (COG) annotations, and Kyoto Encyclopedia of Genes and Genomes (KEGG) enrichments were performed. The GO analysis revealed that the identified DEPs were associated with different molecular and biological processes (Figure 5A). Most of the identified DEPs in both rice genotypes were involved in cellular and metabolic processes (biological process). At the molecular level, most of the identified DEPs were involved in catalytic activity, binding, transporter and carrier activity, and structural molecule activity. Similarly, at the cellular component level, the identified DEPs were linked to the cell (membrane and cytoplasm) and organelles. In addition to that, COG analysis of the DEPs grouped them into 24 specific categories on the basis of their functional annotations (Figure 5B). Most of the DEPs were clustered in the "general functional prediction only" category, whereas the post-translational modifications, translation, energy production, carbohydrate metabolism, and amino acid metabolism clusters were found to be the other abundant ones. Altogether, these results suggest that, under salt stress in the rice, salt-responsive proteins might be involved in different metabolic and cellular processes and localize in different cell parts and organelles.

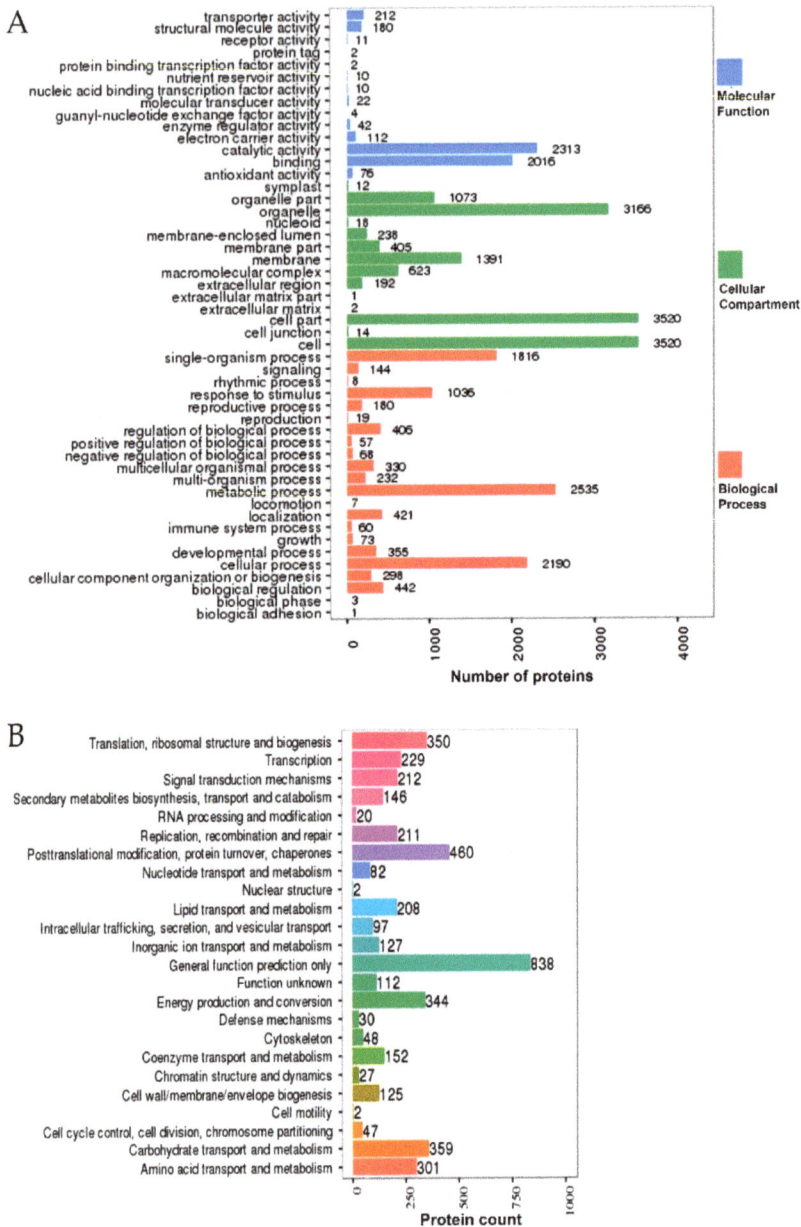

Figure 5. Gene ontology (GO) and Clusters of Orthologous Group (COG) analysis of the differentially responsive proteins in response to salt stress. (**A**) The distribution of number of differentially responsive proteins alongside their corresponding GO terms. Different colors represent different GO categories. (**B**) The distribution of number of differentially responsive proteins alongside their different functions as annotated by COG analysis.

In addition, the KEGG enrichment of the identified DEPs in both rice genotypes revealed their functionality as per the associated pathways. The KEGG pathways, including the metabolic pathway, oxidative phospohorylation, photosynthesis, lysine degradation, glyoxylate metabolism, carbon fixation, photosynthesis-antenna proteins, chlorophyll metabolism, pyruvate metabolism, and ribosomes were found to be the top 10 annotated pathways for the DEPs (Figure 6). From these, the metabolic pathways were found to be the primary enriched pathways in both the rice genotypes. Moreover, analysis of the detail of the KEGG enrichments and associated GO terms revealed that DEPs involved in the salt stress response, redox reactions, photosynthesis, and osmotic stress response were the most abundant in the rice genotypes (Figure 7). For instance, in LYP9, 41 salt-responsive proteins were found to be upregulated under various salt levels, whereas 26 upregulated DEPs were found in the NPBA rice. Similarly, 24 DEPs associated with carbohydrate metabolism were found to be upregulated in LYP9 rice, while 16 DEPs involved with carbohydrate metabolism were found to be upregulated in NPBA. The DEPs from both rice genotypes, with their corresponding fold changes as compared to the controls and their associated physiological pathways, are listed in Table 3. In addition, prediction of the subcellular localizations of the identified DEPs in both the rice genotypes revealed that most of the DEPs localize in the cytoplasm and chloroplasts (Figure 8).

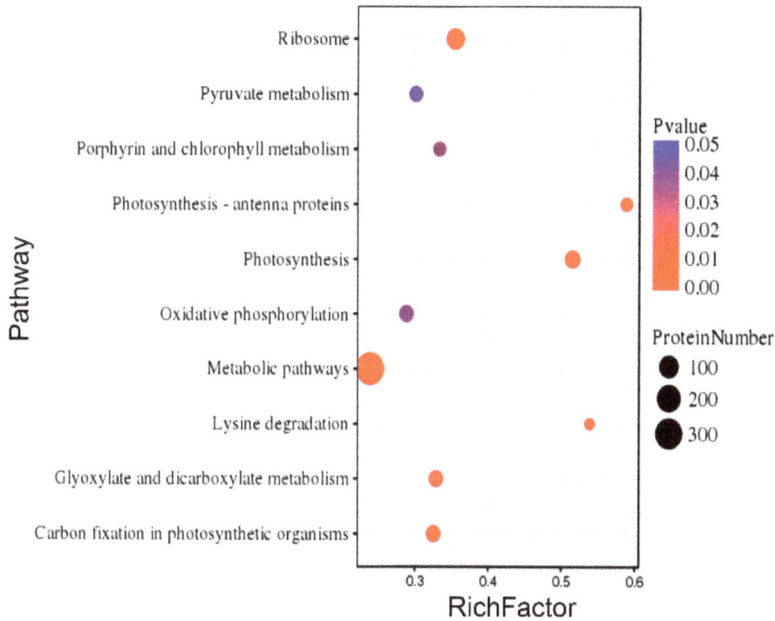

Figure 6. Top 10 pathway enrichments of the identified DEPs in LYP9 and NPBA by KEGG analysis. The corresponding pathways are listed on the Y-axis and the Rich factor values are mentioned along X-axis. Different sized dots represent the distribution of DEPs for a corresponding pathway, whereas, their color represents the *p* value.

A

Metabolic: 1.5%

Ethylene responsive: 2.9%

Omsotic stress: 7.3%

Stress response: 9.5%

Cabohydrate metabolism: 11.7%

Photosyhthesis: 10.9%

Salt responsive: 24.8%

Redox reactions: 31.4%

B

Metabolic: 2.5%

Ethylene responsive: 2.5%

Omsotic stress: 7.5%

Stress response: 5.6%

Cabohydrate metabolism: 14.9%

Photosyhthesis: 9.3%

Salt responsive: 25.5%

Redox reactions: 32.3%

■ Salt responsive ■ Redox reactions ■ Photosyhthesis ■ Cabohydrate metabolism
■ Stress response ■ Omsotic stress ■ Ethylene responsive ■ Metabolic

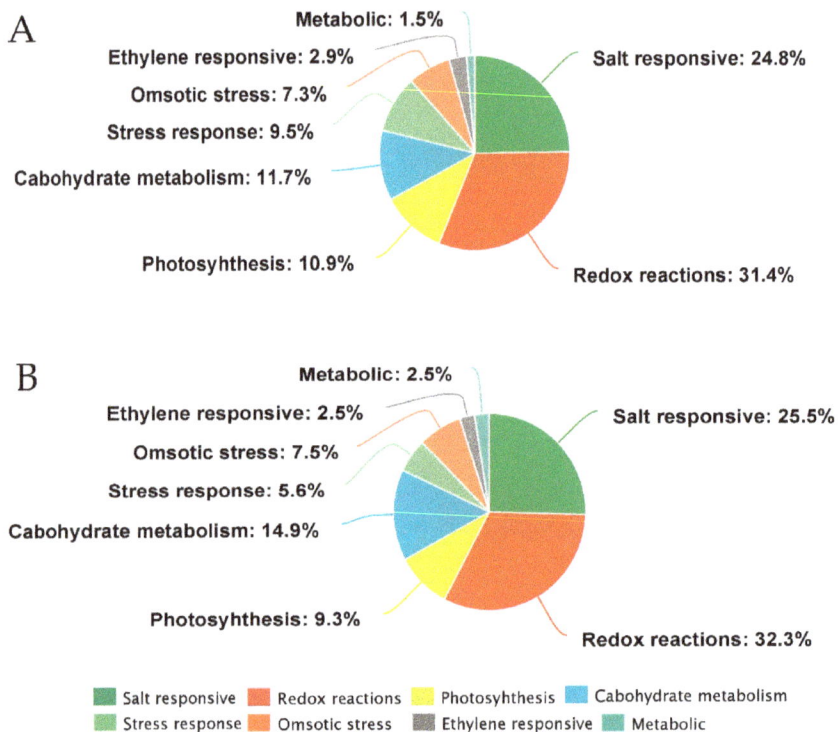

Figure 7. The major pathway annotations of the identified DEPs in LYP9 and NPBA rice. (**A**) Different pathways and their annotated DEP percentages in NPBA rice. (**B**) Different pathways and their annotated DEP percentages in LYP9.

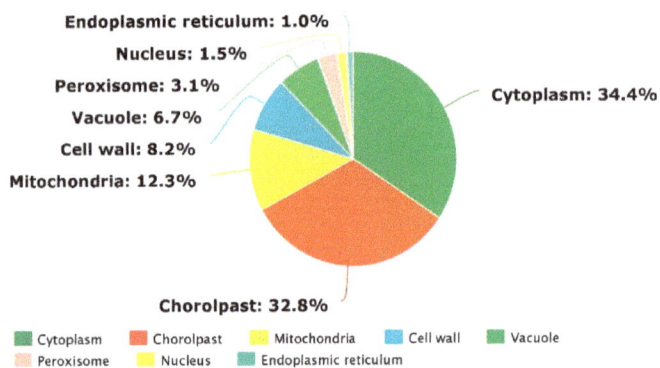

Endoplasmic reticulum: 1.0%

Nucleus: 1.5%

Peroxisome: 3.1%

Vacuole: 6.7%

Cell wall: 8.2%

Mitochondria: 12.3%

Cytoplasm: 34.4%

Chorolpast: 32.8%

■ Cytoplasm ■ Chorolpast ■ Mitochondria ■ Cell wall ■ Vacuole
■ Peroxisome ■ Nucleus ■ Endoplasmic reticulum

Figure 8. The predicted subcellular localization and compartmentation of the identified DEPs in LYP9 and NPBA.

3. Discussion

3.1. Biochemical Responses of Rice Plants to Salt Stress

Salt stress is a major concern in agriculture, affecting crop productivity across the world. Nutrient imbalance, due to the competition of Na^+ and Cl^- with other nutrients, including potassium (K^+), calcium (Ca^{2+}), and nitrate (NO^{3-}) ions, is a result of salt stress that compromises normal plant growth and development [8–12,32]. In addition, salt stress induces early leave-senescence and a decrease in photosynthesis area [33]. Moreover, osmotic imbalance, poor leaf growth, high CMI, and decreased root activity are associated with the typical salt stress responses in plants [31]. In the current study, the subjection of salt stress negatively affected rice growth in the early stages. All four levels of applied salt stress to both rice cultivars resulted in compromised growth parameters along with CMI. The degree of CMI was found to be higher in NPBA as compared with LYP9, suggesting LYP9 has a higher salt tolerance capacity than NPBA (Figure 1). Further, high rice root activity is usually associated with the interaction of the root with rhizosphere soil and the microbial environment [34], changes in physico-chemical status [35], and plant growth [36]. Further, by enhancing the root activity, plants cope better under an unfavorable environment [34] (Figure 3). In this study, the salt tolerance levels of LYP9 were found to be much higher than those of NPBA at high salt conditions (HS). LYP9 plants could survive by significantly increasing their root activities, whereas none of NPBA plants could survive at the same salt concentrations (Figure 2).

3.2. Proteomic Analysis in the Rice Genotypes Under Salt Stress

Both transcriptomic and proteomic dynamics occurring when subjected to salt stress have already been reported in several plants [37]. Further, the availability of substantial sequential information on rice has paved the way for the use of analytical proteomic studies, including iTRAQ analysis. In this study, iTRAQ-based protein identifications in LYP9 and NPBA cultivars revealed their proteome dynamics in response to salt stress. The comparative analysis of the total of identified proteins (5340) revealed that 93, 782, and 1052 proteins were differentially regulated in LYP9 as compared to the control (CK) under LS, MS, and HS salt stress conditions, respectively. On the other hand, in NPBA, 432 and 722 differentially expressed proteins were found as compared to CK under LS and MS salt stress conditions, respectively (Table 3). These results suggest that the numbers of identified proteins are in direct proportion to the increasing salt stress levels. In addition, the finding of increased numbers of differentially expressed proteins in between LS and MS in both cultivars, and in between LS and MS, and MS and HS in LYP9, further strengthens the proposed proportional relationship between differential protein expression and salt stress levels. Moreover, using the iTRAQ identified protein information, we compared the proteins expressed in LYP9 and NPBA, and thereby the biochemical pathways were identified, including salt stress-responsive protein synthesis, redox responses, photosynthesis, and other metabolic processes. Some of these pathways in response to salt stress have been confirmed in some of the previous studies [38,39]; therefore, the functions of the identified DEPs in this study are discussed further below.

The proteome dynamics and the DEPs in NPBA and LYP9 rice genotypes under different salt stress levels were determined by using iTRAQ analysis. Further, to detect and quantify the proteins in the rice genotypes, the high-resolution LC–MS/MS technique was employed. The identified proteins were quantified on automated software called IQuant [40]. Sequences of the identified DEPs were retrieved from the rice protein database based on the GI numbers, and a blastp algorithm was performed against the GO and KEGG databases. GO annotations of the DEPs were performed over three domains—cellular component, molecular function, and biological process—by using *R* software packages. Likewise, the COGs were delineated by using a PERL scripted pipeline. The pipeline of the iTRAQ-based protein identification and the subsequent bioinformatic characterizations are represented in Figure 9.

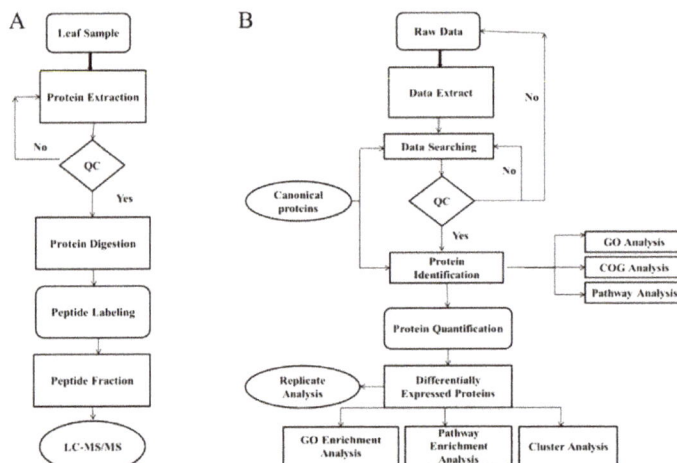

Figure 9. Schematic diagram of the experimental procedures and the complete pipeline for isobaric tags for relative and absolute quantitation (iTRAQ) bioinformatics quantification analysis. (**A**) Steps of the experiment of iTRAQ quantitative proteomics. (**B**) The bioinformatics analysis pipeline for the identified proteins from iTRAQ analysis. All the proteins (FDR < 0.01) proceeded with downstream analysis, including GO, COG, and Kyoto Encyclopedia of Genes and Genomes (KEGG).

3.2.1. Proteins Related to Salt Stress

The comparative proteomics study of both rice genotypes (LYP9 and NPBA) under salt stress revealed new insights into the salt resistance or sensitive mechanisms in rice. In both the rice genotypes, some of the major salt stress-responsive proteins exhibited differential up regulations as compared to the control, including malate dehydrogenase (gi l 115482534), glucanase (gi l 13249140), nascent polypeptide-associated complex (NAC) subunit (gi l 115450217), methyltransferase (gi l 115477769), and chloroplast inorganic pyrophosphatase (gi l 46805452) (Table 3). Plant malate dehydrogenase (MDH) (EC 1.1.1.37) is a member of the oxidoreductase group that catalyzes the inter-conversion of malate and oxaloacetate in a redox reaction [24]. Further, MDH has been shown to play a vital role in regulating the salt stress response in plants [41,42]. Likewise, glucanase and inorganic pyrophosphatases have been associated with salt resistance properties in plants [43,44]. NAC has been reported to be involved in the translocation of newly synthesized proteins from the ribosomes to the endoplasmic reticulum during various physiological conditions, by directly interacting with the signal recognition particles. Further, overexpression of SaβNAC from *Spartina alterniflora* has been reported to enhance the salt tolerance in *Arabidopsis* [45]. In addition, methylation is often utilized by plants under unfavorable conditions as a strategy for gene regulation, protein sorting, and repairs [46]. IbSIMT1, a methyltransferase gene, has been observed to be activated by salt stress, and confers salinity resistance in sweet potato [47]. On the contrary, DEPs associated with salt stress responses, including glutathione peroxidase (GP) (gi l 125540587), fructose-bisphosphate aldolase (FBA) (gi l 218196772), pyruvate dehydrogenase (gi l 125564321), and triosephosphate isomerase (TPI) (gi l 125528336) were found to be significantly upregulated in LYP9, but down regulated in NPBA. Recently, the rice GP gene (*OsGPX3*) has been reported to play a vital role in regulating the salt stress response [48]. Rice plants with silenced *OsGPX3* were found to be highly salt sensitive, confirming the positive role of GP in salinity tolerance. FBA is involved in plant glucose pathways, including glycolysis and gluconeogenesis, and also plays a role in the Calvin cycle [49]. However, the FBA gene has been reported to exhibit induced expressions under salt stress in plants, indicating its role in salt stress. The FBA genes in *Arabidopsis* and *Camellia oleifera* were found to be strongly upregulated under salt stress, conferring salinity tolerance [48,50]. Likewise, the transcription of TPI genes has been reported to become active in rice in response to salt stress [51,52]. The upregulated expression of these salt related proteins in the salt-tolerant genotype LYP9, and their down regulation in the salt-sensitive NPBA, suggests that these genotypes possess a different protein pool in response to salinity. Moreover, the difference in salt tolerance between these two rice genotypes might have resulted due to the differential expression of these key proteins. A functional validation study, such as the Western blot or protein interactions, will add further insights to this hypothesis.

3.2.2. Proteins Related to Redox Reactions

Salt stress in plants induces osmotic imbalances, disrupts ion-homeostasis, and triggers oxidative damage, including the generation of reactive oxygen species (ROS) [53,54]. A fitting response to these adversities caused by salinity stress includes physiological and developmental changes, reprograming of salt-induced gene or proteins, and activation of ROS scavenging pathways [55]. In the current study, the proteomic analysis of LYP9 and NBPA revealed that redox reactions and ROS signaling are involved in the salt stress response in rice. Major enzymes involved in ROS signaling and redox reactions, including peroxidases (POD) (gi | 125525683), superoxide dismutase (SOD) (gi | 125604340), and glutathione s-transferase (GST) (gi | 115459582), were found to be highly expressive in LYP9 and NPBA genotypes under the multiple salt stress levels we investigated (Table 3). Under salt stress, the cell membrane-bound peroxidases like NADPH oxidase and the diamine oxidases present in apoplast are activated, leading to generation of ROS [56,57]. In addition, SOD act as the first line of antioxidant defense in plants under multiple stress responses, and confer enhanced tolerance levels to oxidative stress [54]. Similarly, increased levels of GSTs in response to multiple stimuli have been reported in plants to mitigate oxidative stress [58]. Induced expressions and differential regulation of antioxidant enzymes, including PODs, SODs, and GSTs, have been reported by several studies in rice in response to salt stress [59,60]. Furthermore, comparative proteome analysis has confirmed the involvement of ROS and redox related protein in salt stress in plants, including alfalfa [61], searocket [62], maize [63], barley [64], and wheat [65]. Moreover, as many as 56 DEPs annotated with redox reaction functions were identified in both the rice genotypes under the various salt stress levels, suggesting oxidation and reduction reactions might be the key biochemical changes taking place in rice under salinity.

3.2.3. Proteins Related to Photosynthesis

Photosynthesis is a major physiological process accounting for sustainability and energy production in plants. However, salt stress has adverse effects on the plant photosynthesis process by causing a decrease in the leaf cellular CO_2 levels [7,66]. Additionally, salinity affects the Rubisco activity, retards chlorophyll synthesis, and destabilizes photosynthetic electron transport [66]. The findings from our study revealed that salt stress in rice affects the expression of the proteins involved in the photosynthesis process. These proteins, including the thylakoid lumenal protein (gi | 115477166), psbP domain-containing protein 6 (gi | 115440559), psbP-like protein 1 (gi | 38636895), ferredoxin-thioredoxin reductase (gi | 115447507), photosystem I 9K protein (gi | 218186547), photosystem II oxygen-evolving complex protein 2 (gi | 164375543), and protochlorophyllide reductase B (gi | 75248671), were found to be highly expressed under salt stress conditions (Table 3). Thylakoid luminal protein is required for the functioning of photosystem II (PspB), whereas ferredoxin reductase is a key enzyme that facilitates the conversion of ferredoxin to NADPH in the photosystem I (PSI) complex, and these are also affected by salt stress [67,68]. Moreover, the psbP proteins, thylakoid luminal proteins, and ferredoxin reductase have been reported to be differentially expressed under salt stress [68]. Likewise, differential expression of photosystem proteins was reported in tomatoes in response to salt stress [69]. Similarly, the differential protein expression of protochlorophyllide reductase between the salt stress-induced and control, and its effects on chlorophyll biosynthesis, has been reported in rice [70]. Usually, in salt sensitive plants, salinity causes the down-regulation of photosynthesis proteins, compromising plant sustainability [2,71]. However, the analysis of iTRAQ-based proteomics revealed that the proteins involved in photosynthesis were upregulated in both rice genotypes, which might have aided the rice types to withstand salinity pressures.

3.2.4. Proteins Related to Carbohydrate Metabolism

Apart from being the building blocks in plants, soluble carbohydrates act as osmolytes, and thereby participate in salt tolerance in plants [72]. Besides, the onset of salt stress affects the protein dynamics in plants, resulting in differential protein accumulations [73]. In this study, several carbohydrate metabolism related proteins, including xyloglucan endotransglycosylase/hydrolase protein (XTH) (gi | 115475445), β-glucosidase (gi | 115454825), and polygalacturonase (gi | 115479865), were found to be upregulated in both rice genotypes under various salt stress levels. XTH is known as a cell wall-modifying enzyme, however it also plays a role in salinity resistance responses in plants (Table 3). For instance, the constitutive and heterologous expression of CaXTH3 resulted in increased salt tolerance levels in *Arabidopsis* and tomato plants [74,75]. Similarly, β-glucosidase is a key enzyme in the cellulose hydrolysis process, and has been reported to be involved in the salt stress response. In barley, the activity of an extracellular β-glucosidase was reported to be highly induced in response to salt stress, and cause abscisic acid-glucose conjugate hydrolysis [76]. Further, the overexpression of *Thkel1*, a fungal gene that modulates β-glucosidase activity, improved the salt tolerance levels in transgenic *Arabidopsis* plants [77]. Polygalacturonase, another enzyme capable of hydrolyzing the α-1,4 glycosidic bonds, participates in the salt stress responses in plants. Characterization of the salt stress responses and the associated signal transduction pathways in *Arabidopsis* revealed the elevated transcript accumulation of a polygalacturonase gene (*At1g48100*) under salt stress [78]. However, several proteins related to carbohydrate metabolism, including xylanase inhibitor protein (XIP) (gi | 297605789, gi | 115467998) and MDH (gi | 116310134), were found to be downregulated in the NPBA rice, while being upregulated in the LYP9 rice. MDH is a key enzyme in stress responses and actively participate in the tricarboxylic acid (TCA) cycle [74]. In the current study, upregulated expression of MDH was found in LYP9, however down-regulation in NPBA suggests the inhibition of the TCA cycle in the salt sensitive NPBA, but not in the tolerant LYP9 genotype. Further, OsXIP was reported to be induced under various abiotic stresses, including salt stress, and to take part in the rice defense mechanisms against several biotic and abiotic stresses [79]. Moreover, the induced many-fold expression of the carbohydrate metabolism related proteins in LYP9, but their down regulation in NPBA, indicates that carbohydrate metabolism might be a major physiological process that is affected under salinity in rice, and can show the dynamic changes in protein expression depending on the salt tolerance capacity of a genotype.

3.2.5. Proteins Related to Osmotic Stress

Often, salt stress induces the reduction of cellular water potential, causing osmotic stress to the plant. Osmotic stress responses in plants can be very complex in higher plants, including rice [80]. In this study, 11 osmotic stress related proteins were differentially expressed in both rice genotypes under various salt levels, suggesting salt stress in rice leads to the onset of osmotic stress. For instance, a putative lipid transferase protein (gi | 297612544) identified as a DEP in both the rice genotypes was found to be upregulated under salt stress. The induced expression of *TSW12* and *SiLTP*, coding the lipid transferase proteins in tomato and foxtail millet plants, has been reported under salt stress [80,81]. Conversely, osmotic stress responsive proteins such as sucrose synthase (gi | 125544232) and NADH dehydrogenase (gi | 115473055) were found to exhibit an induced response in LYP9 rice under salt stress, but were not significantly induced in the NPBA rice. Sucrose synthase (Sus) is the major enzyme in sucrose metabolism, however it also plays a part in osmotic stress responses in plants. In *Arabidopsis*, up-regulation of Sus1 has been reported in response to osmotic stresses and water deficit conditions [82]. In addition, involvement of Sus in the osmotic stress response has been reported in *Beta vulgaris* [83]. On the other hand, NADH dehydrogenase facilitates electron transfer from NADH to the mitochondrial respiratory chain [84]. The up-regulation of NADH dehydrogenase under salt stress indicates an increase in the ATP pool in the LYP9 rice, subsequently aiding in sustainable plant growth and salinity tolerance. However, no induced expression of the same in NPBA suggests that, under salt stress, the ATP pool might decrease, resulting in declining plant growth (Table 3).

3.2.6. Proteins Related to Other Metabolic Processes

Salt stress alters the protein pool that contributes to many metabolic mechanisms, such as stress responses, energy metabolism, and phytohormone synthesis [23,85]. In this study, several DEPs have been identified in the rice genotypes under salt stress, with various physiological and metabolic functions. For instance, putative glucan endo-1,3-β-glucosidase 4 (gi|297607511) was found to be up-regulated in both rice types under salt stress conditions. Similar findings were reported in cotton plants, where the subjected salt stress caused an increased accumulation of glucan endo-1,3-β-glucosidase [23]. Further, the strong induced response of a putative zinc finger protein (gi|28971968) was found under salt stress in both rice genotypes. Induced expression of gene finger proteins has been associated with several stresses, including salt stress. Overexpression of a rice zinc-finger protein OsISAP1 in transgenic tobacco resulted in enhanced abiotic stress tolerance levels, including salinity, dehydration, and cold [86]. Recently, OsZFP213 was reported to interact with OsMPK3, conferring salinity tolerance in rice [87]. In addition, many other proteins with annotated functions or which are uncharacterized were found to be differentially regulated at various salt levels in the rice genotypes. Moreover, these results collectively suggest that salinity affects many physiological processes in rice, irrespective of their salt tolerance levels. Furthermore, the protein pool of a salt tolerant and a salt sensitive rice genotype might differ at a specific point of time, which could be the basic reason of their differential salt tolerance responses (Table 3).

4. Materials and Methods

4.1. Plant Material and Growth Conditions

A pot culture experiment was conducted in a greenhouse at China National Rice Research Institute (39°4′49″ N, 119°56′11″ E), Zhejiang Province, China, during the rice growing season (May–November, 2017). Two rice cultivars (origin, China and Japan), Liangyoupiejiu (LYP9, Hybrid, *indica*) and Nipponbare (NPBA, *japonica*) were used as the planting materials. Thirty-day old seedlings were transplanted in pots (45 × 30 cm) with different salt stress levels and 23 kg air-dried soil. The experimental soil was loamy clay with an average bulk density of 1.12 g/cm, 4.7% organic matter, 0.0864 dS/m EC, and 5.95 pH. Each pot contained six rice seedlings with three replications.

Sodium chloride (NaCl) was used in each pot to develop artificial salinity in soil until the maximum tillering stage of the rice seeding was reached (about 45 days). The treatments were comprised of four NaCl levels: 0 (control, CK), 1.5 g NaCl/kg dry soil (low salt stress, LS), 4.5 g NaCl/kg dry soil (moderate salt stress, MS), and 7.5 g NaCl/kg dry soil (high salt stress, HS). After salinity development, the corresponding EC for these levels was 0.086 dS/m (CK), 1.089 dS/m (LS), 3.20 dS/m (MS), and 4.64 dS/m (HS).

Nitrogen was applied in the form of urea (N: 46%), phosphorous as superphosphate (P_2O_5: 12%), and potassium as potassium sulfate (K_2O: 54%). Urea was used at the rate of 4.02 g/pot in two splits: 50% was applied as the basal dose, and 50% was applied at the tillering stage. Potassium sulfate (3.08 g/pot) was applied in two equal splits, as a basal dose and at the tillering stage, while the whole amount of superphosphate (6.93 g/pot) was applied as a basal dose.

4.2. Soil and Plant Sampling

Rice flag leaves were collected at the maximum tillering stage and stored at −80 °C after being frozen in liquid nitrogen. Plants were collected for measurement of Na^+ and Cl^- contents in the roots and leaves at the maximum tillering stage. Soil samples were collected at the transplanting stage and at the maximum tillering stage to check the Na^+ and Cl^- contents in the soil. Five flag leaves with three replicates were collected to measure the cell membrane injury in rice leaves at the maximum tillering stage, while root samples were collected to measure the rice root activity. All these experiments were performed with three independent biological replicates.

4.3. Leaf Proteomics Analysis Pipeline

4.3.1. Protein Extraction

A total of 1–2 g of plant leaves with 10% PVPP were ground in liquid nitrogen and then sonicated on ice for 5 min in Lysis buffer 3 (8M Urea and 40 mM Tris-HCl containing 1 mM PMSF, 2 mM EDTA, 10 mM DTT, and pH 8.5) with 5 mL of samples. After centrifugation, 5 mL of 10% TCA/acetone with 10 mM DTT were added to the supernatant to precipitate the proteins. The precipitation step was repeated with acetone alone until the supernatant became colorless. The proteins were air dried and re-suspended in Lysis buffer 3. Ultra-sonication on ice for 5 min was used to improve protein dissolution with the help of Lysis buffer 3. After centrifugation, the supernatant was incubated at 56 °C for 1 h for reduction, and then alkylated by 55 mM iodoacetamide (IAM) in the dark at room temperature for 45 min. Acetone (5 mL) were used to precipitate the proteins and stored at –80 °C. The quality and quantity of the isolated proteins were estimated by performing Bradford assay and SDS-PAGE [88].

4.3.2. Digestion of Proteins and Peptide Labeling

About 100 μg of the protein solution with 8 M urea was diluted four times with 100 mM TEAB. For the digestion of the proteins, Trypsin Gold (Promega, Madison, WI, USA) was used at a ratio of trypsin: protein of 40:1, at 37 °C, and was put into the samples overnight. After the digestion with trypsin, Strata X C18 column (Phenomenex, Torrance, CA, USA) were used to desalt the peptides and vacuum-dry them according to the manufacturer's protocol. For peptide labeling, the peptides were dissolved in 30 μL 0.5 M TEAB. Then, the peptide labeling was performed by an iTRAQ reagent 8-plex kit. The labeled peptides with different reagents were combined and desalted with a Strata X C18 column (Phenomenex), and vacuum-dried.

4.3.3. Peptide Fractionation and HPLC

The peptide fractionations were performed by using a Shimadzu LC-20AB HPLC pump attached to a high pH RP column. About 2 mL of the reassembled peptides with buffer A (5% ACN, 95% H₂O, pH 9.8) was loaded on a 5 μm particulate column (Phenomenex). The flow rate was adjusted to 1 mL/min with a 5% buffer B (5% H₂O, 95% ACN, pH 9.8) gradient for 10 min, with 5–35% buffer B for 40 min, and with 35–95% buffer B for 1 min, to separate the peptides. An incubation of 3 min in 95% buffer B, and for 1 min in 5% buffer B, followed this, before the final equilibration with 5% buffer B. Each peptide fraction was collected at 1 min time intervals, and OD of the eluted fractions were measured at 214 nm. Twenty fractions were pooled together and vacuum dried. Post drying, the fractions were re-suspended in buffer A solution (2% CAN; 0.1% FA in water) individually and centrifuged. Then, the supernatant was collected and loaded onto a C18 trap column with a rate of 5 μL/min by using a LC-20AD nano-HPLC device (Shimadzu, Kyoto, Japan). Peptide elutions were performed afterwards and separated by using an analytical C18 column with an inner diameter of 75 μm. The gradients were run at 300 nL/min starting from 8 to 35% of buffer B (2% H₂O; 0.1% FA in ACN) for 35 min, with an increase up to 60% in 5 min, then were maintained at 80% buffer B for 5 min before returning to 5% in 6 s, with a final equilibration period of 10 min.

4.3.4. Mass Spectrometer Detection

The spectrometric data were acquired using a TripleTOF 5600 System (SCIEX, Framingham, MA, USA) fit to a Nano-Spray III source (SCIEX, Framingham, MA, USA) and a pulled quartz tip-type emitter (New Objectives, Woburn, MA, USA), which was controlled with the franchise software

Analyst v1.6 (AB-SCIEX, Concord, ON, Canada). The MS data procurements were undertaken as per the following conditions: the ion spray voltage was set to 2300 V, the curtain gas was set to 30, the nebulizer gas was set to 15, and the interface heater temperature was 150 °C. High sensitivity mode was used for the whole data acquisition process. The MS1 accumulation time was set to 250 ms, while 350–1500 Da was the allowed mass range. At least 30 product ion scans were collected based on the MS1 survey intensity, exceeding a threshold of 120 counts/s and a 2^+ to 5^+ charge-state. A value of 1/2 peak width was set for the dynamic exclusion. The collision energy was adjusted to all precursor ions for the collision-induced dissociation for the iTRAQ data acquisition, and the Q2 transmission window for 100 Da was at 100%. Three independent biological replicates were included for each sample in the experiment.

4.4. Cell Membrane Injury

Cell membrane injury (CMI) was determined by using flag leaves. Twenty pieces (1 cm diameter) were cut from these flag leaves, and were submerged into 20 mL distilled water (DI) contained in test tubes. The test tubes were kept at 10 °C in an incubator for 24 h. After 24 h, the samples were kept at 25 °C to warm the samples, and the electrical conductivity (C1) of the samples was measured. These samples were then autoclaved for 20 min at 120 °C and the electrical conductivity (C2) was determined again. Cellular injury was determined by using the following formula [89]:

$$Cellmembraneinjury = \frac{C1}{C2} \times 100 \quad (1)$$

where, 'C' refers to EC 1 and 2. The experiment was performed with three independent biological replicates.

4.5. Rice Root Activity

Rice root activity was analyzed by the triphenyl tetrazolium chloride (TTC) method [90]. Briefly, rice root samples (0.5 g, root tips) were taken, and 5 mL of the phosphate buffer (pH 7) and 5 mL 0.4% TTC (Vitastain, $C_{19}H_{15}N_4Cl$) were added to keep the root activity alive. The samples were kept in an incubator in the dark at 37 °C for 3 h. After 3 h, the samples were taken out and 1 mL 1 mol/L H_2SO_4 was added to stop the reaction. The rice roots were then removed from the test tubes. These root samples were ground by adding a pinch of silica sand, and mixed with 8 mL ethyle acetate. The extract was transferred to test tubes and a 10 mL final volume was reached by adding ethylene acetate. These samples were analyzed using a spectrophotometer (UV-2600, UV-VIS Spectrophotometer Shimadzu) at 485 nm. The formula used for calculation of root activity is as follows:

$$RootActivty = \frac{C}{W/3} \quad (2)$$

where C is the concentration of the samples calculated from a standard curve. W is the weight of the root samples. The experiment was performed with three independent biological replicates.

4.6. Na^+ Concentration in the Soil and Plants

Na^+ was extracted from the soil by ammonium acetate solution using Rihards (1954) method [91]. About 5 g ground (particle size ≤ 2mm) air dried soil was placed in 250 mL plastic bottles and 50 mL ammonium acetate (NH_4OAc, 1 mol/L) was added. These bottles were kept on a shaker for 30 min at 120 rpm. After that, the samples were filtered by using filter paper to obtain the soil solution.

Na^+ in the plants' parts was extracted by digestion with sulfuric acid (H_2SO_4) by following Rihards (1954) method [91] with the necessary modifications. About 0.3 g ground (particle size ≤ 2 mm) root and leaf were taken in 50 mL glass tubes and mixed with 5 mL H_2SO_4. These glass bottles were kept overnight. The samples were put into the fume hood and were incubated at 320 °C for 2 h. After 2 h, hydrogen peroxide solution (H_2O_2) was added drop by drop and the samples were mixed until a whitish or transparent color appeared. Then, the samples were cooled at room temperature before being filtered by using filter paper to get the plant part extracts.

The soil and plant extracts were used to measure the sodium ions (Na^+) by using a flame photometer. The standards used were 0, 2, 4, 6, 8, 10, 15, and 20 mL NaCl. The final soluble sodium (Na) in soil was measured by using the formula:

$$Na\left(\frac{\mu g}{g}\right) = \frac{A \times C}{W} \tag{3}$$

where A is the total volume of the extract (mL), C is the sodium concentration values given by the flame-photometer ($\mu g/mL$), and W is the weight of the air dried soil (g). The experiment was performed with three independent biological replicates.

4.7. Cl^- Concentration in the Soil and Plants

About 10 g air dried soil (particle size $\leq$ 2mm) was placed in 250 mL plastic bottles and mixed with 50 mL deionized water. These bottles were transferred onto a shaker and were shaken for 5 min at 180 rpm. The samples were then filtered by using filter paper to obtain the soil solution extract for Cl^-.

Plant samples weighing approximately 0.1 g were placed in 50 mL glass tubes and mixed with 15 mL deionized water. The tubes were transferred into a hot water bath and kept for 1.5 h. The samples were then diluted with 25 mL deionized water after cooling at room temperature.

The soil and plant extracts were used to measure the chloride (Cl^-) by using a chloride assay kit (QuantiChrom^{TM} Chloride Assay Kit, 3191 Corporate Place Hayward, CA 94545, USA) following the manufacturer's instructions. The standards used were 0, 10, 20, 30, 40, 60, 80, and 100 mL. The final chloride concentration in the solution was measured by the formula:

$$Chloride = \frac{ODsample - ODblank}{Slop} \times n\left(\frac{mg}{dL}\right) \tag{4}$$

where ODsample is the OD 610 nm values of the samples, and ODblank is the OD 610 nm values of the blanks (water). The experiment was performed with three independent biological replicates.

4.8. Statistical Analysis

The statistical software package IBS SPSS Statistics 19.0 was used for the analyses of data. For evaluating the statistical significance of the biochemical parameters, a one-way ANOVA was employed with LSD at the level of $p = 0.05$. For the iTRAQ-based protein quantification, all identified DEPs were required to satisfy the t-test at $p \leq 0.05$, and with a fold change ratio of >1.2 or <0.8.

5. Conclusions

Using comparative iTRAQ-based protein quantification, the proteome dynamics of LYP9 and NPBA rice were explored in this study. The results from the study suggest that rice cell membrane integrity was inversely correlated and root activity was positively correlated with the concentration of salinity. Furthermore, the physiological processes, including carbohydrate metabolism, redox reactions, and photosynthesis, made significant contributions towards the salt tolerance in rice. The number of differentially expressed proteins—salt responsive proteins in particular—suggested that the protein pool in response to salt stress is different in a salt tolerant compared to a susceptible rice genotype. Finally, the *indica* rice LYP9 showed promising results under the subjected salt stress levels, and can be selected over the *japonica* NPBA for salt tolerance. Further works deciphering the functions of some particular proteins of interest will add new insights into their roles in salt tolerance in rice.

Author Contributions: S.H. (Sajid Hussain), J.Z., and Q.J. conceived and designed research. S.H. (Sajid Hussain), J.Z., and Q.J. conducted experiments. J.H., X.C., Z.B., and L.Z. contributed analytical tools. S.H. and C.Z. analyzed

Int. J. Mol. Sci. **2019**, *20*, 547

the data. S.H. (Sajid Hussain), S.N., and C.Z. wrote the manuscript. S.N., S.H. (Saddam Hussain), and A.R. revised the manuscript. Q.L., L.W., and Y.L. help in formal analysis. The manuscript has been read and approved by all authors.

Funding: This work was supported by the National Key Research and Development Program of China (2016YFD0200801), and the Natural Science Foundation of China (31872857, 31771733).

Conflicts of Interest: The authors declare no conflict of interest exists.

Abbreviations

LYP9	Liangyoupeijiu
NPBA	Nipponbare
iTRAQ	Isobaric tags for relative and absolute quantitation
CMI	Cell membrane injury
RRA	Rice root activity
DEPs	Differentially expressed proteins
GO	Gene ontology
KEGG	Kyoto encyclopedia of genes and genomes
PSI	Photosystem I
LS	Low salt stress
MS	Moderate salt stress
HS	High salt stress
COG	Cluster of orthologous groups

References

1. Yamori, W.; Hikosaka, K.; Way, D.A. Temperature response of photosynthesis in C3, C4, and CAM plants: Temperature acclimation and temperature adaptation. *Photosynth. Res.* **2013**, *119*, 101–117. [CrossRef] [PubMed]
2. UNFPA. Linking Population, Poverty and Development. 2014. Available online: http://www.unfpa.org/pds/trends.htm (accessed on 28 January 2019).
3. Ray, D.K.; Mueller, N.D.; West, P.C.; Foley, J.A. Yield trends are insufficient to double global crop production by 2050. *PLoS ONE* **2013**, *8*, e66428. [CrossRef] [PubMed]
4. Nachimuthu, V.V.; Sabariappan, R.; Muthurajan, R.; Kumar, A. *Breeding Rice Varieties for Abiotic Stress Tolerance: Challenges and Opportunities*; Springer: Singapore, 2017.
5. Pushpam, R.; Rangasamy, S.R.S. In vivo response of rice cultivars to salt stress. *J. Ecol.* **2002**, *14*, 177–182.
6. Joseph, B.; Jini, D.; Sujatha, S. Biological and Physiological Perspectives of Specificity in Abiotic Salt Stress Response from Various Rice Plants. *Asian J. Agric. Sci.* **2010**, *2*, 99–105.
7. Hussain, S.; Chu, Z.; Zhigang, B.; Xiaochuang, C.; Lianfeng, Z.; Azhar, H.; Chunquan, Z.; Shah, F.; Allen, B.J.; Junhua, Z.; et al. Effects of 1-Methylcyclopropene on Rice Growth Characteristics and Superior and Inferior Spikelet Development Under Salt Stress. *J. Plant Growth Regul.* **2018**, *37*, 1368–1384. [CrossRef]
8. Islam, T.; Manna, M.; Reddy, M.K. Glutathione peroxidase of Pennisetum glaucum (PgGPx) is a functional Cd21 dependent peroxiredoxin that enhances tolerance against salinity and drought stress. *PLoS ONE* **2015**, *10*, e0143344. [CrossRef] [PubMed]
9. Islam, F.; Yasmeen, T.; Ali, S.; Ali, B.; Farooq, M.A.; Gill, R.A. Priming-induced antioxidative responses in two wheat cultivars under saline stress. *Acta Physiol. Plant.* **2015**, *37*, 153–161. [CrossRef]
10. Islam, F.; Yasmeen, T.; Arif, M.S.; Ali, S.; Ali, B.; Hameed, S.; Zhou, W. Plant growth promoting bacteria confer salt tolerance in Vigna radiata by up-regulating antioxidant defense and biological soil fertility. *Plant Growth Regul.* **2016**, *80*, 23–36. [CrossRef]
11. Islam, F.; Ali, B.; Wang, J.; Farooq, M.A.; Gill, R.A.; Ali, S.; Wang, D.; Zhou, W. Combined herbicide and saline stress differentially modulates hormonal regulation and antioxidant defense system in *Oryza sativa* cultivars. *Plant Physiol. Biochem.* **2016**, *107*, 82–95. [CrossRef] [PubMed]
12. Naeem, M.S.; Jin, Z.L.; Wan, G.L.; Liu, D.; Liu, H.B.; Yoneyama, K.; Zhou, W.J. 5-Aminolevulinic acid improves photosynthetic gas exchange capacity and ion uptake under salinity stress in oilseed rape (*Brassica napus* L.). *Plant Soil* **2010**, *332*, 405–415. [CrossRef]

13. Hussain, S.; Xiaochuang, C.; Chu, Z.; Lianfeng, Z.; Maqsood, A.K.; Sajid, F.; Junhua, Z.; Qianyu, J. Sodium chloride stress during early growth stages altered physiological and growth characteristics of rice. *Chil. J. Agric. Res.* **2018**, *78*, 183–197. [CrossRef]

14. Islam, F.; Ali, S.; Farooq, M.A.; Wang, J.; Gill, R.A.; Zhu, J.; Ali, B.; Zhou, W. Butachlor-induced alterations in ultrastructure, antioxidant, and stress-responsive gene regulations in rice cultivars. *Clean—Soil Air Water* **2017**, *45*, 1500851. [CrossRef]

15. Islam, F.; Farooq, M.A.; Gill, R.A.; Wang, J.; Yang, C.; Ali, B.; Wang, G.X.; Zhou, W. 2,4-D attenuates salinity-induced toxicity by mediating anatomical changes, antioxidant capacity and cation transporters in the roots of rice cultivars. *Sci. Rep.* **2017**, *7*, 10443. [CrossRef] [PubMed]

16. Cui, P.; Liu, H.; Islam, F.; Li, L.; Farooq, M.A.; Ruan, S.; Zhou, W. OsPEX11, a peroxisomal biogenesis factor 11, contributes to salt stress tolerance in *Oryza sativa*. *Front. Plant Sci.* **2016**, *7*, 1357. [CrossRef] [PubMed]

17. Shafi, A.; Chauhan, R.; Gill, T.; Swarnkar, M.K.; Sreenivasulu, Y.; Kumar, S.; Kumar, N.; Shankar, R.; Ahuja, P.S.; Singh, A.K. Expression of SOD and APX genes positively regulates secondary cell wall biosynthesis and promotes plant growth and yield in *Arabidopsis* under salt stress. *Plant Mol. Biol.* **2015**, *87*, 615–631. [CrossRef]

18. Biswas, M.S.; Mano, J.I. Lipid peroxide-derived short-chain carbonyls mediate hydrogen peroxide induced and salt-induced programmed cell death in plants. *Plant Physiol.* **2015**, *168*, 885–898. [CrossRef]

19. Ali, I.; Liu, B.; Farooq, M.A.; Islam, F.; Azizullah, A.; Yu, C.; Su, W.; Gan, Y. Toxicological effects of bisphenol A on growth and antioxidant defense system in Oryza sativa as revealed by ultrastructure analysis. *Ecotoxicol. Environ. Saf.* **2016**, *124*, 277–284. [CrossRef]

20. Ali, I.; Jan, M.; Wakeel, A.; Azizullah, A.; Liu, B.; Islam, F.; Ali, A.; Daud, M.K.; Liu, Y.; Gan, Y. Biochemical responses and ultrastructural changes in ethylene insensitive mutants of *Arabidopsis* thaliana subjected to bisphenol A exposure. *Ecotoxicol. Environ. Saf.* **2017**, *144*, 62–71. [CrossRef]

21. Li, H.; Chang, J.; Chen, H.; Wang, Z.; Gu, X.; Wei, C.; Zhang, Y.; Ma, J.; Yang, J.; Zhang, X. Exogenous Melatonin Confers Salt Stress Tolerance to Watermelon by Improving Photosynthesis and Redox Homeostasis. *Front. Plant Sci.* **2017**, *8*, 295. [CrossRef]

22. Khare, T.; Kumar, V.; Kishor, P.K. Na⁺ and Cl⁻ ions show additive effects under NaCl stress on induction of oxidative stress and the responsive antioxidative defense in rice. *Protoplasma* **2015**, *252*, 1149–1165. [CrossRef]

23. Li, W.; Zhao, F.; Fang, W.; Xie, D.; Hou, J.; Yang, X.; Zhao, Y.; Tang, Z.; Nie, L.; Lv, S. Identification of early salt stress responsive proteins in seedling roots of upland cotton (*Gossypium hirsutum* L.) employing iTRAQ-based proteomic technique. *Front. Plant Sci.* **2015**, *6*, 732. [CrossRef] [PubMed]

24. Wang, Z.Q.; Xu, X.Y.; Gong, Q.Q.; Xie, C.; Fan, W.; Yang, J.L.; Lin, Q.S.; Zheng, S.J. Root proteome of rice studied by iTRAQ provides integrated insight into aluminum stress tolerance mechanisms in plants. *J. Proteom.* **2014**, *98*, 189–205. [CrossRef] [PubMed]

25. Hu, X.; Li, N.; Wu, L.; Li, C.; Li, C.; Zhang, L.; Liu, T.; Wang, W. Quantitative iTRAQ-based proteomic analysis of phosphoproteins and ABA regulated phosphoproteins in maize leaves under osmotic stress. *Sci. Rep.* **2015**, *27*, 15626. [CrossRef] [PubMed]

26. Guo, G.; Ge, P.; Ma, C.; Li, X.; Lv, D.; Wang, S.; Ma, W.; Yan, Y. Comparative proteomic analysis of salt response proteins in seedling roots of two wheat varieties. *J. Proteom.* **2012**, *75*, 1867–1885. [CrossRef] [PubMed]

27. Gong, B.; Zhang, C.; Li, X.; Wen, D.; Wang, S.; Shi, Q.; Wang, X. Identification of NaCl and NaHCO3 stress-responsive proteins in tomato roots using iTRAQ-based analysis. *Biochem. Biophys. Res. Commun.* **2014**, *446*, 417–422. [CrossRef] [PubMed]

28. Kim, D.W.; Rakwal, R.; Agrawal, G.K.; Jung, Y.H.; Shibato, J.; Jwa, N.S.; Iwahashi, Y.; Iwahashi, H.; Kim, D.H.; Shim, I.S.; et al. A hydroponic rice seedling culture model system for investigating proteome of salt stress in rice leaf. *Electrophoresis* **2005**, *26*, 4521–4539. [CrossRef] [PubMed]

29. Dooki, A.D.; Mayer-Posner, F.J.; Askari, H.; Zaiee, A.; Salekdeh, G.H. Proteomic responses of rice young panicles to salinity. *Proteomics* **2006**, *6*, 6498–6507. [CrossRef]

30. Lee, D.G.; Kee, W.P.; Jae, Y.A.; Young, G.S.; Jung, K.H.; Hak, Y.K.; Dong, W.B.; Kyung, H.L.; Nam, J.K.; Byung-Hyun, L.; et al. Proteomics analysis of salt-induced leaf proteins in two rice germplasms with different salt sensitivity. *Can. J. Plant Sci.* **2011**, *91*, 337–349. [CrossRef]

31. Zhang, X.; Huang, G.; Bian, X.; Zhao, Q. Effects of root interaction and nitrogen fertilization on the chlorophyll content, root activity, photosynthetic characteristics of intercropped soybean and microbial quantity in the rhizosphere. *Plant Soil Environ.* **2013**, *59*, 80–88. [CrossRef]
32. Reddy, I.N.B.L.; Kim, S.M.; Kim, B.K.; Yoon, I.S.; Kwon, T.R. Identification of rice accessions associated with K1/Na1 ratio and salt tolerance based on physiological and molecular responses. *Rice Sci.* **2017**, *24*, 36–364. [CrossRef]
33. Amirjani, M.R. Effect of salinity stress on growth, sugar content, pigments and enzyme activity of rice. *Int. J. Bot.* **2011**, *7*, 73–81. [CrossRef]
34. Hinsinger, P.; Betencourt, E.; Bernard, L.; Brauman, A.; Plassard, C.; Shen, J.; Tang, X.; Zhang, F. P for two, sharing a scarce resource: Soil phosphorus acquisition in the rhizosphere of intercropped species. *Plant Physiol.* **2011**, *156*, 1078–1086. [CrossRef] [PubMed]
35. Song, Y.N.; Zhang, F.S.; Marschner, P.; Fan, F.L.; Gao, H.M.; Bao, X.G.; Sun, J.H.; Li, L. Effect of intercropping on crop yield and chemical and microbiological properties in rhizosphere of wheat (*Triticum aestivum* L.), maize (*Zea mays* L.), and faba bean (*Vicia faba* L.). *Biol. Fertil. Soils* **2007**, *43*, 565–574. [CrossRef]
36. Zhang, N.N.; Sun, Y.M.; Li, L.; Wang, E.T.; Chen, W.X.; Yuan, H.L. Effects of intercropping and Rhizobium inoculation on yield and rhizosphere bacterial community of faba bean (*Vicia faba* L.). *Biol. Fertil. Soils* **2010**, *46*, 625–639. [CrossRef]
37. Salekdeh, G.H.; Siopongco, J.; Wade, L.J.; Ghareyazie, B.; Bennett, J. A proteomic approach to analyzing drough and salt responsiveness in rice. *Field Crop Res.* **2002**, *76*, 199–219. [CrossRef]
38. Jiang, Q.; Xiaojuan, L.; Fengjuan, N.; Xianjun, S.; Zheng, H.; Hui, Z. iTRAQ-based quantitative proteomic analysis of wheat roots in response to salt stress. *Proteomics* **2017**, *17*, 1600265. [CrossRef]
39. Abbasi, F.M.; Komatsu, S. A proteomic approach to analyze salt-responsive proteins in rice leaf sheath. *Proteomics* **2004**, *4*, 2072–2081. [CrossRef]
40. Wen, B.; Zhou, R.; Feng, Q.; Wang, Q.; Wang, J.; Liu, S. IQuant: An automated pipeline for quantitative proteomics based upon isobaric tags. *Proteomics* **2014**, *14*, 2280–2285. [CrossRef]
41. Eprintsev, A.T.; Fedorina, O.S.; Bessmeltseva, Y.S. Response of the Malate Dehydrogenase System of Maize Mesophyll and Bundle Sheath to Salt Stress. *Russ. J. Plant Physiol.* **2011**, *58*, 448–453. [CrossRef]
42. Zhang, J.; Jia, W.; Yang, J.; Ismail, A.M. Role of ABA in integrating plant responses to drought and salt stresses. *Field Crops Res.* **2006**, *97*, 111–119. [CrossRef]
43. Su, Y.; Wang, Z.; Liu, F.; Li, Z.; Peng, Q.; Guo, J.; Xu, Q.Y. Isolation and Characterization of ScGluD2, a New Sugarcane beta-1,3-Glucanase D Family Gene Induced by Sporisorium scitamineum, ABA, H2O2, NaCl, and CdCl2 Stresses. *Front. Plant Sci.* **2016**, *7*, 1348. [CrossRef] [PubMed]
44. He, R.; Yu, G.; Han, X.; Han, J.; Li, W.; Wang, B.; Huang, S.; Cheng, X. ThPP1 gene, encodes an inorganic pyrophosphatase in Thellungiella halophila, enhanced the tolerance of the transgenic rice to alkali stress. *Plant Cell Rep.* **2017**, *36*, 1929. [CrossRef] [PubMed]
45. Karan, R.; Prasanta, K.S. Overexpression of a nascent polypeptide associated complex gene (SabNAC) of Spartina alterniflora improves tolerance to salinity and drought in transgenic *Arabidopsis*. *Bioch. Biophys. Res. Commun.* **2012**, *424*, 747–752. [CrossRef] [PubMed]
46. Singh, S.; Singh, C.; Tripathi, A.K. A SAM-dependent methyltransferase cotranscribed with arsenate reductase alters resistance to peptidyl transferase center-binding antibiotics in Azospirillum brasilense Sp7. *Appl. Microbiol. Biotechnol.* **2014**, *98*, 4625–4636. [CrossRef] [PubMed]
47. Paiva, A.L.S.; Passaia, G.; Lobo, A.K.M.; Jardim-Messeder, D.; Silveira, J.A.; Margis-Pinheiro, M. Mitochondrial glutathione peroxidase (OsGPX3) has a crucial role in rice protection against salt stress. *Environ. Exp. Bot.* **2018**, *158*, 12–21. [CrossRef]
48. Zeng, Y.; Tan, X.; Zhang, L.; Long, H.; Wang, B.; Li, Z.; Yuan, Z. A fructose-1,6-biphosphate aldolase gene from Camellia oleifera: Molecular characterization and impact on salt stress tolerance. *Mol. Breed.* **2015**, *35*, 1–17. [CrossRef]
49. Lu, W.; Tang, X.; Huo, Y.; Xu, R.; Qi, S.; Huang, J.; Zheng, C.; Wu, C.A. Identification and characterization of fructose 1,6-bisphosphate aldolase genes in *Arabidopsis* reveal a gene family with diverse responses to abiotic stresses. *Gene* **2012**, *503*, 65–74. [CrossRef]
50. Minhas, D.; Grover, A. Transcript levels of genes encoding various glycolytic and fermentation enzymes change in response to abiotic stresses. *Plant Sci.* **1991**, *146*, 41–51. [CrossRef]

51. Sharma, S.; Mustafiz, A.; Singla-Pareek, S.L.; Shankar, S.P.; Sopory, S.K. Characterization of stress and methylglyoxal inducible triose phosphate isomerase (OscTPI) from rice. *Plant Signal. Behav.* **2012**, *7*, 1337–1345. [CrossRef]
52. Zhu, J.K. *Plant Salt Stress*; John Wiley & Sons: Hoboken, NJ, USA, 2007.
53. Habib, S.H.; Kausar, H.; Saud, H.M. Plant growthpromoting Rhizobacteria enhance salinity stress tolerance in Okra through ROS-scavenging enzymes. *Biomed. Res. Int.* **2016**, *2016*, 6284547.
54. Hossain, M.S.; Dietz, K.J. Tuning of Redox Regulatory Mechanisms, Reactive Oxygen Species and Redox Homeostasis under Salinity Stress. *Front. Plant Sci.* **2016**, *7*, 548. [CrossRef] [PubMed]
55. Sharma, P.; Jha, A.B.; Dubey, R.S.; Pessarakl, M. Reactive oxygen species, oxidative damage, and antioxidative defense mechanism in plants under stressful conditions. *J. Bot.* **2012**, *2012*, 217037. [CrossRef]
56. Rejeb, K.B.; Benzarti, M.; Debez, A.; Bailly, C.; Savoure, A.; Abdelly, C. NADPH oxidase-dependent H2O2 production is required for saltinduced antioxidant defense in *Arabidopsis* thaliana. *J. Plant Physiol.* **2015**, *174*, 5–15. [CrossRef] [PubMed]
57. Chen, J.H.; Han-Wei, J.; En-Jung, H.; Hsing-Yu, C.; Ching-Te, C.; Hsu-Liang, H.; Tsan-Piao, L. Drought and Salt Stress Tolerance of an *Arabidopsis* Glutathione S-Transferase U17 Knockout Mutant Are Attributed to the Combined Effect of Glutathione and Abscisic Acid. *Plant Physiol.* **2012**, *158*, 340–351. [CrossRef] [PubMed]
58. Mishra, P.; Kumari, B.; Dubey, R.S. Differential responses of antioxidative defense system to prolonged salinity stress in salt-tolerant and salt-sensitive Indica rice (*Oryza sativa* L.) seedlings. *Protoplasma* **2013**, *250*, 3–19. [CrossRef] [PubMed]
59. Li, C.R.; Liang, D.D.; Li, J.; Duan, Y.B.; Li, H.; Yang, Y.C.; Qin, R.Y.; Li, L.I.; Wei, P.C.; Yang, J.B. Unravelling mitochondrial retrograde regulation in the abiotic stress induction of rice Alternative oxidase 1 gene. *Plant Cell Environ.* **2013**, *36*, 775–788. [CrossRef]
60. Gao, Y.; Cui, Y.; Long, R.; Sun, Y.; Zhang, T.; Yang, Q.; Kang, J. Salt-stress induced proteomic changes of two contrasting alfalfa cultivars during germination stage. *J. Sci. Food Agric.* **2019**, *99*, 1384–1396. [CrossRef]
61. Belghith, I.; Jennifer, S.; Tatjana, H.; Chedly, A.; Hans-Peter, B.; Ahmed, D. Comparative analysis of salt-induced changes in the root proteome of two accessions of the halophyte Cakile maritime. *Plant Physiol. Biochem.* **2018**, *130*, 20–29. [CrossRef]
62. Soares, A.L.C.; Christoph-Martin, G.; Sebastien, C.C. Genotype-Specific Growth and Proteomic Responses of Maize Toward Salt Stress. *Front Plant Sci.* **2018**, *9*, 661. [CrossRef]
63. Fatehi, F.; Abdolhadi, H.; Houshang, A.; Tahereh, B.; Paul, C.S. The proteome response of salt-resistant and salt-sensitive barley genotypes to long-term salinity stress. *Mol. Biol. Rep.* **2012**, *39*, 6387. [CrossRef]
64. Gao, L.; Yan, X.; Li, X.; Guo, G.; Hu, Y.; Ma, W.; Yan, Y. Proteome analysis of wheat leaf under salt stress by two-dimensional difference gel electrophoresis (2D-DIGE). *Phytochemistry* **2011**, *72*, 1180–1191. [CrossRef] [PubMed]
65. Li, B.; Tester, M.; Gilliham, M. Chloride on the move. *Trends Plant Sci.* **2017**, *22*, 236–248. [CrossRef] [PubMed]
66. Moolna, A.; Bowsher, C.G. The physiological importance of photosynthetic ferredoxin NADP+ oxidoreductase (FNR) isoforms in wheat. *J. Exp. Bot.* **2010**, *61*, 2669–2681. [CrossRef] [PubMed]
67. Ngara, R.; Roya, N.; Jonas, B.J.; Ole, N.J.; Bongani, N. Identification and profiling of salinity stress-responsive proteins in Sorghum bicolor seedlings. *J. Proteom.* **2012**, *75*, 4139–4150. [CrossRef]
68. Bai, J.; Yan, Q.; Jinghui, L.; Yuqing, W.; Rula, S.; Na, Z.; Ruizong, J. Proteomic response of oat leaves to long-term salinity stress. *Environ. Sci. Pollut. Res.* **2017**, *24*, 3387. [CrossRef] [PubMed]
69. Chen, S.; Natan, G.; Bruria, H. Proteomic analysis of salt-stressed tomato (*Solanum lycopersicum*) seedlings: Effect of genotype and exogenous application of glycinebetaine. *J. Exp. Bot.* **2009**, *60*, 2005–2019. [CrossRef] [PubMed]
70. Turan, S.; Baishnab, C.T. Salt-stress induced modulation of chlorophyll biosynthesis during de-etiolation of rice seedlings. *Physiol. Plantarum.* **2015**, *153*, 477–491. [CrossRef] [PubMed]
71. Wang, H.; Wang, H.; Shao, H.; Tang, X. Recent advances in utilizing transcription factors to improve plant abiotic stress tolerance by transgenic technology. *Front. Plant Sci.* **2017**, *7*, 4563. [CrossRef] [PubMed]
72. Boriboonkaset, T.; Theerawitaya, C.; Yamada, N.; Pichakum, A.; Supaibulwatana, K.; Cha-um, S.; Takabe, T.; Protoplasma, C.K. Regulation of some carbohydrate metabolism-related genes, starch and soluble sugar contents, photosynthetic activities and yield attributes of two contrasting rice genotypes subjected to salt stress. *Protoplasma* **2013**, *250*, 1157–1167. [CrossRef] [PubMed]

73. Wu, G.; Jin-Long, W.; Rui-Jun, F.; Shan-Jia, L.; Chun-Mei, W. iTRAQ-Based Comparative Proteomic Analysis Provides Insights into Molecular Mechanisms of Salt Tolerance in Sugar Beet (*Beta vulgaris* L.). *Int. J. Mol. Sci.* **2018**, *19*, 3866. [CrossRef]

74. Cho, S.K.; Jee, E.K.; Jong-A, P.; Tae, J.E. Woo Taek KimConstitutive expression of abiotic stress-inducible hot pepper CaXTH3, which encodes a xyloglucan endotransglucosylase/hydrolase homolog, improves drought and salt tolerance in transgenic *Arabidopsis* plants. *FEBS Lett.* **2006**, *580*, 3136–3144. [CrossRef] [PubMed]

75. Choi, J.Y.; Seo, Y.S.; Kim, S.J.; Kim, W.T.; Shin, J.S. Constitutive expression of CaXTH3, a hot pepper xyloglucan endotransglucosylase/hydrolase, enhanced tolerance to salt and drought stresses without phenotypic defects in tomato plants (*Solanum lycopersicum* cv. Dotaerang). *Plant Cell Rep.* **2011**, *30*, 867–877. [CrossRef] [PubMed]

76. Dietz, K.; Sauter, A.; Wichert, K.; Messdaghi, D.; Hartung, W. Extracellular β-glucosidase activity in barley involved in the hydrolysis of ABA glucose conjugate in leaves. *J. Exp. Bot.* **2000**, *51*, 937–944. [CrossRef] [PubMed]

77. Hermosa, R.; Leticia, B.; Emma, K.; Jesus, A.J.; Marta, M.B.; Vicent, A.; Aurelio, G.-C.; Enrique, M.; Carlos, N. The overexpression in *Arabidopsis* thaliana of a Trichoderma harzianum gene that modulates glucosidase activity, and enhances tolerance to salt and osmotic stresses. *J. Plant Physiol.* **2011**, *168*, 1295–1302. [CrossRef] [PubMed]

78. Liu, J.; Renu, S.; Ping, C.; Stephen, H.H. Salt stress responses in *Arabidopsis* utilize a signal transduction pathway related to endoplasmic reticulum stress signaling. *Plant J.* **2007**, *51*, 897–909. [CrossRef]

79. Takaaki, T.; Muneharu, E. Induction of a Novel XIP-Type Xylanase Inhibitor by External Ascorbic Acid Treatment and Differential Expression of XIP-Family Genes in Rice. *Plant Cell Phys.* **2007**, *48*, 700–714.

80. Upadhaya, H.; Sahoo, L.; Panda, S.K. Molecular Physiology of Osmotic Stress in Plants. In *Molecular Stress Physiology of Plants*; Rout, G., Das, A., Eds.; Springer: India, New Delhi, 2013.

81. Torres-Schumann, S.; Godoy, J.A.; Pintor-Toro, J.A. A probable lipid transfer protein gene is induced by NaCl in stems of tomato plants. *Plant Mol. Biol.* **1992**, *18*, 749–757. [CrossRef]

82. Pan, Y.; Li, J.; Jiao, L.; Li, C.; Zhu, D.; Yu, J. A Non-specific Setaria italica Lipid Transfer Protein Gene Plays a Critical Role under Abiotic Stress. *Front. Plant Sci.* **2016**, *7*, 1752. [CrossRef]

83. Dejardin, A.; Sokolov, L.N.; Kleczkowski, L.A. An *Arabidopsis* Stress-Responsive Sucrose Synthase Gene is UP-Regulated by Low Water Potential. In *Photosynthesis: Mechanisms and Effects*; Garab, G., Ed.; Springer: Dordrecht, The Netherlands, 1998.

84. Vastarelli, P.; Moschella, A.; Pacifico, D.; Mandolino, G. Water Stress in Beta vulgaris: Osmotic Adjustment Response and Gene Expression Analysis in ssp. Vulgaris and maritima. *Am. J. Plant Sci.* **2013**, *4*, 11–16. [CrossRef]

85. Sobhanian, H.; Razavizadeh, R.; Nanjo, Y.; Ehsanpour, A.A.; Jazii, F.R.; Motamed, N.; Komatsu, S. Proteome analysis of soybean leaves, hypocotyls and roots under salt stress. *Proteome Sci.* **2010**, *8*, 19. [CrossRef]

86. Sampath, K.I.; Ramgopal, R.S.; Vardhini, B.V. Role of Phytohormones during Salt Stress Tolerance in plants. *Curr. Trends Biotechnol. Pharm.* **2015**, *9*, 334–343.

87. Mukhopadhyay, A.; Shubha, V.; Akhilesh, K.T. Overexpression of a zinc-finger protein gene from rice confers tolerance to cold, dehydration, and salt stress in transgenic tobacco. *Proc. Natl. Acad. Sci. USA* **2004**, *101*, 6309–6314. [CrossRef] [PubMed]

88. Zhang, Z.; Huanhuan, L.; Ce, S.; Qibin, M.; Huaiyu, B.; Kang, C.; Yunyuan, X. A C2H2 zinc-finger protein OsZFP213 interacts with OsMAPK3 to enhance salt tolerance in rice. *J. Plant Phys.* **2018**, *229*, 100–110. [CrossRef] [PubMed]

89. Jamil, M.; Iqbal, W.; Bangash, A.; Rehman, S.; Imran, Q.M.; Rha, E.S. Constitutive Expression of OSC3H33, OSC3H50 AND OSC3H37 Genes in Rice under Salt Stress. *Pak. J. Bot.* **2010**, *42*, 4003–4009.

90. Wang, X.K.; Zhang, W.H.; Hao, Z.B.; Li, X.R.; Zhang, Y.Q.; Wang, S.M. *Principles and Techniques of Plant Physiological Biochemical Experiment*; Higher Education Press: Beijing, China, 2006; pp. 118–119. (In Chinese)

91. U.S. Salinity Laboratory Staff. *Diagnosis and Improvement of Saline and Alkali Soils*; Richards, L.A., Ed.; U.S. Goverment Publishing Office: Washington, DC, USA, 1954.

MDPI

St. Alban-Anlage 66

4052 Basel

Switzerland

Tel. +41 61 683 77 34

Fax +41 61 302 89 18

www.mdpi.com

International Journal of Molecular Sciences Editorial Office

E-mail: ijms@mdpi.com

www.mdpi.com/journal/ijms

www.ingramcontent.com/pod-product-compliance
Lightning Source LLC
Chambersburg PA
CBHW051706210326
41597CB00032B/5383